T0202190

INTERACTING SYSTEMS FAR FROM EQUILIBRIUM

Interacting Systems far from Equilibrium
Quantum Kinetic Theory

Klaus Morawetz

*Münster University of Applied Sciences, Stegerwaldstraße 39, 48565 Steinfurt,
Germany*
*International Institute of Physics (IIP), Universidade Federal do Rio
Grande do Norte Campus Universitário Lagoa nova,
CEP: 59078-970, Natal, Brazil*
*Max-Planck-Institute for the Physics of Complex Systems,
Nöthnitzer Straße 87, 01187 Dresden, Germany*

OXFORD
UNIVERSITY PRESS

OXFORD

UNIVERSITY PRESS

Great Clarendon Street, Oxford, OX2 6DP,
United Kingdom

Oxford University Press is a department of the University of Oxford.
It furthers the University's objective of excellence in research, scholarship,
and education by publishing worldwide. Oxford is a registered trade mark of
Oxford University Press in the UK and in certain other countries

Published in the United States of America by Oxford University Press
198 Madison Avenue, New York, NY 10016, United States of America

British Library Cataloguing in Publication Data
Data available

Library of Congress Control Number: 2017959075

ISBN 978–0–19–879724–1

DOI 10.1093/oso/9780198797241.001

Printed and bound by
CPI Group (UK) Ltd, Croydon, CR0 4YY

Simplicity is the ultimate sophistication.
Leonardo da Vinci

Preface

With this book a current gap should be filled between standard textbooks on many-body theory and monographs on specific topics. The need for such a summarising overview arises first due to the current interest in describing nonequilibrium processes in different fields and secondly because of the inherent demand of theoretical physics to develop an easy and transparent language to condense many complicated developments into a straight and simple notation. The final aim is to provide a concise textbook which enables students to learn the method of nonequilibrium Green's functions.

Atoms in gases, electrons in crystals or nucleons in nuclear matter behave as independent particles in many aspects so that we can understand the properties of these systems from statistical considerations. We will follow the original question of Ludwig Boltzmann as to how one can describe the motion of many-body systems from individual collisions. In this respect the philosophy is to consider motion versus force and to distinguish between large- and small-angle collisions, where the first one leads to dissipation and collision delays and the second one to the mean field and more complex effects on the drift. In dense systems the motion of particles is not only affected by the surrounding medium but feeds back to the embedding medium which allows us to describe Bose-Einstein condensation and superconductivity on the same footing.

In the course of study there are two ways to become acquainted with a new subject. One consists of developing the theory from phenomenological findings as a bottom-up or inductive way towards the abstract formulation with the upmost simplicity in notation and compression. The second way is to learn the final theory directly and to deduce all different phenomenon as specific applications. Here both possibilities are provided, offering students the ability to learn the quantum kinetic theory in terms of Green's functions using their experience in different fields and at the same time experienced researchers offer a framework to develop and apply the theory straight to the phenomena. Reflecting the versatile applications of nonequilibrium Green's functions, examples are collected ranging from solid state physics (impurity scattering, systems with spin-orbit coupling and magnetic fields, diffraction on organic barriers, semiconductors, superconductivity, Bose-Einstein condensation, ultrafast phenomena, graphene), plasma physics (screening and ultrashort-time modes), quenches of cold atoms in optical lattices up to heavy ion collisions and relativistic transport.

Given the intricate matter and the versatile recent developments, many different views exist on how to present a theory consistently. As the old motto of the University of Rostock said: 'Doctrina multiplex veritas una'. Though preparing collectively to avoid misprints and errors, the alert reader is kindly thanked in advance for bringing any of them to my attention.

Klaus Morawetz

Dresden,
14 May 2017

Acknowledgements

From a common book (Lipavský, Morawetz and Špička, 2001*b*) I could adapt parts into the chapters 3, 13, 14, 23 and appendix C, D for which copyright permission by EDP science is thanked for. I would have liked to convince Pavel Lipavský to be a coauthor, since he contributed main parts of sections 2.1, 2.2.2, 2.4, 2.5.1–2, 4, 5, 6.1, 6.3 and 16.1., 16.5–6. Especially the proof in appendix D has been composed in the book (Lipavský, Morawetz and Špička, 2001*b*) by him. The many enlightening discussions with Michael Männel are acknowledged who has worked out the Bose condensation of section 12.2 and 12.5 within his PhD thesis.

For the long-standing encouragement to start this book project, I would like to thank the Editors Sonke Adlung and Ania Wronski also for continuously giving advice and support to any occurring problems. The carefull proof reading finally has been performed by Marie Felina Francois and Lesley Harris.

Contents

Part V   Selected Applications

16   Diffraction on a Barrier   371

17   Deep Impurities with Collision Delay   392

18   Relaxation-Time Approximation   402

Part I

Classical Kinetic Concepts

The classical concepts of distribution functions are introduced from elementary principles. The motion of single particles is contrasted with the forces and the account of short- and long-range correlations in kinetic theory is explored. The division of collisional correlations into mean-fields and into nonlocal collisions is presented and the concept of quasiparticles is derived. One can learn about virial theorems and viral corrections and entropy production as well as hidden latent heat in collisions.

1
Historical Background

1.1 Introduction

Nonequilibrium phenomena are the essential processes which occur in nature. Any evolution is built up of involved causal networks which may render a new state of quality in the course of time evolution. The steady state or equilibrium is rather the exception in nature, if not a theoretical abstraction. The objective of statistical mechanics is to explain and predict the properties of macroscopic matter from the properties of its microscopic constituents. The equilibrium properties of macroscopic systems, with a given microscopic Hamiltonian, can be obtained as suitable averages in well-defined Gibbs ensembles. Our understanding of nonequilibrium phenomena is much less than satisfactory at the present time (Ernst, 1983). The aim of nonequilibrium statistical mechanics is to determine how macroscopic properties of matter evolve with time, in terms of the laws of quantum mechanics that govern the motion of its constituents—atoms, molecules, etc.

One line of approache extends the idea of the Gibbs ensemble to nonequilibrium systems, explicitly specifying the nonequilibrium density operator by a set of relevant observables (Zubarev, 1971; Röpke, 1987). For a modern overview see Zubarev et al. (1996), (Zubarev, Morozov and Röpke, 1997). Besides the great advantage of this method for maintaining the Gibbs ensemble picture and providing a convenient linear response formalism, one has to know in advance the set of relevant observables. In this sense, one has to specify possible evolutions of the system before actually knowing them. The concept of kinetic theory or equation of motions is different.

Kinetic theory tries to find basic evolution equations for the time-dependence of distribution functions with the help of which variables are determined. Kinetic equations for distribution functions ensure the microscopic formulation of hydrodynamics and thermodynamics. Therefore, it is the link between the nonequilibrium statistical mechanics of many particle systems and macroscopic or phenomenological physics.

The kinetic description, which started with the foundation of Ludwig Boltzmann's famous equation (Boltzmann, 1872), has been rapidly developed from important classical contributions made by Chapman and Cowling (1939), Enskog (1917), Kirkwood (1946), Bogoliubov (1946) and Prigogine (1962) to quantum extensions, where the pioneering work along these lines was performed by Bogoliubov and Gurov (1947) and

Interacting Systems far from Equilibrium. Klaus Morawetz, Oxford University Press (2018).
© Klaus Morawetz. DOI: 10.1093/oso/9780198797241.001.0001

Mori and Ono (1952). The quantum Boltzmann equation, called Boltzmann–Uhling–Uhlenbeck (BUU) equation differs from the classical one in the collision term, which takes into account that the final scattering states can be occupied and consequently blocked by the Pauli exclusion principle. Moreover, the quantum-mechanical transition rate, rather than the classical one, is used. Attempts to extend the kinetic theory to higher densities have been fraught with severe difficulties. One might have imagined being able to develop a power series expansion of the transport coefficients in the same way as one expands the equilibrium equation of state in the virial series. In (Cohen and Dorfman, 1965) and (Cohen and Dorfman, 1972) it was proved that such an expansion does not exist. The Navier–Stokes coefficients are non-analytic functions of the density (Evans and Morriss, 1990). We have to search for another expansion schema.

One general starting point in deriving kinetic equations was the coupled set of equations of motion for the reduced density operators. This set was first derived by Irving and Zwanzig (1951). The formal structure is similar to the so-called Bogoliubov–Born–Green–Kirkwood–Yvon (BBGKY) hierarchy for reduced distribution functions in classical statistical physics, see Bogoliubov (1946), Born and Green (1946), Kirkwood (1946) or (Balescu, 1978). The reduced density of the BBGKY hierarchy expressed in the mixed Wigner representation becomes a function of the position in phase space like the Boltzmann distribution. The Wigner distribution of quantum theory has thus a close analogy to its classical precursor, the Boltzmann distribution. A quantum particle, however, has a property which disturbs such a classical picture: it can tunnel into regions where the local potential exceeds the energy of the particle, and similarly, it can reach momenta which don't correspond to its kinetic energy. It became necessary to distinguish particles having an energy corresponding to the classical expectation (particles on the energy shell) from those which have a different energy (particle off the energy shell). As Köhler and Malfliet (1993) show, the off-shell contributions are necessary to describe the high-momenta tails of the Wigner distribution and to distinguish the mean removal energy from the quasiparticle one. While the classical picture treats the energy as a property of the particle that is determined by the phase space point, the new concept of quantum statistics known as Green's function treats the energy as an independent variable.

The method of the quantum statistical Green's function was introduced by Matsubara (1955). The Green's functions used by him are also called imaginary-time Green's functions because their time arguments stands for the imaginary inverse temperature. For this reason they can only be used to describe equilibrium many-body systems. In 1959 Martin and Schwinger introduced another Green's function technique for the description of many-body systems (Martin and Schwinger, 1959). As starting point they constructed a hierarchy of equations of motion for the imaginary-time Green's functions, nowadays known as Martin–Schwinger hierarchy. The real-time Green's functions (Schwinger, 1961) have been presented in the textbook by their students Kadanoff and Baym (1962).

An important ingredient in the real-time Green's functions is the analytic continuation of the Green's function from imaginary times to real times in order to describe nonequilibrium situations. Both Keldysh (1964) and Craig (1968) used this concept and formulated a Dyson equation on the time contour.

The real-time Green's functions provide a unifying many-body theory. They allow us to consistently investigate properties of many-particle systems from the ground state, over a finite temperature equilibrium state, to nonequilibrium situations. Applications of real-time many-body methods range from problems of quantum chromodynamics (Landsman and van Weert, 1987), through nuclear physics (Botermans and Malfliet, 1990, 1988; Tohyama, 1987; Danielewicz, 1984a), to the theory of liquid helium (Serene and Rainer, 1983), physics of plasmas (Bezzerides and DuBois, 1972), physics of condensed matter (Rammer and Smith, 1986), astrophysics (Keil, 1988) and cosmology (Calzetta and Hu, 1988). Moreover there are many investigations in classical relativistic treatment (Manzke and Kremp, 1979; Groot, a. van Leeuwen and van Weert, 1980) and quantum relativistic treatments can be found in (Brown, Puff and Wilets, 1970; Poschenrieder and Weigel, 1988; de Jong and Malfliet, 1991). Several papers are devoted to the formulation of transport equations in quantum field theory (Botermans and Malfliet, 1990; Ivanov, 1987; Mrowczynski and Danielewicz, 1990; Elze and et. al., 1987). The real-time Green's function technique is indeed a universal method to describe quantum many-particle systems under extreme conditions like nuclear matter or condensed stellar objects. For reviews see (Botermans and Malfliet, 1990; Danielewicz, 1990, 1984b; Itzykson and Zuber, 1980; Kremp, Schlanges and Bornath, 1985).

1.2 Virial corrections to the kinetic equation

The very basic idea of the Boltzmann equation to balance the drift of particles with dissipation is used both in gases and condensed systems like metals or nuclei. In both fields, the Boltzmann equation allows for a number of different improvements that make it possible to describe phenomena far beyond the range of the validity of the original Boltzmann equation.

In the theory of gases, the focus of the improvements was on a finite volume of molecules. The Boltzmann equation now includes a contradiction. The scattering cross sections used within the Boltzmann equation reflect the finite volume of molecules and their other interactions, however, the instant and local approximation of scattering events implies the equation of state of an ideal gas. To achieve consistency and to extend the validity of the Boltzmann equation to moderately dense gases, Clausius and Boltzmann included the space nonlocality of binary collisions (Chapman and Cowling, 1990). After a century, virial corrections won new interest as they can be incorporated into Monte-Carlo simulation methods (Alexander, Garcia and Alder, 1995).

In the theory of condensed systems modifications of the Boltzmann equation are rather different being determined by the quantum mechanical nature of the single-particle motion and by quantum mechanical statistics. A headway in this field is covered by the Landau concept of quasiparticles (Baym and Pethick, 1991). There are three major modifications: the Pauli blocking of scattering channels; the underlying quantum-mechanical dynamics of collisions; and the quasiparticle renormalisation of a single-particle dispersion relation. With all these deep modifications, the scattering integral of the Boltzmann equation still remains local in space and time. In other words, the Landau theory does not include a quantum-mechanical analogy of virial corrections studied in the theory of gases.

Even if one reinterprets the Boltzmann equation in the spirit of Landau's theory of Fermi liquids incorporating all known quasiparticle renormalisations, the Boltzmann equation still does not cover all known basic features of bulk matter. The structure of the quasiparticle kinetic equation is deduced from the system of weakly coupled particles, while dense matter like nuclear matter exhibits implicit correlations due to strong interactions. These correlations demand to include the virial corrections which are absent in Landau's theory.

1.2.1 Classical gas of hard spheres

Beside nonlocalities, Enskog has also introduced static correlations (Enskog, 1972; Chapman and Cowling, 1990; Hirschfelder, Curtiss and Bird, 1964; Schram, 1991) and later Weinstock (1963a; 1963b; 1965) has discussed dynamic correlations. Although our focus will be on the nonlocalities, it is useful to recall the different kinds of virial corrections.

The first kind of correction beyond the original Boltzmann equation is the collisional transfer of momentum and energy. When two particles undergo a collision, the momentum and energy transferred from one particle to another suddenly jump over the distance between both particles. This process is instant and only two particles are involved. Accordingly, the nonlocal corrections are nothing but a more careful description of an isolated two-particle collision.

The second kind of correction takes into account that particles of finite volume scatter more frequently as is apparent when the density approaches a closely packed system. There are two effects that change the frequency of collisions. Firstly, a particle of a finite volume flies over a shorter distance to meet another partner in a point-like model. Secondly, the particles shield each other. These corrections can be taken into account by introducing static statistical correlations into the scattering integral. The shorter mean-free path and shielding effect appear only if collisions are treated as a process in the many-body surrounding. Accordingly, the static statistical correlations are of a three-particle nature.

The third kind of correction takes into account the factm that any collision creates a perturbation in the vicinity of a particle. Indeed, the energy and momentum which a particle lost at a certain collision is carried by particles in its neighbourhood. This excess energy and momentum influence an initial condition for its next collision. Apparently, this is a dynamic process by its nature and also needs more than two particles to make it happen.

For the hard-sphere gas the nonlocal correction is trivial; therefore the main theoretical focus was on the statistical correlations (Enskog, 1972; Chapman and Cowling, 1990; Hirschfelder, Curtiss and Bird, 1964; Schram, 1991; Cohen, 1962; Weinstock, 1963a,b, 1965; Kawasaki and Oppenheim, 1965; Dorfman and Cohen, 1967; Goldman and Frieman, 1967; van Beijeren and Ernst, 1979). It turned out that the treatment of higher-order contributions is far from trivial, as the dynamical statistical correlations result in divergences that are cured only after resummation of an infinite set of contributions. Naturally, this resummation leads to non-analytic density corrections to the scattering integral (Weinstock, 1965; Kawasaki and Oppenheim, 1965; Dorfman and Cohen, 1967; Goldman and Frieman, 1967; van Beijeren and Ernst, 1979). The essence of these hard-sphere studies is that beyond the nonlocal corrections one has to sum up an infinite set of contributions, the plain perturbative expansion leads to incorrect results.

1.2.2 Hard-sphere corrections to the quantum Boltzmann equation

Simulation codes of the Boltzmann equation and similarly the quantum Monte-Carlo dynamics assume the local and instant scattering. Halbert has discussed nonlocal corrections already in 1981, within the cascade model, Malfliet (1983) proposed to include the nonlocal Enskog-type corrections into simulation codes and Kortemeyer et al. (1996) incorporated the nonlocal corrections into the Monte-Carlo codes for the BUU equation using a method developed within the classical molecular dynamics (Alexander, Garcia and Alder, 1995). All these studies treat the nonlocality of two-particle collisions as if particles are classical hard spheres.

The basic idea of nonlocal corrections is that two colliding particles do not occupy the same space point but are displaced by the sum of their radii. For hard spheres, the direction of this displacement is easily identified with the direction of the transferred momentum. The numerical trick of building this displacement into Monte-Carlo simulations consists of a space displacement of the colliding particles after the collision. The classical molecular dynamics (Alexander, Garcia and Alder, 1995) show that the nonlocal corrections might be similarly implemented into quantum-molecular dynamics.

The hard-sphere corrections, however, didn't improve the agreement between numerical simulations and experimental data in nuclear collisions. This is because of their principal natures. As shown by Danielewicz and Pratt (1996) nucleons don't expel each other as hard spheres do, but tend to stay in the collision state as a short-living two-nucleon cluster. This effective 'molecule' lasts for a certain time called the collision delay. As known from gases, the formation of molecules reduces the pressure because it reduces the number of particles which carry a momentum. In contrast, hard-sphere interaction leads to the excluded volume of the van der Waals equation which increases the pressure. For realistic simulations it is thus necessary to leave intuitively formulated nonlocal corrections and derive their value from microscopic theory.

1.2.3 Quantum nonlocal corrections in gases

The microscopic theory of nonlocal corrections to the collision integral has been pioneered within the theory of gases by many authors (Waldmann, 1957, 1958, 1960; Snider, 1960, 1964; Bärwinkel, 1969a; Thomas and Snider, 1970; Snider and Sanctuary, 1971; Rainwater and Snider, 1976; Balescu, 1975; McLennan, 1989; Laloë, 1989; Tastevin, Nacher and Laloë, 1989; Nacher, Tastevin and Laloë, 1989; Loos, 1990a,b; de Haan, 1990b,a, 1991; Laloë and Mullin, 1990; Snider, 1990, 1991; Nacher, Tastevin and Laloë, 1991a,b; Snider, 1995; Snider, Mullin and Laloë, 1995).

Snider (1960, 1964) generalised Waldmann's equation in such a way that it includes the nonlocality of collisions. Since then a number of similar theories have been developed. It was generally believed that this approach covers virial corrections until Laloë, Mullin, Nacher and Tastevin analysed the so-called Waldmann–Snider theory to conclude that Snider's equation in its original form is not consistent with the second-order virial correction to the equation of state (Laloë and Mullin, 1990; Nacher, Tastevin and Laloë, 1991*a*). In a series of papers (Laloë, 1989; Tastevin, Nacher and Laloë, 1989; Nacher, Tastevin and Laloë, 1989; Nacher, Tastevin and Laloë, 1991*a*, 1991*b*) these authors have introduced a new concept of 'free Wigner transform which wipes out the effect of the potential at short distances'. The main idea of their approach is identical to the idea of quasiparticles in that the kinetic equation is not constructed for the Wigner distribution but for a subsidiary function, the free Wigner transform, that is free of the correlated motion. The Wigner function is constructed then from the free Wigner transform in the final step. Since the free Wigner transform includes only the on-shell contributions, as far as we can see, the free Wigner transform is identical to the quasiparticle distribution. The intuitive theory of Laloë, Tastevin and Nacher has been confirmed by de Haan (1990*b*,*a*, 1991) who used Balescu's formal derivation of kinetic equations (Balescu, 1975).

This theory made headway in the field as it allows one to describe quasiparticle renormalisations including off-shell processes. It gives a nonlocal scattering integral as being consistent with second-order virial corrections. It is important that all nonlocal corrections are presented in terms of derivatives of the scattering phase shift. This provides a clear link between the virial corrections and observable quantities. On the other hand, this approach is limited to dilute systems with non-degenerate statistics, which makes it inapplicable to dense Fermi systems. This is because technically the concept of the free Wigner transform as well as Balescu's method become too complicated for degenerated systems, in particular with in-medium effects.

1.3   Quantum nonlocal corrections in dense Fermi systems

A convenient perturbative expansion for degenerate systems is well established within Green's functions. The pioneering Green's function treatment of non-instant and nonlocal corrections to the scattering integral was done by Bärwinkel (1969*a*) who also discussed the thermodynamical consequences of these corrections 1969*b*. His studies are limited to low-density particles. Nevertheless, his work was the most important contribution to Green's functional approach to virial corrections.

The physics behind these virial corrections are strong-interaction correlations. In the weak-coupling limit the scattering rate of two particles is given by the matrix element of the interaction potential between the plain waves of initial and final states. For a strong potential, the wave function cannot completely penetrate a strongly repulsive potential being expelled from the particle core, but contrarily it can be enhanced at moderately short distances by short-range attractive forces. In the strong-coupling case one thus has to take into account the reconstruction of the wave functions by the interaction potential. This reconstruction is usually called the internal dynamics of collisions as it is reflected in the finite duration of collision. The build-up of the wave function means that the particles can be found at a given distance with an increased probability. Within the ergodic interpretation of probability it means that they have to spend a longer time at this distance than would result from an uncorrelated motion. This interpretation corresponds to the concept of dwell time, see (Hauge and Støvneng, 1989). Accordingly, unlike in Landau's theory, the collision has to be treated as non-instant and nonlocal.

These internal dynamics have thermodynamic consequences. For instance, with a finite collision duration, the density of quasiparticles differs from the density of real particles. The difference is the so-called correlated density. The Fermi momentum (of quasiparticles) differs then from the Fermi momentum defined by the density of particles (Schmidt, Röpke and Schulz, 1990; Morawetz and Röpke, 1995; Bornath, Kremp, Kraeft and Schlanges, 1996). We note that within Landau's theory of Fermi liquids the Fermi momentum is not affected by the interaction and these two values are equal according to the Luttinger theorem (Luttinger, 1960). The shift of the Fermi momentum now reflecting the collision delay leads to second-order virial corrections to the law of mass action (Bornath, Kremp, Kraeft and Schlanges, 1996). A discussion about thermodynamic consequences and applications of this mass-action law can be found in (Kraeft, Kremp, Ebeling and Röpke, 1986). Therefore the contradictory predictions of the virial correction to the mass-action law concern the violation of Luttinger's theorem and are traced back to the question of whether a correlated density appears or not. As shown by Farid (1999), Luttinger's proof of the Fermi liquid ground state of any perturbatively treated many-fermion system is a tautology in the sense that the correlated density is neglected as a definition of a Fermi liquid which obeys, then of course the Luttinger theorem. Similarly, the assumption of the Fermi liquid ground state is a part of Craig's approach. We believe that the zero result of Craig (1966a) and a similar study by Ivanov et al. (2000) are not correct in spite of the employed Φ-derivable approximation. Most Fermi systems do not lead to the Fermi liquid ground state, but provide e.g. a superconducting transition. In the superconducting state the shift of the Fermi momentum is experimentally confirmed because it leads to the electrostatic potential known as the Bernoulli potential (Lipavský, Koláček, Morawetz, Brandt and Yang, 2007).

The question is now: what approximation is capable of covering the internal dynamics of the collisions. Using the expansion in small scattering rates, Craig (1966a) found no virial correction to the density of particles. In contrast, this limit was introduced by Stolz and Zimmermann (1979), Kremp et al. (1984) and Zimmermann and Stolz (1985) to study just this type of virial corrections in a non-ideal plasma. Its implementation for nuclear physics has been considered by Röpke et al. (1982a), Röpke et al. (1982b; 1982c) and Schmidt et al. (1990) who have shown that, for equilibrium, the limit of small scattering rates called the generalised Beth–Uhlenbeck approach (Beth and Uhlenbeck, 1937) provides the correct second-order virial corrections.

1.4   Quantum nonlocal corrections to kinetic theory

In the limit of small scattering rates, the transport equation for the Green's function is converted into the Boltzmann-type kinetic equation by the extended quasiparticle approximation corresponding to the $\rho[f]$ functional. The ability of the limit of small scattering rates to recover the kinetic equation with non-instant corrections has been discussed by Bornath et al. (1996) although only for non-degenerated systems. Another model study of non-interacting electrons scattered by neutral impurities (Špička, Lipavský and Morawetz, 1997a) demonstrated that the non-instant corrections have to be studied together with the quasiparticle corrections, because they partially compensate each other in many observable quantities. A quantum-kinetic theory which unifies the achievements of transport in dense gases with the quantum transport of dense Fermi systems was derived starting with the impurity problem (Špička, Lipavský and Morawetz, 1997b) and then for arbitrary Fermi systems (Špička, Lipavský and Morawetz, 1998; Lipavský, Morawetz and Špička, 2001b; Morawetz, Lipavský and Špička, 2001c). The quasiparticle drift of Landau's equation is connected with a dissipation governed by a nonlocal and non-instant scattering integral in the spirit of the Enskog corrections. These corrections are expressed in terms of shifts in space and time that characterise the nonlocality of the scattering process (Morawetz, Lipavský, Špička and Kwong, 1999b). In this way quantum transport can be recast into a quasiclassical picture suited for simulations. The balance equations for the density, momentum, energy and entropy include quasiparticle contributions and the correlated two-particle contributions beyond the Landau theory will also be demonstrated in this book.

As to special limits, this kinetic theory includes the Landau theory as well as the Beth–Uhlenbeck equation of state which means correlated pairs. The medium effects on binary collisions are shown to mediate the latent heat which is the energy conversion between correlation and thermal energy (Lipavský, Špička and Morawetz, 1999; Lipavský, Morawetz and Špička, 2001b). In this respect the seeming contradiction between particle-hole symmetry and time-reversal symmetry in the collision integral was solved (Špička, Morawetz and Lipavský, 2001). Compared to the Boltzmann equation, the presented form of virial corrections only slightly increases the numerical demands in implementations (Morawetz, Špička, Lipavský, Kortemeyer, Kuhrts and Nebauer, 1999c; Morawetz, 2000a; Morawetz, Ploszajczak and Toneev, 2000b; Morawetz, Lipavský, Normand, Cussol, Colin and Tamain, 2001b) since large cancellations in the off-shell motion appear which are usually hidden in non-Markovian behaviours. The molecular quantum dynamics and the quantum Boltzmann (BUU) simulations (Morawetz, Špička, Lipavský, Kortemeyer, Kuhrts and Nebauer, 1999c; Morawetz, Lipavský, Normand, Cussol, Colin and Tamain, 2001b) with nonlocal binary collisions derived from the scattering T-matrix in (Špička, Lipavský and Morawetz, 1998; Morawetz, Lipavský, Špička and Kwong, 1999b) lead to a better agreement of experimental and theoretical distributions of neutrons and protons emitted from reacting nuclei.

Because of the high complexity of the complete quantum-mechanical treatment, the quasiclassical kinetic equation is still beneficial to the study of nonequilibrium many-body systems. In this book we will discuss various aspects of the Green's function approach and very often we will interpret the derived relations and quantities in terms of corresponding features in the quasiclassical picture.

2

Elementary Principles

The many-body theory combines ideas of thermodynamics with those of mechanics. In this introductory chapter, the symbiosis of these two different fields of physics is demonstrated on overly simplified models.

2.1 Motion versus forces

Two of Newton's theories have influenced the early microscopic pictures of fluids, the law of motion and the law of gravity. The law of gravity supplied the idea that molecules can act one on each other by forces that range over long distances. Such repulsive forces can explain a finite compressibility of solids, liquids and dense gases. The law of motion, $\Delta k = F \Delta t$, says that the change of the momentum is caused by forces and vice versa. This law is sufficient to explain the equations of state of rare gases and provides a link between the heat and the kinetic energy of molecules. The first law of thermodynamics thus became identical with the general law of conservation of energy. This standard interpretation was not so obvious when the heat was viewed as an independent substance.

In the first half of the nineteenth century, the concept based on the law of motion and the concept based on the long-ranged forces were felt to be competitive. The pure theory based on long-ranged forces needed a phenomenological assumption that forces depend on the temperature to explain the equation of state of gases. The pure theory of motion was free of phenomenological assumptions but worked only for rare gases. As often happens, it turned out that the two concepts are not competitive but complementary.

To explain the thermodynamical consequences of molecular motion and of the forces between molecules, we first discuss the pressure in a one-dimensional (1D) classical gas.

2.1.1 Principle of motion

Assume a single particle of mass m in a 1D system of volume Ω. The particle flies between points $-\frac{1}{2}\Omega$ and $\frac{1}{2}\Omega$ with velocity $\pm v$. In a single round trip the particle passes the trajectory of length 2Ω which takes a time $\Delta t = 2\Omega/v$. With each encounter of the surface, the momentum of the particle changes by $\Delta k = 2mv$. The mean force

Interacting Systems far from Equilibrium. Klaus Morawetz, Oxford University Press (2018).
© Klaus Morawetz. DOI: 10.1093/oso/9780198797241.001.0001

needed for these changes of momentum represents the pressure the particle exerts on the surface. According to Newton's law of motion, the pressure

$$\mathscr{P} = \frac{\Delta k}{\Delta t} = \frac{mv^2}{\Omega} \tag{2.1}$$

is proportional to the kinetic energy $\frac{1}{2}mv^2$ and inversely proportional to the volume.

If there are N non-interacting particles, they move independently. Their individual contributions to the pressure thus add

$$\mathscr{P} = \frac{m}{\Omega} \sum_{i=1}^{N} v_i^2. \tag{2.2}$$

The sum over single-particle energies divided by volume,

$$\mathscr{E} = \frac{1}{\Omega} \sum_{i=1}^{N} \frac{1}{2} mv_i^2, \tag{2.3}$$

is the density of energy. The pressure in the non-interacting gas is thus proportional to the density of energy,

$$\mathscr{P} = 2\mathscr{E}. \tag{2.4}$$

Later on, we will establish the relation between the density of energy and the temperature of the system by which 1D Bernoulli's relation (2.4) turns into the equation of state.

2.1.2 Principle of finite-range forces

The long-range forces between particles modify the relation between the pressure and the energy. For the purpose of discussion it is sufficient to assume a model box-like potential

$$\mathscr{V}(r) = \mathscr{V}\theta(R-r), \tag{2.5}$$

where r is a distance of particles. The potential is parametrised by range (radius) R and strength $\mathscr{V}$. We assume a repulsive potential $\mathscr{V} > 0$ to avoid bounded states.

To describe the effect of collisions on the pressure, we select two neighbouring particles moving one to the other with velocities v_1 and v_2. When the two particles meet, their velocities change to $u_{1,2}$ with which they move as long as their distance is smaller than R. According to the momentum and energy conservations

$$mv_1 + mv_2 = mu_1 + mu_2, \quad \frac{1}{2}mv_1^2 + \frac{1}{2}mv_2^2 = \frac{1}{2}mu_1^2 + \frac{1}{2}mu_2^2 + \mathscr{V}, \tag{2.6}$$

the velocities of interacting particles read

$$u_{1,2} = \frac{v_1 + v_2}{2} \pm \sqrt{\left(\frac{v_1 - v_2}{2}\right)^2 - \frac{\mathscr{V}}{m}}. \tag{2.7}$$

For a strong potential, $\mathscr{V} > \frac{1}{4}m(v_1 - v_2)^2$, the interaction region is not accessible for particles; therefore they get reflected at the very first touch.

After the collision, particles have final velocities $w_{1,2}$. In the 1D system, the conservation of momentum and energy,

$$mv_1 + mv_2 = mw_1 + mw_2, \qquad \frac{1}{2}mv_1^2 + \frac{1}{2}mv_2^2 = \frac{1}{2}mw_1^2 + \frac{1}{2}mw_2^2, \tag{2.8}$$

determines the final states of particles,

$$
\begin{aligned}
w_{1,2} &= v_{2,1} \qquad \text{for} \quad \mathscr{V} > \frac{1}{4}m(v_1 - v_2)^2 \\[2ex]
w_{1,2} &= v_{1,2} \qquad \text{for} \quad \mathscr{V} < \frac{1}{4}m(v_1 - v_2)^2
\end{aligned}
\tag{2.9}
$$

In the case of strong potentials, particles exchange their momenta. In the case of weak potentials, they end up with their initial momenta.

The two regimes of the collision, reflection or transmission, contribute differently to the pressure. Before we discuss details, it is necessary to note that the pressure corrections appear only for the forces of a finite range. Indeed, for forces of infinitesimal short range, $R \to 0$, all collisions are point-like, i.e. local. The characteristic time of the interaction scales with $R/(v_1 - v_2)$ so that the point collisions are also instant. The value of the internal velocities $u_{1,2}$ thus does not matter and the collision is characterised fully by the local matching of initial and final velocities (2.9), $v_{1,2} \to w_{1,2}$. If all assumed particles are identical, particles 1 and 2 are not distinguishable; therefore the same physical picture is met if they pass each other, as for $\mathscr{V} < \frac{1}{4}m(v_1 - v_2)^2$, or if they interchange their momenta, as for $\mathscr{V} < \frac{1}{4}m(v_1 - v_2)^2$. For $R \to 0$, with no regards whether the interaction is strong, weak or ignored, there are two particles of velocities v_1 and v_2 reaching the point of encounter, and two particles of the velocities v_1 and v_2 leaving this point. Accordingly, all these systems exert the same pressure on the surface. Briefly, the instant and local collisions do not influence the pressure.

Inaccessible volume   For the potential of finite range, let us assume first a strong potential. The collisions are instant, but at the instant of collision the centres of two colliding particles are displaced by R. Similarly, when reflected by the surface, the centre of the particles is displaced by $\frac{1}{2}R$ from the surface. Trajectories of three particles are shown in Fig. 2.1. Such 1D particles behave like hard rods. As can be seen, such a system of

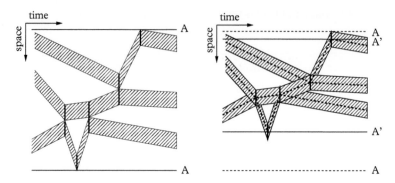

Figure 2.1: *Trajectories of three one-dimensional hard rods colliding three and two times with the wall (left panel). To map this system on a system of point-like particles, we first shift the upper line by $\frac{1}{2}R$ so that the centre of the first particle is reflected on the surface. Now we shift the middle trajectory and the lower line by R so that the centres of the first and second particles at the instant of collision coincide. Next we shift the lowest trajectory by 2R so that centres of the lowest and the middle particle at collision coincide too. In the right panel we thus obtain an effective system of point-like particles in which particles collide and hit the surface with the same frequency.*

hard rods can be mapped on a system of point-like particles in a volume $\Omega - NR$. The pressure in the system of hard rods thus is

$$\mathscr{P} = \frac{1}{\Omega - NR} m \sum_{i=1}^{N} v_i^2 = \frac{2\mathscr{E}}{1 - nR}, \tag{2.10}$$

where $n = \frac{N}{\Omega}$ is the particle density. As (2.4) corresponds to the equation of state of the ideal gas, so (2.10) corresponds to the van der Waals equation of state with the excluded volume NR.

Delay time and internal pressure   Now we turn to collisions in which particles have enough kinetic energy to pass each other. Such weak-potential collisions are non-instant and nonlocal. At the beginning and at the end of collisions, particles are displaced by R but from opposite sides. Accordingly, during interaction particles travel the relative distance $2R$. Since their relative velocity is $u_1 - u_2$, the collision process lasts

$$\tau_C = \frac{2R}{u_1 - u_2} = \frac{R}{\sqrt{\left(\frac{v_1 - v_2}{2}\right)^2 - \frac{\mathscr{V}}{m}}}. \tag{2.11}$$

In the absence of interactions, particles would travel the same distance in the time

$$\tau_0 = \frac{2R}{v_1 - v_2}. \tag{2.12}$$

The collision delay that can be attributed to the interaction thus is

$$\Delta_\tau = \tau_C - \tau_0. \tag{2.13}$$

The collision delay Δ_τ modifies the frequency with which particles hit the surfaces. For simplicity, let us assume for a while that in free motion all particles have the same velocity $\pm v$. In this case, only particles with opposite orientations of velocities can collide being always delayed by the same time

$$\Delta_\tau = \frac{R}{\sqrt{v^2 - \frac{\mathcal{V}}{m}}} - \frac{R}{v}. \tag{2.14}$$

In a single round trip, each particle collides twice with each other; therefore its motion is delayed by $2(N-1)\Delta_t$. Assuming $N \gg 1$, we write the time period as

$$\Delta t = 2\frac{\Omega}{v} + 2N\Delta_\tau. \tag{2.15}$$

The internal pressure caused by particles reads

$$\mathscr{P} = N\frac{2mv}{\Delta t} = \frac{2\mathscr{E}}{1 + nR\left(\left[1 - \frac{\mathcal{V}}{mv^2}\right]^{-\frac{1}{2}} - 1\right)}. \tag{2.16}$$

A particularly simple form of the pressure appears for a very weak interaction potential, $\mathcal{V} \ll mv^2$. The linear correction in $\mathcal{V}$ from (2.16) follows as

$$\mathscr{P} = 2\mathscr{E} - \frac{1}{2}n^2 R\mathcal{V}. \tag{2.17}$$

Unlike the excluded volume, the weak potential correction $-\frac{1}{2}n^2 R\mathcal{V}$ does not depend on the density of energy $\mathscr{E}$ but only on the square of the density of particles n. The internal pressure of the van der Waals equation is exactly of this form.

Transition from weak to strong potential   The effect of the interaction on the pressure depends on the value of the potential. For the potential smaller than the kinetic energy $\mathcal{V} < mv^2$, the interaction decreases the pressure. In the limit $\mathcal{V} \to mv^2$, the pressure goes to zero,

$$\lim_{\mathcal{V} - mv^2 \to 0_-} \mathscr{P} = 0, \tag{2.18}$$

because τ_C becomes infinite. At $\mathcal{V} = mv^2$, the potential dependence of the pressure is non-analytic. From the zero value (2.18), the pressure suddenly jumps to the hard-rod value (2.10). For all strong potentials $\mathscr{V} > mv^2$ the pressure is given by the hard-rod value, i.e. in this region it is independent of the actual value of the potential.

The sharp transition from weak to strong potential regimes follows from non-realistic assumptions of the box-like potential and equal velocities; nevertheless, this discussion shows that the strong and weak forces act principally in a different manner. The strong repulsive forces lead to the inaccessible volume since they simply do not allow two particles to be closer than the sum of their radii. The weak forces modify velocities of particles during their encounter leading to the internal pressure.

2.1.3   Summary

The naive derivation of the van der Waals equation of state shows that the motion of particles is responsible for the basic structure of the equation of state and forces appear only via corrections. The free motion is the dominant and explains why all gases behave identically when sufficiently diluted. On the other hand, the force contribution is the interesting part if we want to learn about molecules composing the gas. The internal pressure and the excluded volume can be used to infer the interaction potential from thermodynamic measurements. For this model, the range R of the interaction is readily obtained from the excluded volume at low temperatures when the thermal velocity does not allow the particles to penetrate each other, and the strength $\mathscr{V}$ is found at high temperatures from the mean-field correction $\sim R\mathscr{V}$. So finally one obtains a relation reminding us of the familiar van der Waals equation of state,

$$\left(\mathscr{P} + a\frac{N^2}{\Omega^2}\right)(\Omega - Nb) = 2\mathscr{E}. \tag{2.19}$$

The (mean-field) corrections to drift result in the internal pressure $a\frac{N^2}{\Omega^2}$. The nonlocal correction to the collisions results in the effect of the molecular volumes Nb. The importance of these two corrections depends on the range of the interaction. The weak interaction dominantly contributes to the drift. The strong interaction leads to nonlocal collisions. Realistic interactions are strong on short distances and weak on long distances. Both groups of corrections are thus needed.

2.2   Random versus deterministic

So far we have assumed that velocities and coordinates of individual particles are known. In reality it is not possible and even not desirable to characterise any many-body system by so many numbers as there are coordinates and velocities of all its particles. From experience we know that in equilibrium all systems of the same density of particles and energy behave so similar that the actual shape of the container and also the actual set of coordinates and velocities are irrelevant. It is sufficient to know the total energy E, the volume Ω and the number of particles N.

The fact that actual coordinates and velocities of particles do not matter can be utilised in two ways. The first one is used within numerical simulations. The system is represented by some randomly selected initial condition that complies with the prescribed total energy. Solving the trajectories of particles numerically with this initial condition one can evaluate e.g. the force on the surface. Although this force fluctuates differently for different initial conditions, its mean value is always the same.

The second way to benefit from the low sensitivity of the system to the actual initial condition is to average it over all possible initial conditions. The averaging wipes out most of the dynamics in the system. For example in equilibrium, any physical quantity averaged over the full set of initial conditions remains time independent. Of course, instead of exact predictions one is able to obtain only the most probable values of observables which are just the mean values independent of initial conditions.

2.2.1 Chaos

The equivalence of different initial conditions is limited to observables which include large numbers of single-particle events. Easily observable physical effects which do not meet this condition are rather rare but exist. A simple example is the Brownian motion of pollen seeds. Each actual initial condition results in a different trajectory of pollen seed.

It is reasonable to assume that each particle in the system moves along a zigzag path as shown in Fig. 2.2. The abrupt changes of direction are caused by collisions. Let us try to understand to what extent the initial condition influences the actual behaviour of the system. The sensitivity of the system to the initial condition can be demonstrated on two particles scattering at three Lenard–Jones impurities with minimal initial displacement shown in Fig. 2.2. The particle is scattered by the walls of the container and by the impurities. The numerical solution shows that two close initial conditions lead after a

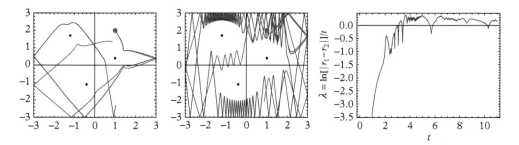

Figure 2.2: *The trajectories of two 2D particles displaced initially by 0.01Å interacting with Lennard–Jones-type impurities (Green, 1952), $V(r) = -\frac{a}{r^6} + \frac{b}{r^{12}}$ with $a = 0.63$ eVÅ^6 and $b = 220$ eVÅ^{12} fitted to helium. The weak attractive long-range potential bends trajectories towards the impurity; the strong repulsive short-range potential bounces particles off in a rather abrupt way. In spite of nearly identical initial conditions, the trajectories become soon very dissimilar. Right: Lyapunov exponent (2.20) for the infinitely mirrored middle figure.*

short time to two very different positions and velocities of the particle. This can be measured by an exponential divergence of the difference in position as

$$\lambda = \frac{1}{t} \log |r_1(t) - r_2(t)| \tag{2.20}$$

called the Lyapunov exponent plotted in Fig 2.2. In other words, two close initial conditions diverge quickly into two well separated ones.

The divergence of the initial conditions is so strong that one can vaguely say that the system can reach any state from any initial condition. If an ensemble of particles starts with initial conditions from a small interval $(x, x + dx) \times (y, y + dx) \times (\phi, \phi + d\phi)$, where ϕ denotes a direction of velocity, they will soon cover the whole space with all possible directions of velocities. For this reason, the averaging over a small interval of initial conditions results in the same observables as the averaging over all initial conditions. The averaging over small intervals is necessary at least because of experimental errors. The averaging over all initial conditions is a powerful theoretical tool. In equilibrium the averaging over time agrees with the ensemble averaging.

It is clear that with such a high sensitivity to the initial condition we cannot solve the actual trajectory of the system, because any small error, either in establishing the initial condition or in numerical treatment of the time evolution, necessarily leads to a completely different trajectory of the particle. Accordingly, it is more reasonable to deal with conserving quantities and to treat the non-conserving degrees of freedom as random. One can see that chaotic evolution leads to a basic paradox of the many-body physics: the evolution of the system is so sensitive to the initial condition of non-conserving degrees of freedom that their actual initial values do not matter.

2.2.2   Velocity distribution

The interaction of particles in the three-dimensional (3D) system results in a randomisation of the single-particle energies. If we assume that the interaction has no other effect than to randomise the positions and velocities of particles, we can treat the system as a system of non-interacting particles; however, averaged over all possible states.

N-particle distribution in a 1D system   To average over all states, assume N particles with the total energy E. Since all states have equal statistical weights, a probability to find the system in one particular state is given only by the restriction that this state corresponds to the energy E,

$$F_N(r_1, \ldots, r_N, v_1, \ldots, v_N) = A_N \delta \left(E - \frac{1}{2} m \sum_{i=1}^{N} v_i^2 \right). \tag{2.21}$$

The norm

$$A_N = (N-1)! \frac{1}{3\Omega^N} \Gamma \left(\frac{3}{2} N + 1 \right) m \left(\frac{2\pi E}{m} \right)^{1 - \frac{3}{2} N} \tag{2.22}$$

is chosen so that the sum over all states is unity[1]

$$\frac{1}{N!} \int dr_1 \ldots dr_N dv_1 \ldots dv_N \, F_N = 1. \tag{2.23}$$

To obtain the analogue for N' particles in 3D system, one takes $N = 3N'$ so that the number of degrees of freedom is the same.

Single-particle distribution Although the N-particle distribution is as simple as the δ function, it escapes our intuition as any N-dimensional function would. Let us take a look at an intuitively simpler function $f(r, v)$, the probability to find at a point r a particle of velocity v. The function f is the local distribution of velocities known as the Boltzmann distribution.

To evaluate the probability that a single particle is at r with v while positions and velocities of the other particles can be any, we fix the 1-th, the 2-nd, $\ldots$, N-th, argument of the N-particle distribution, $r_i = r$ and $v_i = v$, and integrate over the others. Since

---

[1] From (2.21) the norm follows as

$$A_N^{-1} = \frac{1}{N!} \int dr_1 \ldots dr_N dv_1 \ldots dv_N \, \delta \left(E - \frac{1}{2} m \sum_{i=1}^{N} v_i^2 \right).$$

First, we integrate out space variables which yield the factor Ω^N. Secondly, in the velocity space, we substitute into $3N$-dimensional spherical coordinates $w^2 = \sum_{i=1}^{N} v_i^2$,

$$A_N^{-1} = \frac{1}{N!} \Omega^N \int dS_{3N} \int_0^\infty dw w^{3N-1} \delta \left(E - \frac{1}{2} m w^2 \right).$$

The angle element in the $3N$-dimensional sphere S_{3N} does not enter the δ function and can be readily integrated out giving $3N$-times the volume of the $3N$-dimensional unitary sphere [formula (4.632) in (Gradshteyn and Ryzhik, 2014)]

$$\int dS_{3N} = 3N \frac{\pi^{\frac{3}{2}N}}{\Gamma \left(\frac{3}{2}N + 1 \right)}.$$

The δ function reduces as

$$\delta \left(E - \frac{1}{2} m w^2 \right) = \frac{1}{mw} \delta \left(w - \sqrt{\frac{2E}{m}} \right),$$

giving the integral over w

$$\int_0^\infty dw w^{3N-1} \delta \left(E - \frac{1}{2} m w^2 \right) = \frac{1}{m} \left(\frac{2E}{m} \right)^{\frac{3}{2}N-1}.$$

The combination of these formulae yields (2.22).

all particles are identical, the contributions of all particles are equal. Therefore the sum over all particles is N-times the contribution of the 1-th one,[2]

$$f(r,v) = \frac{1}{(N-1)!} \int dr_2 \ldots dr_N dv_2 \ldots dv_N F_N(r, r_2, \ldots, r_N, v, v_2, \ldots, v_N). \tag{2.24}$$

The integration is similar to the evaluation of the norm A_N, see Footnote 1. The space integration is $(N-1)$-dimensional giving the factor of Ω^{N-1},

$$f(r,v) = \frac{\Omega^{N-1}}{(N-1)!} A_N \int dv_2 \ldots dv_N \delta \left(E - \frac{1}{2}mv^2 - \frac{1}{2}m \sum_{i=2}^{N} v_i^2 \right). \tag{2.25}$$

The velocity integration is also similar to the norm A_N except that one has to subtract the kinetic energy of the selected particle from the total energy, because only the energy $E - \frac{1}{2}mv^2$ is left for particles $2, \ldots, N$. The velocity integral is a $3(N-1)$-dimensional one. In parallel with the evaluation of the norm A_N, one finds

$$f(r,v) = A_N \frac{3\Omega^{N-1}}{(N-2)!} \frac{1}{m} \frac{\left(2\pi \left(\frac{E}{m} - \frac{1}{2}v^2 \right) \right)^{\frac{3}{2}N - \frac{5}{2}}}{\Gamma \left(\frac{3}{2}N - \frac{1}{2} \right)}.$$

---

[2] In a more careful formulation one can introduce the probability P to find any of the particles in the interval

$(r, r + dr) \times (v, v + dv) \equiv (x, x + dx) \times (y, y + dy) \times (z, z + dz) \times (v_x, v_x + dv_x) \times (v_y, v_y + dv_y) \times (v_z, v_z + dv_z)$.

For infinitesimal dr and dv, this probability is proportional to $drdv$ so that $P = f(r,v)drdv$.

To evaluate the probability P, we introduce a condition that the i-th particle is from the interval $(r, r + dr)$ by function $\xi_r(r - r_i) = \theta(r_i - r)\theta(r + dr - r_i)$. The function ξ_r equals unity if the i-th particle is in the interval and zero otherwise. A similar function $\xi_v(v - v_i) = \theta(v_i - v)\theta(v + dv - v_i)$ is used for velocities. The probability of finding some of the particles in the given phase-space interval then reads

$$P = \frac{1}{N!} \int dr_1 \ldots dr_N dv_1 \ldots dv_N F_N \sum_{i=1}^{N} \xi_r(r - r_i)\xi_v(v - v_i).$$

The velocity distribution $f = \frac{P}{drdv}$ results sending $dr \to 0$ and $dv \to 0$. Under this limit,

$$\frac{1}{dr}\xi_r(r - r_i) \to \delta(r - r_i) \quad \text{and} \quad \frac{1}{dv}\xi_v(v - v_i) \to \delta(v - v_i),$$

so that

$$f(r,v) = \lim_{drdv \to 0} \frac{P}{drdv} = \frac{1}{N!} \int dr_1 \ldots dr_N dv_1 \ldots dv_N F_N \sum_{i=1}^{N} \delta(r - r_i)\delta(v - v_i)$$

$$= \frac{1}{(N-1)!} \int dr_1 \ldots dr_N dv_1 \ldots dv_N F_N \delta(r - r_1)\delta(v - v_1)$$

which is equivalent to (2.24).

After substitution of the norm A_N from (2.22), we get

$$f(r,v) = \frac{N-1}{\Omega} \left(\frac{m}{2\pi E}\right)^{\frac{3}{2}} \frac{\Gamma\left(\frac{3}{2}N+1\right)}{\Gamma\left(\frac{3}{2}N-\frac{1}{2}\right)} \left(1 - \frac{mv^2}{2E}\right)^{\frac{3}{2}N-\frac{5}{2}}. \tag{2.26}$$

Maxwell distribution For large numbers of particles, $N \to \infty$, the single particle distribution simplifies to Maxwell's distribution. Not to change thermodynamic properties of the system, we have to limit the large number of particles keeping the particle density n and the energy ϵ per particle fixed, i.e. $N, \Omega, E \to \infty$ while $\frac{N}{\Omega} = n$ and $\frac{E}{N} = \epsilon$ are constant. From (2.26) the exponential form of Maxwell's distribution of velocities then results,[3]

$$\lim_{N\to\infty} f(r,v) = n \left(\frac{3m}{4\pi\epsilon}\right)^{\frac{3}{2}} \exp\left(-\frac{3mv^2}{4\epsilon}\right). \tag{2.27}$$

Maxwell's distribution of velocities starts to form for a surprisingly small number of particles. In Fig. 2.3 we show single-particle velocity distributions for systems composed of three, six and an infinite number of particles. At velocities corresponding to the energy per particle, already for the two-particle system, the distribution has the character of a hat. For velocities far exceeding the energy per particle, however, one has to be careful about the number of particles in game; the higher the velocity, the slower the convergence towards Maxwell's distribution. Nevertheless, in the majority of many-body systems, the limit of an infinite number of particles is safely achieved.

---

[3] From Stirling's formula for the Γ function, $\ln \Gamma(z) \approx z \ln z - z - \frac{1}{2}\ln z + \ln\sqrt{2\pi} + o(z^{-2})$, see formula (8.343) in (Gradshteyn and Ryzhik, 2014), we obtain

$$\ln \frac{\Gamma\left(\frac{3}{2}N+1\right)}{\Gamma\left(\frac{3}{2}N-\frac{1}{2}\right)} \approx \left(\frac{3}{2}N+1\right)\ln\left(\frac{3}{2}N+1\right) - \left(\frac{3}{2}N+1\right) - \frac{1}{2}\ln\left(\frac{3}{2}N+1\right)$$

$$- \left(\frac{3}{2}N-\frac{1}{2}\right)\ln\left(\frac{3}{2}N-\frac{1}{2}\right) + \left(\frac{3}{2}N-\frac{1}{2}\right) + \frac{1}{2}\ln\left(\frac{3}{2}-\frac{1}{2}\right)$$

$$\approx \frac{3}{2}\ln\frac{3N}{2}.$$

The Γ factors in the distribution (2.26) thus yield

$$\frac{\Gamma\left(\frac{3}{2}N+1\right)}{\Gamma\left(\frac{3}{2}N-\frac{1}{2}\right)} \approx \left(\frac{3N}{2}\right)^{\frac{3}{2}}.$$

The Gaussian function results from the limit

$$\lim_{N\to\infty} \left(1 - \frac{2}{N}\frac{mv^2}{4\epsilon}\right)^{\frac{3}{2}N-\frac{5}{2}} = \exp\left(-\frac{3mv^2}{4\epsilon}\right).$$

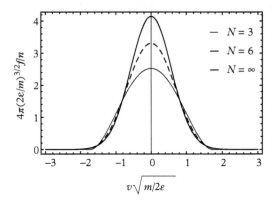

Figure 2.3: *Dimensionless velocity distribution (2.26) for 3, 6 and ∞ particles in the 3D space. With an increasing number of particles the distribution is taller and thinner in the central part and has longer tails. All distributions correspond to the same density and energy per particle.*

Pressure   Having the single-particle distribution one can use probabilistic arguments to evaluate various observables. To connect with the previous discussion we evaluate the pressure.

The force dF exerted by particles on a small area dS of the surface of the container is given by the difference between the momentum flux which particles carry towards the surface and the momentum flux they carry away,

$$dF = \sum_{v_\perp > 0} mv_\perp \frac{dn^{(+)}(v_\perp)}{dt} - \sum_{v_\perp < 0} mv_\perp \frac{dn^{(-)}(v_\perp)}{dt}. \tag{2.28}$$

The number $dn^{(\pm)}(v_\perp)$ of particles which hit/leave the surface during time dt with the perpendicular velocity $v_\perp$ is the sum of the velocity distribution over the parallel velocity,

$$dn^{(\pm)}(v_\perp) = |v_\perp| dt dS \int dv_\parallel f(v). \tag{2.29}$$

The factor $|v_\perp| dt dS$ appears because during dt all particles from this volume cross the surface. Substituting (2.29) into (2.28) the pressure results as

$$\mathcal{P} = \frac{dF}{dS} = \int dv m v_\perp^2 f(v). \tag{2.30}$$

In the 1D system, where $v \equiv v_\perp$, the left-hand side of (2.30) equals twice the density of energy, $\int dv m v_\perp^2 f(v) = 2\mathcal{E}$. Formula (2.30) is thus identical with (2.4). In the 3D system, we can use Maxwell's distribution to obtain the (3D) Bernoulli's formula

$$\mathscr{P} = \int dv m v_{\perp}^2 \, n \left(\frac{3m}{4\pi\epsilon} \right)^{\frac{3}{2}} \exp\left(-\frac{3mv^2}{4\epsilon} \right) = n\frac{2}{3}\epsilon = \frac{2}{3}\mathscr{E}, \qquad (2.31)$$

with the density of energy

$$\mathscr{E} = \int dv \frac{1}{2} m v^2 f(v) = n\epsilon. \qquad (2.32)$$

Temperature   In accordance with our assumption that the interaction of particles only randomises the distribution, the pressure (2.31) has no force corrections; i.e., it corresponds to the equation of state of the ideal gas,

$$\mathscr{P}\Omega = N k_B T, \qquad (2.33)$$

where k_B is Boltzmann's constant and T is a temperature. Comparing with (2.31) we find the energy per particle,

$$\epsilon = \frac{3}{2} k_B T. \qquad (2.34)$$

By this relation, the distribution function f introduced to describe chaotic motion of molecules (2.27) becomes parametrised with a purely thermodynamic quantity, the temperature T. Now the microscopic and macroscopic worlds are connected.

Summary   In principle, the evolution of the many-body system is covered by Newton's equation, but in reality, the exact state of the system is unpredictable because of chaotic evolution of non-conserving degrees of freedom. In the chaotic regime the time evolution scans the phase space in a manner which reminds a random sampling of the whole allowed subspace. Observables thus can be conveniently evaluated as mean values over this allowed subspace.

2.3   Information versus chaos

2.3.1   Entropy as a measure of negative information or disorder

Any randomisation destroys possible information stored in the system in the specific structure of the distribution. Therefore we will measure the dissipation or randomisation by the Shannon measure of information from information theory which is the opposite of the Boltzmann physical entropy. The aim is now to develop a quantitative formula for such measure *entropy S* which is able to count the information stored in the distribution. Amazingly we can derive an explicit mathematical expression for S if we demand only three properties for such a measure:

1. The entropy should be a positive quantity

$$S \geq 0. \tag{2.35}$$

2. S will not judge about the symmetry of distributions or probabilities w_i

$$S(w_1, w_2, w_3, ...) = S(w_2, w_1, w_3, ...) = S(w_3, w_1, w_2, ...) = ... \tag{2.36}$$

3. We divide our distribution into n rough boxes and then further into $m > n$ fine boxes. Then the total entropy should be the entropy of the rough division plus the entropy of the fine division under the condition that we have any given rough division

$$S(w_{nm}) = S(w_n) + \sum_{\bar{n}} w_{\bar{n}} S(w_{m|\bar{n}}). \tag{2.37}$$

Here w_{nm} is the probability to have n and m divisions and $w_{m|\bar{n}}$ is the probability of m under the condition of $\bar{n}$ which occurs with probability $w_{\bar{n}}$.

The usefulness of property 3. is easily seen if we assume that the fine division is independent of the rough division $w_{m,|n} = w_m$. Then the sum in (2.37) about the probability $w_{\bar{n}}$ is unity, as it should be for a probability and one gets simply the sum of the entropies of both divisions $S(w_{nm}) = S(w_n) + S(w_m)$.

Now we hunt for the explicit formula and use the properties 2. and 3. which means that the entropy has to be additive and the sum of an unknown function $s(w)$ of the distribution w_n

$$S(w_n) = \sum_n s(w_n) \tag{2.38}$$

since only in this case the interchangeability of any division n into arbitrary intervals is ensured. One gets from property 3.

$$\sum_{nm} s(w_{nm}) = \sum_n s(w_n) + \sum_{\bar{n}} w_{\bar{n}} \sum_m s(w_{m|\bar{n}}). \tag{2.39}$$

Since one summation occurs on both sides we can choose a specific $\bar{n} = n = 1$ to get

$$\sum_m s(w_1 w_{m|1}) = s(w_1) + w_1 \sum_m s(w_{m|1}) \tag{2.40}$$

where we have employed the formula of Bayes $w_{1m} = w_{m|1} w_1 = w_{1|m} w_m$ as one checks from the definition of the probability to count events 1 or m.

With (2.40) we have arrived at a functional equation for the unknown function s. In fact, the only function which translates an argument of products $w_1 w_{m|1}$ into a sum is the log function. Therefore using the ansatz $s(x) = g(x) \ln x$ in (2.40) leads to

$$\ln w_1 \left[\sum_m g(w_1 w_{m|1}) - g(w_1) \right] + \sum_m \ln w_{m|1} \left[g(w_1 w_{m|1}) - w_1 g(w_{m|1}) \right] = 0. \quad (2.41)$$

Both parts vanish independently if we choose $g(x) = x$. Therefore we have reached our goal and have derived an explicit formula for the entropy obeying the properties 1.–3.:

$$S = -k_B \sum_n w_n \ln w_n. \quad (2.42)$$

Here the minus sign is chosen according to property 1. and the constant k_B turns out to be the Boltzmann constant as the comparison with thermodynamic quantities shows in section 2.3.2.

2.3.2 Maximum entropy and equilibrium thermodynamics

As we will see on each level of kinetic theory throughout this book, the entropy is a quantity which will increase with time. This second law of thermodynamics is one of the most fundamental laws of thermodynamics. In fact the maximum of the entropy (2.42) leads to the different known equilibrium distributions w_n and the thermodynamic quantities.

For micro-canonical ensembles we assume no further restrictions and consider a closed system. Then the maximum of (2.42) is easily seen form $\delta S/\delta w_n = 0$ to be $w_n = 1/N$ which means the equi-partitioning of N events where we used the normalisation

$$1 = \sum_1^N w_n. \quad (2.43)$$

If we allow energy fluctuations between the energy portions ϵ_n in each division and demand constant energy on the ensemble average

$$\sum_1^N w_n \epsilon_n - E = 0, \quad (2.44)$$

we have to add this boundary condition to (2.42) with a Lagrange multiplier β. Rendering this expression extremal one gets the canonical distribution

$$w_n = \frac{e^{-\beta \epsilon_n}}{z_{can}}; \quad z_{can} = \sum_m e^{-\beta \epsilon_m} \quad (2.45)$$

where we have used the normalisation (2.43). The multiplier β turns out to be the inverse temperature $\beta = 1/k_B T$ as one sees from the comparison with the first law of thermodynamics $dE = TdS$ for constant volume and particle number. In fact calculating explicitly the entropy (2.42) with (2.45) one has $S = \beta k_B E + k_B \ln z$ which leads to the desired identification. Furthermore since the free energy is $F = E - TS$ one has direct access to the free energy by calculating the canonical statistical sum $F = -k_B T \ln z_{\text{can}}$.

If one allows density fluctuations between the occupation numbers n_n and demands that the total number conserves,

$$\sum_1^N n_n \epsilon_n - N = 0, \tag{2.46}$$

one has to add this boundary with an additional Lagrange multiplier α and obtains the grand canonical ensemble. By comparison to the first law of thermodynamics, this additional Lagrange multiplier $\alpha = -\beta\mu$ is identified with the chemical potential μ and one has

$$w_n = \frac{e^{-\beta(\epsilon_n - \mu n_n)}}{z_g}; \quad z_g = \sum_m e^{-\beta(\epsilon_m - \mu n_n)} \tag{2.47}$$

and direct access to the Gibbs potential $PV = k_B T \ln z_g$.

This illustrates the importance of the explicit formula for the entropy as the measure of (negative) information or disorder in the system. The kinetic theory as the description of time-dependent distributions will proove the second law of thermodynamics that the entropy (2.42) is a quantity which increase with time due to collisions or dissipation. Consequently, in equilibrium, this provides an extremal principle which allows us to access thermodynamic quantities by microscopic parameters.

2.4   Collisions versus drift

Let us return to Fig. 2.2 which shows a zigzag path of a particle. The path is composed of slightly bent straight lines and sharp curvatures. It is convenient to assume that the collisions are due to centres of fixed positions, such as impurities in semiconductors. The two distinct components of the path follow from the properties of the impurity potential which are strong at short distances and weak at large distances.

The curvature, i.e., the encounter of a large deflection angle, appears only for nearly central events, when the strong part of the potential affects the particle. We will call such an event a collision. In a highly non-central encounter, the particle feels the weak tail of the potential without the strong part. On a short scale, the weak potential of the tails only slightly deviates from a straight motion. These deviations, however, accumulate so that along a longer trajectory the particle slowly changes its velocity or momentum. We will describe the encountering of a low deflection angle as a momentum drift. The momentum drift has to be distinguished from the space drift given by the free motion of particles. Both components together, space and momentum, we call briefly the drift.

2.4.1 Collisions

In Fig. 2.2 one can see that the essential part of the collision happens on the scale which is small compared to the mean distance between impurities. Since the strong potential of the closest impurity dominates, it is possible to solve this short part of the trajectory by neglecting the other impurities. The collision is then approximated by a collision on an isolated impurity.

The dependence of the deflection angle on the impact factor for a Lennard–Jones potential is demonstrated in Fig. 2.4. As expected, large deflection angles appear for impact factors comparable with an atomic radius R. The collision is overly sensitive to an initial condition: nearly the whole sphere of finite directions results from target locations in a region of area $\sim R^2$.

The over-sensitivity of collisions allows us to take the target location as a random quantity. Let us assume that a particle of velocity v somewhere in a cylindrical tube of section $2\pi b^2$ approaches an impurity sitting in the centre of the tube. The target location is given by orientation angle φ of the collision plane and the impact factor b. Accordingly, the particle is in cell $(\varphi, \varphi + d\varphi) \times (b, b + db)$ with the probability $\frac{bd\varphi db}{2\pi b_{cut}^2}$.

From the target location b and φ, the particle is scattered into the final velocity w. Due to the energy conservation, the magnitude of velocity does not change, $|w| = |v|$. Due to the angular momentum conservation, the particle remains in the same collision plane (we assume spherical potential), i.e., φ conserves. The deflection angle θ depends on the impact factor b. The probability $p_{vw}d\varphi \sin\theta d\theta$ that the particle will end up with velocity in the direction cell $(\varphi, \varphi + d\varphi) \times (\theta, \theta + d\theta)$ equals the probability that it has started the collision from the corresponding cell of target locations,

$$p_{vw}d\varphi \sin\theta d\theta = \frac{bd\varphi db}{2\pi b_{cut}^2}. \tag{2.48}$$

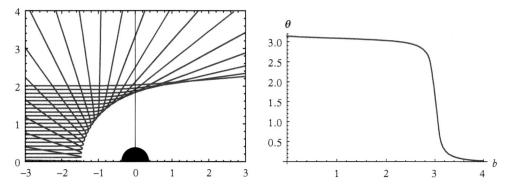

Figure 2.4: *Classical trajectories of a particle with different impact parameter and same velocity (left) and the corresponding deflection angle as a function of the impact factor (right). The interaction potential is of Lennard–Jones type as in Figure 2.2.*

Accordingly,

$$p_{vw} = \frac{b\,db}{\sin\theta\,d\theta}\frac{1}{2\pi b_{\text{cut}}^2}. \tag{2.49}$$

This probability can be derived from the dependence of the deflection angle on the impact factor presented in Fig. 2.4.

Boltzmann equation If one neglects low-angle encounters, the concept of random target location can be used to derive a kinetic equation for a single-particle distribution $f(r, v, t)$. Let us assume a small space and velocity cell $d\Omega_r \times d\Omega_v$ around the point (r, v). The number of particles in this cell is $dN = f(r, v, t)d\Omega_r d\Omega_v$. This number changes with time due to particles which cross the space border of the cell running with velocity v, and due to particles which cross the velocity border being accelerated by the potential of some impurity.

We take a cylindrical space cell, $d\Omega_r = dldS$, with the rounded side of area dS perpendicular to velocity v, i.e. the dl follows the direction $\frac{v}{|v|}$. During time dt, all particles of velocity v which are in the layer of width $|v|dt$ attached to the cylinder from the upwind side enter the cylinder. Their number is

$$dN_{\text{enter}} = |v|dt dSd\Omega_v f\left(r - \frac{1}{2}\frac{v}{|v|}dl, v, t\right). \tag{2.50}$$

On the contrary, during dt, all particles of velocity v in the layer of width $|v|dt$ inside the cylinder on the downwind side leave it. Their number is

$$dN_{\text{leave}} = |v|dt dSd\Omega_v f\left(r + \frac{1}{2}\frac{v}{|v|}dl, v, t\right). \tag{2.51}$$

The acceleration by impurity potential is described as a collision. During dt, each particle will scan the volume $\pi b_{\text{cut}}^2 |v|dt$ reaching some impurity with a probability $\pi b_{\text{cut}}^2 |v|dt\frac{dN_{\text{imp}}}{d\Omega_r}$, where dN_{imp} is the number of impurities in the volume $d\Omega_r$. The probability that during dt some of particles lose their velocity v is

$$dN_{\text{out}} = dN\pi b_{\text{cut}}^2 |v|dt\frac{dN_{\text{imp}}}{d\Omega_r} = f(r, v, t)d\Omega_r d\Omega_v \pi b_{\text{cut}}^2 |v|dt\frac{dN_{\text{imp}}}{d\Omega_r}. \tag{2.52}$$

On the contrary, some particles of velocity w suffer a collision after which they end up in the cell $d\Omega_v$ around v. The number of particles of velocity $|w| = |v|$ scattered into the cell $d\Omega_v$ is

$$dN_{\text{in}} = \int_{d\Omega_v} dv \times \int_0^{2\pi} d\varphi \int_0^{\pi} \sin\theta\,d\theta\,p_{wv} \times \pi b_{\text{cut}}^2 |w|dt\frac{dN_{\text{imp}}}{d\Omega_r} \times d\Omega_r f(r, w, t). \tag{2.53}$$

The symbols × separate components of the formula. The last term counts particles of velocity w. The last but one term specifies which particles undergo some collision. The second term selects collisions which end up with the velocity v. The first term sums over all velocities in an allowed velocity cell. After substitution from (2.49) we obtain

$$dN_{\text{in}} = d\Omega_v d\Omega_r dt \int_0^{2\pi} d\varphi \int_0^{\pi} d\theta \sin\theta P_{wv} f(r, w, t), \qquad (2.54)$$

where we have denoted the transition probability

$$P_{wv} = \frac{dN_{\text{imp}}}{d\Omega_r} |v| \frac{bdb}{\sin\theta \, d\theta}. \qquad (2.55)$$

In terms of the transition probability one can also express the number of particles which leave the cell

$$dN_{\text{out}} = d\Omega_v d\Omega dt \int_0^{2\pi} d\varphi \int_0^{\pi} \sin\theta \, d\theta P_{vw} f(r, v, t), \qquad (2.56)$$

where

$$\int_0^{2\pi} d\varphi \int_0^{\pi} \sin\theta \, d\theta P_{vw} = \int_0^{2\pi} d\varphi \int_0^{\pi} \sin\theta \, d\theta c|v| \frac{bdb}{\sin\theta \, d\theta} = c|v|2\pi \int_0^b bdb = c|v|\pi b^2. \qquad (2.57)$$

Summing all contributions, $dN = dN_{\text{enter}} - dN_{\text{leave}} + dN_{\text{in}} - dN_{\text{out}}$, one finds the time evolution of the distribution. The time derivative results from $\frac{dN}{dtd\Omega d\Omega_v} = \frac{\partial f}{\partial t}$. The space drift is obtained from (2.50) and (2.51) as $\frac{dN_{\text{enter}} - dN_{\text{leave}}}{dtd\Omega d\Omega_v} = v\frac{\partial f}{\partial r}$. The scattering integral results from (2.54) and (2.56) into $\frac{dN_{\text{in}} - dN_{\text{out}}}{dtd\Omega d\Omega_v} = \sum_w P_{wv} f(r, w, t) - \sum_w P_{vw} f(r, v, t)$ and the final form is the kinetic equation,

$$\frac{\partial f}{\partial t} + v\frac{\partial f}{\partial r} = \sum_w P_{wv} f(r, w, t) - \sum_w P_{vw} f(r, v, t), \qquad (2.58)$$

known as the Boltzmann equation. Here $\sum_w \equiv \int_0^{2\pi} d\varphi \int_0^{\pi} \sin\theta \, d\theta$. The left-hand side covers a deterministic space drift, the right-hand side covers a random dissipation.

2.4.2 Drift

Deriving the collision integral we have neglected events with the impact factor larger than b_{cut}, i.e., with the deflection angle smaller than corresponding to θ_{cut}. Formally it seems that low-angle encounters can be included by a straightforward limit $\theta_{\text{cut}} \to 0$.

This approach fails because the actual choice of θ_{cut} determines how many collisions a particle suffers. A mean time τ of straight flight is inversely proportional to the cross section πb^2,

$$\frac{1}{\tau} = \sum_w P_{vw} = c|v|\pi b_{cut}^2, \tag{2.59}$$

therefore with increasing b_{cut} we soon meet a regime in which a particle should scatter again before it leaves the interaction potential of a previous scatterer. To keep binary collisions as independent sequential events, one has to cut collisions at some reasonably selected angle and treat the low-angle encounters in a different manner.

Effect of potential tails on velocity   Let us assume a particle between collisions. If it would be completely free of interactions, its velocity is v. Since it passes through tails of impurity potentials, its motion,

$$m\ddot{r} = -\frac{\partial \phi}{\partial r}, \tag{2.60}$$

is driven by a force from a sum over all potential tails

$$\phi(r) = \sideset{}{'}\sum_i V(r - r_i) \tag{2.61}$$

of impurities placed at positions r_i. The prime on the sum denotes that no impurity is closer than b_{cut}.

Some of impurities push the particle to the left and others to the right. Many of these low-angle deflections compensate after a large number of events. If the impurity concentration, however, has a gradient, the particle is more strongly pushed from the side with the higher concentration. In this case, a distortion of the trajectory accumulates which becomes apparent for long trajectories. These distortions can be described by a mean value of potentials affecting the particle, i.e., approximating the actual potential ϕ by a mean field

$$\bar{\phi} = \int_{|r-y|>b_{cut}} dy c(y) V(r - y). \tag{2.62}$$

Neglecting the short-range deviations from the locally 'straight' trajectory, one can speak only about a mean coordinate $\bar{r}$ and mean velocity $\bar{v} = \frac{d\bar{r}}{dt}$. From an effective Newton equation,

$$m\frac{d\bar{v}}{dt} = -\frac{\partial \bar{\phi}}{\partial r}\bigg|_{\bar{r}}, \tag{2.63}$$

one finds that an effective energy

$$\epsilon = \frac{1}{2}m\bar{v}^2 + \bar{\phi}, \tag{2.64}$$

conserves along the mean trajectory. When the particle enters the impurity-free region, its velocity equals v and the energy is $\epsilon = \frac{1}{2}mv^2$. The mean velocity thus differs from the velocity of a completely isolated particle as

$$|\bar{v}| = \sqrt{v^2 - \frac{2\bar{\phi}}{m}} \approx |v| - \frac{\bar{\phi}}{m|v|}. \tag{2.65}$$

Vlasov equation The mean field is easily incorporated into the kinetic equation. Due to forces caused by the mean field, particles can leave the cell $d\Omega_r \times d\Omega_v$ crossing the velocity border under acceleration $a = -\frac{1}{m}\frac{\partial\bar{\phi}}{\partial r}$. We choose a cubic velocity cell, $d\Omega_v = dv_a dS_v$, with a side of area $dS_v = dv_a dv_a$ perpendicular to a; i.e. the explicit dv_a is the direction of the acceleration $\frac{a}{|a|}$. During dt, all particles from a layer of width $|a|dt$ attached to the cube from the upwind side are driven into the cube, and all particles from a layer of the same width but in the downwind side of cube are driven out of the cube. These two processes change the number of particles as

$$dN_{\text{accel}} = |a| dt dv^2 d\Omega f\left(r, v - \frac{1}{2}\frac{a}{|a|} dv_a, t\right), \tag{2.66}$$

$$dN_{\text{decel}} = |a| dt dv^2 d\Omega f\left(r, v + \frac{1}{2}\frac{a}{|a|} dv_a, t\right). \tag{2.67}$$

Summing all contributions, $dN = dN_{\text{enter}} - dN_{\text{leave}} + dN_{\text{accel}} - dN_{\text{decel}} + dN_{\text{in}} - dN_{\text{out}}$, one finds the kinetic equation,

$$\frac{\partial f}{\partial t} + v\frac{\partial f}{\partial r} - \frac{1}{m}\frac{\partial\bar{\phi}}{\partial r}\frac{\partial f}{\partial v} = \sum_w P_{wv}f(r, w, t) - \sum_w P_{vw}f(r, v, t), \tag{2.68}$$

known as the Vlasov equation (although more common is the Vlasov equation with the collision term neglected).

2.4.3 Equivalence of low-angle collisions and momentum drift

At first glance, the Vlasov equation differs from the Boltzmann equation (2.58) only via the mean field term, the momentum drift. The difference, however, is much larger. The mean field and the collisions are two different manifestations of forces between particles. The separation of the interaction into these components has been defined artificially with an open parameter θ_{cut}. Such a procedure is allowed only if there is a region in which both components are equivalent. Briefly, we have to show that there is a region of small deflection angles which can be described either via mean fields or via collisions, of course, with the same result. It turns out that a more sophisticated approximation of the scattering integral is necessary.

To demonstrate the equivalence of collisions on small angles with the mean field, let us derive the corresponding kinetic equations for a system of particles which interact via a soft potential of finite range, e.g., the 3D box potential, $V(r) = V\theta(R - |r|)$. Since the potential is soft, $V \ll \frac{1}{2}mv_{\text{th}}^2 = \frac{3}{2}k_B T$, all scattering events are on small deflection angles. Accordingly, all interactions can be described via the mean field. Since the range of the potential is finite, the total cross section is finite, $\sigma = \pi R^2$. We assume that the range of interaction is smaller than a mean distance between impurities, $R \ll c^{-1/3}$, so that the interactions can be treated as rare events. Accordingly, all interactions can be described as collisions. To show this equivalency of both pictures we derive the mean-field-less Boltzmann-type kinetic equation and rearrange it into the collision-less Vlasov equation.

Kinetic equation with nonlocal corrections We have seen in section 2.1.2 that the mean field is connected with the time Δ_t by which the particle is delayed in the collision. To introduce the delay time into the kinetic equation, we have to take into account that a particle scattered into the velocity v at time t has started the collision process at $t - \Delta_t$. The number of particles scattered into the cell $d\Omega_r \times d\Omega_v$ thus reads

$$dN_{\text{in}} = dtd\Omega_r d\Omega_v \sum P_{wv} f(r, w, t - \Delta_t). \tag{2.69}$$

The low-angle collisions happen for rather large values of the impact factor b, see Fig. 2.5. Since we identify the position r of the collision with the point on the trajectory where the particle is closest to the impurity, the impurity is displaced at $r + \Delta r$. The number of scattered particles thus has to be modified as

$$dN_{\text{in}} = dtd\Omega_r d\Omega_v \sum P_{wv}(r + \Delta r) f(r, w, t - \Delta_t), \tag{2.70}$$

$$dN_{\text{out}} = dtd\Omega_r d\Omega_v \sum P_{wv}(r - \Delta r) f(r, v, t). \tag{2.71}$$

With these modifications, the kinetic equation reads

$$\frac{\partial f}{\partial t} + v\frac{\partial f}{\partial r} = \sum_w P_{wv}(r + \Delta r) f(r, w, t - \Delta_t) - \sum_w P_{vw}(r - \Delta r) f(r, v, t). \tag{2.72}$$

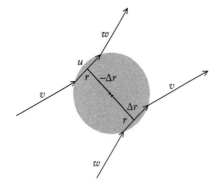

Figure 2.5: *Scattering on the box potential. In the scattering-out (upper trajectory) a particle of initial velocity v is deflected to the velocity w. In the corresponding scattering-in (lower trajectory) a particle goes from w to v. The distance $\pm\Delta r$ of an impurity centre from the point r attributed to the collision event is mutually reversed for -in and -out processes.*

For the soft potential, only small deflection angles and short delay times appear.[4] The free motion of a particle is thus perturbed by two quantities, the deviation of the velocity direction $\Delta v = w - v$ and the collision delay Δ_t. Both these Δ's are small being linear in potential V. The Δr is proportional to the impact factor b which is also small on the scale of gradients in the system.

Expanding (2.72) in Δ's and b, one finds[5]

$$\frac{\partial f}{\partial t} + v\frac{\partial f}{\partial r} = \frac{\partial f}{\partial v}\sum_w \frac{\partial P_{vv}}{\partial r}\Delta r \Delta v - \frac{\partial f}{\partial t}\sum_w \bar{P}_{vv}\Delta_t. \tag{2.73}$$

---

[4] In the collision process on the box-like potential, the particle gains three velocities, v before, u during and w after the collision. Since the collision is symmetric, the change of velocity at matching points, $|r - y| = R$, equal, $u - v = w - u$. (The impurity sits in y.) It is thus sufficient to solve for u as $w = 2u - v$.

When the particle makes a contact, the tangential velocity $v_\parallel = |v|\frac{b}{R}$ conserves, $u_\parallel = v_\parallel$. The normal velocity is given by the energy conservation, $\frac{1}{2}mv_\perp^2 = \frac{1}{2}mu_\perp^2 + V$ as $v_\perp = |v|\sqrt{1 - \frac{b^2}{R^2}}$ and $u_\perp = \sqrt{v_\perp^2 - 2\frac{V}{m}}$. The maximum deflection angle θ_{max} is achieved for the impact factor b_{max} at the edge of allowed penetration, where $u_\perp = 0$. This condition gives

$$1 - \frac{b_{max}^2}{R^2} = \frac{V}{\frac{1}{2}mv^2}, \quad \text{from which} \quad \theta_{max} = 2\arccos\frac{b_{max}}{R} = 2\arcsin\sqrt{\frac{V}{\frac{1}{2}mv^2}} \approx 2\sqrt{\frac{V}{\frac{1}{2}mv^2}}.$$

The collision delay can be estimated in the region of relatively small impact factors from an unperturbed trajectory as

$$\Delta_t = \frac{2\sqrt{R^2 - b^2}}{|u|} - \frac{2\sqrt{R^2 - b^2}}{|v|} \approx \frac{2\sqrt{R^2 - b^2}}{|v|}\frac{V}{mv^2}.$$

The velocity change $\Delta v = w - v = 2(u - v)$ has the dominant contribution in the direction parallel to Δr.

$$\Delta v \approx \frac{\Delta r}{b}\frac{2V}{m|v|}\frac{\frac{b}{R}}{\sqrt{1 - \frac{b^2}{R^2}}} = \frac{\Delta r}{\sqrt{R^2 - b^2}}\frac{2V}{m}.$$

[5] A straightforward expansion of the scattering-in gives

$$\sum_{w=v+\Delta v} P_{wv}(r + \Delta r)f(r, v + \Delta v, t - \Delta_t) = f(r, v, t)\sum_{w=v+\Delta v} P_{wv}(r) + f(r, v, t)\frac{\partial \ln c}{\partial r}\sum_{w=v+\Delta v} P_{wv}(r)\Delta r$$

$$+ \frac{\partial f(r, v, t)}{\partial v}\sum_{w=v+\Delta v} P_{wv}(r)\Delta v - \frac{\partial f(r, v, t)}{\partial t}\sum_{w=v+\Delta v} P_{wv}(r)\Delta_t + \frac{\partial f(r, v, t)}{\partial v}\frac{\partial \ln c}{\partial r}\sum_{w=v+\Delta v} P_{wv}(r)\Delta r \Delta v$$

$$= \frac{f}{\tau} - \frac{\partial f}{\partial t}\sum_{w=v+\Delta v} P_{wv}\Delta_t + \frac{\partial f}{\partial v}\frac{\partial \ln c}{\partial r}\sum_{w=v+\Delta v} P_{wv}\Delta r \Delta v = \frac{f}{\tau} - \frac{\partial f}{\partial t}\sum_{w=v+\Delta v} P_{vv}\Delta_t + \frac{\partial f}{\partial v}\frac{\partial \ln c}{\partial r}\sum_{w=v+\Delta v} P_{vv}\Delta r \Delta v.$$

We have used that P_{wv} depends on the coordinate only via the concentration of impurities. The inverse lifetime is $\frac{1}{\tau} = \sum_w P_{vw}$. Two terms have vanished under the integration over directions: the term $\Delta r = b\sin\varphi$,

$$f\sum_w \left(\frac{\partial P_{wv}}{\partial r} + \frac{\partial P_{vw}}{\partial r}\right)\Delta r = f\frac{\partial \ln c}{\partial r}\sum_w (P_{wv} + P_{vw})\Delta r = f 2\int_0^{2\pi} d\varphi \int_0^\pi d\theta \sin\theta P(|v|, \theta)b\sin\varphi = 0,$$

and from the same reason the term with $\Delta v \propto \sin\varphi$.

A straightforward expansion of the scattering-out gives

$$\sum_{w=v+\Delta v} P_{vw}(r - \Delta r)f(r, v, t) = \frac{f}{\tau}.$$

One can see that the correction terms are small if gradients in time and space are small. The higher powers in Δ's would lead to higher powers in gradients and can be neglected. The linear order is proportional to gradients in the same way as the drift terms on the left-hand side; therefore they cannot be neglected as small in gradients.

Space displacements and direction-changing forces   The space displacement Δr and the velocity change Δv have the same direction which is perpendicular to the velocity v. Accordingly, the gradients are restricted to a plane perpendicular to v which we denote by a projector operator $\mathbf{Q}_{xy} = \delta_{xy} - \frac{v_x v_y}{v^2}$. Clearly, $\mathbf{Q}\Delta r = \Delta r \mathbf{Q} = \Delta r$ and $\mathbf{Q}\Delta v = \Delta v \mathbf{Q} = \Delta v$ while $\mathbf{Q}v = v\mathbf{Q} = 0$. So far we have parametrised the collision by the final velocity w, as it is customary for the scattering integrals. For low-angle collisions when $w \approx v$, this parametrisation becomes clumsy and it becomes more convenient to parametrise the collision by the impact factor b which naturally measures the rate,

$$\sum_w \bar{P}_{wv} = \int_0^{2\pi} d\varphi \int_0^{\pi} d\theta \, \sin\theta \, \bar{P}_{wv} = \int_0^{2\pi} d\varphi \int db \, bc|v|.$$

$$\Delta v \approx \frac{\Delta r}{b} \frac{2V}{m|v|} \frac{\frac{b}{R}}{\sqrt{1 - \frac{b^2}{R^2}}} = \frac{\Delta r}{\sqrt{R^2 - b^2}} \frac{2V}{m}.$$

Then

$$\sum_w \frac{\partial P_{vv}}{\partial r} \Delta r \Delta v = \mathbf{Q}|v| \frac{\partial c}{\partial r} \int_0^R db \, b \int_0^{2\pi} d\varphi b |\Delta v|$$

$$\approx \mathbf{Q} \frac{\partial c}{\partial r} \frac{4\pi V}{m} \int_0^R db \frac{b^3}{\sqrt{R^2 - b^2}} = \mathbf{Q} \frac{1}{m} \frac{\partial c}{\partial r} V \frac{4\pi R^3}{3}. \qquad (2.74)$$

Therefore the effect of the velocity correction is

$$\sum_w \frac{\partial P_{vv}}{\partial r} \Delta r \Delta v = \mathbf{Q} \frac{1}{m} \frac{\partial \bar{\phi}}{\partial r} \qquad (2.75)$$

and we have arrived at a kinetic equation,

$$\left(1 + \sum_w \bar{P}_{vv} \Delta_t\right) \frac{\partial f}{\partial t} + v \frac{\partial f}{\partial r} - \frac{\partial \bar{\phi}}{\partial r} \mathbf{Q} \frac{\partial f}{\partial v} = 0, \qquad (2.76)$$

which reminds us of the Vlasov equation in the absence of a collision. The force perpendicular to v is also the same as in the Vlasov equation. The only peculiar point is that there is no force parallel to v. Instead, we have obtained a strange renormalisation of the time derivative.

Collision delay and energy-changing forces   The renormalisation of the time derivative can be rearranged into the force parallel to the velocity. To this end we evaluate the correction

$$\sum_{w} \bar{P}_{vv} \Delta_t = \int_0^{2\pi} d\varphi \int_0^R db \, bc|v|\Delta_t \approx c \frac{V}{mv^2} \frac{4\pi R^3}{3} = \frac{\bar{\phi}}{mv^2}. \tag{2.77}$$

Now we divide the equation with the renormalisation factor,[6]

$$\frac{\partial f}{\partial t} + \bar{v} \frac{\partial f}{\partial r} - \frac{\partial \bar{\phi}}{\partial r} Q \frac{\partial f}{\partial v} = 0, \tag{2.78}$$

so that the renormalisation moves into the velocity,

$$\bar{v} = \frac{v}{1 + \frac{\bar{\phi}}{mv^2}} \approx v - v \frac{\bar{\phi}}{mv^2}, \tag{2.79}$$

which becomes the mean velocity (2.65). The renormalisation changes the amplitude but not the direction of velocity. Therefore it corresponds to forces parallel to v.

To complete the rearrangement, we substitute the mean velocity instead of the asymptotic velocity v,

$$\bar{f}(r, \bar{v}, t) = f(r, v, t). \tag{2.80}$$

This substitution results into the complete force term[7],

$$\frac{\partial \bar{f}}{\partial t} + \bar{v} \frac{\partial \bar{f}}{\partial r} - \frac{1}{m} \frac{\partial \bar{\phi}}{\partial r} \frac{\partial \bar{f}}{\partial \bar{v}} = 0. \tag{2.81}$$

One can see that $\bar{f}$ obeys the Vlasov kinetic equation. We have thus proved that the low-angle collisions can be rearranged into the mean field.

---

[6] A renormalisation of the perpendicular force term can be neglected being of the second order in potential V.

[7] With substitution (2.79) we have to make an identical substitution of coordinate and time, $\bar{r} = r$ and $\bar{t} = t$. The derivatives transform as

$$\frac{\partial}{\partial t} = \frac{\partial}{\partial \bar{t}} + \frac{\partial |\bar{v}|}{\partial t} \frac{\partial}{\partial |\bar{v}|} = \frac{\partial}{\partial \bar{t}},$$

$$\frac{\partial}{\partial r} = \frac{\partial}{\partial \bar{r}} + \frac{\partial |\bar{v}|}{\partial r} \frac{\partial}{\partial |\bar{v}|} = \frac{\partial}{\partial \bar{r}} - \frac{v}{mv^2} \frac{\partial \bar{\phi}}{\partial r} \frac{\partial}{\partial |\bar{v}|}.$$

We have used that the potential of impurities is stationary, $\frac{\partial \bar{\phi}}{\partial t} = 0$, so that $\frac{\partial \bar{v}}{\partial t} = 0$. The projection of $\bar{v}$ on the derivative in the parallel direction is just $|\bar{v}|$; therefore

$$\bar{v} \frac{v}{mv^2} \frac{\partial \bar{\phi}}{\partial r} \frac{\partial}{\partial \bar{v}} = \frac{1}{m} \frac{\partial \bar{\phi}}{\partial r} (1 - Q) \frac{\partial}{\partial \bar{v}}.$$

Finally, we join both gradient contributions,

$$\frac{\partial \bar{\phi}}{\partial r} Q \frac{\partial \bar{f}}{\partial \bar{v}} + \frac{\partial \bar{\phi}}{\partial r} (1 - Q) \frac{\partial \bar{f}}{\partial \bar{v}} = \frac{\partial \bar{\phi}}{\partial r} \frac{\partial \bar{f}}{\partial \bar{v}}.$$

Energy bottom   The kinetic equation with the nonlocal and non-instant collision integral can be rearranged into the kinetic equation with the mean field for sufficiently high-energy particles. At low energies, however, they differ substantially. Let us assume that the mean potential is repulsive, $\bar{\phi} > 0$. Writing (2.79) in the form of energy conservation,

$$\frac{1}{2}m\bar{v}^2 + \bar{\phi} = \frac{1}{2}mv^2, \tag{2.82}$$

one can see that slow particles, $\frac{1}{2}mv^2 < \bar{\phi}$, correspond to a negative kinetic energy within the mean-field picture. Accordingly, the slow particles are not allowed to penetrate into this region. Within the nonlocal and non-instant picture, no such limitation appears.

For the model assumed, the mean field is an incorrect approximation for low energies and the energy bottom corresponds to $v = 0$. On the other hand, for the Coulomb potential which is truly of long range, there is no space between particles that would be free of the potential of particles around. Accordingly, the mean-field energy bottom is more appropriate to the Coulomb interaction. The optimal separation of the interaction into the mean field and the nonlocal and non-instant collisions thus has to be guided by the actual interaction potential and the density of particles in questions.

2.4.4   Summary

The effects of interaction on the trajectories of particles can be separated into two groups. Firstly, infrequent collisions with a large deflection angle and secondly rather frequent encounters with deflections on small angles. The collisions randomise the trajectory of particles. The low-angle encounters act as a mean field.

These two effects of the interaction naturally appear in the kinetic equation as the mean-field drift and the collision integral. The simplest collision integral of Boltzmann type is local and instant. It describes only the randomisation of directions. In Part IV we discuss the kinetic theory in which collisions are nonlocal and non-instant and thus carry a share of forces.

2.5   Explicit versus hidden forces

Previously we have restricted the deflection angle by an artificial cutoff. In many systematic approaches one never meets the demand to set any cut-offs. In these cases the separation of low- and high-angle collisions happens automatically as a byproduct of binary correlations. Let us demonstrate how binary correlations split the interaction potential into a strong short-range potential core responsible for the high-angle deflections and a weak long-range tail acting like a mean field.

Binary correlation   Naturally, fast particles can penetrate into the repulsive potential deeper than slow ones. The radius of the core thus shrinks with an increasing temperature. Let us evaluate the probability that in a system of N particles interacting by a potential $\mathscr{V}(r-r')$, two particles are found at a distance $|r-r'|$. The N-particle distribution

$$F_N(r_1, \ldots, r_N, v_1, \ldots, v_N) = A_N \delta \left(E - \frac{1}{2}m \sum_{i=1}^{N} v_i^2 - \sum_{i<j}^{N} \mathscr{V}(r_i - r_j) \right), \qquad (2.83)$$

integrated over velocities provides a probability that all N particles are in the prescribed coordinates,

$$n_N(r_1, \ldots, r_N) = \frac{3(2\pi)^{\frac{3}{2}-1} A_N}{(N-1)! \, \Gamma\left(\frac{3}{2}N + 1\right) m^{\frac{3}{2}N}} \left(E - \sum_{i<j}^{N} \mathscr{V}(r_i - r_j) \right)^{\frac{3}{2}N-1}. \qquad (2.84)$$

This is called the N-particle space-correlation function.

To find the probability $n_2(r, r')$ that there are two particles at r and r' with no regards to the coordinates of the others, we associate r with some r_i and r' with some r_j, $j \neq i$, and integrate over the remaining coordinates. There are $N(N-1)$ ways how to select i and j. For identical particles, we can take $i = 1$ and $j = 2$ and multiply them with factor $N(N-1)$; for all other choices,

$$n_2(r, r') = \frac{3(2\pi)^{\frac{3}{2}-1} A_N}{(N-1)! \, \Gamma\left(\frac{3}{2}N + 1\right) m^{\frac{3}{2}N}} N(N-1) \qquad (2.85)$$

$$\times \int dr_3 \ldots dr_N \left(E - \mathscr{V}(r-r') - \sum_{2<i}^{N} \mathscr{V}(r-r_i) - \sum_{2<i}^{N} \mathscr{V}(r'-r_i) - \sum_{2<i<j}^{N} \mathscr{V}(r_i-r_j) \right)^{\frac{3}{2}N-1}.$$

The two-particle probability n_2 cannot be evaluated exactly; however, already a simple approximation provides physically interesting results. If we neglect interactions of three and more particles, the interaction of the remaining particles can be neglected which yields[8]

$$n_2(r, r') = n^2 \exp\left[-\frac{\mathscr{V}(r-r')}{k_B T} \right]. \qquad (2.86)$$

---

[8] From the norm A_N of non-interacting particles (2.22) and the limit of a large number of particles, $N \to \infty$ while, $\frac{N}{\Omega} = n$ and $\frac{E}{N} = \epsilon = \frac{3}{2}k_B T$, one finds

$$n_2(r, r') \approx \frac{3(2\pi)^{\frac{3}{2}-1} A_N}{(N-1)! \, \Gamma\left(\frac{3}{2}N + 1\right) m^{\frac{3}{2}N}} N(N-1) \int dr_3 \ldots dr_N \left(E - \mathscr{V}(r-r') \right)^{\frac{3}{2}N-1}$$

$$= \frac{3(2\pi)^{\frac{3}{2}-1} A_N N \Omega^{N-2}}{(N-2)! \, \Gamma\left(\frac{3}{2}N + 1\right) m^{\frac{3}{2}N}} \left(E - \mathscr{V}(r-r') \right)^{\frac{3}{2}N-1}$$

$$= \frac{N(N-1)}{\Omega^2} \left(1 - \frac{\mathscr{V}(r-r')}{E} \right)^{\frac{3}{2}N-1} = n^2 \exp\left[-\frac{3}{2}\frac{\mathscr{V}(r-r')}{E} \right] = n^2 \exp\left[-\frac{\mathscr{V}(r-r')}{k_B T} \right].$$

As expected, the probability vanishes when the potential exceeds the thermal energy, $n_2 \to 0$ for $\mathscr{V} \gg k_B T$.

2.5.1   Mean field

Here we have derived $n_2(r, r')$ only for a homogeneous system. To extend this formula to slightly inhomogeneous cases, we assume that in a non-interacting system $n_2(r, r') = n(r)n(r')$. The temperature which measures the mutual correlation is centred, $T = T\left(\frac{r+r'}{2}\right)$. For weak gradients, the same result would follow from the algebraic or geometric mean value,

$$T = T\left(\frac{r+r'}{2}\right) = \frac{1}{2}\left(T(r) + T(r')\right) = \sqrt{T(r)T(r')}.$$

We can use the pair-correlation function

$$g(r, r') = \frac{n_2(r, r')}{n(r)n(r')} = \exp\left[-\frac{\mathscr{V}(r-r')}{k_B T\left(\frac{(r+r')}{2}\right)}\right] \tag{2.87}$$

as a filter of the tail part of the potential. A particle feels the potential of the others with the probability of their relative distances; therefore it is natural to define the mean potential as

$$\phi(r) = \int dr'\, n(r')\mathscr{V}(r-r')g(r, r'). \tag{2.88}$$

The binary correlation g naturally eliminates all particles which would be too close to the point r. In the spirit of the decomposition into the tail and core the product with the binary correlation is the tail while the remainder is the core,

$$\mathscr{V}_{\text{tail}}(r-r') = \mathscr{V}(r-r')g(r, r'), \quad \mathscr{V}_{\text{core}}(r, r') = \mathscr{V}(r-r')(1 - g(r, r')). \tag{2.89}$$

The tail and core for He at room temperature are shown in Fig. 2.6. The finite value of the tail at short distances regularises the integration in the mean field (2.88). Without the binary correlation the integral diverges.

Mean field for strong potential   As previously discussed, the tails acting via the mean field lead to effective forces while the dominant effect of the core is to randomise the motion. Besides randomisation, the finite volume of the core results in the excluded volume of the van der Waals equation. The transition between the weak and strong potential limits we demonstrate for the model box-like potential, $\mathscr{V}(r) = \mathscr{V}\theta(R - |r|)$.

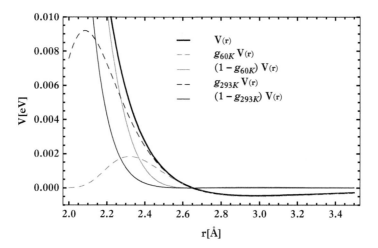

Figure 2.6: *Lennard–Jones potential for He at T = 60K and T = 293K. The tail (dotted) and the core (dashed) are defined via binary correlation (2.89).*

The mean field,

$$\phi = n\mathscr{V} \exp\left[-\frac{\mathscr{V}}{k_B T}\right] \frac{4\pi}{3} R^3,$$ (2.90)

in the weak potential limit $\mathscr{V} \ll k_B T$, approaches an uncorrelated mean field, $\phi \rightarrow n\mathscr{V}\frac{4\pi}{3}R^3$, while in the strong potential limit $\mathscr{V} \gg k_B T$, it vanishes. It shows that the mean field, even in the correlated version, does not cover the excluded volume.[9]

2.5.2 Virial of forces

To describe the excluded volume, we have to evaluate mean forces acting on a particle at r directly,

$$F = -\int dr' n(r') g(r, r') \frac{\partial \mathscr{V}(r - r')}{\partial r}.$$ (2.91)

[9] In plasma physics, it is customary to identify the mean field with its uncorrelated approximation, while all effects of the binary correlations are treated as corrections of a higher order. This choice of the mean field is justified for the Coulomb interaction of electrons and ions because the uncorrelated interaction is usually the dominant part. Moreover, the uncorrelated mean field is identical with the internal or self-consistent field, or Hartree's potential, so that there is a convenient link between the microscopic and phenomenological pictures.

The concept of a correlated mean field is necessary for Lennard–Jones' potentials. The core diverges so strongly that an uncorrelated mean field is not defined,

$$\int dr' \mathscr{V}(r - r') = \int_0^\infty d|r| \left(-\frac{a}{r^6} + \frac{b}{r^{12}}\right) = \lim_{\eta \to 0} \int_\eta^\infty d|r| \left(-\frac{a}{r^6} + \frac{b}{r^{12}}\right) = \infty.$$

For a homogeneous system, forces from the left side compensate forces from the right side; accordingly, the non-zero mean force appears only with a non-zero gradient of the density. Expanding in $r - r'$ around the point r we find[10]

$$F = \frac{1}{2}\frac{\partial nA}{\partial r} + A\frac{1}{2}\frac{\partial n}{\partial r} \qquad A = k_B T\frac{4\pi}{3}R^3\left(1 - \exp\left[-\frac{\mathcal{V}}{k_B T}\right]\right). \qquad (2.92)$$

In the weak potential limit, the mean force agrees with the force found from the (uncorrelated) mean field. In the strong potential limit, however, the mean force does not vanish, while the mean field does,

$$F = \mathcal{V}\frac{4\pi}{3}R^3\frac{\partial n}{\partial r} \qquad\qquad \text{for} \qquad \mathcal{V} \ll k_B T$$

$$\text{(2.93)}$$

$$F = k_B T\frac{4\pi}{3}R^3\frac{\partial n}{\partial r} + n\frac{2\pi}{3}R^3 k_B\frac{\partial T}{\partial r} \qquad \text{for} \qquad \mathcal{V} \gg k_B T$$

This difference follows from the fact that the effective force found from the mean field does not include forces from the collisions but only forces from the low-angle encounters. The mean force includes both.

Pressure The mean force is directly linked to a pressure. Assume a small volume $d\Omega_r = dxdS$ which has $dN = n(r)dxdS$ particles. Particles from the vicinity of $d\Omega_r$ act on each of the particles in this volume by the mean force (2.92); therefore the sum force on the element $d\Omega_r$ is

$$dF = dN\,F = dxdS\frac{1}{2}\frac{\partial}{\partial r_x}n^2 k_B T\frac{4\pi}{3}R^3\left(1 - \exp\left[-\frac{\mathcal{V}}{k_B T}\right]\right). \qquad (2.94)$$

The mean force adds to the momentum transferred by the free motion. During dt, the free motion of particles through area dS on the left/right sides brings momentum $dk^{(\pm)} = dS\,dt\int dv v_x k_x\,f(r \pm \frac{1}{2}dx, v, t)$; therefore the inertial force reads

---

[10] With explicit vector components x and y,

$$F_x = -\int d\bar{r}\,n(r + \bar{r})g(r, r + \bar{r})\frac{\partial\mathcal{V}(\bar{r})}{\partial\bar{r}_x} = -\int d\bar{r}\left(ng + g\sum_{y=1}^{3}\frac{\partial n}{\partial r_y}\bar{r}_y + n\frac{\partial g}{\partial T}\sum_{y=1}^{3}\frac{\partial T}{\partial r_y}\frac{\bar{r}_y}{2}\right)\frac{\partial\mathcal{V}}{\partial\bar{r}_x}$$

$$= -\frac{1}{2}\int d\bar{r}\bar{r}_x\frac{\partial\mathcal{V}}{\partial\bar{r}_x}\left(\frac{\partial ng}{\partial r_x} + g\frac{\partial n}{\partial r_x}\right) = \frac{1}{2}\frac{\partial nA}{\partial r_x} + A\frac{1}{2}\frac{\partial n}{\partial r_x},$$

where A denotes

$$A = -\int d\bar{r}g\bar{r}_x\frac{\partial\mathcal{V}}{\partial\bar{r}_x} = -k_B T\int d\bar{r}\bar{r}_x\frac{\partial}{\partial\bar{r}_x}(1 - g) = k_B T\int d\bar{r}(1 - g) = 4\pi k_B T\int_0^{\infty}d|\bar{r}|\,\bar{r}^2(1 - g).$$

Using explicit g and spherical coordinates one arrives at (2.92).

$$dF_{\text{free}} = \frac{dk^{(+)} - dk^{(-)}}{dt} = dxdS \int dv\, mv_x^2 \frac{\partial f}{\partial x} = dS\, dx \frac{\partial}{\partial x} nk_B T. \tag{2.95}$$

Within the thermodynamic description, the total force on the element $d\Omega_r$ is proportional to a gradient of the pressure, $dF_{\text{free}} + dF = dxdS \frac{\partial \mathcal{P}}{\partial r_x}$, from which one identifies the pressure as

$$\mathcal{P} = nk_B T + \frac{1}{2} n^2 k_B T \frac{4\pi}{3} R^3 \left(1 - \exp\left[-\frac{\mathcal{V}}{k_B T} \right] \right). \tag{2.96}$$

The second term is known as the virial of forces or the virial correction. Again one can see that for a weak potential, the pressure equals the mean-field pressure, $\mathcal{P} = nk_B T + \frac{1}{2} n\phi$, while at the strong potential limit, the pressure gains corrections by the excluded volume, $\mathcal{P} = nk_B T \left(1 - \frac{1}{2} n\frac{4\pi}{3} R^3 \right)^{-1} \approx nk_B T \left(1 + \frac{1}{2} n\frac{4\pi}{3} R^3 \right)$,

$$\mathcal{P} = nk_B T + \frac{1}{2} n^2 \mathcal{V} \frac{4\pi}{3} R^3 \qquad \text{for} \qquad \mathcal{V} \ll k_B T$$

$$\mathcal{P} = nk_B T \left(1 + \frac{1}{2} n\frac{4\pi}{3} R^3 \right) \qquad \text{for} \qquad \mathcal{V} \gg k_B T \tag{2.97}$$

2.5.3 Virial theorem

With the understanding of section 2.5.2 at hand we can now derive the virial theorem. The time derivative of the product of momentum and energy in N-particle systems

$$\frac{d}{dt} \sum_{i=1}^{N} p_i x_i = \sum_{i=1}^{N} \dot{p}_i x_i + \sum_{i=1}^{N} p_i \dot{x}_i = \sum_{i=1}^{N} F_i x_i + 2 \sum_{i=1}^{N} \epsilon_i \tag{2.98}$$

gives the virial of forces $\mathbf{F} \cdot \mathbf{x}$ and twice the kinetic energy. The time averaging of a time derivative vanishes

$$\overline{\frac{d}{dt} A(t)} = \lim_{t \to \infty} \frac{1}{t} \int_0^t dt'\dot{A}(t') = 0 \tag{2.99}$$

and we obtain from (2.98) the virial theorem as

$$\sum_{i=1}^{N} \overline{F_i x_i} = -2 \sum_{i=1}^{N} \epsilon_i. \tag{2.100}$$

This virial theorem is nicely applied if we calculate the virial of the collision with a wall. The pressure exercised on the wall is $\mathcal{P} d A = -\mathbf{F}$ and one obtains

$$\sum_{i=1}^{N} \overline{F_i x_i} = -\int_{\text{wall}} \mathbf{x} \cdot \mathscr{P} \mathbf{d}A = -\mathscr{P} \int \text{div} \mathbf{x} dV = -3\mathscr{P} \int dV = -3\mathscr{P}V \qquad (2.101)$$

and the pressure of the ideal gas appears from the virial theorem in terms of the mean kinetic energy $\mathscr{P} = 2N\bar{\epsilon}/3V$ which is (2.33) and (2.34). If we consider additionally internal forces F_{ij} acting on the place i from all other particles at j, we see that their contribution to the virial theorem is just the potential energy

$$\Phi = \sum_{i,j}^{N} \overline{F_{ij} x_i} = \frac{1}{2} \sum_{i \neq j}^{N} \overline{F_{ij}(x_i - x_j)} \qquad (2.102)$$

and the equation of state becomes the one of an interacting gas

$$\mathscr{P} = \frac{2}{3} n\bar{\epsilon} + \frac{\Phi}{3\Omega}. \qquad (2.103)$$

Now we proceed and consider a thermally distributed system which possesses the mean kinetic energy $\bar{\epsilon} = 3k_B T/2$ due to equi-partitioning. The averaging is now performed over the statistical sum (2.47). We assume the internal forces are given by a potential $F_i = \partial \mathscr{V}/\partial x_i$ such that we can write (2.103) as

$$\mathscr{P}\Omega = Nk_B T - \frac{1}{3} \langle \sum_{ij} r_{ij} \frac{d\mathscr{V}}{dr_{ij}} \rangle = Nk_B T - \frac{N(N-1)}{6} \langle r_{12} \frac{d\mathscr{V}}{dr_{12}} \rangle. \qquad (2.104)$$

In the statistical averaging the kinetic parts cancel and only the integrals over r_1 and r_2 remain. Defining the two-particle correlation function (2.87)

$$g(r_1, r_2) = \Omega^2 \frac{e^{-\mathscr{V}(r_1-r_2)/k_B T}}{\int dr_1 dr_2 e^{-\mathscr{V}(r_1-r_2)/k_B T}} \qquad (2.105)$$

we can finally write from (2.104) the general virial formula for the pressure

$$\mathscr{P} = \frac{N}{\Omega} k_B T - \frac{N(N-1)}{2\Omega^3} \int dr_1 dr_2 r_{12} \frac{d\mathscr{V}}{dr_{12}} g(r_1, r_2) = nk_B T - \frac{2\pi n^2}{3} \int_0^{\infty} \frac{d\mathscr{V}}{dr} g(r) r^3 dr. \qquad (2.106)$$

In lowest approximation $g(r) \approx \exp(-\mathscr{V}/k_B T)$ we obtain the virial expansion

$$\mathscr{P} \approx nk_B T \left[1 + 2\pi n \int_0^{\infty} r^2 \left(1 - e^{-\frac{\mathscr{V}}{k_B T}} \right) dr \right] \qquad (2.107)$$

which leads just to (2.96).

An alternative derivation directly uses the statistical canonical sum (2.45) which separates into a kinetic sum and a configuration sum, $z_{can} = K(T)Q_N(\Omega)$, with

$$Q_N = \frac{1}{N!}\int_V dr_1....dr_N e^{-\frac{\mathscr{V}(r_1....r_n)}{k_B T}} = \frac{\alpha^{3N}}{N!}\int_{V/\alpha^3} dr_1....dr_N e^{-\frac{\mathscr{V}(\alpha r_1....\alpha r_n)}{k_B T}}$$

providing

$$\frac{dQ_N}{d\Omega} = \left(\frac{\frac{\partial Q_N}{\partial \alpha}}{\frac{\partial \Omega}{\partial \alpha}}\right)_{\alpha=1} = \frac{1}{3\Omega}\left(\frac{\partial Q_N}{\partial \alpha}\right)_{\alpha=1}$$

$$= \frac{N}{\Omega}Q_N - \frac{1}{3k_B TN!\Omega}\int_V dr_1....dr_N \sum_{i<j} r_{ij}\frac{d\mathscr{V}(r_{ij})}{dr_{ij}}e^{-\frac{\mathscr{V}(r_{ij})}{k_B T}}. \qquad (2.108)$$

Defining the N-particle distribution $F_N = \exp(-\mathscr{V}/k_B T)/Q_N N!$ one has from the canonical ensemble (2.45) and the relation to the free energy

$$\mathscr{P} = -\frac{\partial F}{\partial \Omega}\bigg|_{T,N} = k_B T\frac{\partial \ln Q_N}{\partial \Omega} = n k_B T - \frac{N(N-1)}{6\Omega}\int dr_1 dr_2 r_{12}\frac{d\mathscr{V}(r_{12})}{dr_{12}}F_2(r_1, r_2).$$

Using the pair-correlation function $g = \Omega^2 F_2$ one obtains just (2.106).

2.5.4 Summary

The binary correlation allows one to distinguish tails and cores of the interaction potential. When the tail is sufficiently regular it allows us to introduce the mean field for deliberately divergent repulsive potentials. An appreciable value of core potential is located at a sphere of a small radius giving a finite total cross section of collisions. The concept of binary correlation is thus behind the intuitive picture of the kinetic equation.

3

Classical Kinetic Theory

The principles outlined in Chapter 2 allow one to formulate a classical microscopic theory of many-body systems applicable to dilute gases. The centre of our interest will be a kinetic equation with a nonlocal and non-instant dissipation. We thus start from parameters which are necessary to describe nonlocal and non-instant features of binary collisions. Finally, we will discuss some thermodynamic properties captured by such a kinetic equation.

The classical nonlocal and non-instant treatment of binary collisions have a long tradition. Pioneering works date back to Enskog (Chapman and Cowling, 1990). A systematic treatment has been developed by Bogoliubov (1946) and Green (1952). This tradition is cultivated in the field of chemical physics being less known among physicists concerned with degenerated systems. There are many reasons why it is useful to view the real scattering event as a chemical reaction where particles form a molecule. Unlike a real (stable) molecule, the scattering state is unstable which means that it decays after a short time Δ_t into the outgoing particles of the collision.

The relationship of scattering states with molecules is particularly strong in thermodynamic properties. For instance, one can derive the law of acting masses for the density of colliding particles, or the partial pressure of colliding particles. In pragmatic formulations of the kinetic equation, the focus shifts from principles to their actual implementations. First of all, it is more convenient to separate low- and high-angle effects on the level of their potential. We thus assume that the interaction potential is composed of two parts

$$\mathscr{V} = \mathscr{V}_{\text{tail}} + \mathscr{V}_{\text{core}}, \tag{3.1}$$

where the tail creates the mean field while the core causes collisions. The core is supposed to be repulsive, i.e. positive, and short range, therefore zero for distances larger than b_{cut}. The tail can be anything, but it is supposed to be sufficiently weak. To keep things simple, we will talk about inert gases only, so that a particle means a spherical atom without internal degrees of freedom. Both parts of the potential thus only depend on the distance between particles.

Interacting Systems far from Equilibrium. Klaus Morawetz, Oxford University Press (2018).
© Klaus Morawetz. DOI: 10.1093/oso/9780198797241.001.0001

3.1   Nonlocal and non-instant binary collisions

The state of particle during the collision, the so-called scattering state, can be characterised by the same internal degrees of freedom as molecules, the mean inter-atomic distance and the rotation. During Δ_t, the molecule flies which makes the collision non-local. The mean inter-atomic distance and the rotation thus contribute to the nonlocality via the internal degrees of freedom while the molecular flight contributes via the standard translational one.

In general, the collision delay, the inter-atomic distance and the rotation are necessary to describe a scattering event. For a special model, some of these parameters are zero. We use this special model to introduce the nonlocal and non-instant corrections step by step.

3.1.1   Hard spheres

A collision of two hard spheres is sketched in Fig. 3.1. Before a collision the particle a follows a trajectory $r_a(t) = r_1 + v_1 t$ and the particle b follows $r_b(t) = r_2 + v_2 t$. At the instant t^{coll} of collision they exchange a part of their momenta and the particle a follows a trajectory $r_a(t) = r_3 + v_3 t$ while the particle b follows $r_b(t) = r_4 + v_4 t$. For the purpose of the scattering integral it is reasonable to relate the trajectories of particles to the instant of collision, $r_a(t) = r_a^{\text{coll}} + v_{1,3}(t - t^{\text{coll}})$ and $r_b(t) = r_b^{\text{coll}} + v_{2,4}(t - t^{\text{coll}})$. At the instant of the event the particles are displaced by $\Delta^{\text{d}} = r_a^{\text{coll}} - r_b^{\text{coll}}$. The amplitude of this displacement equals the sum of radii of particles $|\Delta^{\text{d}}| = R_a + R_b$.

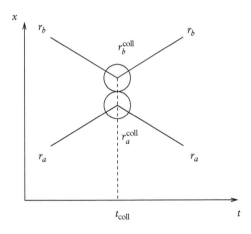

Figure 3.1: *Collision of hard spheres. The coordinates x of two colliding hard spheres are sketched as functions of time t. The collision is instant, i.e. at $t = t^{\text{coll}}$ the straight trajectories break and new straight trajectories begin. Due to the finite diameters of hard spheres, at the instant of collision the two particles are not at the same coordinate but displaced by $r_b^{\text{coll}} - r_a^{\text{coll}}$. The x-axis is taken in the direction of their centres at the instant of the collision to make the displacement clearly visible.*

3.1.2   Sticky point particles

As a second model let us assume point particles which can form unstable molecules. As sketched in Fig. 3.2, at the time instant t^{be} two particles stick into a molecule that

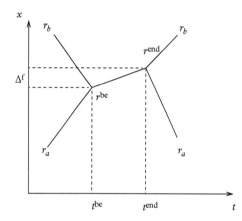

Figure 3.2: *Collision of sticky particles. The co-ordinate x of two colliding sticky-point particles is sketched as a function of time t. At time $t = t^{be}$ the particles form a molecule that breaks apart at $t = t^{end}$. The collision is not instant but lasts over the time interval (t^{be}, t^{end}). During the collision the molecule flies with its centre-of-mass velocity from the coordinate r^{be} to the coordinate r^{end}.*

decomposes at t^{end}. If we describe this two-step process as a single collision event, t^{be} has to be interpreted as the beginning of the collision while t^{end} is interpreted as its end. The length of the time interval, $\Delta_t = t^{end} - t^{be}$, is the collision delay. Due to neglected atomic radii, in this case there is only a displacement between positions of particles at the beginning and at the end of the collision, $\Delta^f = r^{end} - r^{be}$. This displacement reflects the flight of the molecule, therefore the amplitude of this displacement is given by the molecule velocity $v^{mol} = \frac{v_1 m_a + v_2 m_b}{m_a + m_b}$ and the collision delay as $\Delta^f = \Delta_t v^{mol}$.

3.1.3   Realistic particles

The two model collisions in Figs. 3.1 and 3.2 are simple; having trajectories composed from broken straight lines. Such model collisions are fully characterised by a few parameters like Enskog's displacement Δ^d or the collision delay Δ_t. The realistic trajectory, however, is a curve like the one sketched in Fig. 3.3. Although the curve cannot be fully described within few parameters, one can find three parameters which cover its features important for kinetic theory.

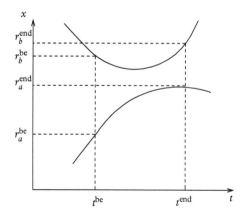

Figure 3.3: *Collision of real particles. The coordinate x of two colliding real particles is sketched as a function of time t. For simplicity, the interaction range is assumed finite; therefore before $t = t^{be}$ and after $t = t^{end}$ particles follow straight lines. During this interval, their trajectory is a general curve. At coordinate r_a^{be} the particle a reaches the potential of the particle b by that being at r_b^{end}. At coordinates $r_{a,b}^{end}$ particles become disconnected.*

As long as we assume a potential of finite range, the asymptotic trajectories are again of the form $r_a(t) = r_{1,3} + v_{1,3}t$ and $r_b(t) = r_{2,4} + v_{2,4}t$. There is a whole time interval, from t^{be} to t^{end}, during which the particles have velocities different from their asymptotic values. Accordingly, as for the sticky point particles, there is no single time instant of the collision but a whole time interval has to be attributed to the collision process. One can see that the collision event links together four space points r_a^{be}, r_b^{be}, r_a^{end} and r_b^{end}. In principle, one could describe the collision by its duration $t^{end} - t^{be}$ and three displacements. Such a set of parameters, however, does not work well realistically. Our choice of the potential range is deliberate since none of the realistic potentials really vanishes at a finite distance and if one uses an artificial cutoff at distance b_{cut}, the non-locality of the collision is not determined by the interaction forces but by the cutoff, $|r_{a,b}^{end} - r_{a,b}^{be}| \sim b_{cut}$. To identify the nonlocal and non-instant features that depend on forces and not on the cutoff, the true collision event has to be mimicked by an effective collision event which provides the same asymptotic states. Let us try to find out what such an effective event should look like and which parameters are necessary to characterise it.

A more detailed picture of the collision can be obtained by plotting the 'trajectory' of the relative distance $r_a - r_b$ in the collision plane, see Fig. 3.4. The matching points A and A′ correspond to $r_a^{be} - r_b^{be}$ and $r_a^{end} - r_b^{end}$, respectively. From parameters seen in Fig. 3.4, the deflection angle θ has a prominent position since it is observable at any macroscopic scale. Other parameters are microscopic.

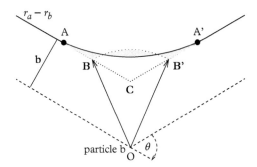

Figure 3.4: *In-plane trajectory of the relative motion of two colliding real particles. The full line shows the trajectory of the relative coordinate $r_a - r_b$ in the (x-y)-plane, where the y-axis is in the direction of their centres at the turning point and the x-axis is the perpendicular one in the collision plane. Points A and A′ correspond to $r_a^{be} - r_b^{be}$ and $r_a^{end} - r_b^{end}$ from Fig. 3.3, respectively. The θ is the barycentric scattering angle. The dashed line shows a trajectory of an instant and local collision as it is used with the Boltzmann equation. The real trajectory differs by the finite impact of factor b and curved trajectory AA′. Point C denotes a turning point within the hard-sphere-like effective trajectory. The dotted lines extrapolate straight lines in the asymptotic regions. Points B and B′ represent the matching point for the dumbbell effective trajectory. The dumbbell effective trajectory includes the arc BB′.*

Sticky hard spheres – 1. reference model   The first microscopic parameter of the collision is the impact factor b which ranges over values of the order of the 'particle diameter' (defined e.g. as a square-root of the total cross section). In the spirit of hard-sphere correction Δ^d, we can include the finite impact factor approximating the real trajectory (curve from A to A') by the broken straight line ACA'. This is exactly the trajectory of the collision of two hard spheres, where the effective Enskog's displacement is $\Delta^d = OC$. Unlike the case for hard spheres, the effective distance of particles $|OC|$ is not constant, but depends on the impact factor b and the deflection angle θ as $b = |OC| \cos \frac{\theta}{2}$.

The second microscopic parameter relates to the duration of the collision. If the particle following the effective trajectory would scatter, as a real hard sphere, it arrives at point A' before t^{end}. To see this from Fig. 3.4, we can use the conservation of the angular momentum (its relative motion component)

$$L = b \frac{m_a m_b}{m_a + m_b} |v_1 - v_2|. \tag{3.2}$$

Let us denote by S the area defined by angle AOA' and the trajectory AA'. According to Kepler's law, S is proportional to the time a particle needs to pass from A to A',

$$t^{end} - t^{be} = \frac{2S}{L} \frac{m_a m_b}{m_a + m_b}. \tag{3.3}$$

The hard-sphere trajectory ACA' encloses a smaller area than AA', it is therefore a faster path.

The collision duration can be fixed by stopping the particle at point C for time Δ_t so that it reaches A' at time t^{end}. The effective collision described by the trajectory ACA' is then described by two parameters, the hard-sphere displacement covering the impact parameter and the collision delay covering the collision duration. Since this approximation provides correct asymptotic trajectories, it represents the most important correction beyond the instant and local approximation of collisions.

Dumbbells – 2. reference model   As we expect from the degrees of freedom of the molecule, the collision should be described by three parameters while only two parameters seem to be sufficient. The third parameter emerges if one handles the angular momentum L during the collision more carefully. Within the hard-sphere-like trajectory ACA', the relative coordinate of the particles does not change during Δ_t. Accordingly, the angular momentum vanishes during this waiting time. This missing contribution to the angular momentum can be included with the help of a more complex effective trajectory ABB'A' shown also in Fig. 3.4. Section BB' corresponds to a rotation of particles with a fixed inter-atomic distance $R = |BO|$ over the angle ϑ. The angular velocity of the rotation is determined by the angular momentum L. The time during which they rotate is interpreted as the collision delay Δ_t.

The collision delay Δ_t is adjusted so that the particles on the effective trajectory from A to A' spend the same time as on the real one, $[r_{ab} = r_a - r_b]$

$$\Delta_t = \int_A^{A'} \frac{d|r_{ab}|}{\left|\frac{dr_{ab}}{dt}\right|} - \frac{|AB|}{|v_2 - v_1|} - \frac{|B'A'|}{|v_4 - v_3|}.$$ (3.4)

Equation (3.4) corresponds to Kepler's geometrical condition that the area of the central shadow element equals the sum of two other elements. The angular velocity $\frac{\vartheta}{\Delta_t}$ is given by the angular momentum,

$$\frac{\vartheta}{\Delta_t} = \frac{L}{R^2} \frac{m_a + m_b}{m_a m_b}.$$ (3.5)

The inter-atomic distance R is given by the impact factor b, the deflection angle θ and the angle of rotation ϑ

$$b = R \cos\left(\frac{\theta + \vartheta}{2}\right),$$ (3.6)

as can be found from Fig. 3.5.

We note that the choice of the effective trajectory is somehow deliberate. In particular the naive picture of two particles rotating like a dumbbell hardly applies to real events. The major aim of this effective picture is to introduce the nonlocal and non-instant features of the collision in a manner that satisfies elementary conservation laws. Moreover, we wanted to demonstrate that concepts like the collision duration are intimately connected with the nonlocal character of the collision. One can see that Δ_t defined within the effective hard-sphere trajectory ACA′ differs from the collision delay defined within the dumbbell effective trajectory, because the free trajectories are of a different length. The three parameters of the dumbbell model correspond to those found for quantum-mechanical collisions.

Finally we have to note that the dumbbell model only works well for repulsive potentials. In the attractive potential any dumbbell-like trajectory is slower than the real

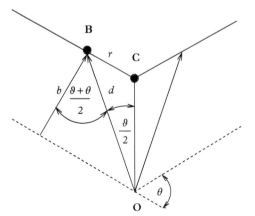

Figure 3.5: *Detail of Fig. 3.4 giving angles needed in triangular relations for the dumbbell effective trajectory.*

trajectory. The hard-sphere-like trajectory leads to peculiarities too having a negative waiting time. Nevertheless if one accepts negative values, both effective trajectories can be used for asymptotic matching without problems.

Vector notation   With respect to the implementation of nonlocalities into the kinetic equation, it is advantageous to express all displacements in the vector notation. Firstly we define generating vectors corresponding to individual degrees of freedom.

The nonlocality of the centre-of-mass motion is independent of the internal degrees of freedom. It depends exclusively on the collision delay,

$$\Delta^{\mathrm{f}} = \Delta_t v^{\mathrm{mol}}. \tag{3.7}$$

We will call this vector the molecular flight.

The internal degrees of freedom are covered by vectors OB and OB'. Due to the time-reversal symmetry of the collision, these vectors are conjugated by a mirror symmetry with the reflection plane being perpendicular to the collision plane and cutting the collision plane at point O with the direction of transferred momentum $q = -m_a(v_3 - v_1) = m_b(v_4 - v_2)$. It is advantageous to represent OB and OB' with the help of two components that are odd and even with respect to the mirror reflection.

The even component $\Delta^{\mathrm{d}} = \frac{1}{2}(\mathrm{OB} + \mathrm{OB}')$ points in the direction of the transferred momentum q,

$$\Delta^{\mathrm{d}} = \frac{q}{|q|} R \cos \frac{\vartheta}{2}, \tag{3.8}$$

where $R = |\mathrm{BO}| = |\mathrm{B'O}|$. This vector has the direction of the hard-sphere displacement to which it reduces for the case of true hard spheres. We will call it Enskog's displacement.

The odd component $\Delta^{\vartheta} = \frac{1}{2}\mathrm{BB}'$ points in a direction that is perpendicular to q and stays in the scattering plane, i.e. in the direction of $v_1 - v_2 + v_3 - v_4$.[1] The amplitude of the odd component can be seen in Fig. 3.4, thus

$$\Delta^{\vartheta} = \frac{v_1 - v_2 + v_3 - v_4}{|v_1 - v_2 + v_3 - v_4|} R \sin \frac{\vartheta}{2}. \tag{3.9}$$

---

[1] The orthogonality of $v_1 - v_2 + v_3 - v_4$ to the transferred momentum q follows from the energy conservation

$$0 = \frac{m_a v_1^2}{2} + \frac{m_b v_2^2}{2} - \frac{m_a v_3^2}{2} - \frac{m_b v_4^2}{2}$$

$$= \frac{m_a v_1^2}{2} + \frac{m_b v_2^2}{2} - \frac{(m_a v_1 - q)^2}{2m_a} - \frac{(m_b v_2 + q)^2}{2m_b}$$

$$= q(v_1 - v_2) - \frac{q^2}{2m_a} - \frac{q^2}{2m_b}$$

$$= \frac{1}{2}q(v_1 - v_2 + v_3 - v_4),$$

where we have used $\frac{q}{m_a} = v_1 - v_3$ and $\frac{q}{m_b} = v_4 - v_2$ to eliminate masses.

We will call this vector the rotation displacement since it vanishes when the molecule does not rotate.

In our further discussion of the effects of nonlocalities within the kinetic equation, it will be more convenient to relate the displacements to one of the end points of the effective collision event. Let us associate the origin of the coordinate system with the position of particle a at the beginning of the effective event. We denote the position of particle b by Δ_2 also at the beginning, and by Δ_3 and Δ_4 the positions of particles a and b at the end, respectively. In the pre-collision asymptotic region the motion of particles thus reads $r_a = r + v_1 t$ and $r_b = r + \Delta_2 + v_2 t$. After the collision we have $r_a = r + \Delta_3 + v_3(t - \Delta_t)$ and $r_b = r + \Delta_4 + v_4(t - \Delta_t)$.

Now we express $\Delta_{2,3,4}$ in terms of the already introduced Δ's. In terms of $\Delta_{2,3,4}$, the Enskog-type displacement reads

$$\Delta_2 + \Delta_4 - \Delta_3 = 2\Delta^d, \tag{3.10}$$

and the rotation displacement is

$$\Delta_4 - \Delta_3 - \Delta_2 = 2\Delta^\vartheta. \tag{3.11}$$

Finally, during the effective collision, the centre of mass moves with constant velocity $v^{mol} = \frac{m_a v_1 + m_b v_2}{m_a + m_b}$. The continuity of the centre-of-mass motion requires

$$m_b(\Delta_4 - \Delta_2) + m_a \Delta_3 = (m_a + m_b)\Delta^f. \tag{3.12}$$

With the help of these conditions one can evaluate all displacements $\Delta_{2,3,4}$ from three scalar parameters Δ_t, R and ϑ.

3.1.4 Kinetic equation without a mean field

The concepts of displacements and the time delay have been used to derive nonlocal and non-instant corrections to the scattering integral of the Boltzmann equation. Taking the Boltzmann equation as a balance equation for the occupation of the phase space, Enskog has shown how to incorporate nonlocal corrections into the Boltzmann equation on the basis of simple intuitive arguments. It is easy to extend his arguments for dumbbell-like effective events.

Let us ignore the mean field for the moment and derive the kinetic equation in heuristic manner. We denote by $f_1 \equiv f_a(k, r, t)$ a probability that at time t a particle of kind a can be found in a unit cell around the phase-space point (k, r), where k is a momentum and r is a coordinate. In this cell, the particle has an energy $\varepsilon_1 = \varepsilon_a(k) = \frac{k^2}{2m_a}$. The probability f_1 changes in time,

$$\frac{\partial f_1}{\partial t} + \frac{\partial \varepsilon_1}{\partial k}\frac{\partial f_1}{\partial r} = \sum_b \int \frac{dp}{(2\pi)^3}\frac{dq}{(2\pi)^3} P 2\pi \delta(\varepsilon_1 + \varepsilon_2 - \varepsilon_3 - \varepsilon_4)\,(f_3 f_4 - f_1 f_2), \tag{3.13}$$

by two mechanisms. The space drift which is the differential term on the left-hand side, and the dissipation covered by the integral terms on the right-hand side.

The dissipation is represented here by the scattering integral which itself has two contributions given by f_3f_4 and f_1f_2. The first one is the scattering-in which counts how many particles end up in the cell (k, r) after they have gone through a collision with some other particle. The second one is the scattering-out which counts how many particles are kicked off the cell (k, r) by a collision with some other particle. The schematic picture of these two processes is shown in Fig. 3.6.

The scattering-out is proportional to the probability f_1 that particle a is at the cell (k, r). The integral over the transferred momentum q sums over all possible final states $k - q$ in which the particle a can end up after the collision. The sum over b and the integral over momentum p sum over all possible partners in the collision. The probability to find such a partner is given by f_2. In this discussion of the binary collision we have seen that particle b is displaced by Δ_2, accordingly we have to look for the partner particle in the cell $(p, r + \Delta_2)$, i.e. $f_2 \equiv f_b(p, r + \Delta_2, t)$.

Other ingredients of the scattering-out relate to properties of an isolated collision event. The differential cross-section $P \equiv P_{ab}(k, p, k - q, p + q)$ measures the probability of two particles a and b with initial momenta k and p to undergo the collision in which the momentum q is transferred from a to b. The δ function shows that the sum energy of the two particles conserves, i.e. the collision is elastic.

Collision processes described by the scattering-in are conjugated to the scattering-out. The integral over q sums over all initial states $k - q$ from which the particle can be scattered into the cell (k, r). As we have seen, the initial and final states of particle a are displaced by Δ_3; therefore we have to look for particle a in the cell $(k - q, r - \Delta_3)$. The collision process lasts for Δ_t, we thus have to find this particle at time $t - \Delta_t$. The

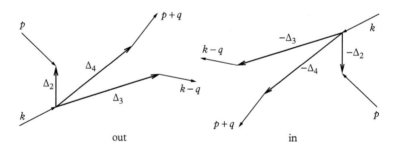

Figure 3.6: *Displacements in the effective collision of real particles. In the scattering-out process, the momenta k and p correspond to the initial states of particles a and b, the momenta $k - q$ and $p + q$ to the final states. In the scattering-in process, the picture is reversed. The momenta k and p correspond to the final states while $k - q$ and $p + q$ to the initial ones. The scattering-in and -out are linked by two symmetry operations, the space inversion and the time reversal. The former results in the opposite orientation of displacements, the latter gives opposite collision delay.*

probability of finding particle a in the suitable initial state is thus given by $f_3 \equiv f_a(k-q, r-\Delta_3, t-\Delta_t)$. At the beginning of the collision, the partner particle b is displaced by Δ_4, i.e. $f_4 \equiv f_b(p+q, r-\Delta_4, t-\Delta_t)$. Since the differential cross section obeys the symmetry $P_{ab}(k, p, k-q, p+q) = P_{ab}(k-q, p+q, k, p)$, we could join both processes into a single scattering integral.

The scattering integral has the same structure as the Boltzmann equation. The only difference appears in arguments of distribution functions that are corrected by displacements $\Delta_{2,3,4}$ and by the collision delay Δ_t. These corrected arguments are

$$f_1 = f_a(k, r, t),$$

$$f_2 = f_b(p, r + \Delta_2, t),$$

$$f_3 = f_a(k - q, r - \Delta_3, t - \Delta_t),$$ (3.14)

$$f_4 = f_b(p + q, r - \Delta_4, t - \Delta_t).$$

The kinetic equation (3.13) has two simple limits. Sending all Δ's to zero one obtains the Boltzmann equation. Using hard-sphere displacements $\Delta_{2,4} = \Delta^d$ and other Δ's equal to zero one recovers the nonlocal corrections of Enskog's equation.

3.1.5 Summary

Collisions of realistic particles are nonlocal and non-instant. A collision delay Δ_t characterises the effective duration of the collision, and three displacements, $\Delta_{2,3,4}$, describe its effective nonlocality. All the three vectors, $\Delta_{2,3,4}$, can be constructed from two scalar parameters and Δ_t. Consequently, the scattering integral of kinetic equation (3.13) is nonlocal and non-instant. Its shifted and delayed initial conditions are given by distributions (3.14).

3.2 Effect of the mean field on collisions

So far we have ignored the mean field. The mean potential acting on a particle of kind a is[2]

$$U_a(x, t) = \sum_b \int dy\, \mathcal{V}_{ab\,\text{tail}}(x - y) n_b(y, t) = \sum_b n_b(x, t) \bar{\mathcal{V}}_{ab},$$ (3.15)

---

[2] The approximate local form $\sum_b n_b(x, t) \bar{\mathcal{V}}_{ab}$ can be used only for a potential which at long distances falls faster than inverse cube, $\mathcal{V}(x - y) < |x - y|^{-3-\eta}$ with $\eta > 0$. The approximation holds up to the linear order in gradients,

$$\int dy\, \mathcal{V}_{ab\,\text{tail}}(x - y) n_b(y, t) \approx \int dy\, \mathcal{V}_{ab\,\text{tail}}(x - y) \left(n_b(x, t) + (y - x)\frac{\partial n_b}{\partial x} \right) = n_b(x, t) \int dy\, \mathcal{V}_{ab\,\text{tail}}(x - y).$$

where n_b is a density of particles b and $\bar{\mathcal{V}}_{ab} = \int dy \mathcal{V}_{ab\,\text{tail}}(y)$. First of all, the force $F_a = -\frac{\partial U_a}{\partial r}$ accelerates particles from the reference cell (k, r) into cells of higher momenta and also accelerates particles of lower momenta into the reference cell. These changes are covered by momentum drift $-\frac{\partial U_1}{\partial r}\frac{\partial f_1}{\partial k}$ which enters the left-hand side of the kinetic equation (3.13).

For instant approximations of collisions, the momentum drift explores the effect of the field on the system. A collision of a finite duration, however, is sensitive to mean-fields $U_{a,b}$. Indeed, the effective molecule is driven by a sum of mean forces on their constituents. The mean field can also influence the relative motion of particles a and b. Since the collision delay represents the microscopic characteristic time that is supposed to be small on the hydrodynamic scale of changes of distributions, the effect of the mean field on the molecule can be evaluated as a correction linear in the collision delay.

In general, the field effect on the scattering is very involved. For simplicity we assume only gravitation-like forces for which $A = \frac{F_a}{m_a} = \frac{F_b}{m_b}$. In this case the effect of the field is explored by the acceleration of the centre-of-mass of colliding particles. This approximation simply neglects the effect of the field on the internal degrees of freedom of effective molecules.

3.2.1   Momentum gains

During its life, length Δ_t, the velocity of the molecule changes by

$$\Delta v^{\text{mol}} = A\Delta_t. \tag{3.16}$$

The momentum gains of individual particles due to the centre-of-mass acceleration is proportional to their masses,

$$\Delta_K^a = m_a \Delta v^{\text{mol}}. \tag{3.17}$$

The momentum gains enter the relation between the initial and final states of the collision. Momenta k, p and q used in kinetic equation (3.13) are related to the time instant t. In the scattering-out process, the initial condition is given at t,[3] therefore no momentum gain enters the arguments of distributions. In the scattering-in process, the initial condition is given at the beginning time $t - \Delta_t$.[4] A reversed momentum gain, $-\Delta_K^a - \Delta_K^b$, is necessary to match finite states k and p, therefore

$$
\begin{aligned}
f_1 &= f_a(k, r, t), \\
f_2 &= f_b(p, r + \Delta_2, t), \\
f_3 &= f_a(k - q - \Delta_K^a, r - \Delta_3, t - \Delta_t), \\
f_4 &= f_b(p + q - \Delta_K^b, r - \Delta_4, t - \Delta_t).
\end{aligned}
\tag{3.18}
$$

---

[3] In the absence of the field by particle distributions $f_1 = f_a(k, r, t)$ and $f_2 = f_b(p, r + \Delta_2, t)$.

[4] In the absence of the field by particle distributions $f_3 = f_a(k - q, r - \Delta_3, t - \Delta_t)$ and $f_4 = f_b(p + q, r - \Delta_4, t - \Delta_t)$.

3.2.2   Energy gain

Similarly, the molecule can gain energy from time-dependent fields. The energy per particle changes along the trajectory of particles by

$$
2\Delta_E = \left(\frac{\partial U_1}{\partial t} + \frac{\partial U_2}{\partial t}\right)\Delta_t + \left(\frac{\partial U_1}{\partial r}\Delta_3 + \frac{\partial U_2}{\partial r}(\Delta_4 - \Delta_2)\right) + (m_a + m_b)v^{\mathrm{mol}}\Delta v^{\mathrm{mol}}. \quad (3.19)
$$

The first term describes the time evolution of the energy bottom. The second term describes potential differences between initial and final positions. The third term is an increase of kinetic energy due to the momentum gain. The second and the third terms cancel as can be seen from $\frac{\partial U_{1,2}}{\partial r} = -Am_{a,b}$ and Eqs. (3.12) and (3.16). This cancellation reflects the fact that forces acting along the trajectory only convert the potential energy into the kinetic one having no effect on the total energy. The energy gain is thus given by the time change of the energy bottom during Δ_t

$$
2\Delta_E = \left(\frac{\partial U_1}{\partial t} + \frac{\partial U_2}{\partial t}\right)\Delta_t. \quad (3.20)
$$

The energy gain appears in the energy-conserving δ function. The scattering-in and scattering-out have to be distinguished. The single-particle energies

$$
\varepsilon_a(k, r, t) = \frac{k^2}{2m_a} + U_a(r, t), \quad (3.21)
$$

in individual initial and final states are
scattering-out:

$$
\begin{aligned}
\varepsilon_1 &\equiv \varepsilon_a(k, r, t), \\
\varepsilon_2 &\equiv \varepsilon_b(p, r + \Delta_2, t), \\
\varepsilon_3 &\equiv \varepsilon_a(k - q + \Delta_K^a, r + \Delta_3, t + \Delta_t), \\
\varepsilon_4 &\equiv \varepsilon_b(p + q + \Delta_K^b, r + \Delta_4, t + \Delta_t),
\end{aligned}
\quad (3.22)
$$

scattering-in:

$$
\begin{aligned}
\varepsilon_1 &\equiv \varepsilon_a(k, r, t), \\
\bar{\varepsilon}_2 &\equiv \varepsilon_b(p, r - \Delta_2, t), \\
\bar{\varepsilon}_3 &\equiv \varepsilon_a(k - q - \Delta_K^a, r - \Delta_3, t - \Delta_t), \\
\bar{\varepsilon}_4 &\equiv \varepsilon_b(p + q - \Delta_K^b, r - \Delta_4, t - \Delta_t).
\end{aligned}
\quad (3.23)
$$

In spite of the complicated arguments of single-particle energies, the energy-conserving δ functions can be rearranged to the (rather misleading) form of kinetic energy conservation known from the Boltzmann equation[5]

scattering-out:

$$2\pi\delta\,(\varepsilon_1 + \varepsilon_2 - \varepsilon_3 - \varepsilon_4 + 2\Delta_E) = 2\pi\delta\left(\frac{k^2}{2m_a} + \frac{p^2}{2m_b} - \frac{(k-q)^2}{2m_a} - \frac{(p+q)^2}{2m_b}\right) \equiv \delta_{1234},$$

(3.24)

scattering-in:

$$2\pi\delta\,(\varepsilon_1 + \bar{\varepsilon}_2 - \bar{\varepsilon}_3 - \bar{\varepsilon}_4 - 2\Delta_E) = 2\pi\delta\left(\frac{k^2}{2m_a} + \frac{p^2}{2m_b} - \frac{(k-q)^2}{2m_a} - \frac{(p+q)^2}{2m_b}\right) \equiv \bar{\delta}_{1234}.$$

(3.25)

The simple form of the uncorrected kinetic energy conservation should not be interpreted as the conservation of kinetic energy since the true asymptotic states are $k - q \pm \Delta_K^a$ and $p + q \mp \Delta_K^b$ not the uncorrected states $k - q$ and $p + q$. In other words, in spite of possible rearrangements (3.24) and (3.25), there is a conversion between the kinetic and the potential energies. The simple form of energy-conserving δ function, however, might be more convenient in implementations.

Kinetic equation with the mean field   Collecting all effects of the field, one finds the kinetic equation

$$\frac{\partial f_1}{\partial t} + \frac{\partial \varepsilon_1}{\partial k}\frac{\partial f_1}{\partial r} - \frac{\partial \varepsilon_1}{\partial r}\frac{\partial f_1}{\partial k} = \sum_b \int \frac{dpdq}{(2\pi)^6} P\delta_{1234}\left(\bar{f}_3\bar{f}_4 - f_1 f_2\right).$$

(3.26)

The subscripts and bars denote variables as in (3.22) and (3.23).

---

[5] The compensation of all Δ's within their linear contributions:

$$\varepsilon_1 + \varepsilon_2 - \varepsilon_3 - \varepsilon_4 + 2\Delta_E = \frac{k^2}{2m_a} + U_a(r,t) + \frac{p^2}{2m_b} + U_b(r+\Delta_2, t) - \frac{1}{2m_a}(k - q + \Delta_K^a)^2$$

$$- U_a(r + \Delta_3, t + \Delta_t) - \frac{1}{2m_b}(p + q + \Delta_K^b)^2 - U_b(r + \Delta_4, t + \Delta_t) + \frac{\partial U_a}{\partial t}\Delta_t + \frac{\partial U_b}{\partial t}\Delta_t$$

$$= \frac{k^2}{2m_a} + \frac{p^2}{2m_b} - \frac{(k-q)^2}{2m_a} - \frac{(p+q)^2}{2m_b} - \frac{\partial U_a}{\partial r}\Delta_3 - \frac{\partial U_b}{\partial r}(\Delta_4 - \Delta_2) - \frac{k-q}{m_a}\Delta_K^a - \frac{p+q}{m_b}\Delta_K^b.$$

For assumed gravitation-like forces, we have

$$\frac{\partial U_{a,b}}{\partial r} = -m_{a,b}A \qquad \text{and} \qquad \Delta_K^{a,b} = m_{a,b}A\Delta_t,$$

from which follows the sum of Δ's as

$$-\frac{\partial U_a}{\partial r}\Delta_3 - \frac{\partial U_b}{\partial r}(\Delta_4 - \Delta_2) - \frac{k-q}{m_a}\Delta_K^a - \frac{p+q}{m_b}\Delta_K^b = A\Big(m_a\Delta_3 + m_b(\Delta_4 - \Delta_2) - (k+p)\Delta_t\Big).$$

Using centre-of-mass continuity (3.12) with $\Delta^f = \Delta_t v^{mol} = \Delta_t \frac{k+p}{m_a+m_b}$, one finds that all Δ-terms cancel.

Note that the scattering-out (3.22) and the scattering-in (3.23) have reversed signs of all Δ's. This follows from the fact that the scattering-in process is the time- and space-reversed counterpart of the scattering-out one. Simply, in the scattering-out particle a faces a collision it will undergo with its partner b in front of it, while in the scattering-in particle a has gone through a collision leaving its partner b behind.

3.2.3 Summary

The interaction of particles via long-range potential tails is approximated by a mean field which acts as an external field. The effect of the mean field on free particles is covered by the momentum drift. The effect of the mean field on the colliding pairs causes the momentum and the energy gains which enter the scattering integral.

3.3 Internal mechanism of energy conversion

The effect of external forces is not the only microscopic mechanism due to which the energy of the colliding pair of particles changes during the collision. In dense systems, the binary collisions are not isolated from the rest of the system but feel the system so that the nominally binary collisions are in fact many-body processes.

To demonstrate such processes, let us assume a model studied by Kortemeyer et al. (1996). They describe the nonlocality of the scattering integral by hard-sphere corrections with the sphere diameter determined from the cross section. Due to in-medium effects, the sphere diameter depends on the density of nucleons, therefore it changes with time following changes in the density. For simplicity, we assume that the diameter of particles $d_a(t)$ is a known function of time with no regard to the mechanism by which it appears. In other words, we discuss a system of breathing hard spheres.

Using the centre-of-mass and the relative coordinates, the collision of two breathing hard spheres corresponds to a scattering of a point particle of the relative mass $\mu = \frac{m_a m_b}{m_a + m_b}$ on the breathing central potential which is infinite at relative distances shorter than $\frac{1}{2}(d_a(t) + d_b(t))$. At the instant of the collision we can decompose the relative velocity $v = \frac{\kappa}{\mu}$ into its tangential and normal components, $v_\parallel$ and $v_\perp$, respectively. The tangential velocity conserves in the collision, the normal does not.

If the central potential does not breath, i.e. for time-independent hard spheres, the normal velocity flips its sign, $v_\perp^{\text{fin}} = -v_\perp$. The final kinetic energy, $\frac{1}{2}\mu v_\parallel^2 + \frac{1}{2}\mu (v_\perp^{\text{fin}})^2$, then equals the initial kinetic energy, $\frac{1}{2}\mu v_\parallel^2 + \frac{1}{2}\mu v_\perp^2$. To make a link with variables used in the kinetic equation we use the fact that the total change of the momentum, $\mu(v_\perp^{\text{fin}} - v_\perp)$, is $-q$, so that $v_\perp = \frac{q}{2\mu}$.

If the central potential breath, i.e. for the time-dependent hard spheres, the normal velocity with respect to the surface is higher by the velocity of the surface, $v_{\text{surf}} = \frac{1}{2}\frac{\partial}{\partial t}(d_a(t) + d_b(t))$. After the collision the normal velocity is given by the flip of the relative motion with respect to the surface, $v_\perp^{\text{fin}} = -v_\perp + 2v_{\text{surf}}$. The final kinetic energy, $\frac{1}{2}\mu v_\parallel^2 + \frac{1}{2}\mu (v_\perp^{\text{fin}})^2 = \frac{1}{2}\mu v_\parallel^2 + \frac{1}{2}\mu v_\perp^2 - 2\mu v_\perp v_{\text{surf}}$, then differs from the initial one. The last term is the corresponding energy gain $-2\Delta_E$. In terms of the transferred momentum the energy gain reads[6]

$$\Delta_E = \mu v_\perp v_{\text{surf}} = \frac{|q|}{4} \frac{\partial}{\partial t} \left(d_a(t) + d_b(t) \right).$$ (3.27)

In the model of breathing hard spheres the total energy of the system equals the kinetic energy. This total energy does not conserve because the mechanism changing the diameter of spheres feeds the energy into the system. In a realistic system the breathing of the effective nucleon diameter is caused by changes in the correlations between particles. The total energy conserves since it is a sum of the kinetic and potential parts. The energy gain Δ_E then represents a mechanism by which the potential energy of binary correlations is converted into the kinetic energy.

3.4   Equation of continuity

The nonlocality and the collision delay produce virial corrections to densities, currents and fluxes. These corrections are important for conservation laws, and in particular, for the equation of state.

3.4.1   Correlated density

Since the particles form effective molecules, the density of particles a is the sum of the density of free particles, the density of non-symmetric molecules $(b \neq a)$ and twice the density of symmetric molecules $(b = a)$. To identify these three contributions, we evaluate the total density of particles.

The total density of particles obeys the equation of continuity

$$\frac{\partial n_a}{\partial t} + \frac{\partial j_a}{\partial r} = 0.$$ (3.28)

Integrating the kinetic equation (3.13) over momentum k, one finds on the left-hand side gradients of

$$n_a^{\text{free}}(r, t) = \int \frac{dk}{(2\pi)^3} f_a(k, r, t), \qquad j_a^{\text{free}}(r, t) = \int \frac{dk}{(2\pi)^3} \frac{k}{m_a} f_a(k, r, t).$$ (3.29)

With the Boltzmann scattering integral (all $\Delta \to 0$), the right-hand side of Eq. (3.13) under the momentum integration vanishes; therefore one can interpret n_a^{free} and j_a^{free} as the density and current of particles, respectively. For the non-instant scattering integral,

---

[6] If the external field is present, contribution (3.20) has to be added.

however, the right-hand side of Eq. (3.13) does not vanish and the density and current pick up virial corrections,

$$n_a = n_a^{\text{free}} + n_a^{\text{corr}}, \qquad j_a = j_a^{\text{free}} + j_a^{\text{corr}}, \tag{3.30}$$

where the correlated density n_a^{corr} is given by the time gradient term and the correlated current is given by the space gradient term remaining from the scattering integral.

The time gradient appears after expansion of distributions (3.14) with respect to the collision delay Δ_t. Since the time dependency enters the scattering integral exclusively via distributions, one finds the correlated density to be[7]

$$n_a^{\text{corr}} = \sum_b \int \frac{dk}{(2\pi)^3} \frac{dp}{(2\pi)^3} \frac{dq}{(2\pi)^3} P_\delta \Delta_t f_a(k-q, r, t) f_b(p+q, r, t), \tag{3.31}$$

The space gradients appear due to displacements in (3.14). Expanding $\Delta_{2,3,4}$ one finds a rather complex expression[8]

$$\frac{\partial j_a^{\text{corr}}}{\partial r} = \sum_b \int \frac{dk\,dp\,dq}{(2\pi)^9} P_\delta$$

$$\times \left(f_a(k) \frac{\partial f_b(p)}{\partial r} \Delta_2 + f_b(p+q) \frac{\partial f_a(k-q)}{\partial r} \Delta_3 + f_a(k-q) \frac{\partial f_b(p+q)}{\partial r} \Delta_4 \right). \tag{3.32}$$

---

[7] We use that $P_\delta = P \, 2\pi \, \delta_{1234}$ and Δ_t are independent of time,

$$\int \frac{dk\,dp\,dq}{(2\pi)^9} P_\delta \left(\frac{\partial f_a}{\partial t} f_b + f_a \frac{\partial f_b}{\partial t} \right) \Delta_t = \frac{\partial}{\partial t} \int \frac{dk\,dp\,dq}{(2\pi)^9} P_\delta f_a f_b \Delta_t.$$

In absence of the mean field, this assumption is justified since an isolated collision does not depend on time in which it happens. The collision under the effect of the mean field might depend on the field, but in reality all time dependencies cancel. For instance, the energy conservation (3.24) is written exclusively in terms of the local kinetic energies, i.e. the time dependency via the mean fields cancel. The same is true for the scattering probability P and for the collision delay Δ_t.

[8] Formal expansion provides also a space gradient due to the drift,

$$\int \frac{dk}{(2\pi)^3} \frac{\partial \varepsilon_1}{\partial r} \frac{\partial f_1}{\partial k} = \frac{\partial U_a}{\partial r} \int \frac{dk}{(2\pi)^3} \frac{\partial f_1}{\partial k} = 0,$$

which vanishes as can be seen integrating by parts over momentum.

An additional vanishing term results from the momentum gain,

$$\int \frac{dk\,dp\,dq}{(2\pi)^9} P_\delta \left(\frac{\partial f_3}{\partial k} \Delta_K^a f_4 + f_3 \frac{\partial f_4}{\partial p} \Delta_K^b \right) = \int \frac{dk\,dp\,dq}{(2\pi)^9} P_\delta \left(\frac{\partial f_3}{\partial k} f_4 + f_3 \frac{\partial f_4}{\partial p} \right) \left(\frac{\partial U_a}{\partial r} + \frac{\partial U_b}{\partial r} \right) \Delta_t$$

$$= \left(\frac{\partial U_a}{\partial r} + \frac{\partial U_b}{\partial r} \right) \int \frac{dk\,dp\,dq}{(2\pi)^9} P_\delta \Delta_t \left(\frac{\partial}{\partial k} + \frac{\partial}{\partial p} \right) f_3 f_4$$

$$= -\left(\frac{\partial U_a}{\partial r} + \frac{\partial U_b}{\partial r} \right) \int \frac{dk\,dp\,dq}{(2\pi)^9} f_3 f_4 \left(\frac{\partial}{\partial k} + \frac{\partial}{\partial p} \right) P_\delta \Delta_t$$

$$= 0.$$

The integral vanishes because P, δ_{1234} and Δ_t depend exclusively on the relative momentum $k-p$.

To show that the right-hand side of Eq. (3.32) reduces to a gradient of a single function, we employ the time inversion from which follows

$$\Delta_2(k, p, q) = \Delta_4(-k+q, -p-q, -q) - \Delta_3(-k+q, -p-q, -q), \tag{3.33}$$

and the symmetry of vectors under space inversion

$$\Delta_i(k-q, p+q, q) = -\Delta_i(-k+q, -p-q, -q). \tag{3.34}$$

These two symmetries allow us to interchange the initial and final states in the first term

$$f_a(k)\frac{\partial f_b(p)}{\partial r}\Delta_2 \to f_a(k-q)\frac{\partial f_b(p+q)}{\partial r}(\Delta_3 - \Delta_4). \tag{3.35}$$

The terms with Δ_4 cancel and one finds

$$j_a^{\text{corr}} = \sum_b \int \frac{dk}{(2\pi)^3}\frac{dp}{(2\pi)^3}\frac{dq}{(2\pi)^3} P_\delta \Delta_3 f_a(k-q, r, t) f_b(p+q, r, t). \tag{3.36}$$

Density of molecules   It is advantageous to view the correlated density in terms of chemical approach as molecules. Let us define densities of *ab*-molecules as

$$n_{ab} = \left(1 - \frac{1}{2}\delta_{ab}\right) \int \frac{dkdpdq}{(2\pi)^9} P_\delta \Delta_t f_a(k-q, r, t) f_b(p+q, r, t). \tag{3.37}$$

Clearly,

$$n_a^{\text{corr}} = 2n_{aa} + \sum_{b\neq a} n_{ab}, \tag{3.38}$$

which confirms that the total density of particles a is the sum of free particles and contributions from molecules. Formula (3.37) then has a straightforward probabilistic interpretation. The product $P_\delta f_a f_b$ provides a rate at which the effective molecules are formed. This rate, weighted with a 'molecular life-time' Δ_t, then gives a molecular density. The probabilistic interpretation of the current (3.36) is similar to the density (3.37). With probability $P_\delta f_a f_b \Delta_t$, particle a moves with the average velocity $\frac{\Delta_3}{\Delta_t}$.

The classical analogy of scattering states with molecules allows us to view formula (3.31) as a law of acting masses. Indeed, for the equilibrium distribution of classical particles

$$f_a = n_a^{\text{free}}\left(\frac{2\pi}{m_a k_B T}\right)^{\frac{3}{2}} e^{-\frac{k^2}{2m_a k_B T}}, \tag{3.39}$$

one finds from Eq. (3.37) that the density of effective molecules obeys the Guldberg—Waage law of acting masses

$$n_{ab} = n_a^{\text{free}} n_b^{\text{free}} K(T). \tag{3.40}$$

The n_a^{free} is a density of free particles a which enters the reaction. Within the chemical picture, its is clear that the density of free particles is lower than the density of all particles. Note that the distribution function f_a is a distribution of free particles a only. In other words, particles just undergoing a collision are excluded from f_a. For the scattering integral this means that particles just undergoing some collision are excluded from initial conditions for all other scattering events. Of course, the kinetic equation (3.13) does not include any chemical reaction, the concept of molecules only serves to describe collisions of finite duration.

3.5 Pressure

The space nonlocality of the scattering leads to corrections to the pressure. For hard spheres these corrections are known as the inaccessible volume in the van der Waals equation of state. The collision delay also modifies the pressure since the effective molecules reduce the number of freely travelling momentum carriers in the system. Also in heavy-ion collisions the equation of state is an important objective studied via kinetic equations. Therefore it is important to remind ourselves about the concept of collision flux, see (Chapman and Cowling, 1990).

The pressure $\mathscr{P}$ is conveniently evaluated from the momentum flux which measures how a momentum is transferred along a selected direction, say along the axis x. The major part of momentum flux is carried by free particles which bring the so-called kinetic part of the pressure

$$\mathscr{P}_{\text{kin}} = \sum_a \int \frac{dk}{(2\pi)^3} \frac{k_x}{m_a} k_x f_a(k) = \frac{1}{3} \sum_a \int \frac{dk}{(2\pi)^3} \frac{k^2}{m_a} f_a(k) = k_B T \sum_a n_a^{\text{free}}. \tag{3.41}$$

Now we re-derive (3.41) from probabilistic arguments. Similar arguments will be then used for the collision flux. Let us assume particle a of momentum k at coordinate $r^{\text{be}} = r$ and time $t^{\text{be}} = t$. After time dt, i.e. at $t^{\text{end}} = t + dt$, this particle is at $r^{\text{end}} = r + \frac{k}{m_a} dt$. The momentum has been transferred by $kr^{\text{end}} - kr^{\text{be}} = k\frac{k}{m_a} dt$. The transferred momentum per time unit can thus be written as $\frac{kr^{\text{end}} - kr^{\text{be}}}{t^{\text{end}} - t^{\text{be}}} = k\frac{k}{m_a}$. This contribution is then summed with probability f_a to find the particle of a given momentum.

The free motion of particles is not the only mechanism to carry the momentum from one part of the system to another. Apparently, molecules carry their share of the momentum in a similar way as reacting gases. Moreover, the transfer momentum q in a collision jumps over a distance Δ^d which for hard spheres gives well-known corrections to the inaccessible volume. Finally, the rotating molecule carries a part of the momentum via its rotation. All these contributions to the momentum flux are called the collision flux as they appear due to internal dynamics during the collision.

In the collision event, the initial momenta are k and p, the final are $k-q$ and $p+q$. The initial and final coordinates are specified in Fig. 3.6. The momentum transfer is given by the positions of particles at matching points of the event $(k-q)r_a^{\text{end}} + (p+q)r_b^{\text{end}} - kr_a^{\text{be}} - pr_b^{\text{be}} = (k-q)\Delta_3 + (p+q)\Delta_4 - p\Delta_2$. The time needed for this transfer is just the collision delay $t^{\text{end}} - t^{\text{be}} = \Delta_t$, therefore the collision transfer time per unit is $\frac{1}{\Delta_t}((k-q)\Delta_3 + (p+q)\Delta_4 - p\Delta_2)$. The probability of finding such a 'molecule' is $P_\delta f_a(k)f_b(p)\Delta_t$. The pressure contribution thus reads

$$\Delta\mathscr{P} = \frac{1}{6}\sum_{a,b}\int \frac{dk\,dp\,dq}{(2\pi)^9} P_\delta f_a(k)f_b(p)\left[(k-q)\Delta_3 + (p+q)\Delta_4 - p\Delta_2\right]. \tag{3.42}$$

The factor $\frac{1}{2}$ compensates for the fact that each event is counted twice as follows from the interchange of $(a,k) \longleftrightarrow (b,p)$.

The pressure correction (3.42) can be simplified for equilibrium distributions (3.39) if we separate the translational and the internal degrees of freedom. Going into centre-of-mass variables, $K = k+p$, $\kappa = \mu\left(\frac{k}{m_a} - \frac{p}{m_b}\right)$, $\mu = \frac{m_a m_b}{m_a+m_b}$, $M = m_a + m_b$, and expressing $\Delta_{2,3,4}$ from microscopic characteristics (3.11)–(3.12), the pressure correction (3.42) can be rearranged as

$$\Delta\mathscr{P} = \frac{1}{6}\sum_{a,b} n_a^{\text{free}} n_b^{\text{free}} \int \frac{dK}{(2\pi)^3}\left(\frac{2\pi}{Mk_BT}\right)^{\frac{3}{2}} e^{-\frac{K^2}{2Mk_BT}} \int \frac{d\kappa}{(2\pi)^3}\left(\frac{2\pi}{\mu k_BT}\right)^{\frac{3}{2}} e^{-\frac{\kappa^2}{2\mu k_BT}}$$

$$\times \int \frac{dq}{(2\pi)^3} P 2\pi\delta\left(\frac{\kappa^2}{2\mu} - \frac{(\kappa-q)^2}{2\mu}\right)\left(\frac{K^2}{M}\Delta_t + |q|R\cos\frac{\vartheta}{2} - |2\kappa-q|R\sin\frac{\vartheta}{2}\right) \tag{3.43}$$

The total momentum K can be directly integrated out, the integration over momenta κ and q can be simplified for isotropic systems. Using the fact that $|q| = 2|\kappa|\sin\frac{\theta}{2}$ and $|2\kappa - q| = 2|\kappa|\cos\frac{\theta}{2}$, where θ is the deflection angle, see Fig. 3.4, one finds $[\kappa \equiv |\kappa|]$

$$\Delta\mathscr{P} = \frac{1}{6}\sum_{a,b} n_a^{\text{free}} n_b^{\text{free}} \frac{\mu}{(2\pi)^3} \int_0^\infty d\kappa\,\kappa^3 \left(\frac{2\pi}{\mu k_BT}\right)^{\frac{3}{2}} e^{-\frac{\kappa^2}{2\mu k_BT}}$$

$$\times \int_0^\pi d\theta \sin\theta P_\delta\left(3k_BT\Delta_t + 2\kappa R\sin\frac{\vartheta-\theta}{2}\right). \tag{3.44}$$

This pressure correction can be decomposed in two parts,

$$\Delta\mathscr{P} = \mathscr{P}_{\text{mol}} + \mathscr{P}_{\text{sp}}, \tag{3.45}$$

where the pressure correction $\mathscr{P}_{mol}$ is given by Δ_t and $\mathscr{P}_{sp}$ is given by the second term of the last integrand in Eq. (3.44).

3.5.1 Partial pressure of effective molecules

Except for factors, the pressure contribution proportional to Δ_t has the same integrand as the molecular densities, i.e. it brings a partial pressure of effective molecules,

$$\mathscr{P}_{mol} = k_B T \left(\sum_a n_{aa} + \frac{1}{2} \sum_{a \neq b} n_{ab} \right). \tag{3.46}$$

This relation between the correlated density and the correction to pressure reflects the centre-of-mass motion of the molecule. It holds thus with no regard to actual interaction forces.

Although for the positive density of molecules the molecular pressure $\mathscr{P}_{mol}$ is also positive, the presence of molecules reduces the total pressure because it reduces the density of free particles. Since the correct parameter of the pressure is the total density $n_a = n_a^{free} + n_a^{corr}$, where n_a^{corr} is reduced by the number of molecules (3.38), the total pressure results in

$$\mathscr{P} = \mathscr{P}_{kin} + \mathscr{P}_{mol} + \mathscr{P}_{sp} = k_B T \sum_a \left(n_a - n_{aa} - \frac{1}{2} \sum_{b \neq a} n_{ab} \right) + \mathscr{P}_{sp}. \tag{3.47}$$

3.5.2 Inaccessible volume

The second term in (3.44) denoted by $\mathscr{P}_{sp}$ brings the part due to space nonlocality following from the relative motion of particles,

$$\mathscr{P}_{sp} = \frac{1}{3} \sum_{a,b} n_a^{free} n_b^{free} \frac{\mu}{(2\pi)^3} \int_0^\infty d\kappa \, \kappa^4 \left(\frac{2\pi}{\mu k_B T} \right)^{\frac{3}{2}} e^{-\frac{\kappa^2}{2\mu k_B T}} \int_0^\pi d\theta \sin\theta \, P_\delta R \sin\frac{\vartheta - \theta}{2}. \tag{3.48}$$

The pressure contribution of the finite inter-atomic distance R and the rotation over the angle ϑ cannot be simplified in a general case, because generally speaking, both parameters are functions of the deflection angle and of the relative momentum of colliding particles, $R \equiv R(\kappa, \theta)$ and $\vartheta \equiv \vartheta(\kappa, \theta)$. The integration, however, can be carried through for the hard-sphere model which we use to drive these corrections into a familiar form.

For hard spheres, the space displacement is a constant, $R = R_a + R_b$, the collision delay and the rotation angle vanish, $\Delta_t = 0$ and $\vartheta = 0$, and the differential cross-section P is proportional to $(R_a + R_b)^2 \sin 2\theta$. Formula (3.48) then gives the correction

$$\mathcal{P}_{sp} \xrightarrow{\text{hard sphere}} k_B T \frac{1}{2} \sum_{a,b} n_a n_b \frac{4\pi}{3} (R_a + R_b)^3 . \qquad (3.49)$$

This is the second-order virial correction to the pressure in the hard-sphere mixture.

For the single-component gas, the pressure correction (3.49) simplifies to the familiar van der Waals form. Hard spheres do not form effective molecules, thus $\mathcal{P}_{\text{mol}} = 0$. The total pressure then reads[9]

$$\mathcal{P} = \mathcal{P}_{\text{kin}} + \mathcal{P}_{sp} = k_B T n + k_B T n^2 \frac{2\pi}{3} R^3 . \qquad (3.50)$$

Within accuracy to linear terms, relation (3.50) is equivalent to the familiar van der Waals equation of state (2.19)

$$\mathcal{P} = \frac{N k_B T}{\Omega - \Omega_{\text{un}}} , \qquad (3.51)$$

where $N = n\Omega$ is the total number of particles, Ω is the sample volume, and

$$\Omega_{\text{un}} = N \frac{2\pi}{3} d^3 \qquad (3.52)$$

is the inaccessible volume Nb with the parameter $b = \frac{2\pi}{3} d^3$.

With respect to the chemical interpretation, it should be noted that real molecules live long enough so that their rotation angle ϑ reaches large random values. In cases where the integrand of Eq. (3.48) rapidly oscillates, the pressure $\mathcal{P}_{sp}$ vanishes.

3.5.3 Mean-field contribution to pressure

All that is missing still is the pressure caused by the mean field. As we have seen on the 1D model, a mean field results in the internal pressure of the van der Waals equation. In the kinetic equation (3.26), the mean field contributes to a drift and also affects dissipation. To evaluate both contributions, we have to approach the pressure directly from the kinetic equation.

The pressure $\mathcal{P}$ results from the momentum balance in the system. The particles in the volume $d\Omega$ have a total momentum $dQ_x = d\Omega \int dv f(r, k, t) k_x$, where subscript x denotes a vector component. The inertial force, $dF_x = \frac{dQ_x}{dt}$, has to be balanced with

---

[9] Here, $R = 2R_a$ is the closest distance of two particles of kind a. The parameter R thus appear in the repulsive potential for two hard spheres. The radius R of the interaction thus equals the diameter $2R_a$ of a particle which makes a notation of radii and diameters sometimes confusing.

a force acting on the surface of the cube (oriented in the direction of dF). At the upwind/downwind side, the force is $F_{\text{up,down}} = \pm dx^2 \mathscr{P}(x \pm \frac{1}{2} dx, t)$. The balance of forces, $dF_x + F_{\text{up}} + F_{\text{down}} = 0$, thus requires

$$\frac{\partial}{\partial t} \sum_a \int \frac{dk}{(2\pi)^3} k_x f_1 + \frac{\partial \mathscr{P}}{\partial x} = 0. \tag{3.53}$$

To evaluate the pressure, we solve Eq. (3.53) with the first term evaluated from the kinetic equation.[10] The space drift results in the kinetic contribution,

$$-\sum_a \int \frac{dk}{(2\pi)^3} k_x \frac{\partial \varepsilon_1}{\partial k} \frac{\partial f_1}{\partial r} = -\sum_{y=1}^{3} \frac{\partial}{\partial r_y} \sum_a \int \frac{dk}{(2\pi)^3} k_x \frac{k_y}{m_a} f_1 = -\frac{\partial \mathscr{P}_{\text{kin}}}{\partial x}. \tag{3.54}$$

The momentum drift results as

$$\sum_a \int \frac{dk}{(2\pi)^3} k_x \frac{\partial \varepsilon_1}{\partial r} \frac{\partial f_1}{\partial k} = -\sum_a \frac{\partial U_a}{\partial r_x} \int \frac{dk}{(2\pi)^3} f_1 = -\sum_a n_a^{\text{free}} \frac{\partial U_a}{\partial r_x}. \tag{3.55}$$

The collision integral contributes via space displacements Δ_{234} which give $\Delta \mathscr{P}$ derived previously from the statistical assumptions, and via the momentum gain which results in[11].

$$\sum_{ab} \int \frac{dk\,dp\,dq}{(2\pi)^9} P\delta_{1234} \left(\bar{f}_3 \bar{f}_4 - f_1 f_2 \right) = -\frac{\partial \Delta \mathscr{P}}{\partial x} - \sum_a n_a^{\text{corr}} \frac{\partial U_a}{\partial x}. \tag{3.56}$$

The contributions of the mean field can be collected,

$$\sum_a n_a^{\text{free}} \frac{\partial U_a}{\partial x} + \sum_a n_a^{\text{corr}} \frac{\partial U_a}{\partial x} = \sum_a (n_a^{\text{free}} + n_a^{\text{corr}}) \frac{\partial U_a}{\partial x}$$

$$= \sum_{ab} n_a \bar{\mathscr{V}}_{ab} \frac{\partial n_b}{\partial x} = \frac{\partial}{\partial x} \frac{1}{2} \sum_{ab} n_a n_b \bar{\mathscr{V}}_{ab}, \tag{3.57}$$

---

[10] We take the time derivative inside the integral and use

$$\frac{\partial f_1}{\partial t} = -\frac{\partial \varepsilon_1}{\partial k} \frac{\partial f_1}{\partial r} + \frac{\partial \varepsilon_1}{\partial r} \frac{\partial f_1}{\partial k} + \sum_b \int \frac{dp\,dq}{(2\pi)^6} P\delta_{1234} \left(\bar{f}_3 \bar{f}_4 - f_1 f_2 \right).$$

[11] The proof that $\Delta \mathscr{P}$ results from Δ_{234} the reader can find in a more sophisticated form in Section 15.5.3. The contribution of Δ_K,

$$\sum_{ab} \int \frac{dk\,dp\,dq}{(2\pi)^9} P\delta k_x \left(\frac{\partial f_3}{\partial k} \Delta_K^a f_4 + f_3 \frac{\partial f_4}{\partial p} \Delta_K^b \right) = -\sum_{ab} \frac{1}{2} \left(\frac{\partial U_a}{\partial x} + \frac{\partial U_b}{\partial x} \right) \int \frac{dk\,dp\,dq}{(2\pi)^9} P\delta \Delta_i f_3 f_4$$

$$= -\sum_a \frac{\partial U_a}{\partial x} n_a^{\text{corr}},$$

can be rearranged in parallel with the vanishing Δ_K contribution to the current. After integration by part over momenta, however, the derivative $\frac{\partial k_x}{\partial k}$ does not cancel.

from which one directly identifies van der Waals pressure caused by the mean field,

$$\mathscr{P}_{mf} = \frac{1}{2}\sum_{ab} n_a n_b \bar{\mathcal{V}}_{ab}. \tag{3.58}$$

The Δ_K contribution shows that the field effect on the collision influences the pressure in the system. Here it resulted in mean-field forces between molecules and particles.

3.6   Latent heat

In systems close to equilibrium, the energy gain Δ_E describes a conversion between the kinetic and the potential energy known as the latent heat. Thermodynamical consequences of this mechanism can be demonstrated on the system of breathing hard spheres introduced above. For simplicity we will assume only the single-component gas which can be compared with the van der Waals equation of state.

Let us start from the thermodynamical picture. An increase of the particle diameter d decreases the accessible volume $\Omega - \Omega_{un}$. For an adiabatically isolated system of constant volume Ω, the total energy W then changes as

$$\frac{\partial W}{\partial t} = -\mathscr{P}\frac{\partial(\Omega - \Omega_{un})}{\partial t} = \mathscr{P}\frac{\partial \Omega_{un}}{\partial t}. \tag{3.59}$$

The right-hand side simply measures the work the hard spheres have to exert to extend the inaccessible volume. Within the lowest-order virial correction we take the pressure $\mathscr{P} = k_B T n$ and the inaccessible volume (3.52). The rate of energy change per unit volume thus reads

$$\frac{1}{\Omega}\frac{\partial W}{\partial t} = k_B T n^2 2\pi d^2 \frac{\partial d}{\partial t}. \tag{3.60}$$

Now we turn to the microscopic picture. Since the only energy in the system of hard spheres is the kinetic one (the extension of hard spheres is treated as the external work),

$$W = \Omega \int \frac{dk}{(2\pi)^3}\frac{k^2}{2m}f(k), \tag{3.61}$$

we can evaluate W from the kinetic equation (3.26) using the energy gain (3.27). If the energy gain is a correct concept, the time derivative of W given by Eq. (3.61) will agree with Eq. (3.60).

For homogeneous systems all space gradients vanish and the energy gain (3.27) remains as the only non-zero correction for breathing hard spheres. The time derivative of kinetic energy from (3.26) results as

$$\frac{1}{\Omega}\frac{\partial W}{\partial t} = \int \frac{dk}{(2\pi)^3}\frac{k^2}{2m}\frac{\partial f(k)}{\partial t}$$

$$= \frac{1}{2}\int \frac{dp'\,dq'\,dk'}{(2\pi)^9}\left(\frac{k'^2}{2m}+\frac{p'^2}{2m}\right)Pf(k'-q')f(p'+q')$$

$$\times 2\pi\delta\left(\frac{k'^2}{2m}+\frac{p'^2}{2m}-\frac{(k'-q')^2}{2m}-\frac{(p'+q')^2}{2m}-|q'|\frac{\partial d}{\partial t}\right)$$

$$-\frac{1}{2}\int \frac{dpdqdk}{(2\pi)^9}\left(\frac{k^2}{2m}+\frac{p^2}{2m}\right)Pf(k)f(p)$$

$$\times 2\pi\delta\left(\frac{k^2}{2m}+\frac{p^2}{2m}-\frac{(k-q)^2}{2m}-\frac{(p+q)^2}{2m}+|q|\frac{\partial d}{\partial t}\right). \tag{3.62}$$

In the second line, we have first used the kinetic equation (3.26) to substitute for $\frac{\partial f(k)}{\partial t}$ and employed the symmetry between momenta k and p to write the mean-kinetic energy in the symmetrical form, $\left\langle\frac{k^2}{2m}\right\rangle = \frac{1}{2}\left\langle\frac{k^2}{2m}+\frac{p^2}{2m}\right\rangle$.

Now we unify the first and the second terms of Eq. (3.62). To this end we make a substitution in the first term, which interchanges the initial and the final states,

$$k = k' - q'$$
$$p = p' + q' \tag{3.63}$$
$$q = -q'.$$

This substitution turns the initial kinetic energies into the final ones, $k'^2 + p'^2 \rightarrow (k-q)^2 + (p+q)^2$. Accordingly, the difference between the first and the second terms of Eq. (3.62) is proportional to the difference between the initial and final energies,

$$\left(\frac{k^2}{2m}+\frac{p^2}{2m}-\frac{(k-q)^2}{2m}-\frac{(p+q)^2}{2m}\right)$$

$$\times 2\pi\delta\left(\frac{k^2}{2m}+\frac{p^2}{2m}-\frac{(k-q)^2}{2m}-\frac{(p+q)^2}{2m}+|q|\frac{\partial d}{\partial t}\right)$$

$$= -|q|\frac{\partial d}{\partial t}2\pi\delta\left(\frac{k^2}{2m}+\frac{p^2}{2m}-\frac{(k-q)^2}{2m}-\frac{(p+q)^2}{2m}\right), \tag{3.64}$$

which is just the energy gain. The rate of energy change thus reads

$$\frac{1}{\Omega}\frac{\partial W}{\partial t} = \int \frac{dpdqdk}{(2\pi)^9}\frac{|q|}{2}\frac{\partial d}{\partial t}Pf(k)f(p)2\pi\delta\left(\frac{k^2}{2m}+\frac{p^2}{2m}-\frac{(k-q)^2}{2m}-\frac{(p+q)^2}{2m}\right). \tag{3.65}$$

This integral can be readily evaluated for the hard-sphere differential cross section giving the expected relation (3.60).

Formula (3.65) has a clear statistical interpretation. At each collision event the breathing of particles feeds the system with the energy $2\Delta_E = |q|\frac{\partial d}{\partial t}$. The number of collision events is given by the standard probability to find two particles to collide times the differential cross section.

The example of breathing hard spheres shows that the energy gain Δ_E describes a mechanism by which the kinetic energy which particles carry between collisions can change during the collisions. Of course, the breathing diameter of hard spheres is just an artificial feature of the model discussed. In fermionic matter, however, the effective particle diameter does depend on time via the density reflecting the correlation with other particles of the system. In particular, the internal dynamics of the collision of two nucleons in a medium of other nucleons is restricted to the unoccupied phase space via the Pauli blocking. Accordingly, the energy gain is made possible by correlations that are of a many-body nature.

Unlike the field effect on scattering, the energy gain in the case of breathing hard spheres has no link to the finite collision duration. As far as we know, such a mechanism has not been discussed in print. This likely follows from the fact that classical collisions of two isolated particles cannot lead to changes of particle diameters. The problem of the energy conversion is treated by Klimontovich (1975). He cuts the BBGKY hierarchy at the three-particle correlation function in agreement with the expectation that the energy conversion is due to the many-body processes. Klimontovich's results have been recovered by Snider (1995) from a purely two-particle treatment which is partially self-consistent by a construction of the initial condition for the scattering integral. Nevertheless, Snider emphasised that a consistent description of the energy conversion requires us to deal with non-instant scattering integrals. In terms of the above intuitive picture, the non-instant approach is necessary to cover the energy conversion due to the field effect on the scattering.

3.7   Entropy production

Any collision dissipates energy and momentum and leads therefore to an increase of disorder and entropy. The entropy (2.42) written for single-particle distributions,

$$S = -k_B \sum_1 f_1 \ln f_1, \tag{3.66}$$

can now be shown to increase with time continuously until it reaches the maximal value in equilibrium. For the purpose of legibility we provide proof of Ludwig Boltzmann without nonlocal corrections at the moment. When we will derive the nonlocal kinetic theory for quantum systems in Chapter 13.3, we will provide the general proof for nonlocal events as well in Chapter 15.5.5.

Calling $s_1 = f_1 \ln f_1$ one has many derivative as $\partial s_1 = \partial f_1 (\ln f_1 + 1)$ and multiplying the kinetic equation (3.26) with s_1 and integrating over k leads to the time derivative of the entropy (3.66)

$$\frac{\partial S}{\partial t} + \sum_1 \left(\frac{\partial}{\partial p} \epsilon_1 \frac{\partial}{\partial r} s_1 - \frac{\partial}{\partial r} \epsilon_1 \frac{\partial}{\partial p} s_1 \right) = \sum_{1234} I_{12|34} (\ln s_1 + 1). \qquad (3.67)$$

Here we abbreviate the corresponding integrations over momenta by the sum and the collision integral by I. Interchanging particles $1 \leftrightarrow 2$ and $3 \leftrightarrow 4$ does not change the collision integral. Interchanging $1 \leftrightarrow 3$ and $2 \leftrightarrow 4$, however provides a minus sign. Adding all 4 possibilities and dividing by 4 provides the balance of entropy which reads

$$\frac{\partial S}{\partial t} - \frac{\partial}{\partial r} k_B \sum_1 \frac{\partial}{\partial p} \epsilon_1 f_1 \ln f_1 = -\frac{k_B}{4} \sum_{1234} I_{12|34} (f_3 f_4 - f_1 f_2) \ln \frac{f_1 f_2}{f_3 f_4}. \qquad (3.68)$$

The left side provides the balance of the entropy density by the time change and a spatial-dependent entropy current. The right side is not zero as in balance equations for the density, momentum or energy, but provides a positive entropy production every time since the appearance of expression $(a - b) \ln b/a < 0$. This confirms the second law of thermodynamics that the entropy increases until it reaches its stationary equilibrium value. It is sometimes called Boltzmann's H-theorem.[12] The extension to nonlocal collision scenario is postponed until Chapter 15.5.4.

One should reflect for a moment on what this H-theorem really means. All basic microscopic equations of motions, like the Hamilton equations in classical systems or the Schrödinger equation in quantum systems, are basically reversible leaving the entropy constant. Letting the particle collide and averaging over such events to form a distribution leads to Boltzmann's collision integral and the entropy production. The mysterious step which makes the theory change from reversible to irreversible appears in the hidden assumption that two particles collide not knowing anything about each other before. This is Boltzmann's hidden chaoticity assumption. It is enough to produce correct irreversibility as we observe in nature. We will see how such asymptotic time-reversal breaking is built in quantum kinetic theory on different levels of approximation.

3.8   Nonequilibrium hydrodynamic equations

Though we have discussed in 3.4, 3.5 and 3.7 Chapters various balance equations we want to summarise here the steps through which one obtains the known hydrodynamic equations from the Boltzmann equation for the sake of completeness. Let us write the

---

[12] Boltzmann signed the entropy in his original papers 1872; 1893 with 'E'. Then Burbury 1890 has changed the notation from E to H which was used by Boltzmann later too 1895. It was speculated that H is the capital of Greek letter *Eta* but this remains a puzzle noticed as early as (Chapman, 1937).

kinetic equation (3.26) without mean fields and nonlocal shifts in the abbreviated form

$$\frac{\partial}{\partial t}f + \frac{\mathbf{k}}{m}\frac{\partial}{\partial \mathbf{r}}f + \mathbf{F}\frac{\partial}{\partial \mathbf{k}}f = I \tag{3.69}$$

where $\mathbf{F}$ represents an external force and the collision integral I vanishes when integrating over moments of $1, \mathbf{k}, k^2/2m$. The momentum integration leads to the continuity equation for the density

$$\frac{\partial}{\partial t}n + \frac{\partial}{\partial \mathbf{r}}\mathbf{j} = 0 \tag{3.70}$$

where the density and mean-mass velocity $\mathbf{u}$ are given by

$$n(\mathbf{r}, t) = \int \frac{dk}{(2\pi)^3}f(\mathbf{k}, \mathbf{r}, t), \qquad \mathbf{j} = \int \frac{dk}{(2\pi)^3}\frac{\mathbf{k}}{m}f(\mathbf{k}, \mathbf{r}, t) = n(\mathbf{r}, t)\mathbf{u}(\mathbf{r}, t) \tag{3.71}$$

Multiplying (3.69) with $\mathbf{k}$ and integrating one obtains

$$m\frac{\partial}{\partial t}j_i + \frac{\partial}{\partial r_j}\left(nmu_iu_j + \Pi_{ij}\right) = nF_i \tag{3.72}$$

where about double occurring indices are summed. We have introduced the relative velocity $\mathbf{v} = \frac{\mathbf{k}}{m} - \mathbf{u}$ and define the stress tensor as

$$\Pi_{ij}(\mathbf{r}, t) = m\int \frac{dk}{(2\pi)^3}v_iv_jf(\mathbf{k}, \mathbf{r}, t). \tag{3.73}$$

Eq. (3.72) can be brought into a more familiar form if we multiply (3.70) with u_i and rewrite it into

$$\frac{\partial}{\partial t}j_i + \frac{\partial}{\partial r_j}(nu_iu_j) = n\left(\frac{\partial}{\partial t} + u_j\frac{\partial}{\partial r_j}\right)u_i. \tag{3.74}$$

Multiplying this with m and subtracting from (3.72) the Navier–Stokes equation appears

$$-\frac{\partial}{\partial r_j}\Pi_{ij} + nF_i = nm\left(\frac{\partial}{\partial t} + u_j\frac{\partial}{\partial r_j}\right)u_i. \tag{3.75}$$

It shows how the external and the internal forces from the stress tensor combined into a substantial derivative of the mean-mass velocity $\mathbf{u}$.

For the energy balance we multiply (3.69) with $k^2/2m$ and obtain

$$\frac{\partial}{\partial t}\int \frac{dk}{(2\pi)^3}\frac{m}{2}(u^2 + v^2)f + \frac{\partial}{\partial r_i}\int \frac{dk}{(2\pi)^3}(u_i + v_i)\frac{m}{2}(v^2 + u^2 + 2\mathbf{u}\cdot\mathbf{v})f = \mathbf{j}\cdot\mathbf{F}. \tag{3.76}$$

Introducing the mean-kinetic energy per particle and the heat flux

$$n(\mathbf{r}, t)\,\bar{\epsilon}(\mathbf{r}, t) = \int \frac{dk}{(2\pi)^3}\frac{m}{2}v^2 f(\mathbf{k}, \mathbf{r}, t), \qquad \mathbf{q}(\mathbf{r}, t) = \int \frac{dk}{(2\pi)^3}\frac{m}{2}v^2 \mathbf{v} f(\mathbf{k}, \mathbf{r}, t) \quad (3.77)$$

and adding (3.70) we get from (3.76) the form

$$\frac{\partial}{\partial t}\left[n\left(\frac{m}{2}u^2 + \bar{\epsilon}\right)\right] + \frac{\partial}{\partial r_i}\left[nu_i\left(\frac{m}{2}u^2 + \bar{\epsilon}\right) + u_j\Pi_{ij} + q_i\right] = \mathbf{j}\cdot\mathbf{F}. \quad (3.78)$$

By multiplying (3.70) now with $\frac{m}{2}u^2 + \bar{\epsilon}$ one gets the relation

$$\frac{\partial}{\partial t}\left[n\left(\frac{m}{2}u^2 + \bar{\epsilon}\right)\right] + \frac{\partial}{\partial r_i}\left[nu_i\left(\frac{m}{2}u^2 + \bar{\epsilon}\right)\right] = n\left(\frac{\partial}{\partial t} + u_j\frac{\partial}{\partial r_j}\right)\left(\frac{m}{2}u^2 + \bar{\epsilon}\right) \quad (3.79)$$

and we recognise the first two terms in (3.78) such that (3.78) can be written as

$$n\left(\frac{\partial}{\partial t} + u_j\frac{\partial}{\partial r_j}\right)\left(\frac{m}{2}u^2 + \bar{\epsilon}\right) + \frac{\partial}{\partial r_i}\left[u_j\Pi_{ij} + q_i\right] = \mathbf{j}\cdot\mathbf{F}. \quad (3.80)$$

In the last step we multiply (3.75) with u_i and add it to (3.80) to see that the external force cancels and the Euler or heat transport equation appears

$$n\left(\frac{\partial}{\partial t} + u_j\frac{\partial}{\partial r_j}\right)\bar{\epsilon} + \frac{\partial}{\partial r_i}q_i = -\Pi_{ij}\frac{\partial}{\partial r_i}u_j. \quad (3.81)$$

All appearing thermodynamic quantities (3.71), (3.73) and (3.77) are given in terms of the nonequilibrium distribution $f(\mathbf{r}, \mathbf{k}, t)$. In this sense we arrive at nonequilibrium thermodynamics without any linearisation. In Chapter 15 we will extend this hydrodynamic equation to the complete quantum version including all nonlocal and non-instant off-sets. If one linearises around a proper local equilibrium one obtains the transport coefficients which we will show in Chapter 18 and the viscosity appearing from the stress tensor in Chapter 18.4.1.

3.9   Two concepts of quasiparticles

The nonlocal picture of collisions described in chapter 3.1 is not customary in the theory of dense Fermi systems. In this field, it is traditional to describe corrections beyond the non-ideal gas with the help of the 'ideal' gas of effective particles called quasiparticles. Since the parameters of the single-particle motion depend on the properties of the system; for example, the mass of effective particles depends on the density and temperature, this seemingly ideal gas can mimic the thermodynamic behaviour of a quite general interacting system.

We introduce the Landau concept of quasiparticles on a one-dimensional system of classical particles interacting via a box potential, $V_r = V$ for $r = |x_1 - x_2| < d$ and $V_r = 0$ elsewhere. In spite of the fact that the Landau concept was proposed for Fermi systems at very low temperatures, it works for this system which is far from the aimed at region of applicability. The simplicity of the selected model allows us to derive the effective mass and potential using simple analytic expressions.

The quasiparticle picture as presented in this chapter is in many aspects very simplified. The reader interested in the complete Landau quasiparticle concept is advised to read the original papers (Landau, 1957a,b; Abrikosov and Khalatnikov, 1959) or some of more recent textbooks (Smith and Hojgaard–Jensen, 1989; Baym and Pethick, 1991).

3.9.1   Virial of forces

Before we describe the system with the help of quasiparticles, it is advantageous to see its properties using an alternative theoretical tool, the virial of forces. We will treat both approaches within the same approximation.

The virial of forces (2.106), see (Hirschfelder et al., 1964), with a simple binary correlation, $g(r) = \exp(-V/k_B T)$, yields the pressure (2.107). For our model box potential the integrand is a non-zero constant for $|r| < d$, therefore

$$\mathscr{P} = nk_B T + n^2 k_B T \frac{2\pi}{3} d^3 \left(1 - e^{-\frac{V}{k_B T}}\right). \tag{3.82}$$

For a weak potential, $V \ll k_B T$, pressure (3.82) simplifies to the mean field,

$$\mathscr{P} = nk_B T + n^2 V \frac{2\pi}{3} d^3. \tag{3.83}$$

This contribution has the form of the internal pressure in the van der Waals equation of state (2.19) with $a = V \frac{2\pi}{3} d^3$.

For a strong repulsive potential, $V \gg k_B T$, the exponential goes to zero and one finds a correction for the excluded volume,

$$\mathscr{P} = nk_B T + n^2 k_B T \frac{2\pi}{3} d^3. \tag{3.84}$$

This formula is restricted to small densities, $n \frac{2\pi}{3} d^3 \ll 1$.

3.9.2   Landau's quasiparticle concept

Now let us assume that the above interacting gas can be simulated by an ideal gas made of effective particles characterised by momentum k. The quasiparticle energy we expect in a form similar to free particles,

$$\epsilon_k = \frac{k^2}{2m^*} + \phi, \tag{3.85}$$

where m^* is an effective mass and ϕ is an effective potential. Both, m^* and ϕ are functions of density and temperature.

The quasiparticle energy represents the total change of the energy $\mathcal{E}$ of the system if one adds a new quasiparticle. This is expressed by the variation of the density of energy,

$$\epsilon_k = \frac{\delta\mathcal{E}}{\delta f_k}, \tag{3.86}$$

where $\mathcal{E}$ depends exclusively on the distribution f.

Since quasiparticles behave as an effective ideal gas, they obey the collision-less Boltzmann equation (3.26),

$$\frac{\partial f}{\partial t} + \frac{\partial \epsilon}{\partial k}\frac{\partial f}{\partial r} - \frac{\partial \epsilon}{\partial r}\frac{\partial f}{\partial k} = 0, \tag{3.87}$$

called the Landau quasiparticle equation.

3.9.3 Quasiparticle pressure

The pressure of the quasiparticle gas can be derived from the momentum balance equation (3.75),

$$\frac{\partial Q}{\partial t} + \frac{\partial \mathcal{P}}{\partial r} = 0. \tag{3.88}$$

We then compare this general equation with the momentum balance obtained from the kinetic equation (3.87).

First we multiply the Landau equation (3.87) with the i-th component of the momentum, k_i, and integrate over momentum,

$$\frac{\partial}{\partial t}\int \frac{dk}{(2\pi)^3} k_i f + \int \frac{dk}{(2\pi)^3} k_i \frac{\partial \epsilon}{\partial k}\frac{\partial f}{\partial r} - \int \frac{dk}{(2\pi)^3} k_i \frac{\partial \epsilon}{\partial r}\frac{\partial f}{\partial k} = 0, \tag{3.89}$$

The first term is the time derivative of the momentum density,

$$Q_i = \int \frac{dk}{(2\pi)^3} k_i f = m j_i \tag{3.90}$$

with (3.71). Accordingly, the second term has to be the gradient of the pressure,

$$\frac{\partial \mathcal{P}}{\partial r_i} = \int \frac{dk}{(2\pi)^3} k_i \frac{\partial \epsilon}{\partial k}\frac{\partial f}{\partial r} - \int \frac{dk}{(2\pi)^3} k_i \frac{\partial \epsilon}{\partial r}\frac{\partial f}{\partial k}. \tag{3.91}$$

We want to solve Eq. (3.91) for pressure. To this end we integrate parts in to the second term,

$$\frac{\partial \mathscr{P}}{\partial r_i} = \sum_j \int \frac{dk}{(2\pi)^3} \left(k_i \frac{\partial \epsilon}{\partial k_j} \frac{\partial f}{\partial r_j} + k_i \frac{\partial^2 \epsilon}{\partial r_j \partial k_j} f + \delta_{ij} \frac{\partial \epsilon}{\partial r_j} f \right), \tag{3.92}$$

and take the derivative in front of the integration,

$$\frac{\partial \mathscr{P}}{\partial r_i} = \sum_j \frac{\partial}{\partial r_j} \int \frac{dk}{(2\pi)^3} \left(k_i \frac{\partial \epsilon}{\partial k_j} + \delta_{ij} \epsilon \right) f - \int \frac{dk}{(2\pi)^3} \epsilon \frac{\partial f}{\partial r_i} \tag{3.93}$$

For an isotropic system, the product of momentum with the velocity contributes only for $i = j$; therefore we can eliminate the sums over direction j,

$$\frac{\partial \mathscr{P}}{\partial r_i} = \frac{\partial}{\partial r_i} \int \frac{dk}{(2\pi)^3} \left(\frac{1}{3} k \frac{\partial \epsilon}{\partial k} + \epsilon \right) f - \int \frac{dk}{(2\pi)^3} \epsilon \frac{\partial f}{\partial r_i}, \tag{3.94}$$

where $k\frac{\partial \epsilon}{\partial k}$ means the scalar product.

In the last term of Eq. (3.94) we use definition (3.86) according to which

$$\int \frac{dk}{(2\pi)^3} \epsilon \frac{\partial f}{\partial r_i} = \int \frac{dk}{(2\pi)^3} \frac{\delta \mathscr{E}}{\delta f} \frac{\partial f}{\partial r_i} = \frac{\partial \mathscr{E}}{\partial r_i}. \tag{3.95}$$

The Landau concept of quasiparticles thus predicts the pressure as

$$\mathscr{P} = \int \frac{dk}{(2\pi)^3} \left(\frac{1}{3} k \frac{\partial \epsilon}{\partial k} + \epsilon \right) f_k - \mathscr{E}. \tag{3.96}$$

See also (Kadanoff and Baym, 1962), where the pressure is given in terms of the pressure tensor $\mathscr{P} = \frac{1}{3} \Pi_{ii}$ with the classical expression (3.73) or the quantum expression in chapter 15.3.3. In the true ideal gas, the second term in brackets cancels with the density of energy and one recovers the pressure of the ideal gas as given by the momentum flux. In the case of an interacting gas, these terms do not cancel giving a non-trivial contribution to the quasiparticle pressure.

3.9.4   Quasiparticle energy

Now we have to find the quasiparticle energy. We start from the density of energy,

$$\mathscr{E} = \frac{3}{2} n k_B T + \frac{1}{2} n^2 \int d^3 r \, g(r) \, V, \tag{3.97}$$

which has the kinetic part (the first term) and the potential contribution which depends on the binary correlation. Within the approximation (2.107) of binary correlations, the energy density reads

$$\mathcal{E} = \frac{3}{2}nk_BT + n^2\frac{2\pi}{3}d^3Ve^{-\frac{V}{k_BT}}.$$

(3.98)

We will evaluate the functional derivative (3.86) using the thermal distribution of quasiparticles $f_k \propto \exp\left(-\frac{k^2}{2m^*k_BT}\right)$, normalised to density,

$$n = \int \frac{dk}{(2\pi)^3}f_k.$$

(3.99)

From the thermal distribution it follows that the effective kinetic energy of quasiparticles has the same form as that of normal particles,

$$\int \frac{dk}{(2\pi)^3}\frac{k^2}{2m^*}f_k = \frac{3}{2}nk_BT.$$

(3.100)

The energy depends on the quasiparticle distribution via the density and the temperature, therefore

$$\epsilon_k = \frac{\delta\mathcal{E}}{\delta f_k} = \frac{\partial\mathcal{E}}{\partial n}\frac{\delta n}{\delta f_k} + \frac{\partial\mathcal{E}}{\partial T}\frac{\delta T}{\delta f_k}.$$

(3.101)

Since the number of particles changes by unity if the occupation of any state changes, one finds

$$\frac{\delta n}{\delta f_k} = 1.$$

(3.102)

This simple formula is the variation of integral (3.99).

The variation of the temperature with respect to the occupation number has to be evaluated indirectly showing the intricate selfconsistencies of the Landau concept. We derive the variation of temperature from the variation of Eq. (3.100). Taking into account that m^* depends on n and T, from Eq. (3.100) we find

$$\frac{\delta}{\delta f_k}\frac{3}{2}nk_BT = \frac{k^2}{2m^*} + \int\frac{dp}{2\pi}p^2f_p\left(\frac{\partial}{\partial n} + \frac{\delta T}{\delta f_k}\frac{\partial}{\partial T}\right)\frac{1}{2m^*}.$$

(3.103)

Using Eq. (3.100) for the p-integral, one has the variation of temperature from relation (3.103),

$$\frac{\delta T}{\delta f_k} = \frac{\frac{2}{3}\frac{k^2}{2m^*k_B} - T - nT\frac{\partial \ln m^*}{\partial n}}{n + nT\frac{\partial \ln m^*}{\partial T}}.$$

(3.104)

With the help of Eq. (3.104) we express the quasiparticle energy (3.101) as

$$
\epsilon_k = \frac{k^2}{2m^*}\frac{\frac{2}{3k_B}\frac{\partial \mathcal{E}}{\partial T}}{n+nT\frac{\partial \ln m^*}{\partial T}} + \frac{\partial \mathcal{E}}{\partial n} - T\left(1+n\frac{\partial \ln m^*}{\partial n}\right)\frac{\frac{\partial \mathcal{E}}{\partial T}}{n+nT\frac{\partial \ln m^*}{\partial T}}.
\tag{3.105}
$$

Comparing expression (3.105) with the expectation (3.85), we find that the coefficient in front of the quasiparticle kinetic energy equals unity, which requires that

$$
\frac{2}{3k_B}\frac{\partial \mathcal{E}}{\partial T} = n + nT\frac{\partial \ln m^*}{\partial T}.
\tag{3.106}
$$

The effective potential then results from the expansion (3.105) as

$$
\phi = \frac{\partial \mathcal{E}}{\partial n} - \frac{3}{2}k_B T\left(1+n\frac{\partial \ln m^*}{\partial n}\right) = n\frac{4\pi}{3}d^3 V e^{-\frac{V}{k_B T}} - n\frac{3}{2}k_B T\frac{\partial \ln m^*}{\partial n}.
\tag{3.107}
$$

Effective mass   Now it remains to solve the differential equation (3.106) for m^*. With the explicit temperature derivative of the energy density (3.98), equation (3.106) reads

$$
\frac{\partial \ln m^*}{\partial T} = n\frac{4\pi}{9}d^3\frac{V^2}{k_B^2 T^3}e^{-\frac{V}{k_B T}}.
\tag{3.108}
$$

At very high temperatures, $k_B T \gg V$, the mass is not renormalised, $m^* \to m$ for $T \to \infty$. This limiting value allows us to evaluate the mass by integration from infinity to T,

$$
\ln m^* = \ln m - n\frac{4\pi}{9}d^3\left(1-e^{-\frac{V}{k_B T}}\right) + n\frac{4\pi}{9}d^3\frac{V}{k_B T}e^{-\frac{V}{k_B T}}.
\tag{3.109}
$$

One can see that at the high temperature or for a weak interaction, $V \ll k_B T$, the mass renormalisation starts from the second order in the powers of interaction,

$$
\frac{m^*}{m} \to 1 - \frac{4\pi}{9}d^3 n\frac{V^2}{k_B^2 T^2} \quad \text{for} \quad V \to 0.
\tag{3.110}
$$

In quantum system, there are two types of the mass renormalisation. The interaction between particles of different kinds also leads to the renormalisation of the mass starting from the second order in the interaction. The interaction between identical particles, however, includes the Fock contribution due to the exchange of particles, which gives a non-zero mass renormalisation already in the first order. When making parallels between classical and quantum systems, one has to keep in mind the exchange contribution which has no classical counterpart.

For hard spheres, $V \to \infty$, the mass renormalisation is also finite,

$$
\frac{m^*}{m} \to 1 - \frac{4\pi}{9}d^3 n \quad \text{for} \quad V \to \infty.
\tag{3.111}
$$

This feature contrasts with the fact that collisions are instant and between them the hard spheres move as free particles with unrenormalised mass. Accordingly, the density of energy of hard spheres is the same as the density of energy of non-interacting particles,

$$\mathcal{E} \to \frac{3}{2} n k_B T \quad \text{for} \quad V \to \infty, \tag{3.112}$$

as one can see from (3.98).

The renormalisation of the mass defined via the variational principle thus has nothing to do with the single-particle motion between collisions. The quasiparticle picture describes the thermodynamic properties (that we will show) but the quasiparticle motion derived from the variational approach cannot be interpreted as a motion of real particles or some collective excitations in the system. The only exception is the limit of weak interaction where trajectories of quasiparticles really follow the mean-field trajectories.

Effective potential   The effective potential follows from Eqs. (3.107) and (3.109) as

$$\phi = n \frac{2\pi}{3} d^3 k_B T \left(1 - e^{-\frac{V}{k_B T}} \right) + n \frac{2\pi}{3} d^3 V e^{-\frac{V}{k_B T}}. \tag{3.113}$$

For weak interaction, $V \ll k_B T$, the potential reduces to the mean field,

$$\phi \to \frac{4\pi}{3} d^3 n V \quad \text{for} \quad V \to 0. \tag{3.114}$$

For hard spheres, $V \to \infty$, the potential remains non-zero,

$$\phi \to \frac{2\pi}{3} d^3 n k_B T \quad \text{for} \quad V \to \infty, \tag{3.115}$$

in spite of the fact that particles never penetrate into the interaction potential. This confirms that the quasiparticle trajectory, given by the variational quasiparticle energy, should not be interpreted as mean trajectory of a real particle.

Pressure   Now we can complete the algebra and evaluate the pressure. Using Eqs. (3.85) and (3.97) in the quasiparticle pressure (3.96) one finds

$$\mathscr{P} = n k_B T + \phi n - n^2 \frac{2\pi}{3} d^3 V e^{-\frac{V}{k_B T}}. \tag{3.116}$$

From the effective potential (3.113) one recovers the pressure (3.82) obtained above from the virial of forces. The quasiparticle picture thus yield the same thermodynamic properties as the corresponding approximation of the virial expansion.

3.9.5 Two forms of quasiparticles

It is striking that the Landau concept of quasiparticles designed for dense Fermi systems at very low temperatures also works perfectly for classical systems at any temperature. Particularly surprising is the limit of strong interaction, $V \gg k_B T$, at which particles behave like hard spheres. Since the contribution of the potential energy vanishes, $\mathscr{E} \to \frac{1}{2} n k_B T$, one might expect that the quasiparticle concept based on variation of $\mathscr{E}$ will result in properties of an ideal gas. As we have seen above, the effective mass, if obtained by the integration from high temperatures, remembers the potential and covers the contribution of the excluded volume.

On the other hand, the effective mass fitted to thermodynamic relations might be misleading when used for physical processes at the microscopic level. Indeed, the classical momentum distribution is given by the Maxwell distribution with the normal mass, $f_k \propto \exp\left(-\frac{k^2}{2mk_B T}\right)$, while the quasiparticle concept results in a distribution with the renormalised mass. In the hard-sphere limit all particles move like free particles since collisions are instantaneous. The renormalised mass then leads to a confusing picture of the single-particle motion.

3.10 Summary

The classical non-ideal gas shows that the two original concepts of the pressure based on the motion and forces have eventually developed into drift and dissipation contributions. The dissipation has to be treated beyond the approximation used within the Boltzmann equation, in particular, the nonlocal and non-instant character of binary collisions have to be taken into account. These nonlocal and non-instant corrections are naturally described by the collision delay Δ_t and displacements $\Delta_{2,3,4}$ of initial and final states. These Δ's enter the scattering integral making it non-instant and nonlocal.

The action of mean fields on particles during the collision requires us to include momentum and energy gains into final states. Although these corrections look at the first glance like the violation of momentum and energy conservation laws, they in fact guarantee their validity.

The non-instant and nonlocal corrections to the scattering integral directly result in virial corrections to the equation of state. The consequences of the finite collision duration can be conveniently described within the dialect of chemical physics as a presence of effective molecules. In particular, one can understand the correlated density in terms of the density of molecules. The nonlocality can be attributed to the finite- mean distance of particles and the rotation degree of freedom of molecules.

The scattering integral of the Boltzmann equation leads to the second law of thermodynamics in that the mean entropy is increasing with time. This irreversibility of the many-body theory is brought about by the chaoticity assumption that two colliding particles have been uncorrelated in the past.

Comparing the results of Chapters 2 and 3, one can see that thermodynamic properties of the system can be covered by two very distinct kinetic equations. Firstly, one can use a simple free drift introducing nonlocal effective collisions. Secondly, one can use simple local collisions and put all corrections into the effective drift. Each of these extreme approaches has its advantages and shortages. Likely, the best quantitative agreement and the widest region of applicability will be achieved with a theory which interpolates between these two extreme cases.

A theory which combines the nonlocal picture with the quasiparticle approach will be derived in Chapter 13. The quasiparticle energy defined from the spectral properties is not identical to Landau's quasiparticle energy obtained by the variation. It is nevertheless customary to call the dispersion relation of a single particle in medium quasiparticle energy. We thus call the single-particle excitations found from the spectrum the spectral quasiparticles, or briefly quasiparticles. Spectral quasiparticles cover some thermodynamic properties but not all. Accordingly, collisions between them have to be treated in the nonlocal picture which covers the rest. To avoid double-counts in such a combined approach, the theory has to be derived as a consistent limit from quantum statistics.

Part II

Inductive Ways to Quantum Transport

We are going to develop the basic ideas of the theory of quantum transport in terms of the scattering of particles on impurities since it reveals the most simple and clear picture of different renormalisation, interferences and the basic physics behind the selfenergy. As the ultimate goal we will obtain the coherent potential approximation which has been shown to be equivalent to the modern dynamical mean-field approximation (Kakehashi, 2002, 2004) which has raised an appreciable renaissance. Since we feel that the original concepts were nearer to the physics we present first step by step the many-body theoretical ingredients to get the transport vertex and the Green's functions. In Part III of the book we will than show how a simple formal theory can cover all these physics in most simple terms.

4

Scattering on a Single Impurity

The evolution of many-body systems consists of permanent collisions among particles. Looking at the motion of a single particle, one can identify encounters through which a particle abruptly changes the direction of flight, these we see as true collisions, and small angle encounters, which in sum act as an applied force rather than randomising collisions. These two basic mechanisms are essential for the concept of the effective medium.

It is intuitively clear that when two colliding particles are hit by a third particle, we have to deal with the three-body process. These processes can often be neglected because of the low probability of three particles meeting at the same point. In contrast, small angle encounters modify initial and final states of colliding particles and have to be assumed in all dense systems. Many-particle processes based on small angle encounters thus contribute even if 'third' particles are displaced at distances exceeding the range of the interaction. It is sufficient if they are comparable to the mean-free path. Apparently, these two types of encounters put two different demands on the self-consistency of the theory.

The two effects of the encounters have been seen already on a simpler model, in which electrons do not interact among themselves but are scattered by randomly distributed impurities. We use the scattering of impurities to introduce the mentioned mechanisms and, in particular, to show how they effect each other. We will assume point impurities, i.e. impurities the potential of which are restricted to a single atomic site of the crystal lattice. In this case interaction potentials never overlap and many-body effects are due to the nonlocal character of the quantum particle.

To introduce elementary components of the formalism, in this chapter we first describe the interaction of an electron with a single impurity.

4.1 Bound state

Assume an ideal crystal with a single impurity. Its Hamiltonian H is composed of the crystal part H^0 and a perturbing impurity potential V. Let us first find the bound state. The bound state $|b\rangle$ satisfies the Schrödinger equation

$$H|b\rangle = E_b|b\rangle. \tag{4.1}$$

Interacting Systems far from Equilibrium. Klaus Morawetz, Oxford University Press (2018).
© Klaus Morawetz. DOI: 10.1093/oso/9780198797241.001.0001

We use Dirac's notation which is free of arguments related to the representation. The wave function $b(\mathbf{r})$ is obtained as a scalar product $b(\mathbf{r}) = \langle \mathbf{r}|b \rangle$. Its conjugate reads $b^*(\mathbf{r}) = \langle b|\mathbf{r} \rangle$. Equation (4.1) is converted to the space form taking the space elements at coordinate $\mathbf{r}$, $\langle \mathbf{r}|H|b \rangle = E_b \langle \mathbf{r}|b \rangle$, and introducing $\mathbf{r}$-decomposition of unitary operator, $1 = \int d\mathbf{r}' |\mathbf{r}' \rangle \langle \mathbf{r}'|$, between H and $|b \rangle$. In the free space,

$$\langle \mathbf{r}|H|\mathbf{r}' \rangle = \left(-\frac{\hbar^2}{2m} \frac{\partial^2}{\partial r^2} + V(\mathbf{r}) \right) \delta(\mathbf{r} - \mathbf{r}').$$

The only question is how to solve (4.1). This problem is very simple and can be solved by many methods. We will follow the iterative procedure based on the resolvent, because we want to prepare the ground for the closely related method of Green's functions.

Let us suppose that we were lucky enough to make a good guess at the eigen-energy, $E = E_b + \Delta E$, where ΔE is small. We also made a guess at the wave function, $|\psi \rangle$, but were not so successful with, $|\psi \rangle = \alpha|b \rangle + \beta|\psi' \rangle$. To improve the wave function, we multiply our guess by $1/(E - H)$ so that

$$\frac{1}{E - H}|\psi \rangle = \frac{\alpha}{\Delta E}|b \rangle + \frac{\beta}{E - H}|\psi' \rangle. \tag{4.2}$$

The closer our guess is to the bound energy, the more enhanced is the component of the bound-state wave function. At the same time, the enhancement of the wave function serves as a criterion for the guess of the energy. The closer we are, the more singular the resolvent $1/(E - H)$ is.

To be more systematic, let us denote as $|n \rangle$ states which diagonalise the Hamiltonian

$$H = \sum_n |n \rangle E_n \langle n|. \tag{4.3}$$

The bound state $|b \rangle$ is one of $|n \rangle$'s. The resolvent,

$$G = \frac{1}{E - H} = \sum_n |n \rangle \frac{1}{E - E_n} \langle n|, \tag{4.4}$$

then diverges for $E \to E_b$. This singularity appears in the matrix elements of the resolvent, say $\langle \psi|G|\psi \rangle$, with the weight $|\langle \psi|b \rangle|^2$,

$$\lim_{E \to E_b} \sum_n \langle \psi|n \rangle \frac{1}{E - E_n} \langle n|\psi \rangle = \frac{|\langle \psi|b \rangle|^2}{E - E_b}. \tag{4.5}$$

The Green's function method identifies eigen energies from the singularities of G.

4.1.1 Secular equation

Unless one is unfortunate making a starting 'guess' of the wave function orthogonal to the desired bound state, any choice of function $|\psi\rangle$ in (4.5) works well. It gives us the freedom to select $|\psi\rangle$ according to a numerical convenience regardless of how closely it approximates the actual bound state $|b\rangle$.

A natural choice of the convenient $|\psi\rangle$ emerges for the Koster–Slater model of neutral impurity, in which the potential affects only a single orbital at the impurity site

$$V = |0\rangle v \langle 0|. \tag{4.6}$$

For convenience we have placed the impurity at the origin of coordinates $\mathbf{r} = 0$. We take $|\psi\rangle = |0\rangle$. The elements

$$G_{00} = \langle 0|G|0\rangle = \langle 0|\frac{1}{E - H^0 - V}|0\rangle, \qquad G_{00}^0 = \langle 0|G^0|0\rangle = \langle 0|\frac{1}{E - H^0}|0\rangle, \tag{4.7}$$

are called local resolvents.

From $(E - H^0 - V)\frac{1}{E-H^0-V} = 1$ one finds $(E - H^0)\frac{1}{E-H^0-V} = 1 + V\frac{1}{E-H^0-V}$ that yields

$$\frac{1}{E - H^0 - V} = \frac{1}{E - H^0} + \frac{1}{E - H^0} V \frac{1}{E - H^0 - V}. \tag{4.8}$$

This is nothing but

$$G = G^0 + G^0 V G. \tag{4.9}$$

The elements are readily solved yielding the local resolvent,

$$G_{00} = \frac{G_{00}^0}{1 - v G_{00}^0}. \tag{4.10}$$

For the bound state the resolvent diverges, $G_{00} \to \infty$. The eigen energy is thus given by roots of the denominator of G_{00},

$$G_{00}^0(E_b) = \frac{1}{v}. \tag{4.11}$$

As one can see, the eigen-state of the Koster–Slater impurity is given by a simple implicit relation (4.11). Due to this simplicity it is ideal for models of multiple scattering processes.

Finally, in the vicinity of a singularity, the limiting form (4.5) provides

$$G_{00} = \frac{G_{00}^0}{1 - G_{00}^0 v} = \frac{1}{\frac{1}{G_{00}^0} - v} \quad \rightarrow \quad \frac{1}{E - E_b} \frac{1}{\frac{\partial}{\partial E} \left(\frac{1}{G_{00}^0} - v \right)_{E_b}} \tag{4.12}$$

and the local amplitude of the eigen function follows from the weight of the pole as

$$|\langle 0|b\rangle|^2 = \frac{1}{\frac{\partial}{\partial E} \frac{1}{G_{00}^0}\Big|_{E_b}}. \tag{4.13}$$

4.1.2   Resolvent of the host crystal

In a homogeneous crystal without any impurity, the eigen functions are Bloch functions normalised to the δ function, $\langle p|k\rangle = (2\pi)^3 \delta(k - p)$, i.e. $\langle r|k\rangle = e^{ik \cdot r}$. The Hamiltonian in this representation reads

$$H^0 = \int \frac{dk}{(2\pi)^3} |k\rangle \epsilon_k \langle k|,$$

where the momentum integration runs over the Brillouin zone. The local resolvent of the host crystal then reads

$$G_{00}^0(E) = \int \frac{dk}{(2\pi)^3} \frac{1}{E - \epsilon_k}. \tag{4.14}$$

This local resolvent can be linked with the experimentally accessible density of states $h^0(E)$ which is defined as a number of eigen-states, $h^0(E)dE$, per site in the energy interval $(E, E + dE)$,

$$h^0(E) = \int \frac{dk}{(2\pi)^3} 2\pi \delta(E - \epsilon_k). \tag{4.15}$$

Comparing (4.15) with (4.14) one finds that the local resolvent is obtained from the density of states via a Hilbert transform,

$$G_{00}^0(E) = \int \frac{dv}{2\pi} \frac{h^0(v)}{E - v}. \tag{4.16}$$

The analytic relation between the local resolvent and the density of states allows us to study the bound state without actual knowledge of the whole Hamiltonian H^0. It is sufficient to know the density of states. For model studies, it is possible to use a model density of states, e.g. the so-called Hubbard bubble,

$$h^0(E) = \frac{4}{w^2}\sqrt{w^2 - (E-w)^2},$$

(4.17)

where $2w$ is the band width. The Hilbert transformation (4.16) yields[1]

$$G_{00}^0(E) = \frac{2}{w^2}\left(E - w - \sqrt{(E-w)^2 - w^2}\right).$$

(4.18)

The solution of secular equation (4.11) for this model function is simple. The energy of the bound state is

$$E_b = w + v + \frac{w^2}{4v} \quad \text{for} \quad v < -\frac{w}{2}.$$

(4.19)

Taking a derivative of E_b with respect to v one can check that the eigen-energy E_b monotonically grows as the binding potential becomes weaker. For $v \to -\frac{w}{2}$, the binding energy touches the energy band, $E_b \to 0$. For $v > -\frac{w}{2}$ and there is no bounded state.

4.2   Scattering and Lippmann–Schwinger equation

The scattering represents a problem rather different from the bound state. The bound state is restricted in space, therefore its wave function is normalised to unity and its phase has no physical meaning. In the scattering, the wave function extends over the whole space, it is thus normalised to the stream of incident particles. The energy is given by the source, we only have to find the wave function. The square of the scattered wave function in the asymptotic region provides us with the scattering cross section which represents changes of direction for large angles. Besides, at low angles the scattered wave interferes with the incident wave. It leads to a shadow representing the depletion of particles from the incident beam. Moreover, the interference of the incident and scattered way leads to a phase shift which reminds us of the effect of the forces.

Phase shifts and scattering rates are mutually connected by the optical theorem as we will see in Chapter 4.4. Since they ought to be described within one scattering process, one should be aware that when choosing an approximation for scattering rates one also specifies the approximation of effective forces and vice versa.

---

[1] One can in fact guess the local resolvent directly from the density of states. Extending the resolvent into the complex plane, $z = E + i0$, from the identity $\text{Im}\frac{1}{E-v} = -\pi\delta(E-v)$, one finds that the density of states is the imaginary part of the complex resolvent, $h^0(E) = -2\text{Im}\,G_{00}^0(E+i0)$. As long as the imaginary part is expressed by an analytic formula, its complete form is given by the same analytic formula except for the polynomials in z (here, the factor $2z/w^2$) which is not visible in the imaginary part along the real energy axis. These polynomials can always be adjusted to give the correct behaviour of G in the asymptotic region $E \to -\infty$, where $G \to 1/E$.

We derive the scattering rate for the Koster–Slater impurity, where the exact result has a simple analytic form. Unless we specify the potential V all relations are general.

Let us assume that there is a source of electrons at infinity, providing the incoming wave

$$\psi_0(\mathbf{r}) = e^{i\mathbf{k}\cdot\mathbf{r}}. \tag{4.20}$$

The total wave is composed of this incoming part and the scattered part ψ which has to be found. The total wave solves the Schrödinger equation

$$(E - H^0 - V)(\psi_0 + \psi) = 0. \tag{4.21}$$

The incoming wave is an eigen-state of the ideal crystal, therefore

$$(E - H^0)\psi_0 = 0, \tag{4.22}$$

which requires the energy to be equal to the kinetic energy, $E = \epsilon_k$. Lippmann and Schwinger used the initial condition (4.22) to rearrange the Schrödinger equation (4.21) as

$$(E - H^0)\psi = V(\psi_0 + \psi). \tag{4.23}$$

This equation is equivalent to (4.21) but it has better properties. Since the incoming wave appears only in the 'source term' of Eq. (4.23), we can easily formulate a boundary condition, which will be necessary for the inversion of operator $E - H^0$. With the incoming wave eliminated, the boundary condition requires that electrons are flying away from the impurity.

To make the inversions of $E - H^0$ well defined, we extend the energy into the complex plane, $z = E \pm i0$. The sign will be selected later in accordance with the boundary condition. Now we introduce the perturbative expansion from equation (4.23) as follows

$$\psi = G^0(z)V(\psi_0 + \psi). \tag{4.24}$$

This relation can be solved by iterations.

4.2.1   Born approximation of the scattered wave

If the impurity potential is weak compared to the kinetic energy of the electron, $|v| \ll \epsilon_k$, the scattered wave at the impurity site is much smaller than the incoming part, $|V\psi| \ll |V\psi_0|$, so that it can be neglected in the source term. This is known as the Born approximation,

$$\psi = G^0 V \psi_0. \tag{4.25}$$

Now we are ready to specify the boundary condition. To this end we evaluate the explicit wave function in the space representation,[2]

$$\psi(\mathbf{r}) = \langle \mathbf{r} | G^0(z) | 0 \rangle v \langle 0 | \mathbf{k} \rangle, \qquad \text{where} \qquad \langle \mathbf{r} | G^0(z) | 0 \rangle = \int \frac{dp}{(2\pi)^3} \frac{e^{i\mathbf{p}\cdot\mathbf{r}}}{z - \epsilon_{\mathbf{p}}}. \tag{4.26}$$

For a parabolic band, $\epsilon_{\mathbf{p}} = \frac{\hbar^2 p^2}{2m}$, Green's function is[3]

$$\langle \mathbf{r} | G^0(z) | 0 \rangle = \frac{2im}{(2\pi)^2 \hbar^2 r} \int_{-\infty}^{\infty} dp \frac{p\, e^{ipr}}{p^2 - \frac{2mz}{\hbar^2}}. \tag{4.27}$$

The value of this integral is obtained by residue analysis. Since $|\mathbf{r}| \equiv r > 0$, we have to close the integration path for $\text{Im}\,p > 0$ with an anticlockwise orientation. The pole with the positive imaginary part is at

$$p = \pm \frac{1}{\hbar} \sqrt{2mE} \qquad \text{for} \qquad z = E \pm i0. \tag{4.28}$$

Positive/negative signs correspond to waves which stream away from/towards the impurity. The scattered wave streams away, we thus keep the solution with the positive sign, $z = E + i0$,

$$\langle \mathbf{r} | G^0(z) | 0 \rangle = -\frac{m}{2\pi \hbar^2 r} e^{ir\frac{1}{\hbar}\sqrt{2mE}}, \qquad \text{and} \qquad \psi(\mathbf{r}) \approx -\frac{m}{2\pi \hbar^2 r} e^{irk} v. \tag{4.29}$$

By extension of the energy into the complex plane one gains two important advantages. Firstly, mathematical operations become covered by the theory of analytic functions. Secondly, the boundary condition can be simply selected by a choice of the complex half plane.

---

[2] Here $|\mathbf{r}\rangle$ is a local orbital at the site $\mathbf{r}$, $|\mathbf{k}\rangle = |\psi_0\rangle$ is a Bloch function with momentum $\mathbf{k}$, and $G^0(z) = 1/(z - H^0)$. The Hamiltonian H^0 is diagonal in the momentum representation,

$$H^0 = \int \frac{dp}{(2\pi)^3} |\mathbf{p}\rangle \epsilon_{\mathbf{p}} \langle \mathbf{p}|, \qquad \text{thus} \qquad G^0(z) = \int \frac{dp}{(2\pi)^3} |\mathbf{p}\rangle \frac{1}{z - \epsilon_{\mathbf{p}}} \langle \mathbf{p}|.$$

[3]

$$\int \frac{dp}{(2\pi)^3} \frac{e^{ipr}}{z - \frac{\hbar^2 p^2}{2m}} = \frac{1}{(2\pi)^2} \int_{-1}^{1} ds \int_0^{\infty} dp\, p^2 \frac{e^{ipsr}}{z - \frac{\hbar^2 p^2}{2m}} = \frac{1}{(2\pi)^2} \int_0^{\infty} dp \frac{p}{ir} \frac{e^{ipr} - e^{-ipr}}{z - \frac{\hbar^2 p^2}{2m}}$$

$$= \frac{2im}{(2\pi)^2 \hbar^2 r} \int_{-\infty}^{\infty} dp \frac{p\, e^{ipr}}{p^2 - \frac{2mz}{\hbar^2}}.$$

4.2.2 T-matrix

The Born approximation fails for strong scattering potentials. For instance, within the Born approximation the wave even penetrates a strongly repulsive impurity potential, while in reality it is expelled from the impurity site. To account for the reconstruction of the wave function at the impurity site we have to solve Eq. (4.24) exactly.

As in the Born approximation, out of the impurity the motion is covered by G^0. At the same time, the scattered wave is linearly proportional to the source $V\psi_0$. It is thus reasonable to write the wave as

$$\psi = G^0 T \psi_0,\tag{4.30}$$

where the operator T is called the T-matrix. To find the T-matrix, we rearrange Eq. (4.23) as $(E - H^0 - V)\psi = V\psi_0$ from which it follows that $\psi = GV\psi_0$. Comparing this form with definition (4.30) one finds that $G^0 T = GV$. Now we substitute identity (4.9) for G and multiply by G_0^{-1} from the left-hand side, which yields

$$T = V + VGV = V + VG^0 T = \frac{V}{1 - G^0 V} = V \sum_{n=0}^{\infty} (G^0 V)^n.\tag{4.31}$$

To evaluate the T-matrix for a realistic potential is the central problem of the theory of collisions (Goldberger and Watson, 1964). For the Koster–Slater potential, the T-matrix is exceptionally simple since it is only non-zero at the central site $|0\rangle$,

$$T = |0\rangle t \langle 0|, \quad t = v + vG^0_{00}v + \ldots = \frac{v}{1 - G^0_{00}v},\tag{4.32}$$

and depends only on the local resolvent of the host crystal G^0_{00}.

The exact scattered wave (4.30) far from the impurity is

$$\psi(\mathbf{r}) = -\frac{m}{2\pi \hbar^2 r} e^{irk} t(\epsilon_k + i0).\tag{4.33}$$

The boundary condition remains the same as in the Born approximation, because it is determined by the behaviour of the wave function in the asymptotic region.

4.3 Collision delay

Due to its complex character the T-matrix describes the finite duration of the collision. To identify the position of the electron, we take the incident particle in the form of a wave packet

$$\psi_0(\mathbf{r}, t) = Z \int d\mathbf{p} \, e^{-\lambda|\mathbf{k}-\mathbf{p}|^2/2} e^{i\mathbf{r}\cdot\mathbf{p} - it\epsilon_\mathbf{p}/\hbar}.\tag{4.34}$$

Each **p**-component has its own energy ϵ_p so that this function evolves in time. In the saddle-point approximation

$$\psi_0(\mathbf{r}, t) \approx Z' e^{i\mathbf{r}\cdot\mathbf{k} - it\epsilon_k/\hbar} e^{|\mathbf{r} - \mathbf{v}_k t|^2/4\lambda}, \tag{4.35}$$

where $\mathbf{v}_k = (1/\hbar)(\partial\epsilon/\partial\mathbf{k})$ is the electron velocity and Z' captures the norm Z and all constant integration factors. At $t = 0$ this wave packet passes the central site at $\mathbf{r} = 0$.

The scattered wave is the same linear combination of waves

$$\psi(\mathbf{r}, t) = -\frac{m}{2\pi\hbar^2 r} Z \int d\mathbf{p}\, e^{-\lambda|\mathbf{k} - \mathbf{p}|^2/2} e^{i\mathbf{r}p} t(\epsilon_p + i0) \tag{4.36}$$

which, in the saddle-point approximation, equals

$$\psi(\mathbf{r}, t) \approx -\frac{Z''}{r} e^{i\mathbf{r}k - it\epsilon_k/\hbar}\, t(\epsilon_k + i0) e^{(r - |\mathbf{v}_k|(t - \Delta_\tau))^2/4\lambda}, \tag{4.37}$$

where the time shift due to the logarithmic derivative of the T-matrix

$$\Delta_\tau = \hbar \text{Im} \left. \frac{\partial \ln t}{\partial\epsilon} \right|_{\epsilon_k} \tag{4.38}$$

is known as Wigner's collision delay.

The sign of the collision delay for a small momentum, $\mathbf{k} \to 0$ depends on the existence of the bound state. For the Koster–Slater impurity

$$\Delta_\tau = -\hbar \text{Im}\, t \frac{\partial t^{-1}}{\partial\epsilon} = \hbar \text{Im} \frac{1}{\frac{1}{v} - G^0} \frac{\partial G^0}{\partial\epsilon}. \tag{4.39}$$

Near the bottom of the energy band the real part of the Green's function G^0 is much bigger than the imaginary part, therefore the collision delay is given by the energy derivative of the density of states

$$\Delta_\tau \approx \frac{\hbar}{\frac{1}{v} - \text{Re}G^0} \frac{\partial \text{Im}G^0}{\partial\epsilon} = -\frac{\hbar}{2} \frac{1}{\frac{1}{v} - \text{Re}G^0} \frac{\partial h}{\partial\epsilon}. \tag{4.40}$$

For energies slightly above the energy bottom, the derivative of the density of states is always positive. The sign of the collision delay is thus determined by the sign of $1/v - \text{Re}G^0$. If there is a bound state, $1/v - \text{Re}G^0 > 0$ for small energies, the collision delay is negative. If the potential is weak or positive so that that bound state is absent, $1/v - \text{Re}G^0 < 0$, the collision delay is positive.

One can understand the sign of the collision delay from the classical picture. If the potential is positive so that it cannot capture a particle, the particle passing through the impurity region loses part of its kinetic energy to overcome the potential. It is thus slow in this region which results in a positive collision delay. On the contrary, passing through the attractive potential, the particle gains kinetic energy so that it crosses this region faster than by free flight. The negative collision delay represents the difference between the free motion and the actual motion.

4.4 Optical theorem

The scattering process can be observed in two different ways: detecting electrons scattered into a selected space angle; and observing how many electrons are scattered out of the incoming beam. As the number of electrons conserves, the total number of electrons scattered out of the beam has to equal the sum over electrons emerging in all final angels. This request puts certain limitations on the approximations of the T-matrix, the identity known as the optical theorem.

To test the conservation of the number of particles, we make a large diameter sphere, $r \gg 1/k$ around the impurity, and evaluate the total current of electrons flowing through it. The current density reads

$$\mathbf{j} = \frac{e\hbar}{m} \mathrm{Im}(\psi_0^* + \psi^*) \frac{\partial}{\partial \mathbf{r}} (\psi_0 + \psi). \tag{4.41}$$

According to Eqs. (4.41) with (4.20) and (4.33), the current density in the vicinity of the Koster–Slater impurity reads

$$\mathbf{j} = \frac{e\hbar\mathbf{k}}{m} + \frac{\mathbf{r}}{r} \frac{e\hbar k}{m} \left| \frac{mt(\epsilon_k + i0)}{2\pi\hbar^2 r} \right|^2 - \left(\mathbf{k} + \frac{\mathbf{r}}{r} k \right) \frac{e}{2\pi\hbar r} \mathrm{Re}\left[t(\epsilon_k + i0) e^{ikr - i\mathbf{k}\cdot\mathbf{r}} \right]. \tag{4.42}$$

The conservation of the number of particles demands the total current flowing through the surface of the sphere to be zero[4]

$$\mathfrak{J} = \int_0^{2\pi} d\phi \int_0^\pi d\vartheta\, r^2 \sin\vartheta\, (j_x \cos\vartheta + j_y \sin\vartheta \cos\phi + j_z \sin\vartheta \sin\phi) = 0. \tag{4.43}$$

The first term of the current density (4.42) does not contribute to the total current, $\mathfrak{J}_1 = 0$. This first term corresponds to the unperturbed incoming beam, $\mathfrak{J}_1 = \frac{e\hbar}{m} \mathrm{Im} \int \psi_0^* \frac{\partial}{\partial \mathbf{r}} \psi_0$. The $\mathfrak{J}_1$ vanishes because the same number of electrons which enters the sphere on one side leaves it on the opposite side. The second term of the current density (4.42) contributes by $\mathfrak{J}_2 = \frac{emk|t|^2}{\pi\hbar^3}$. This term corresponds to particles streaming out of the sphere after they have suffered a large-angle collision with the impurity, $\mathfrak{J}_2 = \frac{e\hbar}{m} \mathrm{Im} \int \psi^* \frac{\partial}{\partial \mathbf{r}} \psi$. The third term describes interference between the incoming and

---

[4] The normal vector of the sphere in cylindrical coordinates is

$$\frac{\mathbf{r}}{r} \equiv (\cos\vartheta, \sin\vartheta \cos\phi, \sin\vartheta \sin\phi).$$

the outgoing waves, $\mathcal{J}_3 = \frac{e\hbar}{m}\operatorname{Im}\int \left(\psi_0^* \frac{\partial}{\partial \mathbf{r}}\psi + \psi^* \frac{\partial}{\partial \mathbf{r}}\psi_0\right)$. Most important is their destructive interference by which the **k**-component of the total wave function is smaller than the incoming beam. The leading terms in $\frac{1}{kr}$ of the third term are

$$\mathcal{J}_3 = \frac{2}{\hbar}\operatorname{Im}t(\epsilon_k + i0). \tag{4.44}$$

Since the total current equals zero, we find from $\mathcal{J}_2 + \mathcal{J}_3 = 0$ the condition on the T-matrix

$$\operatorname{Im}t(\epsilon_k + i0) = -|t(\epsilon_k + i0)|^2 \frac{mk}{2\pi\hbar^2}, \tag{4.45}$$

which is the optical theorem.

The optical theorem (4.45) has been derived for the parabolic band and the Koster–Slater impurity. Before we derive the general optical theorem, let us rearrange (4.45) to the general form. The imaginary part of the local element of Green's function (4.29) for the parabolic band is just $\operatorname{Im}G^0(\epsilon_k + i0) = -mk/2\pi\hbar^2$. Using that $t^*(\epsilon_k + i0) = t(\epsilon_k - i0)$, the optical theorem (4.45) can be expressed as

$$\operatorname{Im}t(\epsilon_{\mathbf{k}} + i0) = t(\epsilon_{\mathbf{k}} + i0)\left[\operatorname{Im}G^0(\epsilon_{\mathbf{k}} + i0)\right]t(\epsilon_{\mathbf{k}} - i0). \tag{4.46}$$

This form is close to the general optical theorem

$$i(T^R - T^A) = T^R i(G_0^R - G_0^A)T^A, \tag{4.47}$$

where the superscript R denotes retarded functions (with energy argument $z = E + i0$) while the superscript A denotes advanced functions (with energy argument $z = E - i0$), and the zero denoting the free function has been moved to the subscript.

Formula (4.47) follows from the expansion series (4.31)

$$T^R(G_0^R - G_0^A)T^A = \sum_{n=0}^{\infty}(VG_0^R)^n V(G_0^R - G_0^A)V\sum_{k=0}^{\infty}(G_0^A V)^k$$

$$= \sum_{n=1}^{\infty}(VG_0^R)^n V\sum_{k=0}^{\infty}(G_0^A V)^k - \sum_{n=0}^{\infty}(VG_0^R)^n V\sum_{k=1}^{\infty}(G_0^A V)^k$$

$$= -V\sum_{k=0}^{\infty}(G_0^A V)^k + \sum_{n=0}^{\infty}(VG_0^R)V$$

$$= T^R - T^A. \tag{4.48}$$

This formula is more general than (4.45) in three ways. Firstly, the band structure can be any, not only the simple parabolic band. Secondly, it holds for any potential V, including cases when V extends to infinity or represents a sum over many impurities. Thirdly, identity (4.48) holds also for energies out of the energy shell $\omega \neq \epsilon_k$.

Let us make the space arguments explicit using the momentum representation. The T-matrix depends on momenta of the incoming wave $\mathbf{k}$ and of the out-going wave $\mathbf{p}$, and the energy ω, so that the optical theorem reads

$$i\left(T^R(\omega, \mathbf{k}, \mathbf{k}') - T^A(\omega, \mathbf{k}, \mathbf{k}')\right)$$

$$= \int \frac{d\mathbf{p}}{(2\pi)^3} T^R(\omega, \mathbf{k}, \mathbf{p}) \, i \left(G_0^R(\omega, \mathbf{p}) - G_0^A(\omega, \mathbf{p})\right) T^A(\omega, \mathbf{p}, \mathbf{k}'). \quad (4.49)$$

The interpretation of the optical theorem applies for $\mathbf{k}' = \mathbf{k}$ and $\omega = \epsilon_k$. The off-diagonal terms, $\mathbf{k}' \neq \mathbf{k}$, and the off-shell terms, $\omega \neq \epsilon_k$, do not have any straightforward interpretation.

4.4.1   Scattering on two impurities

The optical theorem is extremely restrictive. In fact, no approximation of the T-matrix obeys the optical theorem but the exact expression (4.32). Let us first show why this happens. For instance, if we take the Born approximation $t \approx v$, the left-hand side of Eq. (4.46) is zero while the right-hand side is not. This is because the left-hand side is linear in the T-matrix while the right-hand side is quadratic. Apparently, no finite order approximation can satisfy the optical theorem (4.46) since the right-hand side will always be of a two-times higher order. With approximations in game, the conservation of the number of particles has to be tested.

The optical theorem becomes less restrictive if the potential V is composed of two or more individual potentials. Let us assume two impurities of potentials V_1 and V_2. The exact T-matrix is

$$T = \frac{V_1 + V_2}{1 - G^0(V_1 + V_2)}. \quad (4.50)$$

The expansion in powers of the potential,

$$T = V_1 + V_2 + (V_1 + V_2)G^0(V_1 + V_2) + \dots$$
$$= V_1 + V_1 G^0 V_1 + V_2 + V_2 G^0 V_2 + V_1 G^0 V_2 + V_2 G^0 V_1 + \dots \quad (4.51)$$

includes products corresponding to the interaction with a single impurity, like $V_1 G^0 V_1$ or $V_2 G^0 V_2$, and products in which both impurities contribute, like $V_1 G^0 V_2$ or $V_2 G^0 V_1$. If the impurities are displaced one from the other by a distance r, the propagator connecting them is proportional to $1/r$ and for large r it can be neglected compared to the single-impurity contributions. The T-matrix then simplifies as

$$T = V_1 + V_1 G^0 V_1 + V_2 + V_2 G^0 V_2 + \dots. \quad (4.52)$$

In this approximation the T-matrix is a sum of two single-impurity T-matrices,

$$T = T_1 + T_2 \quad \text{with} \quad T_1 = \frac{V_1}{1 - G^0 V_1} \quad \text{and} \quad T_2 = \frac{V_2}{1 - G^0 V_2}. \tag{4.53}$$

This approximation of the T-matrix satisfies the optical theorem (within the approximation $T_1 G^0 T_2 \approx 0$ and $T_2 G^0 T_1 \approx 0$).

4.5 Dissipativeness

In accordance with its natural probabilistic interpretation, the outgoing flow of particles $\frac{\mathcal{I}_2}{e} = \frac{mk|t|^2}{\pi\hbar^3}$ is always positive. The interference leading to the current $\mathcal{I}_3$ thus has to be destructive so that the flow of particles $\mathcal{I}_3/e$ is always negative. From current (4.44) it follows that this condition is satisfied if the imaginary part of the T-matrix is negative

$$\text{Im} t(E + i0) \leq 0. \tag{4.54}$$

The positive sign of the imaginary part of the T-matrix corresponds to a non-physical negative probability of the total scattering rate. In other words, the incoming stream is not reduced after passing such a scattering event but enhanced. The inequality (4.54) is called the condition of dissipativeness.

Let us take a look at expansion (4.32) from the point of dissipativeness. The first-order approximation

$$t \approx v \quad \text{gives} \quad \text{Im} t = 0, \tag{4.55}$$

thus it satisfies (4.54) having no dissipation at all. The second-order approximation

$$t \approx v + v G_{00}^0 v \quad \text{gives} \quad \text{Im} t = v^2 \text{Im} G_{00}^0, \tag{4.56}$$

which satisfies the dissipativeness (4.54) because the local density of states, $h^0(E) = -2\text{Im} G_{00}^0$, see Eq. (4.17), is always positive.

To see the case in which the dissipativeness is violated, let us take look at the third-order approximation for G_{00}^0 given by Eq. (4.18),

$$t \approx v + v G_{00}^0 v + v G_{00}^0 v G_{00}^0 v \quad \text{gives} \quad \text{Im} t = -\frac{2v^2}{w^4}\sqrt{w^2 - (E-w)^2}\left(w^2 + 4v(E-w)\right). \tag{4.57}$$

For $v > 0$, Im t becomes positive for $E < w - \frac{w^2}{4v}$. Since the energy E has to be from the band $E > 0$, the third order approximation (4.57) can violate the dissipativeness for a sufficiently strong potential $v > \frac{w}{4}$.

The dissipativeness is a very important property and we will avoid approximations which violate it in any region of our arguments. The demand of dissipativeness restricts the class of acceptable approximations. It should be noted that approximations which satisfy the optical theorem automatically satisfy dissipativeness.

5

Multiple Impurity Scattering

Furnished with basic ideas about the scattering on a single impurity, we can approach the motion of a particle scattered by many randomly distributed impurities. In spite of having only a single particle, this system already belongs to many-body physics as it combines the randomising effects of high-angle collisions with mean-field effects due to low-angle collisions.

Quantum mechanics treats the two distinct aspects of collisions in a unified manner. This unified approach is enforced by the wave nature of the particle motion. As we have seen when deriving the optical theorem, the scattering out of the incoming beam appears due to the destructive interference of the unperturbed wave with the scattered parts. The same interference is responsible for the low-angle collisions also called zero-angle scattering which we have identified as the mean field.

An approximation made within quantum mechanics thus has to cover two distinct approximations in classical mechanics related to the drift and to the dissipation, respectively. While it might be convenient to solve both problems in a single shot, it puts high demands on the approximation itself. In many cases, an approximation which one selects being concerned with some particular sub-dynamics, has undesirable consequences somewhere else. The question of how to select the right approximation is the main problem of the theory of multiple scattering.

In this section we discuss multiple scattering for a model used to describe properties of alloyed crystals like $GaAl_cAs_{1-c}$. To have a convenient dialect, we will call one of alloy components the host and the other the impurity. For example, the GaAs is then called a pure crystal while AlAs is a crystal made of impurities. The model includes the unperturbed Hamiltonian H^0 of the pure (host) crystal and many Koster–Slater impurities randomly distributed all over the crystal,

$$H = H^0 + \sum_j \eta_j V_j, \tag{5.1}$$

where $V_j = |j\rangle v\langle j|$ is the potential of an impurity at site j.[1] The function η specifies the positions of impurities: η_j equals 1 if the site j is occupied by an impurity and 0 for the host atom.

---

[1] We use site index j instead of the coordinate-like index r used above. The space coordinate relating to site j is denoted by r_j.

Interacting Systems far from Equilibrium. Klaus Morawetz, Oxford University Press (2018).
© Klaus Morawetz. DOI: 10.1093/oso/9780198797241.001.0001

Taking two different sets of η's, one obtains two different configurations of impurities which correspond to two different samples. From experience we know that all samples with the same concentration of impurities reveal the same conductivity, X-ray diffraction, thermopower etc. Clearly, the actual configuration does not matter as long as the concentrations are identical. This equivalence of samples allows us to introduce mean values of physical properties obtained by averaging, $\langle \cdots \rangle$, over a set of macroscopically equivalent samples. Naturally, each site is occupied with a probability equal to the local concentration of impurities $\langle \eta_j \rangle = c_j$. Moreover we assume that there is no correlation between the occupation of individual sites, i.e. for $k \neq j$ we have $\langle \eta_j \eta_k \rangle = c_j c_k$ and similar for all higher correlations.

5.1 Divergence of multiple scattering expansion

Let us try a naive approach first using a straightforward power expansion. We will see that it leads to problems which demand a selfconsistent approach.

After a scattering on the first impurity, for simplicity located at the origin with probability c_0, the out-going wave is the source of a next generation of collisions. The stream of electrons leaving the central impurity is given by the amplitude of the wave function, therefore according to Eq. (4.29) it decays like

$$|\psi_1(r)|^2 \approx \frac{m^2 c_0 v^2}{(2\pi)^2 \hbar^4 r^2}. \tag{5.2}$$

After the second generation of collisions, the stream coming back to the central site from all other impurities will be proportional to

$$|\psi_2(0)|^2 \approx \sum_j \frac{m^4 c_0 c_j v^4}{(2\pi)^4 \hbar^8 r_j^4}. \tag{5.3}$$

The sum over j behaves asymptotically as an integral over r_j, i.e. $\sum_j \approx \frac{4\pi}{a^3} \int_a^\infty d|r_j| r_j^2$, where a is the lattice constant. The second generation gives a convergent contribution. One could hope that this procedure can be extended to any higher order, but this is unfortunately not true. If we consider the third generation, where the electron first gets scattered at site 0, next at site j, and finally after a scattering at site k returns back to the central site, we find that its stream should be

$$|\psi_3(0)|^2 \approx \sum_j \sum_k \frac{m^6 c_0 c_j c_k v^6}{(2\pi)^6 \hbar^{12} r_j^2 r_k^2 |r_j - r_k|^2}. \tag{5.4}$$

For non-vanishing concentration $c_j > c$, we can estimate the wave at the central site from below using the triangular inequality $|r_j - r_k| < |r_j| + |r_k|$

$$|\psi_3(0)|^2 > \int_a^\infty d|r_j| d|r_k| \frac{m^6 c^3 v^6}{4\pi^4 \hbar^{12} a^6 (|r_j| + |r_k|)^2} = \frac{m^6 c^3 v^6}{4\pi^4 \hbar^{12} a^6} \int_a^\infty dR \frac{1}{R^2} \int_{-R}^{R} d|r| \to \infty.$$

$$(5.5)$$

Apparently, the three-impurity scattering is divergent for an arbitrarily small scattering potential and impurity concentration. The divergence at higher orders is even stronger. What is the reason for this embarrassing problem?

This naive approach is missing one important physical ingredient. In the crystal with random impurities, the amplitude of the wave function never decays with a power law, but exponentially due to scattering. To obtain a correct and convergent description of multiple scattering, we have to include the decay of the wave function in a manner which is consistent with the scattering rate evaluated from Eq. (4.29).

5.2   Averaged wave function

According to the Lippmann–Schwinger equation (4.23), the scattered wave is given by $\psi = G_\eta V \psi_0$, where ψ_0 is a wave entering the disordered region of the system from outside. The subscript η denotes that the propagator belongs to a specific configuration of impurities. To write the total wave in a compact form we rewrite the entering wave as $\psi_0 = G^0 \Omega$. In Chapter 16.1.1 we will show that Ω can play the role of a contact by which the electron enters the system. Here we focus on the motion of an electron between individual collisions and the actual character of the source does not matter.

Firstly, we simplify the notation. Individual orders of the expansion (4.24) in the scattering potential are written as

$$\psi_0 = G^0 \Omega,$$

$$\psi_1 = G^0 V G^0 \Omega = \sum_j G^0 \eta_j V_j G^0 \Omega,$$

$$\psi_2 = G^0 V G^0 V G^0 \Omega = \sum_k \sum_j G^0 \eta_k V_k G^0 \eta_j V_j G^0 \Omega,$$

$$\psi_n = (G^0 V)^n G^0 \Omega.$$

$$(5.6)$$

The scattered wave function is the sum over all orders $G_\eta^n = (G^0 V)^n G^0$,

$$\psi = \sum_n \psi_n = \sum_n G_\eta^n \Omega = G_\eta \Omega.$$

$$(5.7)$$

The exact solution of the wave in a given configuration of impurities is inaccessible. Moreover, similarly to the exact trajectory of a classical particle, the exact wave function includes too much information without a clue as to how to read it. To obtain readable information, we want to find properties which are common for the majority of samples.

These properties survive under averaging, $\langle \dots \rangle$, over all samples, i.e. over all configurations of impurities. We will thus start by studying the sample average of Green's function,

$$G = \langle G_\eta \rangle, \qquad \text{with powers} \qquad G^n = \langle G_\eta^n \rangle, \qquad (5.8)$$

which corresponds to the sample average of wave function (5.7).

The random character of the potential does not allow for a simple summation of the series. We have to rely on approximations. Let us take the propagator G from expansion (5.7) term by term. For a better insight into the approximations, we will use the site representation of its elements, $G_{mn} \equiv \langle m|G|n \rangle$.

The average of the first order,

$$G_{mn}^1 = \left\langle \sum_j G_{mj}^0 \eta_j v G_{jn}^0 \right\rangle = \sum_j G_{mj}^0 \langle v \eta_j \rangle G_{jn}^0, \qquad (5.9)$$

can be described by the mean value of the impurity potential

$$\phi = \left\langle V \right\rangle = \sum_j |j\rangle c_j v \langle j| \qquad \text{as} \qquad G^1 = G^0 \phi\, G^0, \qquad (5.10)$$

where we denote the probability or concentration at cite i as $c_i = \langle \eta_i \rangle$. We will call ϕ the mean field.

The second order,

$$G_{mn}^2 = \sum_{jk} G_{mk}^0 \left\langle v \eta_k G_{kj}^0 v \eta_j \right\rangle G_{jn}^0, \qquad (5.11)$$

shows two kinds of contributions. If the sum over intermediate sites j and k runs over different positions, $k \neq j$, the average product of occupational factors equals the product of mean values, $c_j c_k$, but if it enters identical positions, $k = j$, the average product of occupations is simply the probability, c_j, that the site is occupied,

$$\langle \eta_j \eta_k \rangle = \langle \eta_j \rangle \langle \eta_k \rangle = c_j c_k \qquad \text{for} \qquad k \neq j$$
$$\langle \eta_j \eta_k \rangle = \langle \eta_j \eta_j \rangle = \langle \eta_j \rangle = c_j \qquad \text{for} \qquad k = j \qquad (5.12)$$

or together

$$\langle \eta_j \eta_k \rangle = c_j c_k + c_j (1 - c_j) \delta_{jk} \qquad \text{for any } k \text{ and } j. \qquad (5.13)$$

The mean value can then be written in a comprehensive matrix notation as

$$G^2 = G^0\phi\, G^0\phi\, G^0 + G^0\Sigma_B^0 G^0, \quad \text{where} \quad \Sigma_B^0 = \sum_j |j\rangle c_j(1-c_j)vG_{jj}^0 v\langle j|. \quad (5.14)$$

The operator Σ_B^0 describes the second-order process at a single impurity, however, a part of the second-order contribution is included in the square of the mean field which is the first term of expansion (5.14). The subtracted mean-field contribution results in the factor $(1-c)$.

The third order,

$$G_{mn}^3 = \sum_{ijk} G_{mi}^0 \langle v\eta_k G_{ik}^0 v\eta_k G_{kj}^0 v\eta_j\rangle G_{jn}^0, \quad (5.15)$$

has five distinct contributions shown by a combination of Kronecker δ's in the decomposition of the average product of occupation factors,

$$\langle \eta_i\eta_k\eta_j\rangle = c_i c_j c_k + c_i c_j(1-c_j)\delta_{jk} + c_j c_k(1-c_k)\delta_{ik} + c_k c_j(1-c_j)\delta_{ij} + c_j(1-c_j)(1-2c_j)\delta_{ik}\delta_{kj}. \quad (5.16)$$

The first term gives the third-order of the mean field ϕ. The second term results in ϕ at site i and operator Σ_B^0 at site $j = k$. The third term places ϕ at j and Σ_B^0 at $i = k$. The fourth term places ϕ at k, between $i = j$. This corresponds to a second-order process at site $i = j$; however, modified by a first-order process at site k. Finally, the last term describes the third-order process at site $i = j = k$. The matrix notation for these contributions reads

$$G^3 = G^0\phi\, G^0\phi\, G^0\phi\, G^0 + G^0\phi\, G^0\Sigma_B^0 G^0 + G^0\Sigma_B^0 G^0\phi\, G^0 + G^0\Sigma_B^1 G^0 + G^0\Sigma_{(3)}G^0. \quad (5.17)$$

The modified second-order process is included via

$$\Sigma_B^1 = \sum_{jk} |j\rangle c_j(1-c_j)vG_{jk}^0 c_k vG_{kj}^0 v\langle j| = \sum_j |j\rangle c_j(1-c_j)vG_{jj}^1 v\langle j|, \quad (5.18)$$

to be compared with the lowest order (5.14). The irreducible third-order process reads

$$\Sigma_{(3)} = \sum_j |j\rangle c_j(1-c_j)(1-2c_j)vG_{jj}^0 vG_{jj}^0 v\langle j|. \quad (5.19)$$

Let us collect the terms (5.9), (5.14), and (5.17) discussed so far,

$$G^0 + G^1 + G^2 + G^3 = G^0 + G^0\phi\, G^0 + G^0\phi\, G^0\phi\, G^0 + G^0\phi\, G^0\phi\, G^0\phi\, G^0$$

$$+ G^0\Sigma_B^0 G^0 + G^0\phi\, G^0\Sigma_B^0 G^0 + G^0\Sigma_B^0 G^0\phi\, G^0 + G^0\left(\Sigma_B^1 + \Sigma_{(3)}\right)G^0. \quad (5.20)$$

It is already visible that higher orders include products of the mean field and other operators found in lower-order approximations. These products can be collected and summed up.

5.2.1 Mean field

In the expansion (5.20) one can see the four first terms of the geometric series in powers of ϕ. At any n-th power of the random potential V we can split off contributions from different sites and the mean-field contribution $G^0(\phi\, G^0)^n$ hence appears. If one neglects all other terms, the averaged Green's function reads

$$
G \approx G^0 + G^0 \phi\, G^0 + G^0 \phi\, G^0 \phi\, G^0 + \ldots = \frac{G^0}{1 - \phi\, G^0} = \frac{1}{z - H^0 - \phi}. \tag{5.21}
$$

One can see that, within this approximation, the system behaves like a regular system of an effective Hamiltonian $H^0 + \phi$. Being free of disorder, this mean-field approximation is also called the virtual crystal approximation (VCA).[2]

The VCA does not discriminate between the host atoms and the impurities. When all sites are occupied by impurities, $c = 1$, the VCA turns into a Hamiltonian of a pure crystal made of impurity atoms. The VCA hence interpolates the Hamiltonian between two compounds of the system.

In homogeneous systems of Koster–Slater impurities, the mean field is constant so that it merely shifts the energy bottom,

$$
G(z) \approx \frac{1}{z - H^0 - \phi} = G^0(z - \phi). \tag{5.22}
$$

Since a motion of the electron does not depend on the choice of the energy bottom, the mean-field contribution seems to be of no importance. One should keep in mind, however, that the mean field is only an effective picture which does not cover local properties of the wave function. For instance, a particle of energy E radiated from a point-like source placed at the origin of coordinate, $\Omega = |0\rangle$ propagates as in the pure crystal, see (4.29), but with momentum related to the shifted energy bottom,

$$
\langle r|G|0\rangle = -\frac{m}{2\pi\,\hbar^2 |r|} e^{i|r|\frac{1}{\hbar}\sqrt{2m(E-\phi)}}. \tag{5.23}
$$

The mean momentum of the particle is thus $\sqrt{2m(E-\phi)}$, while the local momentum between impurities is $\sqrt{2mE}$.

---

[2] The mean-field approximation is sometimes presented as a dominant contribution in the limit of weak impurity potential, $v \to 0$. The formal argument is as follows. At a given site j, the mean field is linear in v, $\phi_{jj} = c_j v$, while the lowest irreducible term is quadratic, $\Sigma^0_{B\,jj} = c_j(1 - c_j)v^2 G^0_{jj}$, and other terms are of a higher order. As $v \to 0$, the irreducible parts can be neglected. This argument is correct as long as one can neglect dissipative processes.

5.2.2   Dyson equation

Formula (5.23) shows that the mean-field approximation of the wave function does not solve the problem of the divergence of the multiple scattering expansion as $\left|\langle r|G|0\rangle\right|^2$ falls only with $1/r^2$. We have to employ higher-order contributions.

The sum of all contributions (5.20) with orders of selfenergy and mean field is called the selfenergy,

$$\Sigma = \phi + \Sigma_B^0 + \Sigma_B^1 + \Sigma_{(3)} + \dots . \tag{5.24}$$

All orders appear in manifold products in the same way as the mean field (5.21) which allows for their summation in the form of Dyson equation,

$$G = G^0 + G^0 \Sigma G. \tag{5.25}$$

In expansion (5.20), we have not derived a sufficiently high number of terms to see the manifold products of the higher-order contributions. Nevertheless, let us accept for the moment that it is 'evident' that the combination of site indices which lead e.g. to Σ_B^0 will appear also in higher-orders of the expansion and the same happens with all parts of the selfenergy. We will return to the definition of selfenergy later, when it will be clearer what this fundamental concept of Green's functions means.

5.2.3   Born approximation of selfenergy

Already the second-order contribution Σ_B^0, called the Born approximation, differs from the mean field being complex and dependent on the energy. Both properties reflect the properties of the local Green's function G_{jj}^0. For model studies of Born selfenergy in a homogeneous system, one can use the model local Green's function (4.18) with a shifted energy argument

$$G_{00}^0(z) = \frac{2}{w^2}\left(z - w - \sqrt{(z-w)^2 - w^2}\right), \tag{5.26}$$

and directly substitute it into selfenergy (5.14). The imaginary part of the selfenergy follows the shape of a Hubbard bubble being weighted with $c(1-c)v^2$.

The imaginary part of the selfenergy is responsible for the decay of the averaged wave function. From the source at the origin, $\Omega = |0\rangle$, the particle propagates as

$$\langle r|G|0\rangle \approx -\frac{m}{2\pi\hbar^2}e^{i|r|\frac{1}{\hbar}\sqrt{2m(E-\phi-\Sigma_B^0)}}. \tag{5.27}$$

Due to the complex momentum,

$$\hbar k = \sqrt{2m(E - \phi - \Sigma_B^0)} \approx \sqrt{2m(E - \phi - \mathrm{Re}\Sigma_B^0)} + i \frac{\mathrm{Im}\Sigma_B^0}{2\sqrt{2m(E - \phi - \mathrm{Re}\Sigma_B^0)}}, \qquad (5.28)$$

the wave function exponentially decays.

The decay of the wave function restores the convergence of the multiple-scattering theory. From this point of view, we have already found the answer to the multiple-scattering divergence.

5.3   Selfconsistent Born approximation

Before we return to the multiple-scattering problem, we should benefit from the formalism so far introduced for the averaged wave function. For example, the Born selfenergy can be easily improved if we sum it with its higher-order counterpart,

$$\Sigma_B^0 + \Sigma_B^1 = \sum_j |j\rangle c_j (1 - c_j) v^2 \left(G_{jj}^0 + G_{jj}^1 \right) \langle j|. \qquad (5.29)$$

Adding the higher-order contribution, we have corrected the internal propagator G_{jj} from the zeroth order, $G \approx G^0$, to its first order, $G \approx G^0 + G^1$. Again, it is 'evident' also that higher-order corrections to the internal propagator appear, so that the Born approximation eventually arrives at

$$\Sigma_B = \sum_j |j\rangle c_j (1 - c_j) v^2 G_{jj} \langle j|. \qquad (5.30)$$

Unlike the lowest-order approximation (5.14), the selfenergy (5.30) cannot be constructed from properties of the unperturbed crystal, but a selfconsistent evaluation of Green's function G and the selfenergy is needed.

The selfconsistent Green's function in the internal propagator changes the effect of the selfenergy on electrons close to the bottom of the electronic band, i.e. for small momenta $\mathrm{Re}\,k$. Within the mean-field approximation of momentum (5.28), these momenta correspond to the energy, $E \approx \frac{k^2}{2m} + \phi$. For an attractive potential, $v < 0$, the energy can be below the energy bottom of the pure crystal, $E < 0$, where the unperturbed density of state vanishes, $\mathrm{Im}\,G_{jj}^0 = 0$, while the selfconsistent density of states is finite, $\mathrm{Im}\,G_{jj} < 0$. The imaginary part of the lowest-order Born selfenergy Σ_B^0 hence vanishes giving an unphysical result that slow particles do not scatter and thus do not decay exponentially. The imaginary part of the selfconsistent Born approximation Σ_B is non-zero for any small momenta in accordance with the obvious fact that all particles are scattered by impurities.[3]

In the selfconsistent approximation, there is no restriction on the concentration of impurities which can range from zero to unity (per site). As the concentration approaches unity, the Born selfenergy vanishes with $(1-c)$ and the Green's function of a pure crystal made of impurities results from the mean field. At this limit, it is meaningful to interchange host atoms and impurities. A new Hamiltonian then splits into the 'unperturbed' part and 'perturbation' as

$$H = \left(H^0 + \sum_j V_j \right) - \sum_j (1 - \eta_j) V_j. \tag{5.31}$$

The concentration of 'impurities' is then small, $\langle (1 - \eta_j) \rangle = 1 - c_j$. Briefly, we shift the bottom energies by v, switch the sign of the impurity potential $v \to -v$, and substitute the concentration c by $1-c$. Under this interchange, the mean-field Hamiltonian does not change, $H^0 + cv = (H^0 + v) - (1 - c)v$. The selfconsistent Born approximation does not change either, since the interchanged (tilted) and the original selfconsistent conditions are identical,

$$\Sigma_B(z) = c(1 - c)v^2 \langle j | \frac{1}{z - H^0 - cv - \Sigma_B} | j \rangle,$$

$$\tilde{\Sigma}_B(z) = (1 - c)c(-v)^2 \langle j | \frac{1}{z - (H^0 + v) + (1 - c)v - \tilde{\Sigma}_B} | j \rangle. \tag{5.32}$$

As one can see $\Sigma_B = \tilde{\Sigma}_B$.

The non-selfconsistent Born approximation is not invariant with respect to the choice of host atoms and impurities, $\tilde{\Sigma}_B^0 \neq \Sigma_B^0$, as follows from explicit expressions,

$$\Sigma_B^0(z) = c(1 - c)v^2 \langle j | \frac{1}{z - H^0} | j \rangle, \qquad \tilde{\Sigma}_B^0(z) = (1 - c)c(-v)^2 \langle j | \frac{1}{z - (H^0 + v)} | j \rangle. \tag{5.33}$$

---

[3] The imaginary part of the lowest-order Born selfenergy reads

$$\text{Im} \Sigma_B^0(E) = c(1 - c)v^2 \, \text{Im} G_{00}^0(E) = c(1 - c)v^2 \, \text{Im} \int \frac{dp}{(2\pi)^3} \frac{1}{E - \frac{\hbar^2 p^2}{2m} + i0} = -c(1 - c)v^2 \frac{m^{\frac{3}{2}}}{\sqrt{2}\pi\hbar^3} \sqrt{E}.$$

If the impurity potential is attractive, $v < 0$, the mean field is negative, $\phi < 0$, so that for small momenta also $E < 0$. The imaginary part of the lowest-order selfenergy is then zero and corresponding wave functions do not decay exponentially. The imaginary part of the selfconsistent Born selfenergy does not have this problem,

$$\text{Im} \Sigma_B(E) = c(1 - c)v^2 \, \text{Im} \int \frac{dp}{(2\pi)^3} \frac{1}{E - \frac{\hbar^2 p^2}{2m} - \phi - \Sigma_B(E)} \approx -c(1 - c)v^2 \frac{m^{\frac{3}{2}}}{\sqrt{2}\pi\hbar^3} \sqrt{E - \phi}.$$

Selfconsistency is thus a vital improvement to the Born approximation although in some overall pictures of selfenergy it might seem negligible.

5.4 Averaged T-matrix approximation

From selfenergy terms obtained up to the third-order expansion (5.20), let us consider now the three-fold interaction with a single impurity, the $\Sigma_{(3)}$. One can simply name it the third-order Born approximation and add it to the mean field and the (lowest) second-order Born selfenergy, but such an approximation has bad properties. We have seen that the third-order approximation of the T-matrix might violate the dissipativeness. The same problem appears for the third-order selfenergy.

In parallel with the T-matrix, the correct dissipativeness results if one also includes all higher-orders of the interaction with the given impurity. To see how the scattering on a single impurity is embedded in the multiple scattering, let us reorganise the pieces of selfenergy obtained so far, (5.18) and (5.19) so that the processes on a single site become distinguishable,

$$\langle j|(\phi + \Sigma_B^0 + \Sigma_B^1 + \Sigma_{(3)})|j\rangle$$

$$= c_j v + c_j(1 - c_j)v\left(G_{jj}^0 + \sum_k G_{jk}^0 c_k v G_{kj}^0\right)v + c_j(1 - c_j)(1 - 2c_j)v G_{jj}^0 v G_{jj}^0 v$$

$$= c_j\left(v + (1 - c_j)v G_{jj}^0 v + (1 - c_j)^2 v G_{jj}^0 v G_{jj}^0 v\right) + c_j(1 - c_j)\sum_{k \neq j} G_{jk}^0 c_k v G_{kj}^0$$

$$= c_j\left(v + v G_{jj}^0 v + v G_{jj}^0 v G_{jj}^0 v\right) - c_j^2 v G_{jj}^0 v - c_j^2(2 - c_j)v G_{jj}^0 v G_{jj}^0 v + c_j(1 - c_j)\sum_{k \neq j} G_{jk}^0 c_k v G_{kj}^0.$$

$$(5.34)$$

For a low concentration of impurities, $c \ll 1$, we can neglect higher-order terms in c. The three terms linear in c are the three lowest terms, T_j^{1-3}, of the expansion for the T-matrix (4.32). Again, it is 'evident' that all higher-orders of the T-matrix appear too, that suggests an approximation for the low-density limit,

$$\Sigma_{ATA} = \sum_j c_j T_j, \qquad T_j = |j\rangle\frac{v}{1 - v G_{jj}^0}\langle j|, \qquad (5.35)$$

called the averaged T-matrix approximation (ATA).

Naturally, the ATA complies with the optical theorem (4.46) and satisfies the condition of dissipativeness (4.54). The ATA, in parallel with the T-matrix, takes into account the reconstruction of the wave function at the impurity site. Accordingly, it remains regular for infinitely strong impurity potentials, $\frac{v}{1 - v G_{jj}^0} \to -\frac{1}{G_{jj}^0}$.

5.4.1   Double counts in the averaged T-matrix approximation

It is clear that the non-selfconsistent character of the ATA causes similar problems as discussed for the Born selfenergy. According to the optical theorem (4.46), the imaginary part of the selfenergy in the ATA is non-zero only if the local density of states is non-zero, i.e. in the same energy interval as the Σ_B^0. In the limit of high concentrations, $c \to 1$, the ATA does not reproduce the crystal of impurities but results in a peculiar Green's function,

$$G = \frac{1}{z - H^0 - \frac{v}{1 - vG_{00}^0(z)}} \qquad \text{instead of} \qquad G = \frac{1}{z - H^0 - v}. \qquad (5.36)$$

The ATA is not selfconsistent and hence it is compatible with the unperturbed propagation G^0, between individual encounters with impurities. The multiple scattering is then viewed as a sequence of single-impurity collisions, as they appear in the expanded form of Dyson equation (5.25),

$$G = G^0 + G^0 \Sigma_{ATA} G^0 + G^0 \Sigma_{ATA} G^0 \Sigma_{ATA} G^0 + \ldots \qquad (5.37)$$

The sequence of events becomes transparent if one writes the selfenergy in expansion (5.37) explicitly,

$$G = G^0 + G^0 |j\rangle \sum_j c_j T_j \langle j| G^0 + G^0 |k\rangle \sum_k c_k T_k \langle k| G^0 |j\rangle \sum_j c_j T_j \langle j| G^0 + \ldots \qquad (5.38)$$

In the last term, the particle first scatters with the impurity at site j, then at k.

The lack of the selfconsistency is linked with neglected corrections for double counts represented by the higher orders in concentration c. This shortage is already seen from low orders presented explicitly in Eq. (5.38). The first generation of collisions, $G^0 |j\rangle \sum_j c_j T_j \langle j| G^0$, is exact. Indeed, the T-matrix T_j exactly solves the single event which is then weighted with the concentration c_j. The average over the second generation, $G^0 \langle k| \sum_k c_k T_k \langle k| G^0 |j\rangle \sum_j c_j T_j \langle j| G^0$, includes a term in which the electron interacts twice with a single impurity at $k = j$. This interaction is, however, already covered by the first term. To avoid such double counts, the expansion should avoid sequences in which the two next events are located at the same site,

$$G = G^0 + G^0 |j\rangle \sum_j c_j T_j \langle j| G^0 + G^0 |k\rangle \sum_{k \neq j} c_k T_k \langle k| G^0 |j\rangle \sum_j c_j T_j \langle j| G^0 + \ldots \qquad (5.39)$$

Equation (5.39) improves (5.38). On the other hand, the condition that the two next collision centres are at different sites couples the first and second events, therefore one cannot associate a selfenergy with the single-collision process.

Let us try fix the double counts using the approach successful for the Born selfenergy. To make the first and second collisions independent, we extend the summation over all sites and subtract the equal-site component,

$$G = G^0 + G^0|j\rangle \sum_j c_j T_j \langle j| G^0 - G^0|j\rangle \sum_j c_j T_j G^0 c_j T_j \langle j| G^0$$

$$+ G^0|k\rangle \sum_k c_k T_k \langle k| G^0|j\rangle \sum_j c_j T_j \langle j| G^0 + \dots. \tag{5.40}$$

The selfenergy suggested by this rearrangement is a site-diagonal function which at site j reads

$$\Sigma_j = c_j T_j - c_j^2 T_j G^0 T_j. \tag{5.41}$$

This approximation of the selfenergy goes beyond the linear dependence in the impurity concentration, but it does not guarantee the dissipativeness. From this point of view, approximation (5.40) is even worse than the ATA. Due to these double counts, the time for speaking of 'evident' resummations is over.

5.5 Effective medium

The averaged T-matrix and the selfconsistent Born approximation represent two different fundamental approaches to the multiple-scattering problem. A parallel can be drawn with the difference between approaches to neutral gases and plasmas. The averaged T-matrix is close to the early molecular theory of gases modelled by 3D billiards. Indeed, individual collisions of hard spheres have to be treated nonperturbatively to describe abrupt changes of momenta and exclude the possibility that two hard spheres would penetrate into each other, while between collisions, molecules are free as if there were no other molecules around. The Born approximation is close to the theory of plasma where the motion is driven by mean fields, while collisions, if included at all, reflect that real forces acting on a selected electron or ion deviate from the mean field because a surrounding charge is not homogeneous but composed of fluctuating elementary charges. Is there any link between these two distinct approaches? To put this question more pragmatically: how do we unify the advantages of the T-matrix and the selfconsistent expansion of Born's type?

Unlike the selfconsistent expansion, the T-matrix is not symmetric with respect to the interchange of the roles of impurities and host atoms. This symmetry can be achieved using a following simple idea. Imagine two compounds alloyed with nearly equal concentrations. One feels that both kinds of atoms are deviations from 'something between'. It is customary to call this 'something between' the effective crystal.

5.5.1 Averaged T-matrix approximation in the virtual crystal

The simplest effective crystal is the virtual crystal represented by the mean-field Hamiltonian, $H^0 + \phi$. We hence rearrange the Hamiltonian (5.1) with the mean-field (5.10) as

$$H = \left(H^0 + \phi \right) + \sum_j V_j^\phi, \qquad V_j^\phi = \eta_j |j\rangle (1 - c_j) v \langle j| + (\eta_j - 1) |j\rangle c_j v \langle j|. \tag{5.42}$$

The impurity potential, v, deviates from the mean-field, cv, by the potential $(1-c)v$. The potential at a host atom deviates from the mean field by $-cv$.

Starting from the effective Hamiltonian (5.42) one can repeat the expansion. The mean-field propagator, $G^\phi = \frac{1}{z-H^0-\phi}$, will now play the role of G^0 above. Since the mean value of the deviation vanishes,

$$\left\langle V_j^\phi \right\rangle = \left\langle \eta_j |j\rangle (1-c_j)v\langle j| + (\eta_j - 1)|j\rangle c_j v\langle j| \right\rangle = 0, \tag{5.43}$$

no additional mean field to $\langle H^0 + \phi \rangle$ appears. The lowest-order Born selfenergy reads

$$\Sigma_B^\phi = \sum_{jk} \left\langle V_j^\phi G^\phi V_k^\phi \right\rangle = \sum_j \left\langle V_j^\phi G^\phi V_j^\phi \right\rangle + \sum_{j\neq k} \left\langle V_j^\phi G^\phi V_k^\phi \right\rangle. \tag{5.44}$$

For $k \neq j$, the mean value of potentials becomes a product of two mean values, $\left\langle V_j^\phi G_{jk}^\phi V_k^\phi \right\rangle = \left\langle V_j^\phi \right\rangle G_{jk}^\phi \left\langle V_k^\phi \right\rangle = 0$, and vanishes in accordance with (5.43). The remaining part is site diagonal,

$$\Sigma_B^\phi = \sum_j |j\rangle \left\langle \left(\eta_j(1-c_j)v - (1-\eta_j)c_j v \right) G_{jj}^\phi \left(\eta_j(1-c_j)v - (1-\eta_j)c_j v \right) \right\rangle \langle j|. \tag{5.45}$$

Using that $\eta_j(1-\eta_j) = 0$, one arrives at

$$\Sigma_B^\phi = \sum_j |j\rangle \left\langle \eta_j \right\rangle (1-c_j)v G_{jj}^\phi (1-c_j)v\langle j| + \sum_j |j\rangle \left\langle 1-\eta_j \right\rangle c_j v G_{jj}^\phi c_j v\langle j|. \tag{5.46}$$

It covers the $\Sigma_B^0 + \Sigma_B^1$ terms of the Born selfenergy (5.29) having the mean field in the internal propagator,

$$\Sigma_B^\phi = \sum_j |j\rangle c_j(1-c_j)v G_{jj}^\phi v\langle j|. \tag{5.47}$$

But let us remain at the form (5.46). The first term tells us that at site j, the electron is scattered either by an impurity, with probability $\langle \eta_j \rangle$, or by a host atom, with probability $\langle 1-\eta_j \rangle$. Similarly, any order of the deviation potential at site j splits into a sum over these two processes,

$$\left(V_j^\phi G_{jj}^\phi \right)^n V_j^\phi = \eta_j |j\rangle \left((1-c_j)v G_{jj}^\phi \right)^n (1-c_j)v\langle j| + (1-\eta_j)|j\rangle \left(-c_j v G_{jj}^\phi \right)^n (-c_j v)\langle j|. \tag{5.48}$$

This separation is a straight consequence of the fact that site j is occupied either by an impurity or by a host atom, mathematically it follows from $\eta_j = 0$ or 1, i.e. $\eta_j(1-\eta_j) = 0$. Consequently, the T-matrix at site j splits too,

$$T_j^\phi = V_j^\phi + V_j^\phi G_{jj}^\phi V_j + V_j^\phi G_{jj}^\phi V_j^\phi G_{jj}^\phi V_j + \ldots = \eta_j \,^{\mathrm{i}}T_j^\phi + (1 - \eta_j) \,^{\mathrm{h}}T_j^\phi, \qquad (5.49)$$

where the impurity and host T-matrices are

$$^{\mathrm{i}}T_j^\phi = |j\rangle \frac{(1 - c_j)v}{1 - (1 - c_j)vG_{jj}^\phi} \langle j| \qquad ^{\mathrm{h}}T_j^\phi = |j\rangle \frac{-c_j v}{1 - (-c_j v)\,G_{jj}^\phi} \langle j|. \qquad (5.50)$$

The selfenergy defined as the averaged T-matrix then reads

$$\Sigma_{ATA}^\phi = \sum_j \left\langle T_j^\phi \right\rangle = \sum_j c_j \,^{\mathrm{i}}T_j^\phi + (1 - c_j) \,^{\mathrm{h}}T_j^\phi. \qquad (5.51)$$

Approximation (5.51) covers all terms of selfenergy found from the third-order expansion. From the propagator $G = \frac{1}{z - H^0 - \phi - \Sigma_{ATA}^\phi}$ one sees that the selfenergy (5.24) has to be compared with $\phi + \Sigma_{ATA}^\phi$. In fact, the expansion of the selfenergy (5.51) up to the third order in v reads

$$\langle j|\Sigma_{ATA}^\phi|j\rangle = c\frac{(1 - c)v}{1 - (1 - c)vG_{jj}^\phi} + (1 - c)\frac{-cv}{1 - (-cv)\,G_{jj}^\phi}$$

$$= c\left((1 - c)v + (1 - c)v \left(G_{jj}^0 + \sum_k G_{jk}^0 cv G_{kj} \right)(1 - c)v + (1 - c)vG_{jj}^0(1 - c)vG_{jj}^0(1 - c)v \right)$$

$$+ (1 - c)\left(-cv + cv\left(G_{jj}^0 + \sum_k G_{jk}^0 cv G_{kj} \right) cv - cvG_{jj}^0 cv G_{jj}^0 cv \right)$$

$$= c(1 - c)v \left(G_{jj}^0 + \sum_k G_{jk}^0 cv G_{kj} \right) v + c(1 - c)(1 - 2c)vG_{jj}^0 vG_{jj}^0 v = \Sigma_B^0 + \Sigma_B^1 + \Sigma_{(3)}.$$

$$(5.52)$$

The averaged T-matrix approximation (5.51) is symmetric with respect to the interchange of impurities and host atoms. For a low concentration of impurities, $c \to 0$, it reduces to the ATA, $\Sigma_{ATA}^\phi \to c\frac{v}{1 - G_{jj}v} - cv$. For a high concentration of impurities, $c \to 1$, it reduces to the ATA with interchanged roles of impurities and host atoms, $\Sigma_{ATA}^\phi \to (1 - c)\frac{-v}{1 + G_{jj}v} + (1 - c)v$. From these points of view, (5.51) is superior to the ATA.

On the other hand, the selfenergy (5.51) provides an incorrect dissipation. From the optical theorem one does not expect any problems. Being defined via the T-matrix, the imaginary part of the selfenergy is always non-positive,

$$\mathrm{Im}\Sigma_{ATA}^\phi = \sum_j c_j \,^{\mathrm{i}}T_j^\phi \mathrm{Im}G_{jj}^\phi \,^{\mathrm{i}}\bar{T}_j^\phi + (1 - c_j) \,^{\mathrm{h}}T_j^\phi \mathrm{Im}G_{jj}^\phi \,^{\mathrm{h}}\bar{T}_j^\phi \leq 0, \qquad (5.53)$$

because $\bar{T}$ is complex conjugate to T and the imaginary part of the internal propagator is always non-positive, $\mathrm{Im}\,G^{\phi}_{jj} \leq 0$. The problems result from the mean-field approximation of the internal propagator. For a demonstration, we assume a parabolic band, $H^0 = \frac{p^2}{2m}$, so that we can easily associate energy E with momentum,

$$p = \sqrt{2m}\sqrt{E - \phi - \Sigma^{\phi}_{ATA}(E + i0)}. \tag{5.54}$$

For the negative real part of selfenergy, $\mathrm{Re}\,\Sigma^{\phi}_{ATA} < 0$, which is always the case when the interaction is weak,[4] there exists a region of small momenta, $\frac{p^2}{2m} < -\mathrm{Re}\,\Sigma^{\phi}_{ATA}(E + i0)$, where the mean-field density of states,

$$h^{\phi}(E) = -\frac{1}{\pi}\mathrm{Im}\,G^{\phi}_{jj}(E + i0) = \int \frac{dp}{(2\pi\hbar)^3}\delta\left(E - \frac{p^2}{2m} - \phi\right)$$

$$= \int \frac{dp}{(2\pi\hbar)^3}\delta\left(\frac{k^2\hbar^2}{2m} + \mathrm{Re}\,\Sigma^{\phi}_{ATA} - \frac{p^2}{2m}\right) = 0, \tag{5.55}$$

vanishes. The imaginary part of the selfenergy (5.53) is therefore zero, so that electrons of these momenta are not dissipated. To obtain the correct dissipation, the internal propagator has to be the full Green's function, $G = \frac{1}{z-H^0-\Sigma}$, in other words, the selfenergy has to be constructed in a selfconsistent manner.

5.5.2   Averaged T-matrix approximation in the selfconsistent effective crystal

To have the full Green's function in the internal propagation of the T-matrix, the effective medium has to correspond to the averaged propagator, i.e. instead of the mean field, we want to characterise the effective medium by the full selfenergy. Since we do not know the full selfenergy yet, we can take its latest value, Σ^{ϕ}_{ATA}, and introduce an improved effective propagator

$$G^{(2)} = \frac{1}{z - H^0 - \Sigma^{(2)}}, \qquad \Sigma^{(2)} = \phi + \Sigma^{\phi}_{ATA}. \tag{5.56}$$

The deviation from the effective medium at impurities is $v - \Sigma^{(2)}_{jj}$ and at the host atoms we have $-\Sigma^{(2)}_{jj}$. Again, the T-matrix splits into impurity and host parts, $T^{(2)}_j = \eta_j\,{}^i T^{(2)}_j + (1 - \eta_j)\,{}^h T^{(2)}_j$, where

---

[4] The negative value of the selfenergy and the existence of an energy region out of the mean-field band is shown in Eq. (5.73).

$$^i T_j^{(2)} = |j\rangle \frac{v - \Sigma_{jj}^{(2)}}{1 - \left(v - \Sigma_{jj}^{(2)} \right) G_{jj}^{(2)}} \langle j|, \qquad ^h T_j^{(2)} = |j\rangle \frac{-\Sigma_{jj}^{(2)}}{1 + \Sigma_{jj}^{(2)} G_{jj}^{(2)}} \langle j|. \qquad (5.57)$$

The selfenergy defined as the averaged T-matrix then reads

$$\Delta\Sigma^{(3)} = \sum_j \left\langle T_j^{(2)} \right\rangle = \sum_j c_j \, ^i T_j^{(2)} + (1 - c_j) \, ^h T_j^{(2)}. \qquad (5.58)$$

The correction $\Delta\Sigma^{(3)}$ is the first step of an iterative procedure which can be repeated infinite times. With a corrected effective medium, $\Sigma^{(3)} = \Sigma^{(2)} + \Delta\Sigma^{(3)}$, we can again evaluate a higher correction and so on. The iteration relations are simple modifications of the steps (5.56)–(5.58).

After an infinite number of steps, the increments of selfenergy vanish, $\Delta\Sigma^{(\infty)} = 0$, and the selfenergy reaches its selfconsistent value, $\Sigma^{(\infty)}$. The vanishing of the increments provides us, from relation (5.58), with an implicit condition for the asymptotic selfenergy,

$$0 = c_j \, ^i T_j^{(\infty)} + (1 - c_j) \, ^h T_j^{(\infty)}, \qquad (5.59)$$

with the T-matrices defined selfconsistently from the full Green's function, $G = G^{(\infty)} = \frac{1}{z - H^0 - \Sigma^{(\infty)}}$, as

$$^i T_j^{(\infty)} = |j\rangle \frac{v - \Sigma_{jj}^{(\infty)}}{1 - \left(v - \Sigma_{jj}^{(\infty)} \right) G_{jj}} \langle j|, \qquad ^h T_j^{(\infty)} = |j\rangle \frac{-\Sigma_{jj}^{(\infty)}}{1 + \Sigma_{jj}^{(\infty)} G_{jj}} \langle j|. \qquad (5.60)$$

The selfenergy $\Sigma^{(\infty)}$ defined via relations (5.59)–(5.60) satisfies all requirements we have formulated so far. It is symmetric in host atoms and impurities, selfconsistent and given by the T-matrix, i.e. it cover a reconstruction of the wave function at the impurity site. Moreover, condition (5.59) allows one to reach the selfenergy by numerical procedures alternative to this iteration.

5.6 Coherent potential approximation

The previously described approaches to the perturbative expansion show that reasonable approximations of the averaged propagator are restricted by three requests:

1. The decay of the wave function has to be included via Dyson equation, $G = G^0 + G^0 \Sigma G$.

2. Close to the energy bottom, the decay is guaranteed only by selfconsistent selfenergies, $\Sigma = \Sigma[G]$.

3. The selfenergy defined via the T-matrix includes the repulsion of the wave function from the region of a strong potential.

These requests can be used to define a selfenergy for impurity potentials more general than the Koster–Slater model discussed so far as well. We will thus assume a potential $V_{(j)}$ associated with the impurity at the site j but in general non-zero also at sites in the vicinity, including the site-off-diagonal elements.

The first and second requests are fulfilled if the selfenergy is derived by deviations from the effective medium,

$$H = \left(H^0 + \sum_j \Sigma_{(j)} \right) - \sum_j \left(\eta_j (V_{(j)} - \Sigma_{(j)}) - (1 - \eta_j) \Sigma_{(j)} \right), \tag{5.61}$$

where $\Sigma_{(j)}$ is the selfenergy part associated with an impurity at site j. According to the third request, we derive the T-matrix associated with each site, $T_{(j)} = \eta_j\, {}^i T_{(j)} + (1 - \eta_j)\, {}^h T_{(j)}$, where

$$ {}^i T_{(j)} = \frac{V_{(j)} - \Sigma_{(j)}}{1 - (V_{(j)} - \Sigma_{(j)})G}, \qquad {}^h T_{(j)} = -\frac{\Sigma_{(j)}}{1 + \Sigma_{(j)} G}, \tag{5.62}$$

and demand that in a properly chosen effective medium, its mean value vanishes

$$0 = c_j\, {}^i T_{(j)} + (1 - c_j)\, {}^h T_{(j)}. \tag{5.63}$$

The Soven equation (5.63) is an implicit equation for the selfenergy part $\Sigma_{(j)}$. The total selfenergy is a sum over all lattice sites $\Sigma_{CPA} = \sum_j \Sigma_{(j)}$.

In pioneering studies, the Koster–Slater impurities were assumed; where the selfenergy is site-diagonal reminding us of a complex energy-dependent potential. This is the reason why Soven named this method the coherent potential approximation (CPA).

Note that unlike the previously mentioned examples, we did not have to support the approximation by claims that it is 'evident' that some similar terms also appear in all higher orders. The summations leading to the Dyson equation and to the selfconsistency are hidden in the idea of the effective medium. The T-matrix is the exact solution for a single perturbation in the effective medium. The real approximation is thus given by the physical assumption that when we average over the occupation at site j, it is sufficient to describe the rest of the crystal by the mean propagator G.

5.6.1   Lorentz–Lorenz local-field correction

The coherent potential approximation parallels the Lorentz–Lorenz local-field corrections to the dielectric function introduced in 1869 prior to Maxwell's theory of the electromagnetic field. It is useful to show this similarity explicitly, because the approach of Lorentz is better suited for the generalisation of the effective medium to the two-particle T-matrix.

Lorenz studied the optical index of a crystal made of neutral polarisable atoms. Here we discuss the formula in terms of the dielectric function due to Mossotti. The external charges cause the external electric field $\mathbf{E}_0$. Each atom contributes by its dipole field $\mathbf{E}_i$, therefore the total electric field in the systems is $\mathbf{E} = \mathbf{E}_0 + \sum_i \mathbf{E}_i$.

The dipole field of an isolated atom at position $\mathbf{R}_i$ in the external field would be $\mathbf{E}_i(\mathbf{r}) = -\hat{\alpha}(\mathbf{r} - \mathbf{R}_i)\mathbf{E}_0(\mathbf{R}_i)$, where $\hat{\alpha}$ is the tensor of the dipole field. For weakly dielectric materials, $\left|\sum_i \mathbf{E}_i\right| \ll |\mathbf{E}_0|$, the non-selfconsistent approximation represented by the external field is sufficient, $\mathbf{E}(\mathbf{r}) \approx \mathbf{E}_0(\mathbf{r}) - \sum_i \hat{\alpha}(\mathbf{r} - \mathbf{R}_i)\mathbf{E}_0(\mathbf{R}_i)$. After averaging over atom positions for a homogeneous gas and smooth electric fields, the response field becomes local and parallel to the external field $\langle\mathbf{E}(\mathbf{r})\rangle \approx \mathbf{E}_0(\mathbf{r}) - n\alpha\mathbf{E}_0(\mathbf{r})$, where n is the density of atoms and α is a scalar obtained by the space integral over $\hat{\alpha}$.

If the polarisability of the crystal is large, we have to include the field from other atoms. The simplest and most common guess is to use the selfconsistent field instead of the external field, $\mathbf{E}(\mathbf{r}) \approx \mathbf{E}_0(\mathbf{r}) - \sum_i \hat{\alpha}(\mathbf{r} - \mathbf{R}_i)\langle\mathbf{E}(\mathbf{R}_i)\rangle$. The averaging then yields $\langle\mathbf{E}(\mathbf{r})\rangle \approx \mathbf{E}_0(\mathbf{r}) - n\alpha\langle\mathbf{E}(\mathbf{r})\rangle$ so that $\langle\mathbf{E}\rangle = \mathbf{E}_0/(1 + n\alpha)$.

In the selfconsistent choice the field polarising the atom includes the field of this atom itself. A straightforward argument for the neglect of this self-interaction is that a single atom out of 10^{22} or a similarly large number does not play any role, therefore the field generated by all other atoms is the same as the field generated by all atoms. This argument is perfect over all the crystal except for the position of the atom in question. Since our atom is excluded from the field acting on it, there is a cavity in the dielectric medium corresponding to its volume. For cubic materials we can use the spherical cavity, inside which the field relates to the total field as $\mathbf{E}_c = (2/3)\langle\mathbf{E}\rangle + (1/3)\mathbf{E}_0$ so that we find $\langle\mathbf{E}\rangle \approx \mathbf{E}_0 - (2/3)n\alpha\langle\mathbf{E}\rangle - (1/3)n\alpha\mathbf{E}_0$. Therefore

$$\langle\mathbf{E}\rangle = \frac{1 - n\alpha/3}{1 + 2n\alpha/3}\mathbf{E}_0. \tag{5.64}$$

Let us point out one of the interesting consequences of this local correction. The dielectric function gains a simple form

$$\epsilon = \frac{\mathbf{E}_0}{\langle\mathbf{E}\rangle} = \frac{1 + \frac{2n\alpha}{3}}{1 - \frac{n\alpha}{3}} = 1 + \frac{na}{\omega_0 - \omega - i\gamma - \frac{na}{3}} \tag{5.65}$$

for the resonant interaction, $\alpha = a/(\omega_0 - \omega - i\gamma)$ with the field frequency ω, the resonance frequency ω_0 and the linewidth γ. The dielectric function has the resonance frequency shifted by the so-called Lorentz redshift $-na/3$. This experimentally well-established shift is missing in the simple selfconsistent approach without the local correction.

5.6.2 Coherent potential approximation as the local-field correction

The essence of the Lorentz theory is that the field acting on the atom is neither the external field $\mathbf{E}_0$ nor the complete averaged field $\langle\mathbf{E}\rangle$ but a microscopic field $\mathbf{E}_c$, which holds only inside the cavity with the atom and thus it is not observable on the macroscopic level. This cavity field depends on the averaged field. Previously we have asked a similar question. Should the selfenergy be constructed from the bare Green's function G^0 or the full Green's function G? The latter is correct, but when the selfconsistency is used without care, it leads to double-counts similar to the self-interaction of atoms.

To show that the Lorentz–Lorenz local-field correction is equivalent to the idea of the effective medium used here, let us recover the CPA from the Lorentz concept of cavity. Assume an alloy described by the selfenergy as a sum of contributions associated with individual crystal lattice sites $\Sigma = \sum_j \Sigma_{(j)}$. At one site, say $j = 0$, we want to evaluate the interaction with the impurity explicitly keeping the rest in the approximation of the effective medium. First we remove the effective medium from the selected site introducing a 'cavity' propagator at site j

$$G^c_{(j)} = G - G\,\Sigma_{(j)}\,G^c_{(j)}. \tag{5.66}$$

From this propagator we can readily construct the T-matrix

$$T_{(j)} = \frac{V_{(j)}}{1 - V_{(j)}\,G^c_{(j)}} \tag{5.67}$$

and the full propagator is an average over the occupation of site j by the host atom or the impurity

$$G = (1 - c_j)G^c_{(j)} + c_j\left(G^c_{(j)} + G^c_{(j)}\,T_{(j)}\,G^c_{(j)}\right) = G^c_{(j)} + G^c_{(j)}\,c_j\,T_{(j)}\,G^c_{(j)}. \tag{5.68}$$

These equations are equivalent to the Soven equation (5.63) with T-matrices (5.62).[5]

5.7 Energy spectrum

The multiple scattering leads to a reconstruction of the energy spectrum. One aspect of this reconstruction, the shift of the energy bottom, has been already used to formulate the request of selfconsistency, $\Sigma = \Sigma[G]$. Other aspects are important to understand the properties of the system. The energy spectrum belongs to problems studied within the secular equation. In this section we thus demonstrates a use of Green's functions to formulate the secular equation.

---

[5] Substituting $T_{(j)}$ from (5.67) and $G^c_{(j)}$ from (5.66) into (5.68) one obtains

$$cV_{(j)} = \left(1 - (V_{(j)} - \Sigma_{(j)})G\right)\Sigma_{(j)}.$$

The same equation results if one substitutes the T-matrices (5.62) into the Soven equation (5.63) and multiplies it by denominators of the T-matrices.

5.7.1 Densities of states

In a homogeneous system, the spectrum of allowed energies is described by the density of states,

$$h(E) = -\frac{1}{\pi} \mathrm{Im}\, G_{00}(E) = -\frac{a^3}{\pi} \int_{B.z.} \frac{dk}{(2\pi)^3} \frac{\mathrm{Im}\Sigma(E,k)}{(E-\epsilon_k - \mathrm{Re}\Sigma(E,k))^2 + (\mathrm{Im}\Sigma(E,k))^2}. \quad (5.69)$$

Here, $\epsilon_k = \langle k|H^0|k\rangle$ is the band of the host crystal and the momentum integration goes over the Brillouin zone. We remind ourselves that

$$h^0(E) = -\frac{a^3}{\pi} \lim_{\nu\to0} \int_{B.z.} \frac{dk}{(2\pi)^3} \frac{\nu}{(E-\epsilon_k)^2 + \nu^2} = a^3 \int_{B.z.} \frac{dk}{(2\pi)^3} \delta(E-\epsilon_k),$$

with a^3 denoting the volume per single site. To fit the mass m of electrons at the band edge, $\epsilon_k \approx \frac{k^2\hbar^2}{2m}$, the half-band width used as a parameter of the Hubbard bubble has to be

$$w = (2\pi)^{\frac{2}{3}} \left(\frac{\hbar}{a}\right)^2 \frac{1}{m}.$$

The effect of impurities on the density of states is conveniently demonstrated for the host density of state modelled by the Hubbard bubble (4.17) and Koster–Slater impurities (5.1).

All approximations of selfenergy we have discussed are independent of the momentum. We can thus avoid the demanding integration over momenta,

$$G_{00}(E) = G_{00}^0(E-\Sigma) = \frac{2}{w^2}\left(E - \Sigma - w - \sqrt{(E-\Sigma-w)^2 - w^2}\right). \quad (5.70)$$

With this local Green's function we can easily test all approximations of selfenergy

mean field	$\Sigma \approx \phi$	$= cv$
non-selfconsistent Born	$\approx \phi + \Sigma_B^0$	$= cv + c(1-c)v^2 G_{00}^0$
selfconsistent Born	$\approx \phi + \Sigma_B$	$= cv + c(1-c)v^2 G_{00}$
averaged T-matrix approximation	$\approx \Sigma_{ATA}$	$= c\dfrac{v}{1 - vG_{00}^0}$
selfconsistent modification of averaged T-matrix approximation	$\approx \Sigma_{SATA}$	$= c\dfrac{v}{1 - vG_{00}}$
coherent potential approximation	$(1-c)\dfrac{\Sigma_{CPA}}{1 + \Sigma_{CPA}G_{00}}$	$= c\dfrac{v - \Sigma_{CPA}}{1 - (v - \Sigma_{CPA})G_{00}}.$

$$(5.71)$$

For some selected values, the resulting densities of states are compared in Fig. 5.1.

5.7.2 Bottom of allowed energies in the main band

Let us first assume moderately strong repulsive potentials, $0 < v < \frac{w}{2}$, for which the Koster–Slater impurity does not form a bound state. The energy bottom appears at the energy where the imaginary part of the local Green's function sets on. One can see in Figure 5.1 that the non-selfconsistent approximations badly fail. For repulsive potentials, the non-zero density of states starts at $E^{\text{bot}} = 0$ due to a non-zero density of states of the host crystal having nothing to do with the real properties of the system. The other approximations have the energy given by a zero value of the discriminant in the local Green's function (5.70), i.e. $E^{\text{bot}} = \Sigma$. Since the local Green's function has a simple value, $G_{00}(E^{\text{bot}}) = -\frac{2}{w}$, the positions of the energy bottom read

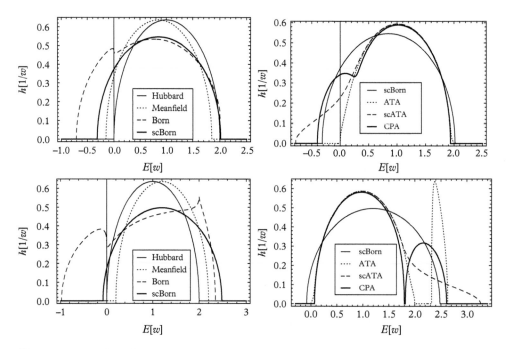

Figure 5.1: *Densities of states in various approximations (5.71) and $c = 0.2$ for the attractive $v = -0.75w$ (above) and repulsive $v = 1w$ (below) impurity potential. The bare density of states h^0 (thin lines) serves as a reference function. The comparison of the non-selfconsistent (dashed) and selfconsistent (solid) lines shows the strong effect of selfconsistency. The Born approximation (left panel) is smooth in selfconsistent approximation while its non-selfconsistent version has non-physical cusps reflecting edges of the non-selfconsistent Green's function inside the selfenergy. The ATA (right panel) with a non-selfconsistent selfenergy shows an impurity-band-split off the main band, while the selfconsistency leads to a broad shoulder at low energies. In the coherent potential approximation (thick line in the right panel) the selfconsistency is locally suppressed and kept around the impurity so that the resulting density of states is between the selfconsistent and non-selfconsistent ATA.*

$$E_{\phi}^{\text{bot}} = cv, \qquad\qquad E_B^{\text{bot}} = cv - 2c(1-c)v^2\frac{1}{w}, \qquad (5.72)$$

$$E_{SATA}^{\text{bot}} = c\frac{v}{1+2\frac{v}{w}}, \qquad E_{CPA}^{\text{bot}} = \frac{w}{4} + \frac{v}{2} - \sqrt{\left(\frac{w}{4}+\frac{v}{2}\right)^2 - \frac{1}{2}cvw}. \qquad (5.73)$$

Note that the mean field always overestimates the value. At the same time, it underestimates the upper limit of the energy spectrum. Differently stated, the mean field does not change the shape and the electron band only shifts its centre. Other approximations deform and widen the band. The wider band confirms that the selfconsistency on the level of the mean field is not sufficient. Being narrower, the unperturbed density of states cannot cover the whole energy spectrum so that regions of un-dissipated states have to appear close to the energy bottom.

5.7.3 Effective mass

In the vicinity of the energy bottom, the density of states has a square-root shape, $h(E) = \alpha\sqrt{E - E^{\text{bot}}}$. In the host crystal, its factor is proportional to the (3/2)-power of electronic mass, $\alpha_0 = \frac{1}{\pi}\left(\frac{2}{w}\right)^{3/2} = \frac{a^3\sqrt{2}}{\hbar^3\pi^2}m^{3/2}$. The mean field does not change the shape of the band, i.e. $\alpha_{\phi} = \alpha_0$. All higher approximations change the density of states at the band edge. We have to evaluate $\text{Im}\,G_{00}(E)$ (5.69) slightly above the energy bottom, $w \gg E - E^{\text{bot}} > |\text{Im}\Sigma|$. The imaginary part of the local Green's function (5.70) is

$$\text{Im}\,G_{00} = -\frac{2}{w^2}\text{Im}\Sigma + \frac{2}{w^2}\text{Im}\sqrt{(E - \Sigma - w)^2 - w^2}. \qquad (5.74)$$

Within the linear approximation $\text{Re}\Sigma(E) = \Sigma(E^{\text{bot}}) + (E - E^{\text{bot}})\frac{\partial}{\partial E}\text{Re}\Sigma$ one finds

$$\text{Im}\,G_{00} = -\frac{2}{w^2}\text{Im}\Sigma + \frac{2}{w^2}\sqrt{2w(E - E^{\text{bot}})}\sqrt{1 - \frac{\partial}{\partial E}\text{Re}\Sigma}, \qquad (5.75)$$

where we have used $E^{\text{bot}} - \Sigma(E^{\text{bot}}) = 0$. This results in the density of states (5.69),

$$h(E) = \alpha_0\sqrt{E - E^{\text{bot}}}\sqrt{1 - \frac{\partial}{\partial E}\text{Re}\Sigma} + \frac{2}{\pi w^2}\text{Im}\Sigma. \qquad (5.76)$$

For the Born approximation,

$$\frac{2}{\pi w^2}\text{Im}\Sigma_B = \frac{2}{\pi w^2}c(1-c)v^2\text{Im}\,G_{00} = -\frac{2}{w^2}c(1-c)v^2h_B(E)$$

$$\approx -\frac{2}{w^2}c(1-c)v^2\alpha_0\sqrt{E - E^{\text{bot}}}\sqrt{1 - \frac{\partial}{\partial E}\text{Re}\Sigma}, \qquad (5.77)$$

where we have used the dominant term of (5.76) as an approximation of h_B. After substitution into (5.76), Born's edge density of states results as $h_B(E) = \alpha_B \sqrt{E - E^{bot}}$ with the factor

$$\alpha_B = \alpha_0 \sqrt{1 - \frac{\partial}{\partial E} \mathrm{Re}\Sigma_B \left(1 - c(1 - c)v^2 \frac{2}{w^2}\right)}. \tag{5.78}$$

The two multiplicative corrections are closely related. The derivative of the selfenergy is

$$\frac{\partial}{\partial E} \mathrm{Re}\Sigma_B = c(1 - c)v^2 \frac{\partial}{\partial E} \mathrm{Re}G_{00} \approx c(1 - c)v^2 \frac{2}{w^2}, \tag{5.79}$$

so that

$$\alpha_B = \alpha_0 \left(1 - c(1 - c)v^2 \frac{2}{w^2}\right)^{\frac{3}{2}} = \alpha_0 \left(1 - \frac{\partial}{\partial E} \mathrm{Re}\Sigma_B\right)^{\frac{3}{2}}. \tag{5.80}$$

The decrease of the density of states at the band edge is related to a change of the effective electronic mass. Due to impurities, the band is wider than the band of the host crystal. Since the width of the band is inversely proportional to the electronic mass, $w \sim (2\pi)^{\frac{2}{3}} \left(\frac{\hbar}{a}\right)^2 \frac{1}{m}$, one has to expect that the interaction makes electrons effectively lighter. From the propagation of the wave, (5.28) or (5.54), we can define an effective mass m^* via $k = \frac{1}{\hbar}\sqrt{2m^*(E - E^{bot})}$ as

$$m^* = m \left(1 - \frac{\partial}{\partial E} \mathrm{Re}\Sigma\right). \tag{5.81}$$

Within Born's approximation, the decrease of the density of states at the band edge is fully covered by the decrease of the effective mass,

$$\alpha_B = \alpha_0 \left(\frac{m^*}{m}\right)^{\frac{3}{2}}. \tag{5.82}$$

5.7.4   Effect of the collision delay on the density of states

For a stronger interaction which requires to use the T-matrix, the density of states reflects a finite duration of the collision process. This can be conveniently demonstrated for the selfconsistent ATA. Using the optical theorem, $\mathrm{Im}t = |t|^2 \mathrm{Im}G_{00}$, to rearrange the imaginary part of the selfenergy, $\mathrm{Im}\Sigma = c\mathrm{Im}t$, from the density of states (5.76) one finds the factor

$$\alpha_{SATA} = \alpha_0 \sqrt{1 - \frac{\partial}{\partial E} \mathrm{Re}\Sigma_{SATA} \left(1 - c|t|^2 \frac{2}{w^2}\right)}. \tag{5.83}$$

Again, the energy derivative of the selfenergy can be linked with the other multiplicative factor,[6]

$$\frac{\partial}{\partial E} \mathrm{Re}\, \Sigma_{SATA} = c|t|^2 \frac{2}{w^2} - 2\mathrm{Im}\, \Sigma_{SATA} \Delta_\tau.$$ (5.84)

In Chapter 4.3 we have seen that Δ_τ measures the time by which the particle is delayed in the collision. For a finite collision delay, $\Delta_\tau \neq 0$, the changes in the density of states are not fully covered by the renormalisation of the mass but they also reflect the finite duration of collision processes,

$$\alpha_{SATA} = \alpha_0 \left(\frac{m^*}{m}\right)^{\frac{3}{2}} + \alpha_0 \left(\frac{m^*}{m}\right)^{\frac{1}{2}} (-2)\mathrm{Im}\, \Sigma_{SATA} \Delta_\tau.$$ (5.85)

As we have seen, for the attractive potential which creates a bound state, the collision delay near the band edge is negative. Since the spectral function of the selfenergy is always positive, $(-2)\mathrm{Im}\, \Sigma > 0$, the density of states are reduced in the main band. This reduction is to be expected because a part of the density of states is carried by the bound states.

5.7.5 Test of approximations in the atomic limit

Electrons can tunnel between bound states of impurities. As a result, impurity states will form a new energy band. Let us inspect how this band is described by individual approximations of the selfenergy.

As the simplest test we start from the system with an infinitely narrow energy band, $w \to 0$. At this limit the crystal becomes a lattice of atoms isolated in the sense that electrons cannot hop from one atom to another. The Hamiltonian is diagonal in the site representation, $H_{ij} = 0$ for $i \neq j$. The local elements have two values: at the host atom $H_{ii} = 0$, and at the impurity $H_{ii} = v$. The exact averaged Green's function thus is

---

[6] From $t = \frac{v}{1 - vG_{00}}$ follows $\frac{\partial t}{\partial E} = t^2 \frac{\partial G_{00}}{\partial E}$, with the help of which one expresses the collision delay in terms of the energy derivative of the local Green's function,

$$\Delta_\tau = \mathrm{Im} \frac{\partial}{\partial E} \ln t = \mathrm{Im} \left[t \frac{\partial}{\partial E} G_{00} \right].$$

The energy derivative of the selfenergy expressed via the energy derivative of the local Green's function reads

$$\frac{\partial}{\partial E} \mathrm{Re}\, \Sigma_{SATA} = c\mathrm{Re}\left[t^2 \frac{\partial}{\partial E} G_{00} \right] = c\mathrm{Re}\left[t(t^* + (t - t^*)) \frac{\partial}{\partial E} G_{00} \right]$$

$$= c|t|^2 \frac{\partial}{\partial E} \mathrm{Re}\, G_{00} - 2c\mathrm{Im}\, t\, \mathrm{Im}\left[t \frac{\partial}{\partial E} G_{00} \right].$$

With $\frac{\partial}{\partial E} \mathrm{Re}\, G_{00} \approx \frac{2}{w^2}$ one arrives at (5.84).

$$G_{ii} = \frac{1-c}{\omega} + \frac{c}{\omega - v}. \tag{5.86}$$

In the mean-field approximation we only obtain a single pole centred between actual poles

$$G_{ii}^{\phi} = \frac{1}{\omega - cv}. \tag{5.87}$$

This result was expected as the mean field cannot describe the bound state.

The non-selfconsistent Born approximation gives

$$G_{ii}^{0B} = \frac{1}{\omega - cv - v^2c(1-c)/\omega} = \frac{\frac{1}{2} + \frac{c}{2\sqrt{c(4-3c)}}}{\omega - \frac{v}{2}\left(c + \sqrt{c(4-3c)}\right)} + \frac{\frac{1}{2} - \frac{c}{2\sqrt{c(4-3c)}}}{\omega - \frac{v}{2}\left(c - \sqrt{c(4-3c)}\right)}. \tag{5.88}$$

These two poles do not originate in the bound state but result from a level repulsion of the mean-field state at cv and the host crystal level in the non-selfconsistent selfenergy. Their positions and weights are thus very different from the exact values.

The selfconsistent Born approximation $G_{ii}^{B} = 1/\left(\omega - cv - v^2c(1-c)G_{ii}^{B}\right)$ leads to a quadratic equation with the roots

$$G_{ii}^{B} = \frac{\omega - cv}{2c(1-c)v^2} \pm \sqrt{\left(\frac{\omega - cv}{2c(1-c)v^2}\right)^2 - \frac{1}{c(1-c)v^2}}. \tag{5.89}$$

This solution corresponds to an energy band centred at cv with the energy width $4v\sqrt{c(1-c)}$. According to expectations we have found a single band, because the Born approximation does not describe the bound state. The finite band width is a consequence of the selfconsistency. As one can see from the equation, any pole of the Green's function has to be a pole of the selfenergy at the same time. This is not possible because functions G and $1/G$ cannot have poles at the same energy. This artificial band is rather wide: for $c = 0.5$ the band width equals $2v$ which twice exceeds the distance between the host and the impurity level.

In the non-selfconsistent averaged T-matrix approximation one finds two poles

$$G_{ii}^{ATA} = \frac{1}{\omega - \frac{cv}{1-v/\omega}} = \frac{\frac{1}{1+c}}{\omega} + \frac{\frac{c}{1+c}}{\omega - (1+c)v}. \tag{5.90}$$

As one expects, the pole weights and positions are correct only in the lowest order of c. If we apply this approximation far beyond its validity region taking the crystal made of impurities, $c = 1$, it predicts two levels at positions $\omega = 0$ and $\omega = 2v$. Similar to the non-selfconsistent Born approximation, the host crystal in the selfenergy leads to an artificial level splitting.

The selfconsistent averaged T-matrix approximation also leads to a quadratic equation with the roots now

$$G_{ii}^{SATA} = \frac{1}{\omega - \frac{cv}{1-vG_{ii}^{SATA}}} = \frac{\omega + (1-c)v}{2v\omega} \pm \sqrt{\left(\frac{\omega + (1-c)v}{2v\omega}\right)^2 - \frac{1}{v\omega}}. \tag{5.91}$$

At $\omega = 0$ it has a pole with the exact weight $1 - c$. The imaginary part due to the negative value of the discriminant represents an impurity band in the energy interval $(1 + c)v + 2v\sqrt{c} < \omega < (1 + c)v - 2v\sqrt{c}$. Again, the selfconsistency leads to the incorrect broadening of the impurity level.

In the coherent potential approximation, the cavity propagator is $G_{ii}^c = 1/\omega$, the T-matrix is $T_{(i)} = v/(1 - v/\omega)$, so that the full propagator

$$G_{ii}^{CPA} = \frac{1}{\omega} + c\frac{1}{\omega^2}\frac{v}{1 - v/\omega} = \frac{1-c}{\omega} + \frac{c}{\omega - v} \tag{5.92}$$

equals the exact solution. As one can see, the cavity propagator is free from the unphysical self-effect of the impurity on itself via the effective medium. The impurity level thus remains sharp.

6

Selfenergy

6.1 Averaged amplitude of wave function

Although the averaged wave function provides valuable information about the energy spectrum of the system, it is not sufficient to evaluate single-electron observables,

$$o = \langle \psi^* O \psi \rangle \neq \langle \psi^* \rangle_\eta O \langle \psi \rangle, \tag{6.1}$$

where O is the operator of the observable o and the conjugated wave function obeys an expansion conjugate to (5.6), e.g. $\psi_n^* = \Omega^* G_0^* (V G_0^*)^n$. To avoid confusion, we specify the boundary condition by superscript

$$G \equiv G^R \quad \text{and} \quad G^* \equiv G^A, \tag{6.2}$$

denoting the retarded and the advanced functions. The advanced Green's function has to be used for expansion of the conjugated function ψ^* as it follows from $(E + i0)^* = E - i0$. To find the mean value of an observable, we have to return to our original problem of multiple scattering, the average of the amplitude of the wave function.

We have already seen that if we evaluate the scattering rate on a single impurity and try to use the outgoing wave as an initial condition for the next generations of scattering, we run into divergences. These divergences are suppressed if we properly combine a perturbative expansion for the amplitude with the perturbative expansion for a single wave function. To explain why there are two 'independent' perturbative expansions for a single physical event, we start with a straightforward perturbative expansion. Then we show how one can use the requirement of consistency to deduce the averaging of the amplitude from the averaging of the wave function and derive simple rules for this procedure.

6.2 Straightforward perturbative expansion

Firstly we denote the sample average of the wave function in the dialect of Green's functions as

$$G^< \equiv \langle \psi \psi^* \rangle, \quad \text{i.e.,} \quad o = \text{Tr}\left[O G^<\right]. \tag{6.3}$$

Interacting Systems far from Equilibrium. Klaus Morawetz, Oxford University Press (2018).
© Klaus Morawetz. DOI: 10.1093/oso/9780198797241.001.0001

Now we expand the amplitude in powers of V. The expansion of $G^<$ combines individual orders of expansions for ψ and ψ^*,

$$G^{n<} = \sum_{m=0}^{n} \langle \psi_m \psi^*_{n-m} \rangle. \tag{6.4}$$

With the explicit forms of ψ_m and ψ^*_{n-m} the n–th order reads

$$G^{n<} = \sum_{m=0}^{n} \left\langle (G^{0R}V)^m G^{0<}(VG^{0A})^{n-m} \right\rangle, \tag{6.5}$$

where $G^{0<} = G^{0R}\Omega\Omega^* G^{0A}$ is used to make the notation more compact.

6.2.1 Transport vertex in the non-selfconsistent Born approximation

Now we will discuss the lowest orders to identify the physical meaning of individual terms. In the first order, there is only a single potential, therefore its average is always the mean field,

$$G^{1<}_{mn} = \sum_{j} G^{0R}_{mj}\langle v\eta_j\rangle G^{0<}_{jn} + \sum_{j} G^{0<}_{mj}\langle v\eta_j\rangle G^{0A}_{jn} \quad \text{briefly} \quad G^{1<} = G^{0R}\phi G^{0<} + G^{0<}\phi G^{0A}. \tag{6.6}$$

This order does not contribute to the scattering because the mean-field potential cannot change the momentum of the electron. In second order we find

$$G^{2<}_{mn} = \sum_{ji} G^{0R}_{mk}\langle v\eta_k G^{0R}_{kj} v\eta_j\rangle G^{0<}_{jn} + \sum_{ji} G^{0R}_{mk}\langle v\eta_k G^{0<}_{kj} v\eta_j\rangle G^{0A}_{jn} + \sum_{ji} G^{0<}_{mk}\langle v\eta_k G^{0A}_{kj} v\eta_j\rangle G^{0A}_{jn}. \tag{6.7}$$

Each term will split into two when we use $\langle \eta_k \eta_j\rangle = c_k c_j + c_k(1-c_k)\delta_{jk}$. The terms with $c_k c_j$ contribute to the mean field, the equal-site elements result in the Born selfenergy,

$$G^{2<} = G^{0R}\phi G^{0R}\phi G^{0<} + G^{0R}\phi G^{0<}\phi G^{0A} + G^{0<}\phi G^{0A}\phi G^{0A}$$
$$+ G^{0R}\Sigma^{0R}_B G^{0<} + G^{0R}\Sigma^{0<}_B G^{0A} + G^{0<}\Sigma^{0A}_B G^{0A}. \tag{6.8}$$

where we have introduced the transport vertex in the non-selfconsistent Born approximation,

$$\Sigma^{0<}_B = \sum_{j} |j\rangle c_j(1-c_j)v^2 G^{0<}_{jj}\langle j|. \tag{6.9}$$

The function $\Sigma^<$ appears for any approximation of the scattering process. Because of its importance, it has several different names: transport vertex, correlation function of the selfenergy, Keldysh selfenergy. We will use the transport vertex which was introduced first within the Kubo formalism. One can go on and write down higher and higher orders and identify individual contributions. In fact, this is what people did in the 1960s and 1970s. There are diagrammatic representations for these kinds of expansions, and Ward identities, to test whether the optical theorem is satisfied.

Inspecting the second order (6.7) one can derive the following recipes to form retarded and advanced functions as well as the vertex from causal functions

$$(BC)^R = B^R C^R, \tag{6.10}$$

$$(BC)^A = B^A C^A, \tag{6.11}$$

$$(BC)^< = B^R C^< + B^< C^A. \tag{6.12}$$

These are the Langreth–Wilkins rules.

The reader might first practise the Langreth–Wilkins rules for this second order. The expansion for the symbolic Green's function reads

$$G = G_0 + G_0 V G_0 + G_0 V G_0 V G_0 + \dots, \tag{6.13}$$

from which the second order is $G_2 = G_0 V G_0 V G_0$. Now we denote $B = G_0 V G_0 V$ and $C = G_0$. According to rule (6.12), $G_2^< = (G_0 V G_0 V)^R G_0^< + (G_0 V G_0 V)^< G_0^A$. The retarded term is already simple because according to rule (6.10) we only have to assign the superscript R to all its functions G_0, $G_2^< = G_0^R V G_0^R V G_0^< + (G_0 V G_0 V)^< G_0^A$. For the correlation part we again use rule (6.12) and find

$$G_2^< = G_0^R V G_0^R V G_0^< + G_0^R V G_0^< V G_0^A + G_0^< V G_0^A V G_0^A. \tag{6.14}$$

which is just expansion (6.7).

Apparently, rule (6.12) works like a derivative, where the correlation part $<$ is the differentiated function. This is how a single symbolic term G_n splits into $n + 1$ terms for the correlation function $G_n^<$.

6.3   Generalised Kadanoff and Baym formalism

The right-hand side of the optical theorem (4.47) reminds one of the transport vertex. This similarity is not accidental. The transport vertex and the right-hand side of the optical theorem describe outgoing particles from collisions. As we have seen, the optical theorem is very restrictive towards approximations of the scattering process. Similar restrictions link the decay of the wave function described by the selfenergy with the transport vertex. These restrictions can be used as a tool to derive the transport vertex.

There are various techniques available as to how to fulfil this task: the cutting of diagrams, the functional derivatives, the analytical continuation and so on. In the early

1970s, Langreth and Wilkins modified the Kadanoff and Baym formalism for a nonequilibrium Green's functions so that instead of complicated analytical continuations they obtained rules which are particularly simple for scattering on impurities. Here, we deduce these rules from the perturbative expansion.

The general order of $G^<$ always has the same structure

$$G_n^< = \sum_{m=0}^{n} \langle (G_0^R V)^m G_0^< (V G_0^A)^{n-m} \rangle. \tag{6.15}$$

Unless we will need a site index, we move the index of the power in V from the superscript to the subscript, $G_n^< \equiv G^{n<}$.

If we ignore the superscripts $R, A, <$, all the $n+1$ terms in the sum (6.15) are identical. We can write them in a symbolic way as

$$G_n = (G_0 V)^n G_0. \tag{6.16}$$

The initial condition $G_0^<$ enters each position of G_0 leaving retarded Green's functions on the left-hand side and advanced Green's functions on the right-hand side.

The term $G_n^<$ can be generated from the symbolic term G_n using the three rules for a product of operators (6.10)–(6.12). We also have to use

$$V^< = 0 \quad \text{and} \quad V^{R,A} = V, \tag{6.17}$$

which reflects that the initial condition cannot enter via the potential. This formalism is a simplified version of the generalised Kadanoff and Baym (GKB) formalism.

To find the perturbative expansion for the amplitude of the wave function, one needs a perturbative expansion for the symbolic Green's functions. According to the rule (6.10), the symbolic expansion is a one-to-one correspondence with the expansion for the retarded Green's functions that we discussed in section 6.2. Therefore we can use the formulae already developed and apply rules (6.10)–(6.12) to obtain single-particle observables.

In section 6.2 we found three important forms of selfconsistency:

- The Dyson equation, $G = G_0 + G_0 \Sigma G$.
- The selfconsistent selfenergy , $\Sigma = \Sigma[G]$.
- The T-matrix, $\Sigma \sim \langle T \rangle$.

The arguments about the decay close to the band edge show that the selfconsistency on the level of the Dyson equation and on the level of the selfenergy is essential for a correct description of the multiple scattering. The selfconsistency based on the T-matrix is only necessary for strong interaction potentials.

6.3.1   GKB equation

Let us take the approximations of the selfenergy, which we derived in section 6.2, and evaluate the equation for the amplitude $G^<$. The amplitude follows from the Dyson equation. If we apply the rule (6.12), we find that

$$G^< = G_0^< + G_0^R \Sigma^R G^< + G_0^R \Sigma^< G^A + G_0^< \Sigma^A G^A. \tag{6.18}$$

Therefore,

$$G^< = \frac{1}{1 - G_0^R \Sigma^R} G_0^< (1 + \Sigma^A G^A) + \frac{1}{1 - G_0^R \Sigma^R} G_0^R \Sigma^< G^A. \tag{6.19}$$

The symbolic Dyson equation is converted into the Dyson equation for the retarded function by rule (6.10),

$$G^R = G_0^R + G_0^R \Sigma^R G^R, \tag{6.20}$$

from which we find by simple algebra that

$$\frac{1}{1 - G_0^R \Sigma^R} = (1 + G^R \Sigma^R). \tag{6.21}$$

The amplitude (6.19) can thus be reorganised as

$$G^< = G^R \Sigma^< G^A + (1 + G^R \Sigma^R) G_0^< (1 + \Sigma^A G^A). \tag{6.22}$$

This equation is called the GKB equation.

The first term in the GKB equation is the most important one. The second term represents the initial or boundary condition. For bulk systems, the second term is irrelevant and one can use a simpler form of the GKB equation

$$G^< = G^R \Sigma^< G^A. \tag{6.23}$$

This approximation simply means that we ignore electrons which did not undergo any scattering event. In bulk samples these electrons are absent.

6.3.2   Transport vertex in the Born approximation

Now we have to use some approximation of Σ. The mean field does not give any transport vertex. Indeed, from

$$\Sigma = \phi = \sum_j |j\rangle v c_j \langle j|, \tag{6.24}$$

we find only $\Sigma^R = \Sigma^A = \phi$ and $\Sigma^< = 0$. This is consistent with the fact that the mean-field approximation does not provide scattering. The Born approximation,

$$\Sigma_B = \phi + \sum_j |j\rangle c_j (1 - c_j) v^2 G_{jj} \langle j|, \tag{6.25}$$

gives us

$$\Sigma_B^R = \phi + \sum_j |j\rangle c_j (1 - c_j) v^2 G_{jj}^R \langle j|$$

$$\Sigma_B^A = \phi + \sum_j |j\rangle c_j (1 - c_j) v^2 G_{jj}^A \langle j|. \tag{6.26}$$

$$\Sigma_B^< = \qquad \sum_j |j\rangle c_j (1 - c_j) v^2 G_{jj}^< \langle j|$$

These equations together with (6.22) and the Dyson equation form a closed set from which we can evaluate $G^<$.

The transport vertex in the non-selfconsistent Born approximation (6.9) describes a single collision event. On the contrary, the selfconsistent approximation (6.26) describes an infinite sequence of collisions which electrons undergo in the crystal. Since the selfconsistent transport vertex $\Sigma^<$ depends on $G^<$ which depends on $\Sigma^<$ via the GKB equation (6.23), this set of equations describes the evolution of the electron as the coherent propagation which take turns due to collisions.

During the coherent propagation the electron stream decays due to imaginary parts of the retarded and advanced selfenergy. This decay is compensated by the electron density fed from the transport vertex. The compensation has to be exact because the number of electrons conserves in time. This is guaranteed if all functions $\Sigma^{R,A,<}$ are used in the same approximation.

6.3.3 Transport vertex in the selfconsistent averaged T-matrix approximation

Similarly one can play with the selfconsistent averaged T-matrix approximation

$$\Sigma_{SATA} = \sum_j |j\rangle c t_{jj} \langle j| \qquad \text{with} \qquad t_{jj} = \frac{v}{1 - v G_{jj}}. \tag{6.27}$$

To find the transport vertex $\Sigma_{SATA}^<$, we rewrite the T-matrix in an equivalent form

$$(1 - v G_{jj}) t_{jj} = v, \tag{6.28}$$

and apply rule (6.12),

$$(1 - v G_{jj}^R) t_{jj}^R = v \qquad \text{and} \qquad (1 - v G_{jj}^R) t_{jj}^< - v G_{jj}^< t_{jj}^A = 0. \tag{6.29}$$

From the symbolic selfenergy (6.27) we find

$$\Sigma^<_{SATA} = \sum_j |j\rangle ct^<_{jj} \langle j| \qquad \text{with} \qquad t^<_{jj} = t^R_{jj} G^<_{jj} t^A_{jj}. \tag{6.30}$$

Note that this formula includes the absolute value of the square of the T-matrix in agreement with the optical theorem.

6.3.4   Coherent potential approximation of the transport vertex

Finally, we can also evaluate the coherent potential approximation (CPA), where the selfenergy is given only by the implicit condition

$$0 = ct_V + (1-c)t_0 = c\frac{v - \Sigma_{CPA}}{1 - (v - \Sigma_{CPA})G_{jj}} - (1-c)\frac{\Sigma_{CPA}}{1 + \Sigma_{CPA}G_{jj}}. \tag{6.31}$$

Again, the retarded selfenergy results if we just put the superscript in to all functions. To evaluate the transport vertex we take the single T-matrix

$$(1 - (v - \Sigma_{CPA})G_{jj})t_V = v - \Sigma_{CPA}, \tag{6.32}$$

which eventually gives us

$$t^<_V = t^R_V G^<_{jj} t^A_V - \frac{1}{1 - (v - \Sigma^R_{CPA})G^R_{jj}}\Sigma^<_{CPA}\frac{1}{1 - G^A_{jj}(v - \Sigma^A_{CPA})}. \tag{6.33}$$

The correlation part of t_0 is similar,

$$t^<_0 = t^R_0 G^<_{jj} t^A_0 - \frac{1}{1 + \Sigma^R_{CPA}G^R_{jj}}\Sigma^<_{CPA}\frac{1}{1 + G^A_{jj}\Sigma^A_{CPA}}. \tag{6.34}$$

The transport vertex is given by the implicit condition,

$$ct^<_V + (1-c)t^<_0 = 0. \tag{6.35}$$

This condition can be rearranged into an explicit formula for a single band. Formula (6.35) holds for a multiband system too, the potential v is then a matrix.

6.4   Optical theorem

As we have seen, the transport vertex is always a linear functional of the local averaged amplitude $\Sigma^< = LG^<_{jj}$. The optical theorem requires that the transport vertex $\Sigma^<$ has the same functional dependence on the averaged amplitude $G^<$ as the scattering cross section $\Gamma = -2\mathrm{Im}\Sigma^R$ has on the local density of states $h = -2\mathrm{Im}G_{jj}^R$, $\Gamma = Lh$. This identity is often used to simplify numerical treatment, because $\Sigma^< = (\Gamma/h)G_{jj}^<$. Within Green's functions, the optical theorem is usually called the Ward identity, see (11.37). Since there are many Ward's identities, we prefer the specific name optical theorem.

We can see that the optical theorem is satisfied for the Born approximation,

$$\Sigma^< = c(1-c)v^2 G_{jj}^<, \qquad -2\mathrm{Im}\Sigma^R = c(1-c)v^2(-2)\mathrm{Im}G_{jj}^R. \tag{6.36}$$

It would be easy to check this property for all the selfenergies above, however, it is not necessary, because this property follows directly from the Langreth–Wilkins rules.

Let us evaluate the imaginary part of a product of two complex numbers,

$$\mathrm{Im}(BC) = \mathrm{Im}[(b+ib')(c+ic')] = b'c+bc' = b'(c-ic')+(b+ib')c' = (\mathrm{Im}B)C^* + B\mathrm{Im}C. \tag{6.37}$$

If we start from the retarded functions and take into account that the advanced functions are their complex conjugate, we find that

$$-2\mathrm{Im}(B^R C^R) = (-2\mathrm{Im}B^R)C^A + B^R(-2\mathrm{Im}C^R). \tag{6.38}$$

Apparently, the imaginary part $\mathrm{Im}(BC)$ obeys the same rules as the correlation functions $(BC)^<$. Therefore, if the retarded selfenergy and the transport vertex are derived from the same symbolic selfenergy, they always satisfy the optical theorem.[1]

6.5 Application scheme

Chapters 6.1–6.3 show the basic structure of the nonequilibrium Green's function approach. To obtain a transport equation all one has to do is to:

1. Use the idea of the effective medium within symbolic Green's functions to find the condition for the selfenergy.
2. Use the LW rules to turn this symbolic condition into a condition for the transport vertex.

This provides a set of equations for non-equilibrium Green's functions with quite sophisticated drift and dissipation. As seen from the density of state at the band edge, the drift includes the mean-field and energy-dependent effective potential responsible for the renormalisation of the electron mass. The scattering includes the finite collision duration. This procedure is quite general and the eventual physical interpretation of individual

---

[1] For readers having some experience with the Keldysh formalism we note that the Keldysh Green's function G^K is not directly related to observables, but it reads $G^K = -2\mathrm{Im}G^R - 2G^<$. The optical theorem has to be used to obtain observables linked to the amplitude of the wave function, $G^<$, at the end of calculations.

elements of the Green's function and the selfenergy depends on the system in question and approximations employed. Nevertheless, under the condition that the system can be described by quasiparticles, there are general relations between the selfenergy and quasiparticle features. To see them we first express the selfenergy in terms of a secular equation and then discuss its interpretation.

6.6   Elimination of surrounding interaction channels-secular equation

Any of the approximations in the previous chapters describe the influence of other particles on the property of a single selected particle. This might be formally considered as separating the world into two channels, the state $\langle 1|$ with the Hamiltonian $\hat{H}_1$ and the rest as remaining channels. Each channel has its Schrödinger equation $\hat{H}_k|\psi_k\rangle = E_k|\psi_k\rangle$. Between the sub-state $\langle 1|$ and the rest we assume an interaction $\hat{V}^{r1}$. Let us consider the many-body Schrödinger equation

$$\left(\omega - \hat{H}_1 - \sum_{k=2} \hat{H}_k - \hat{V}^{r1} \sum_{k=1} |k\rangle\langle k|\right)\left(|\psi_1\rangle + \sum_{k=2}|\psi_k\rangle\right) = 0 \qquad (6.39)$$

where we added a complete set of states $1 = \sum_{k=1}|k\rangle\langle k|$ in the first bracket. We can write it more explicitly

$$(\omega - E_1)|\psi_1\rangle - \hat{V}^{r1}|1\rangle\langle 1|\psi_1\rangle + \sum_{k=2}(\omega - E_k)|\psi_k\rangle - \hat{V}^{r1}\sum_{k=2}|k\rangle\langle k|\psi_k\rangle = 0. \quad (6.40)$$

Multiplying with $\langle 1|$ and $\langle q \neq 1|$ from the left leads to the two coupled equations

$$(\omega - E_1)\Psi_1 - \sum_{k=2} V_{1k}\Psi_k = 0, \qquad \text{and } \Psi_q = \frac{V_{q1}}{(\omega - E_q)}\Psi_1 \qquad (6.41)$$

with the abbreviation $V_{ij} = \langle i|\hat{V}^{r1}|j\rangle$, $\Psi_1 = \langle 1|\psi_1\rangle$ and $\Psi_q = \langle q|\psi_q\rangle$. Introducing the second equation into the first equation, one obtains the secular equation

$$(\omega - E_1 - \Sigma_1)\Psi_1 = 0 \qquad (6.42)$$

where the coupling to the other channels has changed the Schrödinger equation for the problem in channel 1 by the appearance of the selfenergy

$$\Sigma_1 = \sum_{k=2} \frac{V_{1k}V_{k1}}{\omega - E_k}. \qquad (6.43)$$

It is important to note that the selfenergy contains exclusively the interaction of channel 1 with the surroundings and not with itself. Of course, the true selfenergy as defined above results after averaging over the impurity configuration if the impurity potential enters all parts of the Hamiltonian. Here we keep the non-averaged 'selfenergy' for the moment to obtain convenient and easy-to-interpret relations.

Due to the imaginary part of the selfenergy, the secular equation (6.42) does not result in eigen-states in the exact sense. We can introduce such eigen-states approximately neglecting the imaginary part,

$$[\varepsilon - E_1 - \mathrm{Re}\Sigma_1(\varepsilon)]\,\Psi_1 = 0. \tag{6.44}$$

The eigen-energy obtained in this way is the quasiparticle energy ε.

6.6.1 Wave function renormalisation

Considering the norm

$$|\Psi_1|^2 + \sum_k |\Psi_k|^2 = 1 \tag{6.45}$$

we can introduce (6.41) which reads

$$|\Psi_k|^2 = |\Psi_1|^2 V_{1k}\left(\frac{1}{\omega - E_k}\right)^2 V_{k1} = -|\Psi_1|^2 \frac{\partial}{\partial\omega}\left(\frac{V_{1k}V_{k1}}{\omega - E_k}\right) = -|\Psi_1|^2\frac{\partial}{\partial\omega}\Sigma_1 \tag{6.46}$$

such that the norm (6.45) of the wave function becomes normalised to

$$|\Psi_1|^2 = \frac{1}{1 - \frac{\partial}{\partial\omega}\Sigma} = z. \tag{6.47}$$

This function is the weight of the pole of the Green's function. Accordingly, the weight z is called wave-function renormalisation. With the probabilistic interpretation of the wave function, one can say that a particle spends the fraction of time z in its main state $|1\rangle$ and the fraction $1 - z$ in the environment.

6.6.2 Effective mass

The free energy of the Schrödinger equation $\omega = \epsilon_0 = k^2/2m$ becomes shifted by the real part of the selfenergy $\omega = \epsilon = \epsilon_0 + \mathrm{Re}\Sigma(\epsilon)$. Therefore an effective mass appears as one can find from the definition of the velocity

$$v_{\mathrm{quasi}} = \frac{k}{m*} = \partial_k\epsilon = \partial_k\epsilon_0 + \left.\frac{\partial\Sigma}{\partial\omega}\right|_{\omega=\epsilon}\partial_k\epsilon = \frac{\partial_k\epsilon_0}{1 - \left.\frac{\partial\Sigma}{\partial\omega}\right|_{\omega=\epsilon}} = z\frac{k}{m}. \tag{6.48}$$

The same wave function renormalisation leads to the effective mass of quasiparticles, see (5.81). Relation (6.48) has a simple interpretation: a particle moves with the velocity k/m during the time fraction z. In the quasiparticle dialect this reduced motion is expressed effectively in a way as if the quasiparticle moves continuously with a reduced velocity k/m^* given by its effective mass m^*.

6.6.3   Two actions of selfenergy

Now we consider the single-channel Schrödinger equation (6.42) as an impurity problem for the single-particle wave function $\Psi_1 = \langle \psi_1 | 1 \rangle$. The single-channel Hamiltonian consists of the (analogously host crystal) free part and the interaction $\hat{H}_1 = \hat{H}_0 + \hat{V}$. Separating the incoming wave $(E - \hat{H}_0)\Psi_0 = 0$ from the reconstructed one as $\Psi_1 = \Psi_0 + \Psi$ we obtain

$$(E - \hat{H}_0 - \Sigma(E))\Psi = V(\Psi_0 + \Psi) + \Sigma\Psi_0. \tag{6.49}$$

This shows the two-fold meaning of the surrounding medium condensed in Σ. Firstly it normalises the single-particle free propagator in the energy spectrum

$$G_0(z) = [z - H_0 - \Sigma(z)]^{-1} \tag{6.50}$$

and secondly, it provides an additional source for the recursive equation which one gets from (6.49) with (6.50) as

$$\Psi = G_0(z)[V(\Psi_0 + \Psi) + \Sigma\Psi_0]. \tag{6.51}$$

In other words, it is not sufficient to calculate the selfenergy microscopically since it has a twofold effect on the propagator. Firstly, the renormalisation of the spectra, but additionally it changes the recursive property which is an expression of the modification of coherence.

Iterating now (6.51) one obtains

$$\Psi = (1 + G_0 V + G_0 V G_0 V + \ldots)G_0(\Sigma + V)\Psi_0 \equiv G(V + \Sigma)\Psi_0 \tag{6.52}$$

where we introduced the usual Dyson equation for the full propagator $G = G_0 + G_0 V G = [1 - G_0 V]^{-1}G_0$. Alternatively we might summarise the interaction in the irreducible $\mathscr{T}$-matrix as

$$\Psi = G_0\mathscr{T}\Psi_0 \tag{6.53}$$

which provides, by comparison with (6.52),

$$\mathscr{T} = (1 + VG)(V + \Sigma) = V + \Sigma + VG_0\mathscr{T} = [1 - VG_0]^{-1}(V + \Sigma). \tag{6.54}$$

6.7 Summary

The Green's function is a secular equation in which the complicated motion of a particle during interaction is represented by the selfenergy. For long-living excitations, i.e. the small imaginary part of the selfenergy, it is possible to focus on the effective eigen-energy (called the quasiparticle energy) and study the system dynamics in terms of quasiparticles. When doing so one should keep in mind that the quasiparticle spends its full time in the state of momentum k while the real particle stays there only for the fraction of its time z. Also the velocity of the quasiparticle is uniformly reduced while the real particle commutes between the full velocity and the zero velocity during interaction. The quasiparticle picture is thus applicable to long trajectories, where the averaged velocity is of interest. In processes like emission, the Doppler shift depends on the local velocity and one has to return to the microscopic picture.

Part III

Deductive Way to Quantum Transport

The nonequilibrium quantum statistical approaches to dense Fermi systems include a number of alternative techniques. From this large family of techniques we focus on nonequilibrium Green's functions pioneered by Schwinger (1961), Kadanoff and Baym (1962). The convention follows here Langreth and Wilkins (1972) who introduced the so-called generalised Kadanoff and Baym (GKB) formalism. The GKB formalism combines advantages of the original Kadanoff and Baym and Schwinger approach with the matrix formulation due to Keldysh (1964) giving particularly convenient rules for implementation.

7

Nonequilibrium Green's Functions

7.1 Method of equation of motion

The system is described by a Hamiltonian

$$\hat{H}(t) = \sum_a \int dx \hat{\psi}_a^\dagger(x,t) \left(\frac{\nabla_x^2}{2m_a} + U_a(x,t) \right) \hat{\psi}_a(x,t)$$

$$+ \frac{1}{2} \sum_{ab} \int dx_1 dx_2 dx_3 dx_4 \hat{\psi}_a^\dagger(x_3,t) \hat{\psi}_b^\dagger(x_4,t) \mathcal{V}_{ab}(x_1,x_2,x_3,x_4) \hat{\psi}_b(x_2,t) \hat{\psi}_a(x_1,t), \quad (7.1)$$

where $\hat{\psi}^\dagger$ and $\hat{\psi}$ are the creation and annihilation operators, x is the coordinate, t is the time, and a indexes spin and isospin. As a rule, we use hats to denote operators in second quantisation and calligraphic letters for two-particle functions. Mostly we will deal with non relativistic two-particle interactions $\mathcal{V}_{ab}(x_1,x_2,x_3,x_4) = \mathcal{V}_{ab}(x_1,x_2)\delta(x_3-x_1)\delta(x_4-x_2)$, e.g. including the Coulomb interaction. The external potential is U.

We will consider the two distinct double-time correlation functions

$$G^<(1,2) = \text{Tr}\left[\hat{\rho} \hat{\psi}_2^\dagger \hat{\psi}_1 \right] = \langle \hat{\psi}_2^\dagger \hat{\psi}_1 \rangle \quad \text{and} \quad G^>(1,2) = \langle \hat{\psi}_1 \hat{\psi}_2^\dagger \rangle, \quad (7.2)$$

where numbers are cumulative variables, $1 \equiv x_1, t_1, a_1$. The meaning of $G^<(1,2)$ is to re-move first a particle at the space-time point 1 creating a hole and to add then the particle again at the space-time point 2. For $G^>$ we consider the reverse process. By $G^<(1,2)$ we test therefore occupied states by $G^>(1,2)$ the unoccupied states. The disturbance of the system due to such adding/removal of particles is calculated by averaging over the nonequilibrium many-body density operator $\hat{\rho}$. The latter is essentially unknown. Density-operator techniques try to calculate it directly, see Appendix A with the argu-ment in favour of having a single-time function to be treated. However the equations of time-evolution of the density operator also involve propagators of two-time characters such that this is of no real advantage.

Instead, we will follow the equation of motion which has the virtue of avoiding the determination of the nonequilibrium density operator but allows us to determine the

Interacting Systems far from Equilibrium. Klaus Morawetz, Oxford University Press (2018).
© Klaus Morawetz. DOI: 10.1093/oso/9780198797241.001.0001

correlation functions (7.2). In order to specify the propagation of the disturbance we will associate $G^<(1,2)$ with times $t_1 < t_2$ and $G^>(1,2)$ with $t_2 < t_1$ according to the time-points of the removal and the addition of a particle. The causal propagation of the disturbance can then be written in terms of the causal function

$$G^c(1,2) = \frac{1}{i}\Theta(t_1 - t_2)G^>(1,2) \mp \frac{1}{i}\Theta(t_2 - t_1)G^<(1,2) = \frac{1}{i}\langle\hat{T}^c\hat{\psi}_1\hat{\psi}_2^\dagger\rangle, \qquad (7.3)$$

where we account for Fermi/Bose systems according to the sign change of fermions if two operators are interchanged and introduced the time-ordering operator

$$\hat{T}^c\hat{A}\hat{B} = \begin{cases} \hat{A}\hat{B} , & t_A > t_B \\ \mp\hat{B}\hat{A} , & t_B > t_A \end{cases}. \qquad (7.4)$$

Applying the equation of motion for the field operators in the Heisenberg picture one finds

$$i\frac{\partial}{\partial t}\hat{\psi}_1 = -[\hat{H}, \hat{\psi}_1] = \left(-\frac{\nabla_1^2}{2m} + U(1)\right)\hat{\psi}_1 + \int d3\,V(3,1)\hat{\psi}_3^\dagger\hat{\psi}_3\psi_1, \qquad (7.5)$$

where $V(1,3) = \mathscr{V}_{a_1 a_3}(x_1, x_3)\delta(t_1 - t_3)$ and integration over the cumulative variable includes time and the sum over the discrete index, $\int d3 \equiv \sum_{a_3} \int dt_3 dx_3$. We multiply the Heisenberg equation (7.5) with the creation operator $\hat{\psi}_2^\dagger$ from the right and with the time-ordering operator $\hat{T}^c$ from the left and average them with the statistical operator to get

$$\left\langle\hat{T}^c\left(i\frac{\partial}{\partial t_1} + \frac{\nabla_1^2}{2m} - U(1)\right)\psi_1\psi_2^\dagger\right\rangle = \mp\int d3\,V(1,3)\langle\hat{T}^c\hat{\psi}_1\hat{\psi}_3\hat{\psi}_{3+}^\dagger\hat{\psi}_2^\dagger\rangle_{t_3=t_1}$$

$$= \mp i^2\int d3\,V(1,3)G_2(1,3;2,3^+)_{t_3=t_1}. \qquad (7.6)$$

The right-hand side defines the two-particle causal Green's function. We used t_3^+ to be infinitesimally larger than t_3 in order to ensure the correct ordering of operators by the time-ordering operator $\hat{T}^c$.

The spatial derivatives on the left side can be interchanged with the time-ordering operator but the time derivative gives

$$\frac{\partial}{\partial t_1}\langle\hat{T}^c\hat{\psi}_1\hat{\psi}_2^\dagger\rangle = \delta(t_1 - t_2)\langle\hat{\psi}_1\hat{\psi}_2^\dagger\rangle \pm \delta(t_2 - t_1)\langle\hat{\psi}_2^\dagger\hat{\psi}_1\rangle + \left\langle\hat{T}^c\frac{\partial}{\partial t_1}\hat{\psi}_1\hat{\psi}_2^\dagger\right\rangle$$

$$= \delta(t_1 - t_2)\langle[\hat{\psi}_1, \hat{\psi}_2^\dagger]_\pm\rangle + \left\langle\hat{T}^c\frac{\partial}{\partial t_1}\hat{\psi}_1\hat{\psi}_2^\dagger\right\rangle = \delta(1-2) + \left\langle\hat{T}^c\frac{\partial}{\partial t_1}\hat{\psi}_1\hat{\psi}_2^\dagger\right\rangle, \qquad (7.7)$$

where we used the equal-time (anti)commutator relations leading to δ-functions in all indices, e.g. space, spin and sort indices. We obtain from (7.6) the Martin–Schwinger hierarchy (Martin and Schwinger, 1959; Kadanoff and Baym, 1962), where the single-particle Green's function couples to the two-particle one, the two-particle to the three-particle, etc.[1]

$$\left(i\frac{\partial}{\partial t_1} + \frac{\nabla_1^2}{2m} - U(1)\right) G^c(1,2) = \delta(1-2) \mp i \int d3\, V(1,3)\, G_2^c(1,3;2,3^+)$$

$$\left(i\frac{\partial}{\partial t_1} + \frac{\nabla_1^2}{2m} - U(1)\right) G_2^c(1,4;2,5) = \delta(1-2)\,G^c(4,5) - \delta(1-5)\,G^c(4,2)$$

$$\mp i \int d3\, V(1,3)\, G_3^c(1,4,3;2,5,3^+). \qquad (7.8)$$

7.1.1 Enclosure of hierarchy

Our intention is to express the two-particle Green's function in terms of the single-particle Green's function so that the first equation of the hierarchy (7.8) becomes a closed equation.

A formal closure for the single-particle Green's function can be reached with the introduction of the selfenergy

$$\mp i \int d3\, V(1,3)\, G_2^c(13,23^+) = \int_C d3\, \Sigma^c(1,3)\, G^c(3,2), \qquad (7.9)$$

as illustrated in Figure 7.1. The integration $\int_C$ has to be determined in such a way that certain boundary conditions are fulfilled. Let us therefore investigate a first guess by

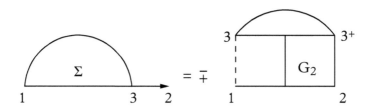

Figure 7.1: *Formal closure of the Martin–Schwinger hierarchy introducing the selfenergy (7.9). At about variable 3 one integrates and 3^+ means an infinitesimal time later that at 3. We introduce diagrammatic rules that a broken line means iV and an arrow line means G^c.*

---

[1] The second equation is obtained by multiplying (7.5) with $\hat{\psi}_4\hat{\psi}_5^\dagger\hat{\psi}_2^\dagger$ from the right. For the interchange of the time-ordering operator and the time derivative one observes

$$\left(\frac{\partial}{\partial t_1}\hat{T}^c - \hat{T}^c\frac{\partial}{\partial t_1}\right)\hat{\psi}_1\hat{\psi}_4\hat{\psi}_5^\dagger\hat{\psi}_2^\dagger$$

$$= \delta(t_1-t_4)[\hat{\psi}_1,\hat{\psi}_4]_\pm\hat{\psi}_5^\dagger\hat{\psi}_2^\dagger \mp \delta(t_1-t_5)\hat{\psi}_4[\hat{\psi}_1,\hat{\psi}_5^\dagger]_\pm\hat{\psi}_2^\dagger + \delta(t_1-t_2)\hat{\psi}_4\hat{\psi}_5^\dagger[\hat{\psi}_1,\hat{\psi}_2^\dagger]_\pm.$$

simply understanding $\int_C$ as a standard integration over time. Breaking the time integrations into pieces and identifying the parts of the causal functions (7.3) would lead to

$$\int_{t_0}^{+\infty} dt_3 \Sigma^c(1,3) G^c(3,2) = \pm \int_{t_0}^{t_1} dt_3 \Sigma^>(1,3) G^<(3,2) - \int_{t_1}^{t_2} dt_3 \Sigma^<(1,3) G^<(3,2)$$

$$\pm \int_{t_2}^{\infty} dt_3 \Sigma^<(1,3) G^>(3,2)$$

$$\rightarrow \pm \int_{t_0}^{\infty} dt_3 \Sigma^<(1,3) G^>(3,2) \quad \text{for } t_1 = t_2 = t_0 \qquad (7.10)$$

for the case $t_1 < t_2$ and the same result one obtains for $t_2 < t_1$. This means a trivial time integration would lead to the unphysical result that in the infinite past there are correlations. It is more physical to assume that at a certain initial time t_0 all higher-order correlations other than meanfield correlations had been absent

$$\lim_{t_1=t_2=t_0} \int_C d3 \Sigma^c(1,3) G^c(3,2) = 0 \qquad (7.11)$$

which is called the weakening of initial correlations. Therefore the simplest way to complete (7.11) is to subtract the necessary part (7.10) from the time integration

$$\int_C d3 \Sigma^c(1,3) G^c(3,2) = \int_{t_0}^{+\infty} d3 \left\{ \Sigma^c(1,3) G^c(3,2) \mp \Sigma^<(1,3) G^>(3,2) \right\}. \qquad (7.12)$$

In such a way we ensure that the asymptotic limit of initial times vanishes.

Please note that we have used an additional asymptotic condition in this way on the time evolution. This breaks the time-symmetry and our equations will become irreversible from now on. This is the analogous condition to Boltzmann's chaoticity assumption. The virtue of the choice (7.12) is that the hierarchy is formally enclosed and the appearing selfenergy can be determined from Figure 7.1 if we specify the two-particle Green's function by an appropriate approximation.

The way of summation (7.10) is equivalent to the Keldysh contour of time integration, see Appendix B. Thus we see that such a Keldysh contour invokes the weakening of initial correlations and therefore introduces the principle of irreversibility. Though most of the readers will be familiar with the Keldysh contour of Appendix B, we prefer this way of presentation making the inherent assumptions transparent.

With the expressions (7.9) and (7.12) we can finally write the first equation of the Martin–Schwinger hierarchy in the following explicit form,

$$\left(i\frac{\partial}{\partial t_1} - \frac{\nabla_1^2}{2m} \right) G^c(1,2) - \int d3 \Sigma^{HF}(1,3) G^c(3,2) \delta(t_1 - t_3)$$

$$= \delta(1-2) + \int_{-\infty}^{+\infty} d3 \left\{ \Sigma^c(1,\bar{1}) G^c(3,2) - \Sigma^<(1,3) G^>(3,2) \right\}. \qquad (7.13)$$

It was genius of (Langreth and Wilkins, 1972) to formulate simple rules for how to handle this seemingly cumbersome time integration. For this purpose we consider the $t_1 < t_2$ part of the causal product (7.12) according to (7.3)

$$C^c(1,2)_{t_1 < t_2} = \pm i C^<(1,2)$$

$$= \pm \int_{t_0}^{t_1} dt_3 \, \Sigma^>(1,3) G^<(3,2) - \int_{t_1}^{t_2} dt_3 \, \Sigma^<(1,3) G^<(3,2) \mp \int_{t_0}^{t_2} d\bar{t}_3 \, \Sigma^<(1,3) G^>(3,2)$$

$$= \pm \int_{t_0}^{t_1} dt_3 \, \Sigma^> G^< + \left(\int_{t_0}^{t_1} dt_3 - \int_{t_0}^{t_2} dt_3 \right) \Sigma^< G^< \mp \int_{t_0}^{t_2} d\bar{t}_3 \, \Sigma^< G^>$$

$$= \pm \int_{t_0}^{t_1} dt_3 \, [\Sigma^> - \Sigma^<] G^< \mp \int_{t_0}^{t_2} d\bar{t}_3 \, \Sigma^< [G^> - G^<]$$

$$= \pm i \int_{t_0}^{\infty} dt_3 \, \Sigma^R(1,3) G^<(3,2) \pm i \int_{t_0}^{\infty} d\bar{t}_3 \, \Sigma^<(1,3) G^A(3,2) \tag{7.14}$$

where we introduced conveniently retarded and advanced functions to be able to extend the time integration again to the complete interval,

$$C^R(t,t') = -i\theta(t-t') \left[C^>(t,t') \pm C^<(t,t') \right], \tag{7.15}$$

$$C^A(t,t') = i\theta(t'-t) \left[C^>(t,t') \pm C^<(t,t') \right]. \tag{7.16}$$

We arrive at the Langreth–Wilkins rules 1972, seen earlier in (6.10) and (6.12), to show how one obtains from a causal product, $C^c = A^c B^c$ the correlation functions

$$C^{\gtrless} = A^{\gtrless} B^A + A^R B^{\gtrless} \tag{7.17}$$

where we summarised also the part $t_1 > t_2$ leading to $> \leftrightarrow <$.

Analogously one obtains the retarded and advanced functions

$$C^{R/A} = A^{R/A} B^{R/A}, \tag{7.18}$$

where again the product of functions are understood as integrations running along the real-time axis and about space. Both correlation functions $C^{\gtrless}$ are now valid on the whole-time axes without restrictions $t_1 \gtrless t_2$. One can convince oneself repeating all steps with the anti-causal function (B.8). Summarising the Langreth–Wilkins rules (7.18) and (7.17) can be considered just as a consequence of the weakening of the initial correlations. Since the LW rules are equivalent to Keldysh's matrix formulation or to the complex path, see Appendix B, all three approaches are equivalent.

7.2   Quantum transport equation

Now we search the equation for the correlation functions itself. The formally closed Martin–Schwinger hierarchy (7.13) reads

$$G_0^{-1}(1,3)G^c(3,2) = \delta(1-2) + \Sigma^c(1,3)G^c(3,2) \tag{7.19}$$

with the Hartree–Fock drift term

$$G_0^{-1}(1,3) = \left(i\frac{\partial}{\partial t_1} - \frac{\nabla_{x_1}^2}{2m} \right)\delta(1-3) - \Sigma^{HF}(1,3)\delta(t_1 - t_3) \tag{7.20}$$

where we subsumed the time-diagonal part of the selfenergy. Double occurring internal variables are integrated over. The Dyson equation (7.19) can be written in integral form

$$G^c(1,3) = G_0^c(1,3) + G_0^c(1,\bar{5})\Sigma^c(\bar{5},\bar{7})G^c(\bar{7},3), \tag{7.21}$$

where G_0^c is the causal function of non-interacting (only meanfield) particles and the selfenergy Σ^c describes all further correlations.

The Dyson equation itself does not specify any approximation. From its differential form (7.19),

$$G_c^{-1}(1,3) = G_0^{-1}(1,3) - \Sigma^c(1,3), \tag{7.22}$$

one can see that it only expresses our determination to develop the perturbation expansion for the inverse Green's function instead of the straightforward expansion for G^c. Accordingly, all physical approximations are given by the approximation of the selfenergy.

The implementation of rule (7.18) is straightforward. The Dyson equation (7.21) for causal functions is converted into the Dyson equation for retarded functions

$$G^R = G_0^R + G_0^R \Sigma^R G^R. \tag{7.23}$$

Similarly one obtains advanced functions which we do not write explicitly as they are obtained by a substitution of the superscript $A \leftrightarrow R$.

A transport equation for the correlation function $G^<$ is obtained from Dyson equation (7.21) by the LW rules (7.17) and (7.18) as

$$\begin{aligned} G^< &= G_0^< + (G_0\Sigma)^< G^A + G_0^R \Sigma^R G^< \\ &= G_0^< + G_0^< \Sigma^A G^A + G_0^R \Sigma^< G^A + G_0^R \Sigma^R G^< \\ &= G^R \Sigma^< G^A + \left(1 + G^R \Sigma^R\right) G_0^< \left(1 + \Sigma^A G^A\right) \end{aligned} \tag{7.24}$$

or analogously $>\longleftrightarrow<$. In the last line we have used the Dyson equation (7.23) to obtain the symmetrical form. The last term represents the asymptotic initial condition and how the uncorrelated single-particle $G_0^<$ evolves from the infinite past. It can be replaced by the more physical boundary condition that $G^<$ has a required value shortly before the collision of the two particles. We adopt this approach and deal with the simpler transport equation

$$G^{\lessgtr} = G^R \Sigma^{\lessgtr} G^A, \tag{7.25}$$

called the GKB equation. The set of equations for nonequilibrium Green's functions on the real-time axis is complete provided we specify the selfenergy.

The question of asymptotic single-particle initial correlations has been transferred in this way to the boundary conditions. They reappear as initial conditions in the final kinetic equation. The inclusion of explicit initial correlations will be postponed until Chapter 19 where higher than one-particle initial correlations will be treated in Chapter 19.4.

With the equation (7.25) we have already given a general transport or kinetic equation. To convert the GKB equation (7.25) into its differential form is a very simple task. One multiplies (7.25) with G_R^{-1} from the left-hand side, which gives $G_R^{-1} G^< = \Sigma^< G^A$, and similarly obtains the adjoined equation $G^< G_A^{-1} = G^R \Sigma^<$. Subtracting both equations, the differential Kadanoff and Baym equation results

$$-i\left(G_R^{-1} G^< - G^< G_A^{-1}\right) = i\left(G^R \Sigma^< - \Sigma^< G^A\right) \tag{7.26}$$

where the retarded and advanced functions are introduced as (7.15) and (7.16). One may also write the selfenergy explicitly,

$$-i\left(G_0^{-1} G^< - G^< G_0^{-1}\right) = i\left(G^R \Sigma^< - \Sigma^< G^A\right) - i\left(\Sigma^R G^< - G^< \Sigma^A\right). \tag{7.27}$$

Here operator notation is used where products are understood as integrations over intermediate variables (time and space). Eq. (7.27) represents the equation derived by Kadanoff and Baym (Kadanoff and Baym, 1962), which reads explicitly

$$\left(i\frac{\partial}{\partial t_1} - \frac{\nabla_1^2}{2m}\right) G^{\gtrless}(1,2) - \int d3\, \Sigma^{HF}(1,3) G^{\gtrless}(3,2)\delta(t_1 - t_3)$$

$$= \int_{-\infty}^{t_1} d3\left[\Sigma^>(1,3) - \Sigma^<(1,3)\right] G^{\gtrless}(3,2) - \int_{-\infty}^{t_2} d3\, \Sigma^{\gtrless}(1,3)\left[G^>(3,2) - G^<(3,2)\right]. $$

$$\tag{7.28}$$

Of course, using the Langreth–Wilkins rules (7.18) and (7.17), from the causal equation (7.13) one obtains also (7.28).

The Kadanoff and Baym equations are treated further in two different ways within the literature. One possibility is to rearrange the terms such that they yield the Landau–Silin kinetic equation for the quasi-particle distribution function when gradient expansion is applied. This will be presented in Chapter 9. The other possibility is to recollect the terms on the right-hand side to a non-Markovian collision integral leading to the Levinson equation for the reduced density matrix (9.76). The connection between both has remained dubious because in the Levinson equation the drift term is only controlled by the Hartree–Fock selfenergy while in the Landau–Silin equation the drift term is represented by the full quasi-particle energy. In Chapter 9 we like to close this gap and demonstrate how the equations can be converted into each other employing a functional between the reduced density matrix and the quasiparticle distribution function.

7.3 Information contained in Green's functions

The single-particle observables are all given by the reduced density matrix,

$$\rho_{ab}(x, y, t) = \mathrm{Tr}\left(\hat{\rho}\hat{\psi}_b^\dagger(y, t)\hat{\psi}_a(x, t)\right),\tag{7.29}$$

where $\hat{\rho}$ is the density operator. Since the Hamiltonian (7.1) does not include any spin flip, we can limit our attention to spin-diagonal reduced density matrices.

The reduced density matrix is simply obtained from the double-time correlation functions $G^<$, see (7.2), on the time diagonal

$$\rho_{ab}(x, y, t) = G^<(x, t, a, y, t, b).\tag{7.30}$$

To point out the analogy of the reduced-density matrix and Boltzmann's distribution, we have to express the reduced-density matrix as a density in phase space. This is achieved by the Wigner mixed representation

$$g_{aa}^<(\omega, k, r, t) = \int dx d\tau e^{i\omega\tau - ikx} G^<\left(r + \frac{x}{2}, t + \frac{\tau}{2}, a, r - \frac{x}{2}, t - \frac{\tau}{2}, a\right),\tag{7.31}$$

where ω is the independent energy. In the mixed representation one can see that the Wigner distribution includes contributions from all independent energies,

$$\rho_{aa}(k, r, t) = \int \frac{d\omega}{2\pi} g_{aa}^<(\omega, k, r, t).\tag{7.32}$$

The energy as an independent variable allows us to distinguish the on-shell contributions, for which the energy is determined by the position of the particle in the phase space, from the off-shell contributions for which no such relation holds. This distinction improves the convergence of perturbation expansions. Briefly, according to the philosophy of Green's functions, the perturbation expansion has to be closed for $G^<$. The desired Wigner distribution is obtained from $G^<$ via (7.32).

The kinetic equation, be it either the Boltzmann equation or its generalisation, is derived as an asymptotic limit of the equation for $G^<$. It turns out that this asymptotic equation is not closed for the Wigner distribution but for the on-shell part of $G^<$ which can be interpreted as the distribution of elementary single-particle excitations, the quasiparticles.

7.3.1 Density and current

Two observables of particular interest are the local density,

$$n_a(x, t) = \rho_{aa}(x, x, t) = \int \frac{dk}{(2\pi)^3} \rho_{aa}(k, x, t) = \int \frac{dk}{(2\pi)^3} \int \frac{d\omega}{2\pi} g_{aa}^<(\omega, k, r, t), \qquad (7.33)$$

and the local current density of particles,

$$j_a(x, t) = \frac{i}{2m_a} (\nabla_x - \nabla_y)\rho_{aa}(x, y, t)\big|_{y=x}$$

$$= \int \frac{dk}{(2\pi)^3} \frac{k}{m_a} \rho_{aa}(k, x, t) = \int \frac{dk}{(2\pi)^3} \frac{k}{m_a} \int \frac{d\omega}{2\pi} g_{aa}^<(\omega, k, r, t). \qquad (7.34)$$

Until now there seems to be no need to use the double-time Green's functions since all variables are expressed by the time-diagonal one which is the reduced density matrix or Wigner function. The reduced density does not provide all desirable observables, however. For instance, the local density of energy and the pressure are not single-particle observables so that they cannot be evaluated from the reduced-density matrix.

The Wigner distribution can be interpreted as a distribution of particles in the phase space. In this sense it is a natural generalisation of the Boltzmann distribution. The similarity of the two distributions makes the impression that a quantum generalisation of the Boltzmann equation should be the kinetic equation for the Wigner distribution. This is unfortunately not true. In the kinetic equation, the energy of a particle is determined by its position (k, r) in the phase space. The Wigner distribution, however, has the high-momenta tail which cannot be interpreted as the presence of particles of correspondingly high-kinetic energies. Until we can understand how to determine the energy of a particle from its position in phase space, we have to keep the energy as an independent variable. This is the very basic idea of Green's functions.

7.3.2 Total energy content

By definition, the density of energy is the mean value of the Hamiltonian, for constant volume

$$\mathscr{E} = \text{Tr}[\hat{\rho}\hat{H}]. \qquad (7.35)$$

Here $\hat{H} = \hat{H}_0 + \hat{V}$ is composed of the kinetic energy $\hat{H}_0$ and the interaction potential $\hat{V}$.

The density of energy is a two-particle observable since the interaction potential $\hat{V}$ is a two-particle operator. Approximations used to derive the kinetic equation are, however, dedicated to the proper description of single-particle properties. To link the single- and two-particle levels, one can use the equations of motion in the backward direction: to express the Hamiltonian in terms of the time derivative. Using the Heisenberg equation (7.5) to evaluate the time derivatives of $G^<$ one finds

$$\frac{i}{4}\left(\frac{\partial}{\partial t_1} - \frac{\partial}{\partial t_3}\right) G^<(1,3)\Big|_{1=3} = \frac{1}{2}\mathrm{Tr}[\hat{\rho}\hat{H}_0] + \mathrm{Tr}[\hat{\rho}\hat{V}]. \tag{7.36}$$

To obtain the mean value of the Hamiltonian (7.35), the time derivative has to be complemented with half of the mean kinetic energy,

$$\mathcal{E} = \frac{1}{2}\mathrm{Tr}[\hat{\rho}\hat{H}_0] + \frac{i}{4}\left(\frac{\partial}{\partial t_1} - \frac{\partial}{\partial t_3}\right) G^<(1,3)\Big|_{1=3}. \tag{7.37}$$

In the Wigner representation (7.31) this relation has the familiar form (Kadanoff and Baym, 1962)

$$\mathcal{E} = \sum_a \int \frac{dk}{(2\pi)^3}\frac{d\omega}{2\pi}\frac{1}{2}\left(\omega + \frac{k^2}{2m}\right) g^<_{aa}(\omega, k, r, t) \tag{7.38}$$

which we use to evaluate the density of energy.

7.3.3 Conservation laws

We could have applied the derivatives in the Martin–Schwinger hierarchy (7.8) also to the other variable and would have obtained

$$\left(-i\frac{\partial}{\partial t_2} - \frac{\nabla_2^2}{2m} - U(2)\right) G^c(1,2) = \delta(1-2) \mp i\int_C d3\, V(2,3) G^c_2(13^-23) \tag{7.39}$$

which corresponds to the reverse ordering in (7.21). In fact both equations (7.8) and (7.39) have to be fulfilled. This requirement was called criterion (A) by (Baym, 1962; Kadanoff and Baym, 1962). In fact it ensures the density conservation since subtracting (7.39) from (7.8) and setting $2 = 1^+$ leads to

$$(i\partial_{t_1} + i\partial_{t_2}) G^c(1,2)_{2=1^+} + \nabla \cdot \left(\frac{\nabla_1 - \nabla_2}{m}\right)_{2=1^+} = 0 \tag{7.40}$$

and using the density (7.33) and the current (7.34) it is just the balance

$$\frac{\partial}{\partial t}n_a + \nabla j_a = 0. \tag{7.41}$$

Next we multiply the first equation of (7.8) with $(\nabla_1 - \nabla_2)/2i$, setting $2 = 1^+$ and integrate over r_1 to obtain

$$\frac{d}{dt_1} \int dr_1 \frac{\nabla_1 - \nabla_2}{2} G^<(1,2) \bigg|_{2=1^+} = \mp \int dr_1 dr_3 \nabla_1 V(r_1 - r_2) G_2(r_1 t_1 r_3 t_1; r_1 t_1^+ r_3 t_1^+)$$

$$- \int dr_1 \nabla_1 U(1) G^<(r_1 t_1 r_1 t_1). \tag{7.42}$$

Now we demand as criterion (B) the symmetry in the two-particle Green's function that the incoming and outgoing particles can be interchanged simultaneously

$$G_2(12, 1^+ 2^+) = G_2(21, 2^+ 1^+). \tag{7.43}$$

With this criterion (B) the double integral in (7.42) vanishes and we obtain

$$\frac{d}{dt_1} \int dr_1 \frac{\nabla_1 - \nabla_2}{2} G^<(1,2) \bigg|_{2=1^+} = - \int dr_1 \nabla_1 U(r_1, t_1) n(r_1 t_1). \tag{7.44}$$

which is just the balance for the mean momentum which is fed by the external force

$$\frac{d}{dt} \langle P(t) \rangle = -\langle \nabla U(r, t) \rangle. \tag{7.45}$$

Using the criteria (A) and (B) one can show analogously that the total energy (7.37) results into the balance

$$\frac{d}{dt} \langle H \rangle = -\langle \nabla U \cdot j \rangle. \tag{7.46}$$

Summarising, if the criteria (A) and (B) of Kadanoff and Baym are fulfilled we have a conserving approximation, i.e. the energy, momentum and density are conserved (Baym and Kadanoff, 1961). A simple way to ensure that any approximation completes criteria (A) and (B) is to represent the Green's functions due to a variational expression as we will do in Chapter 11.

7.3.4   Equilibrium information

In order to learn about the manifold of information contained in the Green's functions, let's inspect them in equilibrium. There we know the density operator, e.g. the grand canonical (2.47) one $\hat{\rho} = \exp\left[-\beta(\hat{H} - \mu \hat{N})\right]/z$, to get for the correlation function

$$G^<(1,2) = \langle \hat{\psi}^\dagger(r_2, t_2) \hat{\psi}(r_1, t_1) \rangle = \text{Tr}\left[e^{-\beta(\hat{H} - \mu \hat{N})} \hat{\psi}^\dagger(r_2, t_2) \hat{\psi}(r_1, t_1) \right]$$

$$= \text{Tr}\left[e^{-\beta(\hat{H} - \mu \hat{N})} \left(e^{\beta(\hat{H} - \mu \hat{N})} \hat{\psi}(r_1, t_1) e^{-\beta(\hat{H} - \mu \hat{N})} \right) \hat{\psi}^+(r_2, t_2) \right] \tag{7.47}$$

where we used the cyclic property of the trace. Further, one shows readily the property

$$\hat{\psi} f(\hat{N}) = f(\hat{N} + 1)\hat{\psi} \tag{7.48}$$

and therefore we have in (7.47)

$$e^{-\beta\mu\hat{N}}\hat{\psi}e^{\beta\mu\hat{N}} = e^{\beta\mu}\hat{\psi} \tag{7.49}$$

and only the exponential of $\hat{H}$ remains in front and after $\hat{\psi}$. These exponentials realise the time propagation of an operator

$$\hat{\Psi}(r, t) = e^{it\hat{H}}\hat{\psi}(r, 0)e^{-it\hat{H}} \tag{7.50}$$

such that we recognise in (7.47) a time-propagation with the imaginary time $t' = -i\beta$. This leads us to a relation between the two correlation functions

$$G^<(1, 2) = \langle \hat{\psi}(r_1, t_1 - i\beta)\hat{\psi}^+(2)\rangle e^{\beta\mu} = G^>(r_1, t_1 - i\beta, 2)e^{\beta\mu}. \tag{7.51}$$

In mixed representations (7.3) this results into

$$g^<(\omega, k, r, t) = e^{-\beta(\omega-\mu)} g^>(\omega, k, r, t) \tag{7.52}$$

known as Kubo–Martin–Schwinger relation. It shows that the two correlation functions are not independent. Indeed, in equilibrium, there should be a fixed relation between the two possibilities first to remove a particle and than to add, or the inverse process.

In the Green's functions there are two basic items of information contained, the spectral and the distribution information. Let's denote the spectral function $A = G^> \pm G^<$ and the rest of the information contained in the correlation function with

$$g^<(\omega, k, r, t) = f(\omega, k, r, t) a(\omega, k, r, t) \tag{7.53}$$

and

$$g^>(\omega, k, r, t) = a(\omega, k, r, t) \mp g^<(\omega, k, r, t) = [1 \mp f(\omega, k, r, t)] a(\omega, k, r, t). \tag{7.54}$$

From the KMS conditions (7.52) one sees then that the ratio of the two correlation functions becomes

$$\frac{g^>}{g^<} = \frac{\text{adding}}{\text{removing}} = \frac{1 \mp f}{f} = e^{\beta(\omega-\mu)}. \tag{7.55}$$

This can be considered as a detailed balance representing the ratio between adding and removing a particle or vice versa. Further one immediately sees from (7.55) that the Fermi and Bose distributions appear

$$f = \frac{1}{e^{\beta(\omega-\mu)} \pm 1}. \tag{7.56}$$

These statistical factors are universal. All information about system properties is in the spectral function a.

Out of equilibrium the situation is often reversed. As we will show, the Boltzmann equation corresponds to the approximation of the spectral function by a single δ-function at the free-particle kinetic energy. The spectral function is then independent of the state of the system and all the rich variety of properties described by the Boltzmann equation follow exclusively from the nonequilibrium distribution $f(r, k, t)$.

The spectral function delivers the excitation weight of the system, i.e. with which statistical weight the energy ω in the distribution is populated. We will discuss these spectral properties in Chapter 8 more extensively. The density (7.33) can be written with the help of (7.53) as

$$n(r) = \int \frac{d\omega}{2\pi} f(\omega) \int \frac{dk}{(2\pi)^3} a(\omega, k, r), \qquad (7.57)$$

such that we see that the spectral function represents the momentum-resolved density of states

$$D(\omega, r) = \int \frac{dk}{(2\pi)^3} a(\omega, k, r). \qquad (7.58)$$

7.3.5 Matsubara technique

Since the equilibrium Green's function is only dependent on the time difference one finds a periodicity for the causal Green's function (7.3)

$$G^c(\tau) = \frac{1}{i}\Theta(\tau)G^>(\tau) \mp \frac{1}{i}\Theta(-\tau)G^<(\tau)$$

$$= \frac{1}{i}\Theta(\tau)G^<(\tau + i\beta)e^{-\beta\mu} \mp \frac{1}{i}\Theta(-\tau)G^>(\tau - i\beta)e^{\beta\mu}$$

$$= \frac{1}{i}\Theta(\tau)G^<(\tau - i\beta)e^{\beta\mu} \mp \frac{1}{i}\Theta(-\tau)G^>(\tau - i\beta)e^{\beta\mu}$$

$$= \mp e^{\beta\mu}G^c(\tau - i\beta). \qquad (7.59)$$

This allows us to cleverly define a discrete Fourier transform

$$G^c(\tau) = c\sum_\nu e^{-iz_\nu\tau} G_\nu = G(\tau), \qquad z_\nu = \frac{i\pi}{\beta}\nu + \mu \qquad (7.60)$$

with the Matsubara frequencies z_ν for even numbers ν describing Bosons and odd numbers describing Fermions. This exactly represents the periodicity (7.59). The constant

c we will determine as convenient in a minute. Therefore we extend the $\Theta(\tau)$ function towards the imaginary axes such that

$$0 \le \tau \le -i\beta : G^c = \frac{1}{i}G^>(\tau) = \frac{1}{i}\int \frac{d\omega}{2\pi}e^{-i\omega\tau}\frac{A(\omega)}{1 \pm e^{-\beta(\omega-\mu)}}$$

$$i\beta \le \tau \le 0 : G^c = \mp\frac{1}{i}G^<(\tau) \tag{7.61}$$

where we used (7.54) with (7.56) for the first line. We will now search for the inverse transformation of (7.60) and integrate

$$\int_0^{-i\beta} d\tau\, G^c(\tau)e^{iz_\nu\tau} = \sum_{\nu'} G_{\nu'}\, c\int_0^{-i\beta} d\tau\, G^c(\tau)e^{i(z_\nu-z_{\nu'})\tau} = G_\nu \tag{7.62}$$

by choosing $c = i/\beta$ since

$$\int_0^{-i\beta} d\tau\, e^{i(z_\nu-z_{\nu'})\tau} = \begin{cases} -\beta\frac{e^{i\pi(\nu-\nu')}-1}{\pi(\nu-\nu')} = 0; & \nu \ne \nu' \\ -i\beta; & \nu = \nu' \end{cases}. \tag{7.63}$$

Using now the same integration in (7.61) we also get

$$\int_0^{-i\beta} d\tau\, G^c(\tau)e^{iz_\nu\tau} = \int \frac{d\omega}{2\pi}\frac{A(\omega)}{z_\nu - \omega} \tag{7.64}$$

such that we find from (7.62)

$$G_\nu = \int \frac{d\omega}{2\pi}\frac{A(\omega)}{z_\nu - \omega}. \tag{7.65}$$

For a free particle with $A(\omega) = 2\pi\delta(\omega - p^2/2m)$ we have therefore

$$G_\nu^0 = \frac{1}{z_\nu - \frac{p^2}{2m}} \tag{7.66}$$

as one can verify by transforming the equation for the free Green's function, $\Sigma = 0$ in (7.13), with respect to (7.60).

Any sum of functions of Matsubara frequencies can be calculated by the following Feynman trick. Since the Matsubara frequencies are just the first-order poles of the

Fermi/Bose distributions $f(z)$ with the residue $\mp 1/\beta$, one can calculate an integral over a function $h(z)$ not having the same poles as

$$\int_c \frac{dz}{2\pi} f(z) h(z) = i \sum_\nu \text{Res}(f, z_\nu) g(z_n u) = \mp \frac{i}{\beta} \sum_\nu h(z_\nu) \tag{7.67}$$

where we assumed a closed loop c including all Matsubara frequencies in the mathematically positive sense. Changing this integration path into the one where all poles z_i of $h(z)$ are encountered, we have a minus sign due to the inversion of the integration direction and can write alternatively

$$\int_c \frac{dz}{2\pi} f(z) h(z) = -i \sum_i f(z_i) \text{Res}(h, z_i). \tag{7.68}$$

Therefore one has

$$\sum_\nu h(z_\nu) = \pm \beta \sum_i f(z_i) \text{Res}(h, z_i). \tag{7.69}$$

This allows us to set up a diagram technique which we will demonstrate in section 10.4 or to calculate conveniently the higher-order equilibrium correlations needed for correlated Bose condensates in Chapter 12.5.1. The rules are:

1. Draw all topological different diagrams you want to consider, each line representing a free Green's function (7.66) and each broken line representing an interaction $iV(q)$ carrying a bosonic Matsubara line.
2. Name all lines with momenta and Matsubara frequencies such that energy (frequency) and momentum is conserved at each point.
3. Sum over all internal lines, i.e. Matsubara frequency summation with a factor i/β and integrate over internal momenta, each closed Fermi loop gets an additional factor of -1 .
4. The sums are calculated according to (7.69) as

$$\sum_\nu \frac{1}{z_\nu - \epsilon} = \pm \beta f^{F/B}(\epsilon). \tag{7.70}$$

7.3.6   Equilibrium pressure

In addition to the evaluation of thermodynamic variables given above one might also express the pressure $\mathscr{P}$ directly by Green's functions. In the grand canonical ensemble (2.47),

$$Z_g = \exp{-\beta(H - \mu N)} = e^{\beta \mathscr{P} V} \tag{7.71}$$

the derivative of pressure with respect to the chemical potential is seen to give the mean-density $V\partial_\mu \mathscr{P} = \langle N \rangle$. Therefore one might use

$$\mathscr{P} = \int_{-\infty}^{\mu} d\bar{\mu}\, n(\bar{\mu}) = \int_{-\infty}^{\mu} d\bar{\mu} \int \frac{d\omega}{2\pi} \sum_a \int \frac{dk}{(2\pi)^3} g_{aa}^<(\omega, k). \tag{7.72}$$

Looking at this relation one has to keep in mind that the pressure is not given by the correlation function $G^<$ of the system in question, but by the integral over all systems of lower-chemical potential.

Similarly, there is another formula which reveals more insight into the physics and which works in fact better. We consider the interaction W coupled to the free Hamiltonian H_0 by a constant $H = H_0 + \lambda W$ to be set $\lambda = 1$ in the end. One gets easily $\partial_\lambda \ln Z_g = -\langle W \rangle$. The pressure (7.71) is therefore given after one integration by

$$\mathscr{P}V = \mathscr{P}_0 V_0 - \int_0^1 \frac{d\lambda}{\lambda}\langle \lambda W \rangle = \mathscr{P}_0 V_0 - \sum_a \int_0^1 \frac{d\lambda}{2\lambda} \int \frac{d\omega}{2\pi} \frac{dk}{(2\pi)^3}\left(\omega - \frac{k^2}{2m}\right) g_{aa}^<(\omega, k, \lambda).$$

$$\tag{7.73}$$

where we subtracted from (7.38) twice the kinetic energy in order to get the mean-interaction energy $\langle W \rangle$. Again, the pressure is not a function $G^<$ of the given state but depends on correlation functions of all (theoretical) states with weaker interaction. The expression (7.73) shows how the pressure is formed in the system when an interaction is switched on. Consequently it is named the charging formula.

7.4   Summary

From all the above equations and definitions, the central one is the GKB equation (7.25) which describes the time evolution of the single-particle correlation function $G^<$. The Dyson equation (7.23) supplies the coherent propagation between individual collisions. The correlation function of the selfenergy, also called the transport vertex, provides the transitions between individual coherent states.

8

Spectral Properties

8.1 Spectral function

8.1.1 Causality and Kramers–Kronig relation

We have seen, from the information contained in the correlation function, that the statistical weight of excitations with which the distributions are populated is given by the spectral function. This momentum-resolved density of state can be found by the retarded (7.15) and advance functions (7.16)

$$A = i\left(G^R - G^A\right) = (G^> \pm G^<) \tag{8.1}$$

called the spectral identity. From these two forms we can identify some general properties of the spectral function.

The retarded functions which propagate only in the future are given by

$$g^R(\tau, k, r, t) = -i\Theta(\tau)a(\tau, k, r, t). \tag{8.2}$$

The step function can be represented as $[\eta \to 0]$

$$\Theta(\tau) = \int \frac{d\omega}{2\pi i} \frac{e^{i\omega\tau}}{\omega - i\eta}. \tag{8.3}$$

With that one gets the Fourier transform

$$g^R(\omega, k, r, t) = \int \frac{d\bar{\omega}}{2\pi} \frac{a(\bar{\omega}, k, r, t)}{\omega + i\eta - \bar{\omega}} = \int \frac{d\bar{\omega}}{2\pi} \frac{a(\bar{\omega}, k, r, t)}{\omega - \bar{\omega}} - \frac{i}{2}a(\omega, k, r, t) \tag{8.4}$$

where we used the Plemelj formula

$$\int_0^\infty dt' \exp\left(ixt'\right) = \frac{i}{x + i\eta} = \pi\delta(x) + i\frac{\wp}{x} \tag{8.5}$$

Interacting Systems far from Equilibrium. Klaus Morawetz, Oxford University Press (2018).
© Klaus Morawetz. DOI: 10.1093/oso/9780198797241.001.0001

to decompose the integral into a principle-value fraction, $\wp$, and a δ-function. Eq. (8.4) shows that the Kramers–Kronig relation exists between real and imaginary parts

$$\text{Reg}^R(\omega, k, r, t) = \int \frac{d\bar{\omega}}{\pi} \frac{\text{Img}^R(\bar{\omega}, k, r, t,)}{\bar{\omega} - \omega}. \tag{8.6}$$

These relations hold, in any case, nonequilibrium and equilibrium. Comparing the retarded and advanced function we find that these functions are complexly conjugated.

8.1.2   Sum rules

Definition (8.1) written in terms of the field operators,

$$A(1,2) = \text{Tr}\left[\hat{\rho} \left(\hat{\psi}_1 \hat{\psi}_2^\dagger \pm \hat{\psi}_2^\dagger \hat{\psi}_1 \right) \right], \tag{8.7}$$

shows that the spectral function is the averaged (anti)commutator for (fermions) bosons with no regards to the Hamiltonian of the system. At the time-diagonal $t_1 = t_2 = t$, this (anti)commutator has a fixed value,

$$\hat{\psi}_a(x, t)\hat{\psi}_b^\dagger(y, t) \pm \hat{\psi}_b^\dagger(y, t)\hat{\psi}_a(x, t) = \delta_{ab}\delta(x - y), \tag{8.8}$$

which determines the spectral function

$$A(1,2)|_{t_1=t_2} = \delta_{a_1 a_2}\delta(x_1 - x_2). \tag{8.9}$$

This relation is often used in the equivalent form

$$A(1,2)\delta(t_1 - t_2) = \delta(t_1 - t_2)\delta_{a_1 a_2}\delta(x_1 - x_2) \equiv \delta(1 - 2) \tag{8.10}$$

or its Wigner transform

$$\int \frac{d\omega}{2\pi} a_{aa}(\omega, k, r, t) = 1. \tag{8.11}$$

Relation (8.11) is called the sum rule for the norm of the spectral function.
    The second sum rule relates to the mean energy

$$\int \frac{d\omega}{2\pi} \omega a = \frac{k^2}{2m_a} + \sigma^{HF}(k, r, t) \tag{8.12}$$

where σ^{HF} is the Hartree–Fock selfenergy. This sum rule follows from the Heisenberg equation of motion for field operators, (7.5),

$$i\frac{\partial}{\partial t_1}\hat{\psi}_1 = -\hat{H}\hat{\psi}_1 + \hat{\psi}_1\hat{H}\int d5 H^0_{15}\hat{\psi}_5 + \frac{1}{2}\int d3 d4 d5\,(\mathcal{V}_{1354} - \mathcal{V}_{3154})\,\hat{\psi}_3^\dagger\hat{\psi}_4\hat{\psi}_5, \quad (8.13)$$

$$-i\frac{\partial}{\partial t_2}\hat{\psi}_2^\dagger = \hat{H}\hat{\psi}_2^\dagger - \hat{\psi}_2^\dagger\hat{H}\int d4 H^0_{42}\hat{\psi}_4^\dagger + \frac{1}{2}\int d3 d4 d5\,(\mathcal{V}_{5423} - \mathcal{V}_{5432})\,\hat{\psi}_5^\dagger\hat{\psi}_4^\dagger\hat{\psi}_3, \quad (8.14)$$

where, the Hamiltonian $\hat{H}$ from definition (7.1) is abbreviated as

$$\hat{H} = \int d4 d5\, H^0_{45}\hat{\psi}_4^\dagger\hat{\psi}_5 + \frac{1}{2}\int d3 d4 d5 d6\,\mathcal{V}_{3456}\hat{\psi}_3^\dagger\hat{\psi}_4^\dagger\hat{\psi}_5\hat{\psi}_6, \quad (8.15)$$

with H^0 being the kinetic energy and $\mathcal{V}$ the interaction potential extended by δ-functions like $\delta(4-5)\delta(3-6)$ to match the definition (7.1). From definition (8.7) and Heisenberg equations (8.13)–(8.14) follows

$$\frac{i}{2}\left(\frac{\partial}{\partial t_1} - \frac{\partial}{\partial t_2}\right)A(1,3)\Big|_{t_1=t_2} = \frac{i}{2}\,\mathrm{Tr}\left(\hat{\rho}\left(\frac{\partial\hat{\psi}_1}{\partial t_1}\hat{\psi}_2^\dagger \pm \hat{\psi}_2^\dagger\frac{\partial\hat{\psi}_1}{\partial t_1} - \hat{\psi}_1^\dagger\frac{\partial\hat{\psi}_2}{\partial t_2} \mp \frac{\partial\hat{\psi}_2^\dagger}{\partial t_2}\hat{\psi}_1\right)\right)\Big|_{t_1=t_2}$$

$$= \frac{1}{2}\int d5\,\left(H^0_{15}A(5,2) + A(1,5)H^0_{52}\right)\Big|_{t_1=t_2} + \int d3 d4\,(\mathcal{V}_{1324} - \mathcal{V}_{3124})\,\rho(4,3). \quad (8.16)$$

In the Wigner representation, the first term results in the kinetic energy $k^2/2m_a$ and the second term gives the Hartree–Fock selfenergy energy σ^{HF} such that the sum rule (8.12) is proved.

8.2 Quasiparticle and extended quasiparticle picture

Knowledge of the retarded Green's function provides the spectral function and therefore the spectral properties of the system. The Dyson equation for the retarded one, (7.15) reads explicitly

$$\left(i\frac{\partial}{\partial t_1} - \frac{\nabla_1^2}{2m}\right)G^R(1,1') = \delta(1-1') + \int d\bar{1}\,\Sigma^R(1,\bar{1})G^R(\bar{1},1'). \quad (8.17)$$

From this equation it is now possible to derive an *exact* expression for the inverse retarded Green's function where we use Wigner coordinates $t = (t_1 + t_2)/2$, $\tau = t_1 - t_2$, $r = (x_1 + x_2)/2$ and $x = x_1 - x_2$

$$(G^R)^{-1}(x,\tau,r,t) = -\left[i\frac{1}{2}\frac{\partial}{\partial t} - i\frac{\partial}{\partial \tau} - \frac{(\frac{1}{2}\nabla_r - \nabla_x)^2}{2m}\right]\delta(\tau)\delta(x) - \Sigma^R(x,\tau,r,t). \quad (8.18)$$

We now add the adjoined equation and use the properties of δ- functions to find, in a mixed representation, the inverse Green's function in *second order* gradient expansion

$$(G^R)^{-1}(k, \omega, r, t) = \left[\omega - \frac{k^2}{2m}\right] - \Sigma^R(k, \omega, r, t). \tag{8.19}$$

The problem is to find the retarded Green's function itself for any approximation of Σ^R, see (Malfliet, 1992; Morawetz, 1994). We can invert the field-free Dyson equation (Mahan, 1987) (8.17) and get, for the spectral function, $a = i(g^R - g^A)$ and for the real part of retarded function $g = \frac{1}{2}(g^R + g^A)$

$$a(k, \omega, r, t) = \frac{\gamma(k, \omega, r, t)}{[\omega - \frac{k^2}{2m} - \sigma(k, \omega, r, t)]^2 + \frac{1}{4}\gamma(k, \omega, r, t)^2} \tag{8.20}$$

$$g(k, \omega, r, t) = \frac{\omega - \frac{k^2}{2m} - \sigma(k, \omega, r, t)}{[\omega - \frac{k^2}{2m} - \sigma(k, \omega, r, t)]^2 + \frac{1}{4}\gamma(k, \omega, r, t)^2} \tag{8.21}$$

with

$$\gamma = 2\mathrm{Im}\sigma^R = i(\sigma^R - \sigma^A), \qquad \sigma = \mathrm{Re}\sigma^R = \frac{1}{2}(\sigma^R + \sigma^A). \tag{8.22}$$

Since we will develop a theory valid in first-order gradient approximation, this form of spectral function is appropriate. For small imaginary parts of the selfenergy and, consequently, a small damping, one can expand this expression following (Zimmermann and Stolz, 1985; Kraeft, Kremp, Ebeling and Röpke, 1986; Špička and Lipavský, 1995)

$$a(k, \omega, r, t) = \frac{2\pi\delta(\omega - \epsilon(k, r, t))}{1 + \frac{\partial \sigma(k, \omega, r, t)}{\partial \omega}\big|_{\omega = \epsilon}} - \gamma(k, \omega, r, t)\frac{\partial}{\partial \omega}\frac{\wp}{\omega - \epsilon(k, r, t)} + o(\gamma^2) \tag{8.23}$$

where $\wp$ denotes the principal value. The quasiparticle energy $\epsilon(prt)$ is the solution of the dispersion relation

$$\omega - \frac{k^2}{2m} - \sigma(k, \omega, r, t) = 0. \tag{8.24}$$

Apparently, the quasiparticle approximation has a wrong norm,

$$\int \frac{d\omega}{2\pi} 2\pi z\delta(\omega - \varepsilon) = z. \tag{8.25}$$

The extended quasiparticle approximation (8.23) has the correct norm,

$$\int \frac{d\omega}{2\pi}\left(1+\frac{\partial\sigma}{\partial\omega}\right)2\pi\delta(\omega-\varepsilon)+\int\frac{d\omega}{2\pi}\gamma\frac{\wp'}{\omega-\varepsilon}=1+\frac{\partial\sigma}{\partial\varepsilon}+\frac{\partial}{\partial\varepsilon}\int\frac{d\omega}{2\pi}\gamma\frac{\wp}{\omega-\varepsilon}=1. \quad (8.26)$$

In the last step we have used the Kramers–Kronig relation

$$\sigma(\varepsilon)=\sigma^{HF}-\int\frac{d\omega}{2\pi}\gamma(\omega)\frac{\wp}{\omega-\varepsilon}. \quad (8.27)$$

The extended quasiparticle spectral function (8.23) fulfils the spectral sum rules (8.11) and the energy-weighted sum rule (8.12) as we will see now. The quasiparticle approximation

$$\int\frac{d\omega}{2\pi}\omega 2\pi z\delta(\omega-\varepsilon)=z\varepsilon=z\frac{k^2}{2m_a}+z\sigma(\varepsilon,k,r,t) \quad (8.28)$$

does not satisfy sum rule (8.12) even if one omits the wave-function renormalisation z, because the full selfenergy σ appears instead of its singular Hartree–Fock part σ^{HF}.

The extended quasiparticle approximation satisfies

$$\int\frac{d\omega}{2\pi}\omega\left(1+\frac{\partial\sigma}{\partial\omega}\right)2\pi\delta(\omega-\varepsilon)+\int\frac{d\omega}{2\pi}\omega\gamma\frac{\wp'}{\omega-\varepsilon}$$

$$=\varepsilon\left(1+\frac{\partial\sigma}{\partial\varepsilon}\right)+\int\frac{d\omega}{2\pi}\omega\gamma\frac{\wp'}{\omega-\varepsilon}$$

$$=\varepsilon\left(1+\frac{\partial\sigma}{\partial\varepsilon}\right)+\int\frac{d\omega}{2\pi}\gamma\frac{\wp}{\omega-\varepsilon}+\varepsilon\int\frac{d\omega}{2\pi}\gamma\frac{\wp'}{\omega-\varepsilon}$$

$$=\varepsilon\left(1+\frac{\partial\sigma}{\partial\varepsilon}\right)-(\sigma-\sigma^{HF})-\varepsilon\frac{\partial\sigma}{\partial\varepsilon}$$

$$=\varepsilon-(\sigma-\sigma^{HF})=\frac{k^2}{2m_a}+\sigma^{HF} \quad (8.29)$$

which is just the sum rule (8.12). The extended quasiparticle approximation is thus superior to the quasiparticle approximation in satisfying this sum rule.

We note that the mean-field approximation,

$$a\approx 2\pi\delta\left(\omega-\frac{k^2}{2m_a}-\sigma^{HF}\right), \quad (8.30)$$

obeys both sum rules too.

8.3 Comparison with equilibrium

Neglecting off-shell terms, one can end up with very different approximations of the spectral function a. It is worthwhile comparing these approximations with the full spectral function (8.20). In the ground state, the spectral function has been studied by Köhler 1992; 1993, and at equilibrium of finite temperature by Alm et al. (1996) where we refer the reader for details.

For the purpose of our discussion it is sufficient to use the spectral function shown in Fig. 8.1 which we have obtained in ladder approximation in Chapter 10.6, with the help of the Yamaguchi potential for nuclear matter, see Appendix E.1. From the plastic map of the spectral function in the $(\omega-k)$-plane one can see that the spectral function behaves as a δ function only in the vicinity of the Fermi energy. At a finite-energy distance, the spectral function broadens reflecting the finite life-time of the quasiparticle states. The spectral function for Fermi momentum is plotted in Fig. 8.2.

For momenta close to the Fermi momentum and energies close to the Fermi energy, the quasiparticle approximation

$$a = 2\pi\delta\left(\omega - \frac{k^2}{2m} - \sigma\right) = 2\pi z\delta\left(\omega - \varepsilon\right), \tag{8.31}$$

becomes well justified, the free-particle approximation and the mean-field approximation

$$a = 2\pi\delta\left(\omega - \frac{k^2}{2m} - \sigma^{HF}\right) \tag{8.32}$$

place the singularity at wrong energies, see Fig. 8.2.

The positions of the singularity of the spectral function are given by the quasiparticle dispersion relation (8.24) shown in Fig. 8.3. While the quasiparticle approximation follows the position of the singularity nearly exactly, the free-particle approximation is badly off by 60 MeV and the mean-field approximation differs by 7 MeV.

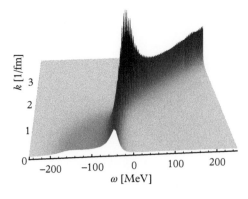

Figure 8.1: *Spectral function $a(\omega, k)$ for nuclear matter at a temperature of $T = 10$ MeV. The energy ω runs from left to right, the momentum $k \equiv |k|$ runs from the front to back. The central massif follows the quasiparticle dispersion $\omega = \varepsilon(k)$ which is crudely parabolic. Close to the Fermi momentum the spectral function reaches the highest values due to the partial Pauli blocking of the collisions. The off-shell satellites appear mainly for energies below the quasiparticle dispersion, $\omega < \varepsilon$, reflecting the off-shell and the correlated motion of particles.*

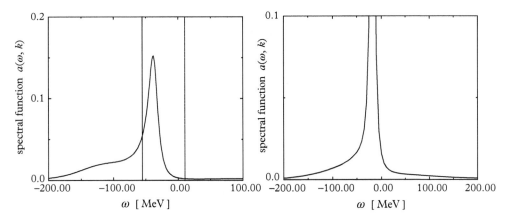

Figure 8.2: *The equilibrium spectral functions $a(\omega, k)$ at $T = 10$ MeV as a function of the energy. Left: for momentum $k = 0.7$ fm^{-1} (about half of the Fermi momentum) and right: momentum $k = 1.3$ fm^{-1} corresponding to the quasiparticle energy on the left side of the Luttinger dip in the spectral function of the selfenergy, see also Fig. 9.3, where the energy dependence is strongest. The sharp peak appears at the quasiparticle energy $\omega = \varepsilon(k)$. Its width is proportional to the inverse life-time of the quasiparticle. The satellite visible in particular below the quasiparticle energy reflects the off-shell motion of particles between individual binary interactions. From the area below the curve is 2π, the presence of satellites is compensated by a reduced weight of the peak. For comparison, the vertical lines denote the free-particle spectral function $a^{free}(\omega, k) = 2\pi\delta\left(\omega - \frac{k^2}{2m}\right)$ (dotted line) and the Hartree–Fock approximation $a^{HF}(\omega, k) = 2\pi\delta\left(\omega - \frac{k^2}{2m} - \sigma^{HF}(k)\right)$, where σ^{HF} is the Hartree–Fock mean field.*

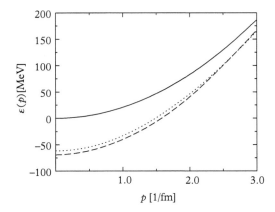

Figure 8.3: *Quasiparticle energy dispersion relation (8.24) at $T = 10$ MeV. The free-particle energy $\frac{k^2}{2m}$ (full line) is corrected mainly by the Hartree-Fock interaction $\varepsilon^{HF} = \frac{k^2}{2m} + \sigma^{HF}$ (long dashed line). The regular (dynamical) part of the interaction brings a smaller correction, $\varepsilon = \frac{k^2}{2m} + \sigma_\varepsilon$ (short dashed line), the shift by about 7 MeV seems to be small but is important on the energy scale of thermal excitations.*

The quasiparticle approximation also provides a non-trivial norm of the singular contribution which equals the wave-function renormalisation z. Simpler approximations are normalised to unity. On the other hand, the quasiparticle approximation does not satisfy the sum rules (8.12) which have to be fulfilled for any value of the momentum k. The quasiparticle approximation only gives the pole contribution z. The free-particle approximation obeys (8.11) but fails on (8.12). The mean-field approximation obeys both sum rules. The failure of the quasiparticle approximation to comply with these sum rules follows from the missing satellites that are clearly seen in Fig. 8.2. These satellites are also visible in Fig. 8.1, in particular for very low momenta.

Since the mean-field approximation obeys both sum rules, it is superior to the quasiparticle approximation on the short time scale. One can conclude that the best approximation would be to use the quasiparticle approximation within the transport theory and the mean-field approximation to derive subsidiary functionals like the quasiparticle energy. Although such a pragmatic approach is very reasonable and works well at equilibrium, in transport it can only be implemented after a careful separation of the short and long time scales.

8.4   The problem of ansatz

Since the time-diagonal Kadanoff and Baym equations (7.28) determine the Wigner function, the problem remains that the collision side is given explicitly by $G^{\gtrless}$. In order to close such kinetic equations, (see 9.76), it is necessary to know the relation between $G^>$ and $G^<$. This problem is known as an ansatz and must be constructed consistently with the required approximation of the selfenergy. The conventional way is to use the spectral function in order to change the correlation functions into an expression of the Wigner function employing the spectral identity $g^> \pm g^< = a$. This is the basis for the next two ways of constructing the ansatz.

Kadanoff and Baym ansatz   Neglecting the off-pole part in the spectral function leading to a δ shape (8.32), one finds the relation between the correlation function and the Wigner distribution function ρ as

$$g^<(k, \tau, r, t) = e^{-i\epsilon(k,r,t)\tau} \rho(k, r, t) \tag{8.33}$$

with ϵ the quasiparticle energy (8.24). This is quite good as long as the quasiparticle picture holds true. This is connected with the neglected off-shell parts in (8.23). Therefore the simple ansatz, denoted above as *the conventional ansatz*, will certainly fail in dense correlated systems. The failure of the quasiparticle approximation is reflected in the failure of the energy-weighted sum rule (8.12).

Another obscure discrepancy is the fact that with the conventional ansatz one has some minor differences in the resulting collision integrals, if one compares it with the results from the density operator technique (Jauho and Wilkins, 1984; Levinson, 1969). With the conventional ansatz one gets just one half of all retardation times in the various time arguments (Jauho and Wilkins, 1984; Jauho, 1991). This annoying difference has been solved by the generalised Kadanoff and Baym ansatz.

Generalised Kadanoff and Baym ansatz   The construction of the generalised ansatz is given in (Lipavský, Špička and Velický, 1986). Here we repeat the derivation for the

Hartree–Fock approximation or white noise model, where the ansatz is exact. Using the time-diagonal Hartree–Fock selfenergy in the equation for the retarded Green's function (8.17) one can see that the semi-group relation holds (Lipavský, Špička and Velický, 1986)

$$iG^R(tt_1) \cdot G^R(t_1 t_2) = G^R(tt_2). \tag{8.34}$$

Now the integral form of the Kadanoff–Baym equation (7.28) reads

$$G^<(11') = G^R(12) \left[(G_0^R(23))^{-1} G_0^<(34)(G_0^A(45))^{-1} + \Sigma^<(25) \right] G^A(51').$$

Here we concentrate on the time evolution only, 1 means t_1, and assume summation over equal indices. First let us set $t_1 > t_1'$. Then, by using the property (8.34), one finds

$$G^<(11') = iG^R(11') \cdot G^<(1'1').$$

Including the opposite case $t_1' > t_1$ one arrives finally at

$$G^<(11') = iG^R(12) \cdot G^<(22) - iG^A(12) \cdot G^<(11) \tag{8.35}$$

or in mixed representation

$$g^<(k, \tau, r, t) = \rho \left(k, r, t - \frac{|\tau|}{2} \right) a(k, \tau, r, t) \tag{8.36}$$

and analogously

$$g^>(k, \tau, r, t) = \left[1 \mp \rho \left(k, r, t - \frac{|\tau|}{2} \right) \right] a(k, \tau, r, t). \tag{8.37}$$

Therefore, the connection between $G^>$ and $G^<$ is given and the equation (9.76) can be closed. This ansatz is superior to the Kadanoff and Baym ansatz in several respects (Morawetz and Jauho, 1994): (i) it has the correct spectral properties, (ii) it is gauge invariant, (iii) it preserves causality, (iv) the quantum kinetic equations derived with Eq. (20.18) coincide with those obtained with the density matrix technique (Levinson, 1965, 1969; Jauho and Wilkins, 1984), and (v) it reproduces the Debye–Onsager relaxation effect (Morawetz and Kremp, 1993).

One serious drawback of this ansatz is the overestimation of wave function renormalisation (Špička and Lipavský, 1995) at large times, see Chapter 19.3.1. This indicates that the high energy tails of the Wigner function prevent the simple picture of scattering. To provide a clear interpretation of scattering into and out of the phase space element, we have to convert the transport equation into the equation for the quasiparticle, which does not have problematic energy tails. Interestingly, the GKB ansatz is consistent with the first gradient corrections corresponding to the extended quasiparticle picture (Špička and Lipavský, 1995) which will be demonstrated in Chapter 9.4.

8.4.1 Further spectral functions

The set of equations for nonequilibrium Green's functions is most conveniently written in terms of retarded and advanced functions. With respect to the physical interpretation, it is sometimes appropriate to work with an alternative representation that decomposes retarded and advanced functions into real and imaginary parts

$$
A = i\left(G^R - G^A\right), \quad G = \frac{1}{2}\left(G^R + G^A\right),
$$

$$
\Gamma = i\left(\Sigma^R - \Sigma^A\right), \quad \Sigma = \frac{1}{2}\left(\Sigma^R + \Sigma^A\right),
$$

$$
\mathscr{A} = i\left(\mathscr{G}^R - \mathscr{G}^A\right), \quad \mathscr{G} = \frac{1}{2}\left(\mathscr{G}^R + \mathscr{G}^A\right),
$$

$$
\mathscr{M} = i\left(\mathscr{T}^R - \mathscr{T}^A\right), \quad \mathscr{T} = \frac{1}{2}\left(\mathscr{T}^R + \mathscr{T}^A\right).
$$

(8.38)

The last two lines we wrote for two-particle and two-time Green's functions and for the T-matrix analogously. The imaginary parts are called spectral functions. The real or imaginary functions are only the diagonal elements of operators. More exactly, one should talk about Hermitian or anti-Hermitian parts.

According to definitions (7.15) and (7.16), the imaginary parts can be alternatively expressed via correlation functions as

$$
A = i\left(G^R - G^A\right) = G^> + G^<, \quad \Gamma = i\left(\Sigma^R - \Sigma^A\right) = \Sigma^> + \Sigma^<,
$$

$$
\mathscr{A} = i\left(\mathscr{G}^R - \mathscr{G}^A\right) = \mathscr{G}^> - \mathscr{G}^<, \quad \mathscr{M} = i\left(\mathscr{T}^R - \mathscr{T}^A\right) = \mathscr{T}^> - \mathscr{T}^<.
$$

(8.39)

These relations are called spectral identities.

8.4.2 Optical theorems

From the spectral identities (8.39) and the transport equations $[G^{\gtrless} = G^R \Sigma^{\gtrless} G^A, \mathscr{T}^{\gtrless} = \mathscr{T}^R \mathscr{G}^{\gtrless} \mathscr{T}^A$ one finds that the spectral functions obey equations similar to transport ones

$$
A = G^R \Gamma G^A, \quad \mathscr{M} = \mathscr{T}^R \mathscr{A} \mathscr{T}^A.
$$

(8.40)

The last relation is known as the optical theorem.

Identities (8.40) can also be derived without using the transport equations. For instance, multiplying $A = i(G^R - G^A)$ by G_R^{-1} from the left-hand side one finds

$$
G_R^{-1} A = i\left(1 - G_R^{-1} G^A\right).
$$

(8.41)

The differential Dyson's equation (7.22) transformed to the retarded and advanced functions reads

$$G_R^{-1} = G_0^{-1} - \Sigma^R = \left(G_A^{-1} + \Sigma^A\right) - \Sigma^R = G_A^{-1} - \left(\Sigma^R - \Sigma^A\right) = G_A^{-1} + i\Gamma. \qquad (8.42)$$

Substituting (8.42) into the right-hand side of (8.41) one directly recovers the equation $A = G^R \Gamma G^A$.

Identities (8.40) have the causal order of operators (retarded–spectral–advanced). The algebraic evaluation of the anti-Hermitian part allows one to find similar identities with the anti-causal order (advanced–spectral–retarded). Let us multiply $A = i(G^R - G^A)$ by G_R^{-1} from the right-hand side

$$A G_R^{-1} = i\left(1 - G^A G_R^{-1}\right). \qquad (8.43)$$

Using (8.42) on the right-hand side, one finds $A = G^A \Gamma G^R$. Similarly one can derive all anti-causal counterparts of relations (8.40),

$$A = G^A \Gamma G^R, \quad \mathcal{M} = \mathcal{T}^A \mathcal{A} \mathcal{T}^R. \qquad (8.44)$$

9

Quantum Kinetic Equations

9.1 Kadanoff and Baym equation in quasiclassical limit

The quasiclassical approximation is achieved turning (7.27) into the Wigner mixed representation (7.31) and neglecting all gradients in time and space but linear gradients. The complex unit turns the anti-Hermitian operators in the brackets into real functions.

9.1.1 Gradient expansion

Let us consider a product of two-particle functions

$$C(1234) = \int d3' d4' A(123'4') B(3'4'34). \tag{9.1}$$

We transform this product into mixed representation (7.31) keeping gradients up to the linear order and using the coordinates

$$\alpha = 1 - 2, \quad \beta = 3 - 4, \quad \tau = \frac{1}{2}(1 + 2 - 3 - 4), \quad x = \frac{1}{4}(1 + 2 + 3 + 4). \tag{9.2}$$

Since the relative coordinates, α and β, obey the standard matrix algebra while only the coordinate τ and x undergo the gradient expansion, we write these arguments in the form $\langle \alpha | C(\tau, x) | \beta \rangle$. In this notation the product (9.1) reads with integration variables $\gamma = 3' - 4', \bar{\tau} = (3 + 4 - 3' - 4')/2$

$$\langle \alpha | C(\tau, x) | \beta \rangle = \int d\gamma \, d\bar{\tau} \langle \alpha | A \left(\tau - \bar{\tau}, x + \frac{1}{2}\bar{\tau} \right) | \gamma \rangle \langle \gamma | B \left(\bar{\tau}, x - \frac{1}{2}(\tau - \bar{\tau}) \right) | \beta \rangle. \tag{9.3}$$

We see a convolution in the first variables which is spoilt by the arguments beyond the centre-of-mass coordinate x which we therefore expand around x. The relative coordinates can be suppressed being independent of this expansion. Any two-point function $C(x_1, x_2) = \int dx_3 A(x_1, x_3) B(x_3, x_2)$ has therefore the same expansion with

Interacting Systems far from Equilibrium. Klaus Morawetz, Oxford University Press (2018).
© Klaus Morawetz. DOI: 10.1093/oso/9780198797241.001.0001

$x = (x_1 + x_2)/2$, and $\tau = x_1 - x_2$. Via the Fourier transformation of coordinate τ, we can establish the gradient expansion up to any order in (9.3)

$$C(\kappa, x) = \int d\tau e^{-i\tau\kappa} \sum_{n,l} \int d\bar{\tau} \frac{A^{(n)}(\tau - \bar{\tau}, x)}{n!} \left(\frac{\bar{\tau} - \tau}{2}\right)^l \frac{B^{(l)}(\bar{\tau}, x)}{l!} \left(\frac{\bar{\tau}}{2}\right)^n$$

$$= \sum_{n,l} \frac{\left(-\frac{i}{2}\partial_\kappa{}^A\partial_x{}^B\right)^l \left(\frac{i}{2}\partial_\kappa{}^B\partial_x{}^A\right)^n}{l!} A(\kappa, x) B(\kappa, x)$$

$$= e^{\frac{i}{2}\left(\partial_x{}^A\partial_\kappa{}^B - \partial_\kappa{}^A\partial_x{}^B\right)} A(\kappa, x) B(\kappa, x) . \tag{9.4}$$

Assuming a slow variation of this centre-of-mass coordinate we can keep gradients in x up to the linear order

$$c = ab - \frac{i}{2}[a, b] = ab\left(1 - \frac{i}{2}[\ln a, \ln b]\right), \tag{9.5}$$

where the (generalised) Poisson bracket

$$[a, b] \equiv \frac{\partial a}{\partial \omega}\frac{\partial b}{\partial t} - \frac{\partial a}{\partial t}\frac{\partial b}{\partial \omega} - \frac{\partial a}{\partial k}\frac{\partial b}{\partial r} + \frac{\partial a}{\partial r}\frac{\partial b}{\partial k} \tag{9.6}$$

applies to the sum coordinate and time. The logarithmic form of (9.5) is merely a convenient notation because a and b are matrices in α and β.

In this way, the anticommutator reduces to a simple product

$$\frac{1}{2}\{A, B\}_+ = \frac{1}{2}(AB + BA) \rightarrow a_a(\omega, k, r, t)b_a(\omega, k, r, t), \tag{9.7}$$

and the commutator is linear in gradients

$$-i[A, B]_- = -i(AB - BA) \rightarrow [a, b] \tag{9.8}$$

with the Poisson bracket (9.6) which is the quasiclassical limit. Please remember as a rule, we use lower cases to denote operators in the Wigner representation (7.31).

9.1.2 Quasiclassical Kadanoff and Baym equation

Rearranging the Kadanoff and Baym equation (7.27) we can write

$$-i\left[G_0^{-1} - \Sigma, G^<\right]_- + i\left[G, \Sigma^<\right]_- = \frac{1}{2}\{A, \Sigma^<\}_+ - \frac{1}{2}\{\Gamma, G^<\}_+, \tag{9.9}$$

where relations $G_{R,A}^{-1} = G_0^{-1} - \Sigma^{R,A} = G_0^{-1} - \Sigma \pm \frac{i}{2}\Gamma$ and $G^{R,A} = G \mp \frac{i}{2}A$ have been used in the rearrangement. This equation is the starting point to the quasiparticle transport equation of the Landau type. The quasiclassical limit of the Kadanoff and Baym equation (9.9) corresponding to the gradient expansion up to the second order (9.5) is

$$[g_0^{-1} - \sigma, g^<] + [g, \sigma^<] = a\sigma^< - \gamma g^<. \tag{9.10}$$

The free-particle inverse Green's function (7.20) in the Wigner representation reads

$$g_0^{-1} = \omega - \frac{k^2}{2m_a} - \sigma^{HF}. \tag{9.11}$$

From the differential Dyson equation $G_R^{-1} = G_0^{-1} - \Sigma^R$ it follows that the propagator is free of explicit gradient contributions,

$$g^R = \frac{1}{\omega - \frac{k^2}{2m_a} - \sigma^{HF} - \sigma^R}. \tag{9.12}$$

The spectral function a and the real part g thus are

$$a = \frac{\gamma}{\left(\omega - \frac{k^2}{2m_a} - \sigma\right)^2 + \frac{1}{4}\gamma^2}, \quad g = \frac{\omega - \frac{k^2}{2m_a} - \sigma}{\left(\omega - \frac{k^2}{2m_a} - \sigma\right)^2 + \frac{1}{4}\gamma^2}. \tag{9.13}$$

where $\sigma = \sigma^{HF} + (g^R + g^A)/2$ is the real part and $\gamma = i(\sigma^R - \sigma^A)$ the imaginary part of the selfenergy.

The Kadanoff and Baym equation (9.10) is the quantum generalisation of the Boltzmann equation. Indeed, the left-hand side describes a drift while the right-hand side describes the scattering. On the other hand, the Kadanoff and Baym equation includes a number of features that cannot be understood within the classical picture, in particular it includes the off-shell motion. To enlighten these quantum features, we discuss various approximations by which the Kadanoff and Baym equation can be reduced to a kinetic equation of the Boltzmann type.

It is striking that nearly all approximations which lead to different kinetic equations are justified by the same physical argument of the smallness of gradient corrections to the scattering integral. This lack of the systematics follows from the fact that it is not clearly specified what the scattering integral really is. If one vaguely identifies the scattering integral with the interaction terms represented by Σ's in the Kadanoff and Baym equation (7.27), one can make various rearrangements which are equivalent but the neglect of gradient corrections leads to different kinetic equations.

9.1.3   Collision-less Landau equation

To fix the wrong position of the pole in the mean-field approximation, Kadanoff and Baym (1962) introduce an approach which is able to recover the collision-less kinetic equation of Landau's theory of normal Fermi liquids. Assuming that close to the Fermi level μ all collisions freeze out, $\sigma^{>,<} \to 0$ for $\omega \to \mu$, they simplify the transport equation (9.10) as

$$[g_0^{-1} - \sigma, g^<] = 0. \tag{9.14}$$

This approximation is not based on the neglect of gradients, therefore it allows one to see the structure of the kinetic equation from an independent angle.

One can prove that the drift resulting from approximation (9.14) differs from the mean-field drift, i.e. that the regular part of the selfenergy does not vanish in the vicinity of the Fermi level, even though the imaginary part does. Indeed, $\sigma^> + \sigma^< > 0$ for $\omega \neq \mu$ guarantees that for $\omega \to \mu$ the energy derivative of the selfenergy does not vanish

$$\frac{\partial \sigma_{\text{reg}}}{\partial \omega} = -\int \frac{d\omega'}{2\pi} \frac{\sigma^>(\omega') + \sigma^<(\omega')}{(\omega - \omega')^2} < 0. \tag{9.15}$$

In the narrow region close to the Fermi level, the spectral function also includes only the quasiparticle pole, as it follows from the limit $\gamma \to 0$ of (9.13),

$$a = 2\pi \delta \left(\omega - \frac{k^2}{2m} - \sigma \right) = 2\pi z \delta \left(\omega - \varepsilon \right), \tag{9.16}$$

where z is the wave-function renormalisation and ε is the quasiparticle energy.

The initial condition $g^< = f_{FD} a$ in the remote past is thus singular and the correlation function close to the Fermi level remains singular for all times. One can then write the correlation function as

$$g^{\lessgtr}(\omega, k, r, t) = \left(\begin{matrix} f(k, r, t) \\ 1 \mp f(k, r, t) \end{matrix} \right) 2\pi \delta \left(\omega - \frac{k^2}{2m} - \sigma \right), \tag{9.17}$$

which is called the quasiparticle approximation. Note that the quasiparticle approximation only holds in the vicinity of the Fermi level. It gives no information about the correlation function $g^<$ for energies far from the Fermi level.

Using the quasiparticle approximation in the collision-less transport equation (9.14), one readily recovers the Landau kinetic equation for quasiparticles,

$$\frac{\partial f}{\partial t} + \frac{\partial \varepsilon}{\partial k} \frac{\partial f}{\partial r} - \frac{\partial \varepsilon}{\partial r} \frac{\partial f}{\partial k} = 0, \tag{9.18}$$

where the quasiparticle energy ε given by

$$\varepsilon = \frac{k^2}{2m_a} + \sigma_\varepsilon \tag{9.19}$$

as a pole value of $g_0^{-1} - \sigma$ in the transport equation (9.14). Since we can use this transport equation to derive the spectral function (9.16), the quasiparticle energy also defines the singularity of the spectral function.

The quasiparticle kinetic equation shows that the mean field does not fully explore the effect of the medium on the drift, i.e. on the asymptotic states of the binary processes. Since the mean-field approximation is well established in many fields of many-body physics, a natural question arises whether the corrections beyond are really worthy of any attention. In metals, Fock's interaction (exchange term of Σ^{HF}) is known to be strongly reduced by the screening of potentials. This screening is hidden in the regular part of the selfenergy Σ_{reg}^R. In nuclear matter, the mean-field interaction is usually fitted, therefore it also includes contributions from correlations of higher order than the pure Hartree–Fock term. Briefly, even if one aims to arrive at a simplified kinetic equation with a mean field, e.g. of Skyrme type, a consistent approach requires us to include the effect of the regular selfenergy on the drift of particles.

9.1.4   Landau equation with collisions

Let us specify which neglect of gradients will lead us to the kinetic equation with a quasi-particle drift as in (9.18) and with a corresponding scattering integral. The selfenergy Σ^R can be viewed as an effective energy-dependent Hamiltonian. Its Hermitian (real) part

$$\Sigma = \frac{1}{2}\left(\Sigma^R + \Sigma^A\right) = \Sigma^{HF} + \frac{1}{2}\left(\Sigma_{reg}^R + \Sigma_{reg}^A\right) \tag{9.20}$$

corrects an undamped motion, while its anti-Hermitian (imaginary) part

$$\Gamma = i\left(\Sigma^R - \Sigma^A\right) = i\left(\Sigma_{reg}^R - \Sigma_{reg}^A\right) \tag{9.21}$$

describes the damping. Moving the Hermitian part into the drift term; the transport equation gains the form

$$-i\left[G_0^{-1} - \Sigma, G^<\right]_- = i\left(G^R\Sigma^< - \Sigma^< G^A\right) - \frac{1}{2}\left[\Gamma, G^<\right]_+ . \tag{9.22}$$

This equation is the starting point for the Landau type quasiparticle transport equation.

Following our convention, we keep gradient terms on the drift side of (9.22) and neglect gradient terms on the scattering side,

$$[g_0^{-1} - \sigma, g^<] = \sigma^< g^> - \sigma^> g^< . \tag{9.23}$$

This is equivalent to neglecting $[g, \sigma^<]$ in the Kadanoff and Baym equation (9.10). By interchanging $> \longleftrightarrow <$ one obtains the approximate equation for $g^>$, from which follows the damping-free approximate equation for the spectral function a solved by (9.16).

Again, from the initial condition and absence of the off-shell contributions in equation (9.23) one can show that the quasiparticle approximation (9.17) follows from (9.23). After a direct substitution one finds that the distribution function obeys the kinetic equation

$$\frac{\partial f}{\partial t} + \frac{\partial \varepsilon}{\partial k}\frac{\partial f}{\partial r} - \frac{\partial \varepsilon}{\partial r}\frac{\partial f}{\partial k} = z\sigma_\varepsilon^< (1 - f) - z\sigma_\varepsilon^> f, \tag{9.24}$$

which is Landau's equation (9.18) extended by the scattering integral called the Landau–Silin equation.

Note that the scattering rate is reduced by the wave-function renormalisation (6.47)

$$z = \frac{1}{1 - \frac{\partial \sigma}{\partial \omega}\big|_\varepsilon}. \tag{9.25}$$

According to the quasiparticle approximation, additional factors z enter the scattering integral with each initial and final state. The Landau–Silin equation (9.24) can explain the reduced relaxation rate found in numerical treatments of Green's functions (Danielewicz, 1984*a*; Köhler, 1995*a*).

9.1.5   Missing satellites

As we have seen in Chapter 8.3, the full spectral function (9.13) possesses a rich structure with satellites besides the pronounced quasiparticle peak. The missing satellites show that the quasiparticle approximation (9.17) works only in a narrow energy region close to the Fermi level. As shown by Köhler and Malfliet (1993), the quasiparticle approximation used to evaluate σ leads to incorrect total binding energy, i.e. it would lead to incorrect quasiparticle energies. Why this failure of the quasiparticle approximation appears one can understand from two different time scales connected with the dynamics of the system. These two time scales can be seen on the propagator $G^R(t, 0)$ shown in Fig. 9.1.

The main features of $G^R(t, 0)$ can be identified by the spectral function shown in Fig. 8.2 to which it is linked by the inverse Fourier transformation ($t > 0$)

$$G^R(t, 0; k) = \int \frac{d\omega}{2\pi} e^{i\omega t} a(\omega, k). \tag{9.26}$$

In the very short time domain, the whole spectrum contributes and the value of the propagator at $t \to 0$ has to equal to the unity in agreement with the sum rule (8.11).

The backward extrapolation of the long time part to $t = 0$ suggests a wave-function renormalisation $z = 0.75$ while the approximate formula provides the value of 1.03, see

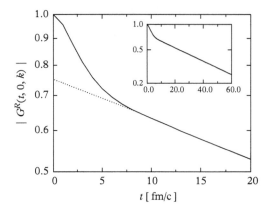

Figure 9.1: *Decay of the absolute value of the retarded propagator $|G^R|$ is plotted to reveal two time scales. On the long time scale presented in the insertion as log-plot, one can see that the propagator decays exponentially. The small deviation from the exponential decay on short time scale of 5 fm/c is the formation of quasiparticles. The momentum $k = 1.3$ fm/c and the temperature $T = 10$ MeV are identical to the spectral function shown in Fig. 8.2.*

Fig. 9.5. This results from the neglect of $\frac{\partial \gamma}{\partial \omega}$ that is extremely strong in this region as seen from the non-symmetric shape of the spectral function in Fig. 8.2.

The width of the peak in $a(\omega, k)$ determines the decay and the satellites determine the quasiparticle formation. After a certain time, called the quasiparticle formation time, the satellite contributions to the propagator vanish due to dephasing and the pole contribution dominates. This will be discussed within the transient time regime in Chapter 19. The kinetic equation operates on the long time scale, where the quasiparticle approximation is nearly exact. The internal dynamics of binary interactions hidden in the selfenergy happen on a short time scale where the quasiparticle approximation fails.

9.1.6   Causality and gradient corrections

We can conclude from Chapters 9.1.1–9.1.5 that neglecting the gradient term $[g, \sigma^<]$, we have lost the satellites of the spectral function, i.e. the off-shell contributions of the particle motion. This neglect has one more serious consequence, the resulting kinetic theory is not causal.

The full Kadanoff and Baym equation (7.27) is strictly causal as it follows from time-step functions hidden in retarded and advanced functions, e.g.

$$i\left(\Sigma^R_{\mathrm{reg}} G^< - G^< \Sigma^A_{\mathrm{reg}}\right)(t_1, t_3) = i\int_{-\infty}^{t_1} d\bar{t}\, \Sigma^R_{\mathrm{reg}}(t_1, \bar{t}) G^<(\bar{t}, t_3) - i\int_{-\infty}^{t_3} d\bar{t}\, G^<(t_1, \bar{t}) \Sigma^A_{\mathrm{reg}}(\bar{t}, t_3).$$

$$(9.27)$$

Note that time integrals extend exclusively to the past only if operators are in an algebraic order retarded–correlation–advanced, as they result from Langreth–Wilkins rules.

Rearrangements of the kinetic equation into commutators and anti-commutators disturb this strictly causal structure leading to integrals extended into the future, e.g.

$$\frac{1}{2}\left[\Gamma, G^<\right]_+(t_1, t_3) = \frac{i}{2}\int_{-\infty}^{t_1} d\bar{t}\,\Sigma_{\text{reg}}^R(t_1, \bar{t})\,G^<(\bar{t}, t_3) - \frac{i}{2}\int_{t_1}^{\infty} d\bar{t}\,\Sigma_{\text{reg}}^A(t_1, \bar{t})\,G^<(\bar{t}, t_3)$$

$$+ \frac{i}{2}\int_{t_3}^{\infty} d\bar{t}\,G^<(t_1, \bar{t})\,\Sigma_{\text{reg}}^R(\bar{t}, t_3) - \frac{i}{2}\int_{-\infty}^{t_3} d\bar{t}\,G^<(t_1, \bar{t})\,\Sigma_{\text{reg}}^A(\bar{t}, t_3). \tag{9.28}$$

The second and the third terms disturb the transparent causality.

The causality violated by the scattering-out (9.28) is restored if the kinetic equation also includes the associated commutator

$$i[\Sigma_{\text{reg}}, G^<]_-(t_1, t_3) = \frac{i}{2}\int_{-\infty}^{t_1} d\bar{t}\,\Sigma_{\text{reg}}^R(t_1, \bar{t})\,G^<(\bar{t}, t_3) + \frac{i}{2}\int_{t_1}^{\infty} d\bar{t}\,\Sigma_{\text{reg}}^A(t_1, \bar{t})\,G^<(\bar{t}, t_3)$$

$$- \frac{i}{2}\int_{t_3}^{\infty} d\bar{t}\,G^<(t_1, \bar{t})\,\Sigma_{\text{reg}}^R(\bar{t}, t_3) - \frac{i}{2}\int_{-\infty}^{t_3} d\bar{t}\,G^<(t_1, \bar{t})\,\Sigma_{\text{reg}}^A(\bar{t}, t_3). \tag{9.29}$$

In other words, the scattering-out integral does not violate the causality if it is complemented by the quasiparticle correction to the drift. From this point of view, the free-particle and the mean-field approximations violate the causality having only the scattering-out integral but no renormalisations.

The quasiparticle approximation, however, incorrectly treats the scattering-in resulting from

$$i\left(G^R\Sigma^< - \Sigma^<G^A\right)(t_1, t_3) = i\int_{-\infty}^{t_1} d\bar{t}\,G^R(t_1, \bar{t})\,\Sigma^<(\bar{t}, t_3) - i\int_{-\infty}^{t_3} d\bar{t}\,\Sigma^<(t_1, \bar{t})\,G^A(\bar{t}, t_3), \tag{9.30}$$

because only the anti-commutator part $\frac{1}{2}[A, \Sigma^<]_+$ is kept while the associated commutator $[G, \Sigma^<]_-$ is neglected. Accordingly, even the quasiparticle approximation does not comply with the requirement of causality.

To have a correct causal description, the commutator term $i[G, \Sigma^<]_-$ has to be kept, i.e. included into the drift part of (9.9) or its quasiclassical form (9.10). In this form, no commutator (gradient term) appears on the scattering side and no anti-commutator (non-gradient term) appears on the drift side. This rearrangement thus represents a correct separation of the drift and the scattering.

The physical meaning of the term $-i[G, \Sigma^<]_-$ can be seen from the decomposition of the propagator (for illustration we take a simple propagator in free space)

$$g^R(\omega, k) = \frac{1}{\omega - \frac{k^2}{2m_a} + i0} = \frac{\wp}{\omega - \frac{k^2}{2m}} - i\pi\delta\left(\omega - \frac{k^2}{2m}\right), \tag{9.31}$$

The principal-value term is G and the singular term is the spectral function a. The latter one describes the on-shell motion allowing the independent energy ω to contribute only for $\omega = \frac{k^2}{2m}$. The real part g describes the off-shell motion with $\omega \neq \frac{k^2}{2m}$. Accordingly, the alien term $-i[g, \sigma^<]_-$ describes the off-shell drift of particles.

9.1.7   Conclusion from off-shell contributions

The broken causality shows that the quasiparticle approximation is not the finite answer to the question of how to construct the kinetic equation. A correct treatment has to also handle the off-shell motion represented by the gradient term $[g, \sigma^<]$. Since the off-shell motion belongs to the short time scale, one cannot expect that it will directly enter the kinetic equation. In fact, Landau–Silin kinetic equation (9.24) as it stands is basically correct. Botermans and Malfliet (1990) and (Malfliet, 1992) have shown how the off-shell motion can be eliminated from the kinetic equation without the unjustified neglect of $[g, \sigma^<]$. This off-shell drift term compensates for the off-shell part of the standard term $[g_0^{-1} - \sigma, g^<]$, while the remaining on-shell parts lead to Landau–Silin kinetic equation (9.24). The same compensation has been found within semiconductor physics by Špička and Lipavský (1994; 1995). They show that the correct treatment of the off-shell motion is achieved using the very idea of the kinetic equation, a consistent separation of the hydrodynamical and the microscopic scales. On the hydrodynamical scale one has to find the kinetic equation that describes a balance between diffusion and scattering of quasiparticles, i.e. the evolution on the energy shell. On the microscopic scale, one has to describe the internal dynamics of interactions which have an off-shell character. Špička and Lipavský (1995) have shown that by this recollection the causality of observable quantities is guaranteed.

  The consistent separation of the hydrodynamic and microscopic scales allows us to regain the off-shell part and to use it in subsidiary relations, in particular to construct the Wigner distribution from the quasiparticle one, i.e. to construct the $\rho[f]$-functional (9.64). The resulting transport theory then has exactly the structure visualised by Landau–Silin (9.24). The time evolution of the system is governed by the kinetic equation for quasiparticle distribution, and observables are evaluated with the help of local functionals resulting from the $\rho[f]$-functional (9.64). The ansatz (9.62) is derived in this second way in a constructive manner.

9.2   Separation of on-shell and off-shell motion

The correlation function $G^<$ includes both on-shell and off-shell contributions. These two contributions can be separated with the help of a rearrangement that basically follows the idea of the Fermi golden rule.

  The GKB equation (7.25), $G^< = G^R \Sigma^< G^A$ describes a coherent and decaying propagation of the particle after its latest dissipative interaction given by $\Sigma^<$. To explain let us focus on the time-diagonal of the correlation function,

$$G^<(t, t) = \int_{-\infty}^{t} d\bar{t} \int_{-\infty}^{t} d\tilde{t}\, G^R(t, \bar{t}) \Sigma^<(\bar{t}, \tilde{t}) G^A(\tilde{t}, t). \tag{9.32}$$

If the latest interaction happened at a sufficiently remote time, $t - \bar{t} \gg \tau_{QF}$ and $t - \tilde{t} \gg \tau_{QF}$, where the quasiparticle-formation time τ_{QF} measures a formation of the asymptotic state, the particle has to be on the energy shell. If the latest interaction happened in the recent past, $t - \bar{t} \sim \tau_{QF}$ or $t - \tilde{t} \sim \tau_{QF}$, the on-shell or off-shell character of the particle motion cannot be decided.

Idea of the Fermi golden rule   Fermi proposed identifying the on-shell contributions by integrating through the collision into a future where only the on-shell contributions survive while the off-shell contributions die out by dephasing. The same idea can be directly implemented into the GKB equation. Let us extend the time integration over $\bar{t}$ in (9.32) into a future using the spectral function $A = i(G^R - G^A)$ instead of the retarded propagator,

$$\Lambda^<_{\text{left}}(t, t) = -i \int\limits_{-\infty}^{\infty} d\bar{t} \int\limits_{-\infty}^{t} d\tilde{t} A(t, \bar{t}) \Sigma^<(\bar{t}, \tilde{t}) G^A(\tilde{t}, t). \tag{9.33}$$

This extended integration wipes out the off-shell contributions so that $\Lambda^<_{\text{left}}$ is dominated by the on-shell ones.

The function $\Lambda^<_{\text{left}}$ is not symmetric with respect to its left and right arguments. One can introduce an adjoined whipping extension as

$$\Lambda^<_{\text{right}}(t, t) = i \int\limits_{-\infty}^{t} d\bar{t} \int\limits_{-\infty}^{\infty} d\tilde{t} G^R(t, \bar{t}) \Sigma^<(\bar{t}, \tilde{t}) A(\tilde{t}, t). \tag{9.34}$$

The symmetric form is obtained as their average,

$$\Lambda^< = \frac{1}{2} \Lambda^<_{\text{left}} + \frac{1}{2} \Lambda^<_{\text{right}}. \tag{9.35}$$

The function $\Lambda^<$ is the on-shell part of the correlation function $G^<$.

On-shell and off-shell parts of the correlation function   Let us assume that the correlation function is composed of the on-shell contribution $\Lambda^<$ and the off-shell part $\Xi^<$,

$$G^< = \Lambda^< + \Xi^<. \tag{9.36}$$

According to (9.35), the on-shell part reads

$$\Lambda^< = \frac{1}{2}(G^R - G^A) \Sigma^< G^A - \frac{1}{2} G^R \Sigma^< (G^R - G^A). \tag{9.37}$$

The off-shell part is the remainder of (9.32),

$$\Xi^< = \frac{1}{2} \left(G^R \Sigma^< G^R + G^A \Sigma^< G^A \right). \tag{9.38}$$

Similarly for $> \longleftrightarrow <$.

Definitions (9.37) and (9.38) are at the same time transport equations for $\Lambda^<$ and $\Xi^<$. This set of two equations is equivalent to the transport equation (7.25) for $G^<$ as one can directly check substituting (9.37) and (9.38) into (9.36). The transport equations (9.37) for $\Lambda^<$ and (9.38) for $\Xi^<$ are mutually coupled via the selfenergy that depends on $G^{>,<}$, i.e. on $\Lambda^{>,<} + \Xi^{>,<}$.

Although the equations (9.37) and (9.38) are equivalent to the GKB equation (7.25), they provide a very different starting point for the asymptotic theory. In any feasible asymptotic theory, one has to employ some approximation that is more or less equivalent to the Fermi golden rule. If the off-shell contributions are not separated in advance, they will be either lost (by neglect of the off-shell drift), or mixed up with the on-shell contributions (by integration over the energy). In the former case, one arrives at a theory which misses some contributions of the off-shell processes, in the latter case, the theory includes unphysical off-shell contributions.

On-shell and off-shell parts of the spectral function   Along with the separation of the on-shell and off-shell parts of the correlation function, one has to separate the on-shell and off-shell parts of the single-particle propagation. This can be done writing the spectral function as

$$A = S + B, \tag{9.39}$$

where

$$S = \frac{1}{2}(G^R - G^A)\Gamma G^A - \frac{1}{2}G^R\Gamma(G^R - G^A) \tag{9.40}$$

includes the on-shell contribution, and

$$B = \frac{1}{2}\left(G^R\Gamma G^R + G^A\Gamma G^A\right) \tag{9.41}$$

includes the off-shell part. Evidently, this separation parallels the separation of the correlation function. Again, substituting (9.40) and (9.41) into (9.39) one arrives at (8.40). From the spectral identities (8.39) one finds spectral identities for S and B,

$$S = \Lambda^> + \Lambda^<, \qquad B = \Xi^> + \Xi^<. \tag{9.42}$$

Accordingly, the functions S and B are spectral functions related to $\Lambda^{>,<}$ and $\Xi^{>,<}$, respectively. The separation of the on-shell and off-shell parts is accomplished.

9.3   Differential transport equation

As in the Kadanoff and Baym approach, the integral transport equation (9.37) for $\Lambda^<$ has to be converted into a differential form and the drift and the dissipation have to be separated. This procedure is a minor modification of the one used above for $G^<$.

As a preliminary step we use (8.38) to write (9.37) as

$$\Lambda^< = -\frac{i}{2}A\Sigma^< G^A + \frac{i}{2}G^R\Sigma^< A. \tag{9.43}$$

The standard procedure to derive the differential form is to multiply (9.43) with G_R^{-1} from the left-hand side,

$$G_R^{-1}\Lambda^< = -\frac{i}{2}\Gamma G^A\Sigma^< G^A + \frac{i}{2}\Sigma^< A, \tag{9.44}$$

by which one obtains a one-sided differential equation. We have used $G_R^{-1}A = \Gamma G^A$ which follows from (8.40). Similarly multiplying from the right-hand side, one finds the adjoined equation $[AG_A^{-1} = G^R\Gamma]$

$$\Lambda^< G_A^{-1} = -\frac{i}{2}A\Sigma^< + \frac{i}{2}G^R\Sigma^< G^R\Gamma. \tag{9.45}$$

The symmetric differential form of the transport equation is obtained subtracting the two one-sided forms,

$$-i(G_R^{-1}\Lambda^< - \Lambda^< G_A^{-1}) = \frac{1}{2}(\Sigma^< A + A\Sigma^<) - \frac{1}{2}\left(\Gamma G^A\Sigma^< G^A - G^R\Sigma^< G^R\Gamma\right). \tag{9.46}$$

Equation (9.46) parallels the differential Kadanoff and Baym equation (7.27). In fact, the crude interpretation of these two equations would hardly make any distinction. On the other hand, one can see that the left-hand side is free of the off-shell contribution being proportional to $\Lambda^<$. The first term on the right-hand side includes the off-shell part being proportional to the spectral function A. The off-shell part of A is compensated by the second term on the right-hand side.

Drift and dissipation Finally, we split the inverse Green's function into free parts and selfenergies, $G_{R,A}^{-1} = G_0^{-1} - \Sigma^{R,A}$, and decompose all retarded and advanced functions into real and imaginary parts,

$$-i[G_0^{-1} - \Sigma, \Lambda^<]_- + \frac{i}{4}[\Gamma, A\Sigma^< G + G\Sigma^< A]_-$$

$$= \frac{1}{2}\left([A, \Sigma^<]_+ - [\Gamma, G\Sigma^< G - \frac{1}{4}A\Sigma^< A]_+\right) - \frac{1}{2}[\Gamma, \Lambda^<]_+. \tag{9.47}$$

Commutators leading to gradient drift terms are collected on the left-hand side and anti-commutators leading to non-gradient scattering terms on the right-hand side.

Equation (9.47) does not include any approximation (except for those given by a choice of the model and the approximation of the selfenergy). Its structure is as close as possible to the kinetic equation, further progress requires approximations.

9.3.1   Quasiclassical limit

Backed by the previously discussed numerical justification of the quasiclassical limit, we apply this limit to the equation for the on- and off-shell parts of the correlation function $G^<$, and to the associated spectral functions.

On-shell quasiclassical transport   Using (9.5) and (9.6) one readily obtains the quasiclassical limit of the transport equation (9.47)

$$[g_0^{-1} - \sigma, \lambda^<] - \frac{1}{2}[\gamma, ag\sigma^<] = \sigma^< s - \gamma \lambda^<. \tag{9.48}$$

We have used that in the quasiclassical limit the symmetrical ternary products $C = ABA$ are free of gradients,

$$c = a^2 b, \tag{9.49}$$

as follows from (9.5) and (9.6).

Equation (9.48) parallels the Kadanoff and Baym equation (9.10). Again, a crude interpretation of both equations is identical. The left-hand side describes the drift of quasiparticles, and the free-particle energy is renormalised by the real part of the selfenergy . The right-hand side describes the dissipation with the first term corresponding to the scattering-in and the second one to the scattering-out. The differences appear in the off-shell parts. While (9.48) is free of the off-shell contributions, the Kadanoff and Baym equation (9.10) includes them both in the drift and the dissipative terms. Note that the complicated first term on the right-hand side of (9.47) has collapsed into $\sigma^< s$. This is because it is free of linear gradients and $a - g^2 + \frac{1}{4}a^2 = s$. The spectral function s shows that scattering-in processes have to end up on the energy shell. The off-shell processes are separated in $\Xi^<$.

Explicit Pauli blocking   The transport equation (9.48) does not show the full Pauli blocking of final scattering states. The explicit Pauli blocking is easily introduced using the spectral identities (8.39) and (9.42),

$$[g_0^{-1} - \sigma, \lambda^<] - \frac{1}{2}[\gamma, ag\sigma^<] = \sigma^< \lambda^> - \sigma^> \lambda^<. \tag{9.50}$$

This equation is a quantum precursor of the kinetic equation.

Off-shell contributions   The off-shell propagation is limited to a small space and time region in the vicinity of the interaction process. Accordingly, the off-shell transport equation (9.38) can be advantageously treated in the integral form. Using (9.49) one finds that the integral equation for the off-shell part $\Xi^<$ in the Wigner representation $\xi^<$ (7.31) is free of explicit gradients

$$\xi^< = \frac{1}{2}(g_R^2 + g_A^2)\sigma^< = \left(g^2 - \frac{1}{4}a^2\right)\sigma^<. \tag{9.51}$$

Spectral functions   As in the Kadanoff and Baym approach, the set of equations is closed by the associated spectral functions. Using (9.49), the off-shell spectral function B from (9.41) is free of gradients. With the help of the known spectral function a given by (9.13) the off-shell spectral function results

$$b = \left(g^2 - \frac{1}{4}a^2\right)\gamma = \frac{\left(\omega - \frac{k^2}{2m_a} - \sigma\right)^2 \gamma - \frac{1}{4}\gamma^3}{\left[\left(\omega - \frac{k^2}{2m_a} - \sigma\right)^2 + \frac{1}{4}\gamma^2\right]^2}. \tag{9.52}$$

Since the spectral function a is also free of gradients and $s = a - b$, see (9.39), one finds

$$s = \frac{1}{2}\gamma a^2 = \frac{\frac{1}{2}\gamma^3}{\left[\left(\omega - \frac{k^2}{2m_a} - \sigma\right)^2 + \frac{1}{4}\gamma^2\right]^2}. \tag{9.53}$$

The quasiclassical limit is accomplished.

Note that the explicit energy dependencies of spectral functions s and b confirm the interpretation of the separated parts as the on-shell and the off-shell parts, respectively. The on-shell part sharply peaks at the quasiparticle energy and falls in the off-shell region as the square of the Lorentzian, $s \to \frac{1}{2}\gamma^3/\omega^4$. The off-shell part changes sign in the vicinity of the quasiparticle energy giving no contribution to the integral over this region, and in the off-shell region falls slowly as the Lorentzian, $b \to \gamma/\omega^2$.

9.4   Extended quasiparticle picture

Now we approach the last step, elimination of the energy. Due to the complex character of the correlations and the finite lifetime of the quasiparticles, the elimination of the independent energy can only be done in an approximate way. Starting from Kadanoff and Baym (1962), such an approximation is traditionally called the ansatz, although in some cases this elimination is based on a well-defined approximation. The approximation we use here is the limit of small scattering rates, $\gamma \to 0$. In this limit one keeps only terms linear in γ.

To perform the limit of small scattering rates we write $\lambda^<$ as an equilibrium-like product

$$\lambda_a^<(\omega, k, r, t) = s_a(\omega, k, r, t)\tilde{f}_a(\omega, k, r, t). \tag{9.54}$$

In equilibrium, the function $\tilde{f}$ equals the Fermi–Dirac distribution, $\tilde{f}_a(\omega, k, r, t) = f_{FD}(\omega)$. Relation (9.54) does not, however restrict the validity of the theory to systems near equilibrium as it simply defines $\tilde{f}$.

$\gamma \to 0$ *limit of the quasiparticle spectral function*   Off the pole, the quasiparticle spectral function falls as γ^3/ω^4, i.e. in the limit of small scattering rates it behaves as a δ function,

$$s = 2\pi\delta\left(\omega - \frac{k^2}{2m} - \sigma\right) = 2\pi z \delta\left(\omega - \varepsilon\right),$$ (9.55)

where ε is the quasiparticle energy given by

$$\varepsilon_a(k, r, t) = \frac{k^2}{2m_a} + \sigma_a(\omega, k, r, t)|_{\omega=\varepsilon_a(k,r,t)},$$ (9.56)

and z is the wave-function renormalisation factor

$$z_a(k, r, t) = 1 + \left.\frac{\partial\sigma_a(\omega, k, r, t)}{\partial\omega}\right|_{\omega=\varepsilon_a(k,r,t)}.$$ (9.57)

This confirms that, in the limit of small scattering rates, the spectral function s depends exclusively on the on-shell contributions. Formula (9.57) is by itself the $\gamma \to 0$ limit of more general (9.25).

$\gamma \to 0$ *limit of the correlation function*   The δ function character of the quasiparticle spectral function allows one to eliminate the independent energy. Within the approximation of small scattering rates one can easily define the quasiparticle distribution as the on-shell part of $\tilde{f}$,

$$f_a(k, r, t) = \tilde{f}_a(\omega, k, r, t)|_{\omega=\varepsilon_a(k,r,t)},$$ (9.58)

and according to (9.54) the function $\lambda^<$ is

$$\lambda^< = f 2\pi z \delta(\omega - \varepsilon).$$ (9.59)

The off-shell part $\xi^<$ does not depend explicitly on the 'distribution' $\tilde{f}$, but it is explicitly proportional to $\sigma^<$ which is of the order of γ. In the limit of small scattering rates, one can thus neglect γ in $g^{R,A}$ so that

$$\xi^< = \sigma^< \frac{\wp'}{\omega - \varepsilon},$$ (9.60)

where $\wp'$ denotes the derived principal value,

$$\frac{\wp'}{\omega - \varepsilon} = \mathrm{Re}\frac{1}{(\omega - \varepsilon + i0)^2}.$$ (9.61)

For small scattering rates the correlation functions $g^{>,<}$ thus read

$$g^< = f 2\pi z \delta(\omega - \varepsilon) + \sigma^< \frac{\wp'}{\omega - \varepsilon},$$

$$g^> = (1-f) 2\pi z \delta(\omega - \varepsilon) + \sigma^> \frac{\wp'}{\omega - \varepsilon} \tag{9.62}$$

which is the extended quasiparticle picture. Formula (9.62) can be compared with the quasiparticle approximation. The first term is exactly the same as the quasiparticle approximation (9.17). The second term extends the quasiparticle approximation by the off-shell contributions. The correlation function given by (9.62) does not include any explicit gradient term.

The ansatz (9.62) has been derived for the first time in (Špička and Lipavský, 1994, 1995). The spectral identity $a = g^> + g^<$ prooves that this ansatz is consistent with the extended quasiparticle picture (8.23). It is wort remarking that (9.62) fulfils the spectral sum rule (8.11) and (8.12) (Kremp, Kraeft and Lambert, 1984). The latter one is violated in the simple quasiparticle picture.

As far as we know, the limit of small scattering rates was first introduced by Craig (1966*b*) for highly degenerated Fermi liquids and later used in (Stolz and Zimmermann, 1979; Kremp, Kraeft and Lambert, 1984) for equilibrium nonideal plasmas. The same approximation, but under the name of the generalised Beth–Uhlenbeck approach, has been used by Schmidt and Röpke (1987) in nuclear matter for studies of the correlated density. Köhler and Malfliet (1993) have used this approximation for the study of the mean-removal energy and high-momenta tails of the Wigner distribution. They call this the extended quasiparticle approximation. The nonequilibrium form was introduced by Špička and Lipavský (1994) as a modified Kadanoff and Baym ansatz and later on rederived from the limit of small scattering rates and thus called the small Γ expansion in (Špička and Lipavský, 1995). In a study of the kinetic equation for nonideal gases, Bornath et al. (1996) came to a similar approximation (differing in gradient terms), but they do not use any specific name for it. From this list of names we prefer to use the original name, the *limit of small scattering rates*, for the approach itself, while the later name, the *extended quasiparticle approximation*, is used for the resulting approximation of the correlation functions (9.62).

Wigner distribution Unlike the quasiparticle approximation, the limit of small scattering rates provides a link between the Wigner distribution and the quasiparticle distribution. The Wigner distribution is simply the energy integral (7.32) over $g^<$ given by (9.62),

$$\rho = fz + \int \frac{d\omega}{2\pi} \sigma^< \frac{\wp'}{\omega - \varepsilon} = fz + \int \frac{d\omega}{2\pi} \frac{\wp}{\omega - \varepsilon} \frac{\partial \sigma^<}{\partial \omega} \tag{9.63}$$

$$= f + \int \frac{d\omega}{2\pi} \frac{\wp}{\omega - \varepsilon} \frac{\partial}{\partial \omega} \left(\sigma^< (1-f) - \sigma^> f \right). \tag{9.64}$$

Form (9.63) provides the Wigner distribution $\rho(k)$ as the sum of the quasiparticle occupation $f(k)$ reduced by the wave-function renormalisation $z(k)$ and the off-shell contribution to the state k from other quasiparticle of momentum $p \neq k$. Form (9.64) provides the Wigner distribution as the sum of the zeroth and the first order in small γ. This form of $\rho[f]$-functional allows for easier control over consistency of further eventual approximations.

Formula (9.63) directly follows from (9.62). Its rearrangement (9.64) deserves a justification. The wave-function renormalisation is linked with the off-shell contribution by the sum rule $\int \frac{d\omega}{2\pi} a = 1$. From the Kramers–Kronig relation (9.65)

$$\sigma_{\text{reg}}(\omega') = -\int \frac{d\omega}{2\pi} \frac{\wp}{\omega - \omega'} \gamma(\omega) \tag{9.65}$$

and the spectral identity $\gamma = \sigma^> + \sigma^<$ follows that the wave-function renormalisation in the limit of small scattering rates (9.57) can be written as

$$z = 1 - \int \frac{d\omega}{2\pi} \frac{\wp}{\omega - \varepsilon} \frac{\partial(\sigma^> + \sigma^<)}{\partial \omega}. \tag{9.66}$$

The second term of the wave-function renormalisation can be joined with the off-shell contribution in (9.63) giving (9.64).

The $\rho[f]$-functional can be also written in terms of the derived principal value,

$$\rho = f + \int \frac{d\omega}{2\pi} \left(\sigma^<(1-f) - \sigma^> f\right) \frac{\wp'}{\omega - \varepsilon}, \tag{9.67}$$

that directly follows from the integration by parts in (9.64). This form we will use most often and will be called extended quasiparticle approximation.

9.4.1 Precursor of kinetic equation

Now we employ the limit of small scattering rates to derive the kinetic equation. Similarly to the Kadanoff and Baym approach we have to separate the spectral function (here s) into a multiplicative factor of the entire transport equation. This separation should be done before the limit of small scattering rates is applied.

To this end we substitute (9.54) into (9.50) which yields

$$s[g_0^{-1} - \sigma, \tilde{f}] - \tilde{f}[g_0^{-1} - \sigma, s] + \frac{1}{2}[\gamma, ag\sigma^<] = s\sigma^< - s\gamma\tilde{f}. \tag{9.68}$$

The second and the third drift terms mutually compensate in the limit of the small scattering rate. Indeed, from (9.53) follows

$$[g_0^{-1} - \sigma, s] = \frac{1}{2}[\gamma, ag\gamma] \tag{9.69}$$

with the help of which the second and the third terms of (9.68) can be rearranged as

$$\frac{1}{2}[\gamma, ag\sigma^<] - \tilde{f}[g_0^{-1} - \sigma, s] = \frac{1}{2}ag\gamma[\gamma,\tilde{f}] + \frac{1}{2}[\gamma, ag(\sigma^< - \tilde{f}\gamma)]. \tag{9.70}$$

As pointed out by Botermans and Malfliet, the term $\sigma^< - \tilde{f}\gamma$ is effectively linear in gradients which can be easily verified from the transport equation (9.68). Accordingly, the second term on the right-hand side of (9.70) vanishes in the quasiclassical limit being quadratic in gradients. Using (9.53) again to rearrange

$$\frac{1}{2}ag\gamma[\gamma,\tilde{f}] = s(g_0^{-1} - \sigma)[\ln\gamma,\tilde{f}], \tag{9.71}$$

one can see the remaining term vanishes in the limit of small scattering rates because the singularity of s is just at $g_0^{-1} - \sigma = 0$ and $[\ln\gamma, .] = \frac{1}{\gamma}[\gamma, .]$ is regular. The transport equation for $\tilde{f}$ thus reads

$$s[g_0^{-1} - \sigma, \tilde{f}] = s\sigma^< - s\gamma\tilde{f}. \tag{9.72}$$

In the limit of small scattering rates the spectral function s goes to the δ function (9.55), therefore only on-shell values of all functions in the transport equation (9.72) contribute. In particular, the auxiliary distribution $\tilde{f}$ contributes only by its on-shell value, where it equals the quasiparticle distribution (9.58). Expressing the drift term with the help of the quasiparticle energy (9.56), $\varepsilon = \frac{k^2}{2m_a} + \sigma_\varepsilon$, the Landau–Silin equation (9.24) appears

$$\frac{\partial f}{\partial t} + \frac{\partial\varepsilon}{\partial k}\frac{\partial f}{\partial r} - \frac{\partial\varepsilon}{\partial r}\frac{\partial f}{\partial k} = z(\sigma^<(1-f) - \sigma^> f)_{\omega=\varepsilon}. \tag{9.73}$$

The limit of small scattering rates thus leads to the Landau–Silin–Boltzmann kinetic equation. However, it is connected with an explicit $\rho[f]$- functional (9.64).

Note that quasiparticle energy ε, which controls the drift in kinetic equation (9.73), is defined from single-particle spectral properties of the system, see (9.56). This energy differs from Landau's definition based on the variation of the total density by the so called rearrangement energy which we will discuss in section 15.7. In the liquid ^{3}He, the rearrangement energy is rather small (Glyde and Hernadi, 1983). In nuclear matter, the rearrangement energy is sufficiently large to be observed. The study of Danielewicz (2000) shows that the dynamics of quasiparticles in nuclear matter are better described by the spectral quasiparticle energy (Brieva and Rook, 1977) than by the variational one (Friedmann and Pandharipande, 1981).

9.5 Numerical examples for equilibrium

The separation of the off-shell motion derived in this section ends up with the same kinetic equation as if we had just neglected it. One might feel that all this algebra is thus merely for the satisfaction of theoretical rigour having no real impact on practical implementations. In fact, this is not the case. The separation of the off-shell motion provides us with a link between the kinetic equation and the Green's function treatment which allows us to understand the numerical results and, in particular, helps us to identify which parameters should not be omitted in phenomenological approaches.

Wigner distribution   As an example let us consider heavy-ion collisions in the low-energy regime which are well described within the Vlasov equation, i.e. within the kinetic equation free of the scattering integral. As long as the mean-field interaction is fitted by momentum-dependent potentials, the Vlasov equation is formally identical to Landau's collision-less kinetic equation (9.18). With respect to the numerical demands of the implementations of these two theories there cannot be any difference as in both cases one solves the same equation. In equilibrium (in particular in the ground state), this equation is solved by Fermi–Dirac distribution shown in Fig. 9.2 as the dashed line.

The difference between the Vlasov and the Landau theories emerges when one interprets the resulting distribution function. Within the Vlasov theory, the resulting function is believed to be the desired Wigner distribution. Within the Landau theory, the resulting function is interpreted as the quasiparticle distribution, and the Wigner distribution is yet to be evaluated from the $\rho[f]$-functional (9.67).

Accordingly, within the kinetic-equation approach to nuclear matter, nucleons have to be viewed as quasiparticles in the spirit of Landau's theory of Fermi liquids. We note that the quasiparticle interpretation is also supported by the renormalisation of nucleonic mass which near the Fermi level reduced to about 70% of the bare mass. The mass

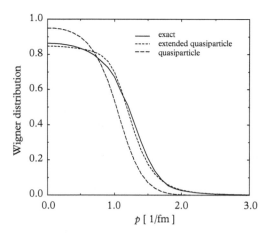

Figure 9.2: *The Wigner distribution ρ (full line) is compared to the quasiparticle distribution f (long dashed line). The extended quasiparticle approximation (short dashed line) almost mach the Wigner distribution. A two-channel Yamaguchi potential of Appendix E.1 is used to calculate the ladder approximation specified in Chapter 10.6. The temperature $T = 10\ MeV$ is safely above the critical temperature $T_c = 5\ MeV$ where the Cooper pairing appears.*

renormalisation is not explicitly visible from Fig. 9.2, but it can be extracted from the slope of the quasiparticle distribution.

In Fig. 9.2, the Wigner distribution as evaluated from (9.67) is compared with its exact value found within the equilibrium Green's functions. One can see that on the scale of the difference between the Wigner distribution and the Fermi–Dirac distribution shown in Fig. 9.2, formula (9.67) provides inevitable and sufficiently precise off-shell corrections.

The quasiparticle distribution exponentially decays at high momenta while the Wigner distribution has a depleted low-momenta region and power-law tails at high momenta. These two differences reflect the off-shell motion of particles between individual binary interactions. The differences between the momentum dependence of the Wigner and the quasiparticle distribution do not matter if one is concerned with the time evolution of local densities, as it is often the case in the simulations of collisions. Indeed, both functions are normalised to the same density. One must however be cautious when the distributions obtained from Vlasov-type simulations are compared with the results of alternative approaches.

In terms of off-shell contributions one can also understand the high-momenta tails found by Danielewicz (1984*a*) and by Köhler 1995*a*; 1996; 1995*b* in direct numerical treatments of the nonequilibrium Green's functions. Whether the system is in equilibrium or not, in regions of sufficiently high momenta, the power-law off-shell tails always dominate over the exponentially falling quasiparticle distribution.

Renormalisation of mass A complementary manifestation of the off-shell motion is the quasiparticle character of the drift. Let us assume that we already know $\varepsilon(k)$, say from its phenomenological fit to experimental data. Is there any practical difference between calling ε the quasiparticle energy instead of the standard mean-field energy? For implementation in the numerical codes of course not. On the other hand being aware of its quasiparticle nature we can estimate the region of validity of this fit.

The difference between the quasiparticle and the mean-field interpretation of the single-particle energy can be seen e.g. in the renormalisation of mass, $\frac{k_F}{m^*} \approx \frac{\partial \varepsilon}{\partial k}\big|_{k_F}$. Let us evaluate the quasiparticle velocity from (9.19),

$$\frac{\partial \varepsilon}{\partial k} = \frac{k}{m} + \frac{\partial \sigma}{\partial k} + \frac{\partial \sigma}{\partial \omega} \frac{\partial \varepsilon}{\partial k} = \frac{\frac{k}{m} + \frac{\partial \sigma}{\partial k}}{1 - \frac{\partial \sigma}{\partial \omega}}. \tag{9.74}$$

Within the mean-field interpretation the denominator equals unity and all renormalisation goes on at the cost of the momentum dependence of the mean-field selfenergy. For most of the mean-field interactions there is no particular reason for their strong temperature dependence, therefore one would expect that a $\varepsilon(k)$ fitted to reproduce the ground state and low-lying excitations can be safely used for such highly excited systems as two colliding nuclei.

Within the quasiparticle interpretation one finds quite a different picture. At low temperatures, the major part of the energy dependence of σ in the vicinity of the Fermi level follows from the deep dip in the imaginary part γ shown in Fig. 9.3. Indeed,

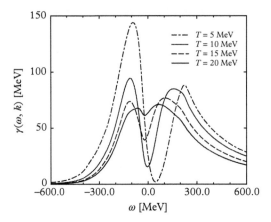

Figure 9.3: *Imagine part of the selfenergy, $\gamma = -2\,Im\sigma^R(\omega, k)$, as a function of temperature for the zero momentum, $k = 0$. The energy dependence shows that the Pauli blocking of collisions strongly affects the values close to the Fermi energy. At low temperatures, there is a dip in agreement with Luttinger's theorem. This dip closes as the temperature increases.*

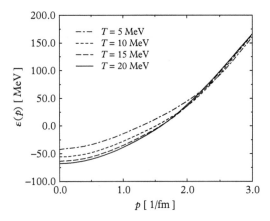

Figure 9.4: *Temperature dependency of the quasiparticle energy. The temperature-dependent dip in the imaginary part of the selfenergy seen in Fig. 9.3 leads, via the Kramers–Kronig relation, to the temperature dependency of the real part of selfenergy close to the Fermi energy which finally results in the temperature dependency of the quasiparticle energy.*

from the Kramers–Kronig relation, one can see that any dip in the imaginary part of an analytic function has to be accompanied by a strong energy dependency in the real part. Alternatively, one can seek for the link between the imaginary part and the energy dependence of the real part from the derived Kramers–Kronig relation (9.15). At higher temperatures, the dip closes up and the effect of the wave-function renormalisation on the effective mass m^* is reduced. The corresponding temperature dependency of the band structure and the wave-function renormalisation are in Figs. 9.4 and 9.5. More details about the temperature and density dependence of the selfenergy are in (Alm, Röpke, Schnell, Kwong and Köhler, 1996).

The temperature dependency of the effective mass is not an accidental property but rather a general feature of the mass renormalisation in Fermi liquids. According to Luttinger's theorem, at low temperatures the dip in the vicinity of the Fermi energy appears with no regards to a particular form of the binary interaction since the Pauli exclusion principle blocks finite states of eventual collisions. At higher temperatures the finite

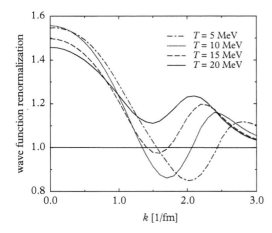

Figure 9.5: *Wave-function renormalisation. For the same reason as the quasiparticle energy, see Fig. 9.4, the wave-function renormalisation depends on the temperature.*

states become accessible and the dip closes up. Briefly, the temperature dependency of the effective mass has to appear in any system except for the non-interacting one.

9.6 Direct gradient expansion of non-Markovian equation

Though we have already discussed various on- and off-shell parts in the Kadanoff–Baym equation and derived the precursor of a kinetic equation by observing compensations of the off-shells, it is instructive to see how the same equation appears if we start with the other line of treatment found in the literature. There all correlation beyond the mean field are collected in the (non-Markovian) collision integral. On the time diagonal, $t_{1,2} = t$, the Kadanoff and Baym equation (7.28) yields an identity

$$-i\big(G_0^{-1}G^< - G^< G_0^{-1}\big)(t, x_1, t, x_2) = \int\limits_{-\infty}^{t} dt' \int dx'$$

$$\times \Big\{ G^>(t, x_1, t', x')\Sigma^<(t', x', t, x_2) + \Sigma^<(t, x_1, t', x')G^>(t', x', t, x_2)$$

$$- G^<(t, x_1, t', x')\Sigma^>(t', x', t, x_2) - \Sigma^>(t, x_1, t', x')G^<(t', x', t, x_2) \Big\}. \tag{9.75}$$

On the left-hand side there is the Hartree–Fock drift and the right-hand side contains a non-Markovian collision integral. Note that correlations beyond the Hartree–Fock field are exclusively in the collision integral.

Now we concentrate on the gradient expansion of the collision integral with the tools of Chapter 9.1.1. The quasiclassical approximation is achieved using the Wigner mixed representation (7.31) and with Fourier transform (9.4), the identity (9.75) reads

$$\left(\frac{\partial}{\partial t} + \frac{k}{m}\frac{\partial}{\partial r}\right)\rho(k,r,t) + 2\sin\left(\frac{1}{2}(\partial_r^1\partial_k^2 - \partial_k^1\partial_r^2)\right)\sigma^{\mathrm{HF}}(k,r,t)\rho(k,r,t)$$

$$= e^{\frac{i}{2}(\partial_r^1\partial_k^2 - \partial_k^1\partial_r^2)} \int_0^\infty d\tau \left(\hat{g}^> \left(k,r,t-\frac{\tau}{2},\tau\right)\hat{\sigma}^< \left(k,r,t-\frac{\tau}{2},-\tau\right)\right.$$

$$+ \hat{\sigma}^< \left(k,r,t-\frac{\tau}{2},\tau\right)\hat{g}^> \left(k,r,t-\frac{\tau}{2},-\tau\right) - \hat{g}^< \left(k,r,t-\frac{\tau}{2},\tau\right)\hat{\sigma}^> \left(k,r,t-\frac{\tau}{2},-\tau\right)$$

$$\left. - \hat{\sigma}^> \left(k,r,t-\frac{\tau}{2},\tau\right)\hat{g}^< \left(k,r,t-\frac{\tau}{2},-\tau\right)\right). \tag{9.76}$$

Here, the superscripts 1,2 denote that the partial derivatives apply only to the first and second function in the product, e.g. $(\partial_r^1\partial_k^2)^3 ab = \frac{\partial^3 a}{\partial^3 r}\frac{\partial^3 b}{\partial^3 k}$.

The power expansion of the goniometric functions in (9.76) defines the expansion in space gradients. This expansion goes to infinite order, but in the following we will restrict our attention to the linear expansion. It is customary to abbreviate the linear gradient terms with the help of the Poisson brackets,

$$2\sin\left(\frac{1}{2}(\partial_r^1\partial_k^2 - \partial_k^1\partial_r^2)\right)ab \approx \{a,b\} = \frac{\partial a}{\partial k}\frac{\partial b}{\partial r} - \frac{\partial a}{\partial r}\frac{\partial b}{\partial k}. \tag{9.77}$$

The linear approximation in space gradients is straightforward, however, one has to be careful about time arguments of functions in the collision integral of (9.76). For example, the first and second product seem to form an anti-commutator in which the linear gradients in space cancel. This is not true because of time arguments.

The gradient approximation in time requires a special treatment because of the lower integration limit. Due to this limit the integral does not define a standard matrix product with respect to the time variables. We will give this expansion an explicit notation.

In equilibrium, the functions $\hat{g}^\gtrless$ and $\hat{\sigma}^\gtrless$ do not depend on the centre-of-mass time. For slowly evolving systems the centre-of-mass dependence is smooth and weak. Accordingly, in the collision integral of (9.76), we can expand the centre-of-mass time dependence around the time t in powers of τ. Since all functions in (9.76) have the same centre-of-mass time, $t-\frac{\tau}{2}$, it is possible to write the linear expansion of (9.76) with respect to time gradients in a compact form

$$\frac{\partial}{\partial t}\rho + \{\epsilon^{\mathrm{HF}}, \rho\} = I + \frac{\partial}{\partial t}R, \tag{9.78}$$

with $\epsilon^{\mathrm{HF}} = \frac{k^2}{2m} + \sigma^{\mathrm{HF}}$. The zero-order gradient term, $I = I^> - I^<$ reads

$$I^< = \left(1 + \frac{i}{2}(\partial_r^1\partial_k^2 - \partial_k^1\partial_r^2)\right)\int_0^\infty d\tau \left(\hat{\sigma}^>(t,\tau)\hat{g}^<(t,-\tau) + \hat{g}^<(t,\tau)\hat{\sigma}^>(t,-\tau)\right). \tag{9.79}$$

$I^>$ is obtained from $I^<$ via the interchange of particles and holes, $> \leftrightarrow <$. In the first-order gradient term, $R = R^> - R^<$ is given by

$$R^< = -\frac{1}{2}\left(1 + \frac{i}{2}\left(\partial_r^1 \partial_k^2 - \partial_k^1 \partial_r^2\right)\right) \int_0^\infty d\tau\, \tau\, \left(\hat{\sigma}^>(t,\tau)\hat{g}^<(t,-\tau) + \hat{g}^<(t,\tau)\hat{\sigma}^>(t,-\tau)\right). \quad (9.80)$$

Now we are ready to turn all functions into the Wigner representation (7.31) for time and energy,

$$\sigma_\omega(k,r,t) = \int d\tau\, e^{i\omega\tau}\, \hat{\sigma}(k,r,t,\tau). \quad (9.81)$$

We will write the energy argument as the subscript and keep other arguments implicit in most of the formulae.

After a substitution of σ's and g's of the Wigner representation (9.81) into (9.79), one can integrate out the difference time τ. The integration over time results in the δ function which describes processes on the energy shell, and the principle value terms which represent off-shell processes,

$$\int_0^\infty d\tau\, e^{i(\omega-\omega')\tau} = \pi\delta(\omega - \omega') + i\frac{\wp}{\omega - \omega'}. \quad (9.82)$$

The δ-function can be readily integrated out leaving both functions with identical energy arguments. The principle value part, $\wp$, is an odd function of energies ω and ω', therefore its non-gradient contributions cancel but the linear gradients survive. Formula (9.79) thus turns into

$$I^< = \int \frac{d\omega}{2\pi}\sigma_\omega^> g_\omega^< + \int \frac{d\omega d\omega'}{(2\pi)^2}\frac{\wp}{\omega' - \omega}\left\{\sigma_{\omega'}^>, g_\omega^<\right\}. \quad (9.83)$$

In the correlated part we express the time integral as

$$\int_0^\infty d\tau\, \tau\, e^{i(\omega-\omega')\tau} = -i\frac{\partial}{\partial\omega}\int_0^\infty d\tau\, e^{i(\omega-\omega')\tau} = -i\pi\delta'(\omega - \omega') + \frac{\wp'}{\omega - \omega'}. \quad (9.84)$$

The meaning of δ' and $\wp'$ is seen from comparison with (9.82). The δ' and $\wp'$ are odd and even functions in $\omega - \omega'$, respectively. The gradient expansion of (9.80) thus reads

$$R^< = -\int \frac{d\omega d\omega'}{(2\pi)^2}\frac{\wp'}{\omega' - \omega}\sigma_\omega^> g_\omega^< + \int \frac{d\omega}{2\pi}\left\{\sigma_\omega^>, \partial_\omega g_\omega^<\right\}. \quad (9.85)$$

The term R contributes to the kinetic equation (9.78) as a gradient correction linear in time. Since the second term of $R^<$ in (9.85) is proportional to space gradients, we neglect this term leading to the second-order contribution in gradients.

9.6.1   Connection between the Wigner and quasiparticle distributions

At very low temperatures, dissipative processes are known to vanish due to the Pauli exclusion principle. It is desirable to reorganise the kinetic equation so that the scattering integral will vanish in this limit too. This can be seen from (9.85), the term $\partial_t R$ remains finite even for very low temperatures. It is possible to formally remove $\partial_t R$ if we shift this term on to the drift side of (9.78) and introduce a new distribution function,

$$f = \rho - R. \tag{9.86}$$

We have denoted the new function as f to anticipate that it is indeed the quasiparticle distribution as will be shown now.

Since we want to arrive at the kinetic equation for f, it is advantageous to write relation (9.86) in the opposite way so that we obtain ρ as a functional of f, $\rho = f + R$. Using (9.85) and the particle–hole conjugated term, $R^>$, one finds

$$\rho = f - \int \frac{d\omega' d\omega}{(2\pi)^2} \frac{\wp'}{\omega' - \omega} \left(\sigma_{\omega'}^> g_\omega^< - \sigma_{\omega'}^< g_\omega^> \right)$$

$$= f + \int \frac{d\omega}{2\pi} \frac{\partial \sigma_\omega}{\partial \omega} g_\omega^< + \int \frac{d\omega' d\omega}{(2\pi)^2} \frac{\wp'}{\omega' - \omega} \sigma_{\omega'}^< a_\omega$$

$$= \left(1 + \left. \frac{\partial \sigma}{\partial \omega} \right|_{\omega=\varepsilon} \right) f + \int \frac{d\omega'}{2\pi} \frac{\wp'}{\omega' - \varepsilon} \sigma_{\omega'}^< \tag{9.87}$$

where σ is the real part of the selfenergy (9.65) and we use in the last line the quasiparticle approximation, $a_\omega = 2\pi \delta(\omega - \varepsilon)$ and $g_\omega^< = f \, 2\pi \delta(\omega - \varepsilon)$ in accordance with the first-order γ expansion. This relation is just the extended quasiparticle approximation of the Wigner distribution (9.64).

9.6.2   Landau–Silin equation

So far we have treated only the time gradients finding that the identity (9.78) can be expressed as

$$\frac{\partial f}{\partial t} + \{\epsilon^{HF}, \rho\} = I. \tag{9.88}$$

Now we rearrange this identity into the Landau–Silin equation for f.

The time-local remainder I of the collision integral still includes the space gradients, $I = I_B - I_\nabla$. The non-gradient part, the first term of (9.83), is the Boltzmann-type scattering integral of the Landau–Silin equation

$$I_B = \int \frac{d\omega}{2\pi} (g_\omega^> \sigma_\omega^< - g_\omega^< \sigma_\omega^>) = z\left((1-f)\sigma_\varepsilon^< - f\sigma_\varepsilon^>\right) \tag{9.89}$$

In the second step we have used (9.62), the off-shell contributions to (9.89) have not been neglected but cancel exactly.

The gradient part which is the second term of (9.83) reads

$$I_\nabla = -\int \frac{d\omega d\omega'}{(2\pi)^2} \frac{\wp}{\omega' - \omega} \left(\{\sigma_{\omega'}^>, g_\omega^<\} - \{\sigma_{\omega'}^<, g_\omega^>\}\right)$$

$$= \int \frac{d\omega}{(2\pi)} \left(\{\sigma_\omega, g_\omega^<\} - \{\sigma_\omega^<, g_\omega\}\right). \tag{9.90}$$

Here (8.27) has been used to remove the ω'-integration for $\sigma^> + \sigma^<$ and a similar Kramers–Kronig relation for the ω-integration over the spectral function $g^> + g^<$ has been applied. The function $g = \text{Reg}^{R,A} = \frac{1}{2}(g^R + g^A)$ is the off-shell part of the propagators.

The mean-field drift $\{\epsilon^{\text{HF}}, \rho\}$ in (9.88) is not compatible with the quasiparticle distribution in the time derivative and the scattering integral. Moreover, the space gradients from the collision integral, I_∇, have to be accounted for. Our aim is now to show that all gradients can be collected into the quasiparticle drift,

$$\{\epsilon^{\text{HF}}, \rho\} + I_\nabla = \{\varepsilon, f\}. \tag{9.91}$$

The quasiparticle energy differs from the mean-field energy by the pole value of the selfenergy, $\varepsilon = \epsilon^{\text{HF}} + \sigma_\varepsilon$.

We will proceed in two steps. Firstly we observe that the right-hand side of (9.91) includes only the on-shell contribution, therefore we show that the on-shell part of the left-hand side equals the right-hand side. Secondly, we show that the off-shell part of the left-hand side of (9.91) vanishes. The on-shell parts of (9.62) used in (9.90) result in

$$I_\nabla^{\text{on}} = \int d\omega \{\sigma_\omega, zf\delta(\omega - \varepsilon)\} = \{\sigma_\varepsilon, zf\} - \sigma_\varepsilon'\{\varepsilon, zf\} + zf\{\sigma_\varepsilon', \varepsilon\}$$

$$= \{\sigma_\varepsilon, zf\} + (1 - z)\{\varepsilon, zf\} + zf\{z - 1, \varepsilon\}. \tag{9.92}$$

Now we add the on-shell part of the commutator $\{\epsilon^{\text{HF}}, \rho\}^{\text{on}} = \{\epsilon^{\text{HF}}, zf\}$. Using a rearrangement

$$\{\epsilon^{\text{HF}}, zf\} = \{\varepsilon - \sigma, zf\} = \{\varepsilon, f\} + (z - 1)\{\varepsilon, f\} + f\{\varepsilon, z\} - \{\sigma, zf\} \tag{9.93}$$

we obtain

$$\{\epsilon^{\mathrm{HF}}, \rho\}^{\mathrm{on}} + I_\nabla^{\mathrm{on}} = \{\varepsilon, f\} - \{\varepsilon, (1-z)^2 f\}. \tag{9.94}$$

The last term is of a higher order than linear in the damping and can be neglected which confirms the relation (9.91) already from the on-shell parts.

It remains for us to prove that all off-shell contributions to the left-hand side of (9.91), $\mathcal{O} = \{\epsilon^{\mathrm{HF}}, \rho\}^{\mathrm{off}} + I_\nabla^{\mathrm{off}}$ cancel, i.e. $\mathcal{O} = 0$. From the off-shell term of (9.62) we find

$$\mathcal{O} = \int \frac{d\omega}{2\pi} \left(\left\{ \epsilon^{\mathrm{HF}} + \sigma_\omega, \frac{\wp'}{\omega - \varepsilon} \sigma_\omega^< \right\} - \{g_\omega, \sigma_\omega^<\} \right). \tag{9.95}$$

The term with ϵ^{HF} results from $\{\epsilon^{\mathrm{HF}}, \rho\}$, while the others are from I_∇. In the last term we use the extended quasiparticle approximation of the real part of the propagator

$$g_\omega = z \frac{\wp}{\omega - \varepsilon} + \frac{\wp'}{\omega - \varepsilon} (\sigma_\omega - \sigma_\varepsilon) \tag{9.96}$$

which directly results from (9.62) and the Kramers–Kronig relation. Using $\left\{ \frac{\wp}{\omega-\varepsilon}, \sigma_\omega^< \right\} = -\left\{ \varepsilon, \frac{\wp}{\omega-\varepsilon} \sigma_\omega^< \right\}$ we can rewrite $\mathcal{O}$ as

$$\mathcal{O} = \int \frac{d\omega}{2\pi} \left(\left\{ \sigma_\omega - \sigma_\varepsilon, \frac{\wp'}{\omega - \varepsilon} \sigma_\omega^< \right\} - \left\{ \frac{\wp'}{\omega - \varepsilon} (\sigma_\omega - \sigma_\varepsilon), \sigma_\omega^< \right\} \right). \tag{9.97}$$

The linear expansion in the vicinity of the pole, $\sigma_\omega - \sigma_\varepsilon = (\omega - \varepsilon)(z - 1) + o(\gamma^2)$, yields

$$\mathcal{O} = \int \frac{d\omega}{2\pi} \left\{ \frac{\wp''}{\omega - \varepsilon}, (z-1)\sigma_\omega^< \right\}. \tag{9.98}$$

The product $(z-1)\sigma^<$ is of second order in γ, i.e. the off-shell contribution $\mathcal{O}$ is negligible within the assumed accuracy. This completes the proof of relation (9.91).

Summary   Considering the causality and large-scale compensations of various off-shell parts leads to the concept of quasiparticles represented by the relation (9.87) between the Wigner (9.64) or the correlation functions (9.62) and the quasiparticle distributions. The space gradients of the collision integral renormalise the mean-field drift into the familiar quasiparticle drift (9.91). The resulting kinetic equation for the quasiparticle distribution is the Landau–Silin equation (9.24). We conclude that the Levinson type of equation (9.76) is equivalent to the Landau–Silin equation (9.24) up to the second order in the damping γ or in the extended quasiparticle picture.

Using different approximations for the selfenergy we obtain all known kinds of kinetic equations with the generalisation that the internal gradients of collision integrals will yield the nonlocal virial corrections. Using ladder summation results in the nonlocal BUU equation, using the random phase approximation leads to the nonlocal Lenard–Balescu equation, etc. As we see, this ansatz provides the correct connection between equations for the Wigner function and the quasiparticle Boltzmann equation (9.24). It has to be remarked that the theory presented here is valid up to first-order gradient corrections. This implies that we will find viral corrections from intrinsic gradients in the scattering integrals (Špička, Lipavský and Morawetz, 1998).

9.7   Alternative approaches to the kinetic equation

The key problem in deriving the kinetic equation from the Kadanoff and Baym equation is to eliminate the independent energy ω. This is the crucial approximation that determines the consistency and the validity of the theory. Beside the approach so far presented, based on the dominant contribution of a singular part of the spectral function, one finds in the literature a number of alternative ways. They differ in details which can be neglected as long as one aims to recover the plain Boltzmann equation. For corrections beyond the Boltzmann equation, however, these details matter. Let us only mention two.

9.7.1   First quasiclassical approximation

For the theory of metals, a convincing way to reduce the Kadanoff and Baym equation to the kinetic equation, has been introduced by Prange and Kadanoff (1964) for the electron-phonon interaction and extended by Prange and Sachs (1967) to the electron-electron and the electron-phonon interactions. Unlike in the above approaches, the independent energy is kept as a parameter of the distribution function while the kinetic energy $\frac{k^2}{2m}$ is integrated out leaving only the dependency on the direction of momentum. This approach, also called the first quasiclassical approximation, allows one to treat systems with strongly energy-dependent selfenergies provided its momentum dependence is weak on the scale of Fermi momentum. This is indeed the case for the electron-phonon interaction since the phonon velocity is far lower than the Fermi velocity. The first quasiclassical approximation turned out to be particularly useful in the theory of nonequilibrium superconductors, where the strong energy dependence reflects the superconducting gap while the momentum dependence remains smooth. This is shown by Prange and Kadanoff resulting in Landau's equation as it is received from the quasiparticle approximation. This shows that Landau's equation is more general than the quasiparticle approximation by which it was derived above.

Let us compare physical pictures behind the quasiparticle approximation and the first quasiclassical approximation. As we have seen above, in deriving the quasiparticle approximation one keeps the momentum fixed and tries to collect and reasonably approximate corrections to the energy, whether it be the mean-field or quasiparticle dispersion relation or the wave-function renormalisation factor. This approach can be described in terms of wave mechanics in a following way. The particle is viewed as an infinite plane wave of a sharp-wave number and one tries to approximate with what frequency it oscillates. Since the wave is damped, this frequency is not sharp as one can see from the width of the pole of the spectral function in Fig. 8.1.

The first quasiclassical approximation approaches the problem from the other side. The energy is fixed and all the corrections and/or renormalisation go into momentum. Within the analogy with the wave mechanics, the particle corresponds to a wave radiated from a point-like source (weak momentum dependency of the selfenergy corresponds to very local scattering events) with a fixed frequency, therefore the interaction with the medium changes the wave length and the damping determines its decay. When the particle reaches a next scattering process, it is characterised by a sharp energy but a broad spectrum of momenta. As long as all the interactions are local, i.e. weakly dependent of momentum, the wide spectrum of momenta does not matter, for the only crucial parameter is the energy.

Which of the two alternative pictures is more relevant depends on the properties of the system. In metals the first quasiclassical approximation is clearly favourable. At low temperatures, any small correction to the energy might appear to be large on the scale of the temperature. Moreover, the selfenergy usually has a sharp energy dependence reflecting thresholds, e.g. of emission of optical phonons. In these cases the quasiparticle approximation may fail.

In the nuclear matter, however, the assumption of a weak momentum dependency is not satisfied since the typical interaction range of 0.7 fm is nearly the same as the wave length of nucleon at the Fermi level at normal density. Moreover, from the first quasiclassical approximation the Landau theory results only in the limit of infinite particle density when the term $(g, \sigma^<)$ vanishes under the integration over kinetic energy. Highly excited nuclear matter in the heavy-ion collision does not fall into this limit. The first quasiclassical approximation thus cannot be easily applied.

The first quasiclassical approximation also has a serious drawback in being an incomplete theory. Due to the indirect way in which the quasiparticle distribution emerges, the first quasiclassical approximation does not provide any recipe for how to evaluate the Wigner distribution. This missing relation reflects the fact that the kinetic equation is free of the off-shell motion and as long as one is only looking for the structure of kinetic equation, the off-shell motion can be ignored. The problem with the off-shell motion emerges if one tries to evaluate observables or the quasiparticle energy.

9.7.2 Thermo field dynamics approach

The two-point correlation function has to be calculated in doubled Hilbert space which is a 2 X 2 matrix in abstract space (Landsman and van Weert, 1987). The reason for this is that in a statistical system the occupation probability of a particle state is not known a priori—and therefore causality of the physical particle propagation requires us to take into account two different temporal boundary conditions for each state (Henning and

Umezawa, 1994). The alternative approach of thermo-field dynamics (Umezawa, 1993; Henning, 1995) starts consequently from two sets of creation and annihilation operators and links these sets to a thermal state by a general Bogoliubov transformation. As a new feature, a free parameter enters the theory and it can be shown that for a special value of this parameter the time-ordered path formalism appears (Henning and Umezawa, 1994). Since the resulting kinetic equations are equivalent to the Kadanoff and Baym equations (Henning, Nakamura and Yamanaka, 1996), we restrict ourselves to GKB Green's function formalism which will lead to the same set of equations of motions.

10

Approximations for the Selfenergy

The simplest and physically most intuitive way to find the appropriate selfenergy is to return to the formal enclosure of the Martin–Schwinger hierarchy in Figure 7.1. Any approximate expression for the two-particle Green's function leads to the corresponding selfenergy. In this way we have the possibility to derive all standard approximations as a clear physical picture of the processes one wants to consider. An alternative way would be to develop a perturbation series expansion directly from complex time paths as outlined in Appendix B. The most simplifying feature of the equation of motion approach is that we can omit that cumbersome steps.

10.1 Hartree–Fock

The simplest approximation for the two-particle Green's function one might think of is the motion of two particles without interaction as illustrated in Figure 10.1.

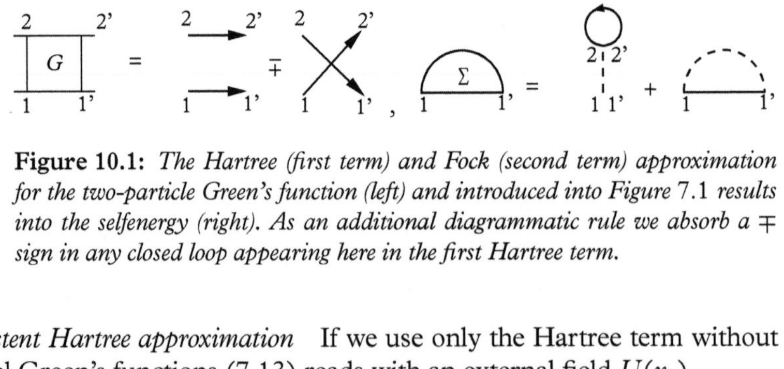

Figure 10.1: *The Hartree (first term) and Fock (second term) approximation for the two-particle Green's function (left) and introduced into Figure 7.1 results into the selfenergy (right). As an additional diagrammatic rule we absorb a $\mp$ sign in any closed loop appearing here in the first Hartree term.*

Selfconsistent Hartree approximation If we use only the Hartree term without exchange the causal Green's functions (7.13) reads with an external field $U(x_1)$

$$\left(i\frac{\partial}{\partial t_1} - \frac{\nabla_1^2}{2m} + U(x_1)\right) G^c(1,1') = \delta(1-1') \mp i \int d2\, V(x_1-x_2) G^c(1,1') G^c(2,2^+)\delta(t_1-t_1').$$

$$(10.1)$$

Interacting Systems far from Equilibrium. Klaus Morawetz, Oxford University Press (2018).
© Klaus Morawetz. DOI: 10.1093/oso/9780198797241.001.0001

Since $iG^c(2, 2^+) = \mp\Theta(t_2^+ - t_2)G^<(2, 2^+) = \mp n(2)$ yields just the density we can write

$$\left[i\frac{\partial}{\partial t_1} + \frac{\nabla_1^2}{2m} - U_{\text{eff}}(x_1, t_1) \right] G^c(1, 1') = \delta(1 - 1') \tag{10.2}$$

which means the effect of all other particles is to produce a mean field

$$U_{\text{eff}}(x_1, t_1) = \int dx_2 \, V(x_1 - x_2)n(x_2, t_1) + U(x_1). \tag{10.3}$$

The equation of motion (10.2) has the same form for the causal and retarded Green's function since the potential is time-diagonal though their solutions are different. It represents the simplest form of time-dependent density functional theory. The effective potential is given by the density and determines the Green's function which provides the density again and so on, such that one has a selfconsistent scheme.

For equilibrium we Fourier transform the time to get

$$\left[\omega + \frac{\nabla_1^2}{2m} - U_{\text{eff}}(x_1, \omega) \right] G^R(x_1, x_1', \omega) = \delta(x_1 - x_1') \tag{10.4}$$

which allows us to represent the retarded Green's function in bilinear expansion

$$G^R(x_1, x_1', \omega) = \sum_i \frac{\phi_i(x_1)\phi_i^*(x_1')}{\omega - E_i + i0} \tag{10.5}$$

where we use the solutions of the effective Schrödinger equation

$$\left[\frac{\nabla_1^2}{2m} + U_{\text{eff}}(x_1) \right] \phi_s(x_1) = E_i\phi_i(x_1). \tag{10.6}$$

Please note that the denominator (10.5) contains an infinitesimal imaginary part ensuring the causality as a retarded function (8.4).

The solution (10.5) provides the spectral function as weighted peaks at the eigen energies

$$A(x_1, x_1', \omega) = 2\pi \sum_i \phi_i(x_1)\phi_i^*(x_1')\delta(\omega - E_i) \tag{10.7}$$

and the density finally as

$$n(x_1) = \int \frac{d\omega}{2\pi} A(x_1, x_1, \omega)f(\omega) = \sum_i |\phi_i(x_1)|^2 f(E_i) \tag{10.8}$$

with the equilibrium Fermi/Bose distribution $f(\omega)$.[1]

---

[1] Actually, any function of the energy $p^2/2m + U_{\text{eff}}$ solves the stationary Hartree equation of motion. The form of Bose or Fermi distribution is approached if additional dissipation is considered such that a collision integral vanishes for these distributions.

Another intuitive form appears for homogeneous systems $n(x) = n$ in equilibrium. Then the effective field reads $U_{\text{eff}} = n \int dx V(x) = nv$,

$$\left(\omega - \frac{p^2}{2m} - nv \right) g^R(p, \omega) = 1 \tag{10.9}$$

which is solved as $g^R = 1/\left(\omega - \frac{p^2}{2m} - nv + i0 \right)$ providing the spectral function

$$a = 2\pi\delta \left(\omega - \frac{p^2}{2m} - nv \right). \tag{10.10}$$

We see that the homogeneous mean field is shifting the free pole of the spectral function at the kinetic energy of the particle by the mean interaction, called the mean field.

In nonequilibrium one has to solve the time-dependent kinetic equation which we will consider now.

Hartree-Fock kinetic equation Returning to Figure 10.1 we consider now both Hartree and exchange (Fock) terms. The selfenergy we obtain using the approximation for G_2 in Figure 7.1 and omitting one Green's function from the right,

$$\Sigma(1, 1') = \delta(t_1 - t_1') \left[\delta(x_1 - x_1') \int dx_2 V(x_1 - x_2)\rho(x_2, x_2, t_1) \right.$$

$$\left. \mp V(x_1 - x_1')\rho(x_1, x_1', t_1) \right] = \delta(t_1 - t_1')U_{\text{eff}}(x_1, x_1', t_1). \tag{10.11}$$

This shows that the Hartree–Fock potential is local in time but compared to the local-in-space Hartree potential the exchange is nonlocal in space. Consequently the reduced density matrix $\rho(t) = G^<(t, t)$ appears.

We might now use the Kadanoff and Baym equation of motions (7.27),

$$-i \left(G_0^{-1} G^< - G^< G_0^{-1} \right) = 0, \tag{10.12}$$

where we need

$$G_0^{-1} G^< = \left(i\partial_{t_1} + \frac{\nabla_1^2}{2m} \right) G^<(1, 2) = \int dx_3 U_{\text{eff}}(x_1, x_3, t_1) G^<(x_3, t_1, x_2, t_2)$$

$$G^< G_0^{-1} = \left(-i\partial_{t_2} + \frac{\nabla_2^2}{2m} \right) G^<(1, 2) = \int dx_3 G^<(x_1, t_1, x_3, t_2) U_{\text{eff}}(x_3, x_2, t_2). \tag{10.13}$$

The right-hand side is a commutator if we consider the time-diagonal part $t_1 = t_2 = t$. In fact, to determine the potential (10.11) one needs the diagonal part of the Green's function. In difference and centre-of-mass coordinates required for mixed representation has $\partial_{t_1} + \partial_{t_2} = \partial_t$ and $\nabla_1^2 - \nabla_2^2 = 2\partial_r \partial_R$ and we get for (10.12)

$$\left(\partial_t - i\frac{\partial_r \partial_R}{m}\right) \rho(r, R, t) = -i\int dx_3 \, [U_{\text{eff}}, \rho]_- . \qquad (10.14)$$

We transform in mixed representation using the gradient expansion (9.4) and see that the commutator in (10.14) becomes

$$\left(\partial_t + \frac{k}{m}\partial_R\right)\rho(k, R, t) = -2\sin\frac{i}{2}\left(\partial_R{}^\rho \partial_k{}^U - \partial_k{}^\rho \partial_R{}^U\right)U(k, R, t)\rho(k, R, t) \qquad (10.15)$$

which is the quantum-Vlasov equation. The quasiclassical limit consists in neglecting higher rather than quadratic orders of gradients which allows us to write for (10.15)

$$\frac{\partial}{\partial t}\rho(k, R, t) + \left[\frac{k}{m} + \frac{\partial}{\partial k}U(k, R, t)\right]\frac{\partial}{\partial r}\rho(k, R, t) - \frac{\partial}{\partial r}U(k, R, t)\frac{\partial}{\partial k}\rho(k, R, t) = 0.$$
$$(10.16)$$

Together with (10.11) we have a complicated selfconsistent kinetic equation for the Wigner function. Due to the time-diagonal character of the selfenergy, the spectral function is a sharply peaked function at the Hartree–Fock energy $k^2/2m + \sigma^{HF}$.

The classical character of (10.16) becomes visible if we consider the Wigner function as a superposition of flow elements $\rho = \sum_i \delta(k - p_i)\delta(R - R_i)$ with velocity R_i and momentum p_i. Then the substantial derivative, i.e. to sit on the flow element reads

$$\frac{d}{dt}\rho = \sum_i \left(\frac{\partial}{\partial t} + \dot{R}_i\frac{\partial}{\partial R} + \dot{p}_i\frac{\partial}{\partial k}\right)\delta(k - p_i)\delta(R - R_i) = 0 \qquad (10.17)$$

which is exactly the Vlasov equation (10.16) if the flow elements obey,

$$\dot{R}_i = \frac{p_i}{m} + \frac{\partial}{\partial p_i}U(R_i, p_i, t), \qquad \dot{p}_i = -\frac{\partial}{\partial R_i}U(R_i, p_i, t), \qquad (10.18)$$

just the classical Hamilton equations of motions for the momentum and the position of the flow element.

10.2 Random phase approximation

10.2.1 Long-range Coulomb interaction

For systems interacting with the Coulomb potential, the interaction is long-ranged with $\sim 1/r$. This means that the perturbation expansion with respect to interactions (see Figure 10.10) as the first Born approximations beyond Hartree–Fock of Figure 10.1, leads to divergent results, since the influence of all other participating particles to a two-particle collision cannot be neglected. A screening cloud is created surrounding a charge. One can see the physics simply from the Poisson equation for the potential ϕ. We assume that the screening cloud around a charge obeys a thermal canonical distribution with the background charge density n_0 and total charge neutrality. This leads to the charge density in the Poisson equation

$$\nabla\phi = \frac{e}{4\pi\epsilon_0}n_0\left(e^{-\frac{e\phi}{k_B T}} - 1\right) \approx -\kappa^2\phi, \qquad \kappa^2 = \frac{n_0 e^2}{4\pi\epsilon_0 k_B T}$$

$$\frac{1}{r}\frac{\partial}{\partial r}^2(r\phi) + \kappa^2\phi = 0 \tag{10.19}$$

where we assume $eV \ll k_B T$ and find the high-temperature definition of the inverse Debye screening length κ. In fact, solving the radial equation (10.19) one gets the screened Debye potential

$$V_D = e\phi = \frac{e^2}{4\pi\epsilon_0 r}e^{-\kappa r}. \tag{10.20}$$

The effective Coulomb potential becomes screened due to the surrounding charges and decays exponentially rather than with $1/r$. This screening renders the theoretical expansions finite. It represents an interaction effect which can be considered as selfenergy. Looking at the place of the charge we might calculate the selfenergy as the limit of the difference of effective Debye and bare Coulomb potential

$$\Sigma = \lim_{r\to 0}\frac{1}{2}(V_D - V_c) = \frac{e^2}{8\pi\epsilon_0}\lim_{r\to 0}\frac{e^{-\kappa r} - 1}{r} = -\frac{\kappa e^2}{8\pi\epsilon_0} \tag{10.21}$$

where we used 1/2 to avoid double counting. One sees that the long-ranged Coulomb interaction leads to screening and a negative selfenergy. The particles in the medium have here a lower energy than the free ones. This might eventually lead to pairing of particles as a prerequisite of superconductivity.

10.2.2   Density-density fluctuations

The screening we saw is an effect of the charge density fluctuations around a moving particle. Therefore we consider the density fluctuation in a nonequilibrium system

$$\delta\rho(11') = \Psi^+(1')\Psi(1) - < \Psi^+(1')\Psi(1) > \tag{10.22}$$

where, as usual, the numbers are cumulative variables $(r_1, t_1, s_1, i_1 ..)$ denoting the space, time, spin, isospin and other required variables. In order to describe these fluctuations in

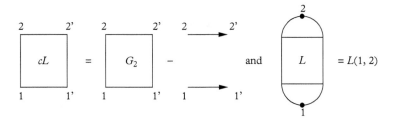

Figure 10.2: *Correlated part of two-particle Green's function (10.23).*

terms of Green's functions we introduce the correlated part of the two-particle Green's function including exchange by

$$cL(121'2') = G_2(121'2') - G^c(11')G^c(22')$$
(10.23)

illustrated by Figure 10.2. Here the factor c can be chosen as convenient. In the literature it ranges from $c = -i$ up to $c = 1$, which we will use in the variational approach in Chapter 11. In order to see how this factor appears or drops out in subsequent formulae, we will keep it here.

The density fluctuations (10.22) can be expressed with the help of two-particle Green's function as

$$\langle \delta\rho(11)\delta\rho(22)\rangle = \langle \Psi_2^+\Psi_2\Psi_1^+\Psi_1\rangle - G^<(11)G^<(22)$$

$$= i^2\Theta(t_1 - t_2)G_{121+2+} + G^c(11^+)G^c(22^+) = icL^>(12) = icL^<(21)$$

$$\langle \delta\rho(22)\delta\rho(11)\rangle = icL^<(12).$$
(10.24)

As seen in Figure 10.2, it can be expressed together as a causal function

$$L^c(12) = \frac{1}{i}\Theta(t_1 - t_2)L^>(12) + \frac{1}{i}\Theta(t_2 - t_1)L^<(12) = L_{121+2+}$$
(10.25)

where we obey the Bose character of fluctuations in the definition (7.3).

With the help of the second equation of the Martin–Schwinger hierarchy (7.8), which couples the two-particle Green's function to the three-particle one, we get the following equation of motion for the correlator L defined in (10.23):

$$\left(i\hbar\frac{\partial}{\partial t_1} - \frac{\nabla_1^2}{2m}\right)L(121'2') = \mp\frac{1}{c}\delta(1 - 2')G^c(21') \pm \frac{i}{c}G^c(11')G^c(22')\int d3V(13)G^c(33^+)$$

$$\pm iG^c(22')\int d3V(13)L(131'3^+) \mp \frac{i}{c}\int d3V(13)G_3(1231'2'3^+).$$
(10.26)

Any further approximation needs a specification on how we want to treat the three-particle Green's functions. Since we are interested in the two-particle density fluctuation we approximate the three-particle Green's function just in terms of all possible two-particle fluctuations (Schlanges and Bornath, 1987),

$$G_3(1231'2'3^+) = G^c(11')G^c(22')G^c(33^+)$$

$$cG^c(11')L(232'3^+) + cG^c(22')L(131'3^+) + cG^c(33^+)L(121'2') \qquad (10.27)$$

illustrated in Figure 10.3.

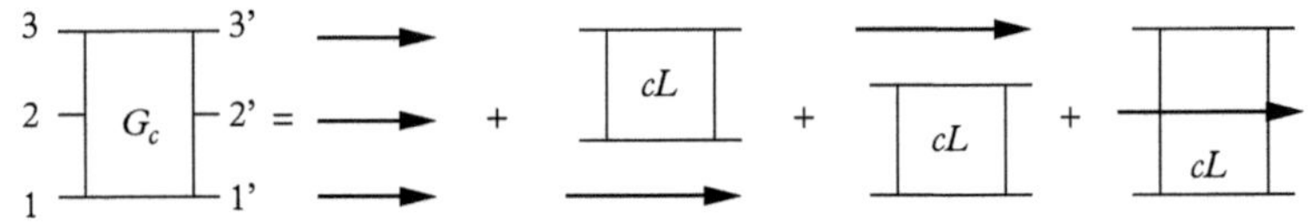

Figure 10.3: *Approximation of three-particle correlations* (10.27).

Using this approximation in (10.26) we obtain the random phase approximation (RPA) in nonequilibrium,

$$\left[i\hbar \frac{\partial}{\partial t_1} - \frac{\nabla_1^2}{2m} - \pm i \int d3 V(13) G^<(33) \right] L(121'2') = \mp \frac{1}{c} \delta(1-2') G^c(21')$$

$$\mp i G^c(11') \int d3 V(13) L(232'3^+). \qquad (10.28)$$

The terms in the bracket of the left-hand side represent just the inverse propagator in Hartree approximation (10.2). We approximate this inverse propagator now by the full one. There is a deep meaning in this neglect. We will return to the more precise treatment of this asymmetry in Chapter 11. The replacement by the full propagator allows us to multiply both sides with $G^c(4, 1)$ and using

$$\mp \frac{1}{c} G^c(12') G^c(21') = L_o(121'2'), \qquad (10.29)$$

we can write (10.28) as integral equation

$$L(421'2') = L_o(421'2') + icL_o(411'1^+)V(13)L(232'3^+) \qquad (10.30)$$

which is given in Figure 10.4.

Introducing (10.23) into the selfenergy (7.9) or graphically in Figure 7.1, we see that besides the Hartree term of Figure 10.1 the selfenergy takes the form presented in Figure 10.5 with the dynamical (screened) potential introduced to take the form of

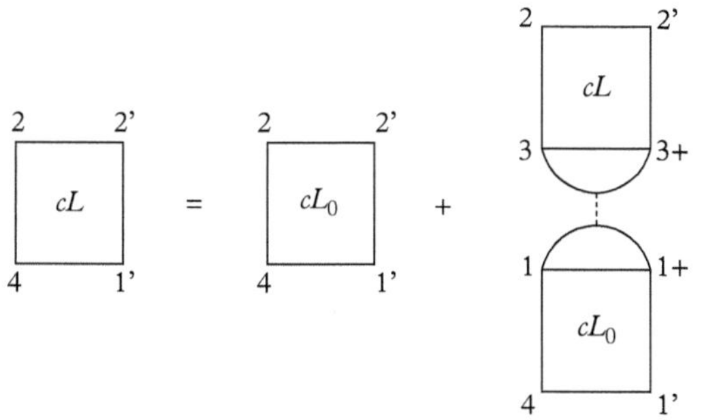

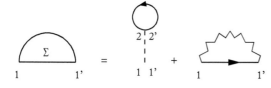

Figure 10.4: *Dynamical (screened) potential approximation.*

Figure 10.5: *Dynamical (screened) potential approximation.*

Fock selfenergy as given in Figure 10.6 where we used the RPA equation (10.30). One might write it explicitly as in Figure 10.7 and see that the effective potential W in the RPA selfenergy is the sum of an infinite order of interactions with particle–hole bubbles as the kernels. Therefore this approximation is known under different names like GW, rainbow, RPA summation. The selfenergy reads:

$$\Sigma^c(1,2) = iG^c(1,2) \cdot W^c(1,2) \tag{10.31}$$

with

$$W^c(1,2) = V(1,2) + ic \int d3d4 V(1,3)L^c(3,4)V(4,2)$$

$$= V(1,2) + ic \int d3d4 V(1,3)L_0^c(3,4)W^c(4,2). \tag{10.32}$$

We see that we need only the two-point function Eq. (10.25) as a special case in Equation (10.30)

$$L_{ab}^c(12) = L_{0aa}^c(12)\delta_{ab} + ic\sum_d \int d3d4 L_{0aa}^c(14)V^{ad}(43)L_{db}^c(32). \tag{10.33}$$

The different kinds of particles have been explicitly marked by Latin letters. Obviously the relation $L_0^{ab} \sim \delta^{ab}$ holds. In the next calculations we will drop this notation for simplicity and restore these notations in the final results.

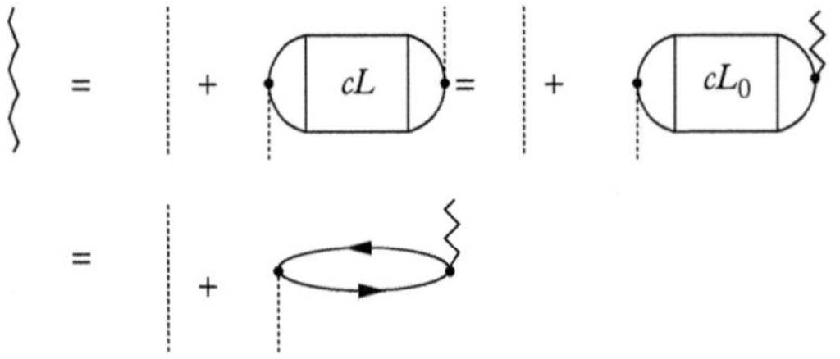

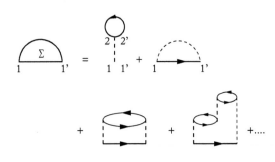

Figure 10.6: *The effective screened potential iW is denoted diagrammatically as a wavy line while the broken line means iV. A closed loop counts as a $\mp$ sign.*

Figure 10.7: *The Hartree–Fock selfenergy (first line) and RPA diagrams (second) line.*

Equation (10.33) is a causal one, which means that all functions entering are causal ones. Therefore we can apply the Langreth–Wilkins rules (7.18) and (7.17) to obtain in operator notation

$$L^{\gtrless} = L_0^{\gtrless} + icL_0^{R}VL^{\gtrless} + icL_0^{\gtrless}VL^{A}$$
$$L^{R/A} = L_0^{R/A} + icL_0^{R/A}VL^{R/A} \tag{10.34}$$

where integrations about inner variables are understood.

The second equation can be written as

$$(1 - icL_0^{R}V)L^{R} = L_0^{R}. \tag{10.35}$$

We could have applied the differential operators in all equations from the right instead from the left. This would have resulted in the conjugated of (10.30)

$$L^{R/A} = L_0^{R/A} + icL^{R/A}VL_0^{R/A} \tag{10.36}$$

which we can write as

$$L^{R} = (1 + icL^{R}V)L_0^{R}. \tag{10.37}$$

Comparing (10.35) and (10.37) we can introduced the operators of retarded and advanced dielectric functions

$$\epsilon^{R} = 1 - L_0^{R}V \equiv (1 + VL^{R})^{-1}$$
$$\epsilon^{A} = 1 - VL_0^{A} \equiv (1 + L^{A}V)^{-1}. \tag{10.38}$$

The first equation of (10.34) leads therefore to the form of an optical theorem

$$L^{\gtrless} = (\epsilon^R)^{-1} L_o^{\gtrless} (\epsilon^A)^{-1}. \tag{10.39}$$

Considering now the screened potential (10.32) we have exactly in the same manner

$$W^{\gtrless} = icVL^{\gtrless}V = icV(\epsilon^R)^{-1}L_0^{\gtrless}(\epsilon^A)^{-1}V = ic(W^R)^{-1}L_0^{\gtrless}(W^A)^{-1} \tag{10.40}$$

where we used (10.39) and introduced the dynamically screened potential as

$$\sum_c V_{ac}(\epsilon_{cb}^R)^{-1} = W_{ab}^R, \qquad \sum_c (\epsilon_{ac}^A)^{-1} V_{cb} = W_{ab}^A. \tag{10.41}$$

Now we can calculate the correlated parts of the selfenergy from (10.31) introducing the definition (7.3)

$$\frac{1}{i}[\Theta(t_1 - t_2)\Sigma^> \mp \Theta(t_2 - t_1)\Sigma^<]$$

$$= i\frac{1}{i}[\Theta(t_1 - t_2)G^> \mp \Theta(t_2 - t_1)G^<]\frac{1}{i}[\Theta(t_1 - t_2)W^> \mp \Theta(t_2 - t_1)W^<] \tag{10.42}$$

from which one reads off

$$\Sigma(12)^{\gtrless} = G(12)^{\gtrless} W(12)^{\gtrless}. \tag{10.43}$$

Exactly in the same way we have from (10.29)

$$L_0^{\gtrless}(12) = -\frac{i}{c} G(12)^{\gtrless} G(21)^{\lessgtr}. \tag{10.44}$$

With (10.44) and (10.40) we have expressed the selfenergy (10.43) completely in terms of the correlation functions $G^{\gtrless}$. The necessary ingredients are the retarded and advanced dielectric functions (10.38) which are given entirely by (10.44).

10.2.3   Lenard–Balescu collision integral

We will use now the Landau–Silin Equation (9.24) to give the collision integral in quasiparticle approximation. We will present the collision integral in quasiparticle approximation which reads in mixed representation

$$g^{\gtrless}(k, \omega, r, t) = 2\pi \delta(\omega - \epsilon(k, r, t)) f_k^{\gtrless}(r, t) \tag{10.45}$$

with with $f^< = f, f^> = 1 \mp f$ and neglecting higher-order gradients. This correspond to the neglect of nonlocal collision scenario which we will discuss in Chapter 13.

The selfenergy (10.43) translates after gradient expansion into

$$\sigma^{\gtrless}(k,\omega,r,t) = \int \frac{d\bar{\omega}}{2\pi} \int \frac{d\bar{p}}{(2\pi)^3} g^{\gtrless}(\bar{p},\omega,r,t) W^{\gtrless}(k-\bar{p},\omega-\bar{\omega}.r.t)$$

$$= \int \frac{d\bar{p}}{(2\pi)^3} W^{\gtrless}(k-\bar{p},\omega-\epsilon_{\bar{p}},r,t) f_{\bar{p}}^{\gtrless}(r,t). \tag{10.46}$$

The correlation function of the screened potential (10.40) becomes

$$W^{\gtrless}(k,\omega,r,t) = ic\frac{V(k)^2}{|\epsilon^R(k,\omega,r,t)|^2} L_0^{\gtrless}(k,\omega,r,t) \tag{10.47}$$

where the free-density fluctuation correlator (10.44) reads

$$L_0^{\gtrless}(k,\omega,r,t) = -\frac{i}{c} \int \frac{d\bar{\omega}}{2\pi} \int \frac{d\bar{p}}{(2\pi)^3} g^{\gtrless}(\bar{p},\bar{\omega},r,t) g^{\lessgtr}(\bar{p}-k,\bar{\omega}-\omega,r,t)$$

$$= -\frac{2\pi i}{c}\delta(\omega + \epsilon_{\bar{p}-k} - \epsilon_{\bar{p}}) \int \frac{d\bar{p}}{(2\pi)^3} f_{\bar{p}}^{\gtrless}(r,t) f_{\bar{p}-k}^{\lessgtr}(r,t). \tag{10.48}$$

Introducing (10.48) in (10.47) and using this in (10.46) the selfenergy becomes

$$\sigma^{\gtrless}(k,\omega) = \int \frac{d\bar{p}}{(2\pi)^3} \frac{V(k-\bar{p})^2}{|\epsilon^R(k-\bar{p},\omega-\epsilon_{\bar{p}},r,t)|^2} \int \frac{dp'}{(2\pi)^3} 2\pi\delta[\omega - \epsilon_{\bar{p}} - \epsilon_{p'} + \epsilon_{p'+\bar{p}-k}]$$

$$\times f_{p'}^{\gtrless}(r,t) f_{\bar{p}}^{\gtrless}(r,t) f_{p'+\bar{p}-k}^{\lessgtr}(r,t). \tag{10.49}$$

Using this in the Landau–Silin equation (9.24) one obtains finally the Lenard–Balescu collision integral

$$\left(\frac{\partial}{\partial t} + \frac{\partial\epsilon_k^a(r,t)}{\partial k}\frac{\partial}{\partial r} - \frac{\partial\epsilon_k^a(r,t)}{\partial r}\frac{\partial}{\partial k}\right) f_k^a(r,t) = \sum_b \int \frac{dpdq}{(2\pi\hbar)^6} 2\pi\delta[\epsilon_k^a + \epsilon_{k-q}^a - \epsilon_p^b - \epsilon_{p+q}^b]$$

$$\times \frac{V(q)^2}{|\epsilon^R(q,\epsilon_k^a - \epsilon_{k-q}^a,r,t)|^2} [f_{k-q}^a f_{p+q}^b(1 \mp f_k^a)(1 \mp f_p^b) - (1 \mp f_{k-q}^a)(1 \mp f_{p+q}^b) f_k^a f_p^b]. \tag{10.50}$$

We see that the bare potential is screened by the dielectric function ϵ^R of (10.38). Without this screening one would have just the Born approximation which one obtains if one takes only the first-order interaction diagram beyond the Hartree–Fock ones in Figure 10.7. The effect of the infinite sum is condensed in the dielectric function. One has in quasiparticle approximation, from (10.38)

$$\epsilon^R(k,\omega,r,t) = 1 - icV(k)L_0^R(k,\omega,r,t) \tag{10.51}$$

and the retarded polarisation function becomes

$$L_0^R(\tau, k, r, t) = -i\theta(\tau)[L_0^>(\tau, k, r, t) - L_0^<(\tau, k, r, t)]$$

$$L_0^R(\omega, k, r, t) = -\frac{i}{c} \int \frac{d\bar{p}}{(2\pi\hbar)^3} \frac{f_{\bar{p}-k}(r, t) - f_{\bar{p}}(r, t)}{\omega + \epsilon_{\bar{p}-k}(r, t) - \epsilon_{\bar{p}}(, r, t)}. \tag{10.52}$$

Fluctuation-dissipation theorem   We might write the Lenard–Balescu collision integral into another form to make the fluctuation properties explicit. For this aim we introduce the symmetrised density fluctuations (10.24) as

$$\frac{ic}{2}[L^>(1, 2) + L^<(1, 2)] = \frac{1}{2}(\langle \delta n_1 \delta n_2 \rangle + \langle \delta n_2 \delta n_1 \rangle) = \overline{\langle \delta n_1 \delta n_2 \rangle} = \frac{\epsilon_0}{V_{12}} \overline{\langle \delta E_1 \delta E_2 \rangle} \tag{10.53}$$

where we have replaced the charge density by the electric field E. Writing the sort indexes explicitly we have for the correlation function of the fluctuation

$$ic \sum_{de} V_{de} L_{de}^{\gtrless} = ic \sum_{de} \left(\frac{L^> + L^<}{2} \pm \frac{L^> - L^<}{2} \right) = \sum_{de} \left(\epsilon_0 \overline{\langle \delta E_d \delta E_e \rangle} \pm \mathrm{Im}(icL^R)_{de} \right)$$

$$= \epsilon \overline{\delta E \delta E} \pm \mathrm{Im}(\epsilon^R)^{-1} \tag{10.54}$$

due to (10.38). Using (10.46) in (10.40) as $W^{\gtrless}(k, \omega, r, t) = icV(k)^2 L^{\gtrless}(k, \omega, r, t)$ we can rewrite the Lenard–Balescu collision integral as

$$\sum_b \int \frac{d\bar{p}}{(2\pi)^3} V_{ab}(k - \bar{p}) \left\{ \epsilon_0 \overline{\langle \delta E \delta E \rangle}(k - \bar{p}, \epsilon_k^a - \epsilon_{\bar{p}}^b)(f_{\bar{p}}^b - f_k^a) \right.$$

$$\left. - \mathrm{Im} \ (\epsilon^R)^{-1}(k - \bar{p}, \epsilon_k^a - \epsilon_{\bar{p}}^b)[f_{\bar{p}}^b(1 \mp f_k^a) + f_k^a(1 \mp f_{\bar{p}}^b)] \right\} \tag{10.55}$$

which clearly shows a fluctuation part and a dissipative part by the dielectric function. In equilibrium the collision integral vanishes and one gets the relation setting $p = \bar{p} - k$ and $\omega = \epsilon_{\bar{p}}^b - \epsilon_k^a - \mu^b + \mu^a$,

$$\epsilon_0 \overline{\langle \delta E \delta E \rangle}(p, \omega) = \mathrm{Im}[\epsilon^R(p, \omega)]^{-1} \coth \frac{\beta\omega}{2} \tag{10.56}$$

equally for Bose or Fermi systems. This represents just the fluctuation-dissipation theorem. It shows the balance between fluctuations and dissipation in equilibrium. Further one sees that $\mathrm{Im}(\epsilon^R)^{-1}$ is the spectral function of the fluctuations.

10.3   Selfenergy and effective mass in quasi two-dimensional systems

We want to model electrons scattering with ions within the quasi two-dimensional gas. Assuming the motion restricted to the $x - y$ plane, the Coulomb potential in this cylindrical Fermi surface is

$$V_{ab}(q_x, q_y) = \frac{e_a e_b \hbar^2}{\epsilon_0} \int\limits_{-\infty}^{\infty} \frac{dq_z}{2\pi\hbar} \frac{1}{q^2 + q_z^2} = \frac{e_a e_b \hbar}{2\epsilon_0 q}. \tag{10.57}$$

10.3.1   Polarisation function in 2D

The equilibrium RPA polarisation function $\Pi^R = icL_0^R$ with (10.48) and (10.52) reads

$$\Pi^R(\omega, q) = \int \frac{dp}{(2\pi\hbar)^2} \frac{f_0\left(\frac{(p+\frac{q}{2})^2}{2m}\right) - f\left(\frac{(p-\frac{q}{2})^2}{2m}\right)}{\frac{p \cdot q}{m} - \omega - i0}. \tag{10.58}$$

The imaginary part is easily rewritten as

$$\text{Im}\Pi^R(\omega, q) = \pi \int\limits_0^\infty \frac{dpp}{(2\pi\hbar)^2} f_0\left(\frac{p^2}{2m}\right) \int\limits_0^{2\pi} d\phi \left\{ \delta\left(\frac{p \cdot q}{m} - \frac{q^2}{2m} - \omega\right) - \delta\left(\frac{p \cdot q}{m} + \frac{q^2}{2m} - \omega\right) \right\}$$

$$= \frac{m^{3/2}}{2^{3/2}\pi\hbar^2 q} \int\limits_0^\infty \frac{d\epsilon}{\sqrt{\epsilon}} \left\{ f_0\left[\epsilon + \left(\frac{q}{2} + \frac{m\omega\hbar}{q}\right)^2 /2m\right] - f_0\left[\epsilon + \left(\frac{q}{2} - \frac{m\omega\hbar}{q}\right)^2 /2m\right] \right\}. \tag{10.59}$$

The energy shifts in the distribution function we absorb into an effective chemical potential which should be positive in order to obtain nonzero contribution at Sommerfeld expansion

$$\mu_{\text{eff}} = \epsilon_f - \frac{\left(\pm\frac{q}{2} + \frac{m\omega\hbar}{q}\right)^2}{2m} \geq 0 \tag{10.60}$$

The low-temperature Sommerfeld expansion reads than with the Fermi–Dirac function $f_0(\epsilon) = f[(\epsilon - \mu_{\text{eff}})/T]$

$$\int\limits_0^\infty \frac{d\epsilon}{\sqrt{\epsilon}} f\left(\frac{\epsilon - \mu_{\text{eff}}}{T}\right) = 2 \int\limits_{\mu_{\text{eff}}/T}^\infty dx f(x) \partial_x \sqrt{Tx + \mu_{\text{eff}}} = 2 \int\limits_{\mu_{\text{eff}}/T}^\infty dx f(x) [1 - f(x)] \sqrt{Tx + \mu_{\text{eff}}}$$

$$= 2\sqrt{\mu_{\text{eff}}} \left(1 - \frac{\pi^2 T^2}{24\mu_{\text{eff}}}\right) = 2\left(1 - \frac{\pi^2 T^2}{12} \frac{\partial}{\partial\epsilon_f}\right) \sqrt{\mu_{\text{eff}}}. \tag{10.61}$$

Therefore it is enough to know the zero-temperature result since the T^2 correction are given simply by a derivative.

Using dimensionless coordinates as in (Hodges, Smith and Wilkins, 1971),

$$x = \frac{q}{2p_f}, \qquad x_0 = \frac{\hbar\omega}{4\epsilon_f}, \tag{10.62}$$

we get finally for the imaginary part of the polarisation function

$$\mathrm{Im}\Pi(\omega,q) = \frac{m}{4\pi\hbar^2 x}\left\{\Theta(x-|x_0+x^2|)\sqrt{1-\left(x+\frac{x_0}{x}\right)^2} - \Theta(x-|x_0-x^2|)\sqrt{1-\left(x-\frac{x_0}{x}\right)^2}\right\}. \tag{10.63}$$

The corresponding real part is given by the Hilbert transform according to (10.58)

$$\mathrm{Re}\Pi(\omega,q) = -2\int \frac{d\omega'}{2\pi}\frac{\mathrm{Im}\Pi(\omega',q)}{\omega-\omega'}. \tag{10.64}$$

Using the integral

$$\eta(a) = \int_{-1}^{1}\frac{\sqrt{1-z^2}}{a-z} = \pi\begin{cases} a, & 1 \geq |a| \\ a - \mathrm{sgn}(a)\sqrt{a^2-1}, & 1 < |a| \end{cases} \tag{10.65}$$

we obtain

$$\mathrm{Re}\Pi = -\frac{m}{4\pi^2\hbar^2 x}\left[\eta\left(\frac{x_0}{x}+x\right) - \eta\left(\frac{x_0}{x}-x\right)\right]. \tag{10.66}$$

The real and imaginary part is plotted in Figure 10.8.

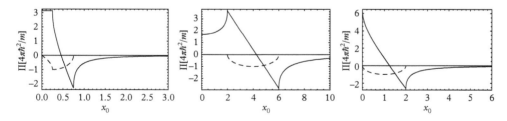

Figure 10.8: *The real (solid) and imaginary (dashed) part of the polarisation versus the dimensionless frequency (10.62) for dimensionless wavelength $x = 0.5$ (left), $x = 1$ (middle) and $x = 2$ (right) according to (10.62).*

10.3.2 Integrals over dielectric functions

Quite often integrals over the dielectric function occur which can be simplified tremendously using a trick given in (Klimontovich, 1975)

$$\int \frac{d\omega}{2\pi} \frac{H(\omega)}{\omega} \mathrm{Im}\,\epsilon^{R-1}(q,\omega) = \frac{H(0)}{2} \mathrm{Re}\left(1 - \frac{1}{\epsilon^R(q,0)}\right) \tag{10.67}$$

for any analytical function $H(\omega)$. To convince us about that relation we consider the following integral including the dielectric function

$$I = \int \frac{d\omega}{2\pi} \frac{H(\omega)}{\omega} \mathrm{Im}\,\epsilon^{R-1}(\omega) = \int \frac{d\omega}{4\pi i}\left(\frac{1}{\omega + i\eta} + \frac{1}{\omega - i\eta}\right) H(\omega)(f^- - f^+) \tag{10.68}$$

where $f^+ = 1 - 1/\epsilon^R$ and $f^- = (f^+)^*$. Assuming that $H(\omega)$ is analytical and since $f^\pm(\omega)$ has no poles in the lower/upper half plane, and vanishes with $\sim \omega^{-2}$ for large ω, we have the identity

$$\int \frac{d\omega}{2\pi i} H(\omega) \frac{f^\pm(\omega)}{(\omega \pm i\eta)} = \mp f^\pm(0) H(0) \tag{10.69}$$

and all other combinations of $f^\pm$ with the denominator vanish. With the help of the relation (10.69) we compute easily for (10.68) the relation (10.67).

10.3.3 Selfenergy and effective mass

For the scattering of electrons on charged ions or holes we consider the dynamically screened approximation (GW) given in Figure 10.7. Due to the big mass difference between electrons and ions the vertex corrections are suppressed. The experimental data we want to compare with are characterised by a relatively low density of electrons such that the electron-electron contributions can be neglected.

The retarded selfenergy of a species noted by the subscript a can be written analogously to the retarded Green's function (8.4) as

$$\Sigma_a^R(q,\omega) = \int \frac{d\omega'}{2\pi} \frac{\wp}{\omega - \omega'} \Gamma_a(q,\omega') - \frac{i}{2}\Gamma_a(q,\omega) \tag{10.70}$$

with the imaginary part of fermionic particles

$$\Gamma_a(q,\omega) = \Sigma_a^>(q,\omega) + \Sigma_a^<(q,\omega). \tag{10.71}$$

Introducing (10.47) into (10.46), the correlation parts of the selfenergy, $\Sigma^{\gtrless}$, in the dynamically screened approximation reads

$$\Sigma_a^{\gtrless}(k,\omega) = \sum_b \int \frac{dq}{(2\pi\hbar)^2} V_{ab}(q)^2 f_{a,k-q}^{\gtrless} \int d\omega' \delta(\epsilon_{k-q} - \hbar\omega - \hbar\omega') \frac{\Pi_b^{\gtrless}(q,\omega')}{|\epsilon^R(q,\omega')|^2}. \quad (10.72)$$

In equilibrium the distributions are Fermi–Dirac ones and we can use $[1 - f(a) - f(b)] = g(a+b)$ with the Bose function $g(x) = 1/(\exp(x/T) - 1)$ and we rewrite the free-density fluctuation or polarisation function

$$\Pi_b^{\gtrless}(q,\omega) = \pm 2 g_b(\pm\hbar\omega)\mathrm{Im}\Pi_b(q,\omega) \quad (10.73)$$

where the imaginary part is

$$\mathrm{Im}\Pi_b(q,\omega) = \pi \int \frac{dp}{(2\pi\hbar)^2} (f_{b,p} - f_{b,p+q})\delta(\epsilon_p - \epsilon_{p+q} - \hbar\omega). \quad (10.74)$$

A final simplification of the selfenergy can be achieved in the limit of heavy masses of the scattering impurities. We are allowed to neglect quantum fluctuations of these impurities which are expressed by $g(\pm\hbar\omega)$ in (10.73) and replace $g(\pm\hbar\omega) \to \pm T/\hbar\omega$. Since $\epsilon^R = 1 - \sum_n V_{bb}\Pi_b^R$ we can split $V_{ab}^2 = V_{aa}V_{bb}$ to realise the $\mathrm{Im}\epsilon^R$ part in (10.72) such that the selfenergy can be written

$$\Sigma_a^{\gtrless}(k,\omega) = 2T \int \frac{dq}{(2\pi\hbar)^2} V_{aa}(q) f_{a,k-q}^{\gtrless} \int \frac{d\omega'}{\omega'} \delta(\epsilon_{k-q} - \hbar\omega - \hbar\omega')\mathrm{Im}\frac{1}{\epsilon^R(q,\omega')}$$

$$= -2\pi T \int \frac{dq}{(2\pi\hbar)^2} V_{aa}(q)\mathrm{Re}\left(1 - \frac{1}{\epsilon^R(q,0)}\right) f_{a,k-q}^{\gtrless}\delta(\epsilon_{k-q} - \hbar\omega). \quad (10.75)$$

In the last step we used (10.67). The real part of the selfenergy (10.70) becomes

$$\Sigma_a(k,\omega) = T \int \frac{dq}{(2\pi\hbar)^2} V_{aa}(q)\mathrm{Re}\left(1 - \frac{1}{\epsilon^R(q,0)}\right) \frac{\wp}{\epsilon_{k-q} - \hbar\omega}. \quad (10.76)$$

In order to evaluate the effective mass according to (10.80), it remains to get an expression for the dielectric function. To this end we use the zero-temperature expansion (Hodges, Smith and Wilkins, 1971; Morawetz, 2003) of the dielectric function $\epsilon = 1 - V\Pi$ in quasi two-dimensions (10.66) since the leading temperature dependence is already in front of the integral. This means we consider the electrons as degenerated but the impurity particle–hole fluctuation as classical here. Changing the integration variables $z = q/2p_f$, we then obtain for the real part (10.76) of the dynamically screened selfenergy [$x = k/2p_f$, $x_0 = \hbar\omega/4\epsilon_f$]

$$
\Sigma_a(k,\omega) = -\frac{e_a^2 T m_a}{2\hbar p_f x}\int_0^\infty \frac{dz}{\frac{xz}{\kappa_a}+1}\int_0^{2\pi}\frac{d\phi}{\frac{x_0}{x^2}-1+2z\cos\phi-z^2} \approx -\frac{e_a^2 T m_a}{2\hbar p_f x}\int_0^\infty dz \int_0^{2\pi}\frac{d\phi}{\frac{x_0}{x^2}-1+2z\cos\phi-z^2}
$$

$$
= \frac{2\pi\, e_a^2 T m_a}{\hbar p_f}
\begin{cases}
\frac{2}{x+\sqrt{x_0}}\mathcal{K}\left[\left(\frac{\sqrt{x_0}-x}{\sqrt{x_0}+x}\right)^2\right] & 0 < x_0 < x^2 \\[2mm]
\frac{\pi}{4x} & x_0 = x^2 \\[2mm]
0 & x_0 > x^2 \\[2mm]
\frac{1}{\sqrt{x^2-x_0}}\mathcal{K}\left[\frac{x^2}{x^2-x_0}\right] & x_0 < 0
\end{cases}
\tag{10.77}
$$

with $\mathcal{K}$ the complete elliptic integral of the first kind. The approximation in the first line concerns large ratios of the inverse screening length to the Fermi momentum which is justified here. In the same way we can evaluate the imaginary part of the selfenergy (10.71) with the result

$$
\Gamma_a(k,\omega) = \frac{2\pi^2 e_a^2 T m_a}{\hbar p_f}\frac{1}{x+\sqrt{x_0}}\mathcal{K}\left[\frac{4\sqrt{x_0}}{(x+\sqrt{x_0})^2}\right].
\tag{10.78}
$$

The real part Σ is related to this imaginary Γ by the Hilbert transform (10.70). The more astonishing is the fact that we find here an additional relation

$$
\Sigma(k,\omega) = -\Gamma\left(k,\frac{k^2}{2m\hbar}-\omega\right)
\tag{10.79}
$$

which is only valid for this specific type of selfenergy besides the Kramers–Kronig relation (10.70). One can see that this first temperature correction has a highly-nontrivial frequency behaviour far from being Fermi-liquid like.

The linear temperature dependence appears here from the neglect of the quantum fluctuations in the ionic particle–hole fluctuation and is due to screening and should not be confused with the standard non-Fermi liquid behaviour in the literature. The latter one is more visible in the divergence at the Fermi energy $x_0 = x^2$ in Figure 10.9. This selfenergy has a finite-zero temperature limit but the imaginary part does not vanish at the Fermi energy though it is not diverging. For a Fermi liquid we would expect a vanishing imaginary part of the selfenergy at the Fermi energy. The linear temperature dependence has been repeatedly reported in the literature both from an experimental and theoretical point of view (Stern, 1980; DasSarma and Hwang, 1999; DasSarma, 1986; Gold and Dolgopolov, 1986; Morawetz, 2003).

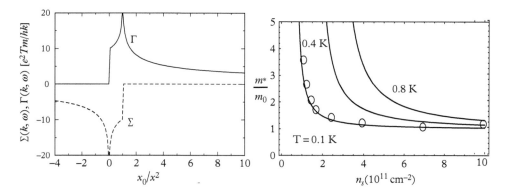

Figure 10.9: *Left: the real and imaginary part of the first temperature correction of the selfenergy (10.78) and (10.77) versus scaled frequency $x_0/x^2 = 2m\hbar\omega/k^2$ from (Morawetz, 2003). Right: the effective mass for temperatures $0.1, 0.4, 0.8K$ from left to right according to (10.81) versus density. The open circles are experimental values (Shashkin, Kravchenko, Dolgopolov and Klapwijk, 2002) at $T = 0.1K$.*

We can now evaluate the effective mass and compare its dependence on the density with the experiments in Figure 10.9. The quasiparticle energy $\hbar\omega = \epsilon_k$ is given by the pole of the spectral function which is the solution of $\omega - \frac{k^2}{2m} - \mathrm{Re}\Sigma(k,\omega) = 0$ and the effective mass m^* follows as

$$\frac{k}{m^*} = \frac{\partial \epsilon_k}{\partial k} = \frac{k}{m} + \frac{\partial \Sigma}{\partial \hbar\omega}\frac{\partial \epsilon_k}{\partial k} + \frac{\partial \Sigma}{\partial k} = \frac{\frac{k}{m} + \frac{\partial \Sigma}{\partial k}}{1 - \frac{\partial \Sigma}{\partial \hbar\omega}} \tag{10.80}$$

where the arguments have to be put on-shell $\hbar\omega = \epsilon_k$ after performing the derivatives. Since the real part of the selfenergy (10.77) scales with the momentum and frequency as $\Sigma \propto 1/k\sigma\,[2m\omega/k^2]$ we see that (10.80) takes exactly the form

$$\frac{m}{m^*} = \frac{1 - \frac{1}{4\epsilon_f}\Sigma_a(p_f, \epsilon_f)}{1 + \frac{1}{4\epsilon_f}\Sigma_a(p_f, \epsilon_f)}. \tag{10.81}$$

The on-shell value of the selfenergy (10.77) simplifies to $\Sigma_a(p_f, \epsilon_f) = \pi^2 T e_a^2 m_a/\hbar p_f$. According to the formulae (10.81) we obtain a sharp increase in the effective mass for lower densities (Figure 10.9), still above the critical one, in good agreement with the measurements. When the density decrease towards the critical value, the electrons become trapped by the charged impurities forming neutral impurities. This is the mechanism of the Mott transition.

10.4   Vertex correction to RPA polarisation

Let us calculate the RPA polarisation in equilibrium with the help of the Matsubara technique, see Chapter 7.3.5, as the lowest-order particle–hole fluctuation (potential leads to an additional i):

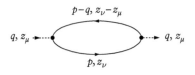

$$= \frac{-(i)^2}{\beta} \sum_{p\nu} \frac{1}{z_\nu - \epsilon_p} \frac{1}{-z_\mu + z_\nu - \epsilon_{p-q}} = \frac{1}{\beta} \sum_{p\nu} \left(\frac{1}{z_\nu - \epsilon_p} - \frac{1}{z_\nu - z_\mu - \epsilon_{p-q}} \right) \frac{1}{\epsilon_p - \epsilon_{p-q} - z_\mu}$$

$$= \sum_p \frac{f^F(\epsilon_p) - f^F(\epsilon_{p-q})}{\epsilon_p - \epsilon_{p-q} - z_\mu} \tag{10.82}$$

and since z_μ are bosonic it has no effect in the Fermi distribution f^F, just the RPA polarisation (10.52) appears, but here in equilibrium, if we extend $z_\mu \to \omega + i0$ in the denominator. As an example of next-order diagrams let us consider now the vertex correction:

$$= \frac{1}{\beta^2} \sum_{kp\nu\iota} V(k) \frac{1}{z_\nu - \epsilon_p} \frac{1}{z_\nu - z_\mu - \epsilon_{p-q}} \frac{1}{z_\nu - z_\mu - \epsilon_{p-k}} \frac{1}{z_\nu - z_\mu - z_\iota - \epsilon_{p-q-k}}$$

$$= \frac{1}{\beta} \sum_{kp\nu} V(k) \frac{1}{z_\nu - \epsilon_p} \frac{1}{z_\nu - z_\mu - \epsilon_{p-q}} \frac{f^B(z_\nu - \epsilon_{p-k}) - f^B(z_\nu - z_\mu - \epsilon_{p-q-k})}{\epsilon_{p-k-q} - \epsilon_{p-k} + z_\mu}$$

$$= \sum_p M(p,q,z_\mu) \frac{f^F(\epsilon_p) - f^F(\epsilon_{p-q})}{\epsilon_p - \epsilon_{q-p} - z_\mu} \tag{10.83}$$

We obtain just the RPA polarisation (10.82) but with an additional form factor

$$M(p,q,z_\mu) = \sum_k V(k) \frac{f^F(\epsilon_{p-k}) - f^F(\epsilon_{p-q-k})}{\epsilon_{p-k-q} - \epsilon_{p-k} + z_\mu}. \tag{10.84}$$

Here we have used $f^B(z_\nu - \epsilon) = -f^F(-\epsilon) = f^F(\epsilon) - 1$ for fermionic Matsubara frequencies z_ν.

10.5   Structure factor and pair-correlation function

The one-time pair-correlation function is defined as

$$n(1)n(2)g(1,2) = \langle \hat{\Psi}_1^+ \hat{\Psi}_2^+ \hat{\Psi}_2 \hat{\Psi}_1 \rangle|_{t_1=t_2} \tag{10.85}$$

and can be obtained if we consider only an infinitesimal propagation $t_2 = t_1 + 0 = t_{1+}$ of the fluctuation function (10.24)

$$-cL(x_1 t_1, x_2 t_{1+}) = icL^<(12) = \langle \Psi_2^+ \Psi_2 \Psi_1^+ \Psi_1 \rangle - n(1)n(2)$$

$$= \langle \Psi_1^+ \Psi_2^+ \Psi_2 \Psi_1 \rangle + \delta_{12} n(1) - n(1)n(2)$$

$$= n(1)n(2)[g(12) - 1] + \delta_{12} n(1) \tag{10.86}$$

which provides the pair-correlation function in terms of the time-diagonal of $L^<$. Changing to difference and centre-of-mass coordinates $r = x_1 - x_2$, $R = (x_1 + x_2)/2$ we can also introduce the structure function

$$[g(r, R) - 1]n(R + \frac{r}{2})n(R - \frac{r}{2}) + \delta(r)n(R) = n(R)(S(r, R) - \delta(r)) \tag{10.87}$$

Neglecting gradients in the density on the left side $n(R \pm r/2) \approx n(R)$ we obtain the known Fourier transform

$$S(R, q, t) - 1 = n(R, t) \int dr e^{-iqr} [g(R, r, t) - 1]. \tag{10.88}$$

The long-wavelength limit in equilibrium

$$S_{q=0} - 1 = \int dr[S(r) - \delta(r)] = \frac{1}{n} \int dr \langle \delta\hat{n}(r)\delta\hat{n}(0) \rangle$$

$$= \frac{1}{n} \frac{\langle N^2 \rangle - \langle N \rangle^2}{V} \equiv \frac{n}{\beta}\kappa = \frac{1}{m\beta v_t} \tag{10.89}$$

determines the isothermal compressibility

$$\kappa = \frac{1}{n}\left(\frac{\partial n}{\partial p}\right)_\beta \tag{10.90}$$

and the isothermal sound velocity $mv_t^2 = (\partial p/\partial n)_\beta$.
Here we use the convenient coordinates (9.2) and its Fourier transform

$$\langle p|L(K)|p' \rangle = \int dr dr' ds\, e^{ipr - ip'r' - iKs} \langle r|L(s, R)|r' \rangle \tag{10.91}$$

where we will drop the centre-of-mass coordinate R. The pair-correlation function reads in these coordinates

$$[g(x_1, x_2) - 1]n(x_1)n(x_2) + \delta_{12}n(1) = ic\langle x_1 - x_2|L_{tt'}^<\left(0, \frac{x_1 + x_2}{2}\right)|x_1 - x_2\rangle$$

$$= ic \int \frac{dp\,dp'\,dK}{(2\pi)^9} e^{-i(p-p')(x_1-x_2)} \langle p|L_{tt}^<(K)|p'\rangle. \tag{10.92}$$

10.5.1   Fock approximation

In lowest-order perturbation theory we calculate the correlation function in Fock approximation which is the second diagram in Figure 10.1. The Hartree term obviously does not contribute to the pair-correlation function since it is diagonal in the spatial coordinates. Neglecting gradients in the centre-of-mass coordinate we obtain the causal function (10.29)

$$c\langle p|L_{tt'}^0(K)|p'\rangle = \mp(2\pi)^3\delta(p+p')G_{tt'}\left(\frac{K}{2}-p\right)\cdot G_{t't}\left(\frac{K}{2}+p\right). \tag{10.93}$$

Here we denote the product explicitly by a dot to distinguish from the operator product. From the causal expression (10.93) the correlation functions (10.44) follows

$$\langle p|L_{tt'}^{0\gtrless}(K)|p'\rangle = -\frac{i}{c}(2\pi)^3\delta(p+p')G_{tt'}^{\gtrless}\left(\frac{K}{2}-p\right)\cdot G_{tt'}^{\lessgtr}\left(\frac{K}{2}+p\right). \tag{10.94}$$

With the Wigner function $\rho(p, R, t) = G_{tt}^<(p, R)$, the structure function (10.88) with the help of (10.92) is computed easily

$$S(q) = 1 + \frac{1}{n}\int\frac{dp}{(2\pi)^3}\rho_p(1 \mp \rho_{p-q}) \tag{10.95}$$

such that the isothermal compressibility (10.90) becomes

$$\frac{n}{\beta}\kappa = \frac{1}{n}\int\frac{dp}{(2\pi)^3}(1 \mp \rho_p)\rho_p \tag{10.96}$$

which is easily checked to be correct by

$$\kappa = \frac{1}{n^2}\frac{\partial n}{\partial \mu} = \frac{1}{n^2}\frac{\partial}{\partial \mu}\int\frac{dp}{(2\pi)^3}\frac{1}{e^{\beta(\epsilon_p-\mu)} \pm 1} = \frac{\beta}{n^2}\int\frac{dp}{(2\pi)^3}\rho_p(1 \mp \rho_p) \tag{10.97}$$

for Fermi/Bose distributions in equilibrium.

10.5.2   Born approximation

Next we want to calculate the correlation function L in Born approximation which is the first interaction in the expansion of the two-particle Green's function illustrated in Figure 10.10.

$$cL(121'2') = i \int d3d4 G^c(14) G^c(41') V(43) G^c(23) G^c(32') \qquad (10.98)$$

which reads in gradient approximation

$$c\langle p_1 | L_{tt'}(K) | p_2 \rangle = i \int \frac{dpdp'}{(2\pi)^6} dt_1 \langle p_1 | \xi_{tt_1}(K) | p \rangle \langle p | V | p' \rangle \langle p' | \xi_{t_1 t'}(K) | p_2 \rangle \quad (10.99)$$

where we suppressed the R coordinate and have introduced

$$\xi(x_1 x_2 t, x_4 x_3 t') = G^c_{tt'}(x_1, x_4) . G^c_{tt'}(x_2, x_3).$$

With the help of

$$\langle p | \xi_{tt'}(k, R) | p' \rangle = (2\pi)^3 \delta(p - p') G^c_{tt'} \left(\frac{k}{2} - p \right) \cdot G^c_{tt'} \left(\frac{k}{2} + p \right) \equiv (2\pi)^3 \delta(p - p') \xi_{tt'}(p)$$
$$(10.100)$$

one obtains

$$c\langle p_1 | L_{tt'}(K) | p_2 \rangle = iV(p_1 - p_2) \int d\bar{t} \xi_{t\bar{t}}(p_1) \xi_{\bar{t}t'}(p_2). \qquad (10.101)$$

Using the GKB ansatz (8.35) $G^{\gtrless}_{t_1,t_2} = iG^R_{t_1,t_2} \cdot \rho^{\gtrless}_{t_2} - i\rho^{\gtrless}_{t_1} \cdot G^A_{t_1,t_2}$ with $\rho^< = \rho, \rho^> = 1 \mp \rho$ we obtain for the time branches of the Hartree correlation function

$$\xi^{\gtrless}_{tt'} = \frac{1}{i} G^{\gtrless}_{tt'} \left(p_1 = \frac{k}{2} + p \right) \cdot G^{\gtrless}_{tt'} \left(p_2 = \frac{k}{2} - p \right)$$

$$= iG^R(1) \cdot G^R(2) \cdot \rho^{\gtrless}_{t'}(1) \cdot \rho^{\gtrless}_{t'}(2) - i\rho^{\gtrless}_t(1) \cdot \rho^{\gtrless}_t(2) \cdot G^A(1) \cdot G^A(2) \equiv \mathscr{G}^R_{tt'} . F^{\gtrless}_{t'} + F^{\gtrless}_t . \mathscr{G}^A_{tt'}$$
$$(10.102)$$

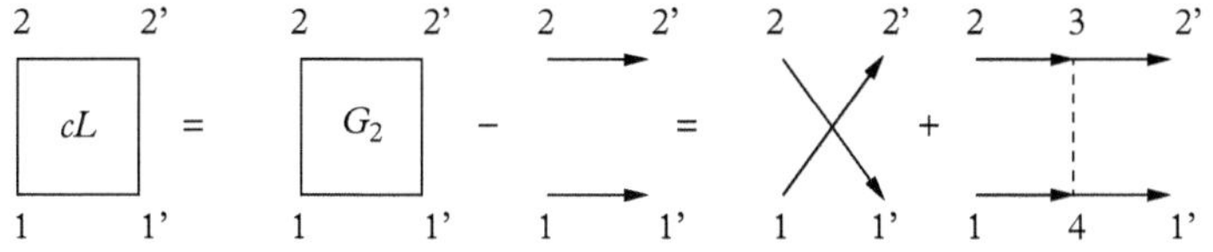

Figure 10.10: *The systematic expansion of the fluctuation function.*

where we introduced $F^{\gtrless}(1,2) = \rho^{\gtrless}(p_1)\rho^{\gtrless}(p_2)$, the two-particle retarded and advanced functions $\mathscr{G}^{R/A}_{tt'} = iG^{R/A}_{tt'}(1)G^{R/A}_{tt'}(2)$ and for the retarded and advanced functions we have

$$\xi^R = -i\theta_{tt'}\mathscr{G}^R.(F^>_{t'} - F^<_{t'}), \qquad \xi^A = i\theta_{t't}(F^>_t - F^<_t).\mathscr{G}^A. \tag{10.103}$$

Now we find the correlation function of (10.101) by applying the Langreth/Wilkens rules

$$cL^< = iV(\xi^<\xi'^A + \xi^R\xi'^<) \tag{10.104}$$

which reads with (10.102) and (10.103)

$$cL^{\gtrless}_{tt'} = -V \int_{t_0}^t d\bar{t}\Big\{\mathscr{G}^R_{t\bar{t}}(1)[F^<_{\bar{t}}(1)F^{\gtrless}_{\bar{t}}(2) - F^{\gtrless}_{\bar{t}}(1)F^<_{\bar{t}}(2)]\mathscr{G}^A_{\bar{t}t'}(2)$$

$$+ F^{\gtrless}_t(1).\mathscr{G}^A_{t\bar{t}}(1).[F^>_{\bar{t}}(2) - F^<_{\bar{t}}(2)].\mathscr{G}^A_{\bar{t}t'}(2) - \mathscr{G}^R_{t\bar{t}}(1).[F^>_{\bar{t}}(1) - F^<_{\bar{t}}(1)].\mathscr{G}^R_{\bar{t}t'}(2).F^{\gtrless}_{t'}(2)\Big\}$$

$$\tag{10.105}$$

We need only the time diagonal part and only the first line survives

$$cL^{\gtrless}_{tt} = -V \int_{t_0}^t d\bar{t}\mathscr{G}^R_{t\bar{t}}.\Big[F^<_{\bar{t}}(1)F^{\gtrless}_{\bar{t}}(2) - F^{\gtrless}_{\bar{t}}(1)F^<_{\bar{t}}(2)\Big].\mathscr{G}^A_{\bar{t}t}(2). \tag{10.106}$$

In quasiparticle approximation,

$$G^{R/A}_{tt'}(p) = \mp i\Theta_{\pm t\mp t'}e^{-i\epsilon_p\left(\frac{t+t'}{2}\right)(t-t')} \tag{10.107}$$

we obtain explicitly with $1 = K/2 - p$, $2 = K/2 + p$, $1' = K/2 - p'$, $2' = K/2 + p'$

$$c\langle p|L^<_{tt}(K)|p'\rangle = V(p-p') \int_{t_0}^t d\bar{t}e^{-i(\epsilon_1+\epsilon_2-\epsilon'_1-\epsilon'_2)(t-\bar{t})}$$

$$\times \Big(F^>_{\bar{t}}.(1,2)F'^<_{\bar{t}}.(1',2') - F^<_{\bar{t}}.(1,2)F'^>_{\bar{t}}.(1',2')\Big) \tag{10.108}$$

where $F^{\gtrless}(12) = \rho^{\gtrless}(1)\rho^{\gtrless}(2)$ as defined in (10.102).

10.5.3   Collision integral in Born approximation

Lets check whether the outlined formalism so far leads to the correct collision integral. The Martin–Schwinger hierarchy (7.8) reads

$$G^{-1}_{HF}(1)G(12) - G(1,2)G^{-1}_{HF}(2) = \mp ic[V(13)L^B(13t_1, 23t_2) - L^B(13t_1, 23t_2)V(2,3)]. \tag{10.109}$$

We obtain with $t_2 = t_1^+ = t$ the equation for the Wigner function $G(12)_{t^+t} = \pm iG^<(x_1t, x_2t) = \pm i\rho_t(x_1, x_2)$ as the kinetic equation

$$\frac{\partial}{\partial t}\rho + \text{Drift} = cVL_{tt}^< - cL_{tt}^< V. \tag{10.110}$$

In explicit notation we have

$$cVL_{tt}^<(r, R) = c \int ds V(s) \langle s|L_{tt}^< \left(\frac{r}{2}, R\right)|s - r\rangle \tag{10.111}$$

after gradient approximation in the centre-of-mass R. Using (10.108) one gets

$$cVL_{tt}^<(k, R) = \int \frac{dpdQ}{(2\pi)^6}|V(q)|^2 \int_{t_0}^{t} d\bar{t}e^{-i(\epsilon_1+\epsilon_2-\epsilon_1'-\epsilon_2')(t-\bar{t})}$$

$$\times \left[F_{\bar{t}}^>.(1,2)F_{\bar{t}}^<.(1',2') - F_{\bar{t}}^<.(1,2)F_{\bar{t}}^>.(1',2') \right] \tag{10.112}$$

where we have renamed $1' = k + q$, $2' = p - q$, $1 = k$, $2 = p$. The other term needed in (10.110), $cL^< V$, is obtained by interchanging $1 \leftrightarrow 1'$ and $2 \leftrightarrow 2'$ which results into a different sign in front of (10.112) and in the exponent. Therefore we obtain from (10.110) finally the Levinson equation

$$\frac{\partial}{\partial t}\rho + \text{Drift} = 2 \int \frac{dpdQ}{(2\pi)^6}|V(q)|^2 \int_0^{t-t_0} d\bar{t} \cos (\epsilon_1 + \epsilon_2 - \epsilon_1' - \epsilon_2')\bar{t}$$

$$\times \left\{[1\pm\rho_1][1\pm\rho_2]\rho_{1'}\rho_{2'} - \rho_1\rho_2[1\pm\rho_{1'}][1\pm\rho_{2'}]\right\}_{(t-\bar{t})}. \tag{10.113}$$

Neglecting the retardation in the distribution functions and setting $t_0 \rightarrow -\infty$ we obtain the standard Boltzmann (BUU) equation in Born approximation. The shortcomings of the retarded Levinson equation will be discussed in Chapter 19.3.1.

10.5.4 Pair-correlation function in equilibrium

In equilibrium we note that

$$F^> F'^< - F^< F'^> = [1\pm\rho_1][1\pm\rho_2]\rho_{1'}\rho_{2'} - \rho_1\rho_2[1\pm\rho_{1'}][1\pm\rho_{2'}]$$

$$= \rho_1\rho_2[1\pm\rho_{1'}][1\pm\rho_{2'}]\left(e^{\beta\Delta E} - 1\right) \tag{10.114}$$

with $\Delta E = \epsilon_1 + \epsilon_2 - \epsilon_1' - \epsilon_2'$ and for long times

$$\lim_{t \to \infty} \int_0^{t-t_0} d\bar{t} e^{-i\Delta E \bar{t}} = \pi \delta(\Delta E) - i \frac{\wp}{\Delta E} \qquad (10.115)$$

such that we obtain from (10.108)

$$c\langle p|L^<(K)|p'\rangle = -iV(p-p')\rho_1\rho_2[1\pm\rho_{1'}][1\pm\rho_{2'}]\left(\frac{e^{\beta\Delta E}-1}{\Delta E}\right). \qquad (10.116)$$

Using (10.92) we have for the pair-correlation function

$$[g(r)-1]n(R+r/2)n(R-r/2) + n(R)\delta(r)$$

$$= \int \frac{dp\,dq\,dk}{(2\pi)^9} e^{-iqr} V(q)\rho_k\rho_p[1\pm\rho_{k-q}][1\pm\rho_{p+q}]\left(\frac{e^{\beta\Delta E}-1}{\Delta E}\right). \qquad (10.117)$$

It is instructive to look at the classical limit, $\beta \to 0$ and neglecting the exchange, one obtains from (10.117)

$$g(r) \approx 1 - \beta V(r). \qquad (10.118)$$

Independent of Fermi or Bosons, the attractive potential will increase, the repulsive potential will expel the particles and increase or diminish the pair-correlation function, respectively. Eq. (10.118) is just the Born approximation, i.e. first-order expansion in V, of the classical pair-correlation function (2.87).

10.6   Ladder approximation

The approximation of the selfenergy must take into account at least two physical properties of the studied system. Firstly, the system is well out of equilibrium and the correlation function $G^<$ undergoes rapid changes in time and space. During and after the collision, the trajectories of non-interacting particles cannot be used even as the crudest approximation of the true state of the system. Accordingly, the perturbative expansion cannot be constructed from the free-particle propagator G_0^c but is strictly required to be self-consistent on the single-particle level, $\Sigma[G]$ not $\Sigma[G_0]$. The reconstruction of the wave function of two interacting particles must be included in solving self-consistently the two-particle problem.

The required self-consistencies on the single- and two-particle levels are satisfied by the selfenergy given by the two-particle T-matrix in the ladder approximation (Kadanoff and Baym, 1962; Danielewicz, 1984b; Botermans and Malfliet, 1990). This approximation guarantees conservation laws for the number of particles, the momentum and the energy, and it obeys the optical theorem (Kadanoff and Baym, 1962). Finally, the ladder approximation gives a non-trivial correlated density (Schmidt, Röpke and Schulz, 1990; Kremp, Kraeft and Lambert, 1984; Zimmermann and Stolz, 1985).

The selfenergy is constructed from the two-particle T-matrix as

$$\Sigma^c(1,3) = \mp i \mathcal{T}^c_{\text{ex}}(1,\bar{2},3,\bar{4}) G^c(\bar{4},\bar{2}). \tag{10.119}$$

This relation still does not specify any approximation but only expresses our determination to specify the approximation on the two-particle level. Relation (10.119) only tells us that the T-matrix is defined as the functional derivative of the selfenergy with respect to Green's function.

The actual approximation is specified by a choice of the T-matrix $\mathcal{T}^c$. Its ladder approximation reads

$$\mathcal{T}^c(12,34) = \mathcal{V}(1234) + \mathcal{V}(12\bar{5}\bar{6})\mathcal{G}^c(\bar{5}\bar{6},\bar{7}\bar{8})\mathcal{T}^c(\bar{7}\bar{8},34), \tag{10.120}$$

where

$$\mathcal{G}^c(56,78) = G^c(5,7) G^c(6,8)\Big|_{\substack{t_7=t_8 \\ t_5=t_6}} \tag{10.121}$$

describes two uncorrelated particles. Missing commas between arguments, e.g. $1\bar{2}$ and $3\bar{4}$ in (10.120), signal that $t_1 = t_2$ and $t_3 = t_4$ due to the instant character of the interaction, $\mathcal{V}(1234) \equiv \mathcal{V}(x_1,x_2,x_3,x_4)\delta_{a_1 a_3}\delta_{a_2 a_4}\delta(t_2-t_1)\delta(t_3-t_1)\delta(t_4-t_1)$.

The ladder approximation (10.120) is a compromise between desirability and tractability. From the diagrammatic representation of its third-order contribution in Fig. 10.11 one can see that it includes only the particle–particle (or hole–hole) channel given by diagram (a), while the particle–hole channel given by diagram (b) is missing. Of course, the interference of the two channels given by diagrams (c–f) is missing too. Some artifacts of the ladder approximation are discussed in Chapter 10.6.4 for a system composed of particles of very unequal masses.

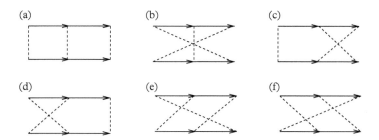

Figure 10.11: *Third-order contributions to the causal T-matrix $\mathcal{T}^c$. Only the diagram (a) is accounted for by the ladder approximation.*

If the 'odd' and the 'even' particles are identical, $a_1 = a_2$, they can mutually exchange. The exchange channel can be easily added yielding the T-matrix

$$\mathscr{T}_{ex}^c(12, 34) = \mathscr{T}^c(12, 34) - \delta_{a_1 a_2} \mathscr{T}^c(12, 43). \tag{10.122}$$

This symmetry also guarantees that the particle does not interact with itself.

Equations (7.21), (10.119)–(10.122) form a closed set. Accordingly, the perturbative expansion is already fully specified.

10.6.1   Analytic pieces

The causal Green's functions allow us to write the transport equation for the correlation function $G^<$ in a compact form (7.21)–(7.28). This formal simplification is counterbalanced by a singular time-dependence of causal functions:

$$G^c(1, 3) = -i\Theta_{13} G^>(1, 3) + i\Theta_{31} G^<(1, 3)$$

$$\Sigma^c(1, 3) = -i\Theta_{13} \Sigma^>(1, 3) + i\Theta_{31} \Sigma^<(1, 3) + \Sigma^{HF}(13)$$

$$\mathscr{T}^c(12, 34) = -i\Theta_{13} \mathscr{T}^>(12, 34) - i\Theta_{31} \mathscr{T}^<(12, 34) + \mathscr{V}(1234)$$

$$\mathscr{G}^c(12, 34) = -i\Theta_{13} \mathscr{G}^>(12, 34) - i\Theta_{31} \mathscr{G}^<(12, 34). \tag{10.123}$$

Here, Θ is the step function. For $t_1 < t_3$ we have $i\Sigma^< = -i(-i\mathscr{T}_{ex}^<)(iG^>)$ which gives

$$\Sigma^<(1, 3) = \mathscr{T}_{ex}^<(1\bar{2}, 3\bar{4}) G^>(\bar{4}, \bar{2}), \tag{10.124}$$

and analogously $> \leftrightarrow <$ for $t_1 > t_3$ There is also no time integration in (10.121). By substitution from (10.123) one finds

$$\mathscr{G}^{\gtrless}(56, 78) = G^{\gtrless}(5, 7) G^{\gtrless}(6, 8) \Big|_{\substack{t_7 = t_8 \\ t_5 = t_6}}. \tag{10.125}$$

Anti-symmetrisation (10.122) in analytical pieces reads

$$\mathscr{T}_{ex}^{\gtrless}(12, 34) = \mathscr{T}^{\gtrless}(12, 34) - \delta_{a_1 a_2} \mathscr{T}^{\gtrless}(12, 43). \tag{10.126}$$

Analogously to the treatment of GW or random-phase approximation Chapter 9 we can apply the Langreth–Wilkins rules to obtain a transport equation for the T-matrix following from the causal equation (10.120) as

$$\mathscr{T}^{\gtrless} = \mathscr{T}^R \mathscr{G}^{\gtrless} \mathscr{T}^A. \tag{10.127}$$

In the T-matrix no initial term appears since the potential $\mathscr{V}$ is singular in time, i.e. $\mathscr{V}^< = 0$. This singular term appears in the retarded T-matrix which follows from (10.120) applying the Langreth–Wilkins rules

$$\mathscr{T}^{R/A} = \mathscr{V} + \mathscr{V}\mathscr{G}^{R/A}\mathscr{T}^{R/A}. \qquad (10.128)$$

Due to the links between retarded, advanced and correlation functions, (7.15) and (7.16), one can write this set in a number of equivalent forms. Some of these rearrangements are useful as they offer an insight into the physical processes.

10.6.2 Selfenergy

According to (7.15), the retarded selfenergy Σ^R is composed of the analytic pieces $\Sigma^<$ and $\Sigma^>$ given by (10.124), and the singular Hartree–Fock term (10.11). Putting all the terms together one finds $[\Theta_{13} \equiv \Theta(t_1 - t_3)]$

$$
\begin{aligned}
\Sigma^R(1,3) &= -i\Theta_{13}\left(\Sigma^>(1,3) + \Sigma^<(1,3)\right) + \Sigma^{HF}(13) \\
&= -i\Theta_{13}\left(\mathscr{T}_{ex}^>(1\bar{2},3\bar{4})G^<(\bar{4},\bar{2}) + \mathscr{T}_{ex}^<(1\bar{2},3\bar{4})G^>(\bar{4},\bar{2})\right) + \mathscr{V}_{ex}(1\bar{2}3\bar{4})G^<(\bar{4}\bar{2}) \\
&= -i\Theta_{13}\left(\mathscr{T}_{ex}^>(1\bar{2},3\bar{4}) - \mathscr{T}_{ex}^<(1\bar{2},3\bar{4})\right)G^<(\bar{4},\bar{2}) + \mathscr{V}_{ex}(1\bar{2}3\bar{4})G^<(\bar{4}\bar{2}) \\
&\quad - \mathscr{T}_{ex}^<(1\bar{2},3\bar{4})i\left(G^>(\bar{4},\bar{2}) + G^<(\bar{4},\bar{2})\right) \\
&= \mathscr{T}_{ex}^R(1\bar{2},3\bar{4})G^<(\bar{4},\bar{2}) - \mathscr{T}_{ex}^<(1\bar{2},3\bar{4})G^A(\bar{4},\bar{2}). \qquad (10.129)
\end{aligned}
$$

Beside a trivial rearrangement we have used that $t_2 = t_1$ and $t_4 = t_3$ so that $\Theta_{13} = \Theta_{24}$, the second form is needed in G^A.

The term $\mathscr{T}_{ex}^R G^<$ has a clear physical interpretation. In the lowest-order approximation $\mathscr{T} \approx \mathscr{V}$, the term $\mathscr{T}_{ex}^R G^<$ turns into the Hartree–Fock potential Σ^{HF}. The term $\mathscr{T}_{ex}^< G^A$ vanishes in the lowest-order limit since $\mathscr{T}^< \approx \mathscr{V}\mathscr{G}^<\mathscr{V}$ is quadratic in $\mathscr{V}$. The term $\mathscr{T}_{ex}^< G^A$ also vanishes at the low-density limit being quadratic in densities while $\mathscr{T}_{ex}^R G^<$ has a dominant linear term.

The quadratic density dependency of $\mathscr{T}_{ex}^< G^A$ shows that the selfenergy includes three-particle contributions, although it has been derived formally on the two-particle level. These three-particle contributions follow from the Pauli blocking of the internal states of the T-matrix. As we will see, they are responsible for the non-trivial conversion of the potential and kinetic energy discussed on the classical level for breathing hard spheres.

Beside the ladder approximation, the physical content of the retarded two-particle T-matrix is determined by the retarded two-particle Green's function $\mathscr{G}^R$ which reads

$$
\begin{aligned}
\mathscr{G}^R(56,78) &= -i\Theta_{57}\left(\mathscr{G}^>(56,78) - \mathscr{G}^<(56,78)\right) \\
&= -i\Theta_{57}\left(G^>(5,7)G^>(6,8) - G^<(5,7)G^<(6,8)\right)_{t_5=t_6}^{t_7=t_8}. \qquad (10.130)
\end{aligned}
$$

One can see that the intermediate propagation includes not only the particle–particle channel with the corresponding Pauli blocking $G^>G^>$, but also the hole-hole channel with the Pauli blocking $G^<G^<$. The propagator $G^>G^> - G^<G^<$ corresponds to the Pauli blocking factor $1 - f - f$.

10.6.3   Scattering T-matrix

The antisymmetrised T-matrix $\mathscr{T}_{\mathrm{ex}}^<$ can be written as a sum over two channels. Substituting (10.127) into (10.126) one finds

$$\mathscr{T}_{\mathrm{ex}}^<(12,34) = \mathscr{T}^R(12,\bar{5}\bar{6})\mathscr{G}^<(\bar{5}\bar{6},\bar{7}\bar{8})\mathscr{T}^A(\bar{7}\bar{8},34) - \delta_{ab}\mathscr{T}^R(12,\bar{5}\bar{6})\mathscr{G}^<(\bar{5}\bar{6},\bar{7}\bar{8})\mathscr{T}^A(\bar{7}\bar{8},43)$$

$$= \frac{1}{2}\delta_{ab}\left[\mathscr{T}^R(12,\bar{5}\bar{6}) - \mathscr{T}^R(12,\bar{6}\bar{5})\right]\mathscr{G}^<(\bar{5}\bar{6},\bar{7}\bar{8})\left[\mathscr{T}^A(\bar{7}\bar{8},34) - \mathscr{T}^A(\bar{7}\bar{8},43)\right]$$

$$+ (1 - \delta_{ab})\, \mathscr{T}^R(12,\bar{5}\bar{6})\mathscr{G}^<(\bar{5}\bar{6},\bar{7}\bar{8})\mathscr{T}^A(\bar{7}\bar{8},34). \tag{10.131}$$

In the rearrangements of the second term we have used the symmetry of two-particle functions, $\mathscr{B}(12,34) = \mathscr{B}(21,43)$. The first term describes a scattering with a particle of different spin or isospin, the second one describes a scattering with an identical particle. Both terms have an evident symmetry with respect to retarded and advanced functions. This symmetry guarantees that one does not arrive at non-physical negative scattering rates. One can unify both terms by introducing the scattering T-matrix

$$\mathscr{T}_{\mathrm{sc}}^R(12,34) = \left(1 - \delta_{a_1 a_2}\right)\mathscr{T}^R(12,34) + \frac{1}{\sqrt{2}}\delta_{a_1 a_2}\left(\mathscr{T}^R(12,34) - \mathscr{T}^R(12,43)\right) \tag{10.132}$$

and the same for the advanced one. This allows us to write $\mathscr{T}_{\mathrm{ex}}^<$ as

$$\mathscr{T}_{\mathrm{ex}}^<(12,34) = \mathscr{T}_{\mathrm{sc}}^R(12,\bar{5}\bar{6})\mathscr{G}^<(\bar{5}\bar{6},\bar{7}\bar{8})\mathscr{T}_{\mathrm{sc}}^A(\bar{7}\bar{8},34). \tag{10.133}$$

10.6.4   Missing particle–hole channels

A validity of the ladder approximation has been questioned e.g. by Botermans and Malfliet (Botermans and Malfliet, 1990). They correctly argue that the particle–hole channels are missing. A description of these channels on a level consistent with the particle–particle and hole–hole channels is desirable but not feasible leading to the so-called parquet diagrams. In spite of number of alternative theoretical derivations (Kadanoff and Baym, 1962; Kremp, Schlanges and Bornath, 1985; Botermans and Malfliet, 1990), the question of the validity of the ladder approximation is still open. We present now a clear and exactly solvable example from which one can see how and why the ladder approximation can fail. This example are electrons scattered by point impurities of a low density.

Ladder approximation versus the exact T-matrix   According to arguments used to justify the ladder approximation, one would expect that the exact T-matrix for elastic impurity scattering will be recovered from the ladder approximation. Indeed, the ladder approximation seems to be based on the exact solution of the motion of two isolated particles, the electron and the impurity. The opposite is true: the ladder approximation appreciably differs from the exact T-matrix.

To perform the impurity limit of the ladder approximation, we designate particles of kind b as impurities and send their mass to infinity so that they become immobile, like impurities in solids. Particles of kind a will play the role of electrons. For simplicity we discuss only a stationary and homogeneous system and assume a contact potential. These simplifications do not touch the structure of the in-medium effects.

Impurity limit of the ladder approximation   Sending the mass of component b to infinity, $m_b \to \infty$, its kinetic energy vanishes, $\frac{k^2}{2m_b} \to 0$, therefore its correlation functions become momentum independent,

$$g_b^<(E) = g_b^<(E, k) = c2\pi\delta(E),$$

$$g_b^>(E) = g_b^>(E, k) = (1 - c)2\pi\delta(E), \tag{10.134}$$

where c is the concentration of the component b.

The limit of infinite mass is directly realised using the limiting correlation functions (10.134) in the T-matrix. One needs the two-particle function

$$\mathscr{G}^<(1, 2) = G_a^<(1, 2)G_b^<(1, 2), \tag{10.135}$$

which in the Wigner representation reads

$$\mathscr{G}^<(\omega) = \int \frac{dE}{2\pi} \int \frac{dk}{(2\pi)^3} g_b^<(E)g_a^<(\omega - E, p - k) = c \int \frac{dk}{(2\pi)^3} g_a^<(\omega, k). \tag{10.136}$$

Differences between the ladder approximation and the exact T-matrix are best visible on the retarded T-matrix. The two-particle retarded Green's functions $\mathscr{G}^R$ corresponding to (10.135) is

$$\mathscr{G}^R(1, 2) = -i\Theta_{12}\left(\mathscr{G}^>(1, 2) - \mathscr{G}^<(1, 2)\right)$$

$$= -i\Theta_{12}\left(G_a^>(1, 2) + G_a^<(1, 2)\right)G_b^>(1, 2) + i\Theta_{12}G_a^<(1, 2)\left(G_b^<(1, 2) + G_b^>(1, 2)\right)$$

$$= G_a^R(1, 2)G_b^>(1, 2) - G_a^<(1, 2)G_b^R(1, 2). \tag{10.137}$$

In the Wigner representation (10.137) reads

$$\mathscr{G}^R(\omega) = \int \frac{dE}{2\pi} \frac{dk}{(2\pi)^3} g_a^R(E,k) g_b^>(\omega-E) - \int \frac{dE}{2\pi} \frac{dk}{(2\pi)^3} g_a^<(E,k) g_b^R(\omega-E)$$

$$= (1-c) \int \frac{dk}{(2\pi)^3} g_a^R(\omega,k) - \int \frac{dE}{2\pi} \frac{dk}{(2\pi)^3} \frac{1}{\omega-E+i0} g_a^<(E,k), \qquad (10.138)$$

where we have used (10.134) and the retarded Green's function of immobile impurities, $g_b^R(\omega-E) = (\omega-E+i0)^{-1}$. From the free-particle approximation for electrons,

$$g_a^<(\omega,p) = f_p 2\pi \delta(\omega-\varepsilon_p), \qquad g_a^R(\omega,p) = \frac{1}{\omega-\varepsilon_p+i0}, \qquad (10.139)$$

one obtains the two-particle propagator,

$$\mathscr{G}^R(\omega) = \int \frac{dk}{(2\pi)^3} \frac{1-c-f_k}{\omega-\varepsilon_k+i0}, \qquad (10.140)$$

corresponding to the ladder approximation.

With the help of this retarded two-particle Green's function one readily evaluates the retarded T-matrix for the contact potential, $\mathscr{V}(x_1,x_2,x_3,x_4) = v\delta(x_1-x_2)\delta(x_1-x_3)\delta(x_1-x_4)$,

$$\mathscr{T}_R^{-1}(\omega) = \frac{1}{v} - \mathscr{G}^R(\omega) = \frac{1}{v} - \int \frac{dk}{(2\pi)^3} \frac{1-c-f_k}{\omega-\varepsilon_k+i0}. \qquad (10.141)$$

In the low-concentration limit, $c \to 0$, the ladder approximation of the T-matrix yields

$$\mathscr{T}_R^{-1} = \frac{1}{v} - \int \frac{dk}{(2\pi)^3} \frac{1-f_k}{\omega-\varepsilon_k+i0}. \qquad (10.142)$$

Exact T-matrix   For the potential of the point impurity at coordinate R,

$$V(x_1,x_2) = v\delta(x_1-R)\delta(x_2-R), \qquad (10.143)$$

the exact T-matrix reads, see (Goldberger and Watson, 1964),

$$T_R^{-1} = \frac{1}{v} - \int \frac{dk}{(2\pi)^3} \frac{1}{\omega-\varepsilon_k+i0}. \qquad (10.144)$$

Now we can compare. Apparently, the ladder approximation (10.142) differs from the exact T-matrix (10.144) by its dependence on the electron distribution f_k. In the ladder approximation (10.142), the intermediate states of electron-impurity interaction are restricted to unoccupied states. In other words, the occupied states are blocked via the Pauli exclusion principle. No such restriction of the intermediate phase-space appears in the exact solution.

Particle–hole channel   Let us identify the origin of the Pauli blocking in the ladder approximation. The ladder approximation includes the particle–particle and hole–hole channels of the intermediate propagator $G_a^> G_b^> - G_a^< G_b^< \propto (1-f)(1-c) - fc$. One can expect that the Pauli blocking vanishes if one adds the particle–hole channels with intermediate propagators $G_a^< G_b^> - G_a^> G_b^< \propto f(1-c) - (1-f)c$.

To include the particle–hole channel we have to construct the two-particle T-matrix in a more elaborate way. Let us write the T-matrix as the ladder summation over the irreducible T-matrix,

$$\mathcal{T} = \mathcal{T}_{ir} + \mathcal{T}_{ir}\mathcal{G}\mathcal{T}. \qquad (10.145)$$

The ladder approximation is just $\mathcal{T}_{ir} \approx \mathcal{V}$. Fig. 10.12(a) shows the sum of maximally crossed diagrams contributing to the irreducible T-matrix. Screwing these diagrams one can see that they describe the ladder approximation of the particle–hole channel, Fig. 10.12(b). This flip interchanges the particle and hole contributions, therefore the irreducible T-matrix describes the particle–hole channel. Within the screwed ladder approximation, diagrams (a–d) from Fig. 10.11 are covered.

In general, the T-matrix given by its irreducible part is badly complicated since genuine four-time functions have to be handled. Due to absence of the momentum and time dependence of the impurity propagators, one can introduce the energy representation coupled to the electronic variables only, and the retarded T-matrix simplifies

$$\mathcal{T}_R^{-1}(\omega) = \mathcal{T}_{Rir}^{-1}(\omega) - \int \frac{dk}{(2\pi)^3} \frac{1 - f_k}{\omega - \varepsilon_k + i0}. \qquad (10.146)$$

The ladder approximation of the particle–hole and hole–particle channels parallels the particle–particle and hole–hole channels. The only difference is that distributions

(a)

(b)

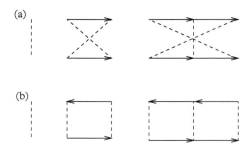

Figure 10.12: *Maximally crossed diagrams contributing to the irreducible T-matrix. The irreducible diagrams (a) are easily mapped on the ladder expansion flipping the orientation of the upper line (b).*

$(1 - c)(1 - f) - cf = 1 - f - c \to 1 - f$ are interchanged by $(1 - c)f - c(1 - f) = f - c \to f$. Accordingly in parallel with (10.141) one finds

$$\mathscr{T}_{\text{Rir}}^{-1}(\omega) = \frac{1}{v} - \int \frac{dk}{(2\pi)^3} \frac{f_k}{\omega - \varepsilon_k + i0}. \qquad (10.147)$$

Substituting (10.147) into (10.146) one recovers the exact result (10.144). We have found that the contribution of the particle–hole channel just compensates the Pauli blocking factor present in the particle–particle channel.

Comment   The failure of the ladder approximation in the impurity limit cannot be taken as a decisive proof against the Galitskii–Feynman approximation. Having no degree of freedom, the impurity is a peculiar collision partner. As known from solids, any degree of freedom of the scattering centre, e.g. the impurity spin, makes the collision sensitive to the distribution of electrons, e.g. magnetic impurities (Fulde, 1991).

   Systems with equal masses do not qualify for the impurity limit. Since the dynamics of both partners in the collision is important, their T-matrix has to depend on the distribution of surrounding particles. On the other hand, one should keep in mind that missing the particle–hole channels, the ladder approximation likely leads to artificial in-medium effects.

10.7   Bethe–Salpeter equation in quasiparticle approximation

We introduce the notation for the known quasiparticle approximation without gradient corrections and use the following representation of the two-particle T-matrix,

$$\langle x_1 x_2 \bar{t} | \mathscr{T} | x_1' x_2' t' \rangle = \int \frac{dkdpdqd\Omega}{(2\pi)^{10}} \mathscr{T}(\Omega, k, p, q, r, t)\, e^{\left[i\left(kx_1 + px_2 - (k-q)x_1' - (p+q)x_2' - \Omega(\bar{t}-t')\right)\right]},$$

$$(10.148)$$

where $r = \frac{1}{4}(x_1 + x_2 + x_1' + x_2')$ and $t = \frac{1}{2}(\bar{t} + t')$ are centre-of-the-mass coordinate and time. In quasiparticle approximation the free two-particle propagator $\mathscr{G}$ (denoted in the mixed Wigner representation as ζ) takes the form

$$\zeta^R(\Omega, p, p', r, t) = \frac{1 \mp f(p, r, t) \mp f(p', r, t)}{\Omega - \varepsilon(p, r, t) - \varepsilon(p', r, t) + i\eta} \qquad (10.149)$$

for fermions/bosons respectively. The selfenergy then reads

$$\sigma^>(\omega, k, r, t) = \int \frac{dp}{(2\pi)^3} t^>(\omega + \varepsilon(p, r, t), k, p, 0)f(p, r, t), \qquad (10.150)$$

and analogously $\sigma^<$. In the rest of this chapter we suppress variables r, t. From (10.128) one finds

$$t^<(\Omega, k, p, 0) = \int \frac{dq}{(2\pi)^3} |t^R|^2 (\varepsilon_{k-q} + \varepsilon_{p+q}, k, p, q) f_{k-q} f_{p+q} 2\pi \delta(\omega - \varepsilon_{p+q} - \varepsilon_{k-q}), \quad (10.151)$$

and $t^>$ is given by interchange $f \leftrightarrow 1 \mp f$.

For the determination of the quasiparticle energy one needs the retarded selfenergy

$$\sigma^R(\omega, k) = -\int \frac{dp}{(2\pi)^3} \frac{d\omega'}{2\pi} \frac{t^>(\omega' + \varepsilon_p, k, p, 0) f_p}{\omega' - \omega - i\eta} - \int \frac{dp}{(2\pi)^3} \frac{d\omega'}{2\pi} \frac{t^<(\omega' + \varepsilon_p, k, p, 0)(1 - f_p)}{\omega' - \omega - i\eta}$$

$$= \int \frac{dp}{(2\pi)^3} t^R(\omega + \varepsilon_p, k, p, 0) f_p - \int \frac{dp dq}{(2\pi)^6} \frac{|t^R(\varepsilon_{k-q} + \varepsilon_{p+q}, k, p, q)|^2 f_{k-q} f_{p+q}}{\varepsilon_{k-q} + \varepsilon_{p+q} - \varepsilon_p - \omega + i\eta}. \quad (10.152)$$

In the real part of the selfenergy the real part of the T-matrix appears,

$$\mathrm{Re} t^R(\Omega, k, p, 0) = \int \frac{dq}{(2\pi)^3} \frac{1 \mp f_{k-q} \mp f_{p+q}}{\Omega - \varepsilon_{k-q} - \varepsilon_{p+q}} |t^R|^2 (\varepsilon_{k-q} + \varepsilon_{p+q}, k, p, q). \quad (10.153)$$

This form follows from the Kramers–Kronig transformation of $t^> - t^<$.

The ladder equation without gradients reads with the notation specified by (9.2)

$$\langle p_1 | t^R(q') | p_2 \rangle = \langle p_1 | \mathcal{V}(q') | p_2 \rangle + \int \frac{dp' \, dp''}{(2\pi)^6} \langle p_1 | \mathcal{V}(q') | p' \rangle \langle p' | \zeta^R(q') | p'' \rangle \langle p'' | t^R_{sc}(q') | p_2 \rangle.$$

$$(10.154)$$

In the coordinates, $k = \frac{q'}{2} + p_1, p = \frac{q'}{2} - p_1, q = p_1 - p_2$, corresponding to (10.148) this equation takes the form

$$t^R(\Omega, k, p, q) = \mathcal{V}(k, p, q) + \int \frac{dp' \, dp''}{(2\pi)^6} \mathcal{V}(k, p, p') \zeta^R(\Omega, k - p', p + p', p' - p'')$$

$$\times t^R(\Omega, k - p'', p + p'', q + p''). \quad (10.155)$$

Since $\zeta^\gtrless(k, p, q) = (2\pi)^3 \delta(q) g^\gtrless(k) g^\gtrless(p)$, in the quasiparticle approximation one obtains the Bethe–Salpeter equation

$$t^R(\Omega, k, p, q) = \mathcal{V}(k, p, q) + \int \frac{dq'}{(2\pi)^3} \mathcal{V}(k, p, q')$$

$$\times \frac{1 \mp f_{k-q'} \mp f_{p+q'}}{\Omega - \varepsilon_{p+q'} - \varepsilon_{k-q'} + i\eta} t^R(\Omega, k - q', p + q', q + q'). \quad (10.156)$$

10.8   Local and instant quantum Boltzmann kinetic equation

The instant and local approximation of the scattering integral is based on the assumption that collisions happen so fast and in such small space region that any gradient in the vicinity of the collision can be neglected. As we have seen in the classical system of Section 2.1.2, the ultimate consequence of this assumption is the neglect of the virial corrections. Let us accept this assumption for a while anyway and derive the non-gradient scattering integral.

Neglecting all gradients, the selfenergy gains the same form as in a locally homogeneous and stationary system. In this case, the integration over space and time variables transforming the selfenergy into the Wigner representation is easy since it is equivalent to a transformation into the standard energy and momentum representation. The scattering integrals follows from the approximation of the selfenergy with (10.124) using (10.133) and (10.125). Writing t_{sc}^R as the Wigner transform of $\mathcal{T}_{\mathrm{sc}}^R$, and $t_{\mathrm{sc}}^R = |t_{\mathrm{sc}}|^{i\phi}$ one has

$$
\sigma_a^{\gtrless}(\omega, k, r, t) = \sum_b \int \frac{dp}{(2\pi)^3} \frac{dE}{2\pi} \frac{dq}{(2\pi)^3} \frac{d\Omega}{2\pi} |t_{\mathrm{sc}}(\omega + E, k, p, q, r, t)|^2
$$

$$
\times g_b^{\lessgtr}(E, p, r, t) g_a^{\gtrless}(\omega - \Omega, k - q, r, t) g_b^{\gtrless}(E + \Omega, p + q, r, t). \quad (10.157)
$$

The binary scattering integral of the Boltzmann equation is obtained from the quasiparticle approximation (9.17), $g^< = zf 2\pi\delta$ and $g^> = z(1-f)2\pi\delta$, as

$$
z_1 \sigma_{1,\varepsilon_1}^< (1 - f_1) - z_1 \sigma_{1,\varepsilon_1}^> f_1 = \sum_b \int \frac{dp dq}{(2\pi)^6} 2\pi\delta \left(\varepsilon_1 + \varepsilon_2 - \varepsilon_3 - \varepsilon_4\right) z_1 z_2 z_3 z_4
$$

$$
\times \left| t_{\mathrm{sc}}\left(\varepsilon_1 + \varepsilon_2, k, p, q, r, t\right) \right|^2 \left((1 \mp f_1)(1 \mp f_2) f_3 f_4 - f_1 f_2 (1 \mp f_3)(1 \mp f_4)\right), \quad (10.158)
$$

where the independent energies have been integrated out with the help of the δ functions. The energy arguments of the quasiparticle energies and distributions are the same as in the Boltzmann equation

$$
\begin{aligned}
\varepsilon_1 &\equiv \varepsilon_a(k, r, t), & \varepsilon_2 &\equiv \varepsilon_b(p, r, t), \\
\varepsilon_3 &\equiv \varepsilon_a(k - q, r, t), & \varepsilon_4 &\equiv \varepsilon_b(p + q, r, t).
\end{aligned} \quad (10.159)
$$

Substituting the scattering integral (10.158) into the kinetic equation (9.73) one obtains the standard BUU equation

$$
\frac{\partial f_1}{\partial t} + \frac{\partial \varepsilon_1}{\partial k} \frac{\partial f_1}{\partial r} - \frac{\partial \varepsilon_1}{\partial r} \frac{\partial f_1}{\partial k} = \sum_b \int \frac{dp}{(2\pi)^3} \frac{dq}{(2\pi)^3} 2\pi\delta \left(\varepsilon_1 + \varepsilon_2 - \varepsilon_3 - \varepsilon_4\right) z_1 z_2 z_3 z_4
$$

$$
\times \left| t_{\mathrm{sc}}\left(\varepsilon_1 + \varepsilon_2, k, p, q, r, t\right) \right|^2 \left[(1 \mp f_1)(1 \mp f_2) f_3 f_4 - f_1 f_2 (1 \mp f_3)(1 \mp f_4)\right]. \quad (10.160)
$$

In the scattering integral on the right-hand side, the distributions of quasiparticles describe the probability of a given initial state for the binary collision, while the hole distribution provides the probability that requested final states are empty. The δ function represents the energy-conservation and the differential cross section given by the amplitude of the T-matrix is reduced by the wave-function renormalisation factors z. In some formulations, the wave-function renormalisation is not assumed. Numerical studies (Danielewicz, 1984a; Köhler, 1995a) show that the wave-function renormalisation should be taken into account.

The link to collisions can be made more visible if we introduce the differential cross section. To this end we use the centre-of-mass momentum $K = k + p$ of the two particles and their difference momenta before, $\bar{p} = \frac{m_b}{M} k - \frac{m_a}{M} p$, and after, $p' = \bar{p} - q$, the collision with the total mass $M = m_a + m_b$ as well as the reduced mass $1/\mu = 1/m_a + 1/m_b$. In this notation the Boltzmann collision integral takes the form of a neglecting wave-function renormalisation

$$\frac{4\pi\mu}{\hbar}\int\frac{dp\,dp'}{(2\pi\hbar)^6}\delta(\bar{p}^2 - p'^2)\left|t_{\bar{p},p'}\left(K,\frac{K^2}{2M}+\frac{\bar{p}^2}{2\mu}\right)\right|^2\left[(1\mp f_1)(1\mp f_2)f_3 f_4 - f_1 f_2(1\mp f_3)(1\mp f_4)\right]$$

(10.161)

where we restored the $\hbar$. Now we can introduce the differential cross section with respect to the spherical angle Ω'

$$\frac{2\pi\mu}{\hbar}\int_0^\infty\frac{dp'}{(2\pi\hbar)^3}p'^2\delta(\bar{p}^2 - p'^2)\left|t_{\bar{p},p'}\left(K,\frac{K^2}{2M}+\frac{\bar{p}^2}{2\mu}\right)\right|^2 = \frac{\bar{p}}{2\mu}\frac{d\sigma}{d\Omega'}$$

(10.162)

to obtain finally

$$\frac{1}{\mu}\int\frac{dp}{(2\pi\hbar)^3}\int d\Omega' p\frac{d\sigma}{d\Omega'}\left[(1\mp f_1)(1\mp f_2)f_3 f_4 - f_1 f_2(1\mp f_3)(1\mp f_4)\right]_{|p'|=|\bar{p}|}.$$

(10.163)

This form shows that indeed the Boltzmann collision integral is given by the differential cross section calculated by the scattering T-matrix .

11

Variational Techniques of Many-Body Theory

11.1 Dyson time ordering and Dirac interaction representation

We consider the time evolution of an interacting many-body state $\langle \psi \rangle$ due to a time-dependent perturbation $U(t)$

$$(i\partial_t - H - U)\psi_t = 0 \tag{11.1}$$

with the interacting Hamiltonian $H = H_0 + V$. Splitting off the time evolution due to the unperturbed Hamiltonian $\hat{\psi}_t = e^{iHt}\psi_t$ the Schrödinger equation (11.1) takes the form of the Dirac picture

$$i\partial\hat{\psi}_t = \hat{U}_t\hat{\psi}_t, \qquad \hat{U}_t = e^{iHt}Ue^{-iHt}. \tag{11.2}$$

The time-evolution operator $\hat{\psi}_t = S(t, t_0)\hat{\psi}_{t_0}$ obeys therefore the equation

$$i\frac{\partial}{\partial t}S(t, t_0) = \hat{U}_t S(t, t_0), \qquad S(t_0, t_0) = 1 \tag{11.3}$$

which can be written compactly into an integral equation

$$S(t, t_0) = 1 - i\int_{t_0}^{t} dt' \hat{U}_{t'} S(t', t_0) = \sum_{n=0}^{\infty} I_n. \tag{11.4}$$

Interacting Systems far from Equilibrium. Klaus Morawetz, Oxford University Press (2018).
© Klaus Morawetz. DOI: 10.1093/oso/9780198797241.001.0001

Here one obtains the infinite sum by iteration with the terms

$$I_n = (-i)^n \int\limits_{t_0}^{t} dt_1 \hat{U}_{t_1} \int\limits_{t_0}^{t_1} dt_2 \hat{U}_{t_2} \dots \int\limits_{t_0}^{t_{n-1}} dt_n \hat{U}_{t_n}$$

$$= (-i)^n \int\limits_{t_0}^{t} dt_1 \hat{U}_{t_1} \int\limits_{t_0}^{t} dt_2 \Theta(t_1 - t_2) \hat{U}_{t_2} \dots \int\limits_{t_0}^{t} dt_n \Theta(t_{n-1} - t_n) \hat{U}_{t_n}$$

$$= \frac{(-i)^n}{n!} \sum\limits_{perm.P_\alpha} \int\limits_{t_0}^{t} dt_1 \dots dt_n P_\alpha \left[\Theta(t_{\alpha_1} - t_{\alpha_2}) \dots \Theta(t_{\alpha_{n-1}} - t_{\alpha_n}) \right] \hat{U}_{t_1} \dots \hat{U}_{t_n}$$

$$= \frac{(-i)^n}{n!} \int\limits_{t_0}^{t} dt_1 \dots dt_n T_c \left[\hat{U}_{t_1} \dots \hat{U}_{t_n} \right] \tag{11.5}$$

where we introduced the time-ordering operator T_c and all permutations of P_α. Therefore the time-evolution operator (11.4) can be written as

$$S(t, t_0) = T_c \mathrm{e}^{-i \int\limits_{t_0}^{t} dt' \hat{U}_{t'}} \tag{11.6}$$

and possesses the property $S(t, t_0) = S^+(t_0, t) = S^{-1}(t_0, t)$.

Now we consider the mean value

$$\langle A_t \rangle = \mathrm{Tr}\, \rho \langle \psi_t | A | \psi_t \rangle = \mathrm{Tr}\, \rho \langle \hat{\psi}_t | \mathrm{e}^{iHt} A \mathrm{e}^{-iHt} | \hat{\psi}_t \rangle = \mathrm{Tr}\, \rho \langle \hat{\psi}_{t_0} | S^+(t, t_0) \hat{A}_t S(t, t_0) | \hat{\psi}_{t_0} \rangle$$

$$= \mathrm{Tr}\, \rho \langle \hat{\psi}_{t_0} | S^+(\infty, t_0) S^+(t, \infty) \hat{A} S(t, t_0) | \hat{\psi}_{t_0} \rangle$$

$$= \mathrm{Tr}\, \rho \langle \hat{\psi}_{-\infty} | S^{-1}(\infty, -\infty) T_c \left[\hat{A}_t S(\infty, -\infty) \right] | \hat{\psi}_{-\infty} \rangle \tag{11.7}$$

where the time of switching-on of the perturbation U is set artificially to $t_0 \to -\infty$ in the last step. This we will understand as an infinitesimally slow switching on of the interaction from times $-\infty$ up to t_0. According to the adiabatic theorem the system will then not be changed which means it can only have obtained a phase L which allows us to write in (11.7)

$$\langle \hat{\psi}_{-\infty} | S^{-1}(\infty, -\infty) = \langle \hat{\psi}_\infty | = \mathrm{e}^{-iL} \langle \psi_{-\infty} | \tag{11.8}$$

and the phase as $\mathrm{e}^{-iL} = \langle \hat{\psi}_\infty | \psi_{-\infty} \rangle$. Therefore we can rewrite (11.7)

$$\langle A_t \rangle = \frac{\mathrm{Tr}\, \rho \langle \hat{\psi}_{-\infty} | T_c \left[\hat{A}_t S(\infty, -\infty) \right] | \hat{\psi}_{-\infty} \rangle}{\mathrm{Tr}\, \rho \langle \hat{\psi}_{-\infty} | T_c S(\infty, -\infty) | \psi_{-\infty} \rangle} = \frac{\mathrm{Tr}\, \rho\, T_c \left[\hat{A}_t S(\infty, -\infty) \right] |}{\mathrm{Tr}\, \rho\, T_c S(\infty, -\infty)}. \tag{11.9}$$

11.2   Representation of Green's functions

Assuming the perturbing potential in mean-field form

$$U = \int d2n(2)u(2) \tag{11.10}$$

where the numbers sign cumulative indices like space-time coordinates, we can present the causal single-particle Green's function according to (11.9) as

$$G(1,1') = \frac{1}{i}\frac{\langle T_c S \hat{\Psi}_1 \hat{\Psi}_{1'}^+ \rangle}{\langle T_c S \rangle} \tag{11.11}$$

and a small variation with respect to $u(2)$ leads to

$$\delta G(1,1') = \frac{1}{i}\left(\frac{\langle T_c \delta S \hat{\Psi}_1 \hat{\Psi}_{1'}^+ \rangle}{\langle T_c S \rangle} - \frac{\langle T_c \delta S \rangle \langle T_c S \hat{\Psi}_1 \hat{\Psi}_{1'}^+ \rangle}{\langle T_c S \rangle^2}\right). \tag{11.12}$$

Since $\delta S = -iS \int d2n(2)\delta u(2)$ we have

$$\delta G(1,1') = \int d2 \left(\frac{\langle T_c S \hat{\Psi}_1 \hat{\Psi}_{1'}^+ \hat{\Psi}_2^+ \hat{\Psi}_2 \rangle}{i^2 \langle T_c S \rangle} - \frac{\langle T_c S \hat{\Psi}_1 \hat{\Psi}_{1'}^+ \rangle}{i\langle T_c S \rangle}\frac{\langle T_c S \hat{\Psi}_2^+ \hat{\Psi}_2 \rangle}{i\langle T_c S \rangle}\right)\delta u(2)$$

$$= \mp \int d2 \left[G_2(121'2^+) - G(1,1')G(2,2^+)\right]\delta u(2) \tag{11.13}$$

such that we can express the two-particle Green's function $i^2 G_{121'2'} = \langle T_c \hat{\Psi}_1 \hat{\Psi}_2 \hat{\Psi}_{2'}^+ \hat{\Psi}_{1'}^+ \rangle$ by a variation of the one-particle Green's functions $iG_{12} = \langle T_c \hat{\Psi}_1 \hat{\Psi}_2^+ \rangle$ with respect to the external potential (Kadanoff and Baym, 1962; Kraeft, Kremp, Ebeling and Röpke, 1986) as

$$G_{121'2'} = G_{11'}G_{22'} \mp \frac{\delta G_{11'}}{\delta U_{2'2}} \tag{11.14}$$

where the upper sign denotes the Fermi and the lower the Bose functions. Using the Dyson equation

$$G^{-1} = G_0^{-1} - \Sigma - U \tag{11.15}$$

we can calculated the derivative in (11.14). With the help of the chain rule and $\delta G = -G\delta G^{-1}G$, one can express the fluctuation function as

$$L_{121'2'} = G_{121'2'} - G_{11'}G_{22'}$$

$$= \mp G_{12'}G_{21'} \mp G_{13}\frac{\delta\Sigma_{34}}{\delta U_{2'2}}G_{41'}$$

$$= \mp G_{12'}G_{21'} + G_{13}\frac{\delta\Sigma_{34}}{\delta G_{56}}L_{5262'}G_{41'}. \tag{11.16}$$

Double occurring indices are understood as integrated over. As introduced in Chapter 10.2.2, we use the convention of $c = 1$. With the definition of the occurring vertex function we can express this graphically:

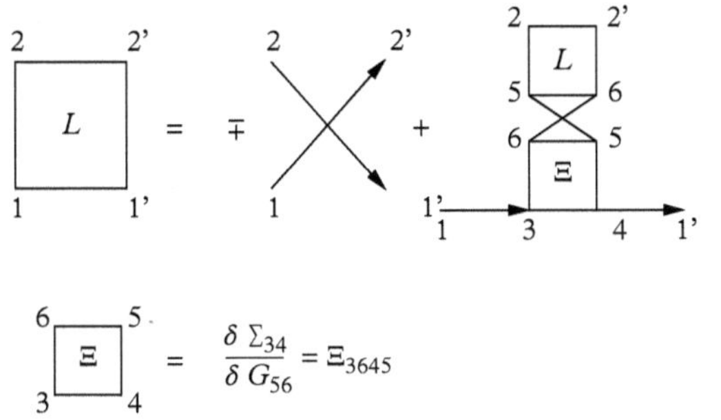

Figure 11.1:

The definition (11.14) reveals that the demand of criterion (B) for conservation laws, see Chapter 7.3.3, requires that

$$\frac{\partial G(11')}{\partial U(2'2)} = \frac{\partial G(22')}{\partial U(1'1)} \tag{11.17}$$

which can only be fulfilled if there exists a function $\mathcal{W}$ such that the single-particle Green's function can be expressed as

$$G(12) = \frac{\partial\mathcal{W}}{\partial U(21)}. \tag{11.18}$$

In equilibrium it is easily given in terms of the grand canonical ensemble

$$\mathcal{W}_{\text{equi}} = \mp\ln\,\mathrm{Tr}e^{-\beta(\hat{H}-\mu\hat{N})}\,Te^{-i\int\hat{\Psi}^+(1)U(1,1')\hat{\Psi}(1')}. \tag{11.19}$$

In nonequilibrium we do not have such explicit functional at hand though it has to exist. Instead we will analyse which demand we can set on the approximation of the selfenergy in order to complete criterion (B). Since the selfenergy itself is given by

$$\Sigma_{13}G_{31'} = \mp iV_{12}G_{121'2^+} \tag{11.20}$$

one finds

$$\Sigma_{13} = \mp i V_{12} \left(\mp \frac{\delta G_{11'}}{\delta U_{22}} + G_{11'} G_{22^+} \right) G_{1'3}^{-1}$$

$$= i V_{13} G_{13} \mp i V_{12} G_{22^+} \delta_{13} + i V_{12} G_{11'} \frac{\delta \Sigma_{1'3}}{\delta U_{22}} \tag{11.21}$$

which is an integral equation to create a consistent selfenergy. The specific forms of selfenergy cannot be arbitrarily chosen since the criterion (B) set sever boundary conditions. In order to see this we multiply (11.16) from both sides with G^{-1} and obtain

$$\int d\bar{2}\bar{2}' \left[G^{-1}(1\bar{2}')G^{-1}(\bar{2}1') - \frac{\partial \Sigma(11')}{\partial G(\bar{2}'\bar{2})} \right] L(\bar{2}'3\bar{2}3') = \mp \delta(1-3')\delta(3-1') \tag{11.22}$$

where the bracket on the left-hand side represents obviously the inverse L. Since criterion (B) of Chapter 7.3.3 demands symmetry if we interchange, in the two-particle Green's functions, the incoming and outgoing particles simultaneously, the same symmetry should be in L and L^{-1}. Therefore we see that it translates into the demand on the selfenergy

$$\frac{\partial \Sigma(11')}{\partial G(2'2)} = \frac{\partial \Sigma(22')}{\partial G(1'1)}. \tag{11.23}$$

This can only be fulfilled if we have a creating functional $\mathscr{S}$ with

$$\Sigma(11') = \frac{\delta \mathscr{S}}{\delta G(1'1)}. \tag{11.24}$$

One can express the density fluctuation function by the T-matrix defined in Figure 11.2 and we write (11.16) of Figure 11.1 as seen in Figure 11.3.

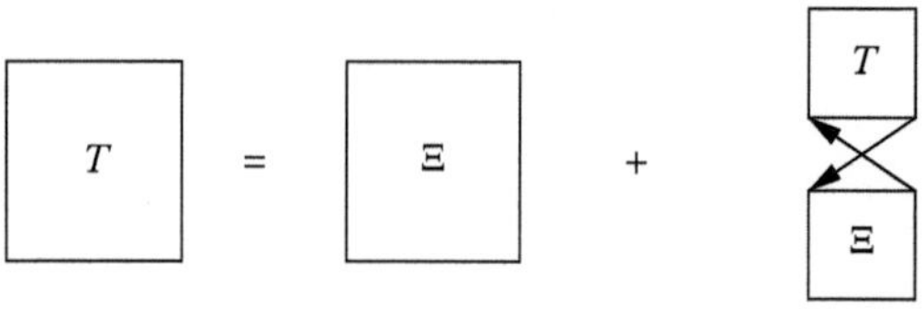

Figure 11.2:

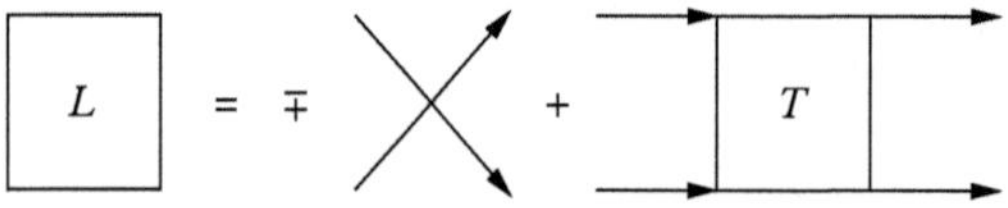

Figure 11.3:

11.2.1 Screening for Coulomb systems

For Coulomb systems it is of advantage to consider the variation with respect to the external and mean field

$$\bar{U}_{11'} = U_{11'} \mp iV_{12}G_{22^+}\delta_{11'}. \tag{11.25}$$

From (11.21) we see now that the screened selfenergy $\bar{\Sigma}$ which is the selfenergy diminished by the Hartree term $\Sigma + U = \bar{\Sigma} + \Sigma^H + U = \bar{\Sigma} + \bar{U}$ is written as

$$\bar{\Sigma}_{14}G_{41'} = iV_{12}\frac{\delta G_{11'}}{\delta U_{22}} = \mp i\Pi_{121'2}W_{12} \tag{11.26}$$

graphically seen in Figure 11.4.

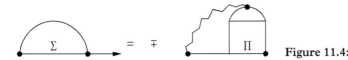

Figure 11.4:

To drive this, we have used the definition of the screened potential of Chapter 10.2.2

$$V_{12}\frac{\delta \bar{U}_{34}}{\delta U_{25}} = W_{12}\delta_{32}\delta_{45} \tag{11.27}$$

and the fact that the polarisation is given as the variation of the Green's functions with respect to the induced potential $\bar{U}$

$$\Pi_{13'1'3} = \mp\frac{\delta G_{11'}}{\delta \bar{U}_{33'}}. \tag{11.28}$$

Calculating the derivative in (11.27) explicitly with (11.25) one obtains

$$\frac{\delta \bar{U}_{33'}}{\delta U_{22'}} = \delta_{23}\delta_{2'3'} + iV_{36}\Pi_{656^+4}\frac{\delta \bar{U}_{45}}{\delta U_{22}}\delta_{33'} \tag{11.29}$$

and the equation for the screened potential becomes

$$W_{13} = V_{13} + iW_{12}V_{34}\Pi_{424^+2}. \tag{11.30}$$

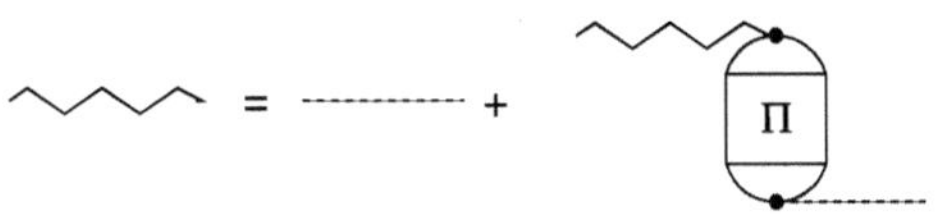

Using $\delta G = -G\delta G^{-1}G$ again we can express the polarisation function

$$\Pi_{12'1'2} = \mp G_{12}G_{2'1'} \mp G_{13}\frac{\delta\bar{\Sigma}_{34}}{\delta\bar{U}_{22'}}G_{41'} \tag{11.31}$$

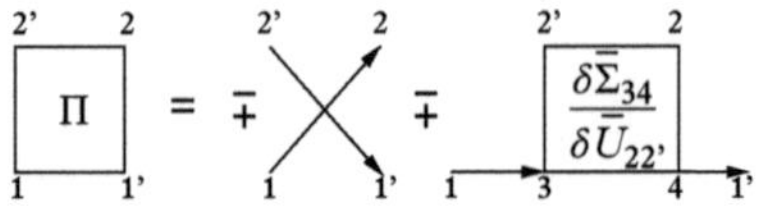

Comparing (11.16) and (11.31) and using the chain rule to express variations with respect to U by variations with respect to $\bar{U}$ we find a relation between L and Π expressed in Figure 11.5.

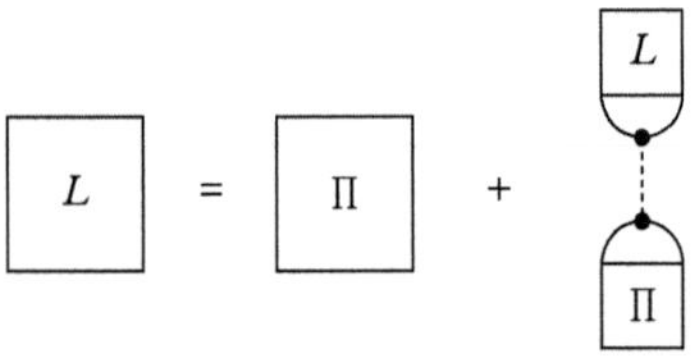

Figure 11.5:

11.3   Hedin equation

We can now define a vertex function

$$\Gamma_{123} = \delta_{12}\delta_{13} + \frac{\delta\bar{\Sigma}_{12}}{\delta\bar{U}_{33}} = \delta_{12}\delta_{13} + \frac{\delta\bar{\Sigma}_{12}}{\delta G_{45}}G_{46}\Gamma_{673}G_{75}. \tag{11.32}$$

Comparing it with (11.31) allows us to express the vertex as written in Figure 11.6.

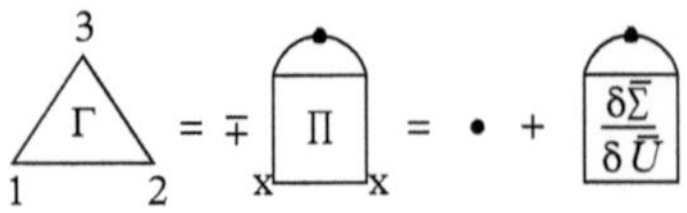

Figure 11.6: *Crosses mean amputated Green's functions.*

With the help of the vertex function (11.32) the selfenergy (11.26) becomes

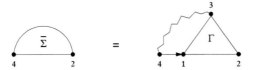

and the special form of the polarisation function needed in (11.31) can be expressed as

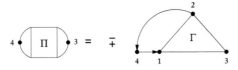

where the rotation of the vortex Γ as signed by the indices should be noted. Finally the screened potential (11.30) is written as

Together with the Dyson equation

where the thin line is G_0, the last five equations form a closed set. These Hedin equations (Hedin, 1965) are used in various applications. This scheme allows a systematic numbering of Feynman diagrams (Molinari, 2005) and has been solved exactly in zero dimensions (Pavlyukh and Hübner, 2007). Approximating the vertex by $\Gamma \approx 1$ one has the GW or RPA approximation. With this scheme one develops systematic vertex corrections beyond the GW approximation (Schindlmayr and Godby, 1998) giving evidence of the convergence of the expansion. It is as well useful to describe spin-dependent interactions (Aryasetiawan and Biermann, 2009). For an overview about recent numerical methods to solve GW approximations and corresponding Bethe-Salpeter equations, see (Leng, Jin, Wei and Ma, 2016).

11.3.1 Under which conditions does RPA become exact in the high-density limit of fermions?

The correlation energy in electron gases are a topic of long-time investigations. In order to avoid divergences in perturbation theory Macke (Macke, 1950) has already summed an infinite series of diagrams (RPA). Later Gell-Man and Brueckner have shown that the RPA at zero temperature becomes exact in a high-density limit (Gell-Mann and Brueckner, 1957) confirmed this up to orders of the logarithm of density (Wang and Perdew, 1991). The corresponding momentum distributions in RPA have been computed already in (Daniel and Vosko, 1960; Kulik, 1961) and recently an improved parametrisation has been presented by cummulant expansions (Gori-Giorgi and Ziesche, 2002). It confirms that the high-density limit is indeed given by the RPA calculation. The analytic expressions of the electron gas can be found in (Ziesche, 2007, 2010). An approximation bridging the low- and high-density expansion of the correlation energy has been provided by Perdew and Wang (1992).

A similar statement contains the Migdal theorem (Migdal, 1958) for an electron-phonon coupling, about the vanishing of higher-order vertex corrections by the order of the ratio of the phonon frequency to the Fermi energy. Violations of this theorem appear if the magnon frequency becomes large (Hertz, Levin and Beal-Monod, 1976) though it was argued that for very large phonon frequencies the Migdal theorem might hold again (Ikeda, Ogasawara and Sugihara, 1992). Strongly coupled electron-phonon systems become essentially non-adiabatic leading eventually to a polaron collapse (Aleksandrov, Grebenev and Mazur, 1987). This has dramatic consequences on the applicability of RPA calculations in high-T_c superconductivity requiring generalised Eliashberg equations (Grimaldi, Pietronero and Strässler, 1995a,b). Basically it was shown that for heavy fermion systems the Migdal theorem is not valid (Wojciechowski, 1999) and near the magnetic boundary the quasiparticle spectra is different from that in the Eliashberg theory (Monthoux, 2003) leading to a pseudogap formation.

Now we can analyse the exact Hedin equations with respect to their high-density dependence. Therefore we scale all momentum and energy variables explicitly in terms of Fermi momentum. Two of the five Hedin equations where the numbers give the space-time variable in general and the letters denote the frequency-momentum ones after Fourier transform in equilibrium read, e.g.

An integration of internal momenta and energy is proportional to $p_f^{d+\beta}$ where d is the dimension of the system and we assume a general relation between Fermi energy and momentum of $\epsilon_f \sim p_f^\beta$. For quadratic dispersion we have of course $\beta = 2$ but one might think of non-Fermi liquids as well. The used potential is supposed to have an $q^{-\alpha}$ dependence which leads to the scaling $V = [\overline{V}] p_f^{-\alpha}$ where we will denote the scaled function with the bracket [] and an over-line. Please note that this scaling is not providing a dimensionless quantity. We only extract the dependence on the Fermi momentum leaving still a dimension quantity. The latter is finite after this Fermi-momentum scaling. The Fermi momentum as the radius of the Fermi sphere is related to the density via $n \sim p_f^d$ where d is the dimensionality of the system. Very often the expansions are given in terms of the Wigner–Seitz radius which describes the radius of a sphere containing one particle. Therefore the scaling with the density is $r_s \sim n^{-1/d}$ and one has $p_f \sim 1/r_s$. The high-density limit is the high-Fermi momentum limit or the small r_s limit.

The propagator scales as inverse energy $G = [\overline{G}] \, p_f^{-\beta}$ and the four Hedin equations scale consequently

$$W = [\overline{V}] \, p_f^{-\alpha} + [\overline{V}] \, \Pi \, W \, p_f^{d+\beta-\alpha}, \qquad\qquad \Pi = [\overline{G}] \, [\overline{G}] \, \Gamma \, p_f^{d-\beta},$$

$$\Sigma = [\overline{G}] \, [\overline{W}] \, \Gamma \, p_f^{d-\alpha}, \qquad\qquad G = [\overline{G}_0] \, p_f^{-\beta} + [\overline{G}_0] \, \Sigma \, [\overline{G}] \, p_f^{d-\beta}. \qquad (11.33)$$

As a consistency check we see that from the first equation we deduce $\Pi = [\overline{\Pi}] \, p_f^{\alpha-\beta-d}$ and with that from the second one $\Gamma = [\overline{\Gamma}] \, p_f^{\alpha-2d}$. Therefore the third equation implies $\Sigma = [\overline{\Sigma}] \, p_f^{-d}$ which fits the fourth equation. Consequently, the equation for the vertex (11.32) itself scales as

$$\Gamma = 1 + \frac{\delta[\overline{\Sigma}][\overline{W}]\Gamma}{\delta[\overline{G}]} \, [\overline{G}] \, \Gamma \, [\overline{G}] \, p_f^{d-\beta-\alpha} = 1 + [...] \, \Gamma^2 \, p_f^{d-\beta-\alpha} = 1 + o\left(p_f^{d-\beta-\alpha}\right). \qquad (11.34)$$

Here we have used in (11.32) the expression for the selfenergy (8) before scaling. This integral equation for the vertex shows how further iterations scale with the orders of the Fermi momentum. In other words, the vertex function becomes unity in the first order of the power of Fermi momentum $p_f^{d-\beta-\alpha}$. In case that $d-\beta-\alpha < 0$ the RPA is exact in the high-density limit.

11.4 Ward identities

There exist variational identities which allow us to check the consistency of the used approximations. These relations discovered by Ward (1950) have soon been found out to be a consequence of gauge invariance and as such are the basis of field theories (Takahashi, 1957). Usually they are used in condensed matter and field theories in equilibrium. Let us derive them here for nonequilibrium situations as has been shown by Velický et al. (2008).

We consider higher-order correlations (11.16) which can be recast into a vertex function analogous to (11.32) defined as the deviation of the two-particle fluctuations from the Fock-term

$$\mp G_{51}^{-1} L_{121'2'} G_{1'6}^{-1}\big|_{2=2'} = \delta_{52}\delta_{26} + \frac{\delta\Sigma_{56}}{\delta U_{22}} = \delta_{52}\delta_{26} + \Lambda(5,6,2). \qquad (11.35)$$

Firstly we consider in the Dyson equation (11.15) a time local and homogeneous to field $U(t_x)$ as a $U(1)$-gauge field. The Dyson equation has to be gauge invariant which means all quantities like selfenergies and Green's functions have to have the form

$$G_U(t, t') = G(t, t') e^{-i \int_{t'}^{t} d\tau\, U(\tau)}. \tag{11.36}$$

This means we have the nonequilibrium Ward identity (Velický, Kalvová and Špička, 2008)

$$\Lambda(t, t', t_x) = \frac{\delta \Sigma(tt')}{\delta U(t_x)} = -i\Sigma(t, t') \int_{t'}^{t} d\tau \, \frac{\partial U(\tau)}{\partial U(t_x)}$$

$$= -i\Sigma(t, t') \Big[\Theta(t - t_x)\Theta(t_x - t') - \Theta(t' - t_x)\Theta(t_x - t) \Big]. \tag{11.37}$$

In fact one can convince oneself that this is indeed the Ward identity by considering the equilibrium where only differences of times occur $\tau = t - t_x$ and $\tau' = t' - t_x$. The Fourier transform leads to the known form

$$\Lambda(\omega, \omega') = \int d\tau e^{i\omega\tau} \int d\tau' e^{-i\omega'\tau} \Lambda(\tau, \tau') = -\frac{\Sigma(\omega) - \Sigma(\omega')}{\omega - \omega'} \tag{11.38}$$

and corresponding derivative forms if $\omega' \to \omega$.

On can now extend this exercise towards inhomogeneous systems and get for (11.37)

$$\Lambda(1, 1', t_2) = \frac{\delta \Sigma(11')}{\delta U(t_2)} = -i\Sigma(1, 1') \Big[\Theta(t - t_2)\Theta(t_2 - t') - \Theta(t' - t_2)\Theta(t_2 - t) \Big]. \tag{11.39}$$

We might integrate over time t_2 to get finally

$$\int dt_2 \Lambda(1, 1', t_2) = \int dt_2 \int d^3 r_2 \Lambda(1, 1', 2) = -i(t - t') \Sigma(1, 1'). \tag{11.40}$$

This means integrating the full inhomogeneous three-point vertex over one time and space variable leading to the selfenergy multiplied with the time difference of the latter. This establishes the Ward identity in nonequilibrium for inhomogeneous systems.

11.5   Asymmetric and cummulant expansion

These variational technique we can use now to derive more closely the structure of selfenergies. We have already seen that an asymmetry in the equation for the density fluctuation in Chapter 10.2.2 appears such that one propagator is only a Hartree and the second one a full propagator. Now we will develop the idea that such asymmetry is a generic feature (Morawetz, 2011). The idea of derivation follows here the centennial overview of many-body approximations by Heinz Puff (1979) which is suited for nonequilibrium Green's function expansions. Therefore all outlined formalism holds in nonequilibrium as well as equilibrium.

The n-particle Green's function can be formally represented by a generating functional

$$G(1, 2, ..., n, 1', 2', ..., n'; U) = \partial_{\eta_{n'}} ... \partial_{\eta_{1'}} \partial_{\lambda_1} ... \partial_{\lambda_n} \mathscr{G}[\lambda, \eta] \big|_{\lambda=\eta=0} \qquad (11.41)$$

with

$$\mathscr{G}[\lambda, \eta] = 1 + \sum_1^n \frac{1}{(n!)^2} \int d1 ... dn \, d1' ... dn' \lambda_n ... \lambda_1 \, G(1...n') \eta_{1'} ... \eta_{n'} \qquad (11.42)$$

where η, λ are Fermi/Bose-commuting auxiliary fields.

The Martin–Schwinger hierarchy (7.8) is expressed in terms of this generating functional by

$$\left[i\partial_{t_1} + \frac{\nabla^2_{r_1}}{2m} - U(1) \right] \partial_{\lambda_1} \mathscr{G} = \eta_1 \mathscr{G} \mp i \int d\bar{1} \, V(1, \bar{1}) \partial_{\eta_{\bar{1}+}} \partial_{\lambda_{\bar{1}}} \partial_{\lambda_1} \mathscr{G}. \qquad (11.43)$$

It is now useful to introduce the correlated n-particle Green's function as the cummulant expansion (Fulde, 1991) due to a new generating functional $\mathscr{G}_c$,

$$\mathscr{G}[\lambda, \eta] = \exp^{\mathscr{G}_c[\lambda, \eta]} \qquad (11.44)$$

with

$$\mathscr{G}_c[\lambda, \eta] = \sum_1^n \frac{1}{(n!)^2} \int d1 ... dn \, d1' ... dn' \lambda_n ... \lambda_1 \, G_c(1...n') \eta_{1'} ... \eta_{n'}. \qquad (11.45)$$

The comparison of (11.44), (11.45) and (11.42) with respect to the orders of λ, η reveals the cummulant expansion. The first order reads

$$G(11') = G_c(11'), \qquad (11.46)$$

i.e. the single-particle Green's function equals its correlated part. The second order shows just the separation of the Hartree–Fock term from the two-particle Green's function

$$G(121'2') = G_c(121'2') + G(11')G(22') \mp G(12')G(21'). \qquad (11.47)$$

The third order is given by all possible exchanges of three one-particle Green's functions together with all possible exchanges of the two-particle correlated Green's function and a one-particle Green's function.

Introducing (11.44) into (11.43) and notating the inverse Hartree–Fock Green's function with

$$
G_{\mathrm{HF}}^{-1}(12) = -iV(12)G(12) + \left[i\partial_{t_1} + \frac{\nabla_{r_1}^2}{2m} - U(1) \pm i \int d\bar{1}\, V(1\bar{1})G(\bar{1}\bar{1}^+) \right] \delta(1-2)
$$

$$(11.48)$$

one can invert the differential equation (11.43) into an integral equation

$$
\partial_{\lambda_1} \mathscr{G}_c = \mathscr{I}(1)
$$

$$(11.49)$$

with

$$
\mathscr{I}(1) = \int d\bar{1}\, G_{\mathrm{HF}}(1\bar{1})\eta_{\bar{1}} \mp i \int d\bar{1}\, d\bar{2}\, G_{\mathrm{HF}}(1\bar{1})V(\bar{1}\bar{2})
$$

$$
\times \left\{ \partial_{\eta_{\bar{2}^+}} \partial_{\lambda_{\bar{2}}} \partial_{\lambda_{\bar{1}}} \mathscr{G}_c + [\partial_{\eta_{\bar{2}^+}} \partial_{\lambda_{\bar{2}}} \mathscr{G}_c - G(2\bar{2}^+)] \partial_{\lambda_{\bar{1}}} \mathscr{G}_c \right.
$$

$$
\left. \mp [\partial_{\eta_{\bar{2}^+}} \partial_{\lambda_{\bar{1}}} \mathscr{G}_c - G(\bar{1}\bar{2}^+)] \partial_{\lambda_{\bar{2}}} \mathscr{G}_c + \partial_{\eta_{\bar{2}^+}} \mathscr{G}_c [\partial_{\lambda_{\bar{2}}} \partial_{\lambda_{\bar{1}}} \mathscr{G}_c + \partial_{\lambda_{\bar{2}}} \mathscr{G}_c \partial_{\lambda_{\bar{1}}} \mathscr{G}_c] \right\}.
$$

$$(11.50)$$

Using (11.45) we see from the definition that the correlated Green's functions can be represented by functional derivatives of (11.49),

$$
G_c(11') = \partial_{\eta_{1'}} \partial_{\lambda_1} \mathscr{G}_c \big|_{\lambda=\eta=0} = \partial_{\eta_{1'}} \mathscr{I}(1) \big|_{\lambda=\eta=0}
$$

$$
G_c(121'2') = \partial_{\eta_{2'}} \partial_{\eta_{1'}} \partial_{\lambda_1} \partial_{\lambda_2} \mathscr{G}_c \big|_{\lambda=\eta=0} = \mp \partial_{\eta_{2'}} \partial_{\eta_{1'}} \partial_{\lambda_2} \mathscr{I}(1) \big|_{\lambda=\eta=0}
$$

$$
G_c(1231'2'3') = \partial_{\eta_{3'}} \partial_{\eta_{2'}} \partial_{\eta_{1'}} \partial_{\lambda_1} \partial_{\lambda_2} \partial_{\lambda_3} \mathscr{G}_c \big|_{\lambda=\eta=0} = \partial_{\eta_{3'}} \partial_{\eta_{2'}} \partial_{\eta_{1'}} \partial_{\lambda_2} \partial_{\lambda_3} \mathscr{I}(1) \big|_{\lambda=\eta=0}
$$

$$(11.51)$$

and so on. We drop the notation of the explicit dependence on the external potential U in the following. The first equation of (11.51) yields for the one-particle correlated Green's function with (11.50)

$$
G_c(11') = G_{\mathrm{HF}}(11') \mp i \int d\bar{1}\bar{2}\, G_{\mathrm{HF}}(1\bar{1})V(\bar{1},\bar{2})G_c(\bar{1}\bar{2}1'\bar{2}^+)
$$

$$(11.52)$$

and is nothing else but the Martin–Schwinger hierarchy (7.8) written for the correlated parts and in integral form.

The integral equation for the two-particle Green's function is more involved and reads from the second equation of (11.51) with (11.50)

$$G_c(121'2') = i \int d\bar{1}d\bar{2}\, G_{\mathrm{HF}}(1\bar{1})V(\bar{1}\bar{2}) \left\{ \mp G_c(\bar{2}1\bar{2}2^+1'2') \right.$$

$$+ G(2\bar{2}^+)\left[G_c(\bar{1}\bar{2}1'2') + G(\bar{2}2')G(\bar{1}1') \mp G(\bar{2}1')G(\bar{1}2') \right]$$

$$+ G_c(\bar{1}2\bar{2}^+2')G(\bar{2}1') + G_c(\bar{1}1'2\bar{2}^+)G(\bar{2}2')$$

$$\left. \mp G_c(\bar{2}2\bar{2}^+2')G(\bar{1}1') \mp G_c(\bar{2}21'\bar{2}^+)G(\bar{1}2') \right\}. \tag{11.53}$$

This equation is graphically presented in Figure 11.7 and is exact so far. Since the following algebra is somewhat involved it is easier to perform it in terms of diagrammatic presentation. Therefore we design a complete selfconsistent one-particle Green's function with a thick full arrow, the interaction with a broken line and the two-, three-, and four-particle Green's functions with a square containing two, three and four legs. The Hartree–Fock Green's function is designed as thin arrow. Corresponding upper signs denote fermions and lower signs bosons. The numbering of in- and outgoing channels are in the direction from bottom to top. Lines without an arrow indicate ends of two- and three-particle Green's functions which are not connected with a one-particle Green's function.

From the derivation above it is not hard to see that the property

$$G_c(121'2') = \mp G_c(211'2') = \mp G_c(122'1') \tag{11.54}$$

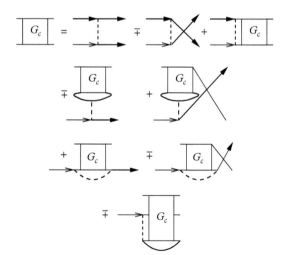

Figure 11.7: *The equation for the correlated two-particle Green's function. The thin open arrows are Hartree–Fock Green's functions, the thick arrows are full Green's functions. The first line represents the ladder approximation and the second and third lines exchange channels.*

is ensured. In other words the criteria (A) and (B) of Chapter 7.3.3 are fulfilled and we have a conserving approximation.

Please note that the corresponding diagrams in the second and third line in Figure 11.7 cancel each other for fermions and contact interaction, i.e. they become small for short-range interactions.

11.5.1 Binary collision approximation

As a first very drastic approximation we want to consider the binary correlation approximation which consists in neglecting the correlated three-particle Green's function in (11.53), i.e. the last diagram in Figure 11.7. Further, we neglect also the second and third line since they vanish for fermions and short-range interaction. This results in the known ladder approximation or Bethe–Salpeter equation. Amazingly we have now obtained that the intermediate propagators of the Bethe–Salpeter equation have to be considered asymmetrically. Indeed, the vertex

$$G_c(121'2') = \int d\bar{1}d\bar{2}d\bar{1}'d\bar{2}'\, G_{\mathrm{HF}}(1\bar{1})G(2\bar{2})\Gamma(\bar{1}\bar{2}\bar{1}'\bar{2}')G(\bar{1}'1')G(\bar{2}'2') \qquad (11.55)$$

can be represented in symmetrized form

$$\Gamma(121'2') = T(121'2') \mp T(122'1') \qquad (11.56)$$

by the T-matrix

$$T(121'2') = iV(12)\delta(1-1')\delta(2-2') + i\int d\bar{1}d\bar{2}V(12)G(2\bar{2})G_{\mathrm{HF}}(1\bar{1})T(\bar{1}\bar{2}1'2'). \qquad (11.57)$$

The asymmetry appears such that a full Green's function G_c is combined with a Hartree–Fock Green's function. This asymmetry is maintained if we go systematically to higher-order approximations neglecting 4-particle correlated Green's functions as we will see in Chapter 12. It is a result of the hierarchical structure of the equations of motion. If one compares it with the linearised parquet approximations (Janis, 2001, 2009), the Hartree–Fock propagator G_{HF} in Figure 11.7 is replaced by the full one G_c which includes nonphysical processes. We will see in Chapter 12 that the result derived here by the hierarchy avoids such nonphysically multiple scatterings with the same channel. Therefore we consider the asymmetric form in Figure 11.7 as superior to the parquet approximation.

11.5.2 Three-particle approximations

We go one step further now and approximate the equation for the three-particle correlation represented by the last line of (11.51). It shows the coupling to the four-particle

Green's function and is quite lengthy. For the sake of legibility we abbreviate the inter-changes of indices of the corresponding foregoing expressions by denoting them in the following formula within the same kind of brackets,

$$
\begin{aligned}
G_c(1231'2'3') = & \mp i \int d\bar{1}\,d\bar{2}\,G_{\mathrm{HF}}(1\bar{1})\,V(\bar{1}\bar{2})\Big(G_c(23\bar{2}\bar{1}\bar{2}^+1'2'3') \\
& \mp G_c(2\bar{2}\bar{1}1'2'3')\,G_c(3\bar{2}^+) \mp (2 \leftrightarrow 3) \\
& + \Big\{ G_c(23\bar{2}\bar{2}^+2'3')\,G_c(\bar{1}1') \mp (\bar{1} \leftrightarrow \bar{2}) \\
& \mp G_c(2\bar{2}\bar{2}^+3')\,G_c(3\bar{1}1'2') \mp (2 \leftrightarrow 3) \\
& + \Big[\mp G_c(3\bar{2}^+)\,G_c(2\bar{2}2'3')\,G_c(\bar{1}1') + (\bar{1} \leftrightarrow \bar{2}) \Big] \mp \Big[2 \leftrightarrow 3 \Big] \\
& + G_c(23\bar{2}^+1')[G_c(\bar{2}\bar{1}2'3') + G_c(\bar{2}2')\,G_c(\bar{1}3') \mp G_c(\bar{2}3')\,G_c(\bar{1}2')] \Big\} \\
& \mp \{1' \leftrightarrow 2'\} + \{1' \leftrightarrow 2' \leftrightarrow 3'\} \Big).
\end{aligned}
\tag{11.58}
$$

Now we consider selected approximations by neglecting the four-particle correlation and selecting special sets of diagrams. First let us show how the known channel approximations appear which are the screened ladder, the maximally crossed diagrams and the ladder diagrams. Then we will consider the pair-pair correlations and their influence on these three channels.

11.5.3 Screened-ladder approximation

We choose, as partial summation from (11.58), the first set of diagrams indicated in Figure 11.8. The diagrams obtained by interchanging $1' \leftrightarrow 2'$ and $1' \leftrightarrow 2' \leftrightarrow 3'$ in

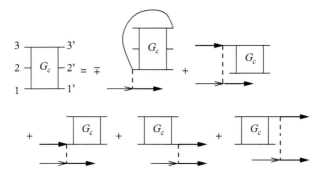

Figure 11.8: *A special set of diagrams for the three-particle correlated Green's function* (11.58).

Figure 11.9: *Definition of the screened potential V_S. Please note the difference to the symmetric one in Figure 10.5.*

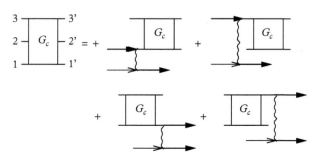

Figure 11.10: *The set of diagrams of Figure 11.8 when introducing the screened potential of Figure 11.9.*

Figure 11.8 are contained in (11.58) as well. Now the iteration of the equation for the three-particle correlated Green's function in Figure 11.8 leads to a repeated sum in the interaction lines which can be summarised by introducing the screened potential of Figure 11.9. This procedure results in the expression for the three-particle correlated Green's function as illustrated in Figure 11.10. Introducing this expression into the last diagram of Figure 11.7 one obtains the diagrams of Figure 11.11.

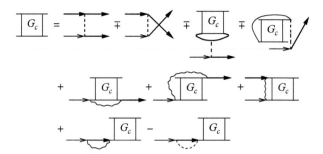

Figure 11.11: *The diagrams of Figure 11.7 when introducing the three-particle Green's function of Figure 11.10.*

We see that the last line normalises the single-particle propagator if brought to the left-hand side. In such a way the equation of two-particle Green's function in Figure 11.11 can be much simplified to which end we introduce the modified propagator

$$G_s = G_{\mathrm{HF}} + G_{\mathrm{HF}}(\Sigma_s - \Sigma_{\mathrm{F}})G_s = G_{\mathrm{H}} + G_{\mathrm{H}}(\Sigma_s)G_s = G_0 + G_0(\Sigma_s + \Sigma_{\mathrm{H}})G_s \qquad (11.59)$$

Figure 11.12: *The screened propagator* (11.59).

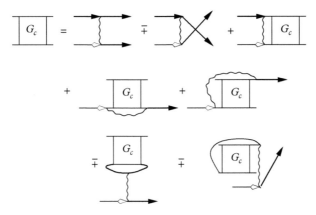

Figure 11.13: *The diagrams of Figure* 11.11 *when introducing the channel-dressed propagator* (11.59) *indicated as thin open arrows.*

where $\Sigma_s(12) = iW(12)G(12)$ is the screened selfenergy and $\Sigma_F(12) = iV(12)G(12)$ the Fock selfenergy as illustrated in Figure 11.12. We will call the following the modified propagator the channel-dressed selfconsistent propagator to distinguish it from the complete-dressed selfconsistent propagator. The channel-dressed propagator is determined by the corresponding selfenergy understood as the lowest selfconsistent diagram in the corresponding channel while the completely-dressed propagator includes all higher-order crossed terms. The introduction of this channel-dressed (screened) propagator results in the final expression of Figure 11.13.

Compared to the Kadanoff–Martin approximation which was represented by diagrams of Figure 11.7 neglecting the three-particle Green's function, we now obtain in Figure 11.13 the same set of diagrams except that the Hartree–Fock propagator has to be replaced by the modified one (11.59) and the bare interaction has to be replaced by the screened interaction of Figure 11.9 in the ladders. Please note that the screened interaction appears with asymmetric propagators.

Again we see that for fermions and short-range screened interaction the second and third line of diagrams in Figure 11.13 are cancelling mutually and only the screened-ladder diagram remains. Together with the screened potential of Figure 11.9 this is the screened-ladder approximation used before (Zimmermann, Kilimann, Kraeft, Kremp and Röpke, 1978) except that now one of the internal propagators has to be replaced by the screened one (11.59). This establishes the asymmetric form.

Please note that (11.59) already looks like the structure of subtracting unphysical repeated collisions (12.33), however, only the Fock term appears to be subtracted from the screened approximation selfenergy in the propagator (11.59) since the scheme accounts for it already in the Bethe–Salpeter equation.

11.5.4 Maximally crossed diagrams

A next set of diagrams included in (11.58) is summarised by the maximally crossed ladders presented in Figure 11.14 where the diagrams interchanging $1' \leftrightarrow 2'$ and $1' \leftrightarrow 2' \leftrightarrow 3'$ are contained in (11.58) as well. We proceed with the same steps as in Chapter 10, by introducing the expression of Figure 11.14 into the last diagram of Figure 11.7 but defining now the channel-dressed propagator

$$G_{\tilde{T}} = G_0 + G_0 \Sigma_{\tilde{T}} G_{\tilde{T}}$$
$$= G_{\mathrm{HF}} + G_{\mathrm{HF}}(\Sigma_{\tilde{T}} - \Sigma_{\mathrm{HF}})G_T \qquad (11.60)$$

to obtain the diagrams in Figure 11.15. Here the selfenergy reads

$$\Sigma_{\tilde{T}}(11') = \mp \int d\bar{1}\bar{2}\,\tilde{\Gamma}(\bar{1}\bar{1}\bar{2}1')G(\bar{2}\bar{1}^+) \qquad (11.61)$$

where the symmetrized vertex $\tilde{\Gamma}(1234) = \tilde{T}(1234) \mp \tilde{T}(1243)$ is expressed via the maximally crossed ladders of Figure 11.16.

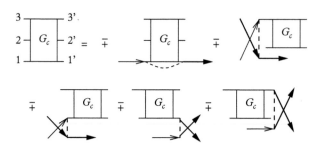

Figure 11.14: *The maximally crossed diagrams contained in* (11.58).

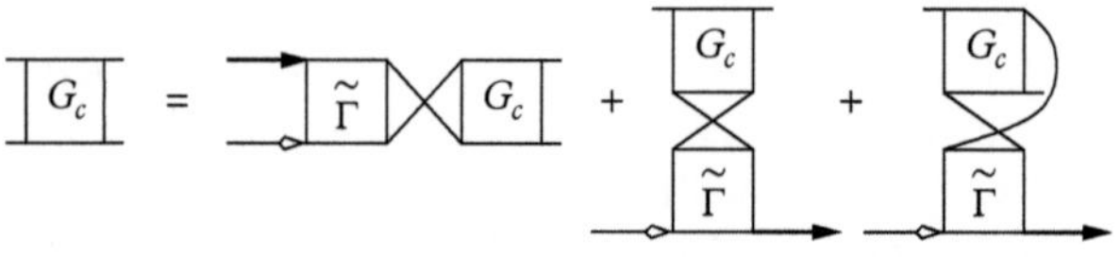

Figure 11.15: *The diagrams introducing Figure 11.14 into Figure 11.7. The thin open arrow marks the channel-dressed propagator* (11.60).

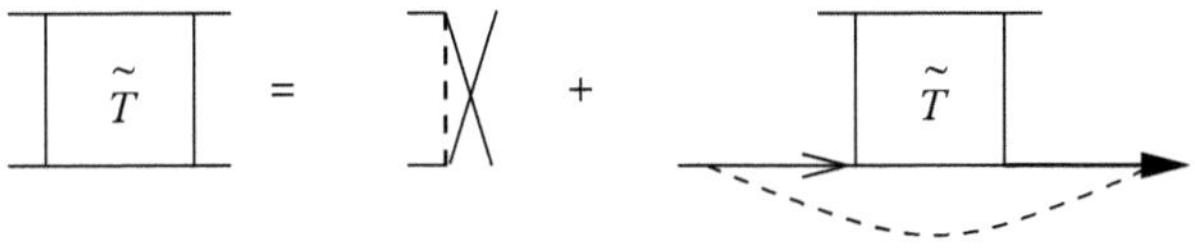

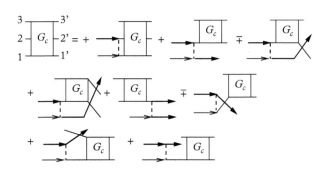

Figure 11.16: *The maximally crossed ladder summation.*

11.5.5 Ladder diagrams

As a last set we select the ladder diagrams included in (11.58) which are collected in Figure 11.17. Also the diagrams interchanging 2 ↔ 3 are contained in (11.58).

Introducing the expression of Figure 11.17 into the last diagram of Figure 11.7 with the channel-dressed propagator

$$G_T = G_0 + G_0 \Sigma_T G_T = G_{\mathrm{HF}} + G_{\mathrm{HF}}(\Sigma_T - \Sigma_{\mathrm{HF}})G_T \qquad (11.62)$$

we obtain the diagrams in Figure 11.18. Here the selfenergy reads

$$\Sigma_T(11') = \int d\bar{1}\bar{2}\, \Gamma(\bar{1}1\bar{2}1')G(\bar{2}\bar{1}^+) \qquad (11.63)$$

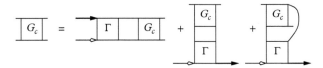

Figure 11.17: *The ladder diagrams contained in* (11.58).

Figure 11.18: *The diagrams introducing the three-particle Green's function of Figure 11.17 into Figure 11.7. The thin open arrows are the channel-dressed propagator (11.62).*

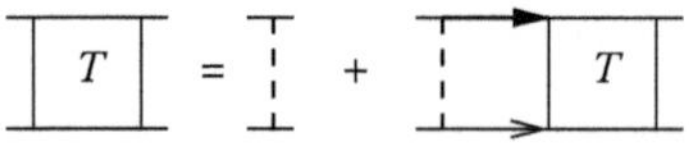

$$T = \cdot + \cdot\quad T$$

Figure 11.19: *The ladder summation.*

and the symmetrized vertex Γ was given by (11.56) in terms of the ladder T-matrix (11.57) presented in Figure 11.19.

11.5.6 Pair-pair correlation

Now that we have seen how the standard channels appear from the cummulant expansion of correlations in terms of Green's functions with special emphasis on the asymmetric propagators, we proceed and investigate the pair-pair correlation. In fact in (11.58) there are diagrams included which describe the interaction of the two two-particle Green's functions outlined in Figure 11.20.

From these diagrams we search only the ones which yield a renormalisation of the lower-left Hartree–Fock propagator after iteration into one of the above three channels, i.e. the first responsible diagram of Figures 11.8, 11.14 or 11.17 respectively. In such a way we can define a channel-dressed propagator as has been done repeatedly. It turns out that only the one diagram written in the second line of Figure 11.20 fulfils this task. All other diagrams give partially repeated iterations included in the above summations which partially lead to new cross diagrams. These diagrams we will not consider here. They have been partially considered in Born approximation named as cluster Hartree–Fock diagrams (Röpke, 1994) and have been applied to exciton problems (Röpke, Seifert, Stolz and Zimmermann, 1980) and to the first-order superfluid phase transition (Röpke, 1995). These diagrams describe the interaction between the cluster and the single particle, while we concentrate here on a genuine cluster–cluster diagram.

We introduce now this single normalising diagram from the last line in Figure 11.20 into the corresponding three-particle ones on the right-hand sides of Figures 11.8, 11.14 or 11.17. This leads to iterations which sum the interactions and result in the channel effective blocks, screened potential, maximally crossed vertex or T-matrix vertex. These diagrams are then again introduced into the last diagram of Figure 11.7 as was done

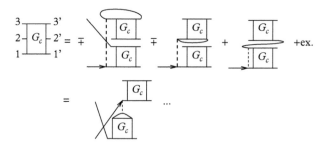

Figure 11.20: *The pair-pair diagrams in* (11.58).

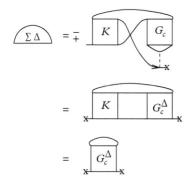

Figure 11.21: *The selfenergy of the normalised propagator due to pair-pair correlations in Figure 11.20 where the kernel K signs $V_s - V$ for the screened channel of Figure 11.13, $i\tilde{T}$ for the maximally crossed channel in Figure 11.15 and $iT + V$ for the ladder diagrams in Figure 11.18. The second line is valid only for separable two-particle correlations, see Figure 11.22, when using the screened diagrams of Figure 11.11. The third line appears using the ladder diagrams of Figure 11.18. Crosses denotes the inverse (amputated) propagator.*

repeatedly before. In this way we obtain an additional renormalisation diagram for the channel-dressed propagators of (11.59), (11.60) and (11.62) which can be written

$$G_\Delta = G_p - G_p \Sigma_\Delta G_\Delta \tag{11.64}$$

with $p = s, \tilde{T}, T$ denoting the channels. The corresponding selfenergy Σ_Δ due to the pair-pair interaction is shown in Figure 11.21 where the kernel K denotes the considered channels $V_s - V$, $i\tilde{T}$ and $iT + V$ correspondingly.

We want to consider specially the particle-particle channel represented by the T-matrix since in this channel in which the pairing appears. There we have

$$G_\Delta = = G_{\text{HF}} + G_{\text{HF}}(\Sigma_T - \Sigma_\Delta - \Sigma_{\text{HF}})G_\Delta$$
$$= G_0 + G_0(\Sigma_T - \Sigma_\Delta)G_\Delta. \tag{11.65}$$

For the singular channel, where the pairing appears, we will now see that the desired specific contribution remains. For this purpose we use the fact that the two-particle correlation separates near the pairing or condensation pole, illustrated in Figure 11.22 and Chapter 12.3. Taking this into account. We see from the screened-ladder diagrams in Figure 11.11 that besides the Hartree and Fock terms which appear to be subtracted in Figure 11.21, only the third term in Figure 11.11 remains as a connected diagram. Therefore we can replace the corresponding form in the second part of Figure 11.21 by a full two-particle propagator which results in the second line of Figure 11.21. Using the iteration of the T-matrix channel of Figure 11.18 once more, we obtain the third line in Figure 11.21.

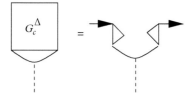

Figure 11.22: *Separation in in- and outgoing channels in the pole of pairing or condensation.*

11.6   Linear response

11.6.1   Basic relations

The response function χ describes the density change δn of the system when an external potential U^{ex} is applied. This disturbance creates an induced potential

$$V^{ind} = (V + f)\delta n + U^{ex} \qquad (11.66)$$

where V is the interaction among the particles and f describes local correlations deviating from the mean field which are called local fields. The density response with respect to this induced potential is the polarisation function Π. Therefore we have the identities

$$\delta n = \chi U^{ex} = \Pi V^{ind} = \frac{\Pi}{1 - (V + f)\Pi} U^{ex} = \chi^s W \qquad (11.67)$$

where we introduced the screened response $\chi^s = \Pi/(1 - f\Pi)$ and the screened potential

$$W = V\delta n + U^{ex}. \qquad (11.68)$$

The external field is given by the external densities $U^{ex} = Vn^{ex}$ and its charges determine the electric displacement field $divD = en^{ex}$ while the sources of the electric field are the present internal charges, i.e. $divE = e(\delta n + n^{ex})$. Therefore one gets the dielectric function as

$$\epsilon = \frac{n^{ex}}{\delta n + n^{ex}} = \frac{U^{ex}}{e\delta n + U^{ex}} = \frac{U^{ex}}{W} = \frac{\chi^s}{\chi} = 1 - V\frac{\Pi}{1 - f\pi} = 1 - V\chi^s. \qquad (11.69)$$

The inverse dielectric function can be written as

$$\frac{1}{\epsilon} = \frac{V\delta n}{U^{ex}} + 1 = 1 + V\chi \qquad (11.70)$$

11.6.2   Connection to diagrammatic expansions

With the help of (11.14), the density response to an external potential can be expressed in terms of the density-fluctuation function L of (11.16). Since the density is given by $\mp iG_{11^+} = <a_1^+ a_1> = <\hat{n}_1> = n_1$ we have from (11.14) and (11.16) for the response function χ to an external potential

$$\chi_{12} = \frac{\delta n_1}{\delta U_{22}} = +iL_{121^+2} = -i < (\hat{n}_1 - n_1)(\hat{n}_2 - n_2) > . \qquad (11.71)$$

The last identity follows from the definition of L and underlines the name-density fluctuation function. We see now that the linear-density variation due to an external potential can be written as

$$\delta n_1 = +iL_{121^+2^+}U_2 = \chi_{12}U_{22}. \qquad (11.72)$$

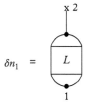

$$\delta n_1 \;=\;$$

Figure 11.23: *Density response* (11.72) *due to external perturbation (cross).*

Graphically we can express it as Figure 11.23 where we will design the external field as a dotted line ending with a cross. From this representation we see that the first-order Green's functions (11.13) with respect to the field and (11.16) is given in Figure 11.24.

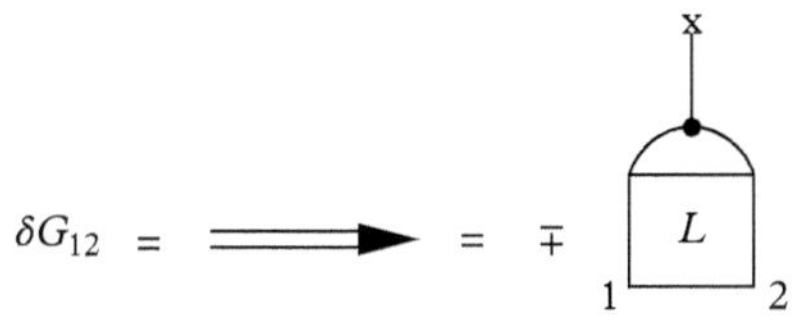

$$\delta G_{12} \;=\; \longrightarrow \;=\; \mp \;\boxed{L}$$

Figure 11.24: *First-order Green's function due to external perturbation (cross).*

This first-order Green's functions can be expressed from the Dyson equation

$$G = G_0^0 + G_0^0(\Sigma + U)G$$

$$G^0 = G_0^0 + G_0^0 \Sigma^0 G^0 \tag{11.73}$$

such that the total Green's function appears as

$$G = G^0 + G^0(\Sigma - \Sigma^0 + U)G. \tag{11.74}$$

Signing the selfenergy in first-order external field as $\Sigma - \Sigma^0 = \delta\Sigma + ...$ we see that the first-order variation of the Green's functions is

$$\delta G = G^0(\delta\Sigma + U)G^0 \tag{11.75}$$

or graphically seen in Figure 11.25.

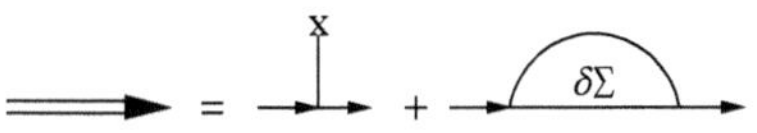

$$\longrightarrow \;=\; \cdot\!\downarrow\!\cdot \;+\; \overset{\frown}{\delta\Sigma}$$

Figure 11.25:

Comparing the last two equivalent representations of Figure 11.24 and Figure 11.25 for δG and using (11.16) we find that the first-order selfenergy with respect to the external field can be expressed via

$$\left\langle \delta\Sigma \right\rangle = \mp \;\boxed{L \;/\; \Xi}^{\,x}$$

11.6.3   Schema for constructing higher-order diagrams

The previously outlined method allows us now to construct higher-order diagrams in the response function employing lower-order diagrams in the selfenergy. One chooses an appropriate approximation for the selfenergy and performs the variation technique to calculate L_0 which involves higher-order diagrams due to the cutting of internal lines during the variation procedure (Kwong and Bonitz, 2000). The explicit schema is now as follows. Using Figs. 11.6 and 11.5 we can write

$$\mp\, L = \Gamma + \frac{L}{\Gamma}$$

which allows to rewrite the first-order Green's functions Figs. 11.25 or 11.24 as (Bonitz, Kwong, Semkat and Kremp, 1999)

$$\Longrightarrow \;=\; \mp\, L^{x} \;=\; \Gamma^{x} \;+\; \frac{L}{\Gamma}^{x}$$

According to (11.72) and Figure 11.23 we can express the first-order response function

$$\chi_{12} = i\, L \Big|_{1}^{2} = i\,() \mp i\, T \Big|_{1}^{2}$$

where we have used the definition of the T-matrix in the u-channel of Figure 11.2 and the vertex can be expressed as well

11.6.4 Linearising kinetic equations

The outline schema above can be used now for obtaining higher-order response functions from kinetic equations. Using a lower-order diagram for the selfenergy in the kinetic equation but formulating all field dependencies carefully allows us to linearise with respect to all field effects. Since the kinetic equation describes the propagator in the used approximation for Σ the linearisation leads to a higher-order of diagrams as given by Figures 11.2 and 11.3.

Let us consider two examples. For the simplest approximation of Hartree and Fock which is (11.21) without the last part, we obtain from definition in Fig. 11.1.

$$\Xi_{1524} = iV_{12}\delta_{25}\delta_{14} \mp iV_{14}\delta_{54}\delta_{12} \tag{11.76}$$

and we have from Figure 11.2

which is the usual ladder approximation in the u-channel. This means linearising the Hartree–Fock kinetic equation (Vlasov) gives a response function of au-channel ladder approximation.

Another example is the screened (GW) approximation which has lead us to the Lenard–Balescu kinetic equation (10.50). Here the order of diagrams created in the response function includes maximally crossed diagrams. The complete scheme is illustrated here as

We conclude that linearising the kinetic equation has the advantage that higher-order diagrams are created in the response function than used in the kinetic equation. This comes from linearisation which cuts diagrams according to the external potential. We explore such procedures further when establishing the link to renormalisation methods in Chapter 11.8.

11.7   Response in finite systems

We want to discuss now the quantum response function in a finite system. We start from the quantum (Hartree) kinetic equation (10.15) where we approximate any collision in the bulk matter in relaxation-time approximation (Kadanoff and Baym, 1962)

$$\frac{\partial}{\partial t}\rho(\mathbf{p}, \mathbf{x}, t) + \frac{\mathbf{p}}{m}\frac{\partial}{\partial \mathbf{x}}\rho(\mathbf{p}, \mathbf{x}, t) - \int \frac{d^3 p' d^3 s}{(2\pi \hbar)^3 i}\left[U\left(\mathbf{x} + \frac{\mathbf{s}}{2}\right) - U\left(\mathbf{x} - \frac{\mathbf{s}}{2}\right)\right]e^{\frac{i}{\hbar}\mathbf{s}(\mathbf{p}' - \mathbf{p})}\rho(\mathbf{p}', \mathbf{x}, t)$$

$$= \frac{\rho_0(\mathbf{p}, \mathbf{r}) - \rho(\mathbf{p}, \mathbf{r}, t)}{\tau} \tag{11.77}$$

with $U = V^{\text{ind}} + U^{\text{ext}}$ containing the external and the mean field. The first-order gradient expansion in U leads to the quasiclassical expression (10.16). For details of relaxation times see Chapter 18.

When the equation (11.77) is linearised with respect to the external perturbation, the selfconsistent potential V^{ind} gives a linear density contribution δn via $\delta V^{\text{ind}} = V\delta n$. As described above in Chapter 11.6.1, the response function is the connection between induced density variation and external perturbation (11.67)

$$\delta n(\mathbf{x}, \omega) = \int d^3 x'\, \chi(\mathbf{x}, \mathbf{x}', \omega)\, U^{\text{ext}}(\mathbf{x}', \omega). \tag{11.78}$$

Therefore one finds the relation between the response function χ including the effect of the selfconsistent potential and the polarisation Π_τ. Without this selfconsistent potential

$$\chi(\mathbf{x}, \mathbf{x}') = \Pi_\tau(\mathbf{x}, \mathbf{x}') + \int d^3\bar{x} d^3\bar{\bar{x}}\, \Pi_\tau(\mathbf{x}, \bar{\mathbf{x}})V(\bar{\mathbf{x}}, \bar{\bar{\mathbf{x}}})\chi(\bar{\bar{\mathbf{x}}}, \mathbf{x}'). \tag{11.79}$$

In other words, it is sufficient to concentrate on the response function Π_τ to an external potential without the selfconsistent potential V^{ind}. The response χ is then given by the solution of the integral equation (11.79). The following derivation of Π_τ is generalising the quasi-classical one by Kirzhnitz et al. (1975).

We introduce the Lagrange picture by following the trajectory $\mathbf{x}(t), \mathbf{p}(t)$ of a particle in phase space. The trajectories are now described instead of the quasi-classical ones (10.18) by the following set

$$\frac{d}{dt}\mathbf{x} = \frac{\mathbf{p}}{m}, \qquad \mathbf{s} \cdot \frac{d}{dt}\mathbf{p} = U\left(\mathbf{r} + \frac{\mathbf{s}}{2}\right) - U\left(\mathbf{r} - \frac{\mathbf{s}}{2}\right) \tag{11.80}$$

where the arbitrary vector $\mathbf{s}$ shows the infinite possibilities of trajectories by quantum fluctuations.[1]

We linearise the kinetic equation equation (11.77) around the stationary state ρ_0 according to $\rho(\mathbf{x}, p, t) = \rho_0(\mathbf{x}, p) + \delta\rho(\mathbf{x}, p, t)e^{-t/\tau}$ and obtain

$$\frac{d}{dt}\delta\rho[\mathbf{x}(t), \mathbf{p}(t), t] = \int d^3x' 2\delta\left[\mathbf{x}' - \mathbf{x}(t) + \mathbf{s}/2\right] \int \frac{d^3p' d^3s}{(2\pi)^3} \sin\left[\mathbf{s}(\mathbf{p}' - \mathbf{p}(t))\right]$$

$$\times \rho\left(\mathbf{p}', \mathbf{x}' - \frac{\mathbf{s}}{2}\right) e^{\frac{i}{\tau}} V^{\text{ext}}(\mathbf{x}', t) \tag{11.81}$$

where we introduced an integration over a dummy variable x'. Integrating over time from $(-\infty, t)$ and about momentum p, multiplying with $\exp(-t/\tau)$, we get the deviation of the density brought by the external field. Conveniently we shift $\tilde{x}(t') = x(t + t')$ and $\bar{t} = t' - t$ to obtain

$$\delta n(\mathbf{x}, t) = \int d^3x' \int_{-\infty}^{0} d\bar{t} 2\delta\left[\mathbf{x}' - \tilde{\mathbf{x}}(\bar{t}) + \mathbf{s}/2\right] \int \frac{d^3p' d^3p d^3s}{(2\pi)^6} \sin\left[\mathbf{s}(\mathbf{p}' - \mathbf{p})\right]\rho(\mathbf{p}', \mathbf{x}' - \mathbf{s}/2)$$

$$\times e^{\frac{i}{\tau}} V^{\text{ext}}(\mathbf{x}', t + \bar{t}) \tag{11.82}$$

where we denote the spatial coordinates at the present time $\mathbf{x} = \mathbf{x}(t) = \tilde{\mathbf{x}}(0)$. After the Fourier transform we can read off the polarisation function for a finite quantum system $\Pi_\tau(\omega) = \Pi(\omega + \frac{i}{\tau})$ with Morawetz (2000b),

$$\Pi(\mathbf{x}, x', \omega) = 2 \int_{-\infty}^{0} d\bar{t} \delta\left[\mathbf{x}' - \tilde{\mathbf{x}}(\bar{t}) + \frac{\mathbf{s}}{2}\right] \int \frac{d^3p' d^3p d^3s}{(2\pi)^6} \sin(\mathbf{s}(\mathbf{p}' - \mathbf{p}))\rho\left(\mathbf{p}', \mathbf{x}' - \frac{\mathbf{s}}{2}\right) e^{-i\omega\bar{t}} \tag{11.83}$$

including the relaxation time.

The knowledge of the evolution of all trajectories is necessary to evaluate this formula. Molecular dynamical simulations can perform this task but it requires an astronomical amount of memory to store all trajectories. Rather, we discuss approximations which will give us more insight into the physical processes behind it.

---

[1] In fact introducing (11.80) into (11.77) one gets, for the integral term,

$$-\dot{\mathbf{p}} \int \frac{d^3s d^3p'}{(2\pi)^2} \partial_{\mathbf{p}'} e^{i\mathbf{s}(\mathbf{p}' - \mathbf{p}(t))} \rho[\mathbf{p}', \mathbf{x}(t), t] = \dot{\mathbf{p}}\partial_{\mathbf{p}}\rho[\mathbf{p}', \mathbf{x}(t), t] \tag{11.84}$$

and together with the drift a total time derivative $\frac{d}{dt}\rho[\mathbf{p}'(t), \mathbf{x}(t), t]$ appears.

If we consider only the first-order of the history of the trajectories

$$\mathbf{x}' - \tilde{\mathbf{x}}(\bar{t}) \approx -(\mathbf{x} - \mathbf{x}') - \bar{t}\dot{\tilde{\mathbf{x}}}(0) \tag{11.85}$$

and neglect higher gradients in the centre-off-mass coordinate $\mathbf{R} = (\mathbf{x} + \mathbf{x}')/2$

$$\rho_0\left(\mathbf{p}', \mathbf{x}' - \frac{\mathbf{s}}{2}\right) = \rho_0\left(\mathbf{p}', \mathbf{R} - \frac{\mathbf{r}+\mathbf{s}}{2}\right) \approx \rho_0(\mathbf{p}', \mathbf{R}) + o(\partial_R) \tag{11.86}$$

we can Fourier transform $\mathbf{x} - \mathbf{x}' \to \mathbf{q}$ to get $\dot{\tilde{\mathbf{x}}}(0) = \mathbf{p}/m$

$$\Pi_\tau(\mathbf{q}, \mathbf{R}, \omega) = 2\int_{-\infty}^{0} d\bar{t}\, e^{-i(\omega + \frac{i}{\tau} - \frac{\mathbf{pq}}{m})\bar{t}} \int \frac{d^3p'd^3p}{(2\pi)^6} \frac{\delta(\mathbf{p}' - \mathbf{q}/2) - \delta(\mathbf{p}' + \mathbf{q}/2)}{2i} \rho(\mathbf{p}' + \mathbf{p}, \mathbf{R})$$

$$= \int \frac{d^3p}{(2\pi)^3} \frac{\rho_0(\mathbf{p} + \mathbf{q}/2, \mathbf{R}) - \rho_0(\mathbf{p} - \mathbf{q}/2, \mathbf{R})}{\omega + \frac{i}{\tau} - \frac{\mathbf{pq}}{m}} \tag{11.87}$$

the quantum Lindhard result in local density approximation. The quasiclassical approximation is given by expanding the distribution in first-order q.

The quasiclassical approximation we could have performed before (11.83) by expanding in the first order of s. If we use Fermi functions at zero temperature the original Kirzhnitz formula (Kirzhnitz, Lozovik and Shpatakovskaya, 1975; Dellafiore, Matera and Brink, 1995) appears

$$\Pi_0(\mathbf{x}, \mathbf{x}', \omega) = -\frac{mgp_f(\mathbf{x})}{4\pi^2\hbar^3}\left[\delta(\mathbf{x}' - \mathbf{x}(0)) + i\omega \int_{-\infty}^{0} dt'\, e^{-it'\omega} \int \frac{d\Omega_\mathbf{p}}{4\pi} \delta(\mathbf{x}' - \mathbf{x}(t'))\right], \tag{11.88}$$

where the angular integration of $d\mathbf{p}$ remains as $d\Omega_p$. This formula represents the ideal free part and a contribution which arises by the trajectories $\mathbf{x}(t)$ averaged over the direction at the present time $\mathbf{n}_p p_f = m\dot{\mathbf{x}}(0)$. We see that due to quantum fluctuations an additional integration $\mathbf{s}$ appears in (11.83) which provides the quantum generalisation of the quasiclassical Kirzhnitz formula (11.88) for the response function in finite systems.

Now we focus on the influence of an additional chaotic scattering which might be caused e.g. by a surface boundary. In order to investigate this effect we add to the regular motion (11.85) the effect of an irregular motion characterised by an additional Lyapunov exponent (2.20)

$$\mathbf{x}' - \mathbf{x}(\bar{t}) \to \left[\mathbf{x}' - \mathbf{x}(\bar{t})\right] e^{-\lambda\bar{t}} \approx \left[-\mathbf{r} - \bar{t}\frac{\mathbf{p}}{m}\right] e^{-\lambda\bar{t}}. \tag{11.89}$$

Using this in (11.83) it is easy to see that we obtain

$$\Pi_\tau(\mathbf{q}, \mathbf{R}, \omega) = \int \frac{d^3p'd^3s}{(2\pi)^6} \rho(\mathbf{p}' + \mathbf{p}, \mathbf{R}) \frac{e^{i\mathbf{s}\cdot(\mathbf{p}'-\mathbf{q}/2)} - e^{-i\mathbf{s}\cdot(\mathbf{p}'+\mathbf{q}/2)}}{\omega + i\left(\frac{1}{\tau} + \lambda\right) - \frac{\mathbf{pq}}{m} + \lambda\frac{\mathbf{s}\cdot\mathbf{q}}{2}} \tag{11.90}$$

which shows how additional chaotic motion influences the quantum response. Since the wave vector q scales typically with the Fermi velocity $v_f = p_f/m$, for the condition

$$\lambda \ll qv_F, \tag{11.91}$$

we can neglect the $\lambda \mathbf{s} \cdot \mathbf{q}$ term in the denominator and obtain the Lindhard polarisation (11.87) but with an additional complex shift

$$\Pi_\lambda(\mathbf{q}, \mathbf{R}, \omega) = \Pi_\tau(\mathbf{q}, \mathbf{R}, \omega + i\lambda). \tag{11.92}$$

This represents the known Matthiessen rule which states that the damping mechanisms are additive in the damping $\Gamma = \frac{1}{\tau} + \lambda$.

We would like to point out that this result has far-reaching consequences. Under the assumption (11.91), we have shown in this way that the linear response behaviour is the same if dissipation comes from the relaxation time via collision processes in many-particle theories or from the concept of chaotic processes characterised by the Lyapunov exponent. We can therefore state that for a small Lyapunov exponent compared to the product of wave vector and Fermi velocity in a many particle system, the largest Lyapunov exponent behaves like the relaxation time in the response function. Application to the damping of giant resonances in deformed nuclei can be found in (Morawetz, Vogt, Fuhrmann, Lipavský and Špicka, 1999*d*).

Since the transport theory is well worked out to calculate the transport coefficients in relaxation time approximation we can express by this way the transport coefficients in terms of the Lyapunov exponent alternatively. This illustrates the mutual equivalence of the concept of Lyapunov exponent and dissipative processes in many-particle theories.

11.8 Renormalisation techniques

11.8.1 Low-energy degrees of freedom

In many applications it is not necessary to know all the information about the complete energy scale but sufficient to control the low-energy scale. Therefore one likes to integrate out high-energy modes. The most effective way is to absorb such unwanted scales into effective interactions. This is a typical renormalisation question, how a cut-off in the energy scale can be transferred into an effective interaction describing the same physics without explicit knowledge of the upper scale. Constructing the theory on the appropriate low-energy degrees of freedom; respecting relevant symmetries, goes back to the idea of effective field theory (Weinberg, 1990). Two-particle correlations are described by the scattering T-matrix or by taking into account the many-body environment the Bethe–Salpeter equation. Using a cut-off in energy scales (Birse, McGovern and Richardson, 1999; Bogner, Schwenk, Kuo and Brown, 2001; Birse, 2011) allows us to construct an effective potential which absorbs this energy cut rendering the T-matrix unchanged and such all physical properties of binary correlations. The method is used to integrate out high-energy modes in nuclei (Bogner, Schwenk, Kuo and Brown, 2001; Bogner, Kuo, Schwenk, Entem and Machleidt, 2003; Hergert, Bogner, Morris, Schwenk and Tsukiyama, 2016).

Let us start with the retarded two-particle Bethe–Salpeter equation (10.156)

$$T_0(\mathbf{k}, \mathbf{k}', \omega) = V_0(\mathbf{k}, \mathbf{k}') + \int \frac{d^3 p}{(2\pi \hbar)^3} V_0(\mathbf{k}, \mathbf{p}) \mathscr{G}_0(\mathbf{p}, \omega) T(\mathbf{p}, \mathbf{k}', \omega) \qquad (11.93)$$

with the two-particle retarded propagator

$$\mathscr{G}_0(\mathbf{p}, \omega) = \frac{1 \mp f_1^{F/B}\left(\mathbf{K} + \frac{\mathbf{p}}{2}\right) \mp f_2^{F/B}\left(\mathbf{K} - \frac{\mathbf{p}}{2}\right)}{\omega - \frac{p^2}{2m}} \qquad (11.94)$$

where $\mathbf{K}$ is the centre-of-mass momentum and we absorbed the two particle threshold in the frequency scale $\omega = \Omega - K^2/2M$ with $M = m_1 + m_2$ and the reduced mass m. The Fermi/Bose distributions $f_i^{F/B}(p) = (\exp(p^2/2m_i - \mu) \pm 1)^{-1}$ are describing the influence of the surrounding medium on the two-particle process for Fermions or Bosons respectively. We see that the centre-of-mass momentum introduces explicit angular dependencies into the argument of distribution functions. In order to facilitate further discussion we will use the angle-averaged Pauli blocking/enhancement factor

$$Q(p) = \left\langle 1 \mp f_1^{F/B}\left(\mathbf{K} + \frac{\mathbf{p}}{2}\right) \mp f_2^{F/B}\left(\mathbf{K} - \frac{\mathbf{p}}{2}\right) \right\rangle \qquad (11.95)$$

which is not essential but serves here for legibility to render the integration in (11.93) one-dimensional. We drop the silent dependence on K in the following.

Now we search for an effective dynamical potential $V_\lambda(k, k', \omega)$ which renders the T-matrix unchanged $T = T_\lambda$ if we use a smooth cut-off function $f_\lambda(p)$ in the propagator

$$\mathscr{G}_\lambda = f_\lambda^2(p) \mathscr{G}_0(p\omega) \qquad (11.96)$$

which has the property to render low momenta unchanged $\lim_{p \ll \lambda} f_\lambda(p) \to 1$ and cut out high momenta $\lim_{p \gg \lambda} f_\lambda(p) \to 0$ corresponding to the interesting momentum scale λ. Such a smooth cut-off has some better properties than the sharp cut-off included here in the form used by (Bogner, Furnstahl, Ramannan and Schwenk, 2007). The Bethe–Salpeter equation for the effective potential reads

$$T_\lambda = V_\lambda + V_\lambda \mathscr{G}_\lambda T_\lambda \qquad (11.97)$$

with the product understood as the integrations in (11.93). In this operator notation we have $V_\lambda = T_\lambda(1 + \mathscr{G}_\lambda T_\lambda)^{-1}$ and since we demand an unchanged T-matrix, $T_\lambda = T_0$, we obtain the renormalisation group (RG) equation $dT_\lambda/d\lambda = 0$ which allows to write

$$\frac{dV_\lambda}{d\lambda} = -T_\lambda (1 + \mathcal{G}_\lambda T_\lambda)^{-1} \frac{d\mathcal{G}_\lambda T_\lambda}{d\lambda} (1 + \mathcal{G}_\lambda T_\lambda)^{-1} = -V_\lambda \frac{d\mathcal{G}_\lambda}{d\lambda} V_\lambda. \tag{11.98}$$

Writing this equation into explicit integrations

$$\frac{d}{d\lambda} V_\lambda(k, k', \omega) = \frac{m}{\pi^2 \hbar^3} \int_0^\infty dp \frac{p^2}{p^2 - 2m\omega} V_\lambda(k, p, \omega) \frac{df_\lambda^2(p)}{dp} V_\lambda(p, k, \omega) \tag{11.99}$$

we have the RG equation for the effective potential. It has been given for reduced functions and free two-particle scattering in (Bogner, Furnstahl, Ramannan and Schwenk, 2007) and using a sharp cut-off $f_\lambda(p) = \Theta(\lambda - p)$ the equation of Birse et al. (1999) appears

$$\frac{d}{d\lambda} V_\lambda(k, k', \omega) = \frac{m}{\pi^2 \hbar^3} \frac{\lambda^2}{\lambda^2 - 2m\omega} V_\lambda(k, \lambda, \omega) V_\lambda(\lambda, k', \omega). \tag{11.100}$$

It is now possible to integrate (11.99). To this aim we use (11.98) as well as the operator identity $dX^{-1} = -X^{-1} dX X^{-1}$ to write

$$-\frac{dV_\lambda^{-1}}{d\lambda} = V_\lambda^{-1} \frac{dV_\lambda}{d\lambda} V_\lambda^{-1} = -\frac{d\mathcal{G}_\lambda}{d\lambda} \tag{11.101}$$

which is easily integrated versus λ from 0 to λ and multiplying from the right with V_λ and from the left with V_0 one obtains

$$V_\lambda = V_0 + V_0 (\mathcal{G}_0 - \mathcal{G}_\lambda) V_\lambda \tag{11.102}$$

or explicitly

$$V_\lambda(k, k', \omega) = V_0(k, k') + \frac{m}{\pi^2 \hbar^3} \int_0^\infty dp \frac{p^2}{2m\omega - p^2} V_0(kp) \left[1 - f_\lambda^2(p)\right] V_\lambda(pk\omega) \tag{11.103}$$

called Bloch–Horowitz equation (Bogner, Furnstahl, Ramannan and Schwenk, 2007). We will use either (11.99) or (11.103) for determining the effective potential explicitly.

The equation (11.99) has the advantage that the fixed points can be determined conveniently. Therefore let us repeat first here the steps of Birse (Birse, McGovern and Richardson, 1999; Birse, 2011; Birse, Epelbaum and Geglia, 2016) for the hard cut-off (11.100). In order to get a real renormalisation group equation one has to scale the arguments dimensionless

$$V_\lambda(k, k', \omega) = V_\lambda(\bar{k}\lambda, \bar{k}'\lambda, \bar{\omega}\lambda^2) = \frac{1}{a\lambda} \bar{V}_\lambda(\bar{k}, \bar{k}', \bar{\omega}) \tag{11.104}$$

with $a = mv_0/\pi^2\hbar^3$ where we absorb the strength of the potential v_0. One obtains

$$
\begin{aligned}
\lambda\frac{d}{d\lambda}\bar{V}(\bar{k},\bar{k}',\bar{\omega}) &= \lambda a V_\lambda(k,k',\omega) + a\lambda^2\frac{d}{d\lambda}V_\lambda(\bar{k}\lambda,\bar{k}'\lambda,\bar{\omega}\lambda^2) \\
&= \bar{V}(\bar{k},\bar{k}',\bar{\omega}) + a\lambda\left(\bar{k}\frac{\partial}{\partial\bar{k}} + \bar{k}'\frac{\partial}{\partial\bar{k}'} + 2\bar{\omega}\frac{\partial}{\partial\bar{\omega}}\right)V_\lambda + \lambda\frac{d}{d\lambda}V_\lambda(k,k',\omega) \\
&= \bar{V} + \left(\bar{k}\frac{\partial}{\partial\bar{k}} + \bar{k}'\frac{\partial}{\partial\bar{k}'} + 2\bar{\omega}\frac{\partial}{\partial\bar{\omega}}\right)\bar{V} + \frac{\bar{V}(\bar{k},\bar{k}',\bar{\omega})V(1,\bar{k}',\bar{\omega})}{1-2m\bar{\omega}} \quad (11.105)
\end{aligned}
$$

where we used (11.100) in the last step. This established a real renormalisation group equation which shows how the different dimensionless variables scale.

11.8.2   Results for separable interaction

In order to discuss it further we assume a separable interaction, see appendix E,

$$
V_0(k,k') = v_0 g_k g'_k \quad (11.106)
$$

which leads to $\bar{V}(\bar{k},\bar{k}',\bar{\omega}) = g_{\bar{k}}g_{\bar{k}'}v(\bar{\omega})$ and (11.105) reduces to

$$
\lambda\frac{d}{d\lambda}v(\bar{\omega}) = v(\bar{\omega}) + 2\bar{\omega}\frac{d}{d\bar{\omega}}v(\bar{\omega}) + \frac{v^2(\bar{\omega})}{1-2m\bar{\omega}}. \quad (11.107)
$$

On the scattering side $\bar{\omega} = \frac{\bar{p}^2}{2m}$ one can now search for the low-energy expansion as $v = b_0 + b_0\bar{p}^2 + o(\bar{p})^4$ and from (11.107) one obtains the differential equation system

$$
\lambda\frac{db_0}{d\lambda} = b_0 + b_2, \qquad \lambda\frac{db_2}{d\lambda} = 3b_2 + 2b_2b_0 + b_0^2 \quad (11.108)
$$

which is easily solved as

$$
b_0(\lambda) = \frac{c_1\lambda}{1-c_1\lambda}, \qquad b_2(\lambda) = \frac{c_2\lambda^3 - c_1^2\lambda^2}{(1-c_1\lambda)^2} \quad (11.109)
$$

with two arbitrary constants. The fixed points are seen from (11.108) to be the unstable one $(b_0,b_2) = (-1,-1)$ and stable one $(b_0,b_2) = (0,0)$. For a range of c_1 and c_2 in Figure 11.26 the flow of $b_2(\lambda)$ versus $b_0(\lambda)$ is plotted.

Now we are going to solve (11.99) and (11.98) explicitly with the help of the separable potential (11.106) For any parametrisation of the form factor g_q the equation (11.93) for the two-particle T-matrix is solved as

$$
T_0(k,k',\omega) = v_0\frac{g_k g_{k'}}{1-\langle g^2\rangle} \quad (11.110)
$$

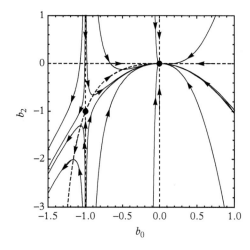

Figure 11.26: *The flow of first two parameter of the low-energy expansion $v = b_0(\lambda) + b_2(\lambda)\bar{p}^2$ when λ is lowered for a couple of different choices of c_1, c_2 in* (11.109). *The critical line connecting the two fixed points (dashed) is given by the zero of the right-hand side of second line of* (11.109) *and eliminating λ becoming $b_2 = -b_0^2/(3 + 2b_0)$.*

with

$$\langle g^2 \rangle = v_0 \mathcal{J}(\omega) = \frac{v_0 m}{\pi^2 \hbar^3} \int\limits_0^\infty dp\, p^2 \frac{Q(p) g_p^2}{2m\omega - p^2 + i\eta} \tag{11.111}$$

with the Pauli blocking/enhancement factors (11.95). The T-matrix possesses a pole at the bound state energy $\omega = -E_b$ given by $1 = v_0 \mathcal{J}(-E_b)$. For details see appendix E

The effective potential is now solved from (11.99) as

$$V_\lambda(k, k', \omega) = v_0\, g_k g_{k'} v_\lambda(\omega),$$

$$v_\lambda(\omega) = \frac{1}{1 - \langle g^2(1 - f_\lambda^2)\rangle} = \left[1 - v_0 \mathcal{J}(\omega) = \frac{v_0 m}{\pi^2 \hbar^3} \int\limits_0^\infty dp\, p^2 \frac{Q(p) g_p^2 [1 - f_\lambda(p)^2]}{2m\omega - p^2 + i\eta} \right]^{-1}.$$

$$\tag{11.112}$$

As a check we have from (11.98)

$$-\frac{dv_\lambda^{-1}}{d\lambda} = -\frac{d}{d\lambda} \langle g^2 f_\lambda^2 \rangle \tag{11.113}$$

which provides after integration indeed (11.112). Please note the cumulative effect of the Pauli blocking /enhancement $Q(p)$, the form factor g_p^2 and the cut-off parameter $f_\lambda(p)$ resulting in an effective restriction of the two-particle scattering seen here as modification of the two-particle retarded propagator.

The Bethe–Salpeter equation with internal cut-off (11.97) becomes

$$T_\lambda(k, k', \omega) = v_0 g_k g_{k'} v_\lambda(\omega) \left[1 + \frac{m v_0}{\pi^2 \hbar^3} \int_0^\infty dp \frac{p^2}{2m\omega - p^2} g_p^2 f_\lambda(p)^2 v_\lambda(\omega) t_\lambda(\omega) \right]$$

$$= v_0 g_k g_{k'} v_\lambda(\omega) t_\lambda(\omega) \tag{11.114}$$

which provides

$$t_\lambda(\omega) = \frac{1}{1 - v_\lambda(\omega) \langle g^2 f_\lambda^2 \rangle}. \tag{11.115}$$

Using the solution (11.112) we have

$$T_\lambda = \frac{v_0 g_k g_{k'} v_\lambda}{1 - v_\lambda \langle g^2 f_\lambda^2 \rangle} = \frac{v_0 g_k g_{k'}}{1 - \langle g^2 (1 - f_\lambda^2) \rangle - \langle g^2 f_\lambda^2 \rangle} = \frac{v_0 g_k g_{k'}}{1 - \langle g^2 \rangle} = T_0 \tag{11.116}$$

which shows that indeed we have transferred the effect of internal cut-off to the effective potential (11.112) rendering the T-matrix unchanged.

The effective potential given by (11.112), we can calculate analytically for a Yamaguchi form factor $g_p = 1/(\beta^2 + p^2)$, see appendix E, and a soft cut-off form $f_\lambda(p) = 1/(1 + p^2/\lambda^2)$. The low momentum expansion $v_\lambda = b_0 + b_2 \bar{p}^2$ reads

$$b_2 = -\frac{4\pi \lambda^2 v_0^2 (\beta + \lambda)^3 \left(3\beta^2 + 6\beta\lambda + 2\lambda^2\right)}{\beta \left(4\beta(\beta + \lambda)^3 + \pi v_0(\beta + 2\lambda)\right)^2}, \quad b_0 = \frac{4\beta\lambda v_0(\beta + \lambda)^3}{4\beta(\beta + \lambda)^3 + \pi v_0(\beta + 2\lambda)}. \tag{11.117}$$

The flow is seen in Figure 11.27 providing a fixed point at zero and a fixed point at the minimal possible interaction strength where a bound state is possible (E.13) which reads in our notation

$$v_0 = -\frac{4}{\pi} (\beta^{3/2} + \sqrt{2m\beta E_b})^2 \leq -\frac{4}{\pi} \beta^3. \tag{11.118}$$

One sees that the fixed point appears exactly at the critical interaction strength where the bound states appear. Comparing both Figures 11.26 and 11.27, one sees clearly the different behaviour. This is partially due to the difference between hard and soft cut-off which tilt the figure. However, the area inside the critical line is not populated. More precisely one obtains values $b_2 > 0$ only for negative scaling λ which is not sensible. Since the general derivation of Figure 11.26 should be valid here as well since we merely used an exactly solvable model, we conclude that the arbitrarily choice of parameters c_1, c_2 in (11.109) are not all representing physically sensible values. As we demonstrate here, only specific parameters are linked to critical interaction strengths and such physically justifiable.

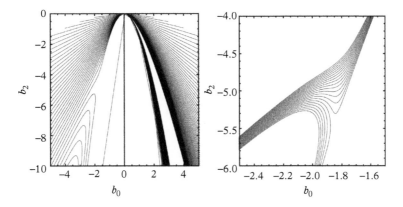

Figure 11.27: *The same flow as in Figure* 11.26 *but for* (11.117) *with soft cut-off and* $\beta = 1.1$ *when* λ *is lowered with* $v_0 = (-5, 5)$ *(left) and details of the fixed point* $v_0 = (-1.7, -1.68)$ *(right) corresponding to the critical strength for the appearance of a bound state* (11.118), $v_0^{\mathrm{crit}} = -\frac{4}{\pi}\beta^3 = -1.695$.

11.8.3   Functional renormalisation

The above illustrated procedure to cut-off energy scales can be realised in general by a smooth- or hard multiplicative cut-off $\Theta(\lambda)$ in front of the free Green's function $G_{0\lambda} = \Theta G_0$ or alternatively by an additive regulator $G_{0\lambda}^{-1} = G_0^{-1} - R$. The relation between both possibilities is easily established writing $G_{0\lambda}^{-1} = G_0^{-1}(1 - G_0 R)$ such that $\Theta^{-1} = 1 - G_0 R$. The scaling-out of high-energy scales requires that the cut-off function should possess the property $\Theta \rightarrow 1$ for $\lambda \rightarrow 0$ and $\Theta \rightarrow 0$ for $\lambda \rightarrow \infty$ and oppositely for R. The idea of functional renormalisation techniques consists now in deriving differential equations for the cut-off dependence of propagators, generating functionals or vertices (Kopietz, Bartosch and Schütz, 2010). Solving these differential equations allow to investigate how the propagator are *flowing* with the cut energy scale. Possible fixed points allow to derive scaling laws and critical exponents when approaching phase transitions (Zinn-Justin, 2007). In general such procedures lead to hierachies of flow equations analogously to the Martin-Schwinger hierarchy. Further approximations are necessary to close such equations which renders ad-hoc methods somewhat uncontrolled. Let us illustrate such a flow equation by differentiating the Dyson equation $G^{-1} = G_0^{-1} - \Sigma$ with respect to λ

$$\frac{\partial}{\partial \lambda} G = -G\left(\frac{\partial}{\partial \lambda} G^{-1}\right) G = -G\partial_\lambda G_0^{-1} G + G\partial_\lambda \Sigma G = G\partial_\lambda R G + G\frac{\delta \Sigma}{\delta G}\partial_\lambda G G. \quad (11.119)$$

or explicitly with double occurring arguments integrated over

$$\partial_\lambda G(1,2) = G(1,3)\partial_\lambda R(3,4) G(4,2) + G(1,3)\frac{\delta \Sigma(3,4)}{\delta G(5,6)}\partial_\lambda G(5,6) G(4,2). \quad (11.120)$$

We recognise an integral equation for the derivative of the propagator $\partial_\lambda G$ with the kernel of Figure 11.1. This kernel requires to use an approximation of the selfenergy which allows to solve the integral equation and then the differential equation for the flow of G with λ. Any underlying many-body approximation is therefore on the level of the used selfenergy. In turn we generate for the propagator G_λ higher-order diagrams by this derivative procedure. Let us illustrate it by starting with the lowest-order Hartree–Fock selfenergy which derivation $\Xi = \delta\Sigma/\delta G$ leads to (11.76). Since it does not contain any propagator we can integrate equation (11.119) for $\partial_\lambda G^{-1}$ from $\lambda = 0$ to λ. In this way we obtain a Dyson equation for the λ-propagator

$$G_\lambda = G + G\Sigma'G_\lambda = G_0 + G_0(\Sigma + \Sigma')G_\lambda \qquad (11.121)$$

with an additional selfenergy in the form of Hartree-Fock

$$\Sigma'(1,2) = R(1,2) + iV(1,2)G'(1,2) \mp iV(1,3)G'(3,3^+)\delta(1,2) \qquad (11.122)$$

but with the closing propagator $G' = G_\lambda - G$. This creates a series of higher-order diagrams beyond the original Hartee-Fock ones.

The flow equation (11.120) allows then to analyse the critical fixed points and critical exponents which provides in principle more information than we have considered so far. Presently there is a great activity in this direction and for an overview see the monographs (Kopietz, Bartosch and Schütz, 2010; Zinn-Justin, 2007).

12

Systems with Condensates and Pairing

12.1 Condensation phenomena in correlated systems

12.1.1 Bose–Einstein condensation in correlated systems

When Einstein (1925) predicted the condensation for an ideal gas of Bosons extending a paper by Bose (1924), it was not foreseeable that it would need 70 years before experimental verification, which was performed in ^{87}Rb by (Anderson, Ensher, Matthews, Wieman and Cornell, 1995), in ^{7}Li by (Bradley, Sackett, Tollett and Hulet, 1995), and in ^{23}Na by (Davis, Mewes, Andrews, van Druten, Durfee, Kurn and Ketterle, 1995) at temperatures between 0.1 and 2 μK. These measurements have encouraged an enormous theoretical activity among which the problem to account adequately for correlations is still unsettled. Specific interesting consequences of correlations are the change of condensation temperature (Andersen, 2004; Baym, Blaizot, Holzmann, Laloë and Vautherin, 2001; Holzmann, Grüter and Laloë, 1999; Braaten and Radescu, 2002; Kneur, Neveu and Pinto, 2004; Hasselmann, Ledowski and Kopietz, 2004; Morawetz, Männel and Schreiber, 2007b; Bala, Srivastava and Pathak, 2015), the occurrence of further phase transitions and even the change of the nature of the Bose–Einstein transition itself. The Bose–Einstein condensation (BEC) is sometimes viewed as a first-order phase transition (Huang, 1995) which seems to be doubtful when attributing a phase transition to interactions and the BEC appearing already in ideal gases. Multiple phase transitions have been reported e.g., in (Stein, Porthun and Röpke, 1998) where the influence of BEC to the liquid-gas phase transition has been calculated.

The multiple-scattering-corrected T-matrix of Section 12.3 will provide a consistent scheme which allows us to describe the situation in and out of the BEC by a common theoretical object. We are therefore in the position to investigate the mutual influence of phase transitions and the BEC due to interactions. This leads to the expectation that interactions and correlations are a proper tool to tune the BEC parameter since they can be controlled fairly well e.g. by Feshbach resonances (Inouye, Andrews, Stenger,

Interacting Systems far from Equilibrium. Klaus Morawetz, Oxford University Press (2018).
© Klaus Morawetz. DOI: 10.1093/oso/9780198797241.001.0001

Miesner, Stamper-Kurn and Ketterle, 1998; Courteille, Freeland, Heinzen, van Abeelen and Verhaar, 1998; Roberts, Claussen, Burke, Greene, Cornell and Wieman, 1998).

12.1.2   Measurement of the quasiparticle spectrum

One possibility to investigate the spectrum and dispersion of excitations in superfluid ^{4}He is the angle-resolved energy-loss spectroscopy of neutrons. In the first-order Born approximation the number of neutrons scattered into a solid angle $d\Omega$ with final energy between E_f and $E_f + dE_f$ is proportional to the double-differential cross section per ^{4}He atom (Glyde, 1994, section 2.1)

$$\frac{d^2\sigma}{d\Omega dE_f} = \frac{a_0^2}{\hbar}\frac{p_f}{p_i}S(\mathbf{q}, \hbar\omega).$$

The energy loss of the neutrons

$$\hbar\omega = E_i - E_f$$

can be obtained by measuring their initial and final energies E_i and E_f. The scattering of the neutrons by the ^{4}He nuclei is described by the scattering length a_0 while the excitations of the ^{4}He medium are covered by the dynamical structure factor $S(\mathbf{q}, \hbar\omega)$, see (10.88). For a given momentum transfer $\mathbf{q}$, excitations are represented by peaks of the dynamical structure factor at the corresponding excitation energies. Fig. 12.1 shows a measurement of the dynamic structure factor where the peak along the phonon-roton dispersion is clearly visible. If there is only a single excitation mode with energy $\epsilon_\mathbf{q}$ and infinite lifetime and the excitation only happens from the macroscopically occupied

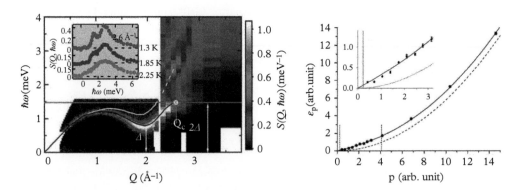

Figure 12.1: *Left: measurement of the dynamic structure factor for liquid ^{4}He by inelastic neutron scattering (Stone et al., 2006). The black solid line corresponds to the phonon-roton dispersion. The white line is the dispersion as obtained by Feynman and Cohen (Feynman and Cohen, 1956). The inset shows the dynamic structure factor for $Q = 2.6$ Å^{-1}. Right: measurement of the Bogoliubov dispersion for ^{87}Rb by Bragg spectroscopy (Steinhauer, Ozeri, Katz and Davidson, 2002). The solid line shows the theoretical Bogoliubov dispersion, the dashed one the quadratic free-particle dispersion. The inset shows the linear phonon regime.*

ground state, the dynamical structure factor will have the form (March and Tosi, 2002, section 14.2.2)

$$S(\mathbf{q}, \omega) = S(\mathbf{q})\delta(\omega - (\epsilon_q - \epsilon_0)),$$

where

$$S(\mathbf{q}) = \int d\omega S(\mathbf{q}, \omega)$$

is the liquid structure factor. The longitudinal sum rule (Pines and Nozieres, 1966)

$$\int d\omega \omega S(\mathbf{q}, \omega) = \frac{Q^2}{2m}$$

with the mass m of a ^{4}He atom yields then the Feynman expression (Feynman, 1954), (Feynman and Cohen, 1956)

$$\epsilon_q = \frac{Q^2}{2mS(\mathbf{q})} + \epsilon_0,$$

which is able to give at least a qualitative description of the phonon-roton dispersion based on the knowledge of the liquid structure factor. For a classical ideal gas one has obviously $S(\mathbf{q}) = 1$.

Another method of measuring the structure factor and the dispersion, which is especially used for ultra-cold alkali gases is Bragg spectroscopy. This method is similar to inelastic light scattering from a laser beam. However, the emission is induced by a second laser beam improving resolution and sensitivity (Stenger, Inouye, Chikkatur, Stamper-Kurn, Pritchard and Ketterle, 1999). The momentum transfer is controlled by the angle between the beams and the energy transfer by their detuning, i.e. first: a photon is absorbed from the beam with higher frequency and second: the emission into the other beam is induced (Steinhauer, Ozeri, Katz and Davidson, 2002).

12.1.3 Quasi-classical approach to superfluidity

One important property of a superfluid is the ability to flow without friction below some critical velocity and temperature. To obtain the upper critical velocity for superfluidity one assumes a fluid of N (quasi)particles with mass m moving with velocity $\mathbf{v}$ past an object. Friction means a transfer of momentum and energy from the fluid to the object that corresponds to the excitation of a quasiparticle, e.g. a phonon, with energy ϵ_p and momentum $\mathbf{p}$ within the fluid. The quasiparticle dispersion ϵ_p is measured in the frame of reference of the fluid. The Galilean transformation yields the corresponding energy in the frame of reference of the object

$$\tilde{\epsilon}_p = \epsilon_p + \frac{m}{2}v^2 + \mathbf{p}\mathbf{v}.$$

In the superfluid state there is a macroscopic number of quasiparticles in the ground state, with $\mathbf{p} = 0$, i.e. Bose condensation. Assuming that all particles are in the ground state the initial energy of the fluid in the frame of reference of the object is

$$E_i = N\epsilon_0 + N\frac{m}{2}v^2.$$

Exciting one quasiparticle with momentum $\mathbf{p}$, the energy of the fluid is (Volovik, 2003, p. 34)

$$E_f = (N-1)\epsilon_0 + (N-1)\frac{m}{2}v^2 + \epsilon_p + \frac{m}{2}v^2 + \mathbf{pv} = E_i + \epsilon_p - \epsilon_0 + \mathbf{pv}.$$

Since the energy transferred to the object has to be positive

$$E_i - E_f = -\mathbf{pv} - (\epsilon_p - \epsilon_0) > 0$$

one obtains for the velocity

$$v \geq -\frac{\mathbf{pv}}{p} > \frac{\epsilon_p - \epsilon_0}{p} \geq c_L = \min_p \frac{\epsilon_p - \epsilon_0}{p}. \tag{12.1}$$

This means that the excitation of a quasiparticle and therefore friction is only possible if $v > c_L$. Below c_L, the critical velocity according to Landau (Landau and Lifschitz, 1980, p. 90), the fluid moves without friction, i.e. it is superfluid. Of course, this argumentation only works if the majority of the quasiparticles are in the ground state, i.e. there has to be a Bose condensation. Since superfluidity is only possible in connection with Bose condensation, it appears below some critical temperatures. The possible excitations are sound waves, i.e. phonons, which according to (12.1) have a linear dispersion $\epsilon_p = c_L p$ with the critical velocity c_L depending on the internal interactions. As Fig. 12.1 shows, the Bogoliubov dispersion can also be measured in the experiment. How the Bogoliubov dispersion or a similar one arises, will be shown in Sections 12.2.2 and 12.5.1.

12.1.4   Quasi-classical approach to superconductivity

The phenomenon of superconductivity has a certain similarity to the phenomenon of superfluidity. In a superconductor the (electrical) conductance diverges below some critical temperature and current, i.e. the (electrical) resistivity vanishes. However the two phenomena differ in the basic excitation mechanisms. Pairs of fermionic quasiparticles are responsible for super conductivity with, opposite spin and zero total momentum, the so-called Cooper pairs. According to the Thouless criterion the energy of a pair is twice the chemical potential (Thouless, 1960). A typical excitation spectrum for the quasiparticles is the BCS dispersion (Bardeen, Cooper and Schrieffer, 1957),

$$\epsilon_p = \sqrt{\left(\frac{p^2}{2m} - \mu\right)^2 + \Delta^2}, \tag{12.2}$$

which has a minimum at the 'Fermi momentum' $p_F = \sqrt{2m\mu}$. The dispersion is given relative to the chemical potential $\mu > 0$. The corresponding energy of the pairs is therefore zero. To break a pair into two quasiparticles at least the energy gap Δ has to be overcome by each particle. It is assumed that breaking is the only possible excitation mechanism for the pairs. Furthermore another a macroscopic occupation of the pairing state is necessary, which only holds up to some critical temperature. The scattering of electrons, e.g. at impurities, connected with a transfer of momentum and energy, is the origin of the resistivity. Since most of the particles are in the pairing state and an energy of at least 2Δ has to be transferred to excite a pair, scattering processes are suppressed at low currents, therefore a current of Cooper pairs feels no resistance. As illustrated in Section 12.1.3, the energy to excite one quasiparticle from the pairing state is

$$\bar{\epsilon}_\mathbf{p} = \epsilon_\mathbf{p} + \mathbf{pv} \geq \epsilon_\mathbf{p} - pv, \tag{12.3}$$

if the condensate of Cooper pairs moves with velocity $\mathbf{v}$ relative to, e.g. an impurity. Since the particles are excited from the pairing state their initial energy is zero. With increasing velocity the minimum of (12.3) is decreased until it reaches zero at the critical velocity v_c. Above v_c it is possible to break the Cooper pairs, i.e. at v_c the resistivity becomes finite. Since the minimum of (12.3) is approximately at p_F the critical velocity can be approximated as $v_c \approx \frac{\Delta}{p_F}$. Again the dispersion and the critical velocity are closely related to inter-particle interactions.

12.2 Cold interacting Bose gas

Before we develop the theory for interacting Bose propagators with gap equations and selfenergies, it is instructive to recap elementary considerations and how these lead to the results we have to expect from more refined microscopic theories.

12.2.1 Ideal Bose gas

The density of an ideal gas of zero-spin Bosons is

$$n = \frac{1}{\Omega} \sum_\mathbf{q} f_B(\epsilon_\mathbf{q}) \tag{12.4}$$

with the free particle dispersion $\epsilon_\mathbf{q} = \frac{\hbar^2 q^2}{2m} - \mu$ and the Bose–Einstein distribution function $f_B(\epsilon) = 1/(\exp(\beta\epsilon) - 1)$. In the thermodynamic limit, i.e. $N, \Omega \to \infty$ with $n = const.$, the contribution of a single state to the density goes to zero-like $\frac{1}{\Omega}$ while the density of states increases proportionally to the volume Ω, and the momentum sum becomes

$$\lim_{\Omega\to\infty} \frac{1}{\Omega} \sum_\mathbf{q} = \int \frac{d^3q}{(2\pi)^3}. \tag{12.5}$$

Bose condensation is an exception to that behaviour. It means a macroscopic occupation of the ground state, i.e. the number of particles in the ground state is of the order of the total number of particles N, its density n_0 staying finite for $\Omega \to \infty$. The chemical potential follows from the condensate density as

$$\mu = -\frac{1}{\beta} \ln \left(1 + \frac{1}{n_0 \Omega} \right) = -\frac{1}{\beta n_0 \Omega} + \mathcal{O}\left(\Omega^{-2}\right) \tag{12.6}$$

and goes to zero in the thermodynamic limit. Bose condensation in an ideal Bose gas with infinite volume appears therefore when the chemical potential μ becomes zero. The density in the thermodynamic limit

$$n = n_0 + \int \frac{d^3q}{(2\pi)^3} f_B(\epsilon_q) = \begin{cases} \left(\frac{m}{2\pi \hbar^2}\right)^{\frac{3}{2}} T^{\frac{3}{2}} P_{\frac{3}{2}} \left(e^{\frac{\mu}{T}}\right) & \text{for } \mu < 0 \\ n_0 + \left(\frac{m}{2\pi \hbar^2}\right)^{\frac{3}{2}} T^{\frac{3}{2}} \zeta \left(\frac{3}{2}\right) & \text{for } \mu = 0 \end{cases} \tag{12.7}$$

is expressed by the polylogarithmic function and zeta function

$$P_k(x) = \frac{1}{\Gamma(k)} \int\limits_0^\infty \frac{t^{k-1} dt}{x^{-1} e^t - 1} = \sum_{j=1}^\infty \frac{x^j}{j^k}, \quad \zeta(k) = P_k(1) = \sum_{j=1}^\infty \frac{1}{j^k} \tag{12.8}$$

with the gamma function Γ. From the conditions $\mu = 0$ and $n_0 = 0$, the critical density and temperature for the Bose condensation (London, 1938) follow

$$n_{\mathrm{id}} = \left(\frac{m}{2\pi \hbar^2}\right)^{\frac{3}{2}} T^{\frac{3}{2}} \zeta \left(\frac{3}{2}\right), \quad T_{\mathrm{id}} = \frac{2\pi \hbar^2}{m} \left(\frac{n}{\zeta\left(\frac{3}{2}\right)}\right)^{\frac{2}{3}}.$$

12.2.2 Bogoliubov transformation for a cold interacting Bose gas

One possibility for obtaining the linear dispersion is a Bogoliubov transformation (Bogoliubov, 1947), (Pethick and Smith, 2004, section 8.1), which shall be illustrated. Considering a Bose gas with contact interaction, the grand-canonical Hamiltonian of the system is

$$\hat{K} = \hat{H} - \mu \hat{N} = \sum_k \left(\frac{\hbar^2 k^2}{2m} - \mu\right) \hat{a}_k^\dagger \hat{a}_k + \frac{U_0}{2\Omega} \sum_{Q,p,p'} \hat{a}_{\frac{Q}{2}+p}^\dagger \hat{a}_{\frac{Q}{2}-p}^\dagger \hat{a}_{\frac{Q}{2}+p'} \hat{a}_{\frac{Q}{2}-p'}.$$

U_0 is the strength of the interaction and μ the chemical potential. In a Bose gas near $T = 0$ with weak interaction most particles will occupy the ground state, i.e. the particles will be in the Bose condensate, only a few particles will be excited. Therefore it is sufficient to concentrate on the contributions of the condensate to the Hamiltonian and keep only those terms up to the second order in $\hat{a}_k^{(\dagger)}$ with $k \neq 0$. With the exception of

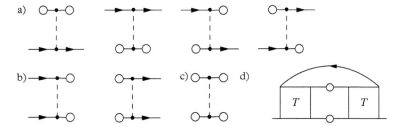

Figure 12.2: *(a)–(c) Scattering processes contributing to the Bogoliubov approximation, broken lines mark the interaction potential, solid lines with arrows show excited particles with finite momentum, dots are vertices where the momentum is conserved, lines with circles mark particles entering or leaving the Bose condensate. (d) Diagram of the bosonic singular selfenergy including the scattering processes (b), which are crucial for obtaining a Bogoliubov-like dispersion.*

the kinetic term and the number operator these contributions come from the interaction part of the Hamiltonian. The relevant scattering processes form three groups: the scattering of an excited particle at a particle from the condensate, shown in Fig. 12.2(a), the scattering of two excited particles into or out of the condensate, shown in Fig. 12.2(b) and the scattering of two particles from the condensate into the condensate, shown in Fig. 12.2(c).

The resulting effective Hamiltonian is therefore

$$\hat{K}_B = \sum_{k \neq 0} \left[\left(\frac{\hbar^2 k^2}{2m} - \mu + \frac{2U_0}{\Omega} \hat{a}_0^\dagger \hat{a}_0 \right) \hat{a}_k^\dagger \hat{a}_k + \frac{U_0}{2\Omega} \left(\hat{a}_0^\dagger \hat{a}_0^\dagger \hat{a}_k \hat{a}_{-k} + \hat{a}_k^\dagger \hat{a}_{-k}^\dagger \hat{a}_0 \hat{a}_0 \right) \right]$$
$$- \mu \hat{a}_0^\dagger \hat{a}_0 + \frac{U_0}{2\Omega} \hat{a}_0^\dagger \hat{a}_0^\dagger \hat{a}_0 \hat{a}_0.$$

Since the ground state is macroscopically occupied with N_0 particles, they are of the order of magnitude of the total particle-number $N \gg 1$ and one can write

$$\hat{a}_0^\dagger |N_0\rangle = \sqrt{N_0 + 1} |N_0 + 1\rangle \approx \sqrt{N_0} |N_0\rangle, \quad \hat{a}_0 |N_0\rangle = \sqrt{N_0} |N_0 - 1\rangle \approx \sqrt{N_0} |N_0\rangle.$$

Neglecting fluctuations one replaces therefore $\hat{a}_0^\dagger$ and $\hat{a}_0$ by the order parameter $\langle \hat{a}_0^\dagger \rangle = \langle \hat{a}_0 \rangle \approx \sqrt{N_0}$ and the Hamiltonian becomes

$$\hat{K}_B = \frac{1}{2} \sum_{k \neq 0} \left[\epsilon_0 \left(\hat{a}^\dagger \hat{a} + \hat{b}^\dagger \hat{b} \right) + \epsilon_1 \left(\hat{a}^\dagger \hat{b}^\dagger + \hat{a}\hat{b} \right) \right] - \mu N_0 + \frac{N_0^2 U_0}{2\Omega} \qquad (12.9)$$

with $\hat{a} \hat{=} \hat{a}_k$, $\hat{b} \hat{=} \hat{a}_{-k}$, $\epsilon_0 \hat{=} \frac{\hbar^2 k^2}{2m} - \mu + 2n_0 U_0$ and $\epsilon_1 \hat{=} n_0 U_0$.

The Bogoliubov transformation introduces new operators $\hat{\alpha} = u\hat{a} + v\hat{b}^\dagger$ and $\hat{\beta} = u\hat{b} + v\hat{a}^\dagger$ and chooses the real numbers u and v such that the Hamiltonian becomes diagonal for $\hat{\alpha}$ and $\hat{\beta}$. Like $\hat{a}$ and $\hat{b}$ also $\hat{\alpha}$ and $\hat{\beta}$ commute with each other and obey the Bose commutation relation

$$1 = [\hat{\alpha}, \hat{\alpha}^\dagger]_- = [\hat{\beta}, \hat{\beta}^\dagger]_- = u^2 - v^2. \tag{12.10}$$

The Hamiltonian (12.9) becomes diagonal if

$$\epsilon_1 (u^2 + v^2) - 2uv\epsilon_0 = 0 \tag{12.11}$$

and condition (12.10) can be fulfilled by $u = \cosh t$ and $v = \sinh t$. The second condition (12.11) yields therefore

$$e^{4t} = \frac{\epsilon_0 + \epsilon_1}{\epsilon_0 - \epsilon_1}.$$

Under the assumption $\epsilon_0 > \epsilon_1 > 0$ one finds

$$u^2 = \frac{\cosh 2t + 1}{2} = \frac{1}{2}\left(\frac{\epsilon_0}{\epsilon} + 1\right), \quad v^2 = \frac{\cosh 2t - 1}{2} = \frac{1}{2}\left(\frac{\epsilon_0}{\epsilon} - 1\right),$$

with the quasiparticle energy $\epsilon = \sqrt{\epsilon_0^2 - \epsilon_1^2}$. Applying the Bogoliubov transformation to the Hamiltonian (12.9) yields

$$\hat{K}_B = \sum_k \epsilon_k \hat{\alpha}_k^\dagger \hat{\alpha}_k - \frac{N_0^2 U_0}{2\Omega} - \sum_{k \neq 0} \epsilon_k v_k^2 \tag{12.12}$$

with

$$v_k^2 = \frac{\frac{\hbar^2 k^2}{2m} + n_0 U_0 - \epsilon_k}{2\epsilon_k}.$$

The Bose gas behaves like a system of quasiparticles with the explicit Bogoliubov dispersion

$$\epsilon_k = \sqrt{\left(\frac{\hbar^2 k^2}{2m} + n_0 U_0\right)^2 - n_0^2 U_0^2}. \tag{12.13}$$

For repulsive interaction, U_0 has to be positive and the condensate density n_0 follows from

$$n = n_0 + \frac{1}{\Omega} \sum_{k \neq 0} \frac{1}{2} \left(\frac{\frac{\hbar^2 k^2}{2m} + n_0 U_0}{\epsilon_k} (1 + 2 f_B(\epsilon_k)) - 1 \right)$$

$$= n_0 + \frac{1}{\Omega} \sum_{k \neq 0} \frac{\frac{\hbar^2 k^2}{2m} + n_0 U_0}{\epsilon_k} f_B(\epsilon_k) + \frac{1}{\Omega} \sum_{k \neq 0} v_k^2. \tag{12.14}$$

The third term on the right-hand side of equation (12.14) represents the depletion of the condensate at $T = 0$, i.e. the density of particles out of the condensate, the term in the middle is the depletion due to excitations at finite temperature (Pethick and Smith, 2004, p. 210 f). Looking for the ground state one assumption would be that all quasiparticles are in the quasiparticle ground state with $k = 0$, i.e. in the condensate, as is the case for the ideal Bose gas. However the expectation value of the energy according to the Hamiltonian (12.12)

$$E = \langle \hat{K}_B \rangle + \mu N \overset{N_0 > 0}{=} \sum_k \epsilon_k f_B(\epsilon_k) + \frac{N_0 U_0}{2\Omega} (2N - N_0) - \sum_{k \neq 0} \epsilon_k v_k^2 \tag{12.15}$$

shows that a finite depletion can lower the energy especially in the ground state. The first contribution to (12.15) is the energy of quasiparticle excitations at finite temperature. The second contribution is a mean field, taking into account the interaction of every particle with the condensate particles, which are assumed to be distributed homogeneously. There is obviously an 'attraction-in-momentum space', i.e. the more particles there are in the ground state the smaller is the mean-field contribution to the energy. The third contribution to (12.15) is the correction to the mean field taking into account correlations, i.e. due to the repulsive interaction the energy can be lowered if the density of particles is lowered locally around the particle under consideration. In contrast to the mean field this third contribution favours a finite depletion.

Assuming that $\langle \hat{\alpha}^\dagger \hat{\beta}^\dagger \rangle = \langle \hat{\beta} \hat{\alpha} \rangle = 0$ and $\langle \hat{\alpha}^\dagger \hat{\alpha} \rangle = \langle \hat{\beta}^\dagger \hat{\beta} \rangle = f_B(\epsilon)$ one can calculate the occupation of the state described by $\hat{a}$

$$\langle \hat{a}^\dagger \hat{a} \rangle = \langle u^2 \hat{\alpha}^\dagger \hat{\alpha} + v^2 (1 + \hat{\beta}^\dagger \hat{\beta}) \rangle = \frac{1}{2} \left(\frac{\epsilon_0}{\epsilon} (1 + 2 f_B(\epsilon)) - 1 \right).$$

The occupation of the ground state is therefore

$$N_0 = \frac{1}{2} \left(\frac{2 n_0 U_0 - \mu}{\sqrt{(2 n_0 U_0 - \mu)^2 - n_0^2 U_0^2}} \left(1 + 2 f_B \left(\sqrt{(2 n_0 U_0 - \mu)^2 - n_0^2 U_0^2} \right) \right) - 1 \right).$$

For a macroscopic occupation of the ground state one can approximate the Bose function near its pole and solve the resulting equation

$$N_0 \approx \frac{2n_0 U_0 - \mu}{\beta((2n_0 U_0 - \mu)^2 - n_0^2 U_0^2)}$$

for μ

$$\mu = 2n_0 U_0 - \frac{1}{2\beta N_0} - \sqrt{n_0^2 U_0^2 + \frac{1}{(2\beta N_0)^2}} = n_0 U_0 + \mathcal{O}(N_0^{-1}) \qquad (12.16)$$

which coincides with (12.6) in the noninteracting limit. The result is that, in the thermodynamic limit, $N_0 \to \infty$, $\mu = n_0 U_0$ (Shi and Griffin, 1998, p. 21). This relation will be used in to eliminate μ.

For weak interaction it can be assumed that at $T = 0$ all particles are in the condensate, i.e. $N = N_0 + \mathcal{O}\left(U_0^2\right)$ and

$$E = \frac{N^2 U_0}{2\Omega} + \mathcal{O}\left(U_0^2\right).$$

From that ground-state energy one can obtain the pressure at $T = 0$

$$\mathscr{P} = -\left.\frac{\partial E}{\partial \Omega}\right|_{N=const.} = \frac{n^2 U_0}{2} + \mathcal{O}\left(U_0^2\right) = \frac{\rho^2 U_0}{2m^2} + \mathcal{O}\left(U_0^2\right) \qquad (12.17)$$

where $\rho = mn$ is the mass density. This relation shows that $U_0 > 0$ guarantees also the stability of the system, i.e. a positive pressure. The pressure for $U_0 = 0$ vanishes at $T = 0$. According to the Nernst theorem (Landau and Lifschitz, 1984, p. 65) the entropy S is zero at $T = 0$. The sound velocity c follows from (Landau and Lifschitz, 1971, p. 284)

$$c^2 = \left.\frac{\partial \mathscr{P}}{\partial \rho}\right|_{S=const.} = \frac{n_0 U_0}{m} + \mathcal{O}\left(U_0^2\right).$$

Obviously the linear approach for small momenta of the Bogoliubov dispersion (12.1) is given by the macroscopic sound velocity at $T = 0$ (Bogoliubov, 1947). When approaching the phase transition from the normal phase, i.e. setting $n_0 = 0$ in (12.14), one finds that the critical temperature and density of the Bogoliubov approximation seem to be the same as for the ideal Bose gas obtained in Section 12.2.1.

Near the phase transition it is convenient to calculate the difference between the density of un-condensed particles and the ideal critical density n_{id} (Shi and Griffin, 1998, p. 33 f.)

$$\Delta n = n - n_0 - n_{id}$$

$$= \frac{1}{\Omega} \sum_{k \neq 0} \frac{1}{2} \left(\frac{\frac{\hbar^2 k^2}{2m} + n_0 U_0}{\epsilon_k} (1 + 2 f_B(\epsilon_k)) - 1 \right) - \frac{1}{\Omega} \sum_{k \neq 0} f_B \left(\frac{\hbar^2 k^2}{2m} \right)$$

$$= \frac{1}{\Omega} \sum_{k \neq 0} \frac{1}{2} \left(\frac{\frac{\hbar^2 k^2}{2m} + n_0 U_0}{\epsilon_k} \coth \frac{\beta \epsilon_k}{2} - \coth \frac{\beta \hbar^2 k^2}{4m} \right).$$

Since both terms in the sum cancel for high momenta, the main contribution to Δn will come from the small k especially if $n_0 U_0$ is small, therefore one can expand the hyperbolic tangents and obtain an analytical solution in the thermodynamic limit

$$\Delta n \overset{\Omega \to \infty}{\approx} \int \frac{d^3 k}{(2\pi)^3} \frac{1}{2} \left(\frac{\frac{\hbar^2 k^2}{2m} + n_0 U_0}{\epsilon_k} \frac{2}{\beta \epsilon_k} - \frac{4m}{\beta \hbar^2 k^2} \right) = -\frac{T}{8\pi} \left(\frac{2m}{\hbar^2} \right)^{\frac{3}{2}} \sqrt{2 n_0 U_0}$$

$$= -\sqrt{\frac{2\pi U_0}{T}} \frac{n_{id}}{\zeta \left(\frac{3}{2} \right)} \sqrt{n_0}.$$

At the expected critical point, i.e. for $n = n_{id}$, one finds that except for the trivial one, $n_0 = 0$, the resulting equation has a second solution

$$n_0 = n_{0c} = \frac{2\pi n_{id}^2 U_0}{T \zeta^2 \left(\frac{3}{2} \right)} = \left(\frac{T}{2\pi} \right)^2 \left(\frac{m}{\hbar^2} \right)^3 U_0, \tag{12.18}$$

when approaching the phase transition from the condensed phase. This coexistence of solutions at the expected critical point indicates a first-order phase transition. According to (12.18), this phase transition remains for any finite interaction strength U_0. However it has to be emphasised that, as mentioned above, (12.18) is only a good approximation for small U_0. Especially, the condition $n_0 \leq n$ is violated if U_0 is too large. Since $\mu = n_0 U_0$, a multivalued condensate density means a multivalued chemical potential.

12.2.3   Popov approximation

At the phase transition the essential contributions to the Hamiltonian discussed in the last section will vanish, therefore it is necessary to include further contributions which remain finite in the normal region too. Assuming that the interaction is weak, the simplest way to do that is to combine the Bogoliubov with the Hartree–Fock approximation.

The Hartree–Fock approximation can also be obtained by a certain neglect of fluctuations and the replacement of a pair of creation and annihilation operators by their expectation value

$$\hat{a}^{\dagger}_{\frac{Q}{2}+p}\hat{a}^{\dagger}_{\frac{Q}{2}-p}\hat{a}_{\frac{Q}{2}+p'}\hat{a}_{\frac{Q}{2}-p'}$$

$$\approx \left(\hat{a}^{\dagger}_{\frac{Q}{2}+p}\hat{a}_{\frac{Q}{2}+p}\left\langle\hat{a}^{\dagger}_{\frac{Q}{2}-p}\hat{a}_{\frac{Q}{2}-p}\right\rangle + \hat{a}^{\dagger}_{\frac{Q}{2}-p}\hat{a}_{\frac{Q}{2}-p}\left\langle\hat{a}^{\dagger}_{\frac{Q}{2}+p}\hat{a}_{\frac{Q}{2}+p}\right\rangle\right)\left(\delta_{pp'}+\delta_{p(-p')}\right).$$

In analogy to Fig. 12.2 there are contributions from four scattering processes, i.e. each two circles have to be associated with an expectation value. The resulting effective Hamiltonian of the Hartree–Fock approximation is therefore

$$\hat{K}_{\mathrm{HF}} = \sum_{\mathbf{k}}\left(\frac{\hbar^2 k^2}{2m} - \mu + \frac{2U_0}{\Omega}\sum_{\mathbf{q}}\langle\hat{a}^{\dagger}_{\mathbf{q}}\hat{a}_{\mathbf{q}}\rangle\right)\hat{a}^{\dagger}_{\mathbf{k}}\hat{a}_{\mathbf{k}} = \sum_{\mathbf{k}}\left(\frac{\hbar^2 k^2}{2m} - \mu + 2nU_0\right)\hat{a}^{\dagger}_{\mathbf{k}}\hat{a}_{\mathbf{k}}.$$

The only difference between the ideal Bose gas and Hartree–Fock approximation is the additional mean field $2nU_0$, i.e. the system behaves like an ideal Bose gas with the effective chemical potential $\mu^* = \mu - 2nU_0$.

Combining the Hartree–Fock and the Bogoliubov approximation, the only difference between the latter and this so-called Hartree–Fock–Bogoliubov or Popov approximation (Shi and Griffin, 1998, p. 23 ff.)

$$\hat{K}_{\mathrm{P}} = \frac{1}{2}\sum_{\mathbf{k}\neq 0}\left[\left(\frac{\hbar^2 k^2}{2m} - \mu + 2nU_0\right)\left(\hat{a}^{\dagger}_{\mathbf{k}}\hat{a}_{\mathbf{k}} + \hat{a}^{\dagger}_{-\mathbf{k}}\hat{a}_{-\mathbf{k}}\right) + n_0 U_0\left(\hat{a}^{\dagger}_{\mathbf{k}}\hat{a}^{\dagger}_{-\mathbf{k}} + \hat{a}_{\mathbf{k}}\hat{a}_{-\mathbf{k}}\right)\right]$$

$$-\mu N_0 + \frac{N_0^2 U_0}{2\Omega}$$

is that now the full density enters the diagonal term. From an analogous argumentation as for the Bogoliubov approximation one obtains therefore a different chemical potential, $\mu = (2n - n_0)U_0$, while the dispersion (12.13) and density (12.14) remain absolutely unchanged (Baym, Blaizot, Holzmann, Laloë and Vautherin, 2001). In contrast to a hard-sphere Bose gas, in the Popov approximation it is not necessary to exclude the exchange in the condensate by hand, since the off-diagonal part of the Hamiltonian takes care of that.

12.3 Generalised Soven scheme

Now that we have made us acquainted with the basic physics of Bose condensation we might proceed and see how the Green's function theory provides a consistent scheme to describe condensation in Bose systems and pairing in Fermi systems.

12.3.1 Missing pole structure

The basis of Bose–Einstein condensation and of superconductivity is the occurrence of anomalous propagators which allow us to describe the Bogoliubov dispersion in BEC or the gap equation in superconductivity. These anomalous propagators are most often assumed ad-hoc and the correct equation of motions appear. Since this approach is connected with particle-non-conserving assumptions, we wish to derive it from a consistent theory. Let us describe how this is possible.

The gap equation does not follow from symmetric T-matrices but from the asymmetric scheme. Let us illustrate this comparing the Dyson equation with symmetric and with asymmetric selfconsistency in Figure 12.3.

The T-matrix has poles at bound states and at the pairing and becomes separable near these poles, $T = \mp \Delta \Delta$ where the upper sign stands for fermions and the lower for Bosons. The two different Dyson equations for the propagator with respect to the symmetry of selfconsistency is illustrated in the next Figure 12.4. The first symmetrical selfconsistent Dyson equation leads to the propagator

$$G(\omega, \mathbf{k}) = \frac{1}{\omega - \epsilon_{\mathbf{k}} \pm \Delta^2 G(-\omega, -\mathbf{k})} \tag{12.19}$$

which shows no pole and no gap equation. Considering the second asymmetrical Kadanoff–Martin approximation of Figure 12.4 we obtain

$$G(\omega, \mathbf{k}) = \frac{1}{\omega - \epsilon_{\mathbf{k}} \mp \frac{\Delta^2}{\omega + \epsilon_{-\mathbf{k}}}} \tag{12.20}$$

which possesses the typical two-pole structure of the BCS gap equation in the case of fermions. Therefore the Kadanoff and Martin approximation is superior to the symmetric

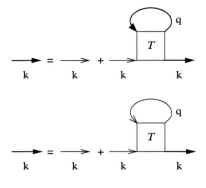

Figure 12.3: *The Dyson equation in ladder approximation for the symmetric selfconsistency (above) and the Kadanoff–Martin asymmetric one derived here (below).*

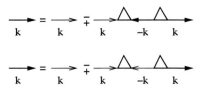

Figure 12.4: *The propagator equations resulting from Figure 12.3, selfconsistency (above), Kadanoff–Martin asymmetric (below). The upper/lower sign stands for Fermi/Bose systems.*

selfconsistent ladder approximation and appears as a consequence of the hierarchical dependencies of correlations (Morawetz, 2011) as derived in Chapter 11.5.6.

In fact this asymmetry has been recognised as being necessary to obtain the gap equation for pairing (Prange, 1960). It was first observed by Kadanoff and Martin (Kadanoff and Martin, 1961) and used later on (Maly, Jankó and Levin, 1999; He, Chien, Chen and Levin, 2007) as an ad-hoc approximation seemingly violating the symmetry of equations and consequently violating conservation laws. This has remained puzzling since a worse approximation obviously leads to better results. Recently it turned out that the repeated collisions (Lipavský, 2008; Šopík, Lipavský, Männel, Morawetz and Matlock, 2011) with the same particle are responsible for this effect as we will explain now.

12.3.2   Link to anomalous propagators

Repeated collisions of two particles in the same state are unphysical since the particles move apart from each other after the collision. Therefore we have to ensure that due to selfconsistency such collisions with the same state do not appear. In fact, as can be seen in Figure 12.5, the selfconsistent Dyson equation does not ensure that the momentum of the repeated collisions, **p**, is unequal to the momentum of the incoming particle **k**. If this would be the case the particle will scatter with a particle in the same state again which is an artifact as outlined in Chapter 6.6. These repeated collisions have to be removed from the T-matrix and the correct gap equation appears and the condensate can be described without asymmetrical ad-hoc assumptions about selfconsistency. The advantage of eliminating only the contributions of single channels as proposed in Refs. (Lipavský, 2008) and (Šopík, Lipavský, Männel, Morawetz and Matlock, 2011; Morawetz, 2010) is that the formation of pairs and their condensation can be described within the same approximation. Explicitly, we split the selfenergy into different channels,

$$\Sigma = \sum_{\mathrm{chann}.j} \Sigma_j \qquad (12.21)$$

where we assume the condensation or pairing to appear in the channel i called the singular channel. Now we define a subtracted propagator

$$G_{\bar{\lambda}} = G - G_{\bar{\lambda}}\Sigma_i G \qquad (12.22)$$

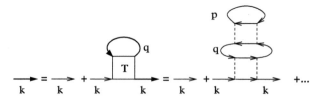

Figure 12.5: *The iteration of the selfconsistent Dyson equation of Figure 12.3 leading to repeated collision with momentum **p**, **q**.*

or $G^{-1} = G_{\bar{k}}^{-1} - \Sigma_i$. Using the standard Dyson equation $G_0^{-1} = G^{-1} + \Sigma$ we obtain the relation

$$G_{\bar{k}} = G_0 + G_0(\Sigma - \Sigma_i)G_{\bar{k}} \qquad (12.23)$$

which shows that in this propagator the own selfenergy channel is subtracted, $\Sigma' = \Sigma - \Sigma_i$. Now we consider a general T-matrix which represents the selfenergy as $\Sigma' = \sum_{j \neq i} T_j \bar{G}$ where the channel T-matrix T_j as the two-particle function is closed by a backward propagator $\bar{G}$. In the singular channel we subtract the repeated interaction within this channel. This is achieved by closing with the subtracted propagator $\Sigma_i = T_i \bar{G}_{\bar{k}}$. Now we can rewrite the Dyson equation as

$$
\begin{aligned}
G^{-1} &= G_0^{-1} - \Sigma = G_0^{-1} - \Sigma' - \Sigma_i \\
&= G_0^{-1} - \Sigma' - T_i \bar{G}_{\bar{k}} \\
&= G_0^{-1} - \Sigma' - T_i \left(\bar{G}_0^{-1} - \bar{\Sigma}' \right)^{-1}
\end{aligned}
\qquad (12.24)
$$

where, in the last step, we have used (12.23). Finally we rewrite (12.24) to obtain the full propagator in momentum-energy (Matsubara) representation $p = (\mathbf{p}, \omega_n)$, or understood as retarded propagator with $p = (\mathbf{p}, \omega)$

$$G(p) = \frac{\bar{G}_0^{-1}(p) - \bar{\Sigma}'(p)}{[G_0^{-1}(p) - \Sigma'(p)][\bar{G}_0^{-1}(p) - \bar{\Sigma}'(p)] - T_i(p)}. \qquad (12.25)$$

Remembering $\Sigma' = \Sigma - T_i \bar{G}_{\bar{k}}$ leads immediately back to the Dyson equation $G = G_0/(1 - \Sigma G_0)$. Therefore it is an exact rewriting so far.

Now we take into account the explicit form of the free propagator $G_0^{-1} = \omega - \epsilon_k$ and $\bar{G}_0^{-1} = -\omega - \epsilon_{-k}$ and call the 'proper' selfenergy

$$\Sigma_{11}(p) \equiv \Sigma'(p). \qquad (12.26)$$

Further, we observe that the T-matrix in the singular channel is separable (Bishop, Strayer and Irvine, 1974, 1975; Gulian and Zharkov, 1999) and can be written $T_i(p) = \mp \Delta(p) \bar{\Delta}(p)$ for fermions/Bosons respectively. Now we can define the 'anomalous' selfenergy as

$$\Sigma_{12}(p) \equiv \Delta(p) \qquad (12.27)$$

such that the propagator (12.25) takes the form

$$G_{11} = \frac{\omega + \epsilon_{-k} + \bar{\Sigma}_{11}}{(\omega + \epsilon_{-k} + \bar{\Sigma}_{11})(\omega - \epsilon_k - \Sigma_{11}) \mp \Sigma_{12}^2} \qquad (12.28)$$

in agreement with (12.20). We add an auxiliary quantity called 'anomalous' Green's function

$$G_{12} \equiv \frac{\pm \Sigma_{12}}{(\omega + \epsilon_{-k} + \bar{\Sigma}_{11})(\omega - \epsilon_k - \Sigma_{11}) \mp \Sigma_{12}^2} \qquad (12.29)$$

and can write in such a way the two equations in matrix form

$$\mathbf{G} = \mathbf{G}^0 + \mathbf{G}^0 \mathbf{\Sigma} \mathbf{G} \qquad (12.30)$$

with

$$\mathbf{G} = \begin{pmatrix} G_{11} & G_{12} \\ \bar{G}_{12} & \mp \bar{G}_{11} \end{pmatrix}, \quad \mathbf{G_0} = \begin{pmatrix} G_0 & 0 \\ 0 & \mp \bar{G}_0 \end{pmatrix}, \quad \mathbf{\Sigma} = \begin{pmatrix} \Sigma_{11} & \Sigma_{12} \\ \bar{\Sigma}_{12} & \mp \bar{\Sigma}_{11} \end{pmatrix} \qquad (12.31)$$

for fermions/Bosons respectively. These are exactly the equations for anomalous propagators derived first by Beliaev (Beliaev, 1958) for Bosons. For fermions these are the Nambu–Gorkov equations (Gorkov, 1958). These equations lead to effective quasiparticle poles in the propagator

$$E_+ = \frac{\epsilon_k + \Sigma_{11} - \epsilon_{-k} - \bar{\Sigma}_{11}}{2} + \sqrt{\left(\frac{\epsilon_k + \Sigma_{11} + \epsilon_{-k} + \bar{\Sigma}_{11}}{2} \right)^2 \pm \Sigma_{12} \bar{\Sigma}_{12}} \qquad (12.32)$$

and E_- with opposite sign of the square root. In the case of symmetric dispersion only one quasiparticle energy appears. One recognises the Bogoliubov dispersion for Bosons (12.13).

In other words, separating a singular channel from the selfenergy and avoiding repeated collision within this channel leads immediately to propagators which have the Beliaev form for Bosons or the Nambu–Gorkov form for fermions. We see that adding the auxiliary quantity (12.29) is not necessary. We have derived all the information without this quantity and it was added here simply to show the same structure of theory appears as provided by the approaches with anomalous functions. In this sense the theory of anomalous functions is overdetermined. We should note, however, that the anomalous propagator G_{12} describes the order parameter. This anomalous propagator appears as a result of the theory here and is not assumed from the beginning as usually done.

Though we have shown above that the Kadanoff–Martin approximation already leads to a two-pole Green's function (12.20) necessary for the gap equation, the alert reader has noticed that the subtraction of unphysical multiple scattering events has assumed a different intermediate propagator to the Kadanoff–Martin approximation though leading to the same gap propagator (12.28). The Kadanoff–Martin approximation requires us to use the Hartree–Fock propagator as derived above by approximating the hierarchy of correlations at the binary level. For the subtraction of repeated collisions we have used the propagator written with the help of (12.22) instead,

$$G_{\bar{A}} = G_{HF} + G_{HF}(\Sigma - \Sigma_{HF} - \Sigma_i)G_{\bar{A}} \tag{12.33}$$

which shows that the subtracted propagator is beyond the Hartree–Fock one.

In Chapter 11.5.6 we have derived an asymmetric cummulant-like expansion of the selfenergy. Due to this procedure we have obtained, with (11.65), the result $\Sigma_{\Delta} = \Sigma_{\bar{A}}$ and comparing (12.33) with (11.65) we concluded that $G_{\Delta} = G_{\bar{A}}$. This is exactly the subtracted propagator (12.33) proposed before which corrects the repeated collisions with the same state. The derived selfenergy diagram in the first line of Figure 11.21 presents the diagrammatic part which leads to the subtraction of unphysical repeated collisions in any channel. For the particle–particle channel we have obtained the correct subtracted propagator and see that a proper collection of pair–pair correlations in the hierarchical expansion of correlations cares for the subtraction of such unphysical processes.

Summarising the theory of superconductivity and Bose–Einstein condensation it is possible to formulate from the asymmetric Bethe–Salpeter equation without the usage of anomalous propagators.

12.3.3 Bogoliubov–DeGennes equation

Now we perform a bilinear expansion of the retarded propagator (12.31) in the same manner as (10.5). The difference is that we have to take two effective wave functions into account

$$G^R(x, x', \omega) = \sum_i \frac{\begin{pmatrix} u_i(x) \\ \pm v_i(x) \end{pmatrix} \otimes \begin{pmatrix} u_i^*(x') \\ \pm v_i^*(x') \end{pmatrix}}{\omega - E_i + i0}. \tag{12.34}$$

This represents the propagator (12.31) exactly, if the wave functions obey the Bogoliubov–DeGennes equations

$$(\epsilon_k + \Sigma_{11})u_i + \Sigma_{12}v_i = E_i u_i$$
$$(-\epsilon_{-k} - \bar{\Sigma}_{11})v_i \pm \bar{\Sigma}_{12}u_i = E_i v_i. \tag{12.35}$$

This can be seen by $\mathbf{G}^{-1}\mathbf{G} = 1$ and from (12.31) one has

$$\mathbf{G}^{-1} = \begin{pmatrix} \omega - \epsilon_k - \Sigma_{11} & -\Sigma_{12} \\ -\bar{\Sigma}_{12} & \pm(\omega + \epsilon_{-k} + \bar{\Sigma}_{11}) \end{pmatrix} \tag{12.36}$$

and applied to (12.34) one gets unity provided the wave function complete

$$\sum_i u_1(x)u_i^*(x') = \delta(x - x'), \qquad \sum_i v_1(x)v_i^*(x') = \delta(x - x') \tag{12.37}$$

and vanishing cross summations between u and v. Equation (12.35) establishes two coupled Schrödinger equations in real space $\epsilon_k = \epsilon_{-i\nabla}$ which can be solved for given boundary conditions. As the reader will have noticed, we assumed a selfenergy as factorised in momentum space. This is only possible exactly for mean field approximations. Therefore the Bogoliubov–DeGennes equations are mean field equations. Any higher-order approximation will introduce dissipation and collision integrals in the appropriate kinetic equation. We will derive these kinetic equation in Chapter 21 since it requires us to extend the kinetic theory to systems with SU(2) structures. An extensive overview on nonequilibrium superconductivity can be found in (Gulian and Zharkov, 1999).

12.4   Gap equations

So far we have used the separability of the singular T-matrix and assumed an ad-hoc gap Δ. In order to see the gap equation for determining Δ we consider the T-matrix (10.131) and the transport vertex reads

$$
T^<(p_1, p_2, K, \omega) = V\mathcal{G}^< T^A + V\mathcal{G}^R T^<
$$

$$
= \int \frac{dp_3}{(2\pi)^3} V(p_1 - p_3)\zeta^<(p_3, K, \omega) T^A(p_3, p_2, K, \omega)
$$

$$
+ \int \frac{dp_3}{(2\pi)^3} V(p_1 - p_3)\zeta^R(p_3, K, \omega) T^<(p_3, p_2, K, \omega) \quad (12.38)
$$

where we introduced the retarded two-particle propagator, see also (10.149),

$$
\mathcal{G}^R(p_1, p_2, K, \omega) = (2\pi)^3 \delta(p_1 - p_2) \int \frac{d\omega_1 d\omega_2}{(2\pi)^2} \frac{g^>(\frac{K}{2}+p_1, \omega_1)g^>(\frac{K}{2}-p_1, \omega_2) - g^< g^<}{\omega - \omega_1 - \omega_2}
$$

$$
\equiv (2\pi)^3 \delta(p_1 - p_2)\zeta(p_1, K, \omega). \quad (12.39)
$$

We know now that the T-matrix shows a pole at twice the chemical potential $\omega = 2\mu$ corresponding to the pairing/condensation instability above (Thouless criterion) (Thouless, 1960). We can therefore assume a separation of wave functions (anomalous) near the pole

$$
T^<(p_1, p_2, K, \omega) = \mp(2\pi)^4 \delta(\omega - 2\epsilon_f)\delta(K)\Delta(p_1)\Delta(p_2). \quad (12.40)
$$

The first part of (12.38) does not contribute to this pole behaviour and we can neglect it if we consider from (12.38) the pole part of equation (12.40). The resulting equation separates in Δ and we get

$$
\Delta(p_1) = \int \frac{dp_3}{(2\pi)^3} V(p_1 - p_3)\zeta(p_3, 0, 2\epsilon_f)\Delta(p_3). \quad (12.41)
$$

It is customary to consider the energy bottom with respect to the chemical potential μ and introducing simply the quasiparticle picture (9.17) into (12.39) and this into (12.41) would result in a no gap equation.

As we have seen in section 12.3.3, the correct gap appears if we introduce only one Green's functions in the intermediate propagator (12.39) selfconsistently. Therefore we look first at the spectral properties given by the Dyson equation for the retarded propagator

$$(G^R)^{-1} = G_0^{-1} - \Sigma^R \tag{12.42}$$

with the selfenergy given in T-matrix approximation $\Sigma^R = T^R G^< - T^< G^A$. Since $T^<$ bears the pole part it makes sense to consider this part separately. Therefore we split (12.42) into

$$(G^R)^{-1} = (G_n^R)^{-1} + T^< G^A \tag{12.43}$$

with the normal Green's functions

$$(G_n^R)^{-1} = G_0^{-1} - T^R G^<. \tag{12.44}$$

Near the pole part we can approximate in (12.43) the normal Green's function (12.44) by the quasiparticle value $G^A \approx G_n^A$ resulting in

$$(G^R)^{-1}(p, \omega) = \omega - \epsilon_p + \int \frac{dp_3 d\omega_1}{(2\pi)^4} \frac{T^< (\frac{p-p_3}{2}, \frac{p-p_3}{2}, p + p_3, \omega_1)}{\omega_1 - \omega - \epsilon_{p_3}}. \tag{12.45}$$

Using the dominant pole part (12.40) we get

$$(G^R)^{-1}(p, \omega) = \omega - \epsilon_p \mp \frac{\Delta(p)^2}{2\mu - \omega - \epsilon_p} \tag{12.46}$$

in agreement with the forms found earlier, (12.20) and (12.28). Therefore the retarded propagator can be written in a fashionable form

$$G^R = \frac{1}{2}\left(1 + \frac{\xi}{E}\right)\frac{1}{\omega - \mu - E} + \frac{1}{2}\left(1 - \frac{\xi}{E}\right)\frac{1}{\omega - \mu + E} \tag{12.47}$$

with $\xi = \epsilon_p - \mu$ and the new quasiparticle energy

$$E = \sqrt{\xi^2 \pm \Delta(p)^2} \tag{12.48}$$

in agreement with (12.32). The spectral function (Ambegaokar, 1969) shows consequently two parts, a particle and 'antiparticle' contribution[1]

$$a = \pi \left(1+\frac{\xi}{E}\right) \delta(\omega-\mu-E) + \pi \left(1-\frac{\xi}{E}\right) \delta(\omega-\mu+E) \qquad (12.49)$$

which fulfils the sum rule and collapses to the normal quasiparticle picture for $\Delta \to 0$. The consequent Green's functions reads

$$g^{<}(p,\omega) = \pi \left(1+\frac{\xi}{E}\right) \delta(\omega-\mu-E) f(E) + \pi \left(1-\frac{\xi}{E}\right) \delta(\omega-\mu+E) f(-E). \qquad (12.50)$$

The spectrum of available states is given by the first part and the spectrum of particles which can be removed are given by the second one. As illustrated in Figure 12.6 the gap for fermions changes the spectra in the characteristic way allowing the population of positive states below the Fermi energy. For a vanishing gap only the negative branch below Fermi energy and the positive branch above Fermi energy bears a non-vanishing weight; such that the normal quasiparticle spectrum $\xi = \epsilon - \epsilon_f$ is approached.

Introducing now one normal quasiparticle Green's functions (9.17) and one anomaly one (12.50) in the two-particle propagator (12.39) we obtain from (12.41) just the correct finite-temperature gap equation

$$\Delta(p_1) = \int \frac{dp_3}{(2\pi)^3} V(p_1-p_3) \frac{f(E)-f(-E)}{2E} \Delta(p_3)$$

$$= \mp \int \frac{dp_3}{(2\pi)^3} V(p_1-p_3) \begin{pmatrix} \tanh\frac{E}{2T} \\ \coth\frac{E}{2T} \end{pmatrix} \frac{\Delta(p_3)}{2E} \qquad (12.51)$$

with (12.48) for fermions/Bosons respectively.

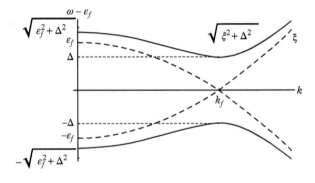

Figure 12.6: *The quasiparticle spectrum for a finite-gap superconductor (fermions) with quadratic dispersion at* $T = 0$.

---

[1] In fact it is the same as in relativistic mean-field models, see (22.32) and (22.33), which analogy is used as equivalent mass approach (Gulian and Zharkov, 1999).

12.4.1   Generalised Soven scheme of coherent potential approximation

Now that we have clarified that the anomalous propagator is an exact rewriting of the Dyson equation, if one corrects a channel of selfenergy for repeated collisions, we might ask what kind of equation such a channel-corrected selfenergy obeys. This will lead us to a generalisation of the Soven equation (Soven, 1967). The Soven equation was proposed to describe impurity scattering in terms of an effective medium and resulted in the coherent potential approximation (CPA) (Velický, Kirkpatrick and Ehrenreich, 1968; Velicky, 1969; Elliott, Krumhansl and Leath, 1974). This CPA improves the averaged T-matrix (Koster and Slater, 1954) with respect to better analytic properties and a wider range of applications. It has turned out that the averaged T-matrix is the uncorrected channel while the CPA is equivalent to a channel-corrected approximation, see Chapter 4. Here we will present the same idea of channel correction but applied to the two-particle scattering. This will lead to a general Soven equation which allows us to improve a chosen approximation scheme in the same manner as the CPA improves the averaged T-matrix approximation for impurity scattering.

Let's assume, in general, the defining equation for the channel T-matrix in terms of the potential V and a block K_j

$$T_j = V + VK_j T_j \tag{12.52}$$

covering both the singular channel $j = i$ as well as the normal channels $j \neq i$. In general, this equation is a two-particle one which is reduced to the one-particle selfenergy by closing the upper line with the backward propagating Green's function $\Sigma' = \sum_{j \neq i} \overrightarrow{T_j G}$ and for the singular channel $\Sigma_i = \overrightarrow{T_i G_{\hat{\imath}}}$. We will now denote explicitly by which function the upper line is closed. All other products are understood as operator products of one-particle functions. For the above mentioned averaged T-matrix approximation of Chapters 5.4 and 5.6, one has $K = G_{\hat{\imath}}$ and a closing by $\bar{G} \to c$ in terms of the impurity concentration c. In the two-particle ladder approximation one would have the form $K_i = G \cdot G_{\hat{\imath}}$. In the following we consider the general block K; such that any more refined approximation can be chosen.

With (12.22) and $\Sigma_i = \overrightarrow{T_i G_{\hat{\imath}}}$ we can write

$$1 = (1-D)G^{-1}G_{\hat{\imath}} + DG^{-1}G_{\hat{\imath}} + G^{-1}G_{\hat{\imath}}\overrightarrow{T_i G_{\hat{\imath}}}G \tag{12.53}$$

where we have added and subtracted an operator D which will be determined later by convenience. Now it easy to proove with the help of the separability of the singular channel $T_i = \mp \Delta \Delta$ that the following relation holds:

$$\overrightarrow{T_i G_{\hat{\imath}}}G = \overrightarrow{T_i G}G_{\hat{\imath}}. \tag{12.54}$$

Indeed, we have with the help of (12.22)

$$\overrightarrow{T_i G} G_\hbar G^{-1} = \overrightarrow{T_i G} - \overrightarrow{T_i G} G_\hbar \overrightarrow{T_i G_\hbar} \tag{12.55}$$

and

$$\overrightarrow{T_i G_\hbar} = \overrightarrow{T_i G} - \overrightarrow{T_i G_\hbar \Sigma_i G}. \tag{12.56}$$

Since we have the identity

$$\overrightarrow{T_i G_\hbar \Sigma_i G} = \overrightarrow{T_i G} G_\hbar \overrightarrow{T_i G_\hbar} \tag{12.57}$$

as shown in Figure 12.7, the equations (12.55) and (12.56) are identical and relation (12.54) is proved.

Therefore we can write for (12.53)

$$1 = (1 - D)G^{-1} G_\hbar + (DG^{-1} + G^{-1} G_\hbar \overrightarrow{T_i G}) G_\hbar \tag{12.58}$$

Multiplying (12.22) from the right with $\overrightarrow{T_i G_\hbar}$ we can find

$$G^{-1} G_\hbar \overrightarrow{T_i G} + \Sigma_i G_\hbar \overrightarrow{T_i G} = \overrightarrow{T_i G} = \overrightarrow{V G} + \overrightarrow{V K T_i G} \tag{12.59}$$

where we have used (12.52) for the second identity. For the last term we now define an effective potential $\tilde{V}_i$

$$\tilde{V}_i G_\hbar \overrightarrow{T_i G} \equiv \overrightarrow{V K T_i G} \tag{12.60}$$

with the help of which we can invert (12.59)

$$G_\hbar \overrightarrow{T_i G} = (G^{-1} + \Sigma_i - \tilde{V}_i)^{-1} \overrightarrow{V G}. \tag{12.61}$$

Using this in (12.58) we arrive at

$$1 = (1 - D)G^{-1} G_\hbar + [1 + (\Sigma_i - \tilde{V}_i) G]^{-1} [(1 + \Sigma_i G)D - \tilde{V}_i GD + \overrightarrow{V G} G] G^{-1} G_\hbar. \tag{12.62}$$

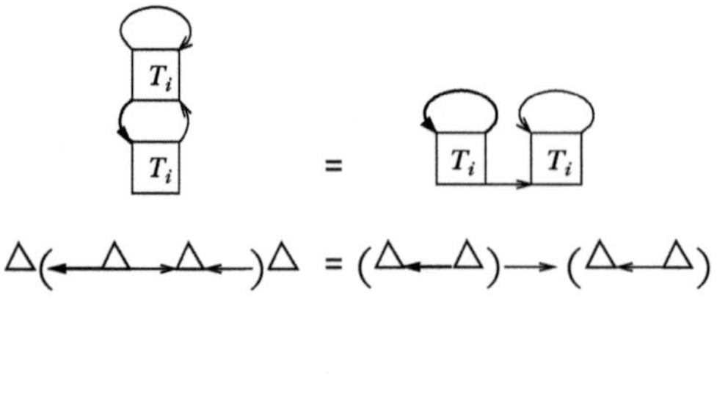

Figure 12.7: *Proof of relation (12.57) in the first line for singular channel T-matrices which become separable as illustrated in the second line.*

Now we choose the operator D conveniently such that the last two terms cancel each other, i.e.

$$\overrightarrow{V}GG \equiv \tilde{V}_i GD. \tag{12.63}$$

Using (12.22) in the form $G^{-1}G_A = (1 + \Sigma_i G)^{-1}$ and subtracting from (12.62) the structure $1 = (1 - D)A^{-1}A + B^{-1}BD$ we obtain finally

$$(1 - D)(1 + \Sigma_i G)^{-1}\Sigma_i = [1 + (\Sigma_i - \tilde{V})G]^{-1}[\tilde{V}_i GDG^{-1} - (1 + \Sigma_i G)D(1 + \Sigma_i G)^{-1}\Sigma_i]. \tag{12.64}$$

Together with the operator (12.63) and the effective potential $\tilde{V}_i$ defined by (12.60) this is the desired generalised Soven equation. It is written in operator form which becomes an algebraic equation in the appropriate representation. In the operator form it is even valid in nonequilibrium and its time ordering can be treated e.g. in the framework of generalised Kadanoff and Baym formalism with the help the Langreth–Wilkins rules.

Though introduced merely for mathematical convenience, the operator D corresponds to the concentration for impurity scattering and the effective potential corresponds to the effective potential in CPA. Therefore D is called the concentration operator hereafter.

Let us illustrate this with the help of some special cases. Choosing the averaged T-matrix approximation we have only one-particle functions, $K = G_A$ and the closing by the concentration as a c-number $\bar{G} \to c$, such that we get from (12.60) $\tilde{V}_i = V$ which gives with (12.63) $D = c$ and from (12.64) the standard Soven equation (Velický et al., 1968) appears

$$(1 - c)\frac{\Sigma_i(p)}{1 + \Sigma_i(p)G(p)} = c\frac{V(p) - \Sigma_i(p)}{1 + [\Sigma_i(p) - V(p)]G(p)} \tag{12.65}$$

which we recognise as the CPA equation (5.62) and (5.63). As a second, so far not known, example we give the explicit expressions for the two-particle T-matrix. Then $K = G \cdot G_A$ is a product in spatial coordinates. The Fourier transform of the different coordinates and gradient expansion then reveals from (12.64) the structure $[p = (\mathbf{p}, \omega_p)]$

$$[1 - D(p)]\frac{\Sigma_i(p)}{1 + \Sigma_i(p)G(p)} = D(p)\frac{\tilde{V}_i(p) - \Sigma_i(p)}{1 + [\Sigma_i(p) - \tilde{V}_i(p)]G(p)}. \tag{12.66}$$

The concentration operator takes the form

$$D(p) = \frac{\sum_{\bar{p},\bar{\omega}} V\left(\frac{\bar{p} - p}{2}\right) G(-\bar{p})}{\tilde{V}_i(p)} = \frac{\Sigma_H(\mathbf{p})}{\tilde{V}_i(p)} \tag{12.67}$$

noting the Hartree selfenergy Σ_H and the effective potential reads

$$\tilde{V}_i(p) = \frac{\overrightarrow{VKT_i\vec{G}}}{\overrightarrow{T_iGG_{\hat{h}}}} = \frac{\overrightarrow{T_i\vec{G} - \vec{V}\vec{G}}}{\overrightarrow{T_iGG_{\hat{h}}}} = \frac{1}{G_{\hat{h}}(p)} - \frac{\Sigma_H(\mathbf{p})}{\Sigma_i(p)G(p)}. \tag{12.68}$$

Here we have used (12.52) for the first equality and (12.54) for the second one. The channel T-matrix reads explicitly

$$\Sigma_i(p) = \sum_{\bar{p}} T_i\left(\frac{\bar{\mathbf{p}}-\mathbf{p}}{2}, \frac{\bar{\mathbf{p}}-\mathbf{p}}{2}, \mathbf{p}+\bar{\mathbf{p}}, \omega_p + \omega_{\bar{p}}\right) G_{\hat{h}}(-\bar{p}). \tag{12.69}$$

It is instructive to see that introducing equations (12.67) and (12.68) into (12.66) leads indeed to an identity. This is due to the fact that we have assumed that all quantities like T-matrix and selfenergy are known exactly. In cases where we start with an approximation for the Green's function or selfenergy we can use the above equation system and the generalised Soven form to iterate and to obtain approximations for the channel-corrected propagators and selfenergies.

12.5   Interacting Bose gas at finite temperatures

We will present a consistent treatment of interactions and condensation in the unified manner with the help of the corrected multiple-scattering T-matrix. Our starting point is a homogeneous gas of interacting Bosons with mass m, temperature T and particle density n. The temperature T scales in energy units so that Boltzmann's constant k_B can be omitted. The interaction is assumed in a separable form, $V_{\mathbf{p},\mathbf{k}} = \lambda g_{\mathbf{p}} g_{\mathbf{k}}$ characterised by the strength λ and the Yamaguchi form factors $g_{\mathbf{p}} = \left(\beta^2 + p^2\right)^{-1}$, see Appendix E.1.

We will give here all formulae in terms of the equilibrium Matsubara–Green's function of Chapter 7.3.5. We absorb the chemical potential as bottom of the single particle energy and the Matsubara frequency reads $z_\nu = -i2\pi\nu T$, $\nu \in \mathbb{Z}$. We use the scheme to eliminate unphysical self-interaction in the T-matrix approximation as presented in Chapter 12.3 and derived in Chapter 11.5.6. This allows us to calculate properties below and above the critical temperature on the same theoretical footing (Lipavský, 2008; Morawetz, 2010; Männel, Morawetz and Lipavský, 2010; Šopík, Lipavský, Männel, Morawetz and Matlock, 2011).

The many-body T-matrix obeys the Bethe–Salpeter equation (10.154) now written in Matsubara frequencies

$$\mathcal{T}_q(\mathbf{p}, \mathbf{k}, z_\nu) = \bar{v}(\mathbf{p}, \mathbf{k}) - \frac{1}{\Omega}\sum_{\mathbf{k}'} \bar{v}(\mathbf{p}, \mathbf{k}')\mathcal{G}_q(\mathbf{k}', z_\nu)\mathcal{T}_q(\mathbf{k}', \mathbf{k}, z_\nu) \tag{12.70}$$

with the symmetrised interaction

$$\bar{v}(\mathbf{p}, \mathbf{q}) = \frac{1}{2}\left[v(\mathbf{p}, \mathbf{q}) + v(-\mathbf{p}, -\mathbf{q})\right] = \bar{v}(-\mathbf{p}, -\mathbf{q}). \tag{12.71}$$

and the two-particle Green's function (10.149) reads

$$\mathscr{G}_{\mathbf{q}}(\mathbf{k}', z_\nu) = \frac{1}{\beta} \sum_{\bar{\nu}} G\left(\frac{\mathbf{q}}{2} + \mathbf{k}', z_{\bar{\nu}}\right) G\left(\frac{\mathbf{q}}{2} - \mathbf{k}', z_\nu - z_{\bar{\nu}}\right).$$

For the selfenergy an exchange of particles has to be considered due to the indistinguishability

$$\Sigma(\mathbf{q}, z_\nu) = -\frac{1}{\beta\Omega} \sum_{\mathbf{p}, \bar{\nu}} \mathscr{T}_{\mathbf{q}+\mathbf{p}}\left(\frac{\mathbf{q}-\mathbf{p}}{2}, \frac{\mathbf{q}-\mathbf{p}}{2}, z_\nu + z_{\bar{\nu}}\right) G(\mathbf{p}, z_{\bar{\nu}})$$

$$-\frac{1}{\beta\Omega} \sum_{\mathbf{p}, \bar{\nu}} \mathscr{T}_{\mathbf{q}+\mathbf{p}}\left(\frac{\mathbf{q}-\mathbf{p}}{2}, \frac{\mathbf{p}-\mathbf{q}}{2}, z_\nu + z_{\bar{\nu}}\right) G(\mathbf{p}, z_{\bar{\nu}}). \tag{12.72}$$

Taking into account the poles of G at the quasiparticle energies $E_{\mathbf{p}}^{(j)}$ as well as the poles of the T-matrix at the energies $\omega_{\mathbf{q}}$ one finds

$$\Sigma(\mathbf{q}, z_\nu) = \sum_{\mathbf{p}, j} n_{\mathbf{p}}^{(j)} \left[\mathscr{T}_{\mathbf{q}+\mathbf{p}}\left(\frac{\mathbf{q}-\mathbf{p}}{2}, \frac{\mathbf{q}-\mathbf{p}}{2}, z_\nu + E_{\mathbf{p}}^{(j)}\right) + \mathscr{T}_{\mathbf{q}+\mathbf{p}}\left(\frac{\mathbf{q}-\mathbf{p}}{2}, \frac{\mathbf{p}-\mathbf{q}}{2}, z_\nu + E_{\mathbf{p}}^{(j)}\right) \right]$$

$$+ \sum_{\mathbf{p}} \left[\Delta_{\mathbf{q}+\mathbf{p}}^2 \left(\frac{\mathbf{q}-\mathbf{p}}{2}\right) + \Delta_{\mathbf{q}+\mathbf{p}}\left(\frac{\mathbf{q}-\mathbf{p}}{2}\right) \Delta_{\mathbf{q}+\mathbf{p}}\left(\frac{\mathbf{p}-\mathbf{q}}{2}\right) \right] G(\mathbf{p}, \omega_{\mathbf{q}+\mathbf{p}} - 2\mu - z_\nu)$$

with

$$\Delta_{\mathbf{q}}(\mathbf{p}) \Delta_{\mathbf{q}}(\mathbf{k}) = \frac{f_B(\omega_{\mathbf{q}} - 2\mu)}{\Omega} \text{Res}\left(\mathscr{T}_{\mathbf{q}}(\mathbf{p}, \mathbf{k}, z), \omega_{\mathbf{q}} - 2\mu\right). \tag{12.73}$$

The occupation numbers are

$$n_{\mathbf{p}}^{(j)} = \frac{f_B\left(E_{\mathbf{p}}^{(j)}\right)}{\Omega} \text{Res}\left(G(\mathbf{p}, z), E_{\mathbf{p}}^{(j)}\right) \tag{12.74}$$

and the density is

$$n = -\frac{1}{\beta\Omega} \sum_{\mathbf{p}, \nu} G(\mathbf{p}, z_\nu) = \sum_{\mathbf{p}, j} n_{\mathbf{p}}^{(j)}. \tag{12.75}$$

If one or several quasiparticle energies touch the chemical potential for $\mathbf{p} = 0$, i.e. $E_0^{(j)} = 0$, the Bose function in (12.74) will diverge and cancel Ω in the thermodynamic limit $\Omega \to \infty$. This means a Bose condensation of quasiparticles in the zero-momentum state. Therefore the condensate density

$$n_0 = \sum_j^{E_0^{(j)}=0} n_0^{(j)}$$

has to be split off from the density and considered separately. Also the singular selfenergy has to be split off from the total one if the Bose function in (12.73) diverges, i.e. if its argument is zero. Assuming that a pair of particles with centre-of-mass momentum $\mathbf{q} = 0$ has the lowest possible energy, the divergence will appear as soon as $\omega_0 = 2\mu$. This condition is identical to the Thouless criterion for fermions (Thouless, 1960).

To avoid self-interaction the loop of the singular selfenergy corresponding to the divergence in (12.73) has to be constructed from reduced Green's functions (12.23)

$$G_{\not{A}} = \frac{1}{G_0^{-1} - \Sigma^{\mathrm{reg}}} = G(1 - \Sigma^{\sin} G_{\not{A}}). \tag{12.76}$$

12.5.1   Condensed phase

In the BEC phase a fraction of particles is condensed, with a condensate density n_0. The T-matrix yields the crucial contribution to obtain a Bogoliubov-like dispersion. In order to obtain the correct dispersion, it is essential to include the scattering of pairs of particles into and out of the condensate.

Calculating the two-particle Green's function

$$\mathscr{G}_{\mathbf{q}}(\mathbf{k}', z_\nu) = -\sum_j f_B\left(E_{\frac{\mathbf{q}}{2}+\mathbf{k}'}^{(j)}\right) \mathrm{Res}\left(G\left(\frac{\mathbf{q}}{2}+\mathbf{k}', z\right), E_{\frac{\mathbf{q}}{2}+\mathbf{k}'}^{(j)}\right) G\left(\frac{\mathbf{q}}{2}-\mathbf{k}', z_\nu - E_{\frac{\mathbf{q}}{2}+\mathbf{k}'}^{(j)}\right)$$

$$+ \sum_j G\left(\frac{\mathbf{q}}{2}+\mathbf{k}', z_\nu - E_{\frac{\mathbf{q}}{2}-\mathbf{k}'}^{(j)}\right) f_B\left(z_\nu - E_{\frac{\mathbf{q}}{2}-\mathbf{k}'}^{(j)}\right) \mathrm{Res}\left(G\left(\frac{\mathbf{q}}{2}-\mathbf{k}', z\right), E_{\frac{\mathbf{q}}{2}-\mathbf{k}'}^{(j)}\right)$$

$$= -\Omega \sum_j \left[n_{\frac{\mathbf{q}}{2}+\mathbf{k}'}^{(j)} G\left(\frac{\mathbf{q}}{2}-\mathbf{k}', z_\nu - E_{\frac{\mathbf{q}}{2}+\mathbf{k}'}^{(j)}\right) + G\left(\frac{\mathbf{q}}{2}+\mathbf{k}', z_\nu - E_{\frac{\mathbf{q}}{2}-\mathbf{k}'}^{(j)}\right) n_{\frac{\mathbf{q}}{2}-\mathbf{k}'}^{(j)} \right]$$

$$- \sum_j G\left(\frac{\mathbf{q}}{2}+\mathbf{k}', z_\nu - E_{\frac{\mathbf{q}}{2}-\mathbf{k}'}^{(j)}\right) \mathrm{Res}\left(G\left(\frac{\mathbf{q}}{2}-\mathbf{k}', z\right), E_{\frac{\mathbf{q}}{2}-\mathbf{k}'}^{(j)}\right)$$

one finds $\mathscr{G}_{\mathbf{q}}\left(\pm\frac{\mathbf{q}}{2}, z_\nu\right) = -\Omega\left[n_0 G(\mathbf{q}, z_\nu) + \mathscr{O}\left(\Omega^{-1}\right)\right]$. We want to split off the condensate contribution to the T-matrix. By inverting the matrix equation (12.70)

$$\bar{v}^{-1}(\mathbf{p}, \mathbf{k}) = \mathscr{T}_{\mathbf{q}}^{-1}(\mathbf{p}, \mathbf{k}, z_\nu) - \frac{1}{\Omega}\mathscr{G}_{\mathbf{q}}(\mathbf{p}, z_\nu)\delta_{\mathbf{pk}}$$

$$= \mathscr{T}_{\mathbf{q}}^{-1}(\mathbf{p}, \mathbf{k}, z_\nu) - \frac{1}{\Omega}\mathscr{G}_{\mathbf{q}}(\mathbf{p}, z_\nu)(1 - \delta_{\frac{\mathbf{q}}{2}\mathbf{p}} - \delta_{-\frac{\mathbf{q}}{2}\mathbf{p}})\delta_{\mathbf{pk}}$$

$$- \frac{1}{\Omega}\mathscr{G}_{\mathbf{q}}\left(\frac{\mathbf{q}}{2}, z_\nu\right) \delta_{\frac{\mathbf{q}}{2}\mathbf{p}}\delta_{\frac{\mathbf{q}}{2}\mathbf{k}} - \frac{1}{\Omega}\mathscr{G}_{\mathbf{q}}\left(-\frac{\mathbf{q}}{2}, z_\nu\right)\delta_{-\frac{\mathbf{q}}{2}\mathbf{p}}\delta_{-\frac{\mathbf{q}}{2}\mathbf{k}}$$

one can define the T-matrix $\mathscr{T}_q^{\prime\prime 0}$

$$\bar{v}^{-1}(\mathbf{p},\mathbf{k}) = \left(\mathscr{T}_q^{\prime\prime 0}\right)^{-1}(\mathbf{p},\mathbf{k},z_\nu) - \frac{1}{\Omega}\mathscr{G}_q(\mathbf{p},z_\nu)(1 - \delta_{\frac{q}{2}\mathbf{p}} - \delta_{-\frac{q}{2}\mathbf{p}})\delta_{\mathbf{pk}},$$

from which the condensate contribution is excluded. Its relation to the full T-matrix

$$\left(\mathscr{T}_q^{\prime\prime 0}\right)^{-1}(\mathbf{p},\mathbf{k},z_\nu) = \mathscr{T}_0^{-1}(\mathbf{p},\mathbf{k},z_\nu) - \frac{1}{\Omega}\mathscr{G}_q\left(\frac{\mathbf{q}}{2},z_\nu\right)\delta_{\frac{q}{2}\mathbf{p}}\delta_{\frac{q}{2}\mathbf{k}} - \frac{1}{\Omega}\mathscr{G}_q\left(-\frac{\mathbf{q}}{2},z_\nu\right)\delta_{-\frac{q}{2}\mathbf{p}}\delta_{-\frac{q}{2}\mathbf{k}}$$

can be inverted to yield finally

$$\mathscr{T}_q^{\prime\prime 0}(\mathbf{p},\mathbf{k},z_\nu) = \bar{v}(\mathbf{p},\mathbf{k}) - \frac{1}{\Omega}\sum_{\mathbf{k}'\neq\pm\frac{q}{2}}\bar{v}(\mathbf{p},\mathbf{k}')\mathscr{G}_q(\mathbf{k}',z_\nu)\mathscr{T}_q^{\prime\prime 0}(\mathbf{k}',\mathbf{k},z_\nu)$$

$$\mathscr{T}_q(\mathbf{p},\mathbf{k},z_\nu) = \mathscr{T}_q^{\prime\prime 0}(\mathbf{p},\mathbf{k},z_\nu) - \frac{1}{\Omega}\mathscr{T}_q^{\prime\prime 0}\left(\mathbf{p},\frac{\mathbf{q}}{2},z_\nu\right)\mathscr{G}_q\left(\frac{\mathbf{q}}{2},z_\nu\right)\mathscr{T}_q\left(\frac{\mathbf{q}}{2},\mathbf{k},z_\nu\right)$$

$$-\frac{1}{\Omega}\mathscr{T}_q^{\prime\prime 0}\left(\mathbf{p},-\frac{\mathbf{q}}{2},z_\nu\right)\mathscr{G}_q\left(-\frac{\mathbf{q}}{2},z_\nu\right)\mathscr{T}_q\left(-\frac{\mathbf{q}}{2},\mathbf{k},z_\nu\right).$$

For separable interaction the symmetry relation (12.71) demands that the potential and the T-matrix are independent of the sign of the relative momenta, i.e.

$$\mathscr{T}_q^{\prime\prime 0}(\mathbf{p},\mathbf{k},z_\nu) = \mathscr{T}_q^{\prime\prime 0}(\mathbf{p},\mathbf{k},z_\nu) + 2n_0\mathscr{T}_q^{\prime\prime 0}(\mathbf{p},\mathbf{k},z_\nu)\,G(\mathbf{q},z_\nu)\,\mathscr{T}_q\left(\frac{\mathbf{q}}{2},\frac{\mathbf{q}}{2},z_\nu\right).$$

The direct part of the selfenergy (12.72) can therefore be approximated by

$$\Sigma^{\text{dir}}(\mathbf{q},z_\nu) \approx -\frac{1}{\beta\Omega}\sum_{\mathbf{p},\bar{\nu}}\mathscr{T}_{\mathbf{q}+\mathbf{p}}^{\prime\prime 0}\left(\frac{\mathbf{q}-\mathbf{p}}{2},\frac{\mathbf{q}-\mathbf{p}}{2},z_\nu+z_{\bar{\nu}}\right)G(\mathbf{p},z_{\bar{\nu}})$$

$$-\frac{2n_0}{\beta\Omega}\sum_{\mathbf{p},\bar{\nu}}\mathscr{T}_{\mathbf{q}+\mathbf{p}}^{\prime\prime 0}\left(\frac{\mathbf{q}-\mathbf{p}}{2},\frac{\mathbf{q}-\mathbf{p}}{2},z_\nu+z_{\bar{\nu}}\right)G(\mathbf{q}+\mathbf{p},z_\nu+z_{\bar{\nu}})$$

$$\times\mathscr{T}_{\mathbf{q}+\mathbf{p}}^{\prime\prime 0}\left(\frac{\mathbf{q}+\mathbf{p}}{2},\frac{\mathbf{q}+\mathbf{p}}{2},z_\nu+z_{\bar{\nu}}\right)G(\mathbf{p},z_{\bar{\nu}}) \tag{12.77}$$

$$=\ldots+2n_0\sum_{\mathbf{p},j}\mathscr{T}_{\mathbf{q}+\mathbf{p}}^{\prime\prime 0}\left(\frac{\mathbf{q}-\mathbf{p}}{2},\frac{\mathbf{q}-\mathbf{p}}{2},E_{\mathbf{q}+\mathbf{p}}^{(j)}\right)n_{\mathbf{q}+\mathbf{p}}^{(j)}$$

$$\times\mathscr{T}_{\mathbf{q}+\mathbf{p}}^{\prime\prime 0}\left(\frac{\mathbf{q}+\mathbf{p}}{2},\frac{\mathbf{q}+\mathbf{p}}{2},E_{\mathbf{q}+\mathbf{p}}^{(j)}\right)G\left(\mathbf{p},E_{\mathbf{q}+\mathbf{p}}^{(j)}-z_\nu\right)$$

$$=\ldots+2n_0^2\mathscr{T}_0^{\prime\prime 0}(\mathbf{q},\mathbf{q},0)\,\mathscr{T}_0^{\prime\prime 0}(0,0,0)\,G(-\mathbf{q},-z_\nu)+2n_0\sum_{\mathbf{p}(\neq-\mathbf{q}),j}\ldots.$$

Obviously, the singular term, i.e. the first one in the last line, corresponds to a double count of the diagram illustrated in Fig. 12.2(a, b). Since inside the condensate we do not have an exchange, this factor of two has to be omitted. Furthermore, to avoid unphysical self interaction like the ones explained in Chapter 12.3, the loop in the singular selfenergy has to be constructed with reduced Green's functions. Splitting off the singular channel in (12.77) one finds

$$\Sigma^{\sin}(\mathbf{q}, z_\nu) = \Delta_0^2(\mathbf{q}) \, G_{\mathbb{A}}(-\mathbf{q}, -z_\nu)$$

with

$$\Delta_0(\mathbf{q}) = \Sigma_{12}(\mathbf{q}, z_\nu) = n_0 \left| \mathcal{T}_0^{\sin}(\mathbf{q}, 0, 0) \right|. \tag{12.78}$$

The latter relation is valid for repulsive as well as for attractive interaction.

The normal selfenergy of (12.77) is related to the regular selfenergy

$$\Sigma_{11}(\mathbf{q}, z_\nu) = \Sigma_{22}(-\mathbf{q}, -z_\nu) = \Sigma^{\text{reg}}(\mathbf{q}, z_\nu).$$

The anomalous selfenergies can be identified as

$$\Sigma_{12}(\mathbf{q}, z_\nu) = \Sigma_{21}(\mathbf{q}, z_\nu) = \Delta_0(\mathbf{q}).$$

From the symmetry relation (12.71) follows $\Delta_0(\mathbf{q}) = \Delta_0(-\mathbf{q})$.

In agreement with (12.28), the full Green's function in the BEC phase, with $n_0 > 0$, becomes (Shi and Griffin, 1998; Maennel, 2011)

$$G(\mathbf{q}, z_\nu) = \frac{z_\nu + \frac{\hbar^2 q^2}{2m} - \mu + \Sigma^{\text{reg}}(-\mathbf{q}, -z_\nu)}{\left(z_\nu + \frac{\hbar^2 q^2}{2m} - \mu + \Sigma^{\text{reg}}(-\mathbf{q}, -z_\nu) \right) \left(z_\nu - \frac{\hbar^2 q^2}{2m} + \mu - \Sigma^{\text{reg}}(\mathbf{q}, z_\nu) \right) + \Delta_0^2(\mathbf{q})}. \tag{12.79}$$

We see that the structure of the resulting Green's function is identical to that of the channel-corrected one (12.28).

Assuming a separable or contact interaction the selfconsistent equation (12.70) can be solved and one obtains

$$\mathcal{T}_\mathbf{q}(\mathbf{p}, \mathbf{k}, z_\nu) = \frac{\bar{v}(\mathbf{p}, \mathbf{k})}{1 + \frac{1}{\Omega} \sum_{\mathbf{k}'} \bar{v}(\mathbf{k}', \mathbf{k}') \mathscr{G}_\mathbf{q}(\mathbf{k}', z_\nu)}. \tag{12.80}$$

In the singular channel, i.e. for $\mathbf{q} = 0$ and $z_\nu = 0$, the two-particle function has to be constructed with one reduced propagator since the upper loop of the singular selfenergy is constructed from these functions. With the Feynman trick (7.69) one finds

$$\mathscr{G}_0^{\mathrm{sin}}(\mathbf{k}',0) = \frac{1}{\beta}\sum_{\bar{\nu}} G\left(\mathbf{k}',z_{\bar{\nu}}\right) G_{\mathbb{A}}\left(-\mathbf{k}',-z_{\bar{\nu}}\right) = -\Omega\sum_j n_{\mathbf{k}'}^{(j)} G_{\mathbb{A}}\left(-\mathbf{k}',-E_{\mathbf{k}'}^{(j)}\right). \qquad (12.81)$$

The pole condition of (12.80) in the singular channel, i.e. the implicit equation that defines $\Delta_0(\mathbf{q})$, is therefore

$$0 = 1 - \bar{v}(0,0)n_0 G_{\mathbb{A}}(0,0) - \sum_{\mathbf{k}'\neq 0,j} \bar{v}(\mathbf{k}',\mathbf{k}')n_{\mathbf{k}'}^{(j)} G_{\mathbb{A}}\left(-\mathbf{k}',-E_{\mathbf{k}'}^{(j)}\right). \qquad (12.82)$$

For attractive interaction the T-matrix can have a pole due to the bound states, i.e. (12.82) can have a solution even if $n_0 = 0$. In that case there is a condensation of bound states. For finite condensate density n_0, condition (12.82) can be rewritten as

$$G_{\mathbb{A}}^{-1}(0,0) = \mu - \Sigma^{\mathrm{reg}}(0,0) = \frac{n_0\bar{v}(0,0)}{1-\sum_{\mathbf{k}'\neq 0,j} \bar{v}(\mathbf{k}',\mathbf{k}')n_{\mathbf{k}'}^{(j)} G_{\mathbb{A}}\left(-\mathbf{k}',-E_{\mathbf{k}'}^{(j)}\right)} = n_0\mathscr{T}_0^{\mathrm{sin}}(0,0,0). \qquad (12.83)$$

In the T-matrix $\mathscr{T}_0^{\mathrm{sin}}$ the condensate contribution is excluded

$$\mathscr{T}_0^{\mathrm{sin}}(\mathbf{p},\mathbf{k},0) = \frac{\bar{v}(\mathbf{p},\mathbf{k})}{1+\frac{1}{\Omega}\sum_{\mathbf{k}'\neq 0}\bar{v}(\mathbf{k}',\mathbf{k}')\mathscr{G}_0^{\mathrm{sin}}(\mathbf{k}',0)}. \qquad (12.84)$$

The singular selfenergy (12.78) can now be written as

$$\Sigma^{\mathrm{sin}}(\mathbf{q},z_\nu) = \mathscr{T}_0^{\mathrm{sin}}(\mathbf{q},0,0)n_0^2\mathscr{T}_0^{\mathrm{sin}}(0,\mathbf{q},0)G_{\mathbb{A}}(-\mathbf{q},-z_\nu).$$

The diagrammatic illustration in Fig. 12.2(d) shows that the singular selfenergy includes exactly those scattering processes crucial for obtaining a Bogoliubov-like dispersion, i.e. the scattering of a pair of particles into and out of the condensate.

For the sake of simplicity, in the following, the regular selfenergy will be assumed to be independent of momentum and energy. Due to Bose statistics the relevant contributions to any property will come from small momenta and energies. One can therefore approximate $\Sigma^{\mathrm{reg}}(\mathbf{q},z_\nu) \approx \Sigma^{\mathrm{reg}}(0,0)$. The reduced Green's function (12.76) becomes

$$G_{\mathbb{A}}(\mathbf{q},z_\nu) = \frac{1}{z_\nu - \epsilon_\mathbf{q}}$$

and the full one (12.79) can then be written as (12.50)

$$G(\mathbf{q},z_\nu) = \frac{1}{2}\left(1+\frac{\epsilon_\mathbf{q}}{E_\mathbf{q}}\right)\frac{1}{z_\nu - E_\mathbf{q}} + \frac{1}{2}\left(1-\frac{\epsilon_\mathbf{q}}{E_\mathbf{q}}\right)\frac{1}{z_\nu + E_\mathbf{q}} \qquad (12.85)$$

with the quasiparticle energies

$$\epsilon_q = \frac{\hbar^2 q^2}{2m} - \mu + \Sigma^{\mathrm{reg}}(0,0) \tag{12.86}$$

and the excitation energy of (12.48)

$$E_q = \sqrt{\epsilon_q^2 - \Delta_0^2(q)}. \tag{12.87}$$

The equation for the density (12.75)

$$n = n_0 + \frac{1}{\Omega} \sum_{p \neq 0} \frac{1}{2} \left[\frac{\epsilon_p}{E_p} \left(1 + 2 f_B(E_p) \right) - 1 \right] = n_0 + \frac{1}{\Omega} \sum_{p \neq 0} \frac{1}{2} \left[\frac{\epsilon_p}{E_p} \coth \frac{\beta E_p}{2} - 1 \right] \tag{12.88}$$

is needed to calculate n_0 and μ. For $E_0 \ll T$ the condensate density can be approximated as $n_0 \approx \epsilon_0 / \Omega \beta E_0^2$ and one finds

$$\mu = \Sigma^{\mathrm{reg}}(0,0) - \frac{1}{2\beta n_0 \Omega} - \sqrt{\Delta_0^2(0) + \frac{1}{(2\beta n_0 \Omega)^2}} = \Sigma^{\mathrm{reg}}(0,0) - \Delta_0(0) + \mathcal{O}(\Omega^{-1}). \tag{12.89}$$

From the dispersion (12.87) and (12.89) follows the generalised Bogoliubov dispersion

$$E_q = \sqrt{\left(\frac{\hbar^2 q^2}{2m} + n_0 \left| \mathcal{T}_0^{\mathrm{sin}}(0,0,0) \right| \right)^2 - n_0^2 \left| \mathcal{T}_0^{\mathrm{sin}}(q,0,0) \right|^2} \tag{12.90}$$

in the form of (12.32). It has also been found in (Shi and Griffin, 1998, pp. 27, 37) and (Bijlsma and Stoof, 1997), that the anomalous selfenergy in T-matrix approximation is given by the product of the condensate density with the T-matrix. In the limit of a weak-repulsive contact interaction, i.e. $\left| \mathcal{T}_0^{\mathrm{sin}}(q,0,0) \right| \to U_0$, the dispersion (12.90) coincides with that obtained by Bogoliubov (12.1). However the more general dispersion (12.90) is also valid for attractive interaction.

Having specified the excitation energy (12.90) and the propagator (12.85), the regular part of the many-body T-matrix (12.80) reads

$$\mathcal{T}(q,0,0) = \lambda g_q \left(1 + \lambda \int \frac{d^3 k}{(2\pi)^3} \frac{g_k^2}{2E_k} (1 + 2 f_B(E_k)) \right)^{-1} \tag{12.91}$$

and $f_B(\epsilon) = 1/\left(e^{\epsilon/T} - 1\right)$ is the Bose distribution function. In the normal phase, for $n_0 = 0$, the dispersion is (12.86)

$$\epsilon_q = \frac{\hbar^2 q^2}{2m} - \mu + 2n\mathcal{T}(0).$$ (12.92)

From (12.90) one sees that the chemical potential μ satisfies the Hugenholtz–Pines relation (Hugenholtz and Pines, 1959) $\mu = 2n\mathcal{T}(0) - n_0\mathcal{T}(0)$ in the condensed phase with the particle density (12.88).

12.6 Comparison of approximations

The expectation value of the total energy density can be calculated from the Green's function (Kadanoff and Baym, 1962), according to (7.38), as

$$\mathcal{E} = \frac{\langle \hat{H} \rangle}{\Omega} = -\frac{T}{\Omega} \sum_{k,\nu} \frac{1}{2} \left(z_\nu + \mu + \frac{\hbar^2 k^2}{2m} \right) G(\mathbf{k}, z_\nu)$$

$$= \underbrace{\int \frac{d^3 k}{(2\pi)^3} E_k f_B(E_k)}_{\mathcal{E}_{qp}} + \underbrace{\mathcal{T}(0) \left(n^2 - n n_0 + \frac{1}{2} n_0^2 \right)}_{\mathcal{E}_{mf}}$$

$$\underbrace{- \int \frac{d^3 k}{(2\pi)^3} E_k v_k^2}_{\mathcal{E}_{cor}} + \underbrace{\int \frac{d^3 k}{(2\pi)^3} \frac{n_0^2 \mathcal{T}^2(\mathbf{k})}{4 E_k} (1 + 2 f_B(E_k))}_{\mathcal{E}_{2p}}.$$ (12.93)

Different levels of approximation can now be distinguished in the way they treat the corresponding contributions to this energy density. The mean-field-like approximation $\mathcal{T}(\mathbf{q}) \approx U_0$ together with $E_q = \epsilon_q \approx \hbar^2 q^2/2m + n_0 U_0$ establishes the Hartree–Fock approximation as proposed by Huang et al. (Huang and Yang, 1957; Huang et al., 1957) and in (12.93) only the contribution of quasiparticles $\mathcal{E}_{qp}$ and the mean field term $\mathcal{E}_{mf}$ survive leading to

$$\mathcal{E} = \int \frac{d^3 k}{(2\pi)^3} \frac{\hbar^2 k^2}{2m} f_B(\epsilon_k) + U_0 \left(n^2 - \frac{1}{2} n_0^2 \right).$$ (12.94)

This energy density shows that, in addition to statistics, BEC is also energetically favoured, since a finite condensate density n_0 lowers the interaction energy. This phenomenon we called 'attraction in momentum space' (Huang, 1964; Leggett, 2001).

Approximating only $\mathcal{T}(\mathbf{q}) \approx U_0$ provides the Hartree–Fock–Bogoliubov or Popov approximation, with the typical Bogoliubov dispersion (12.13). Within this approximation a correlation term $\mathcal{E}_{cor}$ of the energy density (12.93) remains besides the quasiparticle and the mean-field term which favours a finite depletion (Bogoliubov, 1947; Pethick and Smith, 2004). It has to be noted that the original Bogoliubov approximation corresponds to an additional approximation of the chemical potential $\mu \approx n_0 U_0$. In the normal phase the Popov approximation is identical to the Hartree–Fock approximation, with a mean field $2n\lambda$, while the Bogoliubov approximation yields an ideal Bose gas.

For the T-matrix approximation there appears a fourth contribution to the energy density (12.93),

$$\mathscr{E}_{2\mathrm{p}} = -\frac{1}{2} \int \frac{d^3k}{(2\pi)^3} \frac{d^3q}{(2\pi)^3} \lambda g_q g_k C_q C_k - \frac{n_0}{2} \int \frac{d^3k}{(2\pi)^3} \lambda g_k C_k \tag{12.95}$$

which is a two-particle term that can be expressed by the anomalous expectation value of a pair of particles

$$C_k = \langle \hat{a}_k \hat{a}_{-k} \rangle = -\frac{n_0 \mathscr{T}(\mathbf{k})}{2E_k} (1 + 2f_\mathrm{B}(E_k)). \tag{12.96}$$

Please note that this two-particle term appears as a consequence of our theory avoiding unphysical multiple scattering and has not been assumed ad-hoc as has been done in most approaches postulating anomalous functions. A very similar term can be found in the Bardeen–Cooper–Schrieffer (BCS) approximation (Pethick and Smith, 2004) where it describes the contribution of Cooper pairs.

12.6.1 Equation of state

From the set of equations we now calculate the chemical potential μ for different particle densities n, as shown in Fig. 12.8. The discussion follows (Maennel, 2011; Männel et al., 2013). The Hartree–Fock approximation, i.e. the dashed-dotted line, shows a multivalued region near the onset of BEC, where several solutions of the equation of state coexist. The origin of this unphysical behaviour seems to be an overestimation of the

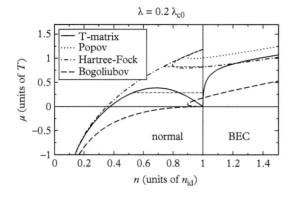

Figure 12.8: *Chemical potential in Bogoliubov, Hartree–Fock, Popov and T-matrix approximation for weak repulsive interaction, the horizontal broken lines correspond to the Maxwell construction, $n_{id} \approx 2.61/\Lambda_{dB}^3$ is the ideal critical density for a Bose condensation with the thermal de Broglie wavelength $\Lambda_{dB} = \hbar\sqrt{2\pi/mT}$. (From (Männel, Morawetz and Lipavský, 2013).)*

attraction in momentum space, which favours a high-condensate fraction. Furthermore Fig. 12.8 shows that there is also a temporary drop of the chemical potential in Hartree–Fock approximation after BEC has set in, which indicates an instability of the gas and a first-order phase transition. It has to be emphasised; however, that this instability has its origin not in the attractive part of the interaction potential but in the BEC and the attraction in momentum space. During the first-order phase transition there is a coexistence of a high- and a low-density phase and according to Gibb's phase rule there is only one free parameter which has to be constant in order to keep the temperature fixed. Therefore, all intensive parameters of the two phases are constant during the phase transition, especially the pressure and chemical potential. In equilibrium, the pressure and the chemical potential have to be equal for both phases and can be obtained via the Maxwell construction illustrated in Fig. 12.8 by the horizontal broken line. As the system follows the curve of constant pressure and chemical potential the unphysical multivalued region is avoided.

As illustrated by the broken curve in Fig. 12.8, the chemical potential in the Bogoliubov approximation shows an unphysical region as well. However, in this approximation the Maxwell construction is not possible. Since the Bogoliubov approximation was developed to describe the system near $T = 0$, it fails near the BEC transition. The approximation can be improved by including the Hartree–Fock mean field, leading to the Popov approximation. Although the unphysical region remains, the Maxwell construction becomes possible. Compared to the Hartree–Fock approximation the width of the unphysical region is reduced with no qualitative change. Therefore in the following it is sufficient to compare only the Hartree–Fock approximation with our T-matrix approximation.

If the repulsive interaction is weak, i.e. $\lambda < 0.23 \, \lambda_{c0}$, the chemical potential in the T-matrix approximation shows no unphysical region. Nevertheless there is still an instability of the gas, i.e. the chemical potential drops down to zero at the onset of BEC, shown by the full line in Fig. 12.8. The reason for the vanishing of the chemical potential is the phenomenon that the many-body T-matrix for zero momentum and energy $\mathscr{T}(0)$ vanish at the critical point (Bijlsma and Stoof, 1997; Shi and Griffin, 1998). In the vicinity of the onset of BEC the repulsive interaction is compensated by the Bose enhancement, leading to the drop of the chemical potential. However, the drop of the chemical potential and the corresponding first-order phase transition might as well be a device for omitting the momentum dependence of the T-matrix in the selfenergy, leading to the dispersion (12.92) (Bijlsma and Stoof, 1997; Holzmann and Baym, 2003; Prokof'ev, Ruebenacker and Svistunov, 2004).

Fig 12.9 shows the chemical potential in the two approximations for a stronger repulsion. In Hartree-Fock approximation, i.e. the dashed-dotted line, there is no qualitative change. However, for the T-matrix approximation (full line) a multivalued region appears for $\lambda > 0.23 \, \lambda_{c0}$, which cannot be avoided by the Maxwell construction. Therefore we attribute a true physical relevance to this behaviour and interpret it as appearance of a hysteresis. Reaching the end of the coexistence region at $n \approx 0.98 \, n_{id}$ from below the chemical potential jumps from $0.65 \, T$ to $3.07 \, T$. Decreasing the density the chemical

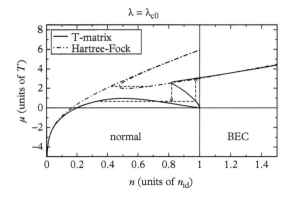

Figure 12.9: *Chemical potential in Hartree–Fock and T-matrix approximation for strong repulsive interaction, the horizontal broken lines correspond to the Maxwell construction, the vertical arrows mark the density hysteresis (from (Männel, Morawetz and Lipavský, 2013)).*

potential decreases and jumps back to $0.65\ T$ at a smaller density near $0.82\ n_{\mathrm{id}}$. This can be understood as hysteresis behaviour.

Strongly correlated systems are connected with a highly nonlinear density dependence of the thermodynamic quantities. Interestingly such nonlinearities can lead to hysteresis behaviour. Besides the known magnetic-field hysteresis there are a number of examples observed in other fields. Optical bistable systems have been reported to show a time hysteresis in the response due to a nonlinear density dependence (Lu et al., 1994). A pressure-induced thermal hysteresis in Kondo lattice systems has been found (Park et al., 2005) and even in plasma discharge systems a density-driven hysteresis is reported (Jiang et al., 2009). A density hysteresis driven by pressure can be found in spin-crossover compounds due to elastic stresses (Stoleriu, Chakraborty, Hauser, Stancu and Enachescu, 2011). Near the BEC of a quantum spin system a peak in sound attenuation was observed (Sherman et al., 2003) and attributed to the hysteresis in magnetic field, which indicates a first-order phase transition. As we have seen here, in strongly correlated Bose systems near BEC, a similar density hysteresis appears (Männel, Morawetz and Lipavský, 2013).

12.7   Superconductivity

Superconductivity results from Bose–Einstein condensation (BEC) of electron (fermion) pairs and the coherence of all pairs in the condensate. We consider that the pairing mechanism (Fröhlich, 1950) is given by phonons creating positive-charge densities due to a moving electron which allows the following electron to feel effectively attracted to the other one. Therefore, this pairing occurs only in the phonon-frequency-ω_c-vicinity of the Fermi energy. For $T = 0$ and homogeneous gap as well as constant attractive potential $V(p) = -V_0$ in the vicinity of $(\epsilon_f, \epsilon_f + \omega_c)$ we get from the gap equation (12.51)

$$1 = \frac{V_0}{2}\int\frac{dp}{(2\pi)^3}\frac{1}{\sqrt{\Delta^2(0)+(\epsilon_p-\epsilon_f)^2}} = \frac{N(0)V_0}{2}\int_0^{\omega_c}\frac{d\xi}{\sqrt{\Delta^2(0)+\xi^2}} = \frac{N(0)V_0}{2}\,\mathrm{arsinh}\frac{\omega_c}{\Delta(0)}$$

$$(12.97)$$

which leads to

$$\Delta(0) = \frac{\omega_c}{\sinh\frac{2}{N(0)V_0}} \approx 2\omega_c\, e^{-\frac{2}{N(0)V_0}} \tag{12.98}$$

showing a non-perturbative result in the interaction.

12.7.1 Critical temperature and density of states

The critical temperature, $\Delta(T_c) = 0$, is (Rees, 1969) derived from (12.51) dividing-out a constant Δ

$$\frac{2}{N(0)V_0} = \int_0^{\omega_c} \frac{d\xi}{\xi} \tanh\frac{\xi_3}{2T_c} = \ln\frac{\omega_c}{T_c}[1 - 2f(\omega_c)] + 2\int_0^{\omega_c/T_c} dx\ln x\frac{d}{dx}\left(\frac{1}{e^x + 1}\right)$$

$$\approx \ln\frac{\omega_c}{T_c}[1 - 2f(\omega_c)] + \gamma + \ln\frac{2}{\pi} \tag{12.99}$$

where we used a constant density of states $N(0)$ at the Fermi energy and one gets the standard BCS result

$$T_c = 1.13\omega_c e^{-\frac{2}{N(0)V_0}} \tag{12.100}$$

and $\Delta(0)/T_c = \pi/\gamma = 1.76$.

The density of states can be calculated from (12.49) considering only $\omega > 0$ states

$$N(\omega - \epsilon_f) = \int \frac{dk}{(2\pi)^3} \frac{1}{2}\left(1 + \frac{\xi}{E}\right)\delta(\omega - \epsilon_f - E) = N(0)\int_{-\epsilon_f}^{\infty} d\xi \frac{1}{2}\left(1 + \frac{\xi}{E}\right)\delta(\omega - \epsilon_f - E)$$

$$= N(0)\frac{\omega - \epsilon_f}{\sqrt{(\omega - \epsilon_f)^2 - \Delta^2}} \tag{12.101}$$

showing the well-known peaked density of states, first experimentally tested by Giaever (1960).

12.8 Stability of the pairing condensate

We might now wonder why only one condensate appears and not two ones in parallel. Let us restrict to conventional superconductors of type I and do not consider recent multi-gaped materials (Machida, Sakai, Izawa, Okuyama and Watanabe, 2011) which have two gaps but in different channels. Within the Bardeen–Cooper–Schrieffer (BCS) theory there are no bounded electron pairs out of the condensate. Consequently a related question is: what scenario do the particles prefer? To go into bound states or into Cooper pairing. This will make it understandable why bounded non-condensed pairs can be neglected in weakly-coupled superconductors.

To answer these two questions we can now compare different T-matrix theories with our multiple scattering-corrected theory of Chapter 12.3. T-matrix approaches of the superconductivity date back to the late 1950s and early 1961 (Wild, 1960; Thouless, 1960; Kadanoff and Martin, 1961). These theories are rooted in the perturbation expansion for the normal state of the system and yield the superconducting phase transition as a bonus. We will compare three versions of the T-matrix theory: the original Galitskii theory (Galitskii, 1958; Fetter and Walecka, 1971) applied to superconductivity by Thouless (Thouless, 1960); the Kadanoff–Martin (KM) theory (Kadanoff and Martin, 1961); and the T-matrix with multiple scattering corrections (Lipavský, 2008; Šopík, Lipavský, Männel, Morawetz and Matlock, 2011).

All these three versions of the T-matrix become singular at the critical temperature that signals the onset of the Cooper pairing (Thouless, 1960; Kadanoff and Martin, 1961; Bishop, Strayer and Irvine, 1975). The predicted critical temperatures differs, however, due to different approximations of single-particle Green's functions from which the T-matrix is constructed. While the KM prediction is close to the BCS result, Haussmann (1994) has shown that the Galitskii theory yields a lower critical temperature with the difference becoming essential for strong interactions. Since the T-matrix with multiple scattering corrections in the normal state is identical to the Galitskii theory (Lipavský, 2008), it gives identical critical temperatures.

Below the critical temperature these three versions of T-matrices significantly differ in the excitation spectra. The T-matrix with multiple-scattering corrections includes the Galitskii as well as Kadanoff–Martin approach as specific simplifications. We thus introduce only the T-matrix with multiple-scattering corrections and eventually simplify it to the other theories.

As we have seen in Section 12.3, multiple-scattering corrections to the T-matrix are equivalent to a reduced self-consistency in the loop of the selfenergy and can be traced back to higher-order cluster-cluster correlations, see Chapter 11.5 and (Morawetz, 2011). These rearrangements are possible due to the symmetry of the T-matrix with respect to the interchange of interacting particles (Šopík, Lipavský, Männel, Morawetz and Matlock, 2011). We note that this symmetry is just the one required by the Baym-Kadanoff criterion (A) for conserving approximations of Chapter 7.3.3. Therefore the multiple-scattering corrected T-matrix conserves energy and momentum.

The formation of bound pairs and their dispersion parallels the formation of molecules in a gas. Unlike the collective motion, this process does not create a space charge or electrical currents, we thus do not include the electromagnetic field in our model. The Hamiltonian assumed has only a single electronic band with parabolic kinetic energy and a separable BCS interaction.

To make a link with the standard notation of the BCS theory we introduce the gap function $|\Delta|$ from the singular term of the T-matrix

$$|\Delta| = \sqrt{\frac{k_{\mathrm{B}} T}{L^3} \mathscr{T}_{0,\mathrm{C}}}, \tag{12.102}$$

where the subscript denotes $Q = (0, \mathbf{C})$. The T-matrix depends on $|\Delta|^2$ being averaged over all thermodynamically allowed phases. On the other hand, its value does not change if we select one preferential phase $\Delta = |\Delta| e^{2i\phi}$ and express the singular element of the T-matrix as

$$\mathscr{T}_{0,\mathbf{C}} = \frac{L^3}{k_\mathrm{B} T} \bar{\Delta} \Delta. \tag{12.103}$$

The **C**-part of the selfenergy

$$\Sigma_{Q\uparrow}(k) = -k_\mathrm{B} T \sum_\Omega \mathscr{T}_Q G_{Q\downarrow}(Q-k), \tag{12.104}$$

is dominated by the condensation at $\Omega = 0$

$$\Sigma_{\mathbf{C}\uparrow}(k) = -k_\mathrm{B} T \sum_\Omega \mathscr{T}_{\Omega,\mathbf{C}} G_{\mathcal{C}\downarrow}(\Omega-\omega, \mathbf{C}-\mathbf{k})$$

$$= -L^3 \bar{\Delta} G_{\mathcal{C}\downarrow}(-\omega, \mathbf{C}-\mathbf{k}) \Delta - k_\mathrm{B} T \sum_{\Omega\neq 0} \mathscr{T}_{\Omega,\mathbf{C}} G_{\mathcal{C}\downarrow}(\Omega-\omega, \mathbf{C}-\mathbf{k})$$

$$\approx -L^3 \bar{\Delta} G_{\mathcal{C}\downarrow}(-\omega, \mathbf{C}-\mathbf{k}) \Delta. \tag{12.105}$$

We denote the spin direction explicitly. The contributions of $\Omega \neq 0$ are regular and become negligible in the thermodynamic limit $L^3 \to \infty$. The selfenergy thus splits into a singular contribution of the **C**-mode and a regular reminder due to all other modes

$$\Sigma_\uparrow(k) = -\bar{\Delta} G_{\mathcal{C}\downarrow}(-\omega, \mathbf{C}-\mathbf{k}) \Delta + \Sigma_{\mathcal{C}\uparrow}(k). \tag{12.106}$$

The **C**-reduced propagator $G_{\mathcal{C}} = G^0 + G^0 \Sigma_{\mathcal{C}} G_{\mathcal{C}}$ does not depend on the singular selfenergy, therefore it has no gap in the energy spectrum. It depends on Δ only indirectly via internal lines of $\Sigma_{\mathcal{C}}$ and plays the role of the normal-state propagator. The energy gap in the full propagator is given by the Gorkov-type equation

$$G_\downarrow(k) = G_{\mathcal{C}\downarrow}(k) - G_\downarrow(k) \bar{\Delta} G_{\mathcal{C}\uparrow}(-\omega, \mathbf{C}-\mathbf{k}) \Delta G_{\mathcal{C}\downarrow}(k), \tag{12.107}$$

which directly follows from (16.49), (12.23) and (12.105).

From (12.107) we can identify the anomalous function $\bar{F}(k) = G_\downarrow(k) \bar{\Delta} G_{\mathcal{C}\uparrow}$ $(-\omega, \mathbf{C}-\mathbf{k})$. Briefly fixing the phase we arrive at the algebra analogous to the Gorkov equations. Note that we have not introduced the anomalous functions as a prerequisite but have obtained them as a result of the algebra in which all interacting pairs were treated equally (Morawetz, 2010).

To be exact, our Δ is rather the off-diagonal selfenergy of the Eliashberg approach and the regular part of the selfenergy is Eliashberg's diagonal selfenergy. As known from the Eliashberg theory, the energy gap in the single-particle spectrum differs from $2|\Delta|$ by a renormalisation factor due to the energy dependence of the regular part of the selfenergy, see (Šopík, Lipavský, Männel, Morawetz and Matlock, 2011). This renormalisation is not essential for our discussion, we thus call Δ the gap for simplicity.

The value of the gap is given by the T-matrix of the condensation mode

$$\mathscr{T}_{0,\mathbf{C}} = V - V\frac{k_\mathrm{B}T}{L^3}\sum_k G_\uparrow(\omega,\mathbf{k})\,G_{\mathscr{C}\downarrow}(-\omega,\mathbf{C}-\mathbf{k})\,\mathscr{T}_{0,\mathbf{C}}. \qquad (12.108)$$

Since the T-matrix diverges, it becomes much larger than the interaction potential, $\mathscr{T}_{0,\mathbf{C}} \gg V$, which allows us to neglect the first term in the right-hand side of (12.108). Substituting $\mathscr{T}_{0,\mathbf{C}}$ from (12.103) and dividing by $L^3\Delta/(k_\mathrm{B}T)$ we obtain the gap equation

$$\bar{\Delta} = -V\frac{k_\mathrm{B}T}{L^3}\sum_k G_\uparrow(\omega,\mathbf{k})\,G_{\mathscr{C}\downarrow}(-\omega,\mathbf{C}-\mathbf{k})\,\bar{\Delta}. \qquad (12.109)$$

In the notation of the T-matrix, the gap equation 12.109 reads $V\mathscr{T}_{0,\mathbf{C}}^{-1}\bar{\Delta} = 0$. Ignoring the normal state solution $\bar{\Delta} = 0$, we can divide (12.109) by $\bar{\Delta}$ and V so that the gap equation reads $\mathscr{T}_{0,\mathbf{C}}^{-1} = 0$. Such form is independent of phase ϕ needed to define Δ.

Near the critical temperature T_c the gap becomes small, $\Delta \to 0$, therefore we can expand the propagator to the lowest order in Δ. For the systematic expansion of the gap equation into the Ginzburg–Landau equation see Ref. (Vagov, Shanenko, Milosevic, Axt and Peeters, 2012). From the Gorkov equation (12.107) follows

$$G_\downarrow(k) \approx G_{\mathscr{C}\downarrow}(k) - G_{\mathscr{C}\downarrow}(k)\bar{\Delta}G_{\mathscr{C}\uparrow}(-\omega,\mathbf{C}-\mathbf{k})\Delta G_{\mathscr{C}\downarrow}(k). \qquad (12.110)$$

At the same time, the critical velocity of the condensate becomes small, therefore the momentum of the condensate $\mathbf{C}$ has to be small. We can thus also expand the sum in the gap equation 12.109 to the second order in $\mathbf{C}$. As a result one obtains the Ginzburg–Landau equation

$$\frac{\hbar^2|\mathbf{C}|^2}{2m^*}\bar{\Delta} + \alpha\bar{\Delta} + \beta|\Delta|^2\bar{\Delta} = 0 \qquad (12.111)$$

with the Ginzburg–Landau parameters

$$\alpha = \chi + \chi V\frac{k_\mathrm{B}T}{L^3}\sum_k G_{\mathscr{C}\uparrow}(\omega,\mathbf{k})\,G_{\mathscr{C}\downarrow}(-\omega,-\mathbf{k}), \qquad (12.112)$$

$$\beta = -\chi V\frac{k_\mathrm{B}T}{L^3}\sum_k G_{\mathscr{C}\uparrow}^2(\omega,\mathbf{k})\,G_{\mathscr{C}\downarrow}^2(-\omega,-\mathbf{k}) \qquad (12.113)$$

and

$$\frac{\hbar^2}{2m^*} = \chi V \frac{\partial^2}{\partial \mathbf{C}^2} \frac{k_B T}{L^3} \sum_k G_{\mathbf{C}\uparrow}(\omega, \mathbf{k}) G_{\mathbf{C}\downarrow}(-\omega, \mathbf{C}-\mathbf{k}) \Bigg|_{\mathbf{C}=0}. \tag{12.114}$$

Replacing $\mathbf{C}$ with the covariant derivative $i\hbar\nabla - e^*\mathbf{A}$ one obtains the customary form of the Ginzburg–Landau equation (de Gennes, 1966).

Gorkov has evaluated the Ginzburg–Landau coefficients (12.112)–(12.114) using the quasiparticle approximation of the normal-state propagator $G_{\mathbf{C}\downarrow}^{-1}(k) \approx \omega - \epsilon(\mathbf{k})$. For the parabolic approximation of the energy dispersion $\epsilon(\mathbf{k}) = \hbar^2|\mathbf{k}|^2/2m$ he obtained $m^* = 2m$,

$$\beta = \frac{3}{2E_F}, \quad \alpha = -\frac{6\pi^2 k_B^2 T_c}{7\zeta_{[3]} E_F}(T_c - T), \quad \chi = \frac{8\pi^2 k_B^2 T_c^2}{7\zeta_{[3]} n} \tag{12.115}$$

where E_F is the Fermi energy and $\zeta_{[3]} = 1.202$ is the Riemann zeta function (12.8). The norm χ cancels in the Ginzburg–Landau equation, with n being the electron density. We restrict our attention to the vicinity of the critical temperature. The limiting form (12.111) will thus be sufficient for our discussion.

Assuming a single condensate, we have in fact assumed that none of the $\mathbf{Q}$-terms of the T-matrix diverges with the volume for $\mathbf{Q} \neq \mathbf{C}$. We will show that if the $\mathbf{Q}$-term was not divergent in the beginning, the presence of the condensate in the $\mathbf{C}$-mode eliminates its chances to become singular.

For the regular mode the $\mathbf{Q}$-reduced propagator approaches the full one except for terms $\propto L^{-3}$,

$$G_{\mathbf{Q}\uparrow} = G_\uparrow. \tag{12.116}$$

The zero-frequency component of the T-matrix of the $\mathbf{Q}$-mode thus satisfies the equation

$$\mathscr{T}_{0,\mathbf{Q}} = V - V \frac{k_B T}{L^3} \sum_k G_\uparrow(\omega, \mathbf{k}) G_\downarrow(-\omega, \mathbf{Q}-\mathbf{k}) \mathscr{T}_{0,\mathbf{Q}}. \tag{12.117}$$

It is more convenient to write it in the inverse form

$$\frac{1}{\mathscr{T}_{0,\mathbf{Q}}} = \frac{1}{V} + \frac{k_B T}{L^3} \sum_k G_\uparrow(\omega, \mathbf{k}) G_\downarrow(-\omega, \mathbf{Q}-\mathbf{k}). \tag{12.118}$$

Near the critical temperature it is sufficient to keep terms quadratic in Δ and $\mathbf{Q}$. From the expansion (12.110) we then find

$$\frac{\chi}{\mathscr{T}_{0,\mathbf{Q}}} = \frac{|\mathbf{Q}|^2}{2m^*} + \alpha + 2\beta|\Delta|^2. \tag{12.119}$$

The factor of two in front of β follows from the fact that for non-condensed pairs both propagators depend on the gap.

The values of Δ are given by the inverse T-matrix in the condensation mode

$$\frac{\chi}{\mathcal{T}_{0,C}} = \frac{|C|^2}{2m^*} + \alpha + \beta|\Delta|^2 = 0 \tag{12.120}$$

which equals zero because of the divergence of $\mathcal{T}_{0,C}$. This equation is identical to the gap equation in the Ginzburg–Landau approximation (12.111).

When equation (12.120) holds, the inverse T-matrix of the **Q**-mode remains non-zero. To see this, let us substitute the gap from (12.120) into the inverse T-matrix (12.119) to get

$$\frac{\chi}{\mathcal{T}_{0,Q}} = \frac{|Q|^2}{2m^*} - \alpha - \frac{|C|^2}{m^*}. \tag{12.121}$$

The right-hand side can reach zero only if $|C|^2$ is sufficiently large to compensate $-\alpha$.

Values of the pair momentum C are limited by the critical current, $|C|^2 < Q_c^2$. The current is proportional to the square of the gap times the momentum, $j \propto C|\Delta|^2$. Using (12.111) one finds $j \propto C\left(-\alpha - |C|^2/2m^*\right)$. The critical current is obtained as the maximum one, $\partial j/\partial C|_{C=Q_c} = 0$, for $Q_c^2 = 2m^*|\alpha|/3$, see Tinkham (1966). Accordingly,

$$\frac{\chi}{\mathcal{T}_{0,Q}} > \frac{|Q|^2}{2m^*} - \alpha - \frac{Q_c^2}{m^*} = \frac{|Q|^2}{2m^*} - \frac{\alpha}{3}. \tag{12.122}$$

Since α is negative below the critical temperature, inequality (12.122) implies that the T-matrix in the **Q**-mode remains finite. This mode thus cannot become singular once the condensation develops in the **C**-mode. Therefore, a parallel condensation in two competitive modes is excluded. Briefly, there is only a single condensate, as it is tacitly assumed in the BCS theory.

12.8.1 Galitskii T-matrix

By omitting the multiple-scattering corrections, the present theory simplifies to the Galitskii T-matrix. This is achieved by approximating $G_{C\downarrow} \approx G_\downarrow$ in Eqs. (12.112)–(12.114) with the help of which one can derive Ginzburg–Landau parameters. The inverse Galitskii T-matrix in the condensation mode thus reads

$$\frac{\chi}{\mathcal{T}_{0,C}} = \frac{|C|^2}{2m^{*\mathrm{Gal}}} + \alpha^{\mathrm{Gal}} + 2\frac{k_B T}{L^3}\beta^{\mathrm{Gal}}\mathcal{T}_{0,C} \to 0 \tag{12.123}$$

with the corresponding (12.112), (12.113) and (12.114).

One can see that $\mathcal{T}_{0,C}$ diverges as L^3 so that the left-hand side of (12.123) goes to zero as indicated by limit $\to 0$. Due to the approximation $G_{C\downarrow} \approx G_\downarrow$ in the selfenergy loop (12.104), the divergence of the T-matrix does not result in the BCS gap, see Wild (Wild, 1960). Apparently, the divergence of the T-matrix is a necessary but not sufficient condition for the BCS gap.

Out of the condensation mode, $\mathbf{Q} \neq \mathbf{C}$, the Galitskii T-matrix is also constructed from the full Green's functions, therefore

$$\frac{\chi}{\mathcal{T}_{0,\mathbf{Q}}} = \frac{|\mathbf{Q}|^2}{2m^{*\mathrm{Gal}}} + \alpha^{\mathrm{Gal}} + 2\frac{k_{\mathrm{B}}T}{L^3}\beta^{\mathrm{Gal}}\mathcal{T}_{0,\mathbf{C}}. \tag{12.124}$$

Subtracting (12.123) we find

$$\frac{\chi}{\mathcal{T}_{0,\mathbf{Q}}} = \frac{|\mathbf{Q}|^2}{2m^{*\mathrm{Gal}}} - \frac{|\mathbf{C}|^2}{2m^{*\mathrm{Gal}}}. \tag{12.125}$$

In contrast to the result for the multiple-scattering corrected T-matrix (12.121), the dispersion (12.125) supports a nucleation of the second condensate at the energy minimum $\mathbf{Q} = 0$ in the presence of the persistent current $\mathbf{C} \neq 0$.

12.8.2 Kadanoff–Martin theory

The analysis of the Kadanoff–Martin theory is very similar. This approximation is obtained from the present one using $G_{\mathbf{Q}\downarrow} \approx G^0$. The inverse T-matrix thus reads

$$\frac{\chi}{\mathcal{T}_{0,\mathbf{C}}} = \frac{|\mathbf{C}|^2}{2m^{*\mathrm{KM}}} + \alpha^{\mathrm{KM}} + \frac{k_{\mathrm{B}}T}{L^3}\beta^{\mathrm{KM}}\mathcal{T}_{0,\mathbf{C}} \to 0 \tag{12.126}$$

with (12.112), (12.113) and (12.114) correspondingly. Out of the condensation mode we find

$$\frac{\chi}{\mathcal{T}_{0,\mathbf{Q}}} = \frac{|\mathbf{Q}|^2}{2m^{*\mathrm{KM}}} + \alpha^{\mathrm{KM}} + \frac{k_{\mathrm{B}}T}{L^3}\beta^{\mathrm{KM}}\mathcal{T}_{0,\mathbf{C}}, \tag{12.127}$$

therefore

$$\frac{\chi}{\mathcal{T}_{0,\mathbf{Q}}} = \frac{|\mathbf{Q}|^2}{2m^{*\mathrm{KM}}} - \frac{|\mathbf{C}|^2}{2m^{*\mathrm{KM}}}. \tag{12.128}$$

Again, like the Galitskii T-matrix (12.125) this dispersion supports a nucleation of the second condensate at the energy minimum $\mathbf{Q} = 0$ in the presence of the persistent current $\mathbf{C} \neq 0$.

To summarise, we have found that only the T-matrix with multiple-scattering corrections excludes a condensation in a second mode, which explains the single-valued condensate tacitly assumed in the BCS theory. In contrast, both the Thouless-like theory based on the Galitskii T-matrix and the Kadanoff–Martin theory allow for a condensation in two (or more) modes, which is in conflict with experimental findings.

12.9   Excitation of Cooper pairs from the condensate

Now we will discuss the possibility of exciting a Cooper pair out of the condensate by an object moving with velocity $\mathbf{v}$ in the static condensate. Going into the running coordinate system, this criterion is used to check the stability of the condensate flowing with velocity $-\mathbf{v}$ around a static obstacle.

Multiple-scattering corrected T-matrix   The right-hand side of (12.119) represents the energy of a non-condensed pair of momentum $\mathbf{Q}$. In the static condensate, $\mathbf{C} = 0$, the gap is given by (12.120) as $|\Delta|^2 = -\alpha/\beta$. From the energy dispersion (12.119 one finds that a Cooper pair can be excited from the condensate into a non-condensed state with the energy cost $\mathscr{E}_{\mathbf{Q}} = |\mathbf{Q}|^2/2m^* + \alpha + 2\beta|\Delta|^2 = |\mathbf{Q}|^2/2m^* - \alpha$. Let us estimate under which conditions Cooper pairs can be excited by an external perturbation.

According to the Landau criterion (Pethick and Smith, 2001), see (12.1), the external perturbation moving with velocity $\mathbf{v}$ can excite the Cooper pair of momentum $\mathbf{Q}$ if the Cherenkov-type condition $(\mathbf{v} \cdot \mathbf{Q}) = \mathscr{E}_{\mathbf{Q}}$ is satisfied, i.e.

$$(\mathbf{v} \cdot \mathbf{Q}) = \frac{|\mathbf{Q}|^2}{2m^*} - \alpha \qquad (12.129)$$

This equation is solved by real $\mathbf{Q}$ only if

$$|\mathbf{v}| > v_{\text{pe}} = \sqrt{\frac{2|\alpha|}{m^*}}. \qquad (12.130)$$

The velocity v_{pe} is the critical velocity for excitation of the Cooper pair from the condensate into a bound pair out of the condensate.

Let us compare the critical velocity of the pair excitation with the critical velocity for the pair breaking $v_{\text{pb}} = \Delta/k_{\text{F}}$, where k_{F} is the Fermi momentum (Dew–Huges, 2001). To this end we use Ginzburg–Landau coefficients derived by Gorkov. From (12.111) follows $\Delta = \sqrt{|\alpha|/\beta} = \sqrt{|\alpha|k_{\text{F}}^2/(3m)}$ so that

$$v_{\text{pe}} = \sqrt{3}v_{\text{pb}}. \qquad (12.131)$$

Since the critical velocity of the pair breaking is lower than the critical velocity of the pair excitation, the stability of the condensate is controlled by the pair breaking.

12.9.1   Galitskii T-matrix and Kadanoff–Martin theory

Within the Galitskii approximation for the system at rest, $\mathbf{C} = 0$, the Cooper pairs can be excited out of the condensate by any small velocity since the Cherenkov-type condition corresponding to dispersion (12.125)

$$(\mathbf{v} \cdot \mathbf{Q}) = \frac{Q^2}{2m^{*\text{Gal}}} \qquad (12.132)$$

can be satisfied for any small velocity **v**. The Galitskii T-matrix thus yields zero critical velocity from the Landau criterion (along with the zero critical velocity from the pair breaking).

For the Kadanoff–Martin approximation and the system at rest, $\mathbf{C} = 0$, the Cooper pairs can also be excited out of the condensate by any small velocity since the Cherenkov-type condition corresponding to dispersion (12.128),

$$(\mathbf{v} \cdot \mathbf{Q}) = \frac{|\mathbf{Q}|^2}{2m^{*\mathrm{KM}}}, \tag{12.133}$$

can be satisfied for any small velocity **v**. We have thus recovered the result of Chen et al. (2005) that the Kadanoff–Martin theory fails to justify superconductivity providing zero critical velocity from the Landau criterion.

12.9.2 Conclusions

We have seen that the T-matrix approach can be used to justify two basic assumptions of the BCS theory: firstly, the condensate is single-valued, i.e. a new condensate cannot nucleate even if the present condensate is driven out of the total energy minim and the new condensate would be thermodynamically favourable. Secondly, excitations of bound electron pairs can be neglected since the critical velocity of their Cherenkov-type generation is higher than the critical velocity of pair breaking.

These conclusions are not general for all T-matrix approaches. Only the T-matrix with multiple-scattering corrections provides this result while the Galitskii T-matrix and the Kadanoff–Martin theory result into a zero-critical velocity of the Cherenkov-type generation of bounded excited pairs. The corresponding differences of the critical velocities are compared in the Table 12.1.

The Galitskii theory fails to reproduce the gap in the single-particle spectrum (Wild, 1960). Since the critical velocity of pair breaking equals the gap divided by the Fermi momentum k_F, the pair-breaking critical velocity is zero. Such a system cannot have permanent supercurrents. The critical velocity of pair excitations is also zero as it follows from the parabolic dispersion obtained e.g. by Haussmann (Haussmann, 1993, 1994), see also the derivation in Section 12.8.1.

The single-particle energy spectrum of the KM theory is very similar to the BCS theory (Wild, 1960; Kadanoff and Martin, 1961). In particular, it has a gap which

Table 12.1 *Comparison of the critical velocities for different T-matrix approaches, Galitskii, Kadanoff–Martin (KM) and the T-matrix corrected by multiple-scattering (TMSC).*

	Critical velocities of:	
	pair breaking	pair excitation
Galitskii	0	0
KM	Δ/k_F	0
TMSC	Δ/k_F	$\sqrt{3}\Delta/k_F$

guarantees the finite pair-breaking critical velocity. Unlike the BCS theory, the KM theory offers excited bound pairs. They have the energy spectrum of an ideal gas, therefore the critical velocity of pair excitation is also zero, see (Chen, Stajic, Tan and Levin, 2005) or Section 12.8.2.

Some approaches combine the anomalous Green's functions with the ladder approximation. In the two-step procedure the T-matrix is used to derive normal-state properties which are than used to study superconductivity on the BCS level (Littlewood, 1990). In the one-step procedure the T-matrix is constructed from Nambu-Gorkov functions (Haussmann, 1993; Dahm and Tewordt, 1995; Keller, Metzner and Schollwöck, 1999). For the contact interaction the evaluation of the T-matrix is feasible even with the screening and exchange channels included, which is known as the fluctuation exchange (FLEX) approximation. The FLEX approximation extended by anomalous functions has been intensively studied for the Hubbard model (Luo and Bickers, 1993; Monthoux and Scalapino, 1994; Pao and Bickers, 1994, 1995; Yanase, Jujo, Nomura, Ikeda, Hotta and Yamada, 2003; Deisz and Slife, 2009). In all these approaches excited pairs eventually appear as poles of the T-matrix, but they are treated differently from Cooper pairs which are covered by a rather complex set of relations for anomalous functions.

The T-matrix with multiple-scattering corrections has a single-particle energy spectrum of BCS type (Lipavský, 2008; Šopík, Lipavský, Männel, Morawetz and Matlock, 2011). We have found that for the Galitskii T-matrix and the Kadanoff–Martin theory the energy spectrum for the motion of the condensate as a whole, i.e. the Ginzburg–Landau equation, is the same as the energy spectrum of excitations out of the condensate. This implies the possibility to nucleate a new condensate and a zero critical velocity for Cherenkov-type processes. Only for the T-matrix with multiple-scattering corrections these two energy spectra differ by a factor of two in the non-linear term (compare β-term in (12.111) and (12.119), this inhibits any new nucleation and gives the finite critical velocity of Cherenkov-type processes.

The factor of two in the non-linear term can be also interpreted in another manner. The non-condensed pairs feel the gap due to the condensate being twice stronger than it is felt by Cooper pairs in the condensate. This reminds the factor of two by which the Bosons out of the Bose–Einstein condensate interact stronger with the condensate than the condensed Bosons interact among themselves, see Chapter 2.3. of Leggett (Leggett, 2006) or the discussion after (12.77) here. In this sense the T-matrix with multiple-scattering corrections seems to be the simplest theory suited to bridge the BCS and Boson picture of the superconductivity (Lipavský et al., 2014). In the strong coupling limit, the multiple-scattering corrections have to be implemented to the composed Boson line. This leads to the expected acoustic spectrum at low energies as outlines in section 12.5, see (Männel, Morawetz and Lipavský, 2010; Männel, Morawetz and Lipavský, 2013).

12.9.3   Relation of pairing density to correlated density

In superconductors, the correlated density which we will find as the consequence of the time-nonlocality of the collision process (15.97) in Chapter 15, becomes visible as the difference between the total and normal density $n_{corr} = n - n_n$. The density n is obtained by the momentum and frequency integral over (12.50) and introducing the density of states $h(\xi) = 2\sum_p 2\pi\delta(\xi - \epsilon_p)$, one has

$$n = \int_{-\bar{\mu}}^{\infty} \frac{d\xi}{2\pi} h(\bar{\mu} + \xi) \left(\frac{1}{2} - \frac{\xi}{2E} \tanh\frac{1}{2}\beta E \right). \tag{12.134}$$

Here we account for a possible electrostatic potential φ and the velocity v of superconducting electrons by $\bar{\mu} = \mu - e\varphi - mv^2/2$. For a vanishing gap we obtain the corresponding density n_n of normal electrons with the chemical potential $\bar{\mu}$ by $n_n = n(\Delta = 0)$. The difference

$$n_{corr} = n - n_n \tag{12.135}$$

describes the correlated density. In the ground state the normal density turns into

$$n_n = 2\sum_p \Theta(\bar{\mu} - \epsilon_p) \approx n_0 - \left(e\varphi + \frac{m}{2}v^2 \right) \frac{h(\mu)}{2\pi} \tag{12.136}$$

where we have expanded $\bar{\mu}$ in first order around the Fermi energy and n_0 describes the number of particles with no motion and no electrostatic potential. The correlated density (12.135) splits into two parts in the zero-temperature limit of (12.134)

$$n_{corr} = \frac{1}{2}\int_0^{\infty} \frac{d\xi}{2\pi} h(\bar{\mu} + \xi) \frac{\sqrt{\xi^2 + \Delta^2} - \xi}{\sqrt{\xi^2 + \Delta^2}} - \frac{1}{2}\int_{-\bar{\mu}}^{0} \frac{d\xi}{2\pi} h(\bar{\mu} + \xi) \frac{\sqrt{\xi^2 + \Delta^2} + \xi}{\sqrt{\xi^2 + \Delta^2}} \tag{12.137}$$

which vanishes for a vanishing gap. Since the gap is only nonzero in the vicinity of the Fermi level given by the Debye frequency ω_D we can restrict the integration to the $\pm\omega_D$-range. Expanding the density of states for $\xi < \omega_D$ we obtain finally (Morawetz, Lipavský, J. Koláček, Brandt and Schreiber, 2007a)

$$n_{corr} = \frac{\partial h}{\partial\mu} \frac{\Delta^2}{4\pi} \left[\ln\left(\frac{\omega_D}{\Delta} + \sqrt{\frac{\omega_D^2}{\Delta^2} + 1} \right) - \frac{1}{1 + \sqrt{1 + \frac{\Delta^2}{\omega_D^2}}} \right] \approx \frac{\partial h}{\partial\mu} \frac{\Delta^2}{4\pi} \ln\left(\frac{2\omega_D}{\sqrt{e}\Delta} \right) \tag{12.138}$$

for $\omega_D \gg \Delta$ in the last step.

Since the total system should stay neutral; we expect $n = n_0$ and the two contributions, $n_n - n_0$ according to (12.136) and n_{corr} of (12.138), should cancel. Therefore the required electrostatic potential must read

$$e\varphi = -\frac{m}{2}v^2 + \frac{\partial\ln h}{\partial\mu} \frac{\Delta^2}{2} \ln\left(\frac{2\omega_D}{\sqrt{e}\Delta} \right). \tag{12.139}$$

This resulting electrostatic potential has the form of a Bernoulli potential. Its purpose is to compensates the contribution inhomogeneities due to diamagnetic currents and the associated inertial and Lorentz forces. It has a part directly linked to the gap.

The great hope was to measure the Bernoulli potential in order to access directly the gap parameter (Rickayzen, 1969). The experimental attempts to measure it; however, have yielded no result (Bok and Klein, 1968; Morris and Brown, 1971). Why no signal of thermodynamic corrections is seen remained a puzzle for nearly 30 years. Recently the solution was found by a modification (Lipavský, Koláček, Mareš and Morawetz, 2001a) of the Budd–Vannimenus theorem (Budd and Vannimenus, 1973) which shows that the surface dipoles cancel the thermodynamical corrections exactly for homogeneous superconductors.

For inhomogeneous superconductors like type-II materials we have, besides the cancellation, a small remaining effect. To see this, the Ginzburg–Landau equation extended to low temperatures due to Bardeen,

$$\frac{1}{2m^*}(-i\hbar\nabla - e^*\mathbf{A})^2\psi + \frac{\gamma T_c^2}{2n}\left(1 - \frac{(T/T_c)^2}{\sqrt{1 - \frac{2}{n}|\psi|}^2}\right)\psi = 0, \qquad (12.140)$$

was solved (Lipavský, Koláček, Morawetz and Brandt, 2002a). Exact solutions can be given for parallel magnetic fields (Lipavský, Morawetz, Koláček, Mareš, Brandt and Schreiber, 2004a) and the numerical solution for perpendicular magnetic fields in type-II superconductors (Lipavský, Koláček, Morawetz and Brandt, 2002a).

The electrostatic potential can leak out of a superconductor by three types of charges: (i) The bulk charge which describes the transfer of electrons from the inner- to the outer-regions of vortices creating a Coulomb force. This force has to balance the centrifugal force by the electrons rotating around the vortex centre; the outward push of the magnetic field via the Lorentz force and the outward force coming from the fact that the energy of Cooper pairs is lower than the one of free electrons, such that unpaired electrons in the vortex core are attracted towards the condensate around the core (Lipavský, Morawetz, Koláček, Mareš, Brandt and Schreiber, 2004a). (ii) The surface charge (Lipavský, Morawetz, Koláček, Mareš, Brandt and Schreiber, 2005) distributed on the scale of the Thomas–Fermi screening length. (iii) The surface dipole which cancels all contributions of pairing forces (Lipavský, Morawetz, Koláček, Mareš, Brandt and Schreiber, 2004b) resulting in an observable surface potential of

$$e\phi_0 = -\frac{f_{\text{el}}}{n}. \qquad (12.141)$$

The latter one gives rise to characteristic features predicted for experimental observations. The quadrupole resonance lines in the high-T_c material YBCO have been measured (Kumagai, Nozaki and Matsuda, 2001) and explained in (Lipavský, Koláček, Morawetz and Brandt, 2002b). More details of the Bernoulli potential can be found in the monography (Lipavský, Koláček, Morawetz, Brandt and Yang, 2007).

Part IV

Nonlocal Kinetic Theory

A quantum kinetic equation which unifies the achievements of transport in dense gases with the quantum transport of dense Fermi systems is presented. The quasiparticle drift of Landau's equation is connected with a dissipation governed by a nonlocal and non-instant scattering integral in the spirit of Enskog corrections. These corrections are expressed in terms of shifts in space and time that characterise the non-locality of the scattering process. In this way quantum transport is possible to recast into a quasiclassical picture. Compared to the Boltzmann equation, the presented form of virial corrections only slightly increase the numerical demands on implementations. In order to achieve this, large cancellations in the off-shell motion have been used which are buried usually in non-Markovian behaviours. The remaining effects are: (i) off-shell tails of the Wigner distribution, (ii) renormalisation of scattering rates and (iii) of the single-particle energy, (iv) collision delay and (v) related nonlocal corrections to the scattering integral. The balance equations for the density, momentum, energy and entropy include quasiparticle contributions and the correlated two-particle contributions beyond the Landau theory. The medium effects on binary collisions are shown to mediate the latent heat, i.e. an energy conversion between the correlation and thermal energy. The two-particle form of the entropy extends the Landau quasiparticle picture by two-particle molecular contributions. The H-theorem is also proved to hold for the nonlocal kinetic equation.

13

Nonlocal Collision Integral

The kinetic equation in its skeleton form (9.24) holds for rather general systems. To obtain its actual form we have to evaluated its ingredients, the scattering integral and the quasiparticle energy at various levels of approximations. Within the same accuracy one has to evaluate the $\rho[f]$-functional. In both cases it means to turn the selfenergy $\Sigma^<$ into a functional of the quasiparticle distribution f. In this section we focus on the scattering integral.

The quasiclassical limit and the limit of small scattering rates explicitly determine how to evaluate the scattering integral from the selfenergy $\sigma^<$. For non-degenerate systems, a very similar scheme was carried through by Baerwinkel 1969a; 1969b. One can see in Baerwinkel's papers, that the scattering integral is troubled by a large set of gradient corrections. This formal complexity seems to be the main reason why most authors either neglect gradient corrections completely (Danielewicz, 1984b; Botermans and Malfliet, 1990) or provide them buried in multi-dimensional integrals (Loos, 1990a,b; Morozov and Röpke, 1995).

To avoid manipulations with long and obscure formulae, the gradient corrections have to be sorted and expressed in a comprehensive form. As we will see, the Enskog-type of non-instant and nonlocal corrections are of such a comprehensive form. The known physical interpretation of Enskog's corrections also provides a guide for reading formulae resulting from the algebra of the quasiclassical limit.

13.1 Gradient expansion

We apply the gradient approximation for the two-particle functions of Chapter 9.1.1. In parallel with the Dyson equation, there are no linear gradients in the ladder equation (10.128). Like the retarded Green function, the T-matrix is given by the matrix inversion, $\mathscr{T}_R^{-1} = \mathscr{V}^{-1} - \mathscr{G}^R$. Since the T-matrix is symmetrical with respect to matrix arguments, $\langle s|\mathscr{T}|s'\rangle = \langle s'|\mathscr{T}|s\rangle$ (s and s' are momenta associated with α and β, respectively), the matrix inversion does not bring any gradients.

The selfenergy, $\Sigma^<(13) = \mathscr{T}^<(1234)G^>(42)$, includes a number of gradient contributions, due to internal gradients in $\mathscr{T}^<$ and a simple nonlocal correction due to the convolution with $G^>$.

Interacting Systems far from Equilibrium. Klaus Morawetz, Oxford University Press (2018).
© Klaus Morawetz. DOI: 10.1093/oso/9780198797241.001.0001

13.1.1   Two-particle matrix products

For the calculation of the selfenergy in (10.124), or its quasiparticle approximation neglecting any gradients (10.150), we need $t^<$ for the zero transferred momentum, $q = 0$, and its infinitesimal vicinity. Making the gradient expansion of (10.127) in terms of the Poisson brackets one finds

$$t^< = t^R \zeta^< t^A \left(1 - \frac{i}{2} \left([\ln t^R, \ln \zeta^<] - [\ln t^A, \ln \zeta^<] + [\ln t^R, \ln t^A] \right) \right)$$

$$= |t|^2 \zeta^< \left(1 - [\phi, \ln \zeta^<] - [\phi, \ln |t|] \right) \tag{13.1}$$

with ζ the two-particle function (10.135). In the second line we have decomposed the T-matrices into amplitudes and phase shifts,

$$t_{\text{sc}}^{R/A} = |t_{\text{sc}}| \, e^{\pm i\phi}. \tag{13.2}$$

Note that all gradient corrections depend exclusively on the derivatives of phase-shift ϕ. We introduce therefore Δ's for these derivatives according to the following list:

$$\Delta_K = \frac{1}{2} \frac{\partial \phi}{\partial r}, \quad \Delta_E = -\frac{1}{2} \frac{\partial \phi}{\partial t}, \quad \Delta_t = \frac{\partial \phi}{\partial \omega},$$

$$\Delta_2 = \frac{\partial \phi}{\partial p} - \frac{\partial \phi}{\partial q} - \frac{\partial \phi}{\partial k}, \quad \Delta_3 = -\frac{\partial \phi}{\partial k},$$

$$\Delta_4 = -\frac{\partial \phi}{\partial q} - \frac{\partial \phi}{\partial k}, \quad \Delta_r = \frac{1}{4} (\Delta_2 + \Delta_3 + \Delta_4). \tag{13.3}$$

Using these definitions and the linear approximation, $a(x)(1 - \Delta \partial_x a) = a(x - \Delta)$, expression (13.1) can be given in the form

$$t^< \left(\Omega, k, p, 0, r - \frac{1}{2}\Delta_2, t \right) = \int \frac{dq}{(2\pi)^3} \frac{dQ}{(2\pi)^3} t^R \left(\Omega - \Delta_\omega, k - \frac{\Delta_k}{2}, p - \frac{\Delta_k}{2}, q, r - \Delta_r, t - \frac{\Delta_t}{2} \right)$$

$$\times \zeta^< \left(\Omega - 2\Delta_\omega, k - q - \Delta_k, p + q - \Delta_k, Q, r - \frac{\Delta_3}{2} - \frac{\Delta_4}{2}, t - \Delta_t \right)$$

$$\times t^A \left(\Omega - \Delta_\omega, k - \frac{\Delta_k}{2}, p - \frac{\Delta_k}{2}, q + Q, r - \Delta_r, t - \frac{\Delta_t}{2} \right). \tag{13.4}$$

Factors of half in the nonlocal corrections to the amplitude $|t^{R,A}|$ result from the fact that this amplitude enters the collision integral in the square.

In the absence of gradients, the two-particle function $\zeta^<$ is singular in the 'transferred' momentum Q giving the only contribution for $Q = 0$. When the gradient corrections are already in the explicit form, one can employ this symmetry as it has been used above. Indeed, the complex conjugacy of the advanced and retarded T-matrices, $t^A = \bar{t}^R$, requires equal arguments of both functions. Now we will show that the infinitesimal vicinity of $Q = 0$ yields additional gradient contributions.

13.1.2 Convolution of initial states

The two-particle correlation function $\zeta^<$ representing initial states of the collision in (13.4) is not a suitable input for the kinetic equation. We have to express $\zeta^<$ in terms of the convolution of two single-particle functions. This convolution provides gradient corrections by which the effective positions of particles entering the collision process become distinct.

With the inverse Fourier transformation to (10.148) we obtain

$$\zeta^<(\Omega, k-p, p+q, Q, r, t)$$

$$= 2^3 \int dx_1 \, dx_2 \, dx_1' \, dx_2' \, d\bar{t} dt' \, e^{iQ(x_1'-x_2')+i\Omega(\bar{t}-t')} e^{-i(k-q)(x_1-x_1')-i(p+q)(x_2-x_2')}$$

$$\times \, \delta(x_1 + x_2 + x_1' + x_2' - 4r) \delta(\bar{t} + t' - 2t) G^<(x_1\bar{t}, x_1' t') G^<(x_2\bar{t}, x_2' t'). \quad (13.5)$$

The substitution, $x_1 = r + \alpha + \beta_1/2$, $x_1' = r + \alpha - \beta_1/2$, $x_2 = r - \tilde{\alpha} + \beta_2/2$, $x_2' = r - \tilde{\alpha} - \beta_2/2$, $\bar{t} = t + \tau/2$, $t' = t - \tilde{\tau}/2$, shows that the δ-functions mean $\tilde{\tau} = \tau$ and $\tilde{\alpha} = \alpha$ and we obtain

$$\zeta^<(\Omega, k-p, p+q, Q, r, t) = 2^3 \int d\beta_1 d\beta_2 d\alpha d\tau e^{2iQ\alpha+i\Omega\tau}$$

$$\times \, e^{-i(k-q+Q/2)\beta_1 - i(p+q-Q/2)\beta_2} g^<(\beta_1, r+\alpha, t, \tau) g^<(\beta_2, r-\alpha, t, \tau), \quad (13.6)$$

where the representation (9.3) has been used.

Now we linearise the centre-of-mass dependence of $g^<$'s in α. The factor α can be represented by a derivative in front of the integral, $\alpha = \frac{i}{2}\partial_Q - \frac{i}{4}\partial_q$. The remaining integration over α results in $\pi^3 \delta(Q)$. When substituted into (13.4), the differential operator in front of the integral is treated, with the help of the integration, by parts giving

$$\alpha \rightarrow -\frac{1}{2}(\Delta_3 - \Delta_4) \quad (13.7)$$

and Q is integrated out. In other words, the α in the arguments of g in (13.6) can be replaced by nonlocal shifts valid up to linear orders in gradients. This leads with (13.4) to

$$t^<\left(\Omega, k, p, 0, r - \frac{1}{2}\Delta_2, t\right) = \int \frac{dq}{(2\pi)^3} \left| t^R\left(\Omega - \Delta_\omega, k - \frac{\Delta_k}{2}, p - \frac{\Delta_k}{2}, q, r - \Delta_r, t - \frac{\Delta_t}{2}\right)\right|^2$$

$$\times \, g^<(k-q, r-\Delta_3, t-\Delta_t, \Omega - 2\Delta_\omega) g^<(p+q, r-\Delta_4, t-\Delta_t, \Omega - 2\Delta_\omega). \quad (13.8)$$

13.1.3   Convolution of T-matrix and hole Green's function

Next we evaluate the convolution of the $\mathscr{T}$-matrix with the hole Green's function which is required for the selfenergy (10.124). Writing the selfenergy in the matrix notation (9.3), $[\tau = 1 - 3$ and $x = \frac{1}{2}(1 + 3)]$

$$\Sigma^<(\tau, x) = \int d\alpha \, d\beta \, G^> \left(\alpha - \tau - \beta, x - \frac{1}{2}(\alpha + \beta) \right) \langle \alpha | \mathscr{T}^< \left(\tau - \frac{1}{2}(\alpha - \beta), x - \frac{1}{4}(\alpha + \beta) \right) | \beta \rangle,$$
(13.9)

we see that the convolution couples matrix arguments α and β with the centre-of-mass variables. By substitution, $\lambda = \frac{1}{2}(\alpha + \beta)$ and $\mu = \alpha - \beta$, and expansion in gradients we obtain

$$
\Sigma^<(\tau, x) = \int d\mu \, G^> (\mu - \tau, x) \int d\lambda \left\langle \lambda + \frac{\mu}{2} \left| \mathscr{T}^< \left(\tau - \frac{\mu}{2}, x \right) \right| \lambda - \frac{\mu}{2} \right\rangle
$$
$$
- \frac{1}{2} \int d\mu \, G^> (\mu - \tau, x) \frac{\partial}{\partial x} \int d\lambda \, \lambda \left\langle \lambda + \frac{\mu}{2} \left| \mathscr{T}^< \left(\tau - \frac{\mu}{2}, x \right) \right| \lambda - \frac{\mu}{2} \right\rangle
$$
$$
- \int d\mu \left(\frac{\partial}{\partial x} G^> (\mu - \tau, x) \right) \int d\lambda \, \lambda \left\langle \lambda + \frac{\mu}{2} \left| \mathscr{T}^< \left(\tau - \frac{\mu}{2}, x \right) \right| \lambda - \frac{\mu}{2} \right\rangle.
$$
(13.10)

The second and third terms are gradient corrections due to the convolution.

Now we can transform the selfenergy (13.10) into the mixed representation. The momentum representation of the matrix algebra is introduced via unity operators, e.g. $1 = \int \frac{ds}{(2\pi)^3} |s\rangle \langle s|$, with $\langle \lambda - \mu/2|s\rangle = e^{is(\lambda - \mu/2)}$ and $\langle s'|\lambda + \mu/2\rangle = e^{-is'(\lambda + \mu/2)}$. For the non-gradient part the integration over λ results in the known fact that only the diagonal element, $s = s'$, contributes. For the gradient contribution, $\lambda \to -(i/2)(\partial_s - \partial_{s'})$, and the diagonal element is taken after the derivatives are performed. Since we have considered the gradient corrections to the $\mathscr{T}$-matrix already in Chapter 13.1.1 and since the shifts are additive, it is sufficient here to use the zero-order approximation of $t^<$,

$$\langle s|t^<|s'\rangle = \int \frac{d\bar{s} \, d\tilde{s}}{(2\pi)^2} \langle s|t^R|\bar{s}\rangle \langle \bar{s}|\zeta^<|\tilde{s}\rangle \langle \tilde{s}|t^A|s'\rangle,$$
(13.11)

where all functions have the centre-of-mass argument (κ, x). From $t^{R,A} = |t|e^{\mp i\phi}$ and condition $s = s'$ we find that $\lambda \to -\partial_s \phi$. By substitution into variables of the kinetic equation, $(q, 0) = s - s'$, $(k, \omega) = \kappa/2 + s$, $(p, 0) = \kappa/2 - s$ and $(r, t) = x$, one confirms that $-\partial_s \phi = \frac{1}{2}\Delta_2$, see (13.3), and the selfenergy reads

$$\sigma_\omega^<(k, r, t) = \int \frac{dp \, d\Omega}{(2\pi)^4} g_{\Omega-\omega}^> (p, r - \Delta_2, t) \left(1 - \frac{1}{2} \frac{\partial \Delta_2}{\partial r} \right) t^< \left(\Omega, k, p, 0, r - \frac{1}{2}\Delta_2, t \right).$$
(13.12)

In this expression, $t^<$ abbreviates the right-hand side of (13.11), because the non-local correction Δ_2 is defined only with respect to the integral over internal states of the collision. The normal term, $1 - \frac{1}{2}\partial_r\Delta_2$ results from the interchange, $\partial_r(\Delta_2 A) = (\partial_r\Delta_2)A + \Delta_2\partial_r A$, one has to make before the derivative is expressed in terms of the displacement, e.g. $g^<(r) - \Delta_2\partial_r g^<(r) = g^<(r - \Delta_2)$.

Together with (13.8) we obtain the result

$$
\sigma_a^<(\omega, k, r, t) = \int \frac{dp\,d\omega'}{(2\pi)^4}\, \mathcal{T}^<\left(kp0, \omega', r + \frac{\Delta_2}{2}\right)g^>(p, \omega' - \omega, r + \Delta_2)
$$

$$
= \sum_b \int \frac{dp\,dq\,dE\,d\Omega}{(2\pi)^8}\left(1 - \frac{1}{2}\frac{\partial\Delta_2}{\partial r}\right)\left|t_{\rm sc}\left(\omega + E - \Delta_E, k - \frac{1}{2}\Delta_K, p - \frac{1}{2}\Delta_K, q, r - \Delta_r, t - \frac{1}{2}\Delta_t\right)\right|^2
$$

$$
\times g_b^>(E, p, r - \Delta_2, t)g_a^<(\omega - \Omega - \Delta_E, k - q - \Delta_K, r - \Delta_3, t - \Delta_t)
$$

$$
\times g_b^<(E + \Omega - \Delta_E, p + q - \Delta_K, r - \Delta_4, t - \Delta_t). \tag{13.13}
$$

This is the resulting comprehensive form of the quasiclassical limit of the selfenergy $\sigma^<$. The selfenergy $\sigma^>$ results in the interchange $> \longleftrightarrow <$. Another direct way is reported in (Lipavský, Morawetz and Špička, 2001b).

Again, sending all Δ's to zero, one recovers the instant and local approximation of the selfenergy, Eq. (10.157). The non-instant and nonlocal corrections given by Δ's appear in three ways: in the arguments of the correlation functions, in the arguments of the T-matrix, and as a prefactor.

The Δ's in the arguments of the correlation functions $g^{>,<}$ in (13.13) remind non-instant and nonlocal corrections in the scattering-in integral for classical particles, see Chapter 3.1.4. The displacements of the asymptotic states are given by $\Delta_{2,3,4}$ where (13.3) are the quantum-mechanical definitions. The time delay enters in an equal way the asymptotic states 3 and 4. The momentum gain Δ_K also appears only in states 3 and 4. Finally, there is the energy gain which will be discussed.

13.2   Binary and ternary collisions

Now we convert the selfenergy $\sigma^<$ as a functional of the correlation functions $g^{>,<}$ into the scattering integral of the quasiparticle distribution f. To be consistent with other steps, we have to use the extended quasiparticle approximation. The extended quasiparticle approximation; however, yields expected binary processes resulting from the on-shell part, and moreover sequential ternary processes resulting from the off-shell parts.

Let us remind ourselves of the scattering integral in terms of the selfenergy and with explicit arguments. According to (9.73) the scattering integral reads

$$I_1 \equiv I_a(k, r, t)$$

$$= z_a(k, r, t) \Big[\sigma_a^<(\omega, k, r, t) \, (1 - f_a(k, r, t)) - \sigma_a^>(\omega, k, r, t) f_a(k, r, t) \Big]_{\omega = \varepsilon_a(k, r, t)}$$

$$= z_1 \sigma_{1,\varepsilon_1}^< (1 - f_1) - z_1 \sigma_{1,\varepsilon_1}^> f_1. \tag{13.14}$$

As in the classical case, an abbreviated notation of the asymptotic state (k, r, t) is the subscript 1. The subscript ε_1 denotes that the selfenergy is taken at the pole value $\omega = \varepsilon_1$. So far, the selfenergies $\sigma^{>,<}$ are functionals of $g^{>,<}$, see (13.13).

The extended quasiparticle approximation (9.62) constructs the correlation function from two parts, $g^< = \lambda^< + \xi^<$, the on-shell quasiparticle contribution $\lambda^<$ and the off-shell correction $\xi^<$. The off-shell contribution is a correction in the limit of small scattering rates, thus one can linearise the selfenergy in ξ's, (in notation of (13.13), $g_2^> = g_b^>$, $g_3^< = g_a^<$, and $g_4^< = g_b^<$)

$$\sigma_1^<[g_2^>, g_3^<, g_4^<] = \sigma_1^<[\lambda_2^>, \lambda_3^<, \lambda_4^<] + \sigma_1^<[\xi_2^>, \lambda_3^<, \lambda_4^<]$$

$$+ \sigma_1^<[\lambda_2^>, \xi_3^<, \lambda_4^<] + \sigma_1^<[\lambda_2^>, \lambda_3^<, \xi_4^<]. \tag{13.15}$$

According to (9.59), the on-shell part $\lambda_i^<$ is proportional to the quasiparticle distribution, e.g. $\lambda_2^> = 2\pi z_2 (1 - f_2) \delta(\omega - \varepsilon_2)$. The first term of (13.15) is thus readily converted to a functional of distributions,

$$\sigma_1^<[\lambda_2^>, \lambda_3^<, \lambda_4^<] \rightarrow \sigma_1^<[1 - f_2, f_3, f_4]. \tag{13.16}$$

The corresponding scattering-in integral $(1 - f_1)\sigma_1^<$ has two-quasiparticle distributions $f_3 f_4$ as an initial condition and two-(quasi)hole distributions $(1 - f_1)(1 - f_2)$ as the Pauli blocking of the final state. This is a typical scattering integral of binary collision outlined by Fig. 13.1 (a).

According to (9.60), the off-shell parts ξ depend on the selfenergy, e.g. $\xi_2^> = \sigma_2^> \wp'/(E - \varepsilon_2)$. In general, the selfenergy $\sigma_2^>$ depends on the full correlation functions $\sigma_2^>[g_5^<, g_6^<, g_7^>]$, i.e. self-consistently on ξ. The off-shell contribution; however, is

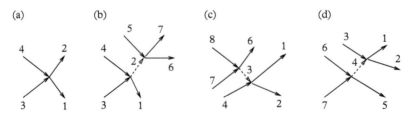

Figure 13.1: *Binary process (a) and ternary processes (b–d). All three ternary processes have the same structure of two binary sub-processes linked by an off-shell propagator (dashed line).*

treated as a correction (being linear in small scattering rates), therefore the lowest-order approximation is sufficient, $\sigma_2^> = \sigma_2^> [\lambda_5^<, \lambda_6^>, \lambda_7^>]$. The second term of (13.15) thus reads

$$\sigma_1^< [\xi_2^>, \lambda_3^<, \lambda_4^<] \rightarrow \sigma_1^< \left[\sigma_2^> [f_5, 1-f_6, 1-f_7] \frac{\wp'}{E-\varepsilon_2}, f_3, f_4 \right]. \qquad (13.17)$$

The corresponding scattering-in integral $(1-f_1)\sigma_1^<$ has three quasiparticle distributions $f_3 f_4 f_5$ at the initial condition, and three-hole distributions $(1-f_1)(1-f_6)(1-f_7)$ at the final one. Accordingly, this is the three-particle process.

One can see that the ternary process given by (13.17) is composed of two sequential binary sub-processes $\sigma_1^<$ and $\sigma_2^>$ with the off-shell single-particle propagation $\wp'/(E-\varepsilon_2)$ between individual sub-processes, see Fig. 13.1 (b). Disregarding their time order, let us call $\sigma_1^<$ the first sub-process and $\sigma_2^>$ the second. In the first sub-process, the particles from states 3 and 4 are scattered into 1 and 2, where state 2 is off-shell so that it cannot become the asymptotic final state. In the second sub-process, the particle is rescued from the off-shell state 2 by the particle from 5 and these two particles end up in 6 and 7.

In a similar manner one can rearrange the third and the fourth terms of (13.15),

$$\sigma_1^< [\lambda_2^>, \xi_3^<, \lambda_4^<] \rightarrow \sigma_1^< \left[1-f_2, \sigma_3^< [1-f_6, f_7, f_8] \frac{\wp'}{\omega-\Omega-\Delta_E-\varepsilon_3}, f_4 \right],$$

$$\sigma_1^< [\lambda_2^>, \lambda_3^<, \xi_4^<] \rightarrow \sigma_1^< \left[1-f_2, f_3, \sigma_4^< [1-f_5, f_6, f_7] \frac{\wp'}{E+\Omega-\Delta_E-\varepsilon_4} \right] \qquad (13.18)$$

which describe ternary processes outlined in Fig. 13.1 (c,d), respectively.

This approximation of the ternary collisions by the sequential binary processes, the so called impulse approximation (Danielewicz and Bertsch, 1991), badly overestimates the actual three-particle scattering rates for energies below 200 MeV/A, where Fadeev equations must be used (Beyer, Röpke and Sedrakian, 1996). The destructive interference leading to a suppression of sequential processes by other three-particle channels cannot be described within the binary T-matrix used in the presented treatment. Nevertheless, we take this experience from the three-body theory and neglect the ternary processes. Let us note that these processes are known as secular divergent and artifacts of the theory (Lifschitz and Pitaevsky, 1981).

13.3 Complete kinetic equation

Now all the approximations needed to complete the kinetic equation with the binary scattering integral are specified. Into the skeleton kinetic equation (9.73) we will substitute the on-shell value of the selfenergy (13.13) where only the on-shell part of the correlation function, $g \rightarrow \lambda$, is used within the limit of small scattering rates. It remains for us to perform the algebra.

In the limit of small scattering rates, the on-shell parts of correlation functions found in (13.13) read:

$$g_b^>(E, p, r - \Delta_2, t) \qquad\qquad \to (1 - \bar{f}_2)\bar{z}_2 2\pi\delta\,(E - \bar{\varepsilon}_2),$$

$$g_a^<(\omega - \Omega - \Delta_E, k - q - \Delta_K, r - \Delta_3, t - \Delta_t) \quad \to \bar{f}_3\bar{z}_3 2\pi\delta\,(\omega - \Omega - \Delta_E - \bar{\varepsilon}_3),$$

$$g_b^<(E + \Omega - \Delta_E, p + q - \Delta_K, r - \Delta_4, t - \Delta_t) \quad \to \bar{f}_4\bar{z}_4 2\pi\delta\,(E + \Omega - \Delta_E - \bar{\varepsilon}_4). \qquad (13.19)$$

where functions without bars have the arguments

$$\varepsilon_1 \equiv \varepsilon_a(k, r, t), \qquad\qquad \varepsilon_2 \equiv \varepsilon_b(p, r + \Delta_2, t),$$
$$\varepsilon_3 \equiv \varepsilon_a(k - q + \Delta_K, r + \Delta_3, t + \Delta_t), \quad \varepsilon_4 \equiv \varepsilon_b(p + q + \Delta_K, r + \Delta_4, t + \Delta_t). \qquad (13.20)$$

Functions with bars have arguments with inverse shifts.

The abbreviated notation parallels (3.23). As in the classical case, the two smaller numbers 1 and 2 denote a pair of particles leaving the collision together and the two higher numbers 3 and 4 denote a pair of particles entering the collision. Upon the interchange $> \longleftrightarrow <$ one must interchange $f \longleftrightarrow (1-f)$.

After the substitution of (13.19) the δ functions from (13.19) can be readily integrated out. Let us first integrate out Ω using the δ function from (13.19). In the second step, one can integrate over E using the δ function from (13.19). In this case, one must take into account that $\bar{\varepsilon}_2$ depends on Δ_2 as a function of the energy E. This E-dependence results in a norm of the singularity

$$\frac{1}{1 - \frac{\partial \bar{\varepsilon}_2}{\partial E}} \approx 1 + \frac{\partial \bar{\varepsilon}_2}{\partial E} = 1 - \frac{\partial \bar{\varepsilon}_2}{\partial r}\frac{\partial \Delta_2}{\partial E} = 1 - \frac{\partial \bar{\varepsilon}_2}{\partial r}\frac{\partial \Delta_2}{\partial \omega}. \qquad (13.21)$$

Putting all the pieces together, the scattering integral results

$$I_1 = \sum_b \int \frac{dp}{(2\pi)^3}\frac{dq}{(2\pi)^3} 2\pi\delta\,(\varepsilon_1 + \bar{\varepsilon}_2 - \bar{\varepsilon}_3 - \bar{\varepsilon}_4 - 2\Delta_E)\, z_1\bar{z}_2\bar{z}_3\bar{z}_4 \left(1 - \frac{1}{2}\frac{\partial \Delta_2}{\partial r} - \frac{\partial \bar{\varepsilon}_2}{\partial r}\frac{\partial \Delta_2}{\partial \omega}\right)$$

$$\times \left| t_{sc}\!\left(\varepsilon_1 + \bar{\varepsilon}_2 - \Delta_E, k - \frac{1}{2}\Delta_K, p - \frac{1}{2}\Delta_K, q, r - \Delta_r, t - \frac{1}{2}\Delta_t\right)\right|^2$$

$$\times \left[(1 - f_1)(1 - \bar{f}_2)\bar{f}_3\bar{f}_4 - f_1\bar{f}_2(1 - \bar{f}_3)(1 - \bar{f}_4)\right]. \qquad (13.22)$$

In Δ's and their derivatives the energy $\omega + E = \varepsilon_1 + \bar{\varepsilon}_2$ is substituted after all derivatives are taken. With the explicit scattering integral (13.22) we can write down the complete kinetic equation

$$\frac{\partial f_1}{\partial t} + \frac{\partial \varepsilon_1}{\partial k}\frac{\partial f_1}{\partial r} - \frac{\partial \varepsilon_1}{\partial r}\frac{\partial f_1}{\partial k} = I_1, \qquad (13.23)$$

Let us summarise the properties of the asymptotic kinetic equation (13.23). The drift is governed by the quasiparticle energy obtained from the single-particle excitation spectrum. The scattering integral is nonlocal and non-instant, including corrections to the conservation of energy and momentum.

In spite of many similarities, the asymptotic kinetic equation (13.23) does not reduce to the intuitive kinetic equation (3.26) in the classical limit, $1 - f \rightarrow 1$. Note that the nonlocal corrections of the scattering-out, the first term in the right-hand side of (13.23), are identical to the nonlocal corrections of the scattering-out. This is in contrast to the intuitive kinetic equation where the nonlocal corrections of the scattering-out are flipped. This disagreement between intuitive approaches in the spirit of Enskog and formal asymptotic theories is not specific to the presented approach but appears in other methods too, see (Bärwinkel, 1969*a*; Tastevin, Nacher and Laloë, 1989; Nacher, Tastevin and Laloë, 1989; Loos, 1990*a,b*; de Haan, 1990*b,a*; de Haan, 1991; Laloë and Mullin, 1990; Snider, 1990; Snider, 1991; Nacher, Tastevin and Laloë, 1991*a,b*; Snider, 1995; Snider, Mullin and Laloë, 1995). We will solve this puzzle in 13.4.

13.4 Particle–hole versus space–time symmetry

Why the asymptotic theory does not lead to the kinetic equation of the Enskog type can be seen from the following consideration. As shown in Chapter 3.1.4, the intuitive Enskog's approach treats the scattering-in as the event which has just ended while treating the scattering-out as the event which is just starting. The many-body perturbation expansion treats both cases as events which have just ended, the scattering-out is, however represented by a collision of two holes while the scattering-in is a collision of two particles. These two pictures are illustrate in Figure 13.2.

To link these two pictures, we need an identity which links the near future with the near past. Such an identity is the optical theorem (8.40) and (8.44) from which follows

$$\mathscr{T}_{\text{sc}}^{R} \mathscr{A} \mathscr{T}_{\text{sc}}^{A} = \mathscr{T}_{\text{sc}}^{A} \mathscr{A} \mathscr{T}_{\text{sc}}^{R}. \tag{13.24}$$

To employ the optical theorem (13.24) requires one to rearrange the scattering integral. In the symbolic notation the related rearrangement reads

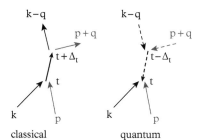

Figure 13.2: *Scattering-out for classical and quantum concept of collisions where the space–time symmetry should be accompanied by particle–hole symmetry.*

$$I_1 = z_1(1-f_1)\sigma_1^< - z_1 f_1 \sigma_1^>$$

$$= z_1(1-f_1)G_2^> \mathscr{T}_{sc}^R \mathscr{G}_{34}^< \mathscr{T}_{sc}^A - z_1 f_1 G_2^< \mathscr{T}_{sc}^R \mathscr{G}_{34}^> \mathscr{T}_{sc}^A$$

$$= \left(z_1(1-f_1)G_2^> - z_1 f_1 G_2^<\right)\mathscr{T}_{sc}^R \mathscr{G}_{34}^< \mathscr{T}_{sc}^A - z_1 f_1 G_2^< \mathscr{T}_{sc}^R \mathscr{A}_{34} \mathscr{T}_{sc}^A. \qquad (13.25)$$

Here, $\mathscr{G}_{34}^< = G_3^< G_4^<$ and $\mathscr{A}_{34} = \mathscr{G}_{34}^> + \mathscr{G}_{34}^<$. In the last term we can apply the optical theorem (13.24) for the interchange $\mathscr{T}_{sc}^R \mathscr{A} \mathscr{T}_{sc}^A \rightarrow \mathscr{T}_{sc}^A \mathscr{A} \mathscr{T}_{sc}^R$, therefore

$$I_1 = \left(z_1(1-f_1)G_2^> - z_1 f_1 G_2^<\right)\mathscr{T}_{sc}^R \mathscr{G}_{34}^< \mathscr{T}_{sc}^A - z_1 f_1 G_2^< \mathscr{T}_{sc}^A \mathscr{A}_{34} \mathscr{T}_{sc}^R. \qquad (13.26)$$

Technically, the interchange of the retarded and the advanced T-matrices is equivalent to the change of sign of the phase shift ϕ, $\mathscr{T}_{sc}^R \rightarrow t_{sc}^R = |t_{sc}|e^{i\phi}$ and $\mathscr{T}_{sc}^A \rightarrow t_{sc}^A = |t_{sc}|e^{-i\phi}$. Accordingly, all shifts which are related to the combination of functions $\mathscr{A}_{34} = G_3^> G_4^> - G_3^< G_4^<$ can be written with all signs inverted. In the kinetic equation (13.23), the combination corresponding to $\mathscr{A}_{34}$ is achieved by regrouping the distributions,

$$(1-f_1)(1-\bar{f}_2)\bar{f}_3\bar{f}_4 - f_1\bar{f}_2(1-\bar{f}_3)(1-\bar{f}_4) = (1-f_1-\bar{f}_2)\bar{f}_3\bar{f}_4 - f_1\bar{f}_2(1-\bar{f}_3-\bar{f}_4). \qquad (13.27)$$

The intermediate-state Pauli-blocking factor $1-\bar{f}_3-\bar{f}_4$ is the occupation number corresponding to $\mathscr{A}_{34}$. Inverting signs of the phase shift, i.e. of Δ's, related to the scattering integral with the Pauli-blocking $1-\bar{f}_3-\bar{f}_4$ one arrives at the Enskog-type collision integral

$$I_1 = \sum_b \int \frac{dp}{(2\pi)^3}\frac{dq}{(2\pi)^3} 2\pi\delta\left(\varepsilon_1 + \bar{\varepsilon}_2 - \bar{\varepsilon}_3 - \bar{\varepsilon}_4 - 2\Delta_E\right) z_1 \bar{z}_2 \bar{z}_3 \bar{z}_4 \left(1 - \frac{1}{2}\frac{\partial\Delta_2}{\partial r} - \frac{\partial\varepsilon_2}{\partial r}\frac{\partial\Delta_2}{\partial\omega}\right)$$

$$\times (1-f_1-\bar{f}_2)\bar{f}_3\bar{f}_4 \left| t_{sc}\left(\varepsilon_1 + \bar{\varepsilon}_2 - \Delta_E, k - \frac{1}{2}\Delta_K, p - \frac{1}{2}\Delta_K, q, r - \Delta_r, t - \frac{1}{2}\Delta_t\right)\right|^2$$

$$- \sum_b \int \frac{dp}{(2\pi)^3}\frac{dq}{(2\pi)^3} 2\pi\delta\left(\varepsilon_1 + \varepsilon_2 - \varepsilon_3 - \varepsilon_4 + 2\Delta_E\right) z_1 z_2 z_3 z_4 \left(1 + \frac{1}{2}\frac{\partial\Delta_2}{\partial r} + \frac{\partial\varepsilon_2}{\partial r}\frac{\partial\Delta_2}{\partial\omega}\right)$$

$$\times f_1 f_2 (1-f_3-f_4) \left| t_{sc}\left(\varepsilon_1 + \varepsilon_2 + \Delta_E, k + \frac{1}{2}\Delta_K, p + \frac{1}{2}\Delta_K, q, r + \Delta_r, t + \frac{1}{2}\Delta_t\right)\right|^2. \qquad (13.28)$$

It is remarkable that the space–time symmetric equation (13.28) is equivalent to the particle–hole symmetric one (13.22) though both symmetries are not visible simultaneously (Špička, Morawetz and Lipavský, 2001).

Please note that the corresponding kinetic equation for Bose systems appear in the same way, just replacing the Pauli blocking $1-f$ by the Pauli enhancement $1+f$ in each place.

In the low-density limit $1-f-f \rightarrow 1$, the kinetic equation (13.28) reduces to the form (3.26) found from intuitive arguments for the classical system. This limit shows that the physical interpretation and consequences of found non-instant and nonlocal corrections are analogous to virial corrections known from the classical statistics.

13.4.1 Kinetic equation for Monte-Carlo simulations

The kinetic equation (13.28) has a peculiar property in that the scattering rates can be negative. This follows from processes in which the negative process can happen if both final states are already occupied. These processes correspond to the combination of occupation factors, $f_q f_2 f_3 f_4$ and $f_1 \bar{f}_2 \bar{f}_3 \bar{f}_4$. In the local approximation these contributions cancel and one does not meet this problem. Within the nonlocal picture, the gradient contributions have opposite signs and do not cancel.

In numerical treatments, the negative scattering rate cannot be simulated by procedures with a statistical interpretation of the scattering events. In this case, one has to neglect these processes using an approximate kinetic equation which is (13.28), but with the replacements

$$(1 - f_1 - \bar{f}_2)\bar{f}_3\bar{f}_4 \longrightarrow (1 - f_1)(1 - \bar{f}_2)\bar{f}_3\bar{f}_4,$$

$$f_1 f_2 (1 - f_3 - f_4) \longrightarrow f_1 f_2 (1 - f_3)(1 - f_4). \tag{13.29}$$

As we have seen in many cases, any neglect of a certain group of gradient corrections results in modified thermodynamic properties implied by the kinetic equation. From the optical theorem (13.24) one can show that the kinetic equation (13.29) corresponds to the approximation of the T-matrix by the Brueckner reaction matrix. The Brueckner reaction matrix differs from the T-matrix in the approximation of the two-particle spectral function which includes only the particle–particle channel, $\mathscr{A}_{34}^{\mathrm{B}} = (G_3^> G_4^>)$, while the Galitskii–Feynman two-particle spectral function (used by us so far) includes also the hole–hole channel, $\mathscr{A}_{34}^{\mathrm{GF}} = (G_3^> G_4^>) - (G_3^< G_4^<)$. The optical theorem for the Brueckner reaction matrix is thus proportional to the occupation factor $(1 - f)(1 - f)$ so that the scattering integral (13.22) directly converts into the scattering integral of (13.29). The virial corrections to the kinetic equation studied with the recent simulation codes are thus limited to the Brueckner approximation. We do not find this limitation serious since the Brueckner approximation used extensively in earlier studies yield satisfactory results.

13.5 Summary

The kinetic equation (13.28) with the nonlocal shifts (13.3) is the final result on our way to derive the kinetic equation with nonlocal corrections. The exclusive dependence of the nonlocal and non-instant corrections on the scattering phase shift confirms the results from the theory of gases (Tastevin, Nacher and Laloë, 1989; Nacher, Tastevin and Laloë, 1989; de Haan, 1990b; Nacher, Tastevin and Laloë, 1991a) obtained by very different technical tools.

With the approximation on the level of the Brueckner reaction matrix, the corresponding the non-instant and nonlocal scattering integral, seen in (13.29) parallels the classical Enskog's equation, therefore it can be treated with Monte-Carlo simulation techniques.

14

Properties of Non-Instant and Nonlocal Corrections

The derivation accomplished in Chapter 13 of the nonlocal scattering integrals is far from being intuitive. We have reached our task, the kinetic equation, being guided by nothing but systematic implementation of the quasiclassical approximation and the limit of small scattering rates. To fill the gap in intuitive understanding of microscopic mechanisms behind the nonlocal corrections, we establish a link of the quantum Δ's to their intuitively clear classical counterparts.

A strong indication that the quantum nonlocal corrections are a straightforward quantum modification of their classical ancestors, is the structure of the scattering integral. Firstly we focus on asymptotic states, say of the scattering-in in the kinetic equation (13.28). They are just the same as in the classical case shown in Fig. 3.6. The collision between particles a and b starts at time $t-\Delta_t$ from states 3 and 4 and lasts till t when both particles are released into their final states 1 and 2. At the beginning, particles a and b do not occupy the same space point but a is at $r-\Delta_3$ while the on-going particle b is at the position $r-\Delta_4$. At the end, the particle a is released at r and the particle b is released at the displaced position $r-\Delta_2$. A single time delay (in contrast to three displacements) follows from the instant character of the interaction due to which both particles have to enter the process at the same time instant $t-\Delta_t$ and are released at the same time instant t.

In the classical introduction of Chapter 3, we have not discussed in detail the possibility that the differential cross section depends on time and space. Certain time and space dependencies appear due to the field effect on the scattering discussed in Chapter 3.2 and for the marginal model of the breathing hard spheres discussed in Chapter 3.3. The actual question of coordinates, however, has not been raised in Chapter 3. In the quantum case, the nonlocal corrections also appeared in the arguments of the cross section. The T-matrix $|t_{\mathrm{sc}}|$ is evaluated in the natural centre of all four asymptotic states. The space coordinate is displaced by $\Delta_r = \frac{1}{4}(\Delta_2 + \Delta_3 + \Delta_4)$ and the time argument is delayed by $\frac{1}{2}\Delta_t$. The momenta k and p are corrected in the T-matrix $|t_{\mathrm{sc}}|$ by $\frac{1}{2}\Delta_K$, and the sum energy is corrected by Δ_E while the sum energy in final states is lifted by $2\Delta_E$.

The centring of the T-matrix reminds Enskog's statistical corrections to the differential cross section. They also depend on time and space via local distributions of

Interacting Systems far from Equilibrium. Klaus Morawetz, Oxford University Press (2018).
© Klaus Morawetz. DOI: 10.1093/oso/9780198797241.001.0001

surrounding particles. Enskog has introduced the centred argument as the most natural one. The hydrodynamic equations derived from the Enskog equation show that the
ceneetring of the differential cross section is inevitable due to conservation laws, see
(Chapman and Cowling, 1990).

14.1 Recovering classical Δ's from quantum formulae

One might argue that the structure of the scattering integral does not tell us much about
the relation between the quantum and classical Δ's. What if conservation laws constrain
the gradient correction so much that it is merely a question of notation whether they look
similar or not? To avoid a suspicion of an artificial link between the quantum and classical
Δ's, we discuss the classical limit, $\hbar \to 0$, of the collision delay and the hard-sphere
displacement. In both cases, the quantum Δ's turn into their classical counterparts.

14.1.1 Collision delay Δ_t

The collision delay Δ_t defined by (13.3) belongs to the family of the collision duration
concept introduced by Wigner (1955), see also (Taylor, 1972; Farina, 1983). Wigner's
concept is based on the asymptotic behaviour of a wave packet. One sends a wave
packet to the scattering centre and monitors when the centre-of-mass of the scattered
wave packet arrives at a detector. The difference compared to the point-like scattering
is Wigner's delay time, also called the time of arrival. Since Wigner's delay time results
from the phase shift ϕ, (as the on-shell derivative for energies at the final state)

$$\Delta_t^W = \frac{\partial \phi(\varepsilon_3 + \varepsilon_4)}{\partial(\varepsilon_3 + \varepsilon_4)}, \tag{14.1}$$

it is also called the phase time. For equilibrium nuclear matter Wigner's collision delay
and its effect on the equation of states have been discussed by Danielewicz and Pratt
(1996).

The energy derivative in Wigner's delay time (14.1) is similar but not identical to the
collision delay Δ_t. Writing (14.1) in terms of the off-shell phase shift

$$\Delta_t^W = \left(\frac{\partial \phi(\omega, k, p, q)}{\partial(\varepsilon_3 + \varepsilon_4)} + \frac{\partial \phi(\omega, k, p, q)}{\partial \omega} \right)_{\omega = \varepsilon_3 + \varepsilon_4}, \tag{14.2}$$

one can see that Wigner's delay time combines the collision delay Δ_t with the time delay
the particle gains due to nonlocal matching in the displaced positions $\Delta_{2,3,4}$. Briefly,
the collision delay Δ_t is defined in the same way as Wigner's delay time Δ_t^W except
for the reference event which is nonlocal while Wigner's time uses the point-like event.
The dependence of the delay time on the reference event has been demonstrated in
Chapters 3.1.2–3.1.3 for sticky hard spheres and dumbbell effective molecules.

Let us return to the original question of how the quantum delay time behaves in the classical limit. The link between Wigner's delay time and the classical delay time defined by the geometrical considerations has been studied at the beginning of 1980s (Bollé, 1981; Martin, 1981; Thirring, 1981). Bollé has shown that the classical geometrical delay time is a direct classical counterpart of Wigner's delay time, in particular, the classical delay time can be defined in terms of a classical T-matrix. Thirring even turns the question around and shows that the classical approach to the delay time can be used to define the classical counterpart of such an exclusively quantum-mechanical concept like the scattering phase shifts. The delay time is then given by the energy derivative of this classical scattering phase shift. A detailed analyses of the time delay and its link to the virial of force has been done by Martin. All these studies thus show that one can conveniently think of the delay time Δ_t in terms of its intuitively more clear classical counterpart.[1]

14.1.2   Hard-sphere displacement

Now we check that space displacements $\Delta_{2,3,4}$ in the proper limit reduce to the known displacement of classical hard spheres. To this end we need the classical limit, $\hbar \to 0$, of the hard-sphere phase shift

$$\phi = \pi - |q| \frac{1}{2}(d_a + d_b),$$ (14.3)

where d_a and d_b are diameters of the spheres. Note that this phase shift depends exclusively on the transferred momentum q. Since the phase shift does not depend on the energy, the collision is instant, $\Delta_t = 0$, as one expects. At the same time, a majority of other Δ's vanish too, $\Delta_3 = 0$, $\Delta_E = 0$ and $\Delta_K = 0$. The only non-trivial displacements read

$$\Delta_2 = \Delta_4 = \frac{q}{|q|} \frac{1}{2}(d_a + d_b).$$ (14.4)

These displacements are classical nonlocal corrections of Enskog's theory. For hard spheres, the classical limit of quantum nonlocal corrections thus results in the expected classical values.

---

[1] Beside Wigner's delay time there are number of other quantum-mechanical concepts of the collision duration. A review of these concepts for a one-dimensional system has been done by Hauge and Støvneng (1989), the three-dimensional systems are discussed by Iannaccone and Pellegrini (1994). Comparison shows that these different concepts are not, in general, equivalent.

The reason for this number of alternative concepts follows from the fact that a physical interpretation of the collision delay is not obvious even for a one-dimensional system with such a simple scattering centre as the rectangular barrier. Since the time appears only as a parameter in the Schrödinger equation and there is no time operator, there are still problems with an interpretation of characteristic times even if the full knowledge of the wave function is at hand. It is based on the impossibility to define the time operator noticed by Pauli (1958). The time is canonical conjugate of the Hamiltonian. The commutator of the time operator with the Hamiltonian thus has to equal $i\hbar$ to yield the standard uncertainty relation $\Delta t \Delta E \sim \hbar$. This relation contradicts the semi-bounded spectrum of Hamiltonian.

Let us also check the energy gain for the model of breathing hard spheres discussed in chapter 3.3. To this end we assume that the hard-sphere diameters depend on time, $d_a(t)$ and $d_b(t)$. The classical limit of the phase shift,

$$\phi = \pi - |q| \frac{1}{2} \left(d_a(t) + d_b(t) \right),$$

(14.5)

is then time dependent which leads to a non-zero energy gain, see (13.3),

$$\Delta_E = \frac{1}{4} |q| \left(\frac{\partial d_a(t)}{\partial t} + \frac{\partial d_b(t)}{\partial t} \right).$$

(14.6)

This energy gain also agrees with the expectation (3.27) from classical mechanics.

The collision delay and the hard-sphere displacement show that the similarity of quantum-mechanical virial corrections with classical virial corrections is not accidental. The nonlocalities given by derivatives of the scattering phase shift are the quantum-mechanical generalisation of the classical concept.

14.2 Invariances of the nonlocal scattering integral

In general, the nonlocal corrections obey point and translational symmetries of the binary collision and general invariances. The point and translational symmetries might be disturbed by medium effects on the collision, since the medium is anisotropic, inhomogeneous, and non-stationary in general. In contrast, the invariances hold always.

Here we discuss only two of many general invariances:

1. the freedom to choose the initial of the single-particle energy which belongs to the gauge invariance, and

2. the invariance under different choices of the inertial network. These two cases are of practical importance. The gauge can be used to remove the effect of the external field on the collision. The Galilean transformation is needed to reach the barycentric framework in which the scattering theory is traditionally formulated.

14.2.1 Gauge invariance

Let us extend the Hamiltonian by gauge potentials $U_a(t)$ which are independent from the space variables. Accordingly, these potentials have no effect on physical processes since they do not cause any forces,

$$\frac{\partial U_a}{\partial r} = 0.$$

(14.7)

We denote by a tilde those quantities (functions) which are evaluated in the presence of the gauge fields. The functions without tilde correspond to the original system in the absence of fields. The gauge potentials lift energy bottoms, therefore the bare-particle energy changes as

$$\epsilon_a(k, r, t) = \frac{k^2}{2m} \qquad \rightarrow \qquad \tilde{\epsilon}_a(k, r, t) = \frac{k^2}{2m} + U_a(t), \tag{14.8}$$

and quasiparticle energy transforms as

$$\tilde{\varepsilon}_a(k, r, t) = \varepsilon_a(k, r, t) + U_a(t). \tag{14.9}$$

The T-matrix depends on U's via single-particle energies (14.8) which appear exclusively in the combination with independent energies ω and E as $\omega + E - \epsilon_a - \epsilon_b$. The effect of the gauge field thus emerges via the energy argument,

$$\tilde{t}_{\text{sc}}^R(\omega + E, k, p, q, r, t) = t_{\text{sc}}^R(\omega + E - U_a(t) - U_b(t), k, p, q, r, t). \tag{14.10}$$

Apparently, the on-shell values of the T-matrix are gauge invariant

$$\tilde{t}_{\text{sc}}^R(\tilde{\varepsilon}_a + \tilde{\varepsilon}_b, k, p, q, r, t) = t_{\text{sc}}^R(\varepsilon_a + \varepsilon_b, k, p, q, r, t). \tag{14.11}$$

The gauge does not thus modify the differential cross section.

From definition (13.3) of Δ's one finds that the $\tilde{\Delta}$'s are identical to the Δ's except for the energy gain which includes the time derivative. Due to the time dependence in the energy argument of $\tilde{t}_{\text{sc}}^R = |\tilde{t}_{\text{sc}}| e^{i\tilde{\phi}}$, the energy gain transforms as

$$\tilde{\Delta}_E = -\frac{1}{2}\frac{\partial \tilde{\phi}}{\partial t} = -\frac{1}{2}\frac{\partial \phi}{\partial t} + \frac{1}{2}\frac{\partial \phi}{\partial \omega}\frac{\partial U_a}{\partial t} + \frac{1}{2}\frac{\partial \phi}{\partial E}\frac{\partial U_b}{\partial t} = \Delta_E + \frac{1}{2}\left(\frac{\partial U_a}{\partial t} + \frac{\partial U_b}{\partial t}\right)\Delta_t. \tag{14.12}$$

The transformation of the energy gain enters the energy-conserving δ function. With the help of (14.9) and (14.12) one can check the gauge invariance of the energy conservation,

$$\tilde{\varepsilon}_1(t) + \tilde{\varepsilon}_2(t) - \tilde{\varepsilon}_3(t - \Delta_t) - \tilde{\varepsilon}_4(t - \Delta_t) - 2\tilde{\Delta}_E$$

$$= \varepsilon_1(t) + U_a(t) + \varepsilon_2(t) + U_b(t) - \varepsilon_3(t - \Delta_t) - U_a(t - \Delta_t) - \varepsilon_4(t - \Delta_t) - U_b(t - \Delta_t)$$

$$- 2\Delta_E - \left(\frac{\partial U_a}{\partial t} + \frac{\partial U_b}{\partial t}\right)\Delta_t$$

$$= \varepsilon_1(t) + \varepsilon_2(t) - \varepsilon_3(t - \Delta_t) - \varepsilon_4(t - \Delta_t) - 2\Delta_E. \tag{14.13}$$

The first and the last lines are formally identical, i.e. the energy-conserving δ function is gauge invariant.

From (14.13) one can see that the energy gain is a necessary ingredient of the energy conservation formulated for the matching of energies at two different times, t and $t - \Delta_t$.

For a collisions of a finite duration, the energy gain is required by the gauge invariance of the theory.

14.2.2   Galilean invariance

Now, variables with a tilde denote the system that runs with velocity u with respect to the laboratory framework while the untilded the one is in the rest. Both systems are identical except for the velocity, therefore any physical quantity of running the system can be expressed in terms of the rest one,

$$\tilde{B}(x_1, t_1, x_2, t_2) = B(x_1 - ut_1, t_1, x_2 - ut_2, t_2). \tag{14.14}$$

In the Wigner representation this relation reads

$$\tilde{b}(\omega, k, r, t) = \int dy d\tau e^{-iky + i\omega\tau} \tilde{B}\left(r + \frac{y}{2}, t + \frac{\tau}{2}, r - \frac{y}{2}, t - \frac{\tau}{2}\right)$$

$$= \int dy d\tau e^{-iky + i\omega\tau} B\left(r + \frac{y}{2} - u\left(t + \frac{\tau}{2}\right), t + \frac{\tau}{2}, r - \frac{y}{2} - u\left(t - \frac{\tau}{2}\right), t - \frac{\tau}{2}\right)$$

$$= b(\omega - uk, k, r - ut, t). \tag{14.15}$$

Firstly we apply the transformation (14.15) to the condition for the quasiparticle energy, $\omega = \frac{k^2}{2m} + \tilde{\sigma}(\omega, k, r, t)$, the solution of which provides

$$\tilde{\varepsilon}_a(k, \tilde{r}, t) = \varepsilon_a(k, r - ut, t) + uk. \tag{14.16}$$

Now we apply the transformation (14.15) to the T-matrix. Since the relative momentum $k - p$ and transferred momentum q do not depend on the velocity of the framework, the transformation applies to the sum momentum $k + p$ only. The T-matrix thus transforms as

$$\tilde{t}^R_{sc}(\omega + E, k, p, q, r, t) = t^R_{sc}(\omega + E - u(k + p), k, p, q, r - ut, t). \tag{14.17}$$

The on-shell values of the T-matrix in both systems are equal, therefore the differential cross section is Galilean invariant.

From (14.17) one can see that derivatives with respect to momenta k and p and the time are modified by the transformation. The transformation of Δ's follows from their definitions (13.3) as

$$\tilde{\Delta}_E = \Delta_E + u\Delta_K, \qquad \tilde{\Delta}_{3,4} = \Delta_{3,4} + u\Delta_t, \tag{14.18}$$

the other Δ's remain unchanged.

The displacements $\Delta_{3,4}$ of the final states are shifted corresponding to the transformation of the flight of the molecule from the running system to the rest one. The momentum corrections Δ_K contributes to the energy gain because the sum momentum of the colliding pair has increased. Substituting into the tilded energy-conserving δ function, one finds that it is identical to the untilded one. The energy-conserving δ function is thus a Galilean invariant.

A practical use of transformations (14.12) and (14.18) is demonstrated in the next subsection. Moreover, these transformations provide vital constrains for eventual *ad hoc* or fitted sets of Δ's. Approximations of Δ's violating (14.12) and (14.18) will likely violate the conservation of energy and momentum.

14.3 Quantum Δ's for isolated particles

In dilute systems, the in-medium effects vanish so that the binary collisions reveal all symmetries following from the spherical potential. As in the classical case, these symmetries and gauge and Galilean transformations allow us to parametrise all the Δ's (four vectors and two scalars) with three scalar parameters, see Chapter 3.

Let us assume that the genuine in-medium effects can be neglected and the colliding particles feel other particles only via mean fields. Under this assumption, one can express the full set of Δ's from the collision delay Δ_t, the mean inter-particle distance d and the rotation angle ϑ in exactly the same way as in the classical system described in Chapter 3. Here we recover these relations for the quantum mechanical non-localities and provide parameters Δ_t, d and ϑ in terms of the barycentric phase shifts.

14.3.1   Gauge transformation to a system free of mean field

We denote by $U_a(r, t)$ the mean field acting on particle a. In the mean-field approximation, the quasiparticle energy reads

$$\varepsilon_a(k, r, t) = \frac{k^2}{2m} + U_a(r, t). \tag{14.19}$$

As the gauge field, the mean field enters the T-matrix exclusively in combinations $\omega + E - \frac{k'^2}{2m} - U_a(r, t) - \frac{p'^2}{2m} - U_b(r, t)$, therefore its effect can be collected in the energy argument

$$t^R_{\text{sc}}(\omega + E, k, p, q, t, r) = \bar{t}^R_{\text{sc}}(\bar{\omega}, k, p, q), \tag{14.20}$$

where

$$\bar{\omega} = E + \omega - U_a(r, t) - U_b(r, t) \tag{14.21}$$

and $\bar{t}^R_{\text{sc}}$ is evaluated in the absence of the field.[2] The T-matrix depends on r and t only via $\bar{\omega}$.

---

[2] In the presence of in-medium effects $\bar{t}^R_{\text{sc}}$ depends on r and t also via the Pauli blocking of internal states which does not vanish when the local energy bottom is rescaled to zero. Compare (14.20) with (14.10).

From the field-free T-matrix $\bar{t}_{sc}^{R}$ we can readily evaluate the energy and momentum gains (13.3)

$$2\Delta_E = \left(\frac{\partial U_a}{\partial t} + \frac{\partial U_b}{\partial t}\right)\Delta_t, \qquad 2\Delta_K = -\left(\frac{\partial U_a}{\partial r} + \frac{\partial U_b}{\partial r}\right)\Delta_t. \tag{14.22}$$

They relate to the delay time Δ_t in the same way as in the classical cases, (3.20) and (3.16).

The field-free T-matrix is the T-matrix of perfectly isolated binary collisions which has been discussed in many details in the extensive theory of collisions. To reduce notational problems, we refer exclusively to the book by Goldberger and Watson (1964).

14.3.2 Galilean transformation to the barycentric coordinate framework

By the gauge transformation we have removed the space dependence of the T-matrix so that the translation invariance of the system has been restored. This allows us to employ the Galilean transformation with the help of which we eliminate the centre-of-mass motion. To this end we transform the system into the barycentric coordinate framework.

It is advantageous to use the centre-of-mass and the relative coordinates. The sum momentum K and the relative momentum κ, and corresponding total mass M and relative mass μ are defined as

$$K = k + p, \qquad \kappa = \frac{1}{2}(k - p), \qquad M = 2m, \qquad \mu = \frac{1}{2}m. \tag{14.23}$$

The sum of particle energies,

$$\frac{k^2}{2m} + \frac{p^2}{2m} = \frac{\kappa^2}{2\mu} + \frac{K^2}{2M}, \tag{14.24}$$

plays a part due to the relative motion and the contribution of the centre-of-mass motion.

During the interaction the sum momentum K conserves. Since the single-particle energies enter the T-matrix $\bar{t}_{sc}^{R}$ exclusively in the form $\bar{\omega} - \frac{k'^2}{2m} - \frac{p'^2}{2m} = \bar{\omega} - \frac{\kappa'^2}{2\mu} - \frac{K^2}{2M}$, it is possible to include the centre-of-mass kinetic energy $\frac{K^2}{2M}$ into the independent energy,

$$\hat{\omega} = \bar{\omega} - \frac{K^2}{2M}. \tag{14.25}$$

This substitution represents the transformation into the barycentric coordinate framework where the sum momentum K equals zero. The field-free T-matrix $\bar{t}_{sc}^{R}$ relates to the barycentric T-matrix $\tilde{t}_{sc}^{R}$ as

$$\bar{t}_{sc}^{R}(\bar{\omega}, k, p, q) = \tilde{t}_{sc}^{R}\left(\hat{\omega}, \kappa, -\kappa, q\right). \tag{14.26}$$

One of the momentum arguments in $\bar{t}^R_{sc}$ is redundant. To reduce the argument and make a link with the notation in (Goldberger and Watson, 1964), we introduce the final relative momentum

$$\kappa_f = \kappa - q \tag{14.27}$$

in terms of which the barycentric T-matrix reads

$$\bar{t}^R_{sc}(\bar{\omega}, k, p, q) = \hat{t}^R_{sc}\left(\hat{\omega}, \kappa, \kappa_f\right). \tag{14.28}$$

From (14.28) and (13.3) one can see that the laboratory and the barycentric delay times equal $[\hat{\phi} = \mathrm{Im}\ln \hat{t}^R_{sc}]$

$$\Delta_t = \frac{\partial \phi}{\partial \omega} = \frac{\partial \hat{\phi}}{\partial \omega} = \frac{\partial \hat{\phi}}{\partial \hat{\omega}}\frac{\partial \hat{\omega}}{\partial \omega} = \frac{\partial \hat{\phi}}{\partial \hat{\omega}} = \hat{\Delta}_t. \tag{14.29}$$

The space displacements (13.3) in terms of the barycentric phase shift read

$$\Delta_2 = -\frac{\partial \hat{\phi}}{\partial \kappa},$$
$$\Delta_{3/4} = \frac{K}{M}\Delta_t - \frac{1}{2}\frac{\partial \hat{\phi}}{\partial \kappa} \mp \frac{1}{2}\frac{\partial \hat{\phi}}{\partial \kappa_f}. \tag{14.30}$$

This transformation combines the Galilean transformation (14.18) for the centre-of-mass velocity $u = \frac{K}{M}$ with the substitution of variables (14.27).

From (14.30) one can see that the space displacements obey the relation

$$\Delta_4 - \Delta_2 + \Delta_3 = 2\frac{K}{M}\Delta_t, \tag{14.31}$$

which is the counterpart of (3.12). According to (14.31), only two independent space displacements are needed. We use

$$\hat{\Delta}_{3/4} = -\frac{1}{2}\frac{\partial \hat{\phi}}{\partial \kappa} \mp \frac{1}{2}\frac{\partial \hat{\phi}}{\partial \kappa_f}, \tag{14.32}$$

which are even/odd with respect to the interchange of the initial and final states.

14.3.3   Rotational symmetry

Before transformations, two vectors determine the special directions in the system: the gradient of the mean-field and the centre-of-mass velocity. By the gauge and Galilean transformation both vectors have been removed so that the full rotational symmetry has been restored. From the rotational symmetry we will show that three scalars are sufficient to parametrise the barycentric displacements $\hat{\Delta}_{3,4}$. These three parameters will be then further reduced to two with the help of the time-reversal symmetry and the energy conservation.

The rotational symmetry restricts the ways in which the barycentric T-matrix depends on its arguments,

$$\hat{t}^R_{\text{sc}}(\hat{\omega}, \kappa, \kappa_{\text{f}}) = \hat{t}^R_{\text{sc}}(\hat{\omega}, \cos\theta, |\kappa|, |\kappa_{\text{f}}|), \qquad (14.33)$$

see (Goldberger and Watson, 1964). It depends only on the amplitude of the initial momentum $|\kappa|$, the amplitude of the final momentum $|\kappa_{\text{f}}|$, and the deflection angle θ given by

$$\cos\theta = \frac{\kappa\kappa_{\text{f}}}{|\kappa||\kappa_{\text{f}}|}. \qquad (14.34)$$

Having only three independent variables, the derivatives of the T-matrix can provide only three independent scalar coefficients.

Restriction (14.33) is sufficient to prove that the barycentric displacements $\hat{\Delta}_3$ and $\hat{\Delta}_4$ are a linear combination of the vectors κ and κ_{f}. Indeed, writing the derivatives as composed ones, all directions are given by

$$\frac{\partial|\kappa|}{\partial\kappa} = \frac{\kappa}{|\kappa|}, \qquad\qquad \frac{\partial\cos\theta}{\partial\kappa} = \frac{\kappa_{\text{f}}}{|\kappa||\kappa_{\text{f}}|} - \frac{\kappa\cos\theta}{|\kappa|^2},$$

$$\frac{\partial|\kappa_{\text{f}}|}{\partial\kappa_{\text{f}}} = \frac{\kappa_{\text{f}}}{|\kappa_{\text{f}}|}, \qquad\qquad \frac{\partial\cos\theta}{\partial\kappa_{\text{f}}} = \frac{\kappa}{|\kappa||\kappa_{\text{f}}|} - \frac{\kappa_{\text{f}}\cos\theta}{|\kappa_{\text{f}}|^2}. \qquad (14.35)$$

Accordingly, the displacements $\hat{\Delta}_{3,4}$ have to be a linear combination of κ and κ_{f}. This condition tells us that the displacements have to be in the collision plane as one expects from the angular-momentum conservation.

14.3.4 Energy conservation and time-reversal symmetry

Now we use the time-reversal symmetry and the energy conservation $|\kappa_{\text{f}}| = |\kappa|$, to show that two linear coefficients are sufficient. The time-reversal symmetry has been restored when we have eliminated the mean fields.

According to the time-reversal symmetry, the T-matrix is symmetric with respect to amplitudes of initial and final momenta[3]

$$\hat{t}^R_{\text{sc}}(\hat{\omega}, \cos\theta, |\kappa|, |\kappa_{\text{f}}|) = \hat{t}^R_{\text{sc}}(\hat{\omega}, \cos\theta, |\kappa_{\text{f}}|, |\kappa|). \qquad (14.36)$$

---

[3] This simple form of the time-reversal symmetry appears only for the T-matrix with independent energy. For T-matrices on the energy shell one has to be more careful. It is customary to use the scattering T-matrix $\mathcal{T}^{(+)}$ in which the energy equals the energy in the initial state, see Chapter 5.6 of (Goldberger and Watson, 1964), as it follows from the Lippmann–Schwinger equation. The time-reversal operation then interchanges with the initial and final energies so that one arrives at another form of the T-matrix $\mathcal{T}^{(-)}$. Although their on-shell values are identical, their derivatives are distinct.

The displacements are evaluated at the energy shell, where (after taking derivatives) the amplitudes of the initial and the final momenta equal each other, $|\kappa_f| = |\kappa|$. From the symmetry (14.36) then follows

$$\left.\frac{\partial \hat{\phi}}{\partial |\kappa_f|}\right|_{|\kappa_f|=|\kappa|} = \left.\frac{\partial \hat{\phi}}{\partial |\kappa|}\right|_{|\kappa_f|=|\kappa|} = \frac{1}{2}\frac{\partial \hat{\phi}'}{\partial |\kappa|}, \tag{14.37}$$

where

$$\hat{\phi}'(\cos\theta, |\kappa|) = \hat{\phi}(\cos\theta, |\kappa|, |\kappa|). \tag{14.38}$$

Accordingly, only one of the derivatives with respect to momentum amplitudes is independent.

On the energy shell, the derivatives of the deflection angle can also be written in a comprehensive form

$$\left.\left(\frac{\partial}{\partial \kappa} - \frac{\partial}{\partial \kappa_f}\right)\cos\theta\right|_{|\kappa_f|=|\kappa|} = (\kappa_f - \kappa)\frac{\cos\theta + 1}{|\kappa|^2},$$

$$\left.\left(\frac{\partial}{\partial \kappa} + \frac{\partial}{\partial \kappa_f}\right)\cos\theta\right|_{|\kappa_f|=|\kappa|} = (\kappa_f + \kappa)\frac{\cos\theta - 1}{|\kappa|^2}. \tag{14.39}$$

Using (14.35), (14.37) and (14.39) in formulae (14.32), one finds the barycentric space displacements to be

$$\hat{\Delta}_3 = -\frac{\kappa_f + \kappa}{2}\left[\frac{\cos\theta - 1}{|\kappa|^2}\frac{\partial \hat{\phi}'}{\partial \cos\theta} + \frac{1}{|\kappa|}\frac{\partial \hat{\phi}'}{\partial |\kappa|}\right], \tag{14.40}$$

$$\hat{\Delta}_4 = -\frac{\kappa_f - \kappa}{2}\left[\frac{\cos\theta + 1}{|\kappa|^2}\frac{\partial \hat{\phi}'}{\partial \cos\theta} - \frac{1}{|\kappa|}\frac{\partial \hat{\phi}'}{\partial |\kappa|}\right]. \tag{14.41}$$

14.3.5   Classical-like parametrisation

One can see that all Δ's can be parametrised by three scalar functions, Δ_t, $\frac{\partial \hat{\phi}'}{\partial \cos\theta}$ and $\frac{\partial \hat{\phi}'}{\partial |\kappa|}$. For implementations, it is advantageous to link the derived parameters with those introduced within the classical mechanics, the mean distance of particles d and the rotation angle ϑ, see Figs. 3.4 and 3.5 and also Formulae (3.8) and (3.9).

The quantum displacements $\hat{\Delta}_{3,4}$ can be identified with their classical counterparts according to their directions. The barycentric displacement $\hat{\Delta}_4$ points into the direction $\kappa - \kappa_f = q$, therefore it is a quantum counterpart of the Enskog-type displacement Δ^d, see (3.8). The barycentric displacement $\hat{\Delta}_3$ points in the direction $\kappa + \kappa_f = 2\kappa - q$ which is perpendicular to q (on the energy shell $q(2\kappa - q) = \kappa^2 - (\kappa - q)^2 = 0$). Accordingly, the $\hat{\Delta}_4$ is a quantum counterpart of the rotation displacement Δ^ϑ introduced in (3.9).

The effective inter-particle distance,

$$d = \sqrt{|\hat{\Delta}_3|^2 + |\hat{\Delta}_4|^2}, \tag{14.42}$$

and rotation angle,

$$\vartheta = 2\arctan\frac{|\hat{\Delta}_3|}{|\hat{\Delta}_4|}, \tag{14.43}$$

parametrise the quantum displacements,

$$\Delta^{\mathrm{d}} = \hat{\Delta}_4 = \frac{\kappa - \kappa_{\mathrm{f}}}{|\kappa - \kappa_{\mathrm{f}}|} d \cos\frac{\vartheta}{2},$$

$$\Delta^{\vartheta} = \hat{\Delta}_3 = \frac{\kappa + \kappa_{\mathrm{f}}}{|\kappa + \kappa_{\mathrm{f}}|} d \sin\frac{\vartheta}{2}, \tag{14.44}$$

as in the classical case, see Fig. 3.4. With the quantum values of the parameters, one can describe the dilute quantum system in the same manner as the classical one.

For implementations, it is an important question under which conditions the above simple free-space form of nonlocal corrections might be a reasonable approximation in dense systems. As already mentioned, in dense matter the anisotropy of medium disturbs the rotational symmetry. There is a common approximation of the medium effect on collisions which neglects the anisotropy, called 'the angle-averaged Pauli blocking', see e.g. (Beyer, Röpke and Sedrakian, 1996). Within this approximation, the phase shift has the symmetry of the isotropic space so that all displacements have to stay in the collision plane. This approximation does not suppress the time and space dependence, therefore the energy and momentum gains are non-zero. Accordingly, $|\kappa_{\mathrm{f}}| \neq |\kappa|$ and the time-reversal symmetry does not apply to this case. The free-space form of displacements thus also implies a neglect of the space and time gradient and corresponding gains.

14.3.6 Representation in partial waves

In spite of their classical interpretation, the virial parameters are quantities derived from a quantum treatment of the binary collision. To this end, one has to evaluate the T-matrix which is commonly done in the representation of partial waves,

$$T = \hat{t}_{\mathrm{sc}}^{R}(\hat{\omega}, \cos\theta, |\kappa|, |\kappa|) = \frac{1}{16\pi} \sum_{l} P_l(\cos\theta) T_l(\hat{\omega}, |\kappa|), \tag{14.45}$$

where P_l is a Legendre polynomial. The coefficients T_l are on the shell $|\kappa_f| = |\kappa|$.

The interaction forces, e.g. among nucleons, might not be perfectly central but might include tensor forces. To circumvent this complication, we approximate the partial

T-matrices by the central-like part given by the trace over singlet and triplet channels (Goldberger and Watson, 1964),

$$T_l = \sum_{I,S=0,1} \sum_{\mathcal{J}=|S-l|}^{S+l} (2\mathcal{J}+1)(2I+1) T_{l,l}^{S,I\mathcal{J}}, \tag{14.46}$$

with the total spin S, isospin I and the total angular momentum $\mathcal{J}$. Due to the summation over projection components m_l, m_s only the diagonal elements in l contribute. The substitution of decomposition (14.45) into (14.29) and (14.40)–(14.41) yields

$$\Delta_t = \mathrm{Im} \frac{1}{T} \sum_l P_l \frac{\partial T_l}{\partial \hat{\omega}},$$

$$\Delta^{\mathrm{d}} = \frac{\sqrt{1-\cos\theta}}{\sqrt{2}|\kappa|} \, \mathrm{Im} \frac{1}{T} \sum_l P_l' T_l (1+\cos\theta) - |\kappa| P_l T_l',$$

$$\Delta^{\vartheta} = \frac{\sqrt{1+\cos\theta}}{\sqrt{2}|\kappa|} \, \mathrm{Im} \frac{1}{T} \sum_l P_l' T_l (1-\cos\theta) + |\kappa| P_l T_l', \tag{14.47}$$

where $P_l'(z) = \frac{\partial}{\partial z} P_l$ is the derivative of the Legendre polynomial and $T_l' = \frac{1}{2} \frac{\partial}{\partial |\kappa|} T_l(|\kappa|)$, where the factor $\frac{1}{2}$ accounts for the fact that on the shell, $|\kappa_f| = |\kappa|$, the derivative with respect to $|\kappa|$ also affects the final momentum.

14.3.7   Numerical results for nuclear matter

The virial parameters for isolated collisions obtained in (Morawetz, Lipavský, Špička and Kwong, 1999b) from (14.47) are shown in Figs. 14.1–14.3. The sum over partial waves is terminated above D-waves but the coupled channels, $^3P_2-{}^3F_2$ and $^3D_3-{}^3G_3$, are included along with $^3S_1-{}^3D_1$. Five approximations of the nucleon-nucleon potential, a set (A–C) of one-Boson-exchange Bonn potentials (Machleidt, 1989; Machleidt, 1993), the Paris potential (Lacombe, Loiseau, Richard, Mau, Côté, Pirès and de Tourreil, 1980) and the separable Paris potential (Heidenberger and Plessas, 1984), are compared.

In the collision delay, Fig. 14.1, one can see three features: (i) at low energies the collision delay reaches large negative values for all deflection angles, (ii) at higher energies the collision delay strongly depends on the deflection angle, and (iii) a singularity at $\theta = 90°$ and energy 80 MeV. Features (ii) and (iii) also appear in Enskog's displacement Fig. 14.2 and the rotation displacement Fig. 14.3.

The large negative values (i) of the collision delay at low energies follow from general properties of the T-matrix. The real part, $\mathrm{Re}\,T$, is regular for $\Omega \to 0$ while the energy dependence of the imaginary part, $\mathrm{Im}\,T$, is proportional to the density of states, $\mathrm{Im}\,T \propto \sqrt{\Omega}$. At low energies, the collision delay hence behaves as $\Delta_t \approx \frac{1}{\mathrm{Re}\,T} \frac{\partial \mathrm{Im}\,T}{\partial \Omega} \propto \frac{1}{\sqrt{\Omega}}$. Using the on-shell condition $\Omega = \frac{|\kappa|^2}{2\mu}$, one can express this singularity as $\Delta_t \propto \frac{1}{|\kappa|}$.

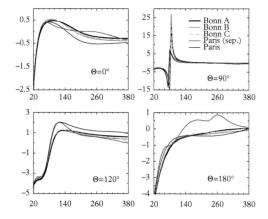

Figure 14.1: *The collision delay* Δ_t *as a function of the deflection angle* θ *and the kinetic energy* $\kappa^2/2\mu$. *A* $\frac{1}{x}$-*singularity at* $\theta = 90°$ *and energies of about 80 MeV, appear for processes of a vanishing scattering rate. All potentials provide nearly identical results, except for the back scattering,* $\theta = 180°$, *where the separable Paris potential yields appreciably different results from the others.*

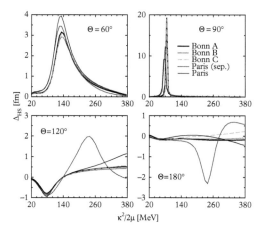

Figure 14.2: *Enskog's displacement* Δ^d *for conditions is identical to Fig. 14.1. The forward scattering is not included because* $\Delta^d = 0$ *for* $\theta = 0°$. *As in Fig. 14.1, processes corresponding to the singularity at* $\theta = 90°$ *and* $\kappa^2/2\mu = 80$ *MeV have a vanishing scattering rate. The separable approximation leads to deviations for large deflection angles* $\theta = 120°$ *and* $\theta = 180°$.

Such a singularity is not dangerous for the kinetic equation since the mean time between collisions also scales with $\frac{1}{|\kappa|}$ being inversely proportional to the velocity.

Enskog's displacement Δ^d shown in Fig. 14.2 and the rotation displacement Δ^φ shown in Fig. 14.3 are regular at low energies going to zero for $|\kappa| \to 0$. This results from the dominant contribution of the S wave in this region. When the S wave dominates, the displacements simplify as $\Delta^\varphi = -\Delta^d = \frac{1}{2} \sin \frac{\theta}{2} \operatorname{Im} \frac{1}{T_0} \frac{\partial T_0}{\partial |\kappa|}$. At small momenta, the T-matrix depends on $|\kappa|^2$, therefore $\frac{\partial T_0}{\partial |\kappa|} \propto |\kappa|$ and vanishes for $|\kappa| \to 0$.

For each of the presented deflection angles (ii), the collision delay has different energy dependence. For the back scattering, $\theta = 180°$, the collision delay monotonically increases with energy. In the forward scattering $\theta = 0°$, there is a single maximum at 100 MeV. In the perpendicular scattering, $\theta = 90°$, appears a $\frac{1}{x}$-singularity at 80 MeV.

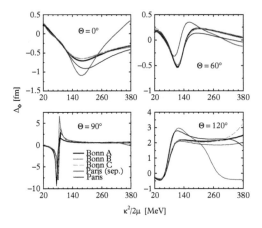

Figure 14.3: *Rotation displacement* Δ^ϑ *for conditions identical to Fig. 14.1. The back scattering is not included because $\Delta^\vartheta = 0$ for $\theta = 180°$. Note the similarities with Fig. 14.1 for deflection angles $\theta = 90°$ and $\theta = 120°$.*

Finally, the scattering at $\theta = 120°$ has a step at 30 MeV which reflects a smoothed $\frac{1}{x}$-singularity at 80 MeV. The angular dependence shows that an interference between S, P, D waves is important. At $\theta = 0°$, $P_{0,1,2} = 1$, therefore $T = (T_0 + T_1 + T_2)/16\pi$. At $\theta = 180°$, $P_{0,2} = 1$ and $P_1 = -1$, therefore $T = (T_0 - T_1 + T_2)/16\pi$. The difference between the collision delay at $\theta = 0°$ and $\theta = 180°$ thus follows from a different interference of the P wave from the other waves. The angular dependence reduces at energies below 30 MeV. For this energy, the de Broglie wave-length $\frac{\hbar}{|\kappa|} \sim 1$ fm is comparable to the range of the interaction potential. With increasing wave length, the interaction potential behaves more and more as a contact potential, therefore the S component of the scattered wave dominates at all angles.

In the singularities (iii) at 80 MeV and $\theta = 90°$, the collision delay and displacements reach values comparable to or exceeding characteristic time and space scales of heavy-ion reactions. For such values of nonlocal corrections the kinetic equation based on the quasiclassical approximation is not justified. In fact, these singularities do not restrict the validity of the kinetic equations, because they appear when the differential cross section vanishes. For $\theta = 90°$, $P_0 = 1$, $P_1 = 0$ and $P_2 = -\frac{1}{2}$, therefore $T = (T_0 - \frac{1}{2}T_2)/16\pi$. At the energy of 80 MeV the S and D waves destructively interfere, $T_2 \to 2T_0$, so that $T \to 0$ while its derivatives remain finite, e.g. $\frac{\partial T}{\partial \Omega} \neq 0$. This causes singularities seen at $\theta = 90°$ and energy 80 MeV in all Δ's. In kinetic equation (13.28), all Δ's are weighted with the cross section as $|T|^2\Delta$. At the point of singularity, $|T|^2\Delta \to 0$. Except for the low-energy region, the rotation displacement has a number of features similar to the collision duration, at least for $\theta = 90°$ and $\theta = 120°$. For the perpendicular scattering, the $1/x$-discontinuity appears at 80 MeV. The step for $\theta = 120°$ at 40 MeV is also similar, including the shoulder at 120 MeV. The velocity of particles in the barycentric framework for 80 MeV is about half of the velocity of light, which crudely corresponds to the coefficient with which the discontinuity of the collision duration scales on the discontinuity of the rotational displacement.

14.4 Summary

At this point we close our discussion of the microscopic mechanism of virial corrections in the kinetic equation. We have seen that, with respect to the microscopic dynamics of collisions and its link to the kinetic equation, one can rely on the classical picture of the collision and corresponding virial corrections to the Boltzmann equation. In particular, the collision delay Δ_t and the displacements $\Delta_{2,3,4}$ are quantum generalisations of their classical counterparts. Eventual complexity of virial corrections does not follow from the quantum mechanics but from the medium effect on the binary collision.

15

Nonequilibrium Quantum Hydrodynamics

15.1 Local conservation laws

We have seen how and what kinetic equation follows from the quantum statistics. Now we derive the consequences of the nonlocal kinetic equation for the thermodynamic properties. Far from equilibrium, the traditional thermodynamic quantities such as the temperature and chemical potential do not capture the time-dependent properties of the system. Accordingly, we want to express the thermodynamic observables as functionals of the time-dependent quasiparticle distribution. Therefore we will multiply the nonlocal kinetic equation (13.28) with a variable $\xi_1 = \delta_{ab}, \mathbf{k}, \varepsilon_1, -k_B \ln[f_1/(1 - f_1)]$, integrate over momentum and sum over species. It results in the equation of continuity, the Navier–Stokes equation, the energy balance and the evolution of the entropy, respectively. All these conservation laws or balance equations for the mean thermodynamic observables

$$\langle \xi \rangle = \int \frac{d^3 k}{(2\pi)^3} \xi_1 f_1 \tag{15.1}$$

will have the form

$$\frac{\partial \langle \xi \rangle}{\partial t} + \frac{\partial}{\partial \mathbf{r}} \mathbf{j}_\xi = \mathscr{I}_{\text{gain}}. \tag{15.2}$$

The additional gain on the right side might be due to an energy or force feed from the outside or the entropy production by collisions. The external potential is absorbed here in the quasiparticle energy and we can concentrate on the internal contributions due to correlations. Then we will obtain a gain only for the entropy while for density, momentum and energy the time change of the density $\langle \xi \rangle$ is exclusively caused by the divergence of the current $\mathbf{j}_\xi$.

We will show from the balances of the kinetic equation that the particle density, momentum flux (pressure), energy and entropy density consist of a quasiparticle part and a correlated contribution $\langle \xi \rangle = \xi^{\text{qp}} + \xi^{\text{mol}}$, respectively. The latter one takes the form of

Interacting Systems far from Equilibrium. Klaus Morawetz, Oxford University Press (2018).
© Klaus Morawetz. DOI: 10.1093/oso/9780198797241.001.0001

a molecular contribution as if two particles form a molecule for a short time, proving the conservation laws (15.2) and showing that also the currents consisting of $j_\xi = j_\xi^{qp} + j_\xi^{mol}$ will be the ultimate goal to convince us about the consistency of the nonlocal kinetic equation.

If our theory is consistent we should obtain the same expressions for ξ starting from Green's functions definitions and applying the extended quasiparticle approximation. We will first integrate the nonlocal kinetic equation (13.28) to prove (15.2) and than show that the observables agree with the extended quasiparticle picture (9.67).

15.2 Symmetries of collisions

Integrating the kinetic equation it will be helpful to perform two transformations, once to interchange incoming and outgoing particles and once to exchange the collision partners *a* and *b*.

15.2.1 Transformation A

The integrated kinetic equation (13.28) is invariant if we interchange particles *a* and *b* or labels $1 \leftrightarrow 2$ and $3 \leftrightarrow 4$. This is realised by the substitution

$$a = \hat{b}, \quad b = \hat{a}$$

$$\mathbf{k} = \hat{\mathbf{p}}, \quad \mathbf{p} = \hat{\mathbf{k}}, \quad \mathbf{q} = -\hat{\mathbf{q}}. \tag{15.3}$$

The local T-matrix obeys this symmetry

$$t_{sc}^R(\omega, \mathbf{k}, \mathbf{p}, \mathbf{q}) = t_{sc}^R(\omega, \mathbf{p}, \mathbf{k}, -\mathbf{q}). \tag{15.4}$$

However, this substitution changes the derivatives of the phase $\phi(\mathbf{k}, \mathbf{p}, \mathbf{q}) = \hat{\phi}(\hat{\mathbf{k}}, \hat{\mathbf{p}}, \hat{\mathbf{q}}) = \hat{\phi}(\mathbf{p}, \mathbf{k}, -\mathbf{q})$ of the T-matrix (13.2) as

$$\frac{\partial \phi}{\partial \mathbf{k}} = \frac{\partial \hat{\phi}}{\partial \hat{\mathbf{p}}}, \quad \frac{\partial \phi}{\partial \mathbf{p}} = \frac{\partial \hat{\phi}}{\partial \hat{\mathbf{k}}}, \quad \frac{\partial \phi}{\partial \mathbf{q}} = -\frac{\partial \hat{\phi}}{\partial \hat{\mathbf{q}}} \tag{15.5}$$

leading to the relation between the displacements (13.3)

$$\mathbf{\Delta}_2 = -\hat{\mathbf{\Delta}}_2, \quad \mathbf{\Delta}_3 = \hat{\mathbf{\Delta}}_4 - \hat{\mathbf{\Delta}}_2, \quad \mathbf{\Delta}_4 = \hat{\mathbf{\Delta}}_3 - \hat{\mathbf{\Delta}}_2 \quad \mathbf{\Delta}_r = \hat{\mathbf{\Delta}}_r - \hat{\mathbf{\Delta}}_2. \tag{15.6}$$

Relations (15.6) merely show that the reference point has been moved to the partner particle and shifts were correspondingly renamed, see Fig. 15.1.

The $\mathbf{\Delta}_t$, $\mathbf{\Delta}_E$ and $\mathbf{\Delta}_K$ remain unchanged as well as the invariant combination

$$\mathbf{\Delta}_{fl} = \frac{1}{2}(\mathbf{\Delta}_3 + \mathbf{\Delta}_4 - \mathbf{\Delta}_2) = \frac{1}{2}\left(\hat{\mathbf{\Delta}}_3 + \hat{\mathbf{\Delta}}_4 - \hat{\mathbf{\Delta}}_2\right) \tag{15.7}$$

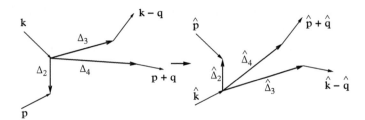

Figure 15.1: *Transform A as interchange of particle a and b leading to (15.6).*

which is the distance between final and initial geometrical centres of the colliding pair. It can be interpreted as the distance on which the particles travel during Δ_t.

The quasiparticle energies (13.20) shift according to

$$\varepsilon_1 \to \varepsilon_b(\mathbf{p}, \mathbf{r}, t) = \varepsilon_2(\mathbf{r} - \mathbf{\Delta}_2)$$

$$\varepsilon_2 \to \varepsilon_a(\mathbf{k}, \mathbf{r} - \mathbf{\Delta}_2, t) = \varepsilon_1(\mathbf{r} - \mathbf{\Delta}_2)$$

$$\varepsilon_3 \to \varepsilon_b(\mathbf{p} + \mathbf{q} + \mathbf{\Delta}_K, \mathbf{r} + \mathbf{\Delta}_4 - \mathbf{\Delta}_2, t + \Delta_t) = \varepsilon_4(\mathbf{r} - \mathbf{\Delta}_2)$$

$$\varepsilon_4 \to \varepsilon_a(\mathbf{k} - \mathbf{q} + \mathbf{\Delta}_K, \mathbf{r} + \mathbf{\Delta}_3 - \mathbf{\Delta}_2, t + \Delta_t) = \varepsilon_3(\mathbf{r} - \mathbf{\Delta}_2) \tag{15.8}$$

where on the right side we denote only changes of the arguments of (13.20) explicitly.

15.2.2 Transformation B

The interchange of initial and final states, $1 \leftrightarrow 3$ and $2 \leftrightarrow 4$, is accomplished by the substitution

$$\hat{\mathbf{k}} = \mathbf{k} - \mathbf{q}, \quad \hat{\mathbf{p}} = \mathbf{p} + \mathbf{q}. \quad \hat{\mathbf{q}} = -\mathbf{q}. \tag{15.9}$$

The general symmetry of the T-matrix with respect to the interchange of the initial and final states,

$$t_{sc}^R(\omega, \mathbf{k}, \mathbf{p}, \mathbf{q}) = t_{sc}^R(\omega, \mathbf{k} - \mathbf{q}, \mathbf{p} + \mathbf{q}, -\mathbf{q}), \tag{15.10}$$

implies that the differential cross section, the collision delay and the energy gain do not change their forms with the substitution. All gradient corrections are explicitly in the form of Δ-corrections. The T-matrix itself thus has the same symmetry as in a homogeneous and stationary medium. Under this substitution, the space displacements effectively behave as if we invert the collision,

$$\mathbf{\Delta}_2 \to \mathbf{\Delta}_3 - \mathbf{\Delta}_4,$$

$$\mathbf{\Delta}_4 \to \mathbf{\Delta}_3 - \mathbf{\Delta}_2,$$

$$\mathbf{\Delta}_r \to \mathbf{\Delta}_3 - \mathbf{\Delta}_r, \tag{15.11}$$

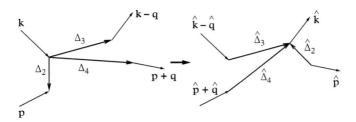

Figure 15.2: *Transform B as an interchange of incoming and outgoing particles leading to (15.11).*

while the other Δ's keeps their values and the combination (15.7) is invariant again. This is illustrated in Figure 15.2.

The distributions of the in-scattering term can be translated into the ones of the out-scattering term if we interchange out- and in-going collisions which means to apply transformation B. Consequently, the arguments of the quasiparticle energies (13.20) transform as

$$\varepsilon_1 \rightarrow \varepsilon_a(\mathbf{k} - \mathbf{q}, \mathbf{r}, t) = \tilde{\varepsilon}_3$$

$$\bar{\varepsilon}_2 \rightarrow \varepsilon_b(\mathbf{p} + \mathbf{q}, \mathbf{r} + \Delta_4 - \Delta_3, t) = \tilde{\varepsilon}_4$$

$$\bar{\varepsilon}_3 \rightarrow \varepsilon_a(\mathbf{k} - \Delta_K, \mathbf{r} - \Delta_3, t - \Delta_t) = \tilde{\varepsilon}_1$$

$$\bar{\varepsilon}_4 \rightarrow \varepsilon_b(\mathbf{p} - \Delta_K, \mathbf{r} + \Delta_2 - \Delta_3, t - \Delta_t) = \tilde{\varepsilon}_2 \qquad (15.12)$$

where the $\tilde{\varepsilon}$ denotes the shift of momentum arguments by $-\Delta_K$, the spatial arguments by $-\Delta_3$, and the time arguments by $-\Delta_t$. If we transform the scattering-in (13.28) with this transformation we obtain the order of distributions as the scattering-out (13.28) and with exactly these shifts inside the different functions. Therefore we will abbreviate in the following

$$D_{in} = \left\{ 1 - \Delta_3 \frac{\partial}{\partial \mathbf{r}} - \Delta_K \left(\frac{\partial}{\partial \mathbf{k}} + \frac{\partial}{\partial \mathbf{p}} \right) - \Delta_t \frac{\partial}{\partial t} \right\} D, \qquad (15.13)$$

where we denote $D = D_{out}$ from the out-scattering of (13.28) with

$$D = |t_{sc}|^2 \, 2\pi \delta(\varepsilon_1 + \varepsilon_2 - \varepsilon_3 - \varepsilon_4 + 2\Delta_E) \, (1 - f_3 - f_4) f_1 f_2. \qquad (15.14)$$

In case where this term D will appear as prefactor to Δs we could ignore the shifts inside D since our theory is linear in Δs. But consistently we will keep the shift as being the one of the out-scattering.

15.2.3   Symmetrisation of collision term

In (13.28), the differential cross section $|t_{sc}|^2$ and the energy argument of the scattering phase shift is based on initial states $\varepsilon_1 + \varepsilon_2$. The transformation B interchanges initial and final states and this sum energy becomes $\varepsilon_3 + \varepsilon_4$. For a convenient implementation, we thus introduce the sum energy at the centre of the collision

$$E = \frac{1}{2}\left(\varepsilon_1 + \varepsilon_2 + \varepsilon_3 + \varepsilon_4\right),$$

$$\bar{E} = \frac{1}{2}\left(\varepsilon_1 + \bar{\varepsilon}_2 + \bar{\varepsilon}_3 + \bar{\varepsilon}_4\right). \tag{15.15}$$

From the energy-conserving δ-functions it follows that on the energy shell the centred energies equal the arguments of the T-matrix,

$$\varepsilon_1 + \varepsilon_2 + \Delta_E = E, \qquad \varepsilon_1 + \bar{\varepsilon}_2 - \Delta_E = \bar{E}. \tag{15.16}$$

The centred energies (15.15) are the physically natural choice and we favour them over $\varepsilon_1 + \varepsilon_2$ resulting from the quasiparticle approximation of Green's functions. The centred energy argument, however, gives a non-trivial contribution to the factor of the energy-conserving δ functions. This comes from the fact that any energy argument in the scattering-in term is given in terms of $\omega = \varepsilon_1 + \bar{\varepsilon}_2$ while the scattering-out has the argument $\omega = \varepsilon_1 + \varepsilon_2$. Changing to the centred energies (15.15) means mathematically

$$\delta[x - \Delta(x)] = \delta[x - \Delta(E)]\left[1 + \frac{\partial \Delta(x)}{\partial x}\left(1 - \frac{\partial E(x)}{\partial x}\right)\right] \tag{15.17}$$

which one gets by comparing

$$\delta[x - \Delta(x)] = \frac{\delta(x - x_0)}{1 - \frac{\partial \Delta(x)}{\partial x}} \approx \delta(x - x_0)\left(1 + \frac{\partial \Delta(x)}{\partial x}\right)$$

with

$$\delta[x - \Delta(E(x))] \approx \delta(x - x_0)\left(1 + \frac{\partial \Delta(x)}{\partial x}\frac{\partial E(x)}{\partial x}\right)$$

in linear order of Δs. Using (15.17) with $x = \varepsilon_1 + \bar{\varepsilon}_1$ and after substitution (15.15) the δ-function of the scattering-in reads

$$\left(1 - \frac{1}{2}\frac{\partial \Delta_2}{\partial \mathbf{r}} - \frac{\partial \bar{\varepsilon}_2}{\partial \mathbf{r}}\frac{\partial \Delta_2}{\partial \omega}\right)\delta\left(\varepsilon_1 + \bar{\varepsilon}_2 - \bar{\varepsilon}_3 - \bar{\varepsilon}_4 - 2\Delta_E\right)\Big|_{\omega = \varepsilon_1 + \bar{\varepsilon}_2}$$

$$= \left[1 - \frac{1}{2}\frac{\partial \Delta_2}{\partial \mathbf{r}} - \frac{1}{2}\left(\frac{\partial \bar{\varepsilon}_3}{\partial \mathbf{k}} + \frac{\partial \bar{\varepsilon}_4}{\partial \mathbf{p}}\right)\frac{\partial \Delta_K}{\partial \omega} - \frac{1}{2}\left(\frac{\partial \bar{\varepsilon}_2}{\partial \mathbf{r}}\frac{\partial \Delta_2}{\partial \omega}\right)\right.$$

$$\left. + \frac{\partial \bar{\varepsilon}_3}{\partial \mathbf{r}}\frac{\partial \Delta_3}{\partial \omega} + \frac{\partial \bar{\varepsilon}_4}{\partial \mathbf{r}}\frac{\partial \Delta_4}{\partial \omega}\right) + \frac{\partial \Delta_E}{\partial \omega} - \frac{1}{2}\frac{\partial \Delta_t}{\partial \omega}\frac{\partial(\bar{\varepsilon}_3 + \bar{\varepsilon}_4)}{\partial t}\right]$$

$$\times \delta\left(\varepsilon_1 + \bar{\varepsilon}_2 - \bar{\varepsilon}_3 - \bar{\varepsilon}_4 - 2\Delta_E\right)\Big|_{\omega = \bar{E}} \tag{15.18}$$

and the scattering-out is given by inverse signs of the Δs at $\omega = E$. Please remember that the quasiparticle energies and distributions have the shifts according to (13.20) and a bar indicates the reversed sign.

15.3   Drift contributions to balance equations

In order to obtain the balance equations for the density, current, energy and entropy, we now multiply the kinetic equation (13.28) with $\xi = \delta_{ab}, \mathbf{k}, \varepsilon_k, -k_B \ln f_k/(1 - f_k)$ respectively. The balance of quantity ξ requires us therefore to evaluate ξ-weighted momentum integrals of the kinetic equation (13.28), i.e.

$$\sum_a \int \frac{d^3k}{(2\pi)^3} \xi_1 \frac{\partial f_1}{\partial t} + \sum_a \int \frac{d^3k}{(2\pi)^3} \xi_1 \left(\frac{\partial \varepsilon_1}{\partial \mathbf{k}} \frac{\partial f_1}{\partial \mathbf{r}} - \frac{\partial \varepsilon_1}{\partial \mathbf{r}} \frac{\partial f_1}{\partial \mathbf{k}} \right) = \sum_a \int \frac{d^3k}{(2\pi)^3} \xi_1 \left(I_1^{\text{in}} - I_1^{\text{out}} \right),$$

$$(15.19)$$

The left-hand side includes terms which have been treated many times within the Boltzmann theory and later extended to the Landau theory of Fermi liquids, see Chapter 3.9. Let us consider them first.

15.3.1   Density balance from drift

For the density $\xi_1 = \delta_{ab}$, the left-hand side of the kinetic equation (15.19) gives

$$\int \frac{d^3k}{(2\pi)^3} \left(\frac{\partial f_1}{\partial t} + \frac{\partial \varepsilon_1}{\partial \mathbf{k}} \frac{\partial f_1}{\partial \mathbf{r}} - \frac{\partial \varepsilon_1}{\partial \mathbf{r}} \frac{\partial f_1}{\partial \mathbf{k}} \right)$$

$$= \frac{\partial}{\partial t} \int \frac{d^3k}{(2\pi)^3} f_1 + \frac{\partial}{\partial \mathbf{r}} \int \frac{d^3k}{(2\pi)^3} \frac{\partial \varepsilon}{\partial \mathbf{k}} f_1 = \frac{\partial n_a^{\text{qp}}}{\partial t} + \frac{\partial \mathbf{j}_a^{\text{qp}}}{\partial \mathbf{r}}. \qquad (15.20)$$

In Landau's theory the integration over the local collision integral of Boltzmann equation is zero and one finds with (15.20) that the divergence of the quasiparticle current

$$\mathbf{j}_a^{\text{qp}} = \int \frac{d^3k}{(2\pi)^3} \frac{\partial \varepsilon_1}{\partial \mathbf{k}} f_1 \qquad (15.21)$$

and the time derivative of the quasiparticle density

$$n_a^{\text{qp}} = \int \frac{d^3k}{(2\pi)^3} f_1 \qquad (15.22)$$

sum to zero in the form of (15.2). The quasiparticle current (15.22) includes the quasiparticle back flow (Lipavský and Špička, 1994) which appears due to a non-symmetry of the quasiparticle energy $\varepsilon(k) \neq \varepsilon(-k)$ resulting from a non-symmetry of the quasiparticle distributions.

15.3.2   Energy balance from the drift

We now integrate the kinetic equation (13.28) multiplied with the energy $\xi_1 = \varepsilon_a(\mathbf{k}, \mathbf{r}, t)$. The drift side

$$\sum_a \int \frac{d^3k}{(2\pi)^3} \varepsilon_1 \frac{\partial}{\partial t} f_1 + \frac{\partial}{\partial \mathbf{r}} \sum_a \int \frac{d^3k}{(2\pi)^3} \varepsilon_1 \frac{\partial \varepsilon_1}{\partial \mathbf{k}} f_1 \qquad (15.23)$$

results in the divergence of the quasiparticle energy current

$$\mathbf{j}_E^{\mathrm{qp}} = \sum_a \int \frac{d^3k}{(2\pi)^3} \varepsilon_1 \frac{\partial \varepsilon_1}{\partial \mathbf{k}} f_1 \qquad (15.24)$$

and the first term of (15.23)

$$\sum_a \int \frac{d^3k}{(2\pi)^3} \varepsilon_1 \frac{\partial}{\partial t} f_1. \qquad (15.25)$$

When Δ's tend to zero the collision integral vanishes after integration over ε and the energy balance is (15.23). Obviously (15.25) has to be rearranged into the time derivative. In the absence of nonlocal collisions which correspond to Landau's concept of quasiparticles, the quasiparticle energy ε equals the functional derivative of the energy density,

$$\varepsilon = \frac{\delta \mathscr{E}^{\Delta=0}}{\delta f}. \qquad (15.26)$$

With the help of (15.26) the drift term (15.23) attains the desired form,

$$\sum_a \int \frac{d^3k}{(2\pi)^3} \varepsilon_1 \frac{\partial}{\partial t} f_1 = \sum_a \int \frac{d^3k}{(2\pi)^3} \frac{\delta \mathscr{E}^{\Delta=0}}{\delta f_1} \frac{\partial f_1}{\partial t} = \frac{\partial \mathscr{E}^{\Delta=0}}{\partial t}. \qquad (15.27)$$

Landau's functional relation (15.26) is consistent with the Boltzmann equation and is particularly useful for phenomenological quasiparticle energies. The variational energy (15.26) makes the conservation laws very convenient and results in correct collective motion. Here we use the quasiparticle energy identified as the pole of the GreenâĂŹs function. Except for simple approximations, these two definition lead to different values of quasiparticle energies. In the theory of liquid $^3_2$He, the difference between these two definitions is know as the rearrangement energy (Glyde and Hernadi, 1983). A relation between these quasiparticle energies and the rearrangement energy has been discussed in (Lipavský, Špička and Morawetz, 1999) and we will present it in Chapter 15.7.

The simplicity of Landaus variational approach makes his concept of quasiparticles very attractive. On the other hand, the Green's function pole represents the true dispersion law of single-particle excitation, therefore the pole definition leads to a better description of the local distribution of particles. Of course, it is at the cost of more complex balance equations. The quasiparticle contributions for all thermodynamical quantities we discuss are complete if we evaluate the collision contributions.

15.3.3   Balance of forces from the drift

Another important thermodynamic quantity is the pressure. In principle, the equilibrium pressure can be obtained from the energy density via the adiabatic derivative with respect to particle density. If one does not know the entropy density, the adiabatic derivative can be evaluated but only for the ground state, where the entropy goes to zero according to the third law of thermodynamics. This approach to the pressure has been used in the search for the pressure in nuclear matter, see (Bertsch and Gupta, 1988; Aichelin, 1991).

Out of equilibrium it is necessary to evaluate the pressure directly from the distribution of quasiparticles. Since the system is anisotropic, one has to deal with the stress tensor Π_{ij}, instead of the scalar pressure $\mathscr{P} = \frac{1}{3} \sum_i \Pi_{ii}$. Here we derive the stress tensor from the kinetic equation (13.28) published in (Špička, Lipavský and Morawetz, 1998).

The stress tensor Π_{ij} represents a flux in the direction i of the momentum density in the direction j. For non-interacting particles the stress tensor reads, see (3.73)

$$\Pi_{ij}^{\mathscr{V}=0} = \sum_a \int \frac{dk}{(2\pi)^3} k_j \frac{k_i}{m} f. \tag{15.28}$$

In an equilibrium non-degenerated system, formula (15.28) results in the familiar equation of state for the ideal gas, $\Pi_{ij}^{\mathscr{V}=0} = \delta_{ij}\mathscr{P} = \delta_{ij}nk_B T$.

The interaction modifies the stress tensor via two kinds of microscopic mechanisms corresponding to quasiparticle renormalisation and to nonlocal collisions. Quasiparticle renormalisations are dominated by the mean fields which attract particles together and contributes like the internal pressure of the van der Waals equation. Besides, due to the mass renormalisation, the quasiparticles carry their momenta with different velocities to the non-interacting particles. Nonlocal collisions have been discussed in Chapter 3. They include the partial pressure of molecules due to the finite collision duration, and the inaccessible volume of the van der Waals equation due to the nonlocal character of binary collisions.

The stress tensor can be derived from the balance between the inertial and the deformation forces

$$\frac{\partial \mathscr{Q}_j}{\partial t} = -\sum_i \frac{\partial \Pi_{ij}}{\partial r_i}. \tag{15.29}$$

The inertial force density is given by the time derivative of the momentum density $\mathscr{D}$. The deformation force density is given by the divergence of the stress tensor.

For the momentum balance one multiplies the kinetic equation (13.28) with the j-th component of momentum $\mathbf{k}$, i.e. $\xi = k_j$, and integrates over momentum k. The $\frac{\partial f}{\partial t}$ results in the time derivative of the momentum density of quasiparticles

$$\mathscr{D}_j^{\text{qp}} = \sum_a \int \frac{d^3k}{(2\pi)^3} k_j \, f_1. \tag{15.30}$$

The other parts of the drift side can be rearranged by integration by parts as

$$\sum_a \int \frac{d^3k}{(2\pi)^3} k_j \left(\frac{\partial \varepsilon_1}{\partial \mathbf{k}} \frac{\partial f_1}{\partial \mathbf{r}} - \frac{\partial \varepsilon_1}{\partial \mathbf{r}} \frac{\partial f_1}{\partial \mathbf{k}} \right)$$

$$= \sum_{i,a} \frac{\partial}{\partial r_i} \int \frac{d^3k}{(2\pi)^3} \left(k_j \frac{\partial \varepsilon_1}{\partial k_i} + \delta_{ij} \varepsilon_1 \right) f_1 - \sum_a \int \frac{d^3k}{(2\pi)^3} \varepsilon_1 \frac{\partial f_1}{\partial r_j}. \tag{15.31}$$

Eq. (15.26) allows us to write the last term of (15.31) as the gradient of the energy density,

$$\frac{\partial \mathscr{E}^{\Delta=0}}{\partial r_j} = \sum_a \int \frac{d^3k}{(2\pi)^3} \frac{\delta \mathscr{E}^{\Delta=0}}{\delta f} \frac{\partial f}{\partial r_j} \equiv \sum_a \int \frac{d^3k}{(2\pi)^3} \varepsilon \frac{\partial f}{\partial r_j}. \tag{15.32}$$

In such a way (15.31) becomes the quasiparticle stress tensor,

$$\Pi_{ij}^{\Delta=0} = \sum_a \int \frac{d^3k}{(2\pi)^3} \left(k_j \frac{\partial \varepsilon_1}{\partial k_i} + \delta_{ij} \varepsilon \right) f_1 - \delta_{ij} \mathscr{E}^{\Delta=0}. \tag{15.33}$$

The quasiparticle momentum-force balance from the drift becomes therefore

$$\sum_a \int \frac{d^3k}{(2\pi)^3} k_j \left(\frac{\partial f_1}{\partial t} + \frac{\partial \varepsilon_1}{\partial \mathbf{k}} \frac{\partial f_1}{\partial \mathbf{r}} - \frac{\partial \varepsilon_1}{\partial \mathbf{r}} \frac{\partial f_1}{\partial \mathbf{k}} \right) = \frac{\partial \mathscr{D}_j^{\text{qp}}}{\partial t} + \sum_i \frac{\partial \Pi_{ij}^{\Delta=0}}{\partial r_i}, \tag{15.34}$$

as the one for the local Boltzmann equation or Landau's theory since without shifts the collision integral vanishes due to momentum conservation. We will obtain additional contributions from the nonlocal collision integral.

15.3.4 Entropy balance

Finally, the single-particle entropy density distribution is given by

$$s_a(\mathbf{k}, \mathbf{r}, t) = -k_B \left[f_1 \ln f_1 + (1 - f_1) \ln(1 - f_1) \right] \tag{15.35}$$

which is the generalisation of the classical expression (2.42) towards quantum effects including the Pauli blocking. The first sum in (15.35) is the entropy of particles but with the quantum quasiparticle distribution. The second sum one can consider as the entropy of holes, $1 - f$, as if they are just added as a second sort of particle.

Since any derivative of (15.35) leads to the derivative of the distribution $\partial s_1 = -k_B \ln[f_1/(1 - f_1)]\partial f_1$ it is advisable to multiply the kinetic equation (13.28) with $\xi_1 = -k_B \ln[f_1/(1 - f_1)]$ and to integrate over **k**. The drift side becomes

$$\sum_a \int \frac{d^3k}{(2\pi)^3} \frac{\partial s_1}{\partial t} + \sum_a \int \frac{d^3k}{(2\pi)^3} \left[\frac{\partial \varepsilon_1}{\partial \mathbf{k}} \frac{\partial s_1}{\partial \mathbf{r}} - \frac{\partial \varepsilon_1}{\partial \mathbf{r}} \frac{\partial s_1}{\partial \mathbf{k}} \right] = \frac{\partial \mathscr{S}^{\mathrm{qp}}}{\partial t} + \frac{\partial \mathbf{j}_S^{\mathrm{qp}}}{\partial \mathbf{r}}. \tag{15.36}$$

It results into the divergence of the quasiparticle entropy current

$$\mathbf{j}_S^{\mathrm{qp}}(\mathbf{r}, t) = \sum_a \int \frac{d^3k}{(2\pi)^3} \frac{\partial \varepsilon_1}{\partial \mathbf{k}} s_a(\mathbf{k}, \mathbf{r}, t)$$

$$= -k_B \sum_a \int \frac{d^3k}{(2\pi)^3} \frac{\partial \varepsilon_1}{\partial \mathbf{k}} \left[f_1 \ln f_1 + (1 - f_1) \ln(1 - f_1) \right] \tag{15.37}$$

and the time derivative of the quasiparticle entropy

$$\mathscr{S}^{\mathrm{qp}}(\mathbf{r}, t) = \sum_a \int \frac{d^3k}{(2\pi)^3} s_a(\mathbf{k}, \mathbf{r}, t)$$

$$= -k_B \sum_a \int \frac{d^3k}{(2\pi)^3} \left[f_1 \ln f_1 + (1 - f_1) \ln(1 - f_1) \right] \tag{15.38}$$

as integral over (15.35). The arguments of f_1, f_2 etc. follow the notation of (13.20).

For the entropy balance, the collision integral does not vanish even neglecting shifts providing an explicit entropy gain. The interesting question is what the molecular part of entropy will look like, what balance we get and whether we can prove Boltzmann's H-theorem, i.e. the second law of thermodynamics. If we manage to derive the expressions including shifts and to prove the H-theorem, this includes, of course, also the simpler case for local Boltzmann equations neglecting shifts. This proof will be given in Chapter 15.5.5.

15.4 Molecular contributions to observables from collision integral

15.4.1 Expansion properties

Now we search for the terms arising from the nonlocal collision integral (13.28). Multiplying the latter one with ξ_1, integrating and applying the B-transform to the in-scattering part, we obtain from (15.18) the structure

$$\int \frac{d^3k d^3p d^3q}{(2\pi)^9} \xi_3(\mathbf{k}-\mathbf{\Delta}_K, r-\mathbf{\Delta}_3, t-\mathbf{\Delta}_t)[1 + \mathscr{T}_{in}]\left[1 - \mathbf{\Delta}_3 \frac{\partial}{\partial r} - \mathbf{\Delta}_K \left(\frac{\partial}{\partial k} + \frac{\partial}{\partial p}\right) - \mathbf{\Delta}_t \frac{\partial}{\partial t}\right] D$$

$$- \int \frac{d^3k d^3p d^3q}{(2\pi)^9} \xi_1[1 + \mathscr{T}_{out}]D \tag{15.39}$$

where we abbreviated (15.14) and adopted the notation (13.20) of the arguments for the observable ξ. One has (15.12) from transform B and in this notation $\xi_3(\mathbf{k}-\mathbf{\Delta}_K, r-\mathbf{\Delta}_3, t-\mathbf{\Delta}_t) = \xi_a(\mathbf{k}-\mathbf{q}, \mathbf{r}, t)$. The factors (15.18) become during this transformation

$$\mathscr{T}_{in} = -\frac{1}{2}\frac{\partial(\mathbf{\Delta}_3 - \mathbf{\Delta}_4)}{\partial r} - \frac{1}{2}\left(\frac{\partial\tilde{\varepsilon}_1}{\partial k} + \frac{\partial\tilde{\varepsilon}_2}{\partial p}\right)\frac{\partial\mathbf{\Delta}_K}{\partial\omega}$$

$$-\frac{1}{2}\left(\frac{\partial\tilde{\varepsilon}_4}{\partial r}\frac{\partial(\mathbf{\Delta}_3 - \mathbf{\Delta}_4)}{\partial\omega} + \frac{\partial\tilde{\varepsilon}_1}{\partial r}\frac{\partial\mathbf{\Delta}_3}{\partial\omega} + \frac{\partial\tilde{\varepsilon}_2}{\partial r}\frac{\partial(\mathbf{\Delta}_3 - \mathbf{\Delta}_2)}{\partial\omega}\right) + \frac{\partial\mathbf{\Delta}_E}{\partial\omega} - \frac{1}{2}\frac{\partial\mathbf{\Delta}_t}{\partial\omega}\frac{\partial(\tilde{\varepsilon}_1 + \tilde{\varepsilon}_2)}{\partial t}$$

$$\mathscr{T}_{out} = \frac{1}{2}\frac{\partial\mathbf{\Delta}_2}{\partial r} + \frac{1}{2}\left(\frac{\partial\tilde{\varepsilon}_3}{\partial k} + \frac{\partial\tilde{\varepsilon}_4}{\partial p}\right)\frac{\partial\mathbf{\Delta}_K}{\partial\omega} + \frac{1}{2}\left(\frac{\partial\tilde{\varepsilon}_2}{\partial r}\frac{\partial\mathbf{\Delta}_2}{\partial\omega} + \frac{\partial\tilde{\varepsilon}_3}{\partial r}\frac{\partial\mathbf{\Delta}_3}{\partial\omega} + \frac{\partial\tilde{\varepsilon}_4}{\partial r}\frac{\partial\mathbf{\Delta}_4}{\partial\omega}\right)$$

$$-\frac{\partial\mathbf{\Delta}_E}{\partial\omega} + \frac{1}{2}\frac{\partial\mathbf{\Delta}_t}{\partial\omega}\frac{\partial(\tilde{\varepsilon}_3 + \tilde{\varepsilon}_4)}{\partial t}. \tag{15.40}$$

The unchanged out-scattering one is just (15.18) with reversed signs and the in-scattering one appears since we have applied transformation B to (15.18). Since our theory is linear in $\mathbf{\Delta}$s we ignore the shifts inside ε_i, i.e. we can use $\tilde{\varepsilon} = \varepsilon = \bar{\varepsilon}$ when they appear as factors with $\mathbf{\Delta}$s.

To start with the treatment of all the following expansions it is very helpful to observe that we can consider the arguments of $\mathbf{\Delta}$s before expansion either being $E = \epsilon_1 + \epsilon_2 + o(\mathbf{\Delta}_E)$ or alternatively $E' = \epsilon_3 + \epsilon_4 + o(\mathbf{\Delta}_E)$ up to the first order in $\mathbf{\Delta}$s, due to the energy conservation in D. Expand the equality $D[\mathbf{\Delta}(E)] = D[\mathbf{\Delta}(E')]$ up to the first order for any $\mathbf{\Delta}$

$$D[0] + (\partial D)\mathbf{\Delta}(E) = D[0] + (\partial D)\mathbf{\Delta}(E') \tag{15.41}$$

with the corresponding derivative ∂. Subtracting the derivative of the equality, $\partial(D\mathbf{\Delta}(E)) = 0 = \partial(D\mathbf{\Delta}(E'))$ due to the δ-function, one gets the useful relation

$$D\partial\mathbf{\Delta}(E) = D\partial\mathbf{\Delta}(E'). \tag{15.42}$$

15.4.2 Correlated observables

In order to make the different parts transparent we concentrate successively on specific terms and collect them together at the end.

For time-derivative and $\mathbf{\Delta}_t$ terms in (15.39) and (15.40), we employ transformation A, added to the original expression and divided by two, resulting in the terms under integration

$$\frac{\xi_3 + \xi_4 - \xi_1 - \xi_2}{2} D - \frac{\Delta_t}{2} \frac{\partial}{\partial t} [(\xi_3 + \xi_4)D] - \frac{\xi_3 + \xi_4}{4} D \left(\frac{\partial \Delta_t}{\partial \omega} \frac{\partial (\varepsilon_1 + \varepsilon_2)}{\partial t} - 2 \frac{\partial \Delta_E}{\partial \omega} \right)$$

$$- \frac{\xi_1 + \xi_2}{4} D \left(\frac{\partial \Delta_t}{\partial \omega} \frac{\partial (\varepsilon_3 + \varepsilon_4)}{\partial t} - 2 \frac{\partial \Delta_E}{\partial \omega} \right). \tag{15.43}$$

The terms in the brackets form the total (on-shell) derivative as follows. From the definition (13.3) we have the identity $\partial_\omega \Delta_E(t, \omega) = -\partial_t \Delta_t(t, \omega)/2$ and therefore for any argument $x(t)$

$$\frac{\partial \Delta_t(t, \omega)}{\partial \omega} \frac{\partial x}{\partial t} - 2 \frac{\partial \Delta_E}{\partial \omega} = \frac{\partial \Delta_t(t, \omega)}{\partial \omega} \frac{\partial x}{\partial t} + \frac{\partial \Delta_t(t, \omega)}{\partial t} = \frac{\partial^{on}}{\partial t} \Delta_t(t, x(t)). \tag{15.44}$$

This means we can write, for (15.43)

$$\frac{\xi_3 + \xi_4 - \xi_1 - \xi_2}{2} D - \frac{\Delta_t(E)}{2} \frac{\partial^{on}}{\partial t} [(\xi_3 + \xi_4)D]$$

$$- \frac{\xi_3 + \xi_4}{4} D \frac{\partial^{on}}{\partial t} \Delta_t(E) - \frac{\xi_1 + \xi_2}{4} D \frac{\partial^{on}}{\partial t} \Delta_t(E') \tag{15.45}$$

Using (15.42) we can add the last two expressions,

$$\frac{\xi_3 + \xi_4 - \xi_1 - \xi_2}{2} D - \frac{\Delta_t(E)}{2} \frac{\partial^{on}}{\partial t} [(\xi_3 + \xi_4)D] - \frac{\xi_3 + \xi_4 + \xi_1 + \xi_2}{4} D \frac{\partial^{on}}{\partial t} \Delta_t(E)$$

$$= \frac{\xi_3 + \xi_4 - \xi_1 - \xi_2}{2} D - \frac{\partial^{on}}{\partial t} \left[\frac{\xi_3 + \xi_4}{2} D \Delta_t(E) \right] - \frac{\partial^{on} \Delta_t(E)}{\partial t} \left[\frac{\xi_3 + \xi_4 - \xi_1 - \xi_2}{4} D \right] \tag{15.46}$$

such that we finally obtain

$$- \frac{\partial}{\partial t} \frac{1}{2} \sum_{ab} \int \frac{d^3k d^3p d^3q}{(2\pi)^9} \Delta_t D(\xi_3 + \xi_4)$$

$$+ \frac{1}{2} \sum_{ab} \int \frac{d^3k d^3p d^3q}{(2\pi)^9} \left(1 + \frac{1}{2} \frac{\partial}{\partial t} \Delta_t \right) D(\xi_3 + \xi_4 - \xi_1 - \xi_2). \tag{15.47}$$

The first term is the negative of the time derivative of a molecular contribution to the observable

$$\xi^{mol} = \frac{1}{2} \sum_{ab} \int \frac{d^3k d^3p d^3q}{(2\pi)^9} D \Delta_t (\xi_3 + \xi_4) \tag{15.48}$$

which will be added to the quasiparticle part from the drift side. It possesses a form which can be statistically understood. With the rate D of (15.14) molecules are formed and multiplied with their lifetime Δ_t to provide the probability with which the observables ξ occur in the molecular state.

The observable gain is the second part of (15.47),

$$\mathscr{I}_{\text{gain}}^{\xi} = \frac{1}{2} \sum_{ab} \int \frac{d^3k d^3p d^3q}{(2\pi)^9} \left(1 + \frac{1}{2} \frac{\partial}{\partial t} \Delta_t\right) D(\xi_3 + \xi_4 - \xi_1 - \xi_2). \tag{15.49}$$

We see that for the density $\xi = 1$ we do not have a gain. For momentum gain $\xi = k_j$ we get from (15.49) linear in Δ

$$\mathscr{F}_j^{\text{gain}} = \sum_{ab} \int \frac{d^3k d^3p d^3q}{(2\pi)^9} D\Delta_{Kj}. \tag{15.50}$$

Dividing and multiplying by Δ_t under the integral we see that the momentum gain is the probability $D\Delta_t$ to form a molecule multiplied with the force Δ_K/Δ_t exercised during the delay time Δ_t from the environment by all other particles. In appendix D it is shown that this momentum gain (15.50) can be exactly recast together with the last term of the drift (15.31) into a spatial derivative

$$\sum_a \int \frac{d^3k}{(2\pi)^3} \varepsilon \frac{\partial f}{\partial r_j} + \mathscr{F}_j^{\text{gain}} = \frac{\partial \mathscr{E}^{\text{qp}}}{\partial r_j} \tag{15.51}$$

of the quasiparticle energy functional (D.19)

$$\mathscr{E}^{\text{qp}} = \sum_a \int \frac{d^3k}{(2\pi)^3} f_a(k) \frac{k^2}{2m} + \frac{1}{2} \sum_{ab} \int \frac{d^3k d^3p}{(2\pi)^6} f_a(k) f_b(p) \mathscr{T}_{\text{ex}}(\varepsilon_1 + \varepsilon_2, k, p, 0) \tag{15.52}$$

instead of the Landau functional (15.32) which was valid only in local approximation.

For the energy gain $\xi = \varepsilon$ we get from (15.49)

$$\mathscr{I}_{\text{gain}}^{E} = \sum_{ab} \int \frac{d^3k d^3p d^3q}{(2\pi)^9} D\Delta_E. \tag{15.53}$$

It represents the mean power Δ_E/Δ_t exerted on the collision multiplied with the probability to form a molecule $D\Delta_t$. As shown in Appendix D, this energy gain combines together with the first term of (15.23) into the total time derivative of the quasiparticle energy functional (D.19)

$$\sum_a \int \frac{d^3k}{(2\pi)^3} \varepsilon \frac{\partial f}{\partial t} - \mathscr{I}_{\text{gain}}^{E} = \frac{\partial \mathscr{E}^{\text{qp}}}{\partial t}. \tag{15.54}$$

The entropy gain (15.49) with $\xi_1 = -k_B \ln f_1/(1 - f_1)$ we will discuss later in Chapter 15.4.4.

Summarising so far, we have arrived from the time parts of the collision integral at the balance equation, almost at the form (15.2)

$$\frac{\partial(\xi^{\text{qp}} + \xi^{\text{mol}})}{\partial t} + \frac{\partial \mathbf{j}_\xi^{\text{qp}}}{\partial \mathbf{r}} = \mathscr{I}_{\text{space}} \tag{15.55}$$

where it remains for us to show that the spatial and momentum derivatives of (15.39), indicated as $\mathscr{I}_{\text{space}}$, can be written as the divergence of the molecular energy current.

15.4.3   Molecular current contributions from collision integral

Firstly we need a guide for what the corresponding molecular currents will look like. Most simply this is seen from the spatial gradients of D in (15.39) and (15.40)

$$\xi_3 D(\mathbf{r} - \boldsymbol{\Delta}_3) - \xi_1 D(\mathbf{r}) = (\xi_3 - \xi_1)D - \boldsymbol{\Delta}_3 \xi_3 \frac{\partial}{\partial \mathbf{r}} D + o(\Delta^2). \tag{15.56}$$

Alternatively, we can apply first the transformation A, and then expand

$$\xi_4 D(\mathbf{r} - \boldsymbol{\Delta}_2 - (\boldsymbol{\Delta}_4 - \boldsymbol{\Delta}_2)) - \xi_2 D(\mathbf{r} - \boldsymbol{\Delta}_2)$$

$$= (\xi_4 - \xi_2)D - \boldsymbol{\Delta}_4 \xi_4 \frac{\partial}{\partial \mathbf{r}} D + \boldsymbol{\Delta}_2 \xi_2 \frac{\partial}{\partial \mathbf{r}} D + o(\Delta^2). \tag{15.57}$$

The shifts inside the ξs can be neglected since we consider only linear orders. We add (15.56) and (15.57) and divide by two to get besides the already counted gain term (15.49) the first-order gradient term

$$\mathscr{I}_{\text{space}}^D = \frac{1}{2}(\boldsymbol{\Delta}_2 \xi_2 - \boldsymbol{\Delta}_3 \xi_3 - \boldsymbol{\Delta}_4 \xi_4)\frac{\partial}{\partial \mathbf{r}} D. \tag{15.58}$$

This suggests what the molecular observable current $\mathbf{j}_\xi$ will look like provided we find the remaining terms such that (15.58) becomes the divergence of the molecular current, $\partial_{\mathbf{r}} \mathbf{j}_\xi^{\text{mol}}$.

We will now consider the frequency derivatives of the spatial shifts in (15.39) and (15.40) and again add the A-transform expression and divide by two. Collecting them together one gets

$$\frac{D}{2}\frac{\partial E}{\partial \mathbf{r}}\left(\xi_2 \frac{\partial \boldsymbol{\Delta}_2}{\partial \omega} - \xi_3 \frac{\partial \boldsymbol{\Delta}_3}{\partial \omega} - \xi_4 \frac{\partial \boldsymbol{\Delta}_4}{\partial \omega}\right) + \frac{D}{4}(\xi_3 + \xi_4 - \xi_1 - \xi_2)\left(\frac{\partial \varepsilon_2}{\partial \mathbf{r}}\frac{\partial \boldsymbol{\Delta}_2}{\partial \omega} + \frac{\partial \varepsilon_3}{\partial \mathbf{r}}\frac{\partial \boldsymbol{\Delta}_3}{\partial \omega} + \frac{\partial \varepsilon_4}{\partial \mathbf{r}}\frac{\partial \boldsymbol{\Delta}_4}{\partial \omega}\right)$$

$$= \mathscr{I}_{\text{space}}^\omega + \mathscr{I}_{\text{gain}}^\omega. \tag{15.59}$$

The first part fits the derivative in (15.58) while the second part obviously counts together with the second part of (15.47), i.e. it is a gain term (15.49) due to the factor $(\xi_3 + \xi_4 - \xi_1 - \xi_2)$.

Next, we collect the spatial derivatives of ξ and the spatial shifts of (15.39) with (15.40). When applying the A-transform, $\xi_3(r - \Delta_3) \rightarrow \xi_4(r - \Delta_2 - (\Delta_4 - \Delta_2))$, one gets

$$
-\Delta_3 \frac{\partial}{\partial \mathbf{r}} \xi_3 - \frac{\xi_3}{2} \frac{\partial}{\partial \mathbf{r}} (\Delta_3 - \Delta_4) - \frac{\xi_1}{2} \frac{\partial}{\partial \mathbf{r}} \Delta_2 = -\Delta_4 \frac{\partial}{\partial \mathbf{r}} \xi_4 + \frac{\xi_4}{2} \frac{\partial}{\partial \mathbf{r}} (\Delta_3 - \Delta_4) + \frac{\Delta_2}{2} \frac{\partial}{\partial \mathbf{r}} \xi_2
$$

$$
= -\frac{\Delta_3}{2} \frac{\partial}{\partial \mathbf{r}} \xi_3 - \frac{\Delta_4}{2} \frac{\partial}{\partial \mathbf{r}} \xi_4 + \frac{\Delta_2}{2} \frac{\partial}{\partial \mathbf{r}} \xi_2 + \frac{\xi_2 - \xi_1}{4} \frac{\partial}{\partial \mathbf{r}} \Delta_2 - \frac{\xi_3 - \xi_4}{4} \frac{\partial}{\partial \mathbf{r}} (\Delta_3 - \Delta_4)
$$

$$
= \mathscr{I}^\xi_{\text{space}} + \mathscr{I}^1_{\text{space}} \tag{15.60}
$$

where we added the first two equations and divided by two to obtain the third equation. The three terms collected in $\mathscr{I}^\xi_{\text{space}}$ will obviously contribute to (15.58).

As remaining parts in (15.39) with (15.40) we now consider the momentum derivatives and Δ_K and have

$$
-\frac{\Delta_K}{2} \left(\frac{\partial}{\partial \mathbf{k}} + \frac{\partial}{\partial \mathbf{p}} \right) [D(\xi_3 + \xi_4)] - \frac{D}{4} \frac{\partial \Delta_K}{\partial \omega}
$$

$$
\times \left[(\xi_1 + \xi_2) \left(\frac{\partial}{\partial \mathbf{k}} + \frac{\partial}{\partial \mathbf{p}} \right) E' + (\xi_3 + \xi_4) \left(\frac{\partial}{\partial \mathbf{k}} + \frac{\partial}{\partial \mathbf{p}} \right) E \right]. \tag{15.61}
$$

Again we have added the A-transformed expression and divided by two. Replacing further

$$
\frac{\partial \Delta_K}{\partial \omega} \left(\frac{\partial}{\partial \mathbf{k}} + \frac{\partial}{\partial \mathbf{p}} \right) E = \left[\left(\frac{\partial}{\partial \mathbf{k}} + \frac{\partial}{\partial \mathbf{p}} \right)^{on} - \left(\frac{\partial}{\partial \mathbf{k}} + \frac{\partial}{\partial \mathbf{p}} \right) \right] \Delta_K(E) \tag{15.62}
$$

by the on-shell derivative one gets

$$
-\frac{\Delta_K}{2} \left(\frac{\partial}{\partial \mathbf{k}} + \frac{\partial}{\partial \mathbf{p}} \right)^{on} [D(\xi_3 + \xi_4)]
$$

$$
-\frac{D}{4} (\xi_1 + \xi_2) \left[\left(\frac{\partial}{\partial \mathbf{k}} + \frac{\partial}{\partial \mathbf{p}} \right)^{on} - \left(\frac{\partial}{\partial \mathbf{k}} + \frac{\partial}{\partial \mathbf{p}} \right) \right] \Delta_K(E')
$$

$$
-\frac{D}{4} (\xi_3 + \xi_4) \left[\left(\frac{\partial}{\partial \mathbf{k}} + \frac{\partial}{\partial \mathbf{p}} \right)^{on} - \left(\frac{\partial}{\partial \mathbf{k}} + \frac{\partial}{\partial \mathbf{p}} \right) \right] \Delta_K(E). \tag{15.63}
$$

Observing (15.42) allows us to add both last terms in (15.63) and we can create the on-shell derivative needed for the first term in (15.61) to find

$$-\left(\frac{\partial}{\partial \mathbf{k}} + \frac{\partial}{\partial \mathbf{p}}\right)^{on}\left[\frac{D}{2}\mathbf{\Delta}_K(\xi_3 + \xi_4)\right]$$

$$+\frac{\xi_3 + \xi_4 - \xi_1 - \xi_2}{4}D\left(\frac{\partial}{\partial \mathbf{k}} + \frac{\partial}{\partial \mathbf{p}}\right)^{on}\mathbf{\Delta}_K$$

$$+\frac{\xi_1 + \xi_2 + \xi_3 + \xi_4}{4}D\left(\frac{\partial}{\partial \mathbf{k}} + \frac{\partial}{\partial \mathbf{p}}\right)\mathbf{\Delta}_K. \tag{15.64}$$

The first term vanishes under integration, the second term obviously counts to wards the gain and the last term can be rewritten as spatial derivatives of the spatial shifts according to the definition (13.3), i.e.

$$\frac{\partial}{\partial \mathbf{r}}(\mathbf{\Delta}_3 + \mathbf{\Delta}_2 - \mathbf{\Delta}_4) = 2\left(\frac{\partial}{\partial \mathbf{p}} + \frac{\partial}{\partial \mathbf{k}}\right)\mathbf{\Delta}_K + 2\frac{\partial}{\partial \mathbf{r}}\mathbf{\Delta}_3. \tag{15.65}$$

The result for (15.64) finally reads

$$\frac{\xi_3 + \xi_4 - \xi_1 - \xi_2}{4}D\left(\frac{\partial}{\partial \mathbf{k}} + \frac{\partial}{\partial \mathbf{p}}\right)^{on}\mathbf{\Delta}_K + \frac{\xi_1 + \xi_2 + \xi_3 + \xi_4}{4}\frac{D}{2}\frac{\partial}{\partial \mathbf{r}}(\mathbf{\Delta}_2 - \mathbf{\Delta}_3 - \mathbf{\Delta}_4)$$

$$= \mathscr{I}^K_{gain} + + \mathscr{I}^2_{space}. \tag{15.66}$$

Now we have all the terms in a form to be combined. The spatial derivatives of Δs in (15.66) and (15.60) can be regrouped together as

$$\mathscr{I}^1_{space} + \mathscr{I}^2_{space} = \frac{\xi_3 + \xi_4 - \xi_1 - \xi_2}{8}D\frac{\partial}{\partial \mathbf{r}}(\mathbf{\Delta}_3 + \mathbf{\Delta}_4 + \mathbf{\Delta}_2)$$

$$-\frac{D}{2}\left(\xi_3\frac{\partial}{\partial \mathbf{r}}\mathbf{\Delta}_3 + \xi_4\frac{\partial}{\partial \mathbf{r}}\mathbf{\Delta}_4 - \xi_2\frac{\partial}{\partial \mathbf{r}}\mathbf{\Delta}_2\right)$$

$$= \mathscr{I}^\Delta_{gain} + \mathscr{I}^\Delta_{space}. \tag{15.67}$$

Collecting the terms from (15.58), (15.59), (15.60) and (15.67) with spatial gradients which have no $\xi_3 + \xi_4 - \xi_1 - \xi_2$ prefactor, we obtain the complete divergence

$$\mathscr{I}^D_{space} + \mathscr{I}^\omega_{space} + \mathscr{I}^\xi_{space} + \mathscr{I}^\Delta_{space} = -\frac{\partial \mathbf{j}^{mol}_\xi}{\partial \mathbf{r}} \tag{15.68}$$

of the molecular current

$$j_{\xi}^{\text{mol}} = -\frac{1}{2}\sum_{ab}\int\frac{d^3k\,d^3p\,d^3q}{(2\pi)^9}D(\xi_2\mathbf{\Delta}_2 - \xi_3\mathbf{\Delta}_3 - \xi_4\mathbf{\Delta}_4) \tag{15.69}$$

which can now be added to the quasiparticle part from the drift side (15.37). Obviously it has again the statistical interpretation as the observable per delay time ξ/Δ_t carried at the points of nonlocal collisions multiplied with the probability to form a molecule $D\Delta_t$.

15.4.4   Remaining gains

The remaining terms with the prefactor $\xi_3 + \xi_4 - \xi_1 - \xi_2$ in (15.49), (15.59), (15.66) and (15.67) are of gain form and read

$$\mathscr{I}_{\text{gain}}^{\xi} + \mathscr{I}_{\text{gain}}^{\omega} + \mathscr{I}_{\text{gain}}^{K} + \mathscr{I}_{\text{gain}}^{\Delta} = \left\{2\left(1 + \frac{1}{2}\frac{\partial}{\partial t}\Delta_t\right)\right.$$

$$+ \left(\frac{\partial}{\partial\mathbf{k}} + \frac{\partial}{\partial\mathbf{p}}\right)^{on}\mathbf{\Delta}_K + \frac{1}{2}\frac{\partial}{\partial\mathbf{r}}(\mathbf{\Delta}_3 + \mathbf{\Delta}_4 + \mathbf{\Delta}_2)$$

$$\left. + \frac{\partial\varepsilon_2}{\partial\mathbf{r}}\frac{\partial\mathbf{\Delta}_2}{\partial\omega} + \frac{\partial\varepsilon_3}{\partial\mathbf{r}}\frac{\partial\mathbf{\Delta}_3}{\partial\omega} + \frac{\partial\varepsilon_4}{\partial\mathbf{r}}\frac{\partial\mathbf{\Delta}_4}{\partial\omega}\right\}\frac{\xi_3 + \xi_4 - \xi_1 - \xi_2}{4}D. \tag{15.70}$$

We see that for both observables, $\xi = k$ for momentum or $\xi = \varepsilon_k$ for energy, the derivative terms are of higher order in Δ since the differences in ξs lead to momentum and energy shift itself, respectively. The zeroth order for momentum and energy gain we had already shown to combine together with a drift part into derivatives of the quasiparticle values, (15.51) and (15.54). The gain for density, $\xi = 1$ vanishes trivially.

Therefore we see that only for the entropy an extra gain term remains from the collision integral. We rewrite the {}-bracket in (15.70) using (15.62) and (15.65) to get

$$\frac{1}{2}\{\} = 1 + \frac{1}{2}\frac{\partial}{\partial t}\Delta_t + \frac{1}{2}\frac{\partial\mathbf{\Delta}_K}{\partial\omega}\left(\frac{\partial}{\partial\mathbf{k}} + \frac{\partial}{\partial\mathbf{p}}\right)E' + \frac{1}{2}\frac{\partial}{\partial\mathbf{r}}\mathbf{\Delta}_2$$

$$+ \frac{1}{2}\frac{\partial\varepsilon_2}{\partial\mathbf{r}}\frac{\partial\mathbf{\Delta}_2}{\partial\omega} + \frac{1}{2}\frac{\partial\varepsilon_3}{\partial\mathbf{r}}\frac{\partial\mathbf{\Delta}_3}{\partial\omega} + \frac{1}{2}\frac{\partial\varepsilon_4}{\partial\mathbf{r}}\frac{\partial\mathbf{\Delta}_4}{\partial\omega} \tag{15.71}$$

where the part $\partial_{\mathbf{r}}\mathbf{\Delta}_2$ vanishes when we add the A-transformed expression and divide by two.

Now we remember that the weight of the energy conserving δ-function has been used for the symmetrised energies (15.18) we had for out-scattering, which is the one contained in D,

$$\delta\big(\varepsilon_1 + \varepsilon_2 - \varepsilon_3 - \varepsilon_4 + 2\Delta_E\big)\big|_{\omega = \varepsilon_1 + \varepsilon_2}$$

$$= \left[1 + \frac{1}{2}\left(\frac{\partial \varepsilon_3}{\partial \mathbf{k}} + \frac{\partial \varepsilon_4}{\partial \mathbf{p}} \right) \frac{\partial \Delta_K}{\partial \omega} + \frac{1}{2}\left(-\frac{\partial \varepsilon_2}{\partial \mathbf{r}} \frac{\partial \Delta_2}{\partial \omega} + \frac{\partial \varepsilon_3}{\partial \mathbf{r}} \frac{\partial \Delta_3}{\partial \omega} \right. \right.$$

$$\left. \left. + \frac{\partial \varepsilon_4}{\partial \mathbf{r}} \frac{\partial \Delta_4}{\partial \omega} \right) - \frac{\partial \Delta_E}{\partial \omega} + \frac{1}{2}\frac{\partial \Delta_t}{\partial \omega}\frac{\partial E'}{\partial t} \right] \delta\left(\varepsilon_1 + \varepsilon_2 - \varepsilon_3 - \varepsilon_4 + 2\Delta_E\right)\big|_{\omega = E} \qquad (15.72)$$

and comparing this with (15.71) we find for the entropy gain (15.70)

$$\left(1 + \frac{\partial \varepsilon_2}{\partial \mathbf{r}} \frac{\partial \Delta_2}{\partial \omega} \right) \frac{\xi_3 + \xi_4 - \xi_1 - \xi_2}{2} D_{\varepsilon_1 + \varepsilon_2} \qquad (15.73)$$

where we used (15.44). If we understand the δ-function as selfconsistent solution with respect to the shifted argument of ε_2 we can absorb the factor

$$\delta\{\omega - \varepsilon_1 - \varepsilon_2[\mathbf{r} + \Delta_2(\omega)]\} = \delta(\omega - \varepsilon_1 - \varepsilon_2)\left(1 + \frac{\partial \varepsilon_2}{\partial \mathbf{r}} \frac{\partial \Delta_2}{\partial \omega} \right) \qquad (15.74)$$

and write finally for the entropy gain

$$\mathscr{I}_{\text{gain}}^S = \frac{1}{2} \sum_{ab} \int \frac{d^3k\, d^3p\, d^3q}{(2\pi)^9} (\xi_3 + \xi_4 - \xi_1 - \xi_2) D_{\varepsilon_1}. \qquad (15.75)$$

Comparing to (15.49) the energy-conserving δ-function is now to be understood as a selfconsistent expression of shifts. This is required in order to have the same shifts in ε inside the δ-function as inside the distributions.

15.5 Balance equations and proof of H-theorem

15.5.1 Equation of continuity

The density balance equation from nonlocal kinetic theory consists of quasiparticle parts and molecular contributions

$$\frac{\partial(n_a^{\text{qp}} + n_a^{\text{mol}})}{\partial t} + \frac{\partial(j_a^{\text{qp}} + j_a^{\text{mol}})}{\partial \mathbf{r}} = 0 \qquad (15.76)$$

with the standard quasiparticle density (15.22) and current (15.21). The correlated or molecular density (15.48)

$$n_a^{\text{mol}} = \sum_b \int \frac{d^3k\, d^3p\, d^3q}{(2\pi)^9} D\Delta_t \qquad (15.77)$$

has the statistical interpretation of the rate of binary processes D of (15.14) weighed with the Δ_t. Please compare with the expressions from the classical nonlocal kinetic theory in Chapter 3.4 and especially (3.37) which had led to the Guldberg–Waage law of acting masses.

The molecular current (15.69) we have obtained as

$$j_a^{\text{mol}} = \sum_b \int \frac{d^3k d^3p d^3q}{(2\pi)^9} D\Delta_3. \tag{15.78}$$

By applying transform A, add and dividing by two we can write equivalently in (15.78) for Δ_3 also Δ_{fl} of (15.7). Again we obtain a statistical interpretation in that the velocity of the molecule $\Delta_{\text{fl}}/\Delta_t$ is multiplied with the rate D to form a molecule and weighted with the duration Δ_t. This quantum expression compares directly with the classical one (3.36).

15.5.2   Energy balance

The energy balance (15.55) we found as

$$\frac{\partial(\mathscr{E}^{\text{qp}} + \mathscr{E}^{\text{mol}})}{\partial t} + \frac{\partial(j_E^{\text{qp}} + j_E^{\text{mol}})}{\partial r} = 0 \tag{15.79}$$

with the quasiparticle energy functional (15.52) having the same structure as the uncorrelated energy functional, the bare interaction potential is, however, replaced by the T-matrix.

The molecular contribution to the energy (15.48)

$$\mathscr{E}^{\text{mol}} = \frac{1}{2} \sum_{ab} \int \frac{d^3k d^3p d^3q}{(2\pi)^9} D\Delta_t E, \tag{15.80}$$

has also a natural statistical interpretation. We saw that the factor $D\Delta_t$ measures the probability of finding two particles in the scattering state. The total energy of these two particles is the mean of $E = \varepsilon_1 + \varepsilon_2$ such that the product gives the mean energy of the molecule living at the time Δ_t.

The energy current is the sum of the quasiparticle current in (15.24) and the molecular current (15.69)

$$j_E^{\text{mol}} = \frac{1}{2} \sum_{ab} \int \frac{d^3k d^3p d^3q}{(2\pi)^9} D(\varepsilon_2 \Delta_2 - \varepsilon_3 \Delta_3 - \varepsilon_4 \Delta_4). \tag{15.81}$$

It is the balance of energies carried by the different spatial off-sets.

15.5.3 Navier–Stokes equation

The inertial-force density is given by the time derivative of the momentum density $\mathscr{Q}$. The deformation-force density is given by the divergence of the stress tensor. The stress tensor we derived from the balance between the inertial and the deformations forces

$$\frac{\partial \left(\mathscr{Q}_j^{\mathrm{qp}} + \mathscr{Q}_j^{\mathrm{mol}} \right)}{\partial t} = -\sum_i \frac{\partial \left(\Pi_{ij}^{\mathrm{qp}} + \Pi_{ij}^{\mathrm{mol}} \right)}{\partial r_i} \tag{15.82}$$

with the momentum density consisting of the quasiparticle (15.30) and molecular part (15.48) with $\xi = k_j$

$$\mathscr{Q}_j^{\mathrm{mol}} = \frac{1}{2} \sum_{ab} \int \frac{d^3k d^3p d^3q}{(2\pi)^9} (k_j + p_j) D \Delta_t \tag{15.83}$$

which gives the mean momentum carried by a molecule formed with the rate D and lifetime Δ_t.

The total stress tensor formed by the quasiparticles reads with (15.51)

$$\Pi_{ij}^{\mathrm{qp}} = \sum_a \int \frac{d^3k}{(2\pi)^3} \left(k_j \frac{\partial \varepsilon}{\partial k_i} + \delta_{ij} \varepsilon \right) f - \delta_{ij} \mathscr{E}^{\mathrm{qp}} \tag{15.84}$$

with (15.52) instead of the local one (15.33). The collision-flux contributions (15.69),

$$\Pi_{ij}^{\mathrm{mol}} = \frac{1}{2} \sum_{ab} \int \frac{d^3k d^3p d^3q}{(2\pi)^9} D \left[(k_j - q_j) \Delta_{3i} + (p_j + q_j) \Delta_{4i} - p_j \Delta_{2i} \right], \tag{15.85}$$

finally possesses a statistical interpretation as well. The two-particle state is characterised by the initial momenta **k** and **p** and the transferred momentum **q**. The momentum tensor is the balance of the momenta carried by the corresponding spatial off-sets weighted with the rate to form a molecule D. From this tensor one can arrive at the pressure contribution which classically limit the van der Waals equation of state we have discussed in Chapter 3.5.

15.5.4 Entropy balance

Finally the entropy balance reads

$$\frac{\partial (\mathscr{S}^{\mathrm{qp}} + \mathscr{S}^{\mathrm{mol}})}{\partial t} + \frac{\partial (j_S^{\mathrm{qp}} + j_S^{\mathrm{mol}})}{\partial \mathbf{r}} = \mathscr{J}_{\mathrm{gain}}^S \tag{15.86}$$

where the entropy consists of the quasiparticle part (15.38) and the molecular part (15.48)

$$\mathscr{S}^{\mathrm{mol}} = -\frac{k_B}{2} \sum_{ab} \int \frac{d^3k d^3p d^3q}{(2\pi)^9} |t_{\mathrm{sc}}|^2 \Delta_t 2\pi \delta(\varepsilon_1 + \varepsilon_2 - \varepsilon_3 - \varepsilon_4)$$

$$\times f_1 f_2 (1 - f_3 - f_4) \ln \frac{f_3 f_4}{(1 - f_3)(1 - f_4)}. \tag{15.87}$$

In the same way, the entropy current has a quasiparticle part (15.37) and a molecular contribution (15.69) with $\xi_1 = -k_B \ln f_1/(1 - f_1)$ reading

$$j_S^{\mathrm{mol}} = \frac{k_B}{2} \sum_{ab} \int \frac{d^3k d^3p d^3q}{(2\pi)^9} D \left[\ln \frac{f_2}{(1 - f_2)} \Delta_2 - \ln \frac{f_3}{(1 - f_3)} \Delta_3 - \ln \frac{f_4}{(1 - f_4)} \Delta_4 \right]. \tag{15.88}$$

The entropy gain (15.75) finally has the form

$$\mathscr{I}_{\mathrm{gain}}^S = -\frac{k_B}{2} \sum_{ab} \int \frac{d^3k d^3p d^3q}{(2\pi)^9} f_1 f_2 (1 - f_3 - f_4)$$

$$\times 2\pi \delta(\varepsilon_1 + \varepsilon_2 - \varepsilon_3 - \varepsilon_4) |t_{\mathrm{sc}}|^2 \ln \frac{f_3 f_4 (1 - f_1)(1 - f_2)}{(1 - f_3)(1 - f_4) f_1 f_2}. \tag{15.89}$$

This entropy gain remains explicit while the momentum gain and energy gain are transferring kinetic into correlation parts and do not appear explicitly.

15.5.5 Proof of H-theorem

Now we are going to proove that the form (15.75) is positive every time . We consider in short-hand notation $\xi = \xi_3 + \xi_4 - \xi_1 - \xi_2$. Then the expansion in Δs reads

$$\mathscr{I}_{\mathrm{gain}}^S = \frac{D}{2} \xi = \left[1 + \Delta_t \left(\frac{\partial^3}{\partial t} + \frac{\partial^4}{\partial t} \right) + \Delta_2 \frac{\partial^2}{\partial \mathbf{r}} + \Delta_3 \frac{\partial^3}{\partial \mathbf{r}} \right.$$

$$\left. + \Delta_4 \frac{\partial^4}{\partial \mathbf{r}} + \Delta_K \left(\frac{\partial^3}{\partial \mathbf{k}} + \frac{\partial^3}{\partial \mathbf{p}} + \frac{\partial^4}{\partial \mathbf{k}} + \frac{\partial^4}{\partial \mathbf{p}} \right) \right] \frac{D_0}{2} \xi_0 \tag{15.90}$$

where we indicate explicitly to which argument $1, 2, 3$ or 4 the derivatives apply.

Firstly we establish a useful relation and focus on the time derivatives. Let's consider an unknown derivative operator $\mathscr{R}$ and apply transform B together with the space-and-time reversal transformation inverting the shifts

$$(1 + \mathscr{R})\frac{D}{2}\xi = \left[1 + \mathscr{R} + \Delta_t\left(\frac{\partial^3}{\partial t} + \frac{\partial^4}{\partial t}\right)\right]\frac{D_0}{2}\xi_0$$

$$= \frac{1}{2}\left[-1 - \tilde{\mathscr{R}}_B - \Delta_t\left(\frac{\partial^1}{\partial t} + \frac{\partial^2}{\partial t}\right)\right](I_0 + D_0)\xi_0 \qquad (15.91)$$

where we denote the symmetrised collision term $I_0 = 2\pi\delta(E - E')|t_{sc}|^2[f_3f_4(1 - f_1 - f_2) - f_1f_2(1 - f_3 - f_4)]$. Subtracting the D_0 part from the left we obtain

$$\left[1 + \frac{\mathscr{R} + \tilde{\mathscr{R}}_B}{2} + \frac{\Delta_t}{2}\left(\frac{\partial^1}{\partial t} + \frac{\partial^2}{\partial t} - \frac{\partial^3}{\partial t} - \frac{\partial^4}{\partial t}\right)\right]\frac{D}{2}\xi = \frac{1}{4}\left[-1 - \tilde{\mathscr{R}}_B - \Delta_t\left(\frac{\partial^1}{\partial t} + \frac{\partial^2}{\partial t}\right)\right]I_0\xi_0.$$

$$(15.92)$$

Now we determine the unknown derivative operator $\mathscr{R} = \Delta_t(a\partial^2 + b\partial^2 + c\partial + d\partial^4)$ and consequently $\tilde{\mathscr{R}}_B = -\Delta_t(c\partial^2 + d\partial^2 + a\partial + b\partial^4)$ such that the left-hand side of (15.92) is unity, which provides $a = b = -1$ and $c = d = 0$ and we obtain finally the identity

$$\mathscr{I}_{\text{gain}}^S = \frac{D}{2}\xi = -\frac{1}{4}\left(1 + \Delta_t\frac{\partial}{\partial t}\right)I_0\xi_0 \qquad (15.93)$$

suited for proving the H-theorem. Replacing the time derivative by the momentum derivative and Δ_t by Δ_K we obtain the analogous expression. Interestingly it remains to show how the $\mathscr{R}$ operator looks like for the spatial derivatives. Analogously to (15.92) we have

$$\left[1 + \frac{\mathscr{R} + \tilde{\mathscr{R}}_B}{2} - \mathbf{\Delta}_2\frac{\partial^2}{\partial\mathbf{r}} + \frac{\mathbf{\Delta}_3}{2}\left(\frac{\partial^1}{\partial\mathbf{r}} + \frac{\partial^2}{\partial\mathbf{r}} - \frac{\partial^3}{\partial\mathbf{r}} + \frac{\partial^4}{\partial\mathbf{r}}\right) - \mathbf{\Delta}_4\frac{\partial^4}{\partial\mathbf{r}}\right]\frac{D}{2}\xi$$

$$= -\left[1 + \tilde{\mathscr{R}}_B - \mathbf{\Delta}_2\frac{\partial^2}{\partial\mathbf{r}} + \mathbf{\Delta}_3\left(\frac{\partial^1}{\partial\mathbf{r}} + \frac{\partial^2}{\partial\mathbf{r}} + \frac{\partial^4}{\partial\mathbf{r}}\right) - \mathbf{\Delta}_4\frac{\partial^4}{\partial\mathbf{r}}\right]\frac{I_0}{4}\xi_0. \qquad (15.94)$$

Again we search for an operator $\mathscr{R}$ which renders the left side unity. A linear equation system provides a manifold of solutions from which we choose one with the final result together with (15.93)

$$\mathscr{I}_{\text{gain}}^S = \frac{D}{2}\xi = -\left[1 + \Delta_t\frac{\partial}{\partial t} - \mathbf{\Delta}_2\frac{\partial^2}{\partial\mathbf{r}} - \mathbf{\Delta}_3\frac{\partial^3}{\partial\mathbf{r}} - \mathbf{\Delta}_4\frac{\partial^4}{\partial\mathbf{r}} + \mathbf{\Delta}_K\left(\frac{\partial}{\partial\mathbf{k}} + \frac{\partial}{\partial\mathbf{p}}\right)\right]\frac{I_0}{4}\xi_0$$

$$= \frac{k_B}{4}\sum_{ab}\int\frac{d^3kd^3pd^3q}{(2\pi)^9}2\pi\delta(\varepsilon_1 + \varepsilon_2 - \varepsilon_3 - \varepsilon_4)|t_{sc}|^2\ln\frac{f_3f_4(1 - f_1)(1 - f_2)}{(1 - f_3)(1 - f_4)f_1f_2}$$

$$\times\left.\left\{f_3f_4(1 - f_1)(1 - f_2) - f_1f_2(1 - f_3)(1 - f_4)\right\}\right|_{1,2,3,4\text{ equally shifted}} \qquad (15.95)$$

where we reestablished the full notation. This entropy gain is every-time positive since with $a = f_3 f_4 (1-f_1)(1-f_2)$ and $b = f_1 f_2 (1-f_3)(1-f_4)$ we have the always positive entropy production density $(a-b)\ln(a/b) > 0$. This is completely analogously to the proof of Boltzmann's H-theorem in the same way as it was proved for classical kinetic theory in Chapter 3.7.

We therefore have shown that the second law of thermodynamics holds also in the nonlocal kinetic theory. We want to emphasise that the molecular contribution to the entropy due to particle interactions as well as the correlated entropy current are new results and show how the two-particle correlations exceed the Landau theory. The single-particle entropy can decrease at the cost of the molecular part of entropy describing the two-particles in a molecular state.

15.6   Equivalence to extended quasiparticle picture

To show the consistency we should obtain the same molecular expressions of observables from direct integration of the correlated part of the Wigner function. This is completed by the extended quasiparticle picture which we present now. In this section we recover the correlated density (15.77) from the $\rho[f]$-functional (9.67). This will confirm the consistency of the extended quasiparticle approximation with the non-instant corrections to the scattering integral. Beside, we show that the correlated density following from the non-instant corrections to the scattering integral is the one known from the generalised Beth–Uhlenbeck approach (Schmidt, Röpke and Schulz, 1990; Morawetz and Röpke, 1995).

According to the definition of the local density (7.33) and relation (9.67) between the Wigner distribution ρ and the quasiparticle distribution f, one finds

$$
n = \int \frac{dk}{(2\pi)^3} \rho = \int \frac{dk}{(2\pi)^3} f - \int \frac{dk}{(2\pi)^3} \frac{d\omega}{2\pi} \left[\sigma^>(\omega) f - \sigma^<(\omega)(1-f) \right] \frac{\wp'}{\omega - \varepsilon}.
$$

$$(15.96)$$

The first term is the free density (15.22), the second term is the correlated density,

$$
n^{\mathrm{corr}} = - \int \frac{dk}{(2\pi)^3} \frac{d\omega}{2\pi} \left[\sigma^>(\omega) f - \sigma^<(\omega)(1-f) \right] \frac{\wp'}{\omega - \varepsilon}.
$$

$$(15.97)$$

15.6.1   Generalised Beth–Uhlenbeck formula

First we show that formula (15.97) is a nonequilibrium modification of the correlated density found within the generalised Beth–Uhlenbeck approach. To this end we express the correlated density in terms of the scattering T-matrix.

The correlated density (15.97) by itself is the linear correction in $\gamma \to 0$, therefore the selfenergy (13.12)

$$\sigma_a^<(\omega, k) = \sum_b \int \frac{dp}{(2\pi)^3} \frac{dE}{2\pi} \frac{dq}{(2\pi)^3} \frac{d\Omega}{2\pi} |t_{sc}(\omega + E, k, p, q)|^2$$

$$\times g_b^>(E, p) g_a^<(\omega - \Omega, k - q) g_b^<(E + \Omega, p + q) \tag{15.98}$$

substituted into (15.97) can be taken in the zeroth order approximation in $\gamma \to 0$. This means that in the functional $\sigma[g^{>,<}]$ we can neglect all off-shell contributions and all gradient contributions given by Δ's.

The correlation functions inside the selfenergy is in the on-shell approximation

$$g_b^>(E, p) = (1 - f_2) 2\pi\delta (E - \varepsilon_2),$$

$$g_a^<(\omega - \Omega, k - q) = f_3 2\pi\delta (\omega - \Omega - \varepsilon_3), \tag{15.99}$$

$$g_b^<(E + \Omega, p + q) = f_4 2\pi\delta (E + \Omega - \varepsilon_4),$$

where the abbreviated arguments are (13.20) without shifts. The on-shell approximation (15.99) allows us to integrate out the energies in (15.98),

$$\sigma_a^<(\omega, k) = \sum_b \int \frac{dpdq}{(2\pi)^6} |t_{sc}(\omega + \varepsilon_2, k, p, q)|^2 (1 - f_2) f_3 f_4 2\pi\delta (\omega + \varepsilon_2 - \varepsilon_3 - \varepsilon_4). \tag{15.100}$$

The selfenergy (15.100) substituted into (15.97) provides the correlated density

$$n_a^{corr} = \sum_b \int \frac{dkdpdq}{(2\pi)^9} |t_{sc}(\varepsilon_3 + \varepsilon_4, k, p, q)|^2 \frac{\wp'}{\varepsilon_3 + \varepsilon_4 - \varepsilon_1 - \varepsilon_2}$$

$$\times [(1 - f_1)(1 - f_2) f_3 f_4 - f_1 f_2 (1 - f_3)(1 - f_4)]. \tag{15.101}$$

Formula (15.101) can be directly compared with equation (53) in (Morawetz and Röpke, 1995). We have thus shown that the correlated density resulting from the $\rho[f]$-functional (9.67) is a nonequilibrium form of the generalised Beth–Uhlenbeck formula.

15.6.2 Correlated density in terms of the collision delay

To check the consistency of the theory, we compare the correlated density (15.101) found from the $\rho[f]$-functional with the correlated density (15.77) resulting from the non-instant corrections to the scattering integral. To rearrange (15.101) into a functional of the collision delay we use the derived optical theorem described in Appendix C.

The particular details of the correlated observables are given in Appendix C with resulting formula (C.30) which for the correlated density reads

$$n_c^{\text{corr}} = \sum_{ab} \int \frac{dk\,dp\,dq}{(2\pi)^9} 2\pi \delta(\varepsilon_1 + \varepsilon_2 - \varepsilon_3 - \varepsilon_4) f_3 f_4 (1 - f_1 - f_2) \Delta_t |t_{\text{sc}}(\varepsilon_1 + \varepsilon_2, k, p, q)|^2 \delta_{ac}.$$

(15.102)

Clearly, (15.102) is identical to (15.77).

The agreement between the correlated density found from the Wigner distribution, and the one found from the equation of continuity documents that the subsidiary relation (9.67) for the Wigner distribution is consistent with the kinetic equation. Moreover, this correlated density is identical to the result of the generalised Beth–Uhlenbeck approach.

In the same way we can derive the molecular energy (15.80) and molecular momentum density (15.83) from the correlated parts of the extended quasiparticle picture. In other words, we have convinced ourselves to interpret the correlated observables as molecular states of two colliding particles.

The correlated density depends exclusively on the collision delay, therefore it provides a sensitive test of the concept of the collision delay. As we have seen, the present concept is just one among many. It differs from Wigner's delay (14.2) promoted by Danielewicz and Pratt (1996), and, of course, from purely space non-local corrections studied by Kortemeyer et al. (1996). According to the classical discussion in chapter 3.1.4–5, the differences in the concepts of the collision delay follow from differences in the reference collision events. Since all concepts of the collision delay allow one to reach the correct asymptotic states of the collision, one might feel that all concepts are equivalent. Apparently, they are not. Physical features connected directly or indirectly with the correlated density require us to use the collision delay (13.3).

15.7   Limit of Landau theory

Now we are able to see what approximations are necessary to obtain the variational concept of the Landau theory. First of all, the scatterings have to be in a local approximation $\Delta_t \to 0$. Then the number of quasiparticles equals the number of particles and we can identify the Landau quasiparticle distribution $\tilde{f}$ with the quasiparticle distribution of the spectral concept f.

If the collisions are nearly instantaneous, one might think that the energy gain is also absent. This is not the case. The energy gain gives the missing link between the quasiparticle energies of the variational approach $\tilde{\varepsilon}$ of (15.26) and the ϵ (8.24) as poles of the spectral function (Lipavský, Špička and Morawetz, 1999). The variational selfenergy indeed mimics the last two terms of the energy balance

$$\frac{d\mathscr{E}}{dt} = \int \frac{dk}{(2\pi)^3} \epsilon_k \frac{\partial f_k}{\partial t} + \frac{d}{dt} \int d\mathscr{D} \frac{\epsilon_k + \epsilon_p}{2} \Delta_t - \int d\mathscr{D} \Delta_E,$$

(15.103)

by the rearrangement contribution to the quasiparticle energy as we will see now. Here we abbreviate $d\mathcal{D} = \frac{d^3k d^3 p d^3 q}{(2\pi)^9} D$ with D of (15.14).

In the instantaneous approximation we have from (15.103)

$$\frac{d\mathcal{E}}{dt} = \int \frac{dk}{(2\pi)^3} \epsilon_k \frac{\partial \tilde{f}_k}{\partial t} - \int d\mathcal{D} \Delta_E. \tag{15.104}$$

The last term can be rewritten using $\Delta_E = -\frac{1}{2}\frac{\partial \phi}{\partial t}$

$$-\int d\mathcal{D} \Delta_E = \frac{1}{2}\int d\mathcal{D}\frac{\partial \phi}{\partial t} = \int \frac{dk}{(2\pi)^3} \int d\mathcal{D}\frac{\delta\phi}{\delta \tilde{f}_k}\frac{\partial \tilde{f}_k}{\partial t} \equiv \int \frac{dk}{(2\pi)^3}\epsilon^{\Delta}_k\frac{\partial \tilde{f}_k}{\partial t}, \tag{15.105}$$

where we have introduced the rearrangement energy

$$\epsilon^{\Delta}_k = \int d\mathcal{D}\frac{\delta\phi}{\delta\tilde{f}_k}. \tag{15.106}$$

Using (15.105) in (15.104) we end up with the variational expression of the Landau theory (15.26)

$$\frac{d\mathcal{E}}{dt} = \int \frac{dk}{(2\pi)^3}(\epsilon_k + \epsilon^{\Delta}_k)\frac{\partial \tilde{f}_k}{\partial t}. \tag{15.107}$$

Therefore we conclude that the relation between the quasiparticle energy of Landau's variational concept (15.26) and the spectral quasiparticle energy (9.56) is just given by the rearrangement energy (15.106), $\tilde{\epsilon} = \epsilon + \epsilon^{\Delta}$.

According to (15.105), the elastic collision integral, $\propto \delta(\tilde{\epsilon}_k + \tilde{\epsilon}_p - \tilde{\epsilon}_{k-q} - \tilde{\epsilon}_{p+q})$ with the variational quasiparticle energy $\tilde{\epsilon} = \epsilon + \epsilon^{\Delta}$, yields the same energy conservation as the non-elastic one, $\propto \delta(\epsilon_k + \epsilon_p - \epsilon_{k-q} - \epsilon_{p+q} - 2\Delta_E)$ with the spectral quasiparticle energy. Without sacrificing energy conservation, one can thus circumvent the inconvenience of non-elastic collision integrals by an incorporation of the rearrangement energy provided the number of correlated pairs can be neglected.

15.8   Summary

In highly excited systems, the phenomenological quasiparticles introduced via variational concepts are not identical to elementary excitations observed in the single-particle spectrum. For collisions of a finite duration, these two pictures result in different densities of quasiparticles and one can hardly establish any connection between them. In the limit of instantaneous collisions the densities become identical, a difference, however, remains in the quasiparticle energy. The variational quasiparticle energy includes the rearrangement energy which mimics the non-elasticity of binary collisions caused by the time-dependence of the Pauli blocking during collisions.

The agreement between the correlated density found from the kinetic equation and from the $\rho[f]$-functional shows that the quasiclassical approximation (used to derive the gradient corrections to the scattering integral) and the limit of small scattering rates (used to separate the off-shell motion) are mutually compatible. Note that these two approximation are of a very different nature and usually are viewed as independent. In fact both approximations are based on the common idea of separation of the microscopic and the hydrodynamic scales.

The structure of the total density and the total current reveals a different character of the quasiparticle and the molecular corrections. The quasiparticle corrections renormalise ingredients of formulae for non-interacting particles, e.g. the velocity in (15.21) has quasiparticle corrections. The molecular corrections are additive changing the structure of thermodynamic relations.

Although the kinetic equation (13.28) wears many features of Landau's quasiparticle theory, thermodynamic consequences of the non-instant scattering show principal contradictions with Landau's postulates. In particular, the postulate of the Luttinger theorem that the density of quasiparticles equals the total density of particles is valid for Landau liquids but not fulfilled due to the correlated density. Only in local approximation the Landau theory can be recovered but with a different quasiparticle energy than the one we obtained from the pole of the spectral function. The Landau quasiparticle energy accounts for the rearrangement energy which represents the energy gain during a collision from all other particles.

The complete set of hydrodynamical balance equations derived from the nonlocal kinetic theory shows that all variables consist of quasiparticle ones and short living molecules describing the correlational state of two particles. It is quite satisfying that this set of balance equations is found to be complete, especially the correlational parts of the entropy and entropy current provide a check of the consistency of the theory. Finally the H-theorem, proven to hold also in nonlocal kinetic theory, is based on the presence of these correlated observables.

Part V

Selected Applications

A selection of various applications is presented which illustrate the quantum-kinetic methods worked out so far. These examples serve as models of how the theory can be used in fields ranging from transport in low-dimensional solid-state structures, impurity scattering, spin-orbit coupled systems up to relativistic transport and heavy-ion collisions. A separate chapter is devoted to short-time dynamics since here the formation of correlations allows some analytical results and provides insight into inherent problems of memory-based approaches due to the double count of correlations.

16

Diffraction on a Barrier

In this section we apply the GKB formalism to the diffraction of electrons on a barrier. The system we study is a planar heterojunction of two ideal semi-infinite crystals or a surface of a crystal. As an initial condition we take a stream of electrons with a sharp momentum $(k_\parallel, k)$. We want to find how many electrons get transmitted through and reflected on a junction. For simplicity, the homogeneous infinite crystals are called leads and the inhomogeneous region between them is called the barrier.

The barrier can be either ideal with a perfect translational symmetry along the junction plane or disordered due to an intentional doping or a diffusion of atoms in a junction. Similarly, a surface roughness or an incomplete adsorbed gas layer have also a random character that can be treated as a disordered barrier.

16.1 One-dimensional barrier

Here we discuss the tight-binding version, know under a number of names: surface Green's functions, recursive method, continued fraction, 1D Bethe lattice, or Cayley tree. We consider the tight-binding Hamiltonian

$$H = \sum_{j=-\infty}^{\infty} \left(|j\rangle v_j \langle j| + |j+1\rangle t_{j+1} \langle j| + |j-1\rangle t_j \langle j| \right) \tag{16.1}$$

describing N molecules in-between the crystal leads at the left, $j \leq 0$, and the right side, $j \geq N + 1$. The energy levels are v_j and the hopping between neighbouring sites are t_j.

Similarly to the scattering on a single impurity, if we want to employ Green's functions for the diffraction on the barrier, we have to separate the incoming wave from the diffracted parts. This is because the incoming wave has a different asymptotic behaviour at infinity.

16.1.1 Convenient formulation of the boundary problem

For the formulation of the scattering one might be tempted to use the Lippmann–Schwinger equation (4.23). To this end one writes the Hamiltonian as a sum of the

Interacting Systems far from Equilibrium. Klaus Morawetz, Oxford University Press (2018).
© Klaus Morawetz. DOI: 10.1093/oso/9780198797241.001.0001

perturbation, $H' = H - H^0$ and the unperturbed crystal, H^0, made by an extension of the left semi-infinite crystal, $t_j = t_0$ and $v_j = v_0$ for $j \leq 0$, to the whole space. An incoming wave ψ^0 is not yet influenced by the barrier of the leads, thus it is the eigenstate of H^0,

$$H^0\psi^0 = E\psi^0. \tag{16.2}$$

Eigenstates of the homogeneous system are plane waves with a momentum k related to the energy by $E = v_0 + 2t_0 \cos(ka)$ with the distance of atoms a. The wave approaching the barrier is $\psi^0(x) = \exp(ikx)$ or discretised $x_j = j\,a$ in the tight-binding representation $\psi_j^0 = \exp(ikaj)$.

The total wave function $\psi^0 + \psi'$ solves the Schrödinger equation

$$(H^0 + H')(\psi^0 + \psi') = E(\psi^0 + \psi'), \tag{16.3}$$

which can be rearranged with the help of (16.2) into the Lippmann–Schwinger equation,

$$(E - H)\psi' = H'\psi^0. \tag{16.4}$$

The boundary condition that the diffracted wave propagates from the barrier to infinity is realised by an infinitesimal shift of the energy E into the complex plane.

The above straightforward implementation of the Lippmann–Schwinger idea becomes numerically extremely inconvenient if the left and right leads are different by hopping, energy levels and external bias. For instance, when the tunnelling junction is biased by the voltage V, the perturbation H' in the right half-space equals the potential of the bias voltage, $H'_{jj} = v_{N+1}$ for all $j \geq N + 1$. The perturbation then extends everywhere except for the left lead, and the source term in the right-hand side of (16.4) is nonzero over this region, too.

In this case, the Lippmann–Schwinger separation of the wave function into an infinite plane wave ψ^0 and correction ψ is not a favourable starting step. This is clearly seen for the case when electrons have an energy at the level of the right lead. All electrons are then reflected at the barrier and only exponential tails of the wave function penetrate into the lead. The finite value of ψ^0 has to be compensated by ψ. This requires a large source term, $H'\psi^0$, and a very accurate treatment of the Lippmann–Schwinger equation. In fact, only identities can guarantee a correct compensation of the large parts of incoming and outgoing waves.

There is a simple modification of the Lippmann–Schwinger idea by which we can circumvent the penetration of the incoming wave into the barrier and the right lead. Let us cut the incoming wave by setting

$$\psi_j^C = \psi_j^0 = \langle j|\Psi^0\rangle = e^{ikaj} \qquad \text{for } j \leq 0, \text{ and}$$

$$\psi_j^C = 0 \qquad\qquad\qquad \text{for } j \geq 1. \tag{16.5}$$

The wave function is now split differently than before, $\psi^0 + \psi' = \psi^C + \psi$, and Schrödinger's equation, $(E - H)(\psi^C + \psi) = 0$, gives

$$(E - H)\psi = \Omega, \qquad \Omega = -(E - H)\psi^C. \tag{16.6}$$

The term Ω is a well-behaved source which is nonzero only at two layers, $j = 0$ and $j = 1$, with

$$\Omega_0 = -t_0\psi_1^0 = -t_0 e^{ika}, \qquad \Omega_1 = t_1\psi_0^0 = t_1, \tag{16.7}$$

which can be easily proved calculating $\langle j|E - H|\psi^C\rangle$ with the help of (16.1). The wave function ψ does not include the incoming part, we can thus use the retarded boundary condition, $z = E + i0$, and write the wave function in terms of Green's functions,

$$\psi = \frac{1}{z - H}\Omega = G\Omega \tag{16.8}$$

or explicitly

$$\psi_j = G_{j1}\Omega_1 + G_{j0}\Omega_0. \tag{16.9}$$

Therefore all what we have to provide is to to solve the Green's functions equation

$$G(z - H) = 1 \tag{16.10}$$

which means the inversion of the matrix $z - H$. The method of surface Green's functions provides an extremely fast numerical inversion method.

16.2   Matrix inversion with surface Green's functions

Let us therefore return to the tight-binding Hamiltonian (16.1) which has the matrix structure

$$H = \begin{pmatrix} & \cdots & & & & & & \\ 0 & t_{j-1} & v_{j-1} & t_j & 0 & 0 & 0 \\ 0 & 0 & t_j & v_j & t_{j+1} & 0 & 0 \\ 0 & 0 & 0 & t_{j+1} & v_{j+1} & t_{j+2} & 0 \\ 0 & 0 & 0 & 0 & t_{j+2} & v_{j+2} & t_{j+3} \\ & & & & & \cdots & \end{pmatrix} \equiv \begin{pmatrix} H^{ll} & H^{lr} \\ \hline H^{rl} & H^{rr} \end{pmatrix} \tag{16.11}$$

where we have cut the matrix at the side j. The cut parts read separately as

$$H^{rr} = \sum_{i=j+1}^{\infty} \Big(|i\rangle v_i \langle i| + |i+1\rangle t_{i+1}|i\rangle + |i-1\rangle t_i |i+1\rangle \Big)$$

$$H^{ll} = \sum_{i=-\infty}^{j} \Big(|i\rangle v_i \langle i| + |i+1\rangle t_{i+1}|i\rangle + |i-1\rangle t_i |i+1\rangle \Big)$$

$$H^{lr} = |j\rangle t_{j+1} \langle j+1|$$

$$H^{rl} = |j+1\rangle t_{j+1} \langle j|. \tag{16.12}$$

Now we employ the general inversion formulae of matrices composed to 2x2 operators, $GB = 1$. The upper-right equation reads

$$G_{11}B_{12} + G_{12}B_{22} = 0 \tag{16.13}$$

from which we obtain

$$G_{12} = -G_{11}B_{12}B_{22}^{-1}. \tag{16.14}$$

This is used in the upper-left equation of $GB = 1$

$$G_{11}B_{11} + G_{12}B_{21} = 1 \tag{16.15}$$

to find finally

$$G_{11} = \left(B_{11} - B_{12}B_{22}^{-1}B_{21} \right)^{-1} \tag{16.16}$$

and from (16.14) the element G_{12} follows. The other two elements of G are given by interchanging $1 \leftrightarrow 2$.

These formulae allows to write the Green's functions (16.10) according to the cutting (16.11) as

$$G_{11} = \left(z - H^{ll} - H^{lr} G^{rr} H^{rl} \right)^{-1}, \quad G_{12} = G_{11}H^{lr}G^{rr},$$

$$G_{21} = G_{22}H^{rl}G^{ll}, \quad G_{22} = \left(z - H^{rr} - H^{rl} G^{ll} H^{lr} \right)^{-1}. \tag{16.17}$$

We consider now the specific energy parts in the diagonal Green's functions with the help of (16.12)

$$H^{lr}G^{rr}H^{rl} = |j\rangle t_{j+1} \langle j+1|G^{rr}|j+1\rangle t_{j+1} \langle j| \equiv t_{j+1}^2 S_{j+1}^r |j\rangle \langle j| \tag{16.18}$$

which defines the right-side surface Green's functions S^r. Analogously we obtain the left-side Green's functions

$$H^{rl} G^{ll} H^{lr} = |j+1\rangle t_{j+1} \langle j | G^{ll} | j \rangle t_{j+1} \langle j+1| \equiv t_{j+1}^2 S_j^l | j+1\rangle \langle j+1|. \qquad (16.19)$$

These surface Green's functions obey simple recursion relations. To see this we cut the left Hamiltonian as

$$H^{ll} = \begin{pmatrix} \ddots & & 0 \\ t_{j-1} & v_{j-1} & t_j \\ 0 & t_j & v_j \end{pmatrix} \equiv \begin{pmatrix} H^{ll}[j-1] & \begin{pmatrix} \vdots \\ 0 \\ t_j \end{pmatrix} \\ \hline (\dots 0\, t_j) & v_j \end{pmatrix} \qquad (16.20)$$

where $H^{ll}[j-1]$ denotes the left-side Hamiltonian when (16.11) is cut at $j-1$. Due to the equation $G^{ll}(z - H^{ll}) = 1$ this partition (16.20) enforces the structure of the Green's functions

$$G^{ll} = \begin{pmatrix} G^{ll}[j-1] & \ddots \\ \ddots & G_{jj}^{ll} \end{pmatrix}. \qquad (16.21)$$

The left-upper part is the Green's functions G^{ll} with the cut at $j-1$. With the help of the inversion formulae above, the lower-right element of (16.21) determines the left-side surface Green's functions

$$S_j^l \equiv G_{jj}^{ll} = \cfrac{1}{z - v_j - (\dots 0, -t_j) G^{ll}[j-1] \begin{pmatrix} \vdots \\ 0 \\ -t_j \end{pmatrix}} = \cfrac{1}{z - v_j - t_j^2 S_{j-1}^l} \qquad (16.22)$$

which establishes the recursion formula for the left-side surface Green's functions. Similarly one obtains the recursion for the right-side surface Green's functions

$$S_{j+1}^r = \cfrac{1}{z - v_{j+1} - t_{j+1}^2 S_{j+2}^r}. \qquad (16.23)$$

If the tight-binding system is in-between two leads with the left material characterised by $t_j = t_0$ and the crystal levels $v_j = v_0$ for $j \le 0$ and the right material characterised by $t_j = t_{N+1}$ and $v_j = v_{N+1}$ for $j \ge N+1$ one finds from (16.22) directly the surface Green's functions of the leads

$$S_0^l = \frac{1}{t_0} \left(\frac{z - v_0}{2t_0} + i\sqrt{1 - \left(\frac{z - v_0}{2t_0}\right)^2} \right) \qquad (16.24)$$

and analogously for S^r_{N+1}. Together with (16.22) this determines all the surface Green's functions completely.

With the knowledge of the surface Green's functions we can now provide the complete Green's functions (16.16). Employing the familiar relation $\frac{1}{A-B} = \frac{1}{A} + \frac{1}{A}B\frac{1}{A} + ...$ we can write the matrix elements of

$$\langle k|G_{11}|m\rangle = \langle k|\frac{(1)}{(G^{ll})^{-1} - a|j\rangle\langle j|}|m\rangle$$

$$= \langle k|G^{ll} + G^{ll}a|j\rangle\langle j|G^{ll} + G^{ll}a|j\rangle\langle j|G^{ll}a|j\rangle\langle j|G^{ll} + ...|m\rangle$$

$$= G^{ll}_{km} + G^{ll}_{kj}a\left(1 + S^l_j a + (S^l_j a)^2 + ...\right)G^{ll}_{jm} = G^{ll}_{km} + \frac{G^{ll}_{kj}aG^{ll}_{jm}}{1 - aS^l_j} \qquad (16.25)$$

where we used the abbreviation $a = S^r_{j+1} t^2_{j+1}$. Analogously we obtain

$$\langle k|G_{12}|m\rangle = \frac{G^{ll}_{kj} t_{j+1} G^{rr}_{j+1m}}{1 - aS^l_j} \qquad (16.26)$$

$$\langle k|G_{22}|m\rangle = G^{rr}_{km} + \frac{G^{rr}_{kj+1} b G^{rr}_{j+1m}}{1 - bS^r_{j+1}} \qquad (16.27)$$

$$\langle k|G_{21}|m\rangle = \frac{G^{rr}_{kj+1} t_{j+1} G^{ll}_{jm}}{1 - bS^r_{j+1}} \qquad (16.28)$$

with the abbreviation $b = t^2_{j+1} S^l_j$.

Now we remember the structure of the complete Green's functions following (16.11)

$$\begin{pmatrix} \begin{array}{c|c} \begin{array}{c} m \\ \hline k \\ \downarrow \\ \vdots \\ j \\ j+1 \\ \vdots \end{array} & \begin{array}{c} \rightarrow ..j\,j+1... \\ \hline \\ G_{11} \quad G_{12} \\ \\ \\ G_{21} \quad G_{22} \end{array} \end{array} \end{pmatrix}. \qquad (16.29)$$

Comparing G_{21} of (16.28) for $k = j + 1$ and $m \le j$ in (16.29) with G_{11} of (16.25) for $k = j$ and $m \le j$ in (16.29) we obtain the recursive relation

$$G_{j+1,m} = t_{j+1} S^r_{j+1} G_{j,m}, \qquad m \le j \qquad (16.30)$$

which allows us to construct the Green's functions on one side from its diagonal part. Analogously we obtain the other diagonal part by comparing G_{22} of (16.27) for $k = j+1$ and $m \ge j + 1$ in (16.29) with G_{12} of (16.25) for $k = j$ and $m \ge j + 1$ in (16.29) as

$$G_{j,m} = t_{j+1} S_j^l G_{j+1,m}, \qquad m \geq j+1. \tag{16.31}$$

The diagonal form is found from G_{11} of (16.25) easily to be

$$\langle k|G_{11}|k\rangle = \frac{S_k^l}{1 - t_{k+1}^2 S_{k+1}^r S_k^l} = \frac{1}{z - v_k - t_{k+1}^2 S_{k+1}^r - t_k^2 S_{k-1}^l} \tag{16.32}$$

where we used (16.22) once more. This completes the recursive construction of the Green's functions.

16.2.1   Summary of the method

For computational purposes let us collect the important steps and formulae.

1. Calculate the surface Green's functions of the left and right lead according to (16.24)

$$S_{N+1}^r = \frac{1}{t_{N+1}} \left(\frac{z - v_{N+1}}{2 t_{n+1}} + i \sqrt{1 - \left(\frac{z - v_{N+1}}{2 t_{N+1}} \right)^2} \right)$$

$$S_0^l = \frac{1}{t_0} \left(\frac{z - v_0}{2 t_0} + i \sqrt{1 - \left(\frac{z - v_0}{2 t_0} \right)^2} \right). \tag{16.33}$$

2. Determine the surface Green's functions due to the recursive relations (16.22) and (16.23)

$$S_j^l = \frac{1}{z - v_j - t_j^2 S_{j-1}^l}$$

$$S_{j+1}^r = \frac{1}{z - v_{j+1} - t_{j+1}^2 S_{j+2}^r}. \tag{16.34}$$

3. Compute the required elements of the Green's functions matrix according to (16.30), (16.31) and (16.32)

$$G_{kk} = \frac{1}{z - v_k - t_{k+1}^2 S_{k+1}^r - t_k^2 S_{k-1}^l}$$

$$G_{j,m} = t_{j+1} S_j^l G_{j+1,m}, \qquad m \geq j+1$$

$$G_{j+1,m} = t_{j+1} S_{j+1}^r G_{j,m}, \qquad m \leq j. \tag{16.35}$$

If there are equal hopping elements $t_k = u$ we can express the off-diagonal element in terms of its diagonal element,

$$G_{ji} = \prod_{m=n-1}^{j} S_m^l u \times G_{nn} \times \prod_{m=n+1}^{i} u S_m^r, \qquad j \le n \le i \qquad (16.36)$$

and

$$G_{ji} = \prod_{m=n+1}^{j} S_m^r u \times G_{nn} \times \prod_{m=n-1}^{i} u S_m^l, \qquad j \ge n \ge i. \qquad (16.37)$$

16.3   Reflection and transmission

When considering the left lead, the incoming wave shows the reflected part characterised by the reflection coefficient r

$$\psi_{-1} = r e^{ika} = t_0 S_{-1}^l (G_{01} \Omega_1 + G_{00} \Omega_0) \qquad (16.38)$$

where we used the recursion formula (16.31) for the Green's functions. Since the energy conservation reads $\cos(ka) = (E - v_0)/2t_0$ we employ the explicit from of the surface Green's functions in the lead (16.24) to derive the relation

$$t_0 S_0^l = e^{ika}. \qquad (16.39)$$

This allows us to identify from (16.38) the reflection coefficient

$$r = G_{01} \Omega_1 + G_{00} \Omega_0. \qquad (16.40)$$

Considering the right lead, we have the transmission coefficient according to

$$\begin{aligned}
\psi_{N+1} &= t e^{i\kappa a(N+1)} \\
&= G_{N+1,1} \Omega_1 + G_{N+1,0} \Omega_0 \\
&= t_{N+1} S_{N+1}^r (G_{N,1} \Omega_1 + G_{N,0} \Omega_0) \qquad (16.41)
\end{aligned}$$

where we used the recursion relation (16.30). Since the energy conservation on the right lead means $\cos(\kappa)a = (z - v_{N+1})/2t_{N+1}$ we employ the explicit form of the surface Green's functions in the lead (16.33) to obtain the relation

$$t_{N+1} S_{N+1}^r = e^{i\kappa a} \qquad (16.42)$$

and from (16.41) follows the transmission coefficient

$$
\begin{aligned}
t &= e^{-i\kappa a N}\left(G_{N,1}\Omega_1 + G_{N,0}\Omega_0\right)\\
&= -2it_1 e^{-i\kappa a N} G_{N,0}\,\sin\left(ka\right)\\
&= -2it_1 e^{-i\kappa a(N+1)} G_{N+1,0}\,\sin\left(ka\right)
\end{aligned}
\tag{16.43}
$$

representing different equivalent forms. The reflection $R = |r|^2$ and transmission $T = |t|^2$ obey $T + R = 1$, of course.

16.3.1   Current and current fluctuations

For the tight-binding Hamiltonian (16.1) the velocity operator $\dot{x} = \frac{i}{\hbar}[H, x]$ is easily computed from the position operator $x = a\sum_j j|j\rangle\langle j|$ as

$$
\dot{x} = \frac{ia}{\hbar}\sum_j t_{j+1}\left(|j\rangle\langle j+1| - |j+1\rangle\langle j|\right).
\tag{16.44}
$$

The mean current $j = \langle e\dot{x}\rangle$ is conserved. Therefore it is sufficient to calculate the current which runs from the atoms to the lead on the right side

$$
j = \frac{iea}{\hbar}t_{N+1}\left(\langle\psi_N^*\psi_{N+1}\rangle - \langle\psi_{N+1}^*\psi_N\rangle\right) = \frac{iea}{\hbar}t_{N+1}\left(G_{N,N+1}^< - G_{N+1,N}^<\right)
\tag{16.45}
$$

where we introduced the definition of the correlated Green's functions $G^<$. The Green's functions provided in the Appendix A are causal functions. According to the Langreth-Wilkins rules (7.18), (7.17) and the recursion relation of the causal Green's functions (16.30) we can write

$$
\begin{aligned}
G_{N,N+1}^< &= t_{N+1}\left(G_{NN}^R S_{N+1}^< + G_{NN}^< S_{N+1}^A\right)\\
G_{N+1,N}^< &= t_{N+1}\left(S_{N+1}^R G_{NN}^< + S_{N+1}^< G_{NN}^A\right)
\end{aligned}
\tag{16.46}
$$

and repeatedly applying (16.22) the current (16.45) reads

$$
j = \frac{ae}{\hbar}\prod_{j=1}^{N+1} t_j^2 \prod_{k=1}^{N-1} |S_k^l|^2 \left(S_{N+1}^{r<}S_0^{l>} - S_{N+1}^{r>}S_0^{l<}\right).
\tag{16.47}
$$

Further simplification can be achieved by rewriting the diagonal Green's functions G_{NN} in (16.47) into the Green's functions $G_{N+1,0}$ with the help of (16.31). We observe that the surface-correlated Green's function $S_0^{l<}$ is the one on the left lead. It can be expressed in terms of the spectral function $A_l = i(S_0^{lR} - S_0^{lA})$ and the Fermi–Dirac function f_l describing the occupation of the left lead via $S_0^{l<} = A^l f_l$ and $S_0^{l>} = A^l(1-f_l)$ and analogously for the right lead. One finally obtains for the current $\mathcal{J} = j/a$

$$\mathcal{J}(z) = \frac{e}{\hbar}|G_{N+1,0}(z)|^2 t_0^2 t_{N+1}^2 A_{N+1}^r(z) A_0^l(z) \left[f_r\left(\frac{z-eV}{T} \right) - f_l\left(\frac{z}{T} \right) \right]. \quad (16.48)$$

In order to obtain the conductance we have to integrate over all energies and to divide by the voltage $V = U/e$

$$G = \frac{e}{U}\int \frac{dz}{2\pi}\mathcal{J}(z) \equiv \frac{e^2}{U}\int \frac{dz}{2\pi\hbar} T_{N+1,0}\,[f_r - f_l] \quad (16.49)$$

which is the known Landauer–Büttiker form of conductance (Blanter and Büttiker, 2000; Kohler, Lehmann and Hänggi, 2005).

The time-averaged current autocorrelation function (Camalet, Kohler and Hänggi, 2004)

$$\bar{\mathcal{J}}^2 = \lim_{T\to\infty} \frac{1}{T}\int\limits_0^T dt \int\limits_{-\infty}^{\infty} d\tau \langle \Delta\mathcal{J}(t)\Delta\mathcal{J}(t-\tau)\rangle \quad (16.50)$$

describes the zero-frequency noise of the current (16.48). The result with (16.48) reads (Büttiker, 1992; Blanter and Büttiker, 2000)

$$\bar{\mathcal{J}}^2 = \frac{e^2}{2\pi\hbar U}\int dz T_{N+1,0}\left[T_{N+1,0}(f_r - f_l)^2 + (1-f_r)f_r + (1-f_l)f_l \right]. \quad (16.51)$$

As a measure for the fluctuations one uses the Fano factor

$$F = \frac{\bar{\mathcal{J}}^2}{e\mathcal{J}}. \quad (16.52)$$

16.4   Transport coefficients: parallel stacked organic molecules

Now we have developed the tools, it is very easy to compute all transport coefficients, even bias-voltage dependent. We assume a different temperature on the right lead $T_r = T + \Delta T/2$ and on the left lead $T_r = T - \Delta T/2$. Linearising with respect to the temperature gradient, but not with respect to the voltage, the particle current (16.48) reads

$$\mathcal{J} = L^{11}V + L^{12}\Delta T \quad (16.53)$$

with the nonlinear Onsager coefficients

$$L^{11} = \frac{e}{\hbar V} \int dz\, T_{N+1,0} \left[f_r \left(\frac{z - eV}{T} \right) - f_l \left(\frac{z}{T} \right) \right]$$

$$L^{12} = -\frac{e}{2\hbar} \int dz\, T_{N+1,0} \left[\frac{z - eV}{T^2} f_r' - \frac{z}{T^2} f_l' \right] \tag{16.54}$$

with $f' = -f(1-f)$. With absent temperature gradients we have the conductance (16.49)

$$G = L^{11}. \tag{16.55}$$

The voltage compensating the current due to the temperature gradient, $V = -Q\Delta T$, such that $\mathcal{J} = 0$ determines the thermoelectric power (Seebeck coefficient) as

$$Q = \frac{L^{12}}{L^{11}}. \tag{16.56}$$

Analogously we can give the heat current $\mathcal{J}_q$ which is just (16.48) but with an additional energy factor z under the integrand. With small temperature gradients the heat current takes the form

$$\mathcal{J}_q = L^{21} V + L^{22} \Delta T \tag{16.57}$$

with the nonlinear Onsager coefficients

$$L^{21} = \frac{e}{\hbar V} \int dz\, z\, T_{N+1,0} \left[f_r \left(\frac{z - eV}{T} \right) - f_l \left(\frac{z}{T} \right) \right]$$

$$L^{22} = -\frac{e}{2\hbar} \int dz\, z\, T_{N+1,0} \left[\frac{z - eV}{T^2} f_r' - \frac{z}{T^2} f_l' \right]. \tag{16.58}$$

The thermoelectric conductance is measured if we keep the particle current zero, $\mathcal{J} = 0$, which results from (16.53) and (16.57) into

$$\mathcal{J}_q = K\Delta T \tag{16.59}$$

with the thermoelectric conductance

$$K = L^{22} - \frac{L^{21} L^{12}}{L^{11}}. \tag{16.60}$$

The ratio between the thermoelectric conductance and the conductance is called the Wiedemann–Franz law and should be linearly proportional to the temperature

$$\frac{K}{G} = \frac{L^{22}}{L^{11}} - \frac{L^{12}L^{21}}{(L^{11})^2}. \tag{16.61}$$

As a measure for the effectiveness of the thermoelectric devices the dimensionless figure of merit or ZT factor is often presented (Hicks and Dresselhaus, 1993) as

$$ZT = \frac{TQ^2 G}{K} = \frac{T(L^{12})^2}{L^{11}L^{22} - L^{21}L^{12}} \tag{16.62}$$

which has become an important quantity for thin-film thermoelectric devices (Venkata-subramanian, 2000; Venkatasubramanian, Silvola, Colpitts and O'Quinn, 2001).

In Figure 16.1 we present the results for the transport coefficients dependent on temperature and the external bias of parallel-stacked thiophene molecules biased in between two leads (Morawetz, Gemming, Luschtinetz, Kunze, Lipavský, Seifert and Eng, 2008). With increasing applied voltage the conductance is lowered since the overlap between left- and right-conductance channel shrinks. The Wiedemann–Franz law is seen to be

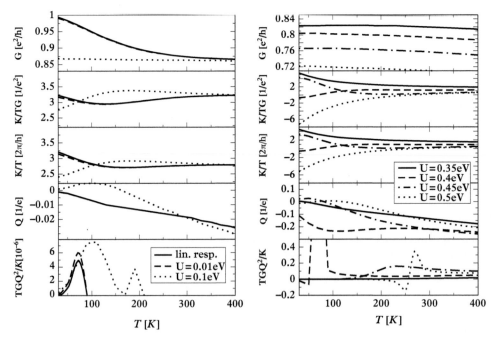

Figure 16.1: *The conductance G, thermal heat conductance K, Wiedemann–Franz law, thermopower Q and figure of merit of thiophene molecules versus temperature for different applied-voltage bias near the linear response (left) and far beyond linear response (right).*

not fulfilled strictly. Instead we observe a deviation from the linear temperature behaviour of up to 10% for lower temperatures. The figure of merit for low-applied voltages is negligible but plotted for completeness. The situation changes if we apply higher voltages. For low temperatures the thermal conductance even changes sign if we approach an external voltage comparable to the highest occupied molecular orbital (HOMO) levels. This has dramatic consequences on the figure of merit. At the temperature where the thermal conductance is changing its sign we observe a resonance structure enhancing the figure of merit dramatically. Interestingly it can be observed that the thermopower is changing its sign for certain temperatures and becomes completely negative for voltages equal to the HOMO levels. This shows that for such voltages the thermal current is reversing its sign which can be seen analogously to the refrigerator effect described in (Rey, Strass, Kohler, Hänggi and Sols, 2007) by an energy-selective transmission of electrons through a spatially asymmetric resonant structure subject to a.c. driving.

16.5 Disordered barrier

Real barriers are never ideal. A diffusion of atoms between the lead and barrier results in a disordered occupations of sites at layers close to the surface of the barrier. A stronger disorder appears when the barrier is made of an alloy, e.g. $Ga_{1-c}Al_cAs$. The concentration c used to control properties of the barrier varies from layer to layer. For simplicity we will call Al atoms impurities.

The disorder complicates the diffraction in two essential points. Firstly, an electron can lose or gain a parallel momentum via collisions with impurities. The diffraction then cannot be reduced to the 1D system but it is of the 3D nature. Secondly, the averaging over random configurations of impurities requires us to employ the multiple scattering theory. From the above methods, Green's functions are best suited to handle this task.

16.5.1 Configurational averaging

In the spirit of Chapter 5, we make the configurational averaging over the square of the scattered wave function,

$$G^< = \left\langle \psi' \psi'^\dagger \right\rangle_S = \left\langle G^R \Omega \Omega^\dagger G^A \right\rangle_S. \tag{16.63}$$

The superscripts R and A denote retarded and advanced boundary conditions, $z = E \pm i0$. Our aim is to formulate an equation for $G^<$.

The source Ω and its conjugate $\Omega^\dagger$ are independent of the barrier, because the cut is before the disordered region. Since the configuration averaging relates only to propagators, we can implement the GKB formalism developed in Chapters 4.6.1 and 7.1. If we denote by V the random part of the impurity potential and by H^0 the regular part, equation (16.63) can be expressed in terms of the power expansion,

$$G^< = \left\langle \left(G^{0R} + G^{0R} V G^{0R} + G^{0R} V G^{0R} V G^{0R} + \ldots \right) \Omega \Omega^\dagger \left(G^{0A} + G^{0A} V G^{0A} + \ldots \right) \right\rangle_S. \tag{16.64}$$

If one denotes the unperturbed part as

$$G^{0<} = G^{0R}\Omega\Omega^{\dagger}G^{0A}, \tag{16.65}$$

the n-th order of (16.64) becomes identical to (6.5). The $G^<$ defined by (16.63) hence obeys the GKB equation,

$$G^< = G^R\Sigma^<G^A + (1 + G^R\Sigma^R)G^{0<}(1 + \Sigma^AG^A). \tag{16.66}$$

As in the general case discussed in Section 16.5, the GKB equation is consistent with any approximation of the selfenergy.

Coherent diffraction   Some electrons keep their initial parallel momentum, $k_{\parallel}$, i.e. they are diffracted coherently, as on the ideal barrier. These processes are covered by the second term of (16.66),

$$\hat{G}^< = (1 + G^R\Sigma^R)G^{0<}(1 + \Sigma^AG^A) = G^R\Omega\Omega^{\dagger}G^A. \tag{16.67}$$

With explicit site indices, the coherent part reads

$$\hat{G}^<_{jr,is} = \sum_{t,w\in\text{plane}} \sum_{l,m=0,-1} G^R_{jr,lt}\Omega_{lt}\Omega^*_{mw}G^A_{mw,is}. \tag{16.68}$$

After the averaging, the system is again a translational invariant so that it is advantageous to transform the in-plane site indexes into the momentum representation. From the Fourier transformation of the source term we find that the source is singular in momentum,

$$\sum_{t\in\text{plane}} \Omega_{lt}\Omega^*_{mw}e^{-ip_{\parallel}(r_t-r_w)} = |\Omega_l)(\Omega_m|(2\pi)^2\delta(p_{\parallel}-k_{\parallel}). \tag{16.69}$$

Here, $|\Omega_{0,1})$ are from (16.91), and $(\Omega_{0,1}|$ are Hermitian conjugate to them. Since propagators, $G^{R,A}$ are regular in $p_{\parallel}$, the distribution of coherently diffracted particles $\hat{G}^<$ is singular, as one expects. We remind ourselves that the conservation of the parallel momentum applies only within the Brillouin zone. Processes with momentum transfer by multiples of the Brillouin vectors are included. The approximation of $G^<$ by $\hat{G}^<$ is called the non-vertex approximation.

16.5.2   Open system

One can take the barrier as a subsystem which is embedded into a reservoir formed by leads. If the barrier is very thick, its properties are dominated by the barrier region itself while the connection to the reservoir represents a perturbation. With the help of the diffraction we can thus formulate the contact between an open system and its reservoir.

To this end we make the Fourier transform of the GKB equation (16.66),

$$G_{ji}^{<}(p_{\parallel}) = \sum_{l,m=1}^{N} G_{jl}^{R}(p_{\parallel}) \Sigma_{lm}^{<}(p_{\parallel}) G_{mi}^{A}(p_{\parallel}) + (2\pi)^2 \delta(p_{\parallel}-k_{\parallel}) \sum_{l,m=-1,0} G_{jl}^{R}(k_{\parallel}) |\Omega_l\rangle \langle\Omega_m| G_{mi}^{A}(k_{\parallel}).$$

(16.70)

In the barrier region, $1 \leq j, i \leq N$, we express the off-diagonal elements in the source term from (16.36) and (16.37)

$$\sum_{l,m=-1,0} G_{jl}^{R} |\Omega_l\rangle \langle\Omega_m| G_{mi}^{A} = G_{j1}^{R} \times tS_0^{lR} \left(|\Omega_0\rangle + tS_{-1}^{lR}|\Omega_{-1}\rangle \right) \left(\langle\Omega_0| + \langle\Omega_{-1}|S_{-1}^{lA}t \right) S_0^{rA}t \times G_{1i}^{A}.$$

(16.71)

With respect to its effect on the barrier region, the source term can be replaced by a source acting exclusively on the 1-th layer,

$$\Sigma^{l<} = tS_0^{lR} \left(|\Omega_0\rangle + tS_{-1}^{lR}|\Omega_{-1}\rangle \right) \left(\langle\Omega_0| + \langle\Omega_{-1}|S_{-1}^{lA}t \right) S_0^{rA}t. \quad (16.72)$$

It is clear that if there would be electrons approaching the barrier from the right-hand side, a similar source, $\Sigma^{r<}$, results acting on the N-th layer. In the barrier region, the GKB equation with suppressed arguments reads

$$G^{<} = G^{R}(\Sigma^{<} + \Sigma^{l<})G^{A}. \quad (16.73)$$

According to its position in the GKB equation (16.73), we have denoted the source, $\Sigma^{l<}$, by the symbol Σ reserved otherwise for the selfenergy. There are similarities and distinctions in their interpretations. Firstly, according to the Lippmann–Schwinger equation (4.23), the transport vertex $\Sigma^{<}$ describes an electrons sourced at a site where the crystal potential is perturbed by an impurity. Similarly, according to the Lippmann–Schwinger equation (16.6), the source term $\Sigma^{l<}$ describes electrons passing through the contact between the lead and barrier. Secondly, all electrons sourced by $\Sigma^{<}$ have been stolen from the system by the decay. The consistency between the transport vertex, $\Sigma^{<}$, and the decay rate $\propto -2\text{Im}\Sigma^{R}$, guarantees that all stolen electrons are re-emitted into the system, i.e. the number of particles conserves. In contrast, new electrons enter via the source $\Sigma^{l<}$. This arrival of electrons is compensated by the emission into leads, however, there is no consistency requirement between the two processes since emitted electrons never return into the barrier. The rate of emissions into leads reminds us that the decay is proportional to the imaginary part of the selfenergy-like surface terms,[1]

---

[1] In equilibrium the number of emitted particles equals the number of entering particles,

$$-2f(E)\text{Im}\Sigma^{lR} = \Sigma^{l<}, \qquad -2f(E)\text{Im}\Sigma^{rR} = \Sigma^{r0}.$$

This relation parallels the equilibrium relation between the selfenergy and the transport vertex, $-2f(E)\text{Im}\Sigma^{R} = \Sigma^{<}$.

$$\Sigma^{\mathrm{l}} = tS_0^{\mathrm{l}}t, \qquad \Sigma^{\mathrm{r}} = tS_{N+1}^{\mathrm{r}}t. \tag{16.74}$$

The sources, $\Sigma^{\mathrm{l}<}$ and $\Sigma^{\mathrm{r}<}$, and the surface terms, Σ^{l} and Σ^{l}, allow one to describe the contact between the open system and reservoir in the form which parallels the selfenergy. In this sense, the contact can be described as a perturbation.[2]

16.5.3   Dissipative diffraction: example GaAs/AlAs layer

To complete the description of diffraction we evaluate the incoherent part. Firstly we have to specify the approximation of the selfenergy. We use the virtual crystal approximation (VCA) of Chapter 5.2.1 for the hopping elements, and the coherent potential approximation (CPA) of Chapter 5.6 for the local elements.

The VCA results into an effective hopping

$$t = (1 - c)t_{\mathrm{GaAs}} + ct_{\mathrm{AlAs}}, \tag{16.75}$$

which varies from layer to layer following the concentration of anionic layers it connects.

The selfenergy in the CPA is site diagonal, i.e. it is layer diagonal and independent of the parallel momentum, $p_\parallel$,

$$\Sigma_{ij}(p_\parallel) = \delta_{ij}\Sigma_j. \tag{16.76}$$

Via the CPA condition,

$$c_j\frac{v_j^{\mathrm{Ga}} - \Sigma_j}{1 - (v_j^{\mathrm{Ga}} - \Sigma_j)g_j} + (1 - c)\frac{v_j^{\mathrm{Al}} - \Sigma_j}{1 - (v_j^{\mathrm{Al}} - \Sigma_j)g_j} = 0, \tag{16.77}$$

the selfenergy depends only on the site-diagonal element of Green's function,

$$g_j = \int_{B.Z.} \frac{dp_\parallel}{(2\pi)^2} G_{jj}(p_\parallel), \tag{16.78}$$

obtained from the layer-diagonal element by integration over the surface Brillouin zone.

---

[2] Except for contact with a very small hopping between 0-th site and 1-th site, the contact has to be treated unperturbatively. For instance, the isolated slab of N layers has discrete quantum levels in the perpendicular direction, while the slab in contact with the reservoir has a continuous spectrum of a band. This principle difference between the two energy spectra is independent from the thickness of the slab.

One might expect that with thicker and thicker slabs, $Na \to \infty$, the contact will play a smaller and smaller role. Indeed, the of wave function of the m-th state, $\psi^m(x) = \sqrt{\frac{2}{Na}}\sin m\frac{\pi}{Na}x$, has the amplitude on the affected first layer, $|\psi_1^m|^2 = \frac{2\pi^2 m^2}{N^3 a}$, which vanishes in this limit. On the other hand, the energy scale on which corrections appreciably change the character of the spectra follows the scale of energy levels, $E_m = 2t\left(1 - \cos m\frac{\pi}{N}\right) = t\frac{m^2\pi^2}{N^2}$, and the number of states with which a selected state couples is proportional to N. The claim that for the thick slabs the contact does not play an important role thus has to be stated in a modified manner. For a thick layer, the discrete energy levels are so close that they well approximate the continued spectrum.

According to rules of Langreth and Wilkins (6.12), the retarded and advanced forms of (16.76)–(16.78) are trivially obtained by adding superscripts R and A. The function $G_{jj}^{R,A}(p_\parallel)$ is evaluated from (16.35) except that the hopping is given by (16.75) and the local potential v_j is substituted by $\Sigma_j^{R,A}$, respectively.

The CPA condition for the transport vertex results from (6.33)–(6.34) and (16.77) as

$$c_j \frac{1}{1-(v_j^{Ga}-\Sigma_j^R)g_j^R}\Sigma_j^< \frac{1}{1-g_j^A(v_j^{Ga}-\Sigma_j^A)} + (1-c_j)\frac{1}{1-(v_j^{Al}-\Sigma_j^R)g_j^R}\Sigma_j^< \frac{1}{1-g_j^A(v_j^{Al}-\Sigma_j^A)}$$

$$= c_j \frac{v_j^{Ga}-\Sigma_j^R}{1-(v_j^{Ga}-\Sigma_j^R)g_j^R}g_j^< \frac{v_j^{Ga}-\Sigma_j^A}{1-g_j^A(v_j^{Ga}-\Sigma_j^A)} + (1-c_j)\frac{v_j^{Al}-\Sigma_j^R}{1-(v_j^{Al}-\Sigma_j^R)g_j^R}g_j^< \frac{v_j^{Al}-\Sigma_j^A}{1-g_j^A(v_j^{Al}-\Sigma_j^A)}.$$

$$(16.79)$$

This linear matrix relation we abbreviate as $M_j\left[\Sigma_j^<\right] = Q_j\left[g_j^<\right]$. From the GKB equation (16.73), the site-diagonal element of the correlation function results as

$$g_j^< = \sum_{n=1}^N \int_{B.Z.} \frac{dp_\parallel}{(2\pi)^2} G_{jn}^R(p_\parallel)\Sigma_n^< G_{nj}^A(p_\parallel) + \hat{g}_j^<, \qquad (16.80)$$

with the coherent contribution,

$$\hat{g}_j^< = G_{j1}^R(k_\parallel)\Sigma^{1<}G_{1j}^A(k_\parallel). \qquad (16.81)$$

We abbreviate (16.80) as $g_j^< = \sum_n P_{jn}\left[\Sigma_n^<\right] + \hat{g}_j^<$. Substituting (16.81) into (16.79) one finds a linear matrix equation for the transport vertex,

$$M_j\left[\Sigma_j^<\right] - Q_j\left[\sum_n P_{jn}\left[\Sigma_j^<\right]\right] = Q_j\left[\hat{g}_j^<\right]. \qquad (16.82)$$

It remains to identify the probabilities of the diffused electron, $p_\parallel \neq k_\parallel$. Let us take the reflected part, the tunnelling part is similar. In the asymptotic region, $s, j \leq 0$ and add for simplicity, $s = -2n_s$ and $j = -2n_j$, the retarded and advanced Green's functions in the GKB equation can be expressed in terms of surface Green's functions,

$$G_{sj}^<(p_\parallel) = \prod_{n=s}^0 S_n^{lR}(p_\parallel)t \times G_{11}^<(p_\parallel) \times \prod_{m=0}^j tS_m^{lA}(p_\parallel)$$

$$= \left(S_a^{lR}\tilde{t}_{ac}S_c^{lR}t_{ca}\right)^{-n_s} \times G_{11}^< \times \left(t_{ac}S_c^{lA}\tilde{t}_{ca}S_a^{lA}\right)^{-n_j}$$

$$= \sum_{\tau\zeta=1}^5 |\psi_a^{-\tau})e^{-ip_\tau n_s a}(\bar{\psi}_a^{-\tau}|G_{11}^<|\psi_a^{-\zeta})e^{-ip_\zeta n_j a}(\bar{\psi}_a^{-\zeta}|, \qquad (16.83)$$

where we have used the decomposition (16.90) for the parallel momentum $p_\parallel$.

To identify formula (16.83), we consider in general a probability, w_ϕ to find an electron after a diffraction in same state ϕ is given by the matrix element $w_\phi = |\langle\phi|\psi'\rangle|^2$, where ψ' is the diffracted wave before the averaging over the configuration of impurities. Using $G^<$ for the averaged square of the wave function, see (16.63), one finds that the averaged probability reads $w_\phi = \langle\phi|G^<|\phi\rangle$. Taking ϕ's as plane waves in the band τ and of momentum $(p_\tau, p_\parallel)$, we obtain from (16.83) that the probability to find a reflected electron in band τ with momentum $(p_\tau, p_\parallel)$ is given by the matrix element,

$$w_\tau(p_\tau, p_\parallel) = (\bar\psi_a^{-\tau}|G_{11}^<|\psi_a^{-\tau}).\tag{16.84}$$

16.6   Three-dimensional barrier in a multi-band crystal

In real crystals, the homogeneity implies that the elements of the Hamiltonian are periodic, $v_{j-N^l} = v_j$ and $t_{j-N^lj\pm1-N^l} = t_{jj\pm1}$, for $j < 0$. The period N^l depends on the material and on the crystallographic direction of the junction. To avoid the number of parameters needed to describe the general situation, we assume an AlAs junction in GaAs grown on the (100) surface. In this case, the period includes two layers, $N^l = 2$, and each layer has either anions (associated to odd layers) or cations (even layers). Let us take the case in which bonds on the left-hand side from anions point in the (111) and $(1\bar1\bar1)$ directions, while the bonds on the right side from anions point in one of the two complementary directions, $(\bar111)$ and $(\bar11\bar1)$. The hopping then has four unequal blocks, t_{ac} and t_{ca} connecting anions and cations along the (111) and $(1\bar1\bar1)$ bonds, and $\tilde t_{ac}$ and $\tilde t_{ca}$ for the complementary bonds. Since the Hamiltonian is Hermitian conjugate to itself, $t_{ac} = t_{ca}^\dagger$ and $\tilde t_{ac} = \tilde t_{ca}^\dagger$. Values of hopping in the left and right leads are in general different.

For all equivalent layers the surface Green's functions equal, $n = 1, 2, 3, \ldots$,

$$S_{1-2n}^l = S_a^l, \qquad S_{-2n}^l = S_c^l,\tag{16.85}$$

because they always describe the same physical quantity, the anion/cation surface layer of a homogeneous semi-infinite crystal. The right semi-infinite crystal is analogous, (barrier width is taken as even, $N = 2k$)

$$S_{N+2n-1}^r = S_a^r, \qquad S_{N+2n}^r = S_c^r.\tag{16.86}$$

These periodic conditions uniquely determine the starting values of recursive relations (16.34). The resulting (5×5)-matrix equation, e.g.

$$S_a^l = \cfrac{1}{z - v_a^l - t_{ac}\cfrac{1}{z - v_c^l - \tilde t_{ca}S_a^l\tilde t_{ac}}t_{ca}},\tag{16.87}$$

however, does not have a sufficient support in numerical tools for matrix quadratic equations.

It is more convenient to construct the surface Green's function from the eigen functions of the infinite crystal. Let us assume an infinite crystal from the material of the left or right lead. For given energy E and parallel momentum, $k_{\parallel}$, its Schrödinger equation is solved by ten independent functions, $\psi^{\pm\tau}$. Each of these functions depends on the layer index via an exponential factor and on the orbital index. Writing the layer index explicit and the orbital dependence as in five-component columns, $\psi_a^{\pm\tau}$ and $\psi_c^{\pm\tau}$, the solutions of Schrödinger equation read $n = \ldots, -1, 0, 1, 2, \ldots$,

$$\psi_{2n+1}^{\pm\tau} = \mathrm{e}^{\pm ik_\tau r_{2n+1}} \psi_a^{\pm\tau}, \qquad \psi_{2n}^{\pm\tau} = \mathrm{e}^{\pm ik_\tau r_{2n}} \psi_c^{\pm\tau}. \tag{16.88}$$

The layer positions are $r_j = ja$, the τ indexes complex bands, $\tau = 1, 2, 3, 4, 5$.

To find vectors $\psi^{\pm\tau}$ and related momenta $\pm k_\tau$, one can use a tight-binding version of the transfer matrix. Schrödinger's equation at two neighbouring layers,

$$(E - v_a)\psi_{2n+1} - \tilde{t}_{ac}\psi_{2n+2} - t_{ac}\psi_{2n} = 0, \qquad (E - v_c)\psi_{2n} - t_{ca}\psi_{2n+1} - \tilde{t}_{ca}\psi_{2n-1} = 0,$$

can be rearranged into a from $\Psi_{n+1} = R\Psi_n$,

$$\begin{pmatrix} \psi_{2n+1} \\ \psi_{2n+2} \end{pmatrix} = \begin{pmatrix} -t_{ca}^{-1}\tilde{t}_{ca} & t_{ca}^{-1}(E - v_c) \\ -\tilde{t}_{ac}^{-1}(E - v_a)t_{ca}^{-1}\tilde{t}_{ca} & \tilde{t}_{ac}^{-1}(E - v_a)t_{ca}^{-1}(E - v_c) - \tilde{t}_{ac}^{-1}t_{ac} \end{pmatrix} \begin{pmatrix} \psi_{2n-1} \\ \psi_{2n} \end{pmatrix}.$$

By a diagonalisation of the 10×10 transfer matrix R we obtain ten vectors Ψ^μ and ten eigen values α_μ which satisfy $R\Psi^\mu = \alpha_\mu \Psi^\mu$. From the time-reversal symmetry follows that to each α_μ exists a conjugated eigen value of the reciprocal value, $\alpha_\nu = \frac{1}{\alpha_\mu}$, so that we can label the eigen values by $\pm\tau$ and write them as $\alpha_{\pm\tau} = \mathrm{e}^{\pm 2ik_\tau a}$. Here, $a = r_{j+1} - r_j$ is a distance of neighbouring layers which equals one quarter of the lattice constant. For the energy shifted into the complex plane, $z = E + i0$, all momenta are complex. The sign + denotes those with a positive imaginary part, i.e. those decaying as n increases.

Apparently, functions $\Psi_n^{\pm\tau} = \mathrm{e}^{\pm 2ik_\tau na}\Psi^{\pm\tau}$ are the desirable solutions of Schrödinger's equation. Comparing with (16.86) one can see that

$$\Psi^{\pm\tau} = \begin{pmatrix} \mathrm{e}^{\mp ik_\tau a}\psi_a^{\pm\tau} \\ \psi_c^{\pm\tau} \end{pmatrix},$$

from which $\psi_{a,c}^{\pm\tau}$ readily follows.

For each layer and each sign we introduce reciprocal vectors $\bar{\psi}$ by orthogonality relations,

$$(\bar{\psi}_a^{+\tau}|\psi_a^{+\zeta}) = \delta_{\tau\zeta}, \qquad (\bar{\psi}_a^{-\tau}|\psi_a^{-\zeta}) = \delta_{\tau\zeta}, \qquad (\bar{\psi}_c^{+\tau}|\psi_c^{+\zeta}) = \delta_{\tau\zeta}, \qquad (\bar{\psi}_c^{-\tau}|\psi_c^{-\zeta}) = \delta_{\tau\zeta}.$$

Being convergent for $n \to -\infty$, the left surface Green's function propagates only solutions decaying as n decreases, i.e. denoted by minus. From (16.36) and (16.37), then follows a matrix generalisation of (16.39), $n = 1, 2, 3, \ldots$,

$$S_a^l \bar{t}_{ac} \psi_c^{-\tau} = e^{ik_\tau a} \psi_a^{-\tau}, \qquad\qquad S_c^l t_{ca} \psi_a = e^{ik_\tau a} \psi_c. \qquad (16.89)$$

With the help of the reciprocal vectors the left surface Green's functions read

$$S_a^l = \sum_{\tau=1}^{5} |\psi_a^{-\tau}\rangle e^{ik_\tau a} (\bar{\psi}_c^{-\tau}|\bar{t}_{ac}^{-1}, \qquad\qquad S_c^l = \sum_{\tau=1}^{5} |\psi_c^{-\tau}\rangle e^{ik_\tau a} (\bar{\psi}_a^{-\tau}|t_{ca}^{-1}. \qquad (16.90)$$

The right surface is analogous.

16.6.1   Diffraction in multi-band crystals

In the multi-band crystal, the diffraction includes processes in which an electron is reflected or transmitted into a different band than the one in which it has approached the barrier. Let us take as the incoming wave function the eigen function $\psi^{+\tau}$. Cutting it on in front of the barrier we find the source terms,

$$|\Omega_0\rangle = -e^{ik_\tau a} t_{ca} |\psi_a^{+\tau}\rangle, \qquad\qquad |\Omega_1\rangle = t_{ac} |\psi_c^{+\tau}\rangle. \qquad (16.91)$$

The reflection into a channel $-\zeta$ is given by a projection on the function $\psi^{-\zeta}$. A matrix modification of (16.40) then reads

$$\beta_{\text{ref}}^{\zeta\tau} = (\bar{\psi}_c^{-\zeta}|G_{01} t_{ac}|\psi_c^{+\tau}\rangle - e^{ik_\tau a}(\bar{\psi}_c^{-\zeta}|G_{00} t_{ca}|\psi_a^{+\tau}\rangle. \qquad (16.92)$$

Although the parallel momentum conserves, different bands of parabolic dispersion relation folded into the Brillouin zone correspond to parallel momenta which differ by multiples of the Brillouin vector. In the LEED, interband reflection thus corresponds to the measured diffracted electrons.

Similarly, the transmission couples all accessible channels. Using κ_ζ for eigen-momenta and $\phi^{\pm\zeta}$ for eigen vectors in the right lead, the transmission coefficient reads

$$\beta_{\text{trn}}^{\zeta\tau} = e^{-i\kappa_\zeta Na}(\bar{\phi}_c^{+\zeta}|G_{N1} t_{ac}|\psi_c^{+\tau}\rangle - e^{-i\kappa_\zeta Na} e^{ik_\tau a}(\bar{\phi}_c^{+\zeta}|G_{N0} t_{ca}|\psi_a^{+\tau}\rangle. \qquad (16.93)$$

Clearly, from the formal point of view, the extension of layered Green's functions to multi-band systems does not represent any difficulty, except for the unavoidable inconvenience given by band indices and the internal structure of Bloch waves.

Numerical demands of layered Green's functions are favourably small compared to other approaches. The comparison should be done with straightforward numerical studies of Schrödinger's equation. In this approach one has to terminate the layer index somewhere, i.e. to replace the semi-infinite crystals by some finite slabs. To reduce artificial size effects, the slabs have to be made of a large number of layers. Most of the computing time is thus wasted by the description of leads, the boring region from the physical point of view. Within the layered Green's functions, the leads are covered by the surface functions of the semi-infinite crystals. The computer time is invested exclusively into the barrier region. As one incorporates more and more realistic features, numerical demands increase only moderately. The size of matrices used in matrix operations depends exclusively on the number of local orbitals included. It is rather difficult to find a more realistic parametrisation of the Hamiltonian than those exceeding the capacity evenof small computers. The number of recursive steps increases only linearly with the width of the barrier. If of interest, one can study barriers with enormously high numbers of layers.

17

Deep Impurities with Collision Delay

17.1 Transport properties

The elastic scattering of electrons by impurities is the simplest, but still a very interesting, dissipative mechanism in semiconductors. Its simplicity follows from the absence of impurity dynamics, so that individual collisions are described by the motion of an electron in a fixed potential. On the other hand, due to the large variety of impurities and their accessible concentrations, impurity-controlled transport regimes span from a simple response characterised by a mean-free path to weak localisation. We will now apply nonlocal kinetic theory (Špička, Lipavský and Morawetz, 1997a).

The rate of scattering by Koster–Slater impurities of concentration c (a probability that impurity occupy a site) follows from the Fermi golden rule as

$$P_{pk} = c|t^R(\epsilon_k)|^2 2\pi \delta(\epsilon_k - \epsilon_p). \qquad (17.1)$$

This scattering rate does not depend on the scattering angle, thus it can also be expressed in terms of the lifetime τ

$$P_{pk} = \frac{1}{\tau} \frac{2\pi^2}{k^2} \delta(|p| - |k|), \qquad (17.2)$$

where τ is conveniently evaluated from the T-matrix

$$\frac{1}{\tau} = c(-2)\mathrm{Im}t^R(\epsilon_k). \qquad (17.3)$$

For the Koster–Slater potential, the T-matrix is also restricted to the selected orbital, $T^R = |0\rangle t^R \langle 0|$, and reads (4.32)

$$t^R = v + v\langle 0|G_0^R|0\rangle t^R = \frac{v}{1 - v\langle 0|G_0^R|0\rangle}. \qquad (17.4)$$

Interacting Systems far from Equilibrium. Klaus Morawetz, Oxford University Press (2018).
© Klaus Morawetz. DOI: 10.1093/oso/9780198797241.001.0001

To obtain the collision delay, we place the impurity in the initial coordinates and express the wave function in the time representation

$$\psi(r, t) = e^{ikx - i\epsilon_k t} - \frac{m}{2\pi |r|} t^R(\epsilon_k) e^{ik|r| - i\epsilon_k t}. \tag{17.5}$$

We have used an asymptotic Green's function for the large r, see (Mesiah, 1961),

$$\langle r | G_0^R(\epsilon_k) | 0 \rangle = -\frac{m}{2\pi |r|} e^{ik|r|}, \tag{17.6}$$

to evaluate the outgoing wave

$$\psi_{\text{out}} = G_0^R(\epsilon_k) T^R(\epsilon_k) \psi_{\text{in}}, \tag{17.7}$$

where

$$G_0^R(\omega) = \frac{1}{\omega - H_0 + i0}, \tag{17.8}$$

is the retarded Green's function of the host crystal. This approximation holds for energies ϵ_k in the parabolic region of the band structure, $\epsilon_k = k^2/2m$. The first term in (17.5) is the incoming wave ψ_{in} and the second one is the outgoing part ψ_{out}.

To see the time delay, we take a linear combination of wave functions ψ, so that the incoming part ψ_{in} forms a wave packet of a narrow momentum width $\kappa \to 0$,

$$\psi_{\text{in}}(r, t) = \frac{1}{\sqrt{\pi\kappa}} \int dp e^{-\frac{(p-k)^2}{\kappa^2}} e^{ipx - i\epsilon_p t} \approx e^{ikx - i\epsilon_k t} \exp\left\{ -\frac{\kappa^2}{4}(x - ut)^2 \right\}, \tag{17.9}$$

where $u = k/m$ is an electron velocity. This wave packet passes the initial coordinates at $t = 0$. A corresponding outgoing wave ψ_{out} reads

$$\psi_{\text{out}}(r, t) = -\frac{m}{2\pi |r|} \frac{1}{\sqrt{\pi\kappa}} \int dp e^{-\frac{(p-k)^2}{\kappa^2}} t^R(\epsilon_p) e^{ip|r| - i\epsilon_p t}$$

$$\approx -\frac{m}{2\pi |r|} t^R(\epsilon_k) e^{ik|r| - i\epsilon_k t} \exp\left\{ -\frac{\kappa^2}{4} \left(|r| - u \left(t + \frac{i}{t^R} \frac{\partial t^R}{\partial \omega} \Big|_{\omega=\epsilon_k} \right) \right)^2 \right\}. \tag{17.10}$$

The outgoing wave passes the initial coordinates with the collision delay

$$\Delta_t = \text{Im} \frac{1}{t^R} \frac{\partial t^R}{\partial \omega} \Big|_{\omega=\epsilon_k} \tag{17.11}$$

which is exactly the definition of our delay time.

Now we use a model local Green's function (Špička, Lipavský and Velický, 1992)

$$\langle 0|G_0^R(\omega)|0\rangle = \frac{2}{W}\left(-\frac{b_1}{2} - \frac{b_3}{8} + z + \left(b_1 - \frac{b_3}{2}\right)z^2 + b_3 z^4\right)$$

$$+ \theta(1 - z^2)\frac{2}{W}(1 + b_1 z + b_3 z^3)\sqrt{1 - z^2}\bigg|_{z=\frac{\omega}{W}-1}. \qquad (17.12)$$

Here, $W = 6$ eV is a half-width of a conductivity band, and parameters $b_1 = 1.2$ and $b_3 = -0.4$ serve to model the local density of state with a shape resembling III–V semiconductors, see Fig. 17.1.

The collision delay is very sensitive to the value of the impurity potential v. Using (17.4), one can rearrange the collision delay (17.11) as

$$\Delta_t = -\text{Im}\left[t^R \frac{\partial}{\partial\omega}\frac{1}{t^R}\right] = \text{Im}\frac{v\frac{\partial}{\partial\omega}\langle 0|G_0^R|0\rangle}{1 - v\langle 0|G_0^R|0\rangle}. \qquad (17.13)$$

Apparently, the collision delay will be large for potentials for which the denominator $1 - v\langle 0|G_0^R|0\rangle$ approaches zero. For these values of potential v, the impurity behaves like a resonant level close to the conductivity band edge.

For a model function (17.12), the real part of the local Green's function at the band edge $\omega = 0$ equals to -0.185 1/eV. For potentials $v < -5.4$ eV, the impurity has a bound state. For $v > -5.4$ eV, there is a resonant level. In our calculations we use the value $v = -5.35$ eV.

The strong dependence of the collisional delay on the position of the resonant level leads to a strong dependence of virial corrections on the impurity potential, see Fig. 17.1. Such changes of the impurity potential can be achieved, for instance, by a hydrostatic

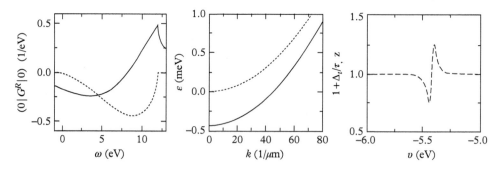

Figure 17.1: *Left: the imaginary part (dotted line) and real part (full) of local Green function. Middle: quasiparticle energy (full line) for resonant levels, $v = -5.35$ eV, of concentration $c = 10^{-6}$ and bare kinetic energy ϵ_k (dotted line). Right: virial (dashed line) and wave function renormalisation (dotted) for electron energy 100 meV above the band edge both curves differing less than 0.8%.*

pressure (Zeman, Šmíd, Krištofik and Mareš, 1993). The impurity concentration and the hydrostatic pressure can thus be used to control the magnitude of virial corrections.

The nonlocal kinetic equation that includes both corrections reads

$$\frac{\partial f}{\partial t} + z \frac{k}{m} \frac{\partial f}{\partial r} - \frac{\partial U}{\partial r} \frac{\partial f}{\partial k} = -\frac{f}{\tau} + \frac{1}{\tau} \frac{2\pi^2}{k^2} \int \frac{dp}{(2\pi)^3} \delta(|p| - |k|) f(p, r, t - \Delta_t).$$

(17.14)

Although this equation has the form of a classical Boltzmann equation, its components z, τ and Δ_t are given by quantum-mechanical microscopic dynamics (13.3). One can also view (17.14) as a phenomenological equation with momentum-dependent parameters z, Δ_t and τ.

From the balance equation of the density (15.76) one finds that the physical density includes only virial corrections,

$$n = \int \frac{dk}{(2\pi)^3} \left(1 + \frac{\Delta_t}{\tau}\right) f,$$

(17.15)

while the density of the particle current has only quasiparticle corrections

$$j = \int \frac{dk}{(2\pi)^3} z \frac{k}{m} f = \int \frac{dk}{(2\pi)^3} \frac{\partial \varepsilon}{\partial k} f.$$

(17.16)

One finds the energy conservation from the balance of the energy (15.79) and it includes both corrections

$$E = \int \frac{dk}{(2\pi)^3} \left(1 + \frac{\Delta_t}{\tau}\right) (\varepsilon + U) f.$$

(17.17)

17.2   Dielectric function

The virial corrections influence the response of the system to perturbations. The time non-locality of the scattering integral emerges in non-stationary processes. The simplest but important process is the linear screening of an external field described by the dielectric function ϵ_r.

The virial corrections enter the dielectric function in two ways, from the transport equation (17.14) and from the functional (17.15). To demonstrate both mechanisms, we evaluate ϵ_r from its definition.

An electrostatic external potential

$$U_0(r, t) = U_0 e^{iqx - i\omega t}$$

(17.18)

creates a perturbation in the electron density $\delta n(r, t) = \delta n e^{iqr - i\omega t}$. The perturbation in density creates a Coulomb potential $\delta U = e^2 \delta n / \epsilon_0 q^2$ that adds to the external one so that the internal field reads

$$U = U_0 + \delta U = U_0 + \frac{e^2}{\epsilon_0 q^2}\delta n = \frac{U_0}{\epsilon_r}. \tag{17.19}$$

Here, ϵ_0 is the permittivity of the host crystal and the dielectric function is

$$\epsilon_r = 1 - \frac{e^2}{\epsilon_0 q^2}\frac{\delta n}{U}. \tag{17.20}$$

For details, see Chapter 11.6.1. To evaluate the perturbation δn of the physical density n, we have to find the linear perturbation of the quasiparticle distribution,

$$\delta f(k, r, t) = \delta f(k) e^{iqr - i\omega t}, \tag{17.21}$$

caused by the potential U. To this end, we use the linearised transport equation (17.14)

$$\left(-i\omega + iz\frac{kq}{m} + \frac{1}{\tau}\right)\delta f(k) - iqU\frac{\partial f_0(k)}{\partial k} = \frac{1}{\tau}\frac{2\pi^2}{k^2}\int\frac{dp}{(2\pi)^3}\delta(|p| - |k|)e^{i\omega\Delta_t}\delta f(p). \tag{17.22}$$

The momentum derivative of the equilibrium distribution $f_0(k) = f_{FD}(\varepsilon_k)$ reads

$$\frac{\partial f_0(k)}{\partial k} = z\frac{k}{m}\frac{\partial f_{FD}(\varepsilon_k)}{\partial \varepsilon_k}. \tag{17.23}$$

The perturbation δf only depends on the absolute value of the momentum $|k|$ and the angle between momentum k and wave vector q. We denote $s = \frac{kq}{|k||q|}$ and $s' = \frac{pq}{|p||q|}$, and integrate over the energy-conserving δ function. With abbreviations $z\frac{|k|}{m} \equiv u$, $|q| \equiv q$, and skipping argument $|k|$ in the distributions, equation (17.22) reads

$$\left(-i\omega + isqu + \frac{1}{\tau}\right)\delta f(s) - isquU\frac{\partial f_{FD}}{\partial \varepsilon} = \frac{1}{2\tau}e^{i\omega\Delta_t}\int_{-1}^{1}ds'\delta f(s'). \tag{17.24}$$

The distribution is easily found from (17.24)

$$\delta f(s) = \frac{isquU\frac{\partial f_{FD}}{\partial \varepsilon} + \frac{1}{\tau}e^{i\omega\Delta_t}\delta F}{-i\omega + isqu + \frac{1}{\tau}}, \quad \text{with } \delta F = \frac{1}{2}\int_{-1}^{1}ds\delta f(s) \tag{17.25}$$

where we have to find its angle-averaged distribution. The latter is found by integrating (17.25) over s

$$\delta F = (1 + (1 - i\omega\tau)\mathcal{F})U\frac{\partial f_{FD}}{\partial \varepsilon} - e^{i\omega\Delta_t}\mathcal{F}\delta F, \tag{17.26}$$

where

$$\mathcal{J} = \frac{i}{2qu\tau} \ln \left(\frac{\omega + \frac{i}{\tau} - qu}{\omega + \frac{i}{\tau} + qu} \right). \tag{17.27}$$

Since the Boltzmann equation only holds for slowly varying fields, we can linearise it in Δ_t, $e^{i\omega\Delta_t} \approx 1 + i\omega\Delta_t$. The angle-averaged distribution from (17.26) then results (Špička et al., 1997b) as

$$\delta F = U \frac{\partial f_{FD}}{\partial \varepsilon} \frac{1 + (1 - i\omega\tau)\mathcal{J}}{1 + (1 + i\omega\Delta_t)\mathcal{J}}. \tag{17.28}$$

Now the perturbation of the quasiparticle distribution is fully determined by (17.25) and (17.28). The perturbation of the electron density is found from (17.15)

$$\delta n = 2 \int \frac{dk}{(2\pi)^3} \delta f(k) = \frac{1}{\pi^2} \int_0^\infty dk k^2 \delta F(k) \left(1 + \frac{\Delta_t}{\tau} \right). \tag{17.29}$$

The factor of two stands for the sum over spins. For simplicity we assume the limit of low temperature

$$\frac{\partial f_{FD}}{\partial \varepsilon} \rightarrow -\delta(\varepsilon - E_F) = -\frac{m}{zk} \delta(k - k_F), \tag{17.30}$$

where one can easily integrate out the momentum

$$\delta n = - U \frac{mk}{\pi^2 z} \left(1 + \frac{\Delta_t}{\tau} \right) \frac{1 + (1 - i\omega\tau)\mathcal{J}}{1 + (1 + i\omega\Delta_t)\mathcal{J}} \bigg|_{k=k_F}. \tag{17.31}$$

Using (17.31) in (17.20) one directly obtains the dielectric function.

17.3 Compensation of virial and quasiparticle corrections

17.3.1 Long wave-length limit

Now we focus on the long wave-length limit, $q \rightarrow 0$. To evaluate this limit from (17.31) we first rearrange (17.27) as

$$\mathcal{J} = -\frac{1}{1 - i\omega\tau} \sum_{x=\pm i\frac{qu\tau}{1-i\omega\tau}} \frac{1}{x} \ln(1 + x) \rightarrow -\frac{1}{1 - i\omega\tau} \left[1 - \frac{1}{3} \left(\frac{qu\tau}{1 - i\omega\tau} \right)^2 \right]. \tag{17.32}$$

In the long wave-length limit, the dielectric function reads therefore

$$\epsilon_r = 1 + \frac{\frac{e^2 m k_F}{\epsilon_0 \pi^2 z}\left(1 + \frac{\Delta_t}{\tau}\right)}{q^2 \frac{1+i\omega\Delta_t}{1-i\omega\tau} - 3\omega\left(\omega + \frac{i}{\tau}\right)\left(1 + \frac{\Delta_t}{\tau}\right)\frac{m^2}{k_F^2 z^2}}, \tag{17.33}$$

where z, τ and Δ_t are values at the Fermi level.

In the static case $\omega = 0$, the dielectric function is of the form $\epsilon_r = 1 + \frac{q_s^2}{q^2}$. From (17.33) one finds that the Thomas–Fermi screening length $1/q_s$ is

$$q_s^2 = \frac{e^2 m k_F}{\epsilon_0 \pi^2}\frac{1}{z}\left(1 + \frac{\Delta_t}{\tau}\right). \tag{17.34}$$

The part $\frac{e^2 m k_F}{\epsilon_0 \pi^2}$ gives the standard Thomas–Fermi screening, the factors $\frac{1}{z}$ and $1 + \frac{\Delta_t}{\tau}$ provide quasiparticle and virial corrections, respectively. As one can see from Fig. 17.1, quasiparticle and virial corrections are nearly equal, therefore they mutually compensate in the Thomas–Fermi screening length (17.34)

$$q_s^2 \approx \frac{e^2 m k_F}{\epsilon_0 \pi^2}. \tag{17.35}$$

For homogeneous perturbations, $q = 0$, the dielectric function is of the form $\epsilon_r = 1 - \frac{\omega_p^2}{\omega(\omega+i/\tau)}$. From (17.33), the plasma frequency ω_p results

$$\omega_p^2 = \frac{e^2 k_F^3}{3\epsilon_0 \pi^2 m} z \tag{17.36}$$

where only the quasiparticle correction z appears.

The virial corrections to the Thomas–Fermi screening q_s and to the plasma frequency ω_p appear in a rather paradoxical way. While the static screening has virial corrections due to $n[f]$, Eq. (17.29), the plasma frequency describing non-stationary behaviour has none. This is because in the homogeneous case, $q = 0$, the virial corrections from $n[f]$ and the scattering integral mutually cancel due to the particle conservation law.

17.3.2   Virial correction to Fermi momentum

The Thomas–Fermi screening length (17.34) and the plasma frequency (17.36) are expressed in terms of the Fermi momentum. Additional virial corrections to those quantities appear if one rewrites them in terms of the physical density n.

In general, the Fermi momentum is a parameter of the quasiparticle distribution f, therefore it is always related to the free density which is (17.15) without virial correction. For the parabolic band, the Fermi momentum results in

$$k_F = \sqrt[3]{3\pi^2 n_{\text{free}}}.$$ (17.37)

For a sufficiently low density n, the ratio Δ_t/τ slightly changes the Fermi momentum. In our case, Δ_t/τ changes by 7% for a density of $n = 10^{16}$ cm^{-3}. This weak dependence allows us to take $k \approx k_F$. From (17.15) one finds that the free and physical densities relate as

$$n = n_{\text{free}} \left(1 + \frac{\Delta_t}{\tau} \right),$$ (17.38)

and the Fermi momentum reads

$$k_F = \sqrt[3]{\frac{3\pi^2 n}{1 + \frac{\Delta_t}{\tau}}}.$$ (17.39)

In terms of the physical density, the Thomas–Fermi screening length (17.35) regains the corrections

$$q_s^2 \approx \frac{e^2 m}{\epsilon_0 \pi^2} \sqrt[3]{\frac{3\pi^2 n}{1 + \frac{\Delta_t}{\tau}}}.$$ (17.40)

Oppositely, the plasma frequency in terms of the physical density regains its free-particle value

$$\omega_p^2 = \frac{e^2 n}{\epsilon_0 m} \frac{z}{1 + \frac{\Delta_t}{\tau}} \approx \frac{e^2 n}{\epsilon_0 m}.$$ (17.41)

Please note that (17.39) clearly violates the Luttinger theorem stating that the Fermi momentum is exclusively determined by the quasiparticle density. Since the Luttinger theorem is valid only for Fermi liquids we have to state that including collision delays describes the correlations beyond Fermi liquids. The appearance of the correlated density here or systems with bound states or pairing all are not Fermi liquids.

17.3.3   dc conductivity

The compensation of virial and quasiparticle corrections also appears for the dc conductivity σ_{dc}. Although this compensation is a direct consequence of the dielectric function, we discuss it in detail for its experimental importance.

The conductivity is related to the dielectric function ϵ_r as

$$\sigma_{\text{dc}} = -i \lim_{q,\omega \to 0} \omega \epsilon_0 (\epsilon_r - 1).$$ (17.42)

This known relation can be recovered from the balance equation of the density $i\omega \delta n - iqj = 0$, where j is the flow of particles. The electric field E results from the electrostatic potential as $eE = iqU$. The conductivity then reads

$$\sigma_{dc} = -\frac{ej}{E} = i\frac{e^2\omega}{q^2}\frac{\delta n}{U}.$$

(17.43)

Comparing (17.43) with (17.20) one recovers (17.42).

For the long wavelength $q \to 0$ and static $\omega \to 0$ limit one finds the standard relaxation time formula with the quasiparticle correction

$$\sigma_{dc} = \frac{e^2 k_F^3 \tau}{3\pi^2 m} z.$$

(17.44)

In terms of the physical density,

$$\sigma_{dc} = \frac{e^2 n\tau}{m}\frac{z}{1 + \frac{\Delta_t}{\tau}} \approx \frac{e^2 n\tau}{m},$$

(17.45)

since virial and quasiparticle corrections mutually compensate, as illustrates in Figure 17.1. Although virial and quasiparticle corrections describe different features of quasiparticle transport, both of them are linked to the energy derivatives of the T-matrix. From this link follows the similarity of their magnitudes as we will see now. One can re-arrange formula (17.11) in a way that reveals the relation of virial correction $1 + \frac{\Delta_t}{\tau}$ to wave-function renormalisation z. Writing (17.11) as $\Delta_t = \mathrm{Im}\frac{1}{t^R}\frac{\partial t^R}{\partial \omega} = \frac{1}{2i}\left(\frac{1}{t^R}\frac{\partial t^R}{\partial \omega} - \frac{1}{t^A}\frac{\partial t^A}{\partial \omega}\right)$, the inverse lifetime (17.3) as $\frac{1}{\tau} = 2c\mathrm{Im}t^R = ic(t^R - t^A)$ with t^A being the complex conjugate to t^R, one obtains

$$\frac{\Delta_t}{\tau} = ic(t^R - t^A)\frac{1}{2i}\left(\frac{1}{t^R}\frac{\partial t^R}{\partial \omega} - \frac{1}{t^A}\frac{\partial t^A}{\partial \omega}\right) = \frac{\partial}{\partial \omega}\frac{c}{2}(t^R + t^A) - \frac{c}{2}\left(\frac{t^A}{t^R}\frac{\partial t^R}{\partial \omega} - \frac{t^R}{t^A}\frac{\partial t^A}{\partial \omega}\right)$$

$$= \frac{\partial \mathrm{Re}\sigma^R}{\partial \omega} - \frac{c}{2}\left(\frac{t^A}{t^R}\frac{\partial t^R}{\partial \omega} - \frac{t^R}{t^A}\frac{\partial t^A}{\partial \omega}\right),$$

(17.46)

where we have used that $\mathrm{Re}\sigma^R = \frac{c}{2}(t^R + t^A)$. From (17.4) one finds

$$\frac{\partial t^R}{\partial \omega} = t_R^2\frac{\partial}{\partial \omega}\langle 0|G_0^R|0\rangle,$$

(17.47)

which substituted into (17.46) provides

$$\frac{\Delta_t}{\tau} = \frac{\partial \mathrm{Re}\sigma^R}{\partial \omega} - c|t^R|^2\mathrm{Re}\frac{\partial}{\partial \omega}\langle 0|G_0^R|0\rangle.$$

(17.48)

Formula (17.48) presents the connection between virial and quasiparticle corrections. One can see that at least two limiting regimes can be distinguished according to the relative values of the first and the second terms in (17.48).

For weak potentials when the selfenergy can be treated in the Born approximation $t^R \approx v$, i.e. $\sigma^R = cv^2 \langle 0|G_0^R|0\rangle$, the virial corrections vanish because the first and the second terms mutually cancel. In contrast, the quasiparticle corrections remain. Since most of the quasiclassical transport equations have been derived within the Born approximation (or single-loop approximation for particle–particle interaction), it is quite natural that they do not include virial corrections.

The scattering by resonant levels is far from the Born approximation, for model and parameters we consider here $|t^R| \sim 100 \times |v|$. In this case, the second term of (17.48) is of the order of 10^{-3} while the first one is of the order of 10^{-1}. Accordingly, the second term can be neglected, i.e. virial and quasiparticle corrections are of the same magnitude.

18

Relaxation-Time Approximation

18.1 Local equilibrium

Transport coefficients connect the induced currents with the perturbing external fields. Therefore in most cases the linear response $\sim \delta f = f - f^0$ is sufficient, see Chapter 11.6. One might be tempted to approximate the collision integral by a simple relaxation time deviation from the equilibrium $\sim (f^0 - f)/\tau$. In fact such expectation leads to conflicting results since we are violating conservation laws. Instead we have to consider relaxation towards a proper local equilibrium, ensuring the corresponding conservation laws. We will illustrate the problem with the example of the electric conductivity and particles with $\epsilon = k^2/2m$ dispersion. One has two possibilities to calculate this conductivity, by the mean current and by the polarisation function. The latter one follows from the Maxwell equation $\mathrm{rot}\mathbf{H} = \mathbf{j} + \epsilon_0 \epsilon_r \dot{\mathbf{E}} = (\epsilon_r + i\sigma/\omega\epsilon_0)(-i\omega\epsilon_0 \mathbf{E})$ where we assumed Ohmic behaviour $\mathbf{j} = \sigma \mathbf{E}$. We see that the effect of the medium can be represented by a complex dielectric function and the conductivity

$$\epsilon = \epsilon_r + i\frac{\sigma}{\omega\epsilon_0}, \qquad \sigma = \omega\epsilon_0 \mathrm{Im}\,\epsilon = -\omega\epsilon_0 V_q \mathrm{Im}\,\Pi \qquad (18.1)$$

with the polarisation function according to (11.69). We consider the long-wavelength limit $q \to 0$ such that the quasiclassical-kinetic equation is sufficient. The quantum-kinetic results are derived completely analogously resulting in the quantum response function instead of the quasiclassical one. The linear response of the kinetic equation with respect to the electric field $(-e)\mathbf{E} = -\partial_r U$,

$$\partial_t \delta f + \frac{\mathbf{k}}{m}\partial_r \delta f - \partial_r U \partial_k f^0 = -\frac{\delta f}{\tau}, \qquad (18.2)$$

is solved by Fourier transform $t \to \omega$ and $r \to q$ and provides the density variation and polarisation function

$$\delta n = \Pi\left(\omega + \frac{i}{\tau}\right)U, \qquad \Pi(\omega) = \int \frac{d^3 k}{(2\pi\hbar)^3}\frac{\mathbf{q}\partial_k f^0}{\frac{\mathbf{k}\mathbf{q}}{m} - \omega} = \frac{nq^2}{m\omega^2} + o(q^4). \qquad (18.3)$$

Interacting Systems far from Equilibrium. Klaus Morawetz, Oxford University Press (2018).
© Klaus Morawetz. DOI: 10.1093/oso/9780198797241.001.0001

Using this in (18.1) with the Coulomb potential $V_q = e^2/\epsilon_0 q^2$ we obtain a vanishing static conductivity

$$\sigma = \frac{ne^2\tau}{m}\frac{2\omega^2}{(1+(\omega\tau)^2)^2} \to 0, \text{ for } \omega \to 0 \tag{18.4}$$

which is obviously a wrong result. The mistake we made is that we have to relax not towards a total equilibrium f^0 but towards a local one f^l which observes the density conservation. This can be achieved by a small deviation of the chemical potential

$$\delta f = f - f^0 = f - f^l + f^l - f^0 = f - f^l + \partial_\mu f^0 \delta\mu \tag{18.5}$$

which is determined by demanding the density conservation $\int d^3k\, f^l = \int d^3k\, f$ such that the deviation of the chemical potential can be expressed in terms of the density variation

$$\delta\mu = \frac{\delta n}{\int \frac{d^3k}{(2\pi\hbar)^3}\partial_\epsilon f^0} = \frac{\delta n}{\Pi(0)}. \tag{18.6}$$

Considering in (18.2) the relaxation towards the local equilibrium on the right side one obtains the Mermin polarisation function (Mermin, 1970; Das, 1975)

$$\Pi^M(\omega) = \frac{\Pi(\omega + i/\tau)}{1 - \frac{1 - \Pi(\omega+i/\tau)/\Pi(0)}{1 - i\omega\tau}}. \tag{18.7}$$

It can be extended to include more conservation laws (Morawetz, 2002, 2013). The conductivity (18.1) becomes now

$$\sigma = -\omega\frac{e^2}{q^2}\text{Im}\frac{nq^2}{m\left(\omega + \frac{i}{\tau}\right)^2\left(1 - \frac{i/\tau}{\omega+i/\tau}\right)} = \frac{ne^2\tau}{m\left[1 + (\omega\tau)^2\right]} \tag{18.8}$$

which provides the correct Drude conductivity and a finite static limit. This can be verified by the charge current

$$\mathbf{j} = -e\mathbf{j}_n = -e\int\frac{d^3k}{(2\pi\hbar)^3}\delta f = -e^2\int\frac{d^3k}{(2\pi\hbar)^3}\frac{\mathbf{k}}{m}\frac{\mathbf{E}\cdot\partial_k f^0}{-i\omega + \frac{1}{\tau} + \frac{i\mathbf{k}\mathbf{q}}{m}}$$

$$= \left[\frac{ne^2\tau}{m\left[1 + (\omega\tau)^2\right]}(1 + i\omega\tau) + o(q)\right]\mathbf{E} \tag{18.9}$$

which for both cases, absolute and local equilibrium leads to the same result. We conclude that the relaxation towards the local equilibrium is the choice of the day.

18.2 Transport coefficients

We now consider Fermi liquids where the quasiparticle energy $\varepsilon(k)$ can have a complicated momentum dependence due to the band structure. We will approximate the collision integral by a momentum-dependent relaxation time $\tau(k)$ towards a local equilibrium

$$f_l = f_0 \left(\frac{\varepsilon(k) - \mu(\mathbf{r})}{T(\mathbf{r})} \right) \tag{18.10}$$

characterised by a space-dependent chemical potential $\mu(\mathbf{r})$ and a temperature $T(\mathbf{r})$. The equilibrium distribution f_0 is assumed to be the Fermi function. If there is additionally a mean momentum $u(r)$ we will obtain the viscosity to be considered later. Also the time dependence of these local quantities can be considered analogously to the following steps. The kinetic equation with an external electric field $\mathbf{E} = \partial_r U/e$ absorbed in $\epsilon = \varepsilon - U$ reads

$$[\partial_t + \partial_k \epsilon \partial_r - \partial_r \epsilon \partial_k] f = -\frac{f - f_l}{\tau} \tag{18.11}$$

Adding f_l on both sides, one can write

$$\left[\partial_t + \partial_k \epsilon \partial_r - \partial_r \epsilon \partial_k + \frac{1}{\tau} \right] (f - f_l) = -[\partial_t + \partial_k \epsilon \partial_r - \partial_r \epsilon \partial_k] f_l. \tag{18.12}$$

In linear response the derivatives of the left side represent higher orders in the deviation of $f - f_l$ and can be neglected. This leads to

$$f - f_l = -\tau \, [\partial_t + \partial_k \epsilon \partial_r - \partial_r \epsilon \partial_k] f_l = -\tau \frac{\partial \varepsilon}{\partial k} \frac{\partial f_0}{\partial \varepsilon} \left[-e\mathscr{E} - \frac{\varepsilon - \mu}{T} \partial_r T \right] \tag{18.13}$$

where we introduced the electrochemical potential $\mathscr{E} = \mathbf{E} + \partial_r \mu/e$.

The deviation from the local equilibrium determines the currents. The charge $\mathbf{j}$, the particle $\mathbf{j}_n$, the energy $\mathbf{j}_\epsilon$, as well as the heat $\mathbf{j}_q$ and entropy current $\mathbf{j}_S$ takes the form

$$\mathbf{j} = -e\mathbf{j}_n = -e \int \frac{dk}{(2\pi\hbar)^3} \frac{\partial \varepsilon}{\partial k} (f - f_l) = L^{11}\mathscr{E} + L^{12}(-\partial_r T)$$

$$\mathbf{j}_q = T\mathbf{j}_S = \mathbf{j}_\epsilon - \mu\mathbf{j}_n = \int \frac{dk}{(2\pi\hbar)^3} \frac{\partial \varepsilon}{\partial k} (\varepsilon - \mu)(f - f_l) = L^{21}\mathscr{E} + L^{22}(-\partial_r T) \tag{18.14}$$

where the Onsager coefficients (matrices) can be written in compact form

$$L^{11} = \mathscr{L}^{(0)}, \quad L^{21} = TL^{12} = -\frac{1}{e}\mathscr{L}^{(1)}, \quad L^{22} = \frac{1}{e^2 T}\mathscr{L}^{(2)} \tag{18.15}$$

with

$$\mathscr{L}^{(\alpha)} = - \int d\omega \, \partial_\omega f_0(\omega)(\omega - \mu)^\alpha \sigma(\omega) \qquad (18.16)$$

and the dynamical conductivity matrix

$$\sigma(\omega) = e^2 \tau(\omega) \int \frac{dk}{(2\pi\hbar)^3} \delta(\omega - \varepsilon) \frac{\partial\varepsilon}{\partial\mathbf{k}} \otimes \frac{\partial\varepsilon}{\partial\mathbf{k}} \qquad (18.17)$$

which one confirms by direct inspection.

18.3   Transport coefficients for metals

For temperatures much lower than the Fermi energy $T \ll \epsilon_f$ which is justified in metals and solid states, we have

$$- \partial_\omega f_0 = \delta(\epsilon_f - \omega) + o\left(e^{-\epsilon_f/T}\right) \qquad (18.18)$$

and therefore $\mathscr{L}^{(0)} = \sigma(\epsilon_f)$. For the integrals over moments of frequency one gets quadratic temperature corrections by the Sommerfeld expansion

$$\int d\omega \phi(\omega) \left(-\frac{\partial f_0}{\partial \omega}\right) = \int_{-\infty}^{\infty} d\omega \left[\phi(\epsilon_f) + \phi'(\epsilon_f)(\omega - \epsilon_f) + \frac{\phi''(\epsilon_f)}{2}(\omega - \epsilon_f)^2\right] \frac{e^{(\omega - \epsilon_f)/T}}{T\left(e^{(\omega - \epsilon_f)/T} + 1\right)^2}$$

$$= \phi(\epsilon_f) + \frac{\phi''(\epsilon_f)}{2} T^2 \int_{-\infty}^{\infty} dx \frac{x^2}{(e^x + 1)(e^{-x} + 1)} = \phi(\epsilon_f) + \frac{\pi^2}{6} T^2 \phi''(\epsilon_f). \qquad (18.19)$$

This procedure we use now for the necessary integration in (18.16) to obtain

$$[(\omega - \varepsilon)\sigma]'' = 2\sigma' + (\omega - \varepsilon)\sigma'' \to 2\sigma'(\epsilon_f)$$

$$\left[(\omega - \varepsilon)^2\sigma\right]'' = 2\sigma + 4(\omega - \varepsilon)\sigma' + (\omega - \varepsilon)^2\sigma'' \to 2\sigma(\epsilon_f) \qquad (18.20)$$

where all other terms integrate to zero. This results into

$$\mathscr{L}^{(0)} = \sigma(\epsilon_f), \quad \mathscr{L}^{(1)} = \frac{\pi^2}{3} T^2 \sigma'(\epsilon_f), \quad \mathscr{L}^{(2)} = \frac{\pi^2}{3} T^2 \sigma(\epsilon_f) \qquad (18.21)$$

and the Onsager coefficients (18.15) take the form

$$L^{11} = \sigma(\epsilon_f), \quad L^{21} = TL^{11} = -\frac{\pi^2}{2e}(k_B T)^2 \sigma'(\epsilon_f), \quad L^{22} = \frac{\pi^2}{3e^2} k_B^2 T \sigma(\epsilon_f). \qquad (18.22)$$

If one has gradients in the temperature and the chemical potential, one gets from (18.14)

$$\mathbf{j} = L^{11}\mathbf{E} \tag{18.23}$$

which justifies calling $\sigma(\epsilon_f)$ the conductivity tensor. For the thermal conductivity we suppress the charge current $\mathbf{j} = 0$, which means

$$\mathscr{E} = -\frac{L^{12}}{L^{11}}(-\partial_\mathbf{r} T) \tag{18.24}$$

and the heat current becomes

$$\mathbf{j}_q = \left(L^{22} - \frac{L^{21}L^{12}}{L^{11}}\right)(-\partial_\mathbf{r} T) \tag{18.25}$$

which allows us to extract the heat conductivity $\kappa \approx L^{22}$. Dividing by the conductivity we obtain the Wiedemann–Franz law

$$\frac{\kappa}{\sigma} = \frac{\pi^2}{3e^2}k_B^2 T. \tag{18.26}$$

The thermopower appears as well by the situation without charge current. Then the gradient in the temperature (18.24) creates an electrochemical potential $\mathscr{E} = Q\partial_\mathbf{r} T$ with the Seebeck coefficient

$$Q = \frac{L^{12}}{L^{11}} = -\frac{\pi^2}{3e}k_B^2 T\frac{\sigma'(\epsilon_f)}{\sigma(\epsilon_f)}. \tag{18.27}$$

The Seebeck effect converts heat gradients into electric voltage at the junction of different types of wires. The opposite Peltier effect shows that a current flowing through the junctions creates a heat gradient. The heating or cooling of a current-carrying conductor created by a temperature gradient is named the Thomson effect.

18.4   Transport coefficients in high-temperature gases

For high temperatures where we can assume a Maxwell distribution $f_0 = e^{(\mu-\epsilon)/k_B T}$ and quadratic dispersion $\epsilon = k^2/2m$, it is easy to see that

$$\int \frac{d^3 k}{(2\pi\hbar)^3}k_i k_j f_0 = \delta_{ij}\frac{2m}{3}\int \frac{d^3 k}{(2\pi\hbar)^3}\frac{k^2}{2m}f_0 = \frac{1}{2}nmk_B T \tag{18.28}$$

with the density $n = \int d^3 k f_0/(2\pi\hbar)^3$. The chemical potential $n\lambda^3 = e^{\mu/k_B T}$ is given by the thermal de Broglie wave length $\lambda^2 = 2\pi\hbar^2/mk_B T$. For the thermal conductivity we only assume gradients in the temperature and isobaric conditions. Then it is useful to call the enthalpy $H = U + PV = TS + \mu N = c_p NT$ with $c_p = \frac{5}{2} k_B$ for the ideal gas and we get, in (18.13)

$$\partial_r \mu - \frac{\mu}{T} \partial_r T = \left[\left(\frac{\partial \mu}{\partial T} \right)_{N,p} - \frac{\mu}{T} \right] \partial_r T = - \left[\frac{\mu}{T} + \left(\frac{\partial S}{\partial N} \right)_{T,p} \right] \partial_r T = \frac{1}{T} \left(\frac{\partial H}{\partial N} \right)_{T,p} \partial_r T = c_p \partial_r T.$$
(18.29)

The particle and energy currents (18.14) are easily integrated using

$$\int \frac{d^3 k}{(2\pi\hbar)^3} \left(\frac{k^2}{2m} \right)^\alpha f_0 = (k_B T)^\alpha \frac{\Gamma\left(\alpha + \frac{3}{2}\right)}{\Gamma\left(\frac{3}{2}\right)}$$
(18.30)

to obtain

$$\mathbf{j}_n = -\frac{ne\tau}{m} \mathbf{E}, \quad \mathbf{j}_\epsilon = -\frac{k_B T}{m} n\tau \left(\frac{5}{2} k_B \partial_r T + \frac{5}{2} e\mathbf{E} \right).$$
(18.31)

We see that no particle current is created due to a temperature gradient. The charge current $j = -ej_n$ and entropy (heat) current $\mathbf{j}_q = \mathbf{j}_\epsilon - (\mu - p\frac{V}{N})\mathbf{j}_n$ leads therefore to the conductivity and the heat conductivity

$$\sigma = \frac{ne^2 \tau}{m}, \quad \kappa = \frac{5}{2} k_B \frac{k_B T}{m} n\tau = \frac{1}{3} c_p v_{\text{th}}^2 n\tau$$
(18.32)

with the thermal velocity $v_{\text{th}}^2 = 3 k_B T / m$.

The diffusivity we can simply obtain assuming only a density gradient $n(r)$ which by (18.13) leads to the particle current

$$\mathbf{j}_n = -D\partial_r n, \quad D = \frac{k_B T}{m} \tau = \frac{1}{3} v_{\text{th}}^2 \tau = \frac{1}{3} v_{\text{th}} l$$
(18.33)

where we have introduced the corresponding mean-free scattering length l and the diffusion constant D.

The shear viscosity one derives assuming a spatial-dependent mean velocity $\mathbf{u}(\mathbf{r})$ in the local equilibrium (18.10). Let's say, it points in x-direction and is dependent on the y coordinate. Then the stress tensor (3.73) gives the pressure exercised (3.75) and the linearisation defines the shear viscosity η in terms of the gradient of the mean velocity

$$\Pi_{xy} = \int \frac{dk}{(2\pi\hbar)^3} \frac{k_x}{m} \frac{k_y}{m} (f - f_l) = -\eta \frac{\partial u_x}{\partial y}.$$
(18.34)

With the linearised Boltzmann equation (18.13) in relaxation-time approximation

$$f - f_l = \tau k_x \frac{k_y}{m} \frac{\partial f_0}{\partial \epsilon} \frac{\partial u_x}{\partial y},$$
(18.35)

the shear viscosity becomes

$$\eta = n\tau k_B T = \frac{1}{3} n m v_{\text{th}}^2 \tau. \tag{18.36}$$

Considering the three balance equations for heat current, the particle current and viscosity, they all have the form of a diffusion equation with the thermal diffusivity and kinematic viscosity,

$$D_T = \frac{\kappa}{n c_p}, \quad D_\eta = \frac{\eta}{n m} \tag{18.37}$$

respectively. This means in relaxation-time approximation one finds

$$D_T = D_\eta = D = \frac{1}{3} v_{\text{th}} l \tag{18.38}$$

and the Prandl number $P_r = D_\eta/D_T = 1$. The experimental values for mono-atomic gases are in contrast

$$\frac{D_\eta}{D} = \frac{\eta}{Dnm} \sim 0.75, \quad P_r = \frac{D_\eta}{D_T} = \frac{\eta c_p}{\kappa n m} \sim \frac{2}{3} \tag{18.39}$$

which shows the missing accuracy of the constant relaxation-time approximation to account for correlations properly. It could be much improved by constructing an energy-dependent relaxation time for each observable specifically. Corresponding methods like the Chapman-Enskog one (Jäckle, 1978; Smith and Hojgaard–Jensen, 1989) or Grad's 13 moments method (Kraeft, Kremp, Ebeling and Röpke, 1986) can be found in these excellent textbooks on transport. In Chapter 18.5 we will solve the Boltzmann equation exactly to derive such a specific energy-dependent relaxation time for each observable.

18.4.1   Stress or momentum current-density tensor

Previously with (18.34) we have considered the shear viscosity since only velocity gradients perpendicular to the mean velocity have been taken into account. In general one can expand the spatial dependence of the mean velocity and obtain for the stress tensor (3.73) by linearisation

$$\Pi_{ik} = \mathscr{P} \delta_{ik} - \eta \left(\frac{\partial u_i}{\partial x_k} + \frac{\partial u_k}{\partial x_i} \right) - \eta' \frac{\partial u_i}{\partial x_i} \delta_{ik} \tag{18.40}$$

with the pressure $\mathscr{P}$, the shear viscosity η and a further viscosity η' describing the gradient in the direction of the velocity. In order to separate the uniform expansion one regroups (18.40)

$$\Pi_{ik} = \mathscr{P} \delta_{ik} - \eta \left(\frac{\partial u_i}{\partial x_k} + \frac{\partial u_k}{\partial x_i} - \frac{2}{3} \frac{\partial u_i}{\partial x_i} \delta_{ik} \right) - \xi \frac{\partial u_i}{\partial x_i} \delta_{ik} \tag{18.41}$$

where $\xi = \frac{2}{3}\eta + \eta'$ are called second or volume viscosity. For a uniformly expanding gas with quadratic dispersion $\epsilon \sim p^2$ this volume viscosity vanishes as one can see in the following expressions. Let us assume a spherical container $V = 4\pi R(t)^3/3$ and a uniform expansion velocity $u_i = ax_i$. This leads to

$$\frac{1}{V}\frac{dV}{dt} = \frac{3}{R}\frac{dR}{dt} = 3a = \text{div } \mathbf{u}. \tag{18.42}$$

When the expansion happens to be isentropic, $VT^{3/2} = \text{const}$, we have

$$\frac{1}{T}\frac{dT}{dt} = -\frac{2}{3}\frac{1}{V}\frac{dV}{dt} = -\frac{2}{3}\text{div }\mathbf{u}. \tag{18.43}$$

The deviation from the local equilibrium becomes

$$f - f_l \sim \partial_\epsilon f^0 \left(-\frac{\epsilon}{T}\frac{dT}{dt} - \mathbf{p}\cdot\frac{d\mathbf{u}}{dt} \right) \tag{18.44}$$

and since $u_i = ax_i = \frac{1}{3}x_i\text{div }\mathbf{u}$ we can write, with the help of (18.43)

$$f - f_l \sim \partial_\epsilon f^0 \left[\frac{2}{3}\text{div }\mathbf{u} \left(\epsilon - \frac{1}{2}\mathbf{p}\cdot\mathbf{v} \right) \right] \tag{18.45}$$

which vanishes for quadratic dispersions. Mono-atomic gases remain in local equilibrium under uniform adiabatic expansion.

18.5   Exact solution of a linearised Boltzmann equation

Especially for low-temperature Fermi systems there exists a method to solve the linearised Boltzmann equation exactly and to obtain energy-dependent relaxation times for each observable. We follow here (Smith and Hojgaard–Jensen, 1989; Baym and Pethick, 1991) and consider only deviations near the Fermi energy $f = f^0 - T\partial_\epsilon f^0 \Psi = f^0 + f^0(1 - f^0)\psi$. The linearised BUU collision integral (10.160) becomes

$$\delta I(\mathbf{p}_1) = \int \frac{d^3p_2 d^3p'_1 d^3p'_2}{(2\pi\hbar)^3} W f_1^0 f_2^0 (1 - f_{1'}^0)(1 - f_{2'}^0)(\psi'_1 + \psi'_2 - \psi_1 - \psi_2) \tag{18.46}$$

where we understand momentum $\mathbf{p}_1 + \mathbf{p}_2 = \mathbf{p}'_1 + \mathbf{p}'_2$ and energy conservation invoked in the integrations. The matrix element $W = \frac{2\pi}{\hbar}|T(p_1, p_2, p'_1, p'_2)|^2$ is given by the T-matrix. Since all moments have the absolute value of Fermi momentum we have only to care about the angular integrations. With respect to the z-axes of $p_1 + p_2$, the integration reads

$$dp_{1'}^3 = m p_f d\epsilon_1' \sin\theta_1' d\theta_1' d\varphi = \frac{m^2}{p_f} d\epsilon_1' d\epsilon_2' \frac{\sin\frac{\theta}{2}}{\sin\theta} d\varphi \qquad (18.47)$$

since $\angle\theta_{1'2'} \approx \theta$ we have replaced here $d\theta_1'$ with the help of $dp_2' = p_f \sin\theta_{1'2'} d\theta_1'$ and $\theta_1' \sim \theta/2$ as illustrated:

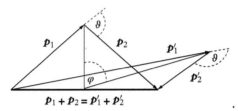

$$p_1 + p_2 = p_1' + p_2'$$

The integration of p_2 we perform with p_1 as z-axes. In such a way we have separated the angular integration from the energy integrations $x = (\epsilon - \mu)/k_B T$ such that the collision integral can be written

$$\delta I(\mathbf{p}_1) = -\frac{2}{\tau(0)\pi^2\langle W\rangle} \int\limits_{-\frac{\mu}{k_B T}}^{\infty} dx_2 dx_1' dx_2' f_1^0 f_2^0 (1-f_{1'}^0)(1-f_{2'}^0)\delta(x_1 + x_2 - x_1' - x_2')$$

$$\times \left\langle \frac{W(\theta,\varphi)}{\cos\frac{\theta}{2}}(\psi_1 + \psi_2 - \psi_1' - \psi_2') \right\rangle \qquad (18.48)$$

where it was useful to abbreviate the energy-independent relaxation time

$$\frac{1}{\tau(0)} = \frac{m^3(k_B T)^2}{16\pi^2\hbar^6}\left\langle \frac{W(\theta,\varphi)}{\cos\frac{\theta}{2}} \right\rangle \qquad (18.49)$$

and the angular integration

$$\langle ... \rangle = \int \frac{d\Omega}{4\pi} \int\limits_0^{2\pi} \frac{d\varphi_2}{2\pi}(...) \qquad (18.50)$$

about the spherical angle $d\Omega = \sin\theta\, d\theta\, d\varphi$ as well as about $d\varphi_2$.

The back scattering can be simplified further by applying a trick expanding in terms of spherical harmonics $Y_{ml}(\hat{p})$

$$\left\langle \frac{W(\theta,\varphi)}{\cos\frac{\theta}{2}} (\psi_2 + \psi_3 - \psi_4) \right\rangle = \left\langle \frac{W(\theta,\varphi)}{\cos\frac{\theta}{2}} \int d\Omega_p \psi_p [\delta(\hat{2}-\hat{p}) - \delta(\hat{3}-\hat{p}) - \delta(\hat{4}-\hat{p})] \right\rangle$$

$$= \left\langle \frac{W(\theta,\varphi)}{\cos\frac{\theta}{2}} \int d\Omega_p \sum_{\bar{m}\bar{l}} \psi_{\bar{m}\bar{l}} Y_{\bar{m}\bar{l}}(\hat{p}) \sum_{ml} Y_{ml}(\hat{p}) \left[Y_{ml}(\hat{2}) - Y_{ml}(\hat{3}) - Y_{ml}(\hat{4}) \right] \right\rangle$$

$$= \sum_{ml} \left\langle \frac{W(\theta,\varphi)}{\cos\frac{\theta}{2}} \psi_{ml}(\eta_2) [Y_{ml}(\hat{2}) - Y_{ml}(\hat{3}) - Y_{ml}(\hat{4})] \right\rangle \qquad (18.51)$$

where we used the completeness and orthogonality relation of the spherical harmonics. With η_2 we indicate the sign of even or odd symmetry by reversing $\hat{p}$. We can now integrate φ_2 with respect to the p_1 direction observing

$$\int \frac{\varphi_2}{2\pi} Y_{ml}(\hat{2}) = Y_{ml}(\hat{1}) P_l(\hat{1}\hat{2}) \qquad (18.52)$$

to get the Legendre polynomials P_l as

$$\sum_{ml} \psi_{ml}(\eta_2) \left\langle \frac{W(\theta,\varphi)}{\cos\frac{\theta}{2}} Y_{ml}(\hat{1}) [P_l(\hat{1}\hat{2}) - P_l(\hat{1}\hat{3}) - P_l(\hat{1}\hat{4})] \right\rangle. \qquad (18.53)$$

The energy integrals in (18.48) can be performed, except one, with the help of

$$\int_{-\infty}^{\infty} dx_2 dx_1' dx_2' f_2^0 (1 - f_{1'}^0)(1 - f_{2'}^0) \delta(x_1 + x_2 - x_1' - x_2') = \frac{1 - f_1^0}{2} (\pi^2 + x_1^2),$$

$$\int_{-\infty}^{\infty} dx f^0 (1 - f^0) = 1, \qquad \int_{-\infty}^{\infty} dx x^2 f^0 (1 - f^0) = \frac{\pi^2}{3},$$

$$\int_{-\infty}^{\infty} dz dw f_x^0 f_w^0 (1 - f_y^0)(1 - f_z^0) \delta(x + w - y - z) = \frac{f_x^0 (1 - f_x^0)}{4} \frac{\cosh\frac{x}{2}}{\cosh\frac{y}{2}} \frac{y - x}{\sinh\frac{y-x}{2}}. \qquad (18.54)$$

Altogether it leads to the practical form of the linearised collision integral

$$\delta I(1) = -\frac{1}{\tau(0)} \left\{ f_1^0 (1 - f_1^0) \left(1 + \frac{x_1^2}{\pi^2} \right) \psi_1 \right.$$

$$\left. + \frac{1}{2\pi^2 \cosh\frac{x_1}{2}} \int dy \frac{F(y - x_1)}{2 \cosh\frac{y}{2}} \sum_{ml} \Psi_{ml}(y) Y_{ml}(\hat{1}) a_l(\epsilon) \right\} \qquad (18.55)$$

with

$$F(x) = \frac{x}{2\sinh\frac{x}{2}}, \quad a_l = \frac{1}{\langle W\rangle}\left\langle \frac{W(\theta,\varphi)}{\cos\frac{\theta}{2}}\eta_2[P_l(-\hat{1}\hat{2}) - P_l(\hat{1}'\hat{1}) - P_l(\hat{2}'\hat{1})]\right\rangle. \quad (18.56)$$

Dependent on the form of the drift for the corresponding observable we obtain the needed l and m components in this expansion. As an example let us consider the thermal conductivity which provides the drift in the Boltzmann equation

$$\partial_{\mathbf{p}_1}\cdot\partial_{\mathbf{r}}f^0 = \frac{\mathbf{p}_1}{m}f^0(1-f^0)\frac{\partial_{\mathbf{r}}T}{T}x = \frac{x}{4\cosh^2\frac{x}{2}}\frac{\mathbf{p}_1\cdot\partial_{\mathbf{r}}T}{mT}. \quad (18.57)$$

We see that it is sufficient to choose $l = 1$ and $m = 0$ and uneven $\eta = -1$ in the expansion. Therefore we choose as ansatz

$$\psi_1 = \psi_{10}(x)Y_{10}(\hat{p}_1) = -\pi^2\tau(0)\cosh\left(\frac{x}{2}\right)\frac{\mathbf{p}_1\cdot\partial_{\mathbf{r}}T}{mT}Q(x) \quad (18.58)$$

where the unknown energy-dependent function $Q(x)$ obeys the linearised Boltzmann equation

$$\frac{x_1}{\cosh^2\frac{x_1}{2}} = (\pi^2 + x_1^2)Q(x_1) + a_1\int dyF(y-x_1)Q(y) \quad (18.59)$$

with

$$a_1 = \frac{-1}{\langle W\rangle}\int\frac{d\Omega}{4\pi}\frac{W(\theta,\varphi)}{\cos\frac{\theta}{2}}\left(\cos\theta + 2\cos^2\frac{\theta}{2}\right) = \frac{-1}{\langle W\rangle}\int\frac{d\Omega}{4\pi}\frac{W(\theta,\varphi)}{\cos\frac{\theta}{2}}(1 + 2\cos\theta).$$

$$(18.60)$$

The inhomogeneous integral equation (18.59) can be solved analytically. Therefore we consider the homogeneous problem and solve the eigenwert equation

$$(\pi^2 + x^2)\phi_n(x) = \lambda_n\int dyF(y-x)\phi_n(y). \quad (18.61)$$

By Fourier transformation

$$\phi_n(x) = \frac{1}{2\pi}\int dz\tilde{\phi}_n(z)e^{\frac{i}{\pi}xz} \quad (18.62)$$

it becomes a Schrödinger equation

$$\left(1 - \partial_z^2 - \frac{\lambda_n}{\cosh^2 z}\right)\tilde{\phi}_n(z) = 0 \quad (18.63)$$

with the known solution (Flügge, 1994)

$$\tilde{\phi}_n(z) = \frac{(-1)^{n+1}}{\pi^2} \frac{2n+1}{n^2(n+1)^2} P_n(\tanh z), \qquad \lambda_n = n(n+1) \tag{18.64}$$

which are orthogonal as

$$\int dx(\pi^2 + x^2)\phi_n(x)\phi_m(x) = \delta_{nm}. \tag{18.65}$$

The inhomogeneous integral equation is then solved by the expansion

$$Q(x) = \frac{X(x)}{\pi^2 + x^2} + \sum_n c_n \phi_n(x), \qquad X(x) = \frac{x}{\cosh \frac{x}{2}} \tag{18.66}$$

which introduced into (18.59) leads to

$$\sum_n \left(1 + \frac{a_1}{\lambda_n}\right) c_n(\pi^2 + x^2)\phi_n(x) = -a_1 \int dy F(y-x) \frac{X(y)}{\pi^2 + y^2}. \tag{18.67}$$

Multiplying with $\phi_m(x)$, integrating and using the homogeneous equation (18.61), one obtains the expansion coefficients in (18.66) as

$$c_n = \frac{-a_1}{\lambda_n + a_1} \int dy X(y)\phi_n(y). \tag{18.68}$$

The integral can be done and the final result for the thermal conductivity reads (Smith and Hojgaard–Jensen, 1989)

$$\kappa = \frac{1}{3} C_v \tau_k v_f^2 \tag{18.69}$$

with the capacity $C_v = Tv_f/2\hbar^2$ and the energy-dependent relaxation time

$$\tau_k = \tau(0)F(a_1), \qquad F(a_1) = \frac{12-\pi^2}{4} - 6a_1 \sum_{n=2,4,\dots} \frac{2n+1}{n^2(n+1)^2(n(n+1)+2a_1)} \tag{18.70}$$

where the energy-dependent a_1 is given by (18.60). Similarly one finds the shear viscosity (Smith and Hojgaard–Jensen, 1989)

$$\eta = \frac{1}{5} nm\tau_\eta v_f^2 \tag{18.71}$$

with a different relaxation time

$$\tau_\eta = \tau(0)F(a_\eta), \quad F(a_\eta) = \frac{\pi^2}{12} - 2a_\eta \sum_{n=1,3,\ldots} \frac{2n+1}{n^2(n+1)^2(n(n+1)+2a_\eta)} \tag{18.72}$$

and

$$a_\eta = \frac{\left\langle \frac{W}{\cos\frac{\theta}{2}} \left(3\sin^2\varphi \sin^4\frac{\theta}{2} - 1\right)\right\rangle}{\left\langle \frac{W}{\cos\frac{\theta}{2}}\right\rangle}. \tag{18.73}$$

With this outlined procedure one can determine the transport coefficients for any observable. This low-temperature form of the collision integral for Fermions allows us to solve the linearised Boltzmann equation exactly giving rise to a specific relaxation time for each variable. Especially the zero sound damping (Abrikosov and Khalatnikov, 1959; Lifschitz and Pitaevsky, 1981) can be calculated. This requires us to consider dynamical relaxation times (Ayik and Boilley, 1992) for which different values by thermal averaging are discussed in (Morawetz, Fuhrmann and Walke, 2000a) and applied to the damping of giant resonances (Morawetz, Fuhrmann and Walke, 1999a). The treatment of collective modes in nuclear matter is documented in an enormous literature starting from their discovery (Bothe and Gentner, 1937), see (Bertsch, Bortignon and Broglia, 1983; Leoni, Døssing and Herskind, 1996) and citations therein. The damping described via a Fermi liquid approach can be found e.g. in (Kamerdzhiev, 1969; Gervais, Thoennessen and Ormand, 1998; Fuhrmann, Morawetz and Walke, 1998; DiToro, Kolomietz and Larionov, 1999).

19

Transient Time Period

Nowadays lasers allow one to create high-density plasmas within a few femtoseconds and allow us to observe their time evolution on a comparable scale (Haug and Jauho, 1996; Theobald, Häßner, Wülker and Sauerbrey, 1996). In semiconductors like GaAs (Huber, Tauser, Brodschelm and Leitenstorfer, 2002) or InP (Huber, Kübler, Tübel, Leitenstorfer, Vu, Haug, Köhler and Amann, 2005) the formation of collective modes and quasiparticles have been observed with the help of femtosecond spectroscopy, see Figure 19.1. Naturally, this plasma is highly excited at the beginning and relaxes towards equilibrium by various mechanisms that might be dominant at some stage and sub-dominant in another one.

The best known regimes are the fast-local equilibration of electron and hole distributions due to binary collisions, and the slow-global relaxation via diffusion, recombination, dissipation of energy into the host crystal, etc. There are, however, even faster processes than the local equilibration. These processes dominate during the very first stage of the relaxation, the so-called transient regime. The same situation of fast disturbance is present in the early stage of heavy-ion collisions. Two nuclei in the first moment of collision create a fast change in the interaction which can be characterised as well as the transient time regime.

The experimental progress with cold atoms triggered by Bose–Einstein condensation has led to enormous activity to understand the time-dependent formation of correlations. Besides the pump and probe experiments, it is now possible to measure the time-dependent occupation of Hubbard-like set ups, see Figure 19.2, and to observe the formation of correlations (Trotzky, Chen, Flesch, McCulloch, Schollwöck, Eisert and Bloch, 2012). The ultra-fast excitations in semiconductors, clusters, or plasmas by ultra-short laser pulses are characterised by long-range Coulomb interactions reflected in the time-dependence of the dielectric function (Huber, Tauser, Brodschelm, Bichler, Abstreiter and Leitenstorfer, 2001; Huber, Tauser, Brodschelm and Leitenstorfer, 2002; Huber, Kübler, Tübel, Leitenstorfer, Vu, Haug, Köhler and Amann, 2005) in the tera-hertz regime (Lloyd-Hughes, Castro-Camus, Fraser, Jagadish and Johnston, 2004). For an overview of theoretical and experimental work see (Axt and Kuhn, 2004). Calculating nonequilibrium Green's functions (Bányai, Vu, Mieck and Haug, 1998; Gartner, Banyai and Haug, 1999) allows one to describe the formation of collective modes (Vu and Haug, 2000; Huber, Kübler, Tübel, Leitenstorfer, Vu, Haug, Köhler and Amann, 2005),

Interacting Systems far from Equilibrium. Klaus Morawetz, Oxford University Press (2018).
© Klaus Morawetz. DOI: 10.1093/oso/9780198797241.001.0001

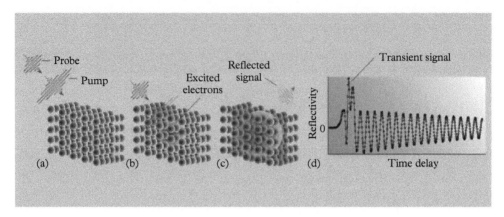

Figure 19.1: *(a) Before excitation by a short, intense laser pulse (the pump), a pure silicon crystal contains few free electrons, and the atoms in the crystal lattice are virtually at rest. (b) The pump hits the sample and creates a localised cloud of excited electrons, which interacts with the surrounding crystal lattice and causes characteristic vibrations and the generation of quasiparticles. (c) A delayed test pulse (the probe) is reflected and modified according to the instantaneous electronic conditions at the surface. (d) By varying the time delay between pump and probe pulses, the excitation dynamics at the femtosecond timescale are measured. (From (Leitenstorfer, 2003))*

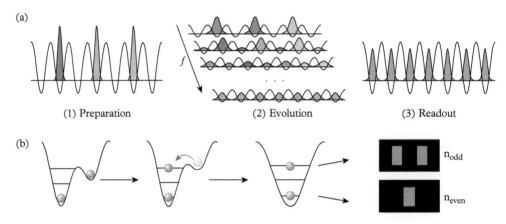

Figure 19.2: *Quench experiment on cold atoms: a (1) after having prepared the density wave $|\Psi(t = 0)\rangle$, (2) the lattice depth was rapidly reduced to enable tunnelling. (3) The properties of the evolved state were read out after all tunnelling was again suppressed. b. Even-odd resolved detection: particles on sites with an odd index were brought to a higher Bloch band. (From (Trotzky, Chen, Flesch, McCulloch, Schollwöck, Eisert and Bloch, 2012).)*

screening (Bányai, Vu, Mieck and Haug, 1998) and even exciton population inversions (Kira and Koch, 2004).

The formation of particles is a single-particle process which has been numerically studied within Green's functions (Bonitz and et. al., 1996). A very similar transient behaviour has been observed for the nuclear matter (Danielewicz, 1984*a*; Köhler, 1995*a*; Morawetz and Koehler, 1999), i.e. the formation/decay of correlation is a rather general phenomenon. If one 'measures' all states of the systems by their distance from equilibrium, instead of *formation* of correlations, one has to talk about their *decay*. Our presumption that the transient period is appreciably shorter than the local relaxation is thus just Bogoliubov's principle of the decay of initial correlations. This decay of correlations is definitely a two-particle process.

The quantity which allows us to follow the two-particle and single-particle pictures in a unified manner is the energy of the system. It is composed from the kinetic energy E_{kin} and the correlation energy $E_{corr} = \langle V_D \rangle - \langle V_C \rangle$, where $\langle V_C \rangle$ subtracts the background. Of course, the total energy conserves,

$$E_{corr}(t) = E_{kin}^0 - E_{kin}(t), \tag{19.1}$$

where E_{kin}^0 is the initial value of the kinetic energy. We will monitor the time dependency of the transfer of the correlation energy into the kinetic one.

It is more convenient to calculate the kinetic energy than the correlation energy because the kinetic one is a single-particle observable. To this end we can use the kinetic equation, of course, which leads to the total energy conservation (19.1). It is immediately obvious that the ordinary Boltzmann equation cannot be appropriate for this purpose because the kinetic energy is an invariant of its collision integral and thus constant in time. It is possible to study this transient time regime within the Levinson equation for the Wigner function.

19.1   Formation of correlations

A first guess of the time-dependent formation of correlations can be found from the time-dependent Fermi golden rule

$$P_{nn'} = \frac{1}{\hbar^2} V_{nn'}^2 \left(\frac{\sin w_{nn'} \frac{t}{2}}{\omega_{nn'} \frac{1}{\hbar}} \right)^2 = V(q)^2 \frac{1 - \cos\left(\Delta E \frac{t}{\hbar}\right)}{(\Delta E)^2} \tag{19.2}$$

expressing the transition probability between states n and n' which we consider as the state before and after the collision and $\Delta E = \epsilon_k + \epsilon_p - \epsilon_{k+q}' - \epsilon_{p-q}'$ denotes the energy difference between initial and final states. Taking the occupation factors into account, the time-dependent formation of kinetic energy is expected to have the form

$$E_{kin}(t) = \int \frac{dkdpdq}{(2\pi\hbar)^9} V(q)^2 \frac{1 - \cos\left(\Delta E \frac{t}{\hbar}\right)}{\Delta E} \rho_{k+q}\rho_{p-q}(1 - \rho_k)(1 - \rho_p). \tag{19.3}$$

Let us examine how this expected expression appears from kinetic theory.

19.1.1   Levinson equation

Though we have seen already how the Levinson equation (10.113) appears to let us shortly summarise the main steps, we substitute of the GKB ansatz (8.35) for $G^<$ into the time diagonal part (9.76) of the Kadanoff and Baym equation (7.28). It is sufficient to restrict our attention to the Born approximation, $\mathscr{T}^{RA} = V\delta(\tau)$ since over short times higher-order correlations have no time to build up as we will see later. The precursor of the Levinson equation for spatial homogeneous systems then reads

$$\frac{\partial}{\partial t}\rho(k,t) = \frac{2}{\hbar^2}\,\text{Re}\int\frac{dpdq}{(2\pi\hbar)^6}V_q^2\int_0^\infty d\tau\, S\left(t-\frac{\tau}{2},\tau\right)D_{2p}(t-\tau). \tag{19.4}$$

All distributions are collected in

$$D_{2p}(t) =[1-\rho_k(t)][1-\rho_p(t)]\rho_{k-q}(t)\rho_{p+q}(t) - \rho_k(t)\rho_p(t)[1-\rho_{k-q}(t)][1-\rho_{p+q}(t)], \tag{19.5}$$

and all propagators are collected in

$$S(t,\tau) = g^R(k,t,\tau)g^R(p,t,\tau)g^A(k-q,t,-\tau)g^A(p+q,t,-\tau). \tag{19.6}$$

Within a simple approximation of the propagator, $[t_1 > t_3]$,

$$g^R(k,t_1,t_3) \approx -i\exp\left\{-i\frac{k^2}{2m\hbar}(t_1-t_3) - \frac{t_1-t_3}{2\tau}\right\}, \tag{19.7}$$

the final Levinson equation (Levinson, 1965, 1969) reads

$$\frac{\partial\rho_k(t)}{\partial t} = \frac{2}{\hbar^2}\int_{t_0}^t d\bar{t}\int\frac{dpdq}{(2\pi\hbar)^6}|V(q)|^2 e^{\frac{\bar{t}-t}{\tau}}\cos\left[\frac{\Delta}{\hbar}E(t-\bar{t})\right]\left\{\rho_{k-q}(\bar{t})\rho_{p+q}(\bar{t})\left[1-\rho_k(\bar{t})\right]\left[1-\rho_p(\bar{t})\right]\right.$$

$$\left. - \rho_k(\bar{t})\rho_p(\bar{t})\left[1-\rho_{k-q}(\bar{t})\right]\left[1-\rho_{p+q}(\bar{t})\right]\right\}, \tag{19.8}$$

where t_0 denotes the time at which the interaction has been switched on and we abbreviate

$$\Delta E = \left(\frac{k^2}{2m} + \frac{p^2}{2m} - \frac{(k-q)^2}{2m} - \frac{(p+q)^2}{2m}\right). \tag{19.9}$$

The solution of the Levinson equation (19.8) in the short-time region $t \ll \tau$ can be written down analytically. In this time domain we can neglect the time evolution of distributions, $\bar{\rho}_a(\bar{t}) \approx \rho_a(0)$, and the life-time factor, $\exp\left\{-\frac{t-\bar{t}}{\tau}\right\} \approx 1$. Therefore the deviation of the Wigner distribution from its initial value, $\rho_a(t) = \rho_a(0) + \delta\rho_a(t)$, reads

$$\delta\rho_a(t) = 2\sum_b\int\frac{dpdq}{(2\pi\hbar)^6}V_D^2(q)\frac{1-\cos\{t\Delta E\}}{\Delta E^2}\left\{\rho_a'\rho_b'(1-\rho_a)(1-\rho_b) - \rho_a\rho_b(1-\rho_a')(1-\rho_b')\right\}. \tag{19.10}$$

This formula shows how the two-particle and the single-particle concepts of the transient behaviour are combined in the kinetic equation. The right-hand side describes how two particles correlate their motion to avoid strong interaction regions. Since the process is very fast, the on-shell contribution to $\delta\rho_a$, proportional to t/τ, can be neglected in the assumed time domain and the $\delta\rho$ has the pure off-shell character as can be seen from the off-shell factor $(1 - \cos\{t\Delta E\})/\Delta E^2$. The off-shell character of mutual two-particle correlations is thus reflected in the single particle Wigner distribution. We refer to Eq. (19.10) as the *finite duration time* approximation. It carries the most important features of the build up of correlations after the interactions are switched on in the initially uncorrelated system. This we will demonstrate by some numerical examples.

Once formed, the off-shell contributions change in time with the characteristic time τ, i.e. following the relaxation (on-shell) processes in the system. Accordingly, the formation of the off-shell contribution signals that the system has reached the state where the further evolution can be described by the Boltzmann equation, i.e. the transient period has been accomplished. From the Wigner distribution (19.10) one can readily evaluate the increase of the kinetic energy,

$$E_{kin}(t) - E_{kin}^0 = \sum_a \int \frac{dk}{(2\pi\hbar)^3} \frac{k^2}{2m_a} \delta\rho_a(t). \tag{19.11}$$

After substitution $\delta\rho_a$ from (19.10) we symmetrize in k and p and anti-symmetrize in the initial and final states and obtain exactly the expected result (19.3).

19.1.2 Formation of correlations in plasma

Starting with a sudden switching approximation, due to Coulomb interaction, the screening is formed during the first transient time period. This can be described by the non-Markovian Lenard–Balescu equation (Morawetz, 1994) instead of the static screened equation (19.8). With the same discussion as above but using the RPA approximation of Chapter 10.2.2 we end up instead of (19.3) with the dynamical expression of the correlation energy (for details, see (Morawetz, Špička and Lipavský, 1998),

$$E_{corr}^{dynam}(t) = -\sum_{ab} \int \frac{dkdpdq}{(2\pi\hbar)^9} \frac{V_C^2(q)}{|\epsilon^R\left(q, \frac{(p+q)^2}{2m_b} - \frac{p^2}{2m_b}\right)|^2}$$

$$\times \frac{1 - \cos\left\{\frac{1}{\hbar}t\Delta E\right\}}{\Delta E^2} \left(\frac{(k-q)^2}{2m_a} - \frac{k^2}{2m_a}\right) \rho_a' \rho_b' (1-\rho_a)(1-\rho_b) \tag{19.12}$$

with the potential screened by the dielectric function ϵ^R. To demonstrate its results and limitations, we discuss (19.3) and (19.12) for equilibrium initial distributions. We assume a one-component plasma which possesses the Maxwellian velocity distribution during this formation time. From (19.12) we find analytically (Morawetz, Špička and Lipavský, 1998) the quantum result of the time derivative of the formation of the correlation for statically as well as dynamically screened potentials as

$$\frac{\partial}{\partial t}\frac{E_{\mathrm{corr}}^{\mathrm{static}}(t)}{n} = -\frac{e^2 \kappa T}{2\hbar}\mathrm{Im}\left[(1+2z^2)e^{z^2}(1-\mathrm{erf}(z)) - \frac{2z}{\sqrt{\pi}}\right]$$

$$\frac{\partial}{\partial t}\frac{E_{\mathrm{corr}}^{\mathrm{dynam}}(t)}{n} = -\frac{e^2 \kappa T}{\hbar}\mathrm{Im}\left[e^{z_1^2}(1-\mathrm{erf}(z_1))\right] \qquad (19.13)$$

where we used $z = \omega_p\sqrt{t^2 - it\frac{\hbar}{T}}$ and $z_1 = \omega_p\sqrt{2t^2 - it\frac{\hbar}{T}}$. For the classical limit one can integrate expression (19.13) with respect to times

$$E_{\mathrm{corr}}^{\mathrm{static}}(t) = -\frac{1}{4}e^2 n\kappa\left\{1 + \frac{2\omega_p t}{\sqrt{\pi}} - \left(1 + 2\omega_p^2 t^2\right)\exp\left(\omega_p^2 t^2\right)\left[1 - \mathrm{erf}(\omega_p t)\right]\right\}$$

$$E_{\mathrm{corr}}^{\mathrm{dynam}}(t) = -\frac{1}{2}e^2 n\kappa\left\{1 - \exp\left(\frac{\omega_p^2}{2}t^2\right)\left[1 - \mathrm{erf}(\frac{\omega_p}{\sqrt{2}}t)\right]\right\}.$$

$$(19.14)$$

In Fig. 19.3 this formulae are compared with molecular dynamic simulations (Zwick-nagel, Toepffer and Reinhard, 1995) for two values of the plasma parameter $\Gamma = 0.1$ and 1. This plasma parameter

$$\Gamma = \frac{e^2}{a_e T}, \qquad (19.15)$$

where $a_e = (\frac{3}{4\pi n})^{1/3}$ is the inter-particle distance or Wigner–Seitz radius, measures the strength of the Coulomb coupling. Ideal plasmas are found for $\Gamma \ll 1$. In this region the static formula (19.14) well follows the major trend of the numerical result, see Fig. 19.3. The static result underestimates the dynamical long-time result of Debye–Hückel $\sqrt{3}/2\Gamma^{3/2}$ by a factor of two. The explanation for this fact is that we can prepare the initial configuration within our kinetic theory such that sudden switching of inter-action is fulfilled. However, in the simulation experiment we have initial correlations which are due to the set up within quasiperiodic boundary condition and Ewald sum-mations. This obviously results into an effective statically screened Debye potential, or at least the simulation results allow for this interpretation. For $\Gamma = 1$ non-ideal effects become important and the formation time is underestimated within (19.14).

The characteristic time of formation of correlations in the high-temperature limit is given by the time where (19.14) shows a saturation. This is reached at about the time of the inverse plasma frequency $\tau_c \approx \frac{1}{\omega_p} = \sqrt{2}/v_{\mathrm{th}}\kappa$ indicating that the dominant role is played by long-range fluctuations. On the other hand, we also see that the correlation time is found to be given by the time a particle needs to travel through the range of the potential with a thermal velocity v_{th} and is not given by the time between successive collisions as one might have thought.

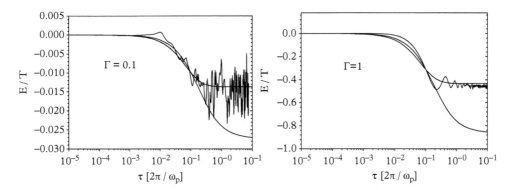

Figure 19.3: *The formation of correlation energy due to molecular dynamic simulations (Zwicknagel et al., 1995) together with the result of (19.14) for a plasma parameter* $\Gamma = 0.1$ *(left) and* $\Gamma = 1$ *(right). The upper curve is the static and the lower the dynamical calculation. The latter one approaches the Debye–Hückel result, from (Morawetz, Špička and Lipavský, 1998).*

19.1.3   Formation of correlations in nuclear matter

As another example, Figure 19.4 shows the time development of the kinetic energy as well as correlation energy for two initially counter-flowing streams of nuclear matter, where the nucleons interact via a Gauß type of potential. The initial temperatures and densities of the colliding beams are $T_1 = 10$ MeV $n_1 = n_o/60$ and $T_2 = 5$ MeV $n_2 = n_o/10$ respectively. The relative momentum is $1.5\hbar/fm$, which corresponds to a colliding energy of 45 MeV/n. Fig. 19.4 shows a build up of correlations during the initial $3-4fm/c$. Total energy is conserved, see Appendix A, and the kinetic energy is increased by the same amount as the correlation energy is decreased. This is because the system is initially prepared to be uncorrelated at $t_0 = 0$. If the time t_0, i.e. the time when the system is uncorrelated is shifted to the infinite past $t_0 = -\infty$, we would not observe any build up of correlations. The equation (19.10) would then in fact reduce to the Boltzmann equation. In Figure 19.4 we compare the results with the exact solution of Kadanoff and Baym equations (Köhler, 1995c, 1996). We see that the finite duration approximation reproduces the exact result quite nicely. The small deviation is due to higher-order effects (Köhler and Morawetz, 2001).

This build-up of correlations is independent of the initial distribution form. If for example we choose a Fermi distribution as the initial distribution, a build up of correlations will occur as well. This is due to the fact that the spatial correlations relate in momentum space to excitations, which results in a distribution looking somewhat like a Fermi distribution but with a temperature higher than the initial uncorrelated Fermi distribution (Köhler, 1995c; Köhler and Morawetz, 2001). One finds the build-up time where the correlation energy reaches its first maximum as the inverse Fermi energy $\tau_c = \hbar/\epsilon_f$ in agreement with the quasiparticle formation time known as Landau's criterion. Indeed, as argued above, the quasiparticle formation and the build up of correlations are two alternative views of the same phenomenon.

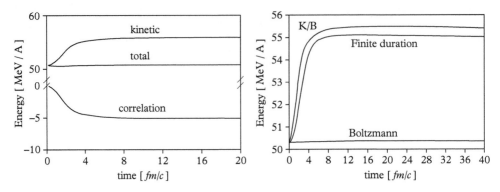

Figure 19.4: *The time dependent kinetic and correlation energy vs. time (left) for a counter-flowing streams of nuclear matter. and the kinetic energy from a solution of the Kadanoff–Baym equation (KB) together with the results from the finite duration approximation (19.10) and the Boltzmann equation (right).*

19.2   Quantum quenches and sudden switching

Special preparation of cold atoms in optical lattices allows us to study the local relaxation (Flesch, Cramer, McCulloch, Schollwöck and Eisert, 2008; Trotzky, Chen, Flesch, Mc-Culloch, Schollwöck, Eisert and Bloch, 2012) and to explore dissipation mechanisms (Syassen, Bauer, Lettner, Volz, Dietze, Garca-Ripoll, Cirac, Rempe and Drr, 2008). At intermediate time scales a quasistationary state had been found during thermalization process (Eckstein, Kollar and Werner, 2009). Both different physical systems, the long-range Coulomb (Morawetz, Lipavský and Schreiber, 2005a) as well as short-range Hubbard systems (Morawetz, 2014) can be described by a common theoretical approach leading even to a unique formula to describe the formation of correlations at short-time scale as we will demonstrate. This could be of interest since normally the formation is explained by numerically demanding calculations solving Green's functions (Bányai, Vu, Mieck and Haug, 1998; Gartner, Banyai and Haug, 1999) or renormalisation group equations (Trotzky, Chen, Flesch, McCulloch, Schollwöck, Eisert and Bloch, 2012). Why do such different systems show similar features? It is just the fact that correlations need time to be formed. In other words higher-order correlations need more time to be build up than low-order ones. Though this statement is strictly only valid for weakly correlated systems, we adopt it here to see that it leads to good results at short-time scales even in the strong-correlated regime. Therefore it is suggested here as a conjecture that the lowest level, the mean-field approximation, is sufficient to describe the basic features of short-time formation of correlations. Therefore we will consider the short-time mean-field approximation with an additional relaxation time mimicking two-particle correlation processes.

We consider the time evolution of the reduced density matrix $\langle p + \frac{1}{2}q|\delta\rho|p - \frac{1}{2}q\rangle = \delta f(p, q, t)$ which is given by linearisation $\delta[H, \rho] = [\delta H, \rho_0] + [H_0, \delta\rho]$ of the kinetic equation [$\hbar = 1$ in the following]

$$\dot{\rho} + i[H, \rho] = \frac{\rho^{\text{l.e.}} - \rho}{\tau} \qquad (19.16)$$

with respect to an external perturbation δV^{ext}. The effective Hamiltonian consists of the quasiparticle energy, the external and induced mean-field $\langle p + \frac{1}{2}q | \delta H | p - \frac{1}{2}q \rangle = \delta V^{\text{ext}} + V_q \delta n_q$ given by the interaction potential V_q and the density variation δn_q. As a possible confining potential we assume a harmonic trap $V^{\text{trap}} = \frac{1}{2}Kx^2$ which leads to $\langle p + \frac{1}{2}q | \delta[V^{\text{trap}}, \rho] | p - \frac{1}{2}q \rangle = -K\partial_p\partial_q\delta f(p, q, t)$.

The kinetic equation (19.16) is assumed to relax towards a local equilibrium of Fermi/Bose distribution with an allowed variation of the chemical potential

$$\langle p + \frac{q}{2} | \rho^{\text{l.e.}} - \rho | p - \frac{q}{2} \rangle = \langle |\rho^{\text{l.e.}} - \rho^0| \rangle - \delta f(p, q, t) = -\frac{\Delta f}{\Delta \epsilon}\delta\mu(q, t) - \delta f(p, q, t). \qquad (19.17)$$

Here we use the short-hand notation $\Delta f = f_0(p + \frac{q}{2}) - f_0(p - \frac{q}{2})$ and $\Delta\epsilon = \epsilon_{p+\frac{q}{2}} - \epsilon_{p-\frac{q}{2}}$. This variation of the chemical potential allows us to enforce the density conservation $n = \sum_p f = \sum_p f^{\text{l.e.}}$ (Mermin, 1970; Das, 1975) leading to the Mermin correction, i.e. a relation between density variation $\delta n(q, t) = \tilde{\Pi}(t, \omega = 0)\delta\mu(q, t)$ and the polarisation in RPA

$$\Pi(t, t') = i\sum_p [f_{p+\frac{q}{2}}(t') - f_{p-\frac{q}{2}}(t')] e^{\left(i\epsilon_{p+\frac{q}{2}} - i\epsilon_{p-\frac{q}{2}} + \frac{1}{\tau} \right)(t'-t)},$$

$$\tilde{\Pi}(t, \omega) = \int d(t - t') e^{i\omega(t-t')} \Pi(t, t'). \qquad (19.18)$$

See also the discussion in Chapter 18.1.

The linearised kinetic equation (19.16) reads therefore for $\delta f_t = \delta f(p, q, t)$

$$\dot{\delta f}_t + \frac{\delta f_t}{\tau} + i\Delta\epsilon\delta f_t = i\Delta f\delta V_t^{\text{ext}} + i\Delta f V_q \delta n_t + \frac{\Delta f}{\Delta\epsilon}\frac{\delta n_t}{\tau\Pi(q, 0, t)} + iK\partial_p\partial_q\delta f_t. \qquad (19.19)$$

The last term describes the confining harmonic trap and the term before comes from the Mermin correction due to density conserving relaxation time approximation.

Neglecting the time derivative of the homogeneous part $f_0(p, t)$ compared to $\delta f(p, q, t)$ we can solve this kinetic equation (19.19) considering the momentum derivatives of the last term as perturbation to obtain (Morawetz, 2014)

$$\delta f(p, q, t) - \delta f(p, q, 0) = i\int_{t_0}^t dt' e^{\left[\left(i\Delta\epsilon + \frac{1}{\tau} \right)(t'-t) \right]} \left\{ \Delta f(t') \left[V_q\delta n(q, t') + V_q^{\text{ext}}(t') \right] \right.$$

$$\left. + \frac{1}{i\tau\tilde{\Pi}(t', 0)}\frac{\Delta f(t')}{\Delta\varepsilon}\delta n(q, t') + K\partial_p\partial_q\delta f(p, q, t') \right\}. \qquad (19.20)$$

The further evaluation is very much dependent on the physical setup and leads to different solutions.

19.2.1   Atoms in a lattice after sudden quench

For cold atoms each occupying second place on a lattice $f_k = (1 + (-1)^k)n/2$ we have a Fourier transform to the momentum distribution

$$f_0(p) = a \sum_{k=-N}^{N} e^{\frac{i}{\hbar}kap} f_k = na \frac{\sin(2N+1)\frac{ap}{\hbar}}{\sin(\frac{ap}{\hbar})} \rightarrow \pi\hbar n\delta(p) = \frac{n}{2}\delta_p, \qquad (19.21)$$

for a large total number of atoms N and the lattice spacing a. We can now Laplace transform the time $t \rightarrow s$ in the kinetic equation (19.19) or (19.20) to get

$$\delta f_s = \frac{\delta f_0}{s+i\Delta\epsilon+\frac{1}{\tau}} + \frac{in}{2}\left(\frac{\delta_{p+q/2}}{s-ib+\frac{1}{\tau}} - \frac{\delta_{p-q/2}}{s+ib+\frac{1}{\tau}}\right)(\delta V^{\text{ext}} + V_q\delta n_s)$$

$$+ \frac{1}{2\tau}\left(\frac{\delta_{p+q/2}}{s-ib+\frac{1}{\tau}} + \frac{\delta_{p-q/2}}{s+ib+\frac{1}{\tau}}\right)\delta n_s + \frac{iK}{s+i\Delta\epsilon+\frac{1}{\tau}}\partial_p\partial_q\delta f_s \qquad (19.22)$$

where we introduced $\Delta\epsilon|_{p=\pm q/2} = \pm 4\mathcal{J}\sin^2\frac{aq}{2\hbar} = \pm b$. The initial time disturbance of the distribution δf_0 is determined according to the different physical preparations.

In case of a sudden quench the interaction is switched on suddenly and no external perturbation will be assumed $\delta V^{\text{ext}} = 0$. Let's consider the time evolution of an empty place in the lattice if each second place was initially populated. The density $n_t = \frac{n}{2} + \delta n_t$ starts with $n_0 = 0$ which means $\delta n_0 = -n/2$ as an initial condition. If we first look at the quench without interaction ($V = 0$) we can solve (19.19)

$$\delta f_t^{V=0} = \delta f(0)e^{-i\Delta\epsilon t - \frac{t}{\tau}} \qquad (19.23)$$

with the choice of $\delta f(0) = -n/2$ that the density

$$\delta n_t^{V=0} = \sum_p \delta f_t = -\frac{n}{2}j_0(\sqrt{4\mathcal{J}bt})e^{-\frac{t}{\tau}} \qquad (19.24)$$

starts with $\delta n_0 = -n/2$ as desired since the Bessel function $j_0(0) = 1$. Please note that $\sum_p = a/2\pi\int_0^{2\pi/a} dp$ according to the finite band.

Now we can integrate (19.22) over momentum including the interaction to get the equation for the density. Using the La Place transform $j_0(\sqrt{4\mathcal{J}bt}) \bullet\!\!-\!\!\circ 1/\sqrt{s^2 + 4\mathcal{J}b}$ let's first inspect the solution without a confining trap ($K = 0$)

$$\delta n_s = -\frac{n}{2}\frac{\left(s+\frac{1}{\tau}\right)^2+b^2}{\sqrt{\left(s+\frac{1}{\tau}\right)^2+4\mathcal{J}b\left(s^2+\frac{s}{\tau}+nbV_q+b^2\right)}} \quad \circ\!\!-\!\!\bullet \quad \delta n_t = -\frac{n}{2}j_0(\sqrt{4\mathcal{J}b}t)\mathrm{e}^{-\frac{t}{\tau}}$$

$$-\frac{n}{4\gamma\tau^2}\int_0^t dx j_0(\sqrt{4\mathcal{J}b}x)\mathrm{e}^{-\frac{t+x}{2\tau}}\left(2\gamma\tau\cos\gamma(t-x)+(1-2bnV_q\tau^2)\sin\gamma(t-x)\right) \quad (19.25)$$

where $\gamma^2 = nbV + b^2 - 1/4\tau^2$. Besides the interaction-free result (19.24) we obtain an additional contribution due to the interaction and dissipation presented by the relaxation time. Without interaction $V = 0$ and damping $1/\tau \to 0$ we obtain the exact result of (Flesch, Cramer, McCulloch, Schollwöck and Eisert, 2008).

In Figures 19.5 we compare (19.25) with the experimental data (Trotzky, Chen, Flesch, McCulloch, Schollwöck, Eisert and Bloch, 2012) with (19.25) where we plot the interaction-free evolution together with the interaction one. The main effect of interaction is the damping which brings the curves nearer to the experimental data. Here we use the parameter for the lattice constant given by half of the short laser wave length $a = \lambda/2 = 765$nm which provides a wave vector of $q = \pi\hbar/a$, and an initial density $n = 1/2a$ with each second place filled. The relaxation time characterises the dissipative processes which we assume to arise due to polaron scattering. These lattice-deformation processes are dominated by hopping transport at high temperatures and band regime

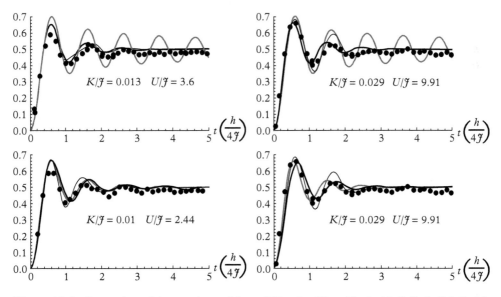

Figure 19.5: *Comparison of the experimental data of (Trotzky, Chen, Flesch, McCulloch, Schollwöck, Eisert and Bloch, 2012) (dots) with the RG calculation (thin line) (Flesch, Cramer, McCulloch, Schollwöck and Eisert, 2008) from (Morawetz, 2014). Left: Mermin's correction of conserving relaxation time $\tau = 0.6\hbar/\mathcal{J}$ approximation (19.25) without (gray) and with interaction (black). Right: with (black) and without (gray) the influence of the trapping potential $K/\mathcal{J}$.*

transport at low temperature with the transition given by $\hbar/\tau = 2\mathcal{J}\exp(-S)$ where S describes the ratio of polaron binding to optical phonon energy. This quantity is generally difficult to calculate (Jones and March, 1973) but in the order of one. We will use it as fit parameter and find a common value $\tau = 0.6\hbar/\mathcal{J}$ for the results in the figures presented here. We see that the analytic result (19.25) describes the data slightly better and we can give the time evolution up to more oscillations than possible by numerical renormalisation group techniques.

19.2.2   Femtosecond laser response

Now we are interested in the short-time response of the system to an external perturbation V^{ext}. This is different from sudden quench since here we have initially $\delta f(p, q, 0) = 0$ and the system is driven out of equilibrium by V^{ext}. As the result we will obtain the dielectric response which gives microscopic access to optical properties. Integrating (19.20) over momentum, one obtains the time-dependent density response

$$\delta n(q, t) = \int_{t_0}^{t} dt' \chi(t, t') V_q^{\text{ext}}(t') \tag{19.26}$$

describing the response of the system with respect to the external field in contrast to the polarisation function (19.18) which is the response to the induced field. One obtains the equation for $\chi(t, t')$ from (19.26) by interchanging integrations in (19.20)

$$\chi(t, t') = \Pi(t, t') + \int_{t'}^{t} d\bar{t} \left\{ \left[\Pi(t, \bar{t}) V_q + I(t, \bar{t}) \right] \chi(\bar{t}, t') + R(t, \bar{t}) \right\} \tag{19.27}$$

with the polarisation (19.18) and Mermin's correction

$$I(t, t') = \sum_p \frac{f_{p+\frac{q}{2}}(t') - f_{p-\frac{q}{2}}(t')}{\varepsilon_{p+\frac{q}{2}} - \varepsilon_{p-\frac{q}{2}}} \frac{e^{\left(i\varepsilon_{p+\frac{q}{2}} - i\varepsilon_{p-\frac{q}{2}} + \frac{1}{\tau} \right)(t'-t)}}{\tau \tilde{\Pi}^{\text{RPA}}(t', 0)}. \tag{19.28}$$

For cold atoms on the lattice we have obtained the solution already (19.22) which we can use here with $\delta f(0) = 0$ and we have also

$$\Pi(t, t') = n e^{\frac{t'-t}{\tau}} \sin[b(t' - t)], \qquad I(t, t') = \frac{1}{\tau} e^{\frac{t'-t}{\tau}} \cos[b(t' - t)]. \tag{19.29}$$

This will lead to the same response formula as a gas of particles with the thermal Fermi/Bose distribution for f_p. For the latter one we work in the limit of long wave lengths $q \to 0$ and the leading terms are $\Pi(t, t') \approx \frac{q^2 n(t')}{m}(t' - t)e^{\frac{t'-t}{\tau}}$ and $I(t, t') \approx \frac{1}{\tau} e^{\frac{t'-t}{\tau}}$ with the time-dependent density $n(t)$.

We introduce the collective mode of plasma/sound-velocity oscillations for Coulomb gas and for the Hubbard models respectively

$$
\omega_p^2 = \begin{cases} \frac{ne^2}{m\varepsilon_0} & \text{for } V_q = \frac{e^2\hbar^2}{\varepsilon_0 q^2}, \epsilon_p = \frac{p^2}{2m} \\[2mm] bnaU & \text{for } V_q = Ua, \epsilon_p = 2\mathcal{J}(1 - \cos pa/\hbar) \end{cases} \tag{19.30}
$$

where we had already used $b = 4\mathcal{J}\sin^2\frac{aq}{2\hbar}$. For Coulomb interactions one has an optical mode while for atoms on the lattice the mode it is acoustic.

For freely moving particles it is convenient to transform (19.27) into a differential equation

$$
\ddot{\chi}(tt') + \frac{1}{\tau}\dot{\chi}(tt') + \omega_p^2\chi(tt') = 0
$$

$$
\chi(t, t) = 0, \dot{\chi}(t, t')|_{t=t'} = -\omega_p^2/V_q \tag{19.31}
$$

where the influence of the trap can be considered as well (Morawetz, 2014).

Interestingly, both solutions, the one for the Hubbard lattice (19.22) and the one for the freely moving particles (19.31) lead to the same result of the integral equation (19.27) via (19.31) for the two-time response function

$$
V\chi(t, t') = -\frac{\omega_p^2}{\gamma}e^{-\frac{t-t'}{2\tau}}\sin\gamma(t - t') \tag{19.32}
$$

but with a different collective mode $\gamma = \sqrt{\omega_p^2 - \frac{1}{4\tau^2}}$ for the Coulomb gas and $\gamma = \sqrt{\omega_p^2 + b^2 - \frac{1}{4\tau^2}}$ for cold atoms. In this sense we consider (19.32) as a universal short-time behaviour.

In the further analysis we will follow the experimental way of analysing the two-time response function closely (Huber, Kübler, Tübel, Leitenstorfer, Vu, Haug, Köhler and Amann, 2005). The pump pulse is creating charge carriers in the conduction band and the probe pulse is testing the time evolution of this occupation. The time delay after this probe pulse $T = t - t_0$ is Fourier transformed into frequency. Similarly we start the half-empty lattice of cold atoms to relax at t_0. The frequency-dependent inverse dielectric function associated with the actual time t is then given by

$$
\frac{1}{\varepsilon(\omega, t)} = 1 + \int_0^{t-t_0} dTe^{i\omega T}V\chi(t, t - T) \tag{19.33}
$$

which is exactly the one-sided Fourier transform introduced in Ref. (ElSayed, Schuster, Haug, Herzel and Henneberger, 1994). The integral (19.33) with (19.32) can be expressed in terms of elementary functions. Without the last term due to the

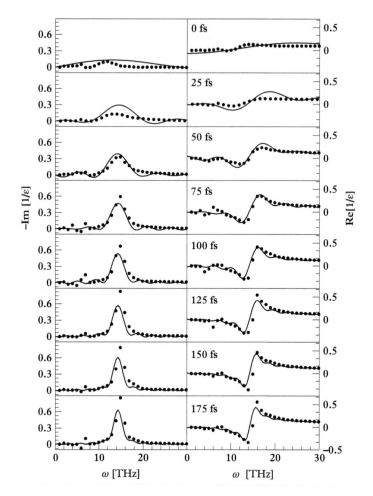

Figure 19.6: *The time evolution of the inverse-dielectric function in GaAs. The labels from top to bottom denote the time t. The pump pulse was at $t_0 = -40$ fs and the probe pulse has a full-width at half-maximum of 27fs. Circles are data from (Huber, Tauser, Brodschelm, Bichler, Abstreiter and Leitenstorfer, 2001; Huber, Tauser, Brodschelm and Leitenstorfer, 2002), solid lines show the electronic part (19.32) published in (Morawetz, Lipavský and Schreiber, 2005a). The plasma frequency is given by $\omega_p = 14.4$ THz and the relaxation time is $\tau = 85$ fs.*

confining potential, it is the solution derived for Coulomb systems in (Morawetz, Lipavský and Schreiber, 2005a).

As Figure 19.6 illustrates, Eq. (19.33) describes the experiments quite accurately. A further virtue is that the long time limit yields the Drude formula correctly,

$$\lim_{t \to \infty} \frac{1}{\varepsilon} = 1 - \frac{\omega_p^2}{\gamma^2 - \omega(\omega + \frac{i}{\tau})} \qquad (19.34)$$

leading to the Drude conductivity (18.8) which is not easy to achieve within short-time expansions (ElSayed, Schuster, Haug, Herzel and Henneberger, 1994) and which had provided the wrong long-time limit $1 - \omega_p^2/[\omega_p^2 - (\omega + i/\tau)^2]$ before.

19.3   Failure of memory-kinetic equations

We saw that the short-time expansion of the mean-field-free examples in Section 19.1.2 and examples of quantum quench of mean-fields in Section 19.2 work quite well. Beyond the very short-time expansion, however, there are double counts in the Levinson equation following from the additional retardation resulting from the GKB ansatz. For slow perturbations one can expand (19.4) to the lowest order in the memory, $D_{2\mathrm{p}}(t{-}\tau) = D_{2\mathrm{p}}(t) - \tau \partial_t D_{2\mathrm{p}}(t)$, so that the collision integral splits into the Boltzmann-like scattering integral and the gradient correction,

$$\frac{\partial}{\partial t}\rho = \mathscr{I} + \frac{\partial}{\partial t}\mathscr{R}, \tag{19.35}$$

with

$$\mathscr{I} = \int \frac{dpdq}{(2\pi)^6}\, V_q^2 D_{2\mathrm{p}}(t) 2\,\mathrm{Re}\int_0^\infty d\tau\, S(\tau), \text{ and } \mathscr{R} = \int \frac{dpdq}{(2\pi)^6}\, V_q^2 D_{2\mathrm{p}}(t) 2\,\mathrm{Re}\int_0^\infty d\tau\, \tau\, S(\tau). \tag{19.36}$$

Please note that $\mathscr{R}$ and $\mathscr{I}$ are different from the extended quasiparticle picture (9.78). We can define a new function to remove the retardation, $\mathscr{F} = \rho - \mathscr{R}$, and try to redefine the transport in terms of $\mathscr{F}$. This will us lead now to unavoidable double counts.

19.3.1   Double counts

The integrals in (19.36) are controlled by two mechanisms, the de-phasing given by momentum integrals and the decay of propagators. Let us assume for a while that the de-phasing is the dominant part leaving the decay aside. Within the quasiparticle approximation, $g^R(k,\tau) \approx -ie^{-i\varepsilon_k\tau}$, which is free of the decay, the time integrals over propagators yield

$$2\,\mathrm{Re}\int_0^\infty d\tau\, S_{\mathrm{qp}}(\tau) = 2\pi\delta\left(\varepsilon_k + \varepsilon_p - \varepsilon_{k-q} - \varepsilon_{p+q}\right), \tag{19.37}$$

$$2\,\mathrm{Re}\int_0^\infty d\tau\, \tau\, S_{\mathrm{qp}}(\tau) = -2\frac{\wp'}{\varepsilon_k + \varepsilon_p - \varepsilon_{k-q} - \varepsilon_{p+q}}. \tag{19.38}$$

After approximation (19.37), $\mathscr{I}$ from (19.36) turns into the Boltzmann-type scattering integral I_B of the Landau–Silin equation (9.24). Naturally, the on-shell contributions are identical since they do not depend on the retardation.

Using (19.38) in (19.36) one finds

$$\mathscr{R} = 2 \int \frac{dpdq}{(2\pi)^6} \mathscr{V}_q^2 D_{2p}(t) \frac{\wp'}{\varepsilon_k + \varepsilon_p - \varepsilon_{k-q} - \varepsilon_{p+q}}. \tag{19.39}$$

The function $\mathscr{R}$ is twice the off-shell contribution R to the Wigner distribution. To show this, we express (19.39) in terms of the Born selfenergy and the approximate correlation functions, $g^< = 2\pi\delta(\omega - \varepsilon)\rho$, as

$$\mathscr{R} = -2 \int \frac{d\omega'd\omega}{(2\pi)^2} \frac{\wp'}{\omega - \omega'} \left(\sigma_{\omega'}^> g_\omega^< - \sigma_{\omega'}^< g_\omega^> \right). \tag{19.40}$$

Comparing with the second term of (9.87) of the extended quasiparticle picture one can see that the additional retardation results in the double count of the correlated part, $\mathscr{R} = 2R$. Due to this double count, the function $\mathscr{F} = \rho - \mathscr{R} = \rho - 2R = f - R$ cannot be interpreted as the distribution of excitations because it has large regions of negative values. Please note this commonly met pitfall in the literature. If one follows this double count the correlated density would show a wrong sign and twice as large. In fact, for the balance equation of the density, the wrong assumption would be used that the correlated part of the collision integral, $\tilde{n}_c = \int dkR$, combines with the left-hand side of (19.35), $\tilde{n}_f = \int dk\rho$, to establish the density conservation, $\frac{\partial}{\partial t}(\tilde{n}_f + \tilde{n}_c) = 0$. The correlated density derived in this way reminds us of the result known from equilibrium (Stolz and Zimmermann, 1979; Schmidt and Röpke, 1987; Beth and Uhlenbeck, 1937; Zimmermann and Stolz, 1985; Schmidt, Röpke and Schulz, 1990). Note that, in this picture, the Wigner distribution on the left-hand side of the Levinson equation is treated as the quasiparticle distribution which is a mistake usually made in density operator studies (Klimontovich, 1975). The wrong sign follows from this last misinterpretation. By definition the momentum integral over the Wigner distribution already yields the full density, $\int dk\rho = n$. The conservation law thus tells us that either $\int dkR = 0$ or the Levinson equation does not conserve the number of particles which demonstrates the failure of interpreting $\mathscr{F} = \rho - \mathscr{R} = \rho - 2R = f - R$ as a quasiparticle distribution function. Instead of the results in the extended quasiparticle picture one has to choose $\rho - R = f$.

In other words, the virial corrections do not appear due to expansion of memory in the non-Markovian collision integral. These terms are exactly compensated by off-shell terms in the Wigner function. Instead, the virial corrections appear after cancellation of the off-shell parts as *internal* nonlocalities of the collision process.

The Levinson equation neglecting or underestimating the decay, also leads to a numerical instability of the equation, as has been reported by Haug (Banyai, Thoai, Remling and Haug, 1992). Other numerical solutions show a continuous increase of kinetic energy, e.g. Figure 4 of (Kremp, Bonitz, Kraeft and Schlanges, 1997) or pulsation modes (Popyrin, 1998). The instability of the solution is a problem which shows

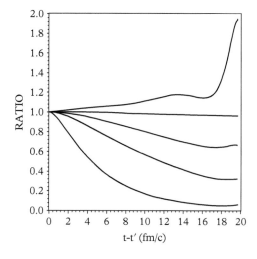

Figure 19.7: $G^<(\mathbf{p}, t, t')/G^<(\mathbf{p}, t, t)$ *vs* $t - t'$ *for different interaction strengths, from bottom to top* 2., 1., 0.5 *and* 0.125 *times normal strength and Levinson (Uppermost curve), momentum* $\mathbf{p} = 0$. *A detailed description of numerical details are in* (Köhler, Kwong and Yousif, 1999).

that the Levinson equation is a difficult tool to handle. The double count of the off-shell contribution resulting in negative effective occupation factors is likely one of the reasons for this problem. As illustrated in Figure 19.7 this double count results in too much off-shell correlations in the Levinson equation (Köhler and Morawetz, 2001).

The double count of correlations is a serious mistake for theories which aim to go beyond the Boltzmann equation. In (Morawetz, Lipavský and Špička, 2001c) it was shown that the decay of propagators during the collision removes this double count, while the original Levinson equation (Levinson, 1965) does not include the decay of propagators in the scattering integral.

19.3.2 Double count of correlation energy explained by extended quasiparticle picture

We consider now homogeneous matter and the total energy of the system reads (7.37)

$$E_{\text{tot}}(t) = \frac{1}{2}K_\rho(t) + \frac{1}{4}\int \frac{d^3k}{(2\pi)^3}\left(\frac{\partial}{\partial t_1} - \frac{\partial}{\partial t_2}\right)G^<(k, t_1, t_2) \tag{19.41}$$

where the kinetic energy K_ρ for the correlated medium is

$$K_\rho(t) = \int \frac{d^3k}{(2\pi)^3}\frac{k^2}{2m}\rho(k, t) \tag{19.42}$$

and the correlation energy is defined by

$$E_{\text{corr}}(t) = E_{\text{tot}}(t) - K_\rho(t). \tag{19.43}$$

Note that the mean or Hartree–Fock field is not included here and the total energy therefore contains only the correlated energy. We also define the *uncorrelated* kinetic energy by

$$K_f(t) = \int \frac{d^3k}{(2\pi)^3} \frac{k^2}{2m} f(k, t). \tag{19.44}$$

The relation between the reduced density matrix ρ and the quasiparticle distribution f will become essential.

We start with an initially uncorrelated system and a momentum-distribution $f(k, t = 0)$ specified by a density ρ and temperature T. The system is then time-evolved beyond equilibrium. The selfenergies are conserving so that the total energy is conserved. Therefore we have during the time evolution

$$K_f^i = E_{\text{tot}}(t) = E_{\text{tot}}^{\text{eq}} \tag{19.45}$$

where K_f^i is the kinetic (and total) energy of the initial unperturbed and uncorrelated system and $E_{\text{tot}}^{\text{eq}}$ is the total energy after equilibration ($t \to \infty$) including correlations.

For the Markovian–Boltzmann equation the kinetic energy is conserved, while the potential energy is zero. The Levinson equation conserves the total energy (Morawetz, 1995), see Appendix A.2. The correlation energy one gets by multiplying the Levinson equation (19.8) with $k^2/2m$ and integrating. In order to see the algebraic step let us shorten the notation and use (19.5) and (19.9) as well as $\langle ... \rangle$ for momentum integration. We have

$$\frac{\partial}{\partial t} \langle \frac{k}{2m} \rho \rangle = \langle \frac{V^2}{2} \int_0^{t-t_0} d\tau \, \Delta E \cos\left(\Delta E\tau\right) D_{2p}(t-\tau)\rangle = \langle \frac{V^2}{2} \int_{t_0}^{t} d\tau \, \Delta E \cos\left(\Delta(t-\tau)\right) D_{2p}(\tau)\rangle$$

$$= \langle \frac{V^2}{2} \int_{t_0}^{t} d\tau \frac{\partial}{\partial t} \sin\left(\Delta E(t-\tau)\right) D_{2p}(\tau)\rangle = \frac{\partial}{\partial t} \langle \frac{V^2}{2} \int_{t_0}^{t} d\tau \sin\left(\Delta E(t-\tau)\right) D_{2p}(\tau)\rangle \tag{19.46}$$

which provides us with the correlation energy

$$E_{\text{corr}}(t) = -\frac{1}{2} \int \frac{d^3k\, d^3p\, d^3q}{(2\pi)^9} V(q)^2 \int_{t_0}^{t} d\tau \sin\left(\Delta E(t-\tau)\right)$$

$$\times \left\{ \rho_{k-q}(t)\rho_{p+q}(t) \left[1 - \rho_k(\tau)\right]\left[1 - \rho_p(\tau)\right] - \rho_k(\tau)\rho_p(\tau)\left[1 - \rho_{k-q}(\tau)\right]\left[1 - \rho_{p+q}(\tau)\right] \right\}. \tag{19.47}$$

For longer times Eq. (19.47) reduces to (Morawetz and Koehler, 1999)

$$E_{\text{corr}}^{\text{eq}} = -\int \frac{d^3k\, d^3p\, d^3q}{(2\pi)^9} V(q)^2 \frac{\wp}{\Delta E} \rho_{eq}(k+q)\rho_{eq}(p-q)[1-\rho_{eq}(k)][1-\rho_{eq}(p)] \qquad (19.48)$$

where $\wp$ denotes the principal value and where ρ_{eq} indicates the equilibrium large-time correlated densities. This energy resembles the second-order Born estimate of the potential energy but with two important differences.

The first is that the densities ρ_{eq} are correlated densities, in the long-time equilibrated limit. A Born estimate would however be done with an uncorrelated distribution f. For weak interactions and/or low density for which the Levinson equation and the Born approximation are certainly valid, the difference between initial uncorrelated and final correlated densities is negligible. Around nuclear matter values, this difference is however important.

The second difference between Eq. (19.48) and the second-order Born estimate is that upon closer inspection a factor of one-half appears missing. This will now be clarified. With the correlation energy given by Eq. (19.48) the total energy after equilibration is using Eq. (19.43)

$$K_f^i = E_{\text{tot}}^{\text{eq}} = K_\rho^{\text{eq}} + E_{\text{corr}}^{\text{eq}}. \qquad (19.49)$$

The second-order Born approximation for the total energy is on the other hand known from perturbation theory

$$E_{\text{tot}}^{\text{eq}} = E_{Born} = K_f^{eq} + \frac{1}{2}E_{\text{corr}}^{\text{eq}}. \qquad (19.50)$$

One should note that in the process of equilibration the system is excited and the correlation energy does in both expressions (19.49) and (19.50) refer to the excited but not to the initial ground-state of matter. Also note that in the process of excitation the uncorrelated kinetic energy K_f^i has increased to K_f^{eq}.

In order to resolve the apparent disagreement between Eqs. (19.49) and (19.50) we first note that Eq. (19.49) results from a time-evolution of the Levinson equation starting from an uncorrelated system with a kinetic energy K_f^i. The correlated and uncorrelated kinetic energies at the end of the time-evolution must therefore be related by

$$K_\rho^{\text{eq}} = K_f^{\text{eq}} - \frac{1}{2}E_{\text{corr}}^{\text{eq}}. \qquad (19.51)$$

To prove Eq. (19.51) we have to discuss the difference between the reduced density matrix ρ and the quasiparticle distribution f. This is performed in the extended quasiparticle picture (9.64) providing the relation between the uncorrelated and correlated energies. Multiplying (9.64) with the kinetic energy $k^2/2m$ and integrating over k, one finds the needed relation (19.51). We can conclude that the extended quasiparticle picture explains the seemingly double count of correlation energy of the non-Markovian Levinson kinetic equation.

19.4   Initial correlations

The Levinson equation discussed so far is incomplete for two reasons: (i) it does not include correlated initial states. (ii) When the evolution of the system starts from the equilibrium state, the collision integral does not vanish, but gives rise to spurious time evolution. The latter point has been addressed by Lee et al. (1970) showing that from initial correlations there have to appear terms in the kinetic equation which ensure that the collision integral vanishes in thermal equilibrium. We have demonstrated that there is a large cancellation of the off-shell effects resulting in a nonlocal kinetic theory. For transient times the Levinson equation has nevertheless its merits and the decay of initial correlations can be studied. We will restrict ourselves to the Born approximation which allows us to present the most straightforward derivation. The inclusion of higher-order correlations can be found in (Bonitz, 1998; Danielewicz, 1984b; Semkat, Kremp and Bonitz, 1999). The effect of initial correlations becomes particularly transparent from our analytical results which may also serve as a bench-mark for numerical simulations.

We start with the RPA approximation outlined in Chapter 10.2.2. The resulting equation of motions for the density fluctuations (10.28) can be inverted into an integral equation but taking into account the initial correlations L_0 as

$$L(1,2,1',2') = L_0(1,2,1',2') - G_H(1,2')G(2,1')$$

$$+ \int d4\, G_H(1,4)G(4,1') \int d3\, V(4,3)L(2,3,2',3^+) \qquad (19.52)$$

where G_H is the Hartree causal Green's function and we have taken into account the boundary condition

$$(G_H^R)^{-1}L_0 = 0. \qquad (19.53)$$

In the case that all times in (19.52) approach t_0, the right-hand side vanishes except L_0 which represents, therefore, the contribution from the initial correlations. They propagate in time according to the solution of (19.53) (Semkat, Kremp and Bonitz, 1999)

$$L_0(121'2') = \int d\bar{x}_1 d\bar{x}_2 d\bar{x}_1' d\bar{x}_2'\, G_H^R(1,\bar{x}_1 t_0)G_H^R(2,\bar{x}_2 t_0)$$

$$\times L_{00}(\bar{x}_1, \bar{x}_2, \bar{x}_1', \bar{x}_2', t_0)G_H^A(\bar{x}_1' t_0, 1')G_H^A(\bar{x}_2' t_0, 2'). \qquad (19.54)$$

Here L_{00} is the initial two-particle correlation function.

Inserting (19.52) into the first equation of the Martin–Schwinger hierarchy (7.8), restricting to the Born approximation and using the inverse Hartree–Fock Green's function $G_{HF}^{-1}(1,2) = G_H^{-1}(1,2) + V(1,2)G^<(1,2)$, we obtain for the causal function

$$G_{HF}^{-1}(1,3)G(3,2) = \delta(1-2) + S_{init}(1,2) + \int_C d4\,\{\Sigma_0(1,4) + \Sigma(1,4)\}\,G(4,2), \quad (19.55)$$

with the selfenergy in Born approximation

$$\Sigma(1,2) = \int d3d5\,V(1,3)G_H(1,2)V(2,5)G_H(3,5^{++})G(5,3^+). \quad (19.56)$$

The initial correlations create two new terms,

$$\Sigma_0(1,2) = \int d3d5\,V(1,3)G_H(1,2)V(2,5)L_0(3,5,3^+,5^{++}),$$

$$S_{init}(1,2) = \int d3\,V(1,3)L_0(1,3,2,3^+). \quad (19.57)$$

The integral form of (19.55) is given in Figure 19.8 from which the definitions (19.57) are obvious.

With the help of the LW rules (7.17), the equation for the causal Green's function (19.55) leads to the Kadanoff and Baym equation with initial correlations

$$(G_{HF}^{-1} - \Sigma^R)G^< - G^<(G_{HF}^{-1} - \Sigma^A) = (\Sigma + \Sigma_0)^<G^A - G^R(\Sigma + \Sigma_0)^< + S_{init} - S_{init}^*. \quad (19.58)$$

Please note that $\Sigma_0^{R/A} = 0$ since $G_H(1,2) \sim \delta(t_1 - t_2)$. Using the generalised Kadanoff and Baym ansatz (8.35) we obtain from (19.58) the kinetic equation for the reduced

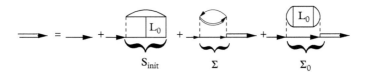

Figure 19.8: *The Dyson equation, including density fluctuation up to a second Born approximation. Besides the initial correlation term S_{init} discussed in (Danielewicz, 1984b; Semkat et al., 1999), a new type of selfenergy Σ_0 appears which is induced by initial correlations. Since the latter one contains an interaction by itself, this term is of the next order Born approximation, from (Morawetz, Bonitz, Morozov, Röpke and Kremp, 2001a).*

density matrix $\rho(t) = G^<(t, t)$

$$\frac{\partial}{\partial t}\rho(k, t) = I(k, t) + S_0(k, t) + S_1(k, t) \tag{19.59}$$

with the Levinson collision integral (10.113) or (19.8)

$$I(k, t) = \frac{2}{\hbar^2}\text{Re}\int_{t_0}^{t} dt_1 \int \frac{dq\,dp}{(2\pi\hbar)^6} V^2(q)\, G_{k-q}^R(t, t_1)\, G_k^A(t_1, t)\, G_{p+q}^R(t, t_1)\, G_p^A(t_1, t)$$

$$\times \left\{ \rho_{k-q}(t_1)\rho_{p+q}(t_1)[1 - \rho_p(t_1)][1 - \rho_k(t_1)] - \rho_k(t_1)\rho_p(t_1)[1 - \rho_{p+q}(t_1)][1 - \rho_{k-q}(t_1)] \right\},$$

$$\tag{19.60}$$

and two new terms arising from the initial correlations, S_{init} and Σ_0,

$$S_0(k, t) = \frac{2}{\hbar}\text{Im}\int \frac{dq\,dp}{(2\pi\hbar)^6} V(q)\, G_{k-q}^R(t, t_0)\, G_k^A(t, t_0)\, G_{p+q}^R(t, t_0)\, G_p^A(t, t_0)$$

$$\times \langle \frac{k-p}{2} + q | L_{00}(p + k, t_0) | \frac{k-p}{2} \rangle, \tag{19.61}$$

$$S_1(k, t) = \frac{2}{\hbar^2}\text{Re}\int_{t_0}^{t} dt_1 \int \frac{dq}{(2\pi\hbar)^3} L_0(q, t, t_1) V^2(q)\, G_{k-q}^R(t, t_1)\, G_k^A(t_1, t)\, \left[\rho_{k-q}(t_1) - \rho_k(t_1)\right]$$

$$\tag{19.62}$$

where we used the matrix notation (9.2). We would like to note that the equation (19.59) is valid up to second order gradient expansion in the spatial variable. In inhomogeneous systems, this variable has to be added simply in all functions and in this way on the left side of (19.59) also the standard mean-field drift would appear.

The first part (19.60) is just the precursor of the Levinson equation in the second Born approximation $\sim V^2$. The S_1 term (19.62) coming from Σ_0 leads to corrections to the third Born approximation since it is $\sim V^2 L_0$. A more general discussion of higher-order correlation contribution within the T-matrix approximation can be found in (Kremp, Bonitz, Kraeft and Schlanges, 1997) and of general initial conditions in (Zubarev, Morozov and Röpke, 1997). The S_0 part, (19.61), following from S_{init} gives just the correction to the Levinson equation at the same order of V, which will guarantee the cancellation of the collision integral for an equilibrium initial state.

Multiplying the kinetic equation (19.59) with a momentum function $\xi(k)$ and integrating over k, one derives the balance equations

$$\langle \dot{\xi}(k) \rangle = \int \frac{dk}{(2\pi\hbar)^3}\xi(k)I + \int \frac{dk}{(2\pi\hbar)^3}\xi(k)S_0. \tag{19.63}$$

For the standard collision integral follows

$$\langle \xi(k)I \rangle = \frac{\mathrm{Re}}{\hbar^2} \int \frac{dkdqdp}{(2\pi\hbar)^9} \int_{t_0}^{t} dt_1 \, V^2(q) G_{k-q}^R(t,t_1) G_k^A(t_1,t) G_{p+q}^R(t,t_1) G_p^A(t_1,t)$$

$$\times \rho_{k-q}(t_1)\rho_y(t_1)[1-\rho_p(t_1)][1-\rho_k(t_1)]\left\{ \xi(k)+\xi(p)-\xi(k-q)-\xi(p+q) \right\}, \quad (19.64)$$

from which it is obvious that density and momentum ($\xi = 1, k$) are conserved, while a change of kinetic energy $\xi = k^2/2m$ is induced which exactly compensates the two-particle correlation energy and, therefore assures total energy conservation of a correlated plasma (Morawetz, 1995). Initial correlations, Eq. (19.61) give rise to additional contributions to the balance equations (Bonitz, 1998; Semkat, Kremp and Bonitz, 1999). We get

$$\langle \xi(k)S_0 \rangle = \frac{\mathrm{Im}}{2\hbar} \int \frac{dkdqdp}{(2\pi\hbar)^9} V(q)(\frac{p-k}{2} + q|L_0(p+k)|\frac{p-k}{2}) \, \{ \xi(k)+\xi(p)-\xi(k-q)-\xi(p+q) \}$$

$$(19.65)$$

which keeps the density and momentum also unchanged and only a correlated energy is induced. The selfenergy corrections from initial correlations which correct the next Born approximation, (19.62), would lead to

$$\langle \xi(k)S_1 \rangle = \frac{2\mathrm{Re}}{\hbar^2} \int \frac{dkdq}{(2\pi\hbar)^6} \int_{t_0}^{t} dt_1 \, V^2(q) L_0(q,t,t_1)\rho_k(t_1) G_{k-q}^R(t,t_1) G_k^A(t_1,t) \, [\xi(k-q)-\xi(k)]$$

$$(19.66)$$

which shows that the initial correlations induce a flux besides an energy in order to equilibrate the correlations imposed initially towards the correlations developed during dynamical evolution if higher than $\sim V^2$ orders are considered.

We will consider this in the following only second Born approximation $\sim V^2$ and have therefore to use

$$G^R(t_1, t_2, k) \approx -i\Theta(t_1 - t_2)e^{i\frac{k^2}{2m\hbar}(t_2-t_1)}, \quad (19.67)$$

and for the initial two-particle (causal) correlation function L_{00} in the first Born approximation one gets from (10.108)

$$\left(\frac{k-p}{2} |L_{00}(k+p)| \frac{k-p}{2} - q \right) = \frac{\wp}{\Delta E} V_0(q) \left\{ \rho_0(k)\rho_0(p)[1-\rho_0(k-q)][1-\rho_0(p+q)] \right.$$

$$\left. - [1-\rho_0(k)][1-\rho_0(p)]\rho_0(k-q)\rho_0(p+q) \right\}. \quad (19.68)$$

where $\wp$ denotes the principal value, ΔE of (19.9), and ρ_0 is the initial Wigner distribution. Then the explicit collision integral (19.60) is just the Levinson equation (19.8) without damping and the new term due to initial correlations (19.61)

$$S_0(k,t) = -\frac{2}{\hbar^2} \int_{t_0}^{t} dt_1 \int \frac{dqdp}{(2\pi\hbar)^6} V(q) V_0(q) \cos\left[\left(\frac{\Delta E}{\hbar}\right)(t-t_1)\right]$$

$$\times \{\rho_0(k-q)\rho_0(p+q)[1-\rho_0(p)-\rho_0(k)] - \rho_0(k)\rho_0(p)[1-\rho_0(p+q)-\rho_0(k-q)]\}\,.$$
$$(19.69)$$

Comparing with (19.8) or (19.60) one sees that with (19.69) exactly the same collision integral appears as in the Levinson equation (19.8) but with initial occupation factors remaining unchanged and the product between the initial interaction and switched-on interaction. This balances the spurious time evolution of the Levinson equation and has been first described by (Lee, Fujita and Wu, 1970) who show that from initial correlations there must appear terms in the kinetic equation which ensure that the collision integral vanishes in thermal equilibrium.

To show the interplay between collisions and correlations, we have calculated the initial two-particle correlation function in the ensemble, where the dynamical interaction $V(q)$ is replaced by some arbitrary function $V_0(q)$. Therefore the initial state deviates from thermal equilibrium except when $V(q) = V_0(q)$ and $\varrho(t_0) = \varrho_0$.

The additional collision term, S_0 cancels exactly the Levinson collision term in the case that we have initially the same interaction as during the dynamical evolution ($V_0 = V$) and if the system starts from the equilibrium $\rho(t) \equiv \rho_0$. Therefore we have completed our task and derived a correction of the Levinson equation which ensures the cancellation of the collision integral in the thermal equilibrium. It is interesting to note that the corrections to the next Born approximation (19.62) due to initial correlations is of the type found in impurity scattering. Therefore the initial correlations higher than $\sim V^2$ are governed by another type of dynamics than the build up of correlations involved in S and S_0.

19.4.1   Formation of correlations with initial correlations

On very short time scales we can neglect the change in the distribution function. Assuming a Maxwellian initial distribution with temperature T and neglecting degeneracy, we can calculate explicitly the collision integrals. We choose as a model interaction a Debye potential $V_i(q) = 4\pi e^2 \hbar^2/[q^2 + \hbar^2 \kappa_i^2]$ with fixed parameter $\kappa_i = \kappa_D$ and for the initial correlations $\kappa_i = \kappa_0$. We obtain, for the change of kinetic energy, over short times from (19.64) and (19.65)

$$\frac{\partial}{\partial t} E_{\text{kin}}(t) = \varepsilon[V(q)^2](t) - \varepsilon[V_0(q)V(q)](t),\qquad (19.70)$$

which can be integrated (Morawetz, Špička and Lipavský, 1998) to yield

$$E_{\text{kin}}(t) = E_{\text{total}} - E_{\text{init}}(t) - E_{\text{coll}}(t).\qquad (19.71)$$

For the classical limit we obtain explicitly the time-dependent kinetic energy (Morawetz, Bonitz, Morozov, Röpke and Kremp, 2001a)

$$\frac{E_{\text{coll}}(t)}{nT} = -\frac{\sqrt{3}\Gamma^{3/2}}{4x} \partial_y(y\mathscr{F}(y))_{y=x\tau}, \tag{19.72}$$

where $\mathscr{F}(y) = 1 - e^{y^2}\text{erfc}(y)$, $\tau = t\omega_p/\sqrt{2}$, $x = \kappa_D/\kappa$ and $\kappa^2 = 4\pi e^2 n/T = \omega_p^2 T/m$ and the plasma parameter Γ of (19.15).

In Fig. 19.9, upper panel, we compare the analytical results of (19.72) with MD simulations using the Debye potential V_i as a bare interaction. The evolution of kinetic energy is shown for three different ratios x. The agreement between theory and simulations is quite satisfactory, in particular, the short-time behaviour for $x = 2$. The faster initial increase of kinetic energy observed in the simulations at $x = 1$ may be due to the finite size of the simulation box which could more and more affect the results for an increasing range of the interaction.

Now we include the initial correlations choosing the equilibrium expression (19.68) which leads to (Morawetz, Bonitz, Morozov, Röpke and Kremp, 2001a)

$$\frac{E_{\text{init}}(t)}{nT} = -\frac{\sqrt{3}\Gamma^{3/2}}{2(x_0^2 - x^2)} [x\mathscr{F}(x\tau) - x_0\mathscr{F}(x_0\tau)], \tag{19.73}$$

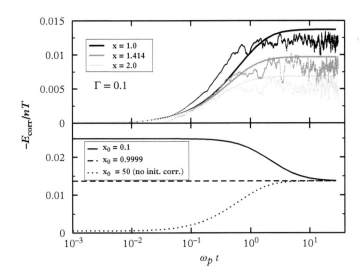

Figure 19.9: *The formation of correlation energy* $-E_{corr} = E_{total} - E_{init} - E_{coll} = E_{kin}$ *in a plasma with Debye interaction* V_i. *The upper panel compares the analytical results (19.72) with MD simulations from (Zwicknagel, 1999) for three different ratios of* κ_D *to the inverse Debye length* $x = \kappa_D/\kappa$. *In the lower panel we compare theoretical predictions for the inclusion of Debye initial correlations characterised by* $x_0 = \kappa_0/\kappa$ *where* $x = \kappa_D/\kappa = 1$.

where $x_0 = \kappa_0/\kappa$ characterising the strength of the initial Debye correlations (19.68) with the Debye potential V_0 which contains κ_0 instead of κ_D. Besides the kinetic energy (19.73) from initial correlations, the total energy E_{total} (19.71) now includes the initial correlation energy which can be calculated from the long time limit of (19.72) leading to

$$\frac{E_{total}}{nT} = \frac{\sqrt{3}\Gamma^{3/2}}{2(x + x_0)}. \qquad (19.74)$$

The result (19.71) is seen in Fig. 19.9, lower panel. We observe that if the initial correlation is characterised by a potential range larger than the Debye screening length, $x_0 < 1$, the initial state is over-correlated, and the correlation energy starts at a higher-absolute value than without initial correlations relaxing towards the correct equilibrium value. If, instead, $x_0 = 1$ no change of correlation energy is observed, as expected. Similar trends have been observed in numerical solutions (Semkat, Kremp and Bonitz, 1999).

19.5   Summary

The formation of binary correlations is very fast on the time scale of dissipative processes. With respect to dissipative regimes, the binary correlations can be treated as instant functionals of the single-particle distribution and thus included into the Boltzmann equation via various renormalisations of its ingredients. They would then be the screened Coulomb potential in the scattering rate or the quasiparticle corrections. Under extremely fast external perturbations, like the massive femtosecond laser pulses, cold atom quenches or heavy ion collisions, the dynamics of binary correlations becomes experimentally accessible. The short-time description of formation of correlations is possible to achieve by the Levinson equation even analytically for the formation of correlations after a sudden quench and for the density response to an external perturbing field. This is possible since the system has had no time to develop higher-order correlations yet. The initial correlations can be treated analytically as well in the lowest-order many-body approximations and it can be shown how their decay is balanced with the formation of correlations. For later times, the Levinson equation fails by double-counting correlations. The delicate balance of propagator decay and correlations is better accounted for by explicit cancellations of the off-shell terms; as performed in the extended quasiparticle picture of Chapter 9.4.

Initial correlations lead to correction terms on every level of perturbation theory correcting the Levinson kinetic equation properly in a way that the collision integral vanishes if the evolution starts from a correlated equilibrium state. Furthermore, the conservation laws of a correlated plasma are proved including the contributions from initial correlations. It is shown that besides the appearance of correlation energy a correlated flux appears if higher than Born correlations are considered.

20

Field-Dependent Transport

20.1 Gauge invariance

In order to get an unambiguous way of constructing approximations we have to formulate the theory in a gauge-invariant way. This can be done following a procedure known from field theory (Itzykson and Zuber, 1990). This method has been applied to high-field problems in (Bertoncini and Jauho, 1991). With the help of the Fourier transform of an arbitrary function $G(x, X)$ for the relative coordinates $x = (\mathbf{r}_2 - \mathbf{r}_1, t_2 - t_1) = (\mathbf{s}, \tau)$ with the centre-of-mass coordinates $X = ((\mathbf{r}_2 + \mathbf{r}_1)/2, (t_2 + t_1)/2) = (\mathbf{r}, t)$, one can introduce a gauge-invariant Fourier-transform of the difference coordinates x

$$\bar{G}(k, X) = \int dx\, G(xX)\, e^{\frac{i}{\hbar} \int_{-\frac{1}{2}}^{\frac{1}{2}} d\lambda x_\mu [k^\mu + \frac{e}{c} A^\mu (X + \lambda x)]} . \tag{20.1}$$

For constant electric fields, the scalar potential can be chosen as zero and $A^\mu = (0, -c\mathbf{E}t)$, and one obtains a generalised Fourier-transform

$$\bar{G}(k, X) = \int dx\, e^{\frac{i}{\hbar} [x_\mu k^\mu + e\mathbf{s} \cdot \mathbf{E}t]} G(x, X),$$

and we have the following rule in formulating the kinetic theory gauge invariantly:

1. Fourier transformation of the four-dimensional difference-variable x to canonical momentum p.
2. Shifting the momentum to kinematic momentum according to $\mathbf{p} = \mathbf{k} - e\mathbf{E}t$.
3. The gauge-invariant functions $\bar{G}$ are given by

$$G(\mathbf{p}, t) = G(\mathbf{k} - e\mathbf{E}t, t) = \bar{G}(\mathbf{k}, t) = \bar{G}(\mathbf{p} + e\mathbf{E}t, t). \tag{20.2}$$

For inhomogeneous fields we consider the general gauge-invariant Fourier transform for $A_\mu = (\Phi(\mathbf{r}, t), \mathbf{A}(\mathbf{r}, t))$ in gradient expansion

Interacting Systems far from Equilibrium. Klaus Morawetz, Oxford University Press (2018).
© Klaus Morawetz. DOI: 10.1093/oso/9780198797241.001.0001

$$\mathbf{p} = \mathbf{k} + e \int_{-\frac{1}{2}}^{\frac{1}{2}} d\lambda \mathbf{A}(\mathbf{r} + \lambda \mathbf{s}, t + \lambda \tau) = \mathbf{k} + e \int_{-\frac{1}{2}}^{\frac{1}{2}} d\lambda e^{\lambda s \partial_r^A} e^{\lambda \tau \partial_t^A} \mathbf{A}(\mathbf{r}, t)$$

$$\rightarrow \mathbf{k} + e \int_{-\frac{1}{2}}^{\frac{1}{2}} d\lambda e^{-i\hbar\lambda \partial_p \partial_r^A} e^{i\hbar\lambda \partial_\omega \partial_t^A} \mathbf{A}(\mathbf{r}, t) = \mathbf{k} + e \, \mathrm{sinc}\left(\frac{\hbar}{2}\partial_p \partial_r^A\right) \mathrm{sinc}\left(\frac{\hbar}{2}\partial_\omega \partial_t^A\right) \mathbf{A}(\mathbf{r}, t)$$

$$(20.3)$$

with $\mathrm{sinc}(x) = \sin x/x = 1 - x^2/3! + - \dots$ and analogously

$$\Omega = \omega + e \, \mathrm{sinc}\left(\frac{\hbar}{2}\partial_p \partial_r^\Phi\right) \mathrm{sinc}\left(\frac{\hbar}{2}\partial_\omega \partial_t^\Phi\right) \Phi(\mathbf{r}, t). \qquad (20.4)$$

One sees that up to second-order gradients we have correctly used these rules for gauge invariant formulation (Bertoncini and Jauho, 1991; Morawetz, 2000c).

20.1.1   Equation for Wigner distribution

With the help of this gauge-invariant formulation of Green's function, we can write the time-diagonal part (9.76) of the Kadanoff and Baym equation (7.28) in mixed representation finally in the following gauge-invariant form

$$\frac{\partial}{\partial t}\rho(\mathbf{k}, t) + e\mathbf{E}\nabla_\mathbf{k}\rho(\mathbf{k}, t) = \int_0^{t-t_0} d\tau \left[\left\{ G^> \left(\mathbf{k} - \frac{e\mathbf{E}}{2}\tau, \tau, t - \frac{\tau}{2}\right), \Sigma^< \left(\mathbf{k} - \frac{e\mathbf{E}}{2}\tau, -\tau, t - \frac{\tau}{2}\right) \right\}_+ \right.$$

$$\left. - \left\{ G^< \left(\mathbf{k} - \frac{e\mathbf{E}}{2}\tau, \tau, t - \frac{\tau}{2}\right), \Sigma^> \left(\mathbf{k} - \frac{e\mathbf{E}}{2}\tau, -\tau, t - \frac{\tau}{2}\right) \right\}_+ \right]. \qquad (20.5)$$

This kinetic equation is exact in time convolutions which is necessary because gradient expansions in time are connected with linearisation in electric fields and consequently fail (Mahan, 1987).

20.1.2   Spectral function and ansatz

The spectral properties of the system are described by the Dyson equation for the retarded Green's function (8.19). A free particle in a uniform electric field, where the field is represented by a vector potential $\mathbf{E}(t) = -\frac{1}{c}\dot{\mathbf{A}}(t)$ leads with (8.17) to the following equation

$$\left[i\hbar\frac{\partial}{\partial t} - \epsilon\left(\mathbf{p} - \frac{e}{c}\mathbf{A}(t)\right) \right] G_0^R(\mathbf{p}, tt') = \delta(t - t'). \qquad (20.6)$$

This equation is easily integrated (Jauho, 1991; Khan, Davies and Wilkins, 1987)

$$G_0^R(\mathbf{p}, tt') = -i\Theta(t-t') e^{\frac{i}{\hbar} \int_t^{t'} du\, \epsilon[\mathbf{p} - \frac{e}{c}\mathbf{A}(u)]}. \tag{20.7}$$

For free particles and parabolic dispersions, the gauge invariant spectral function (Jauho, 1991; Khan, Davies and Wilkins, 1987) follows

$$A_0(\mathbf{k}, \omega) = 2\int_0^\infty d\tau \cos\left(\omega\tau - \frac{k^2}{2m\hbar}\tau - \frac{e^2 E^2}{24 m\hbar}\tau^3\right) = \frac{2\pi}{\epsilon_E} Ai\left(\frac{k^2/2m - \hbar\omega}{\epsilon_E}\right) \tag{20.8}$$

where $\epsilon_E = (\hbar^2 e^2 E^2/8m)^{1/3}$ and $Ai(x)$ is the Airy function (Abramowitz and Stegun, 1984). It is instructive to verify that (20.8) satisfies the frequency sum rule $\int d\omega A_0(\omega) = 2\pi$. The interaction-free but field-dependent retarded Green's function G_o^R is thus obtained from the interaction-free and field-free Green's function by a simple Airy transformation. This is an expression of the fact that the solutions of the Schrödinger equation with constant electric field are Airy functions. The retarded functions can therefore be diagonalised within those eigen-solutions (Bertoncini, Kriman and Ferry, 1989; Bertoncini and Jauho, 1991). It can be shown that (20.8) remains valid even within a quasiparticle picture (Morawetz, 1994), where we have to simply replace the free dispersion $k^2/2m$ by the quasiparticle energy ϵ_k.

In order to close kinetic equation (20.5), it is necessary to know the relation between $G^>$ and $G^<$ and the Wigner function, as we have discussed in Chapter 8.4. With the Kadanoff/Baym ansatz, one gets just one half of all retardation times in the various time arguments (Jauho and Wilkins, 1984; Jauho, 1991). This annoying discrepancy remained obscure until the generalised–Kadanoff–Baym (GKB) ansatz (8.35). Together with the requirement of gauge invariance of Chapter 20.1 and using the quasiparticle spectral function (20.8) with quasiparticle energies ϵ_k instead of $k^2/2m$, the GKB ansatz finally reads

$$G^{\gtrless}(\mathbf{k}, \tau, \mathbf{r}, t) = e^{-\frac{i}{\hbar}\left(\epsilon_k \tau + \frac{e^2 E^2}{24m}\tau^3\right)} \rho^{\gtrless}\left(\mathbf{k} - \frac{e\mathbf{E}|\tau|}{2}, \mathbf{r}, t - \frac{|\tau|}{2}\right) \tag{20.9}$$

with $\rho^< = \rho$ and $\rho^> = 1 \mp \rho$ suited in the case of high external fields. Other choices of ansatz can be appropriate for other physical situations. For a more detailed discussion see (Morawetz, Lipavský and Špička, 2001c).

20.2 Kinetic equation in dynamically screened approximation

Let us repeat here the main formulae for screening of Chapter 10.2.3 with $c = 1$ in the time domain since we want to use them with the gauge-invariant ansatz. The selfenergy (10.43) is given in terms of the dynamical potential W

$$\Sigma_a^< (\mathbf{k}, t, t') = \int \frac{d\mathbf{q}}{(2\pi\hbar)^3} W_{aa}^< (\mathbf{q}, t, t') G_a^< (\mathbf{k} - q, t, t') \qquad (20.10)$$

where the dynamical potential (10.40) is expressed within Coulomb potentials $V_{ab}(\mathbf{q})$

$$W_{aa}^< (\mathbf{q}, t, t') = \frac{i}{\hbar} \sum_{dc} V_{ad}(\mathbf{q}) L_{dc}^< (\mathbf{q}, t, t') V_{ca}(\mathbf{q}) \qquad (20.11)$$

via the density–density fluctuation (10.39)

$$L_{ab}^< (\mathbf{q}, t, t') = \delta_{ab} \int d\bar{t} d\bar{\bar{t}} \left(\epsilon^R \right)^{-1} (\mathbf{q}, t, \bar{t}) \Pi_{aa}^< (\mathbf{q}, \bar{t}, \bar{\bar{t}}) \left(\epsilon^A \right)^{-1} (\mathbf{q}, \bar{\bar{t}}, t'). \qquad (20.12)$$

Here $\Pi = -iL_0$ is the free-density fluctuation (10.44) or polarisation function

$$\Pi_{aa}^< (\mathbf{q}, t, t') = \int \frac{d\mathbf{p}}{(2\pi\hbar)^3} G_a^< (\mathbf{p}, t, t') G_a^> (\mathbf{p} - q, t', t) \qquad (20.13)$$

and $\epsilon^{R/A}$ the retarded/advanced dielectric function

$$\epsilon^{R/A}(\mathbf{q}, t, t') = \delta(t - t') \pm \frac{i}{\hbar} \Theta[\pm(t - t')] \sum_b V_{bb}(\mathbf{q}) [\Pi^> (\mathbf{q}, t, t') - \Pi^< (\mathbf{q}, t, t')]. \qquad (20.14)$$

One easily convince oneself that this set of equations (20.10)–(20.14) is gauge invariant.
We can directly introduce this set of equations into the equation for the Wigner function (20.5) and using (20.9) we obtain some algebra for the in-scattering part of the collision integral

$$I_a^{\text{in}}(\mathbf{k}, t) = \frac{2}{\hbar^2} \sum_b \int \frac{d\mathbf{q}}{(2\pi\hbar)^3} V_{ab}^2(\mathbf{q}) \int_0^\infty d\tau \int \frac{d\omega}{2\pi} \cos\left[(\epsilon_{k-q}^a - \epsilon_k^a - \hbar\omega)\frac{\tau}{\hbar} + \frac{e_a \mathbf{E} \cdot \mathbf{q}\tau^2}{2m_a\hbar} \right]$$

$$\times \rho_a(\mathbf{k} - q - e_a \mathbf{E}\tau, t - \tau)[1 - \rho_a(\mathbf{k} - e_a \mathbf{E}\tau, t - \tau)] \frac{\Pi_{bb}^< (\mathbf{q}, \omega, t - \frac{1}{2}\tau)}{\left| \epsilon^R (\mathbf{q}, \omega, t - \frac{1}{2}\tau) \right|^2} \qquad (20.15)$$

with the polarisation vertex (20.13)

$$\Pi_{bb}^< (\mathbf{q}, \omega, t) = -2 \int \frac{d\mathbf{p}}{(2\pi\hbar)^3} \int_0^\infty d\tau \cos\left[(\omega\hbar - \epsilon_p^b + \epsilon_{p+q}^b)\frac{\tau}{\hbar} + \frac{e_b \mathbf{E} \cdot \mathbf{q}\tau^2}{2m_b\hbar} \right]$$

$$\times \rho_b \left(\mathbf{p} + q, t - \frac{\tau}{2} \right) \left[1 - \rho_b \left(\mathbf{p}, t - \frac{\tau}{2} \right) \right]. \qquad (20.16)$$

The out-scattering term I^{out} is given by $\rho \leftrightarrow 1-\rho$. We have employed the approximation $t \pm \frac{1}{2}\tau \approx t$ in the density fluctuation (20.12) which corresponds to a gradient approximation in time for the density fluctuations. Since the centre-of-mass-time dependence is carried only by the distribution functions in (20.12), this approximation is exact in the quasi-stationary case. All internal time integrations remain exact. Summarising, with

$$\left(\frac{\partial}{\partial t} + e\mathbf{E}\nabla_{\mathbf{k}}\right)\rho_a(\mathbf{k}, t) = I_a^{\text{in}}(\mathbf{k}, t) - I_a^{\text{out}}(\mathbf{k}, t) \tag{20.17}$$

and (20.15) we have the field-dependent Lenard–Balescu kinetic equation (Morawetz, 1994). Other standard approximations, like the T-matrix approximation, result in a field-dependent Bethe–Salpeter (Morawetz and Röpke, 1996).

Kinetic equation in statically screened approximation   Using the static approximation for the dielectric function $\epsilon^R(\mathbf{q}, 0, t)$ in (20.15), the kinetic equation for statically screened Coulomb potentials in high-electric fields appears (Morawetz and Kremp, 1993; Jauho and Wilkins, 1984; Morawetz, 1994) for spin-s_b particles

$$\frac{\partial}{\partial T}\rho_a + e\mathbf{E}\frac{\partial}{\partial \mathbf{k}_a}\rho_a = \sum_b I_{ab}$$

$$I_{ab} = \frac{2(2s_b + 1)}{\hbar^2}\int \frac{d\mathbf{p}d\mathbf{q}}{(2\pi\hbar)^6}\left\{\rho_{a'}\rho_{b'}(1-\rho_a)(1-\rho_b) - \rho_a\rho_b(1-\rho_{a'})(1-\rho_{b'})\right\}$$

$$\times V_s^2(\mathbf{q})\int_0^\infty d\tau \cos\left\{(\epsilon_a + \epsilon_b - \epsilon_a' - \epsilon_b')\frac{\tau}{\hbar} + \frac{\mathbf{E}\cdot\mathbf{q}\tau^2}{2\hbar}\left(\frac{e_a}{m_a} - \frac{e_b}{m_b}\right)\right\} \tag{20.18}$$

with $\rho_b = \rho_b(\mathbf{p} - e_b\mathbf{E}\tau, t - \tau)$, $\rho_b' = \rho_b(\mathbf{p} - \mathbf{q} - e_b\mathbf{E}\tau, t - \tau)$, $\rho_a = \rho_a(\mathbf{k} - e_a\mathbf{E}\tau, t - \tau)$, $\rho_a' = \rho_b(\mathbf{k} + \mathbf{q} - e_a\mathbf{E}\tau, t - \tau)$. The potential is the static Debye one

$$V_s(p) = \frac{V(p)}{\epsilon^R(p, 0)} = \frac{4\pi e_a e_b \hbar^2}{p^2 + \hbar^2\kappa^2} \tag{20.19}$$

and the static screening length κ is given by

$$\kappa^2 = \sum_c \frac{e_c^2 n_c}{\epsilon_0 k_B T_c} \tag{20.20}$$

in the equilibrium and non-degenerated limit. Here T_c is the temperature of specie c, charge e_c, spin s_c and mass m_c respectively.

The gauge-invariant and field-dependent kinetic equation (20.18) of Chapter 19 is suited for a linear response only. As we have seen, the Levinson equation possesses artifacts at larger time scales. Therefore only small short-time derivations from equilibrium can be treated with (20.18).

We are now interested in corrections to the particle flux, and therefore obtain from (20.18) the balance equation for the momentum

$$\frac{\partial}{\partial t} < \mathbf{k}_a > -n_a e_a \mathbf{E} = \sum_b < \mathbf{k}_a I_B^{ab} > .$$

(20.21)

20.3   Feedback and relaxation effects

If there is a motion of particles due to the external field, the two-particle Green's function describing the correlation of the particles with the surroundings will be deformed due to the external field, known as Debye–Onsager relaxation effect. If this deformation is considered selfconsistently it is called the feedback effect. This is the idea. Let us assume the two-particle correlation function is deformed in the first order of the electric field as

$$g_{ab} = g_{ab}^0 + \mathbf{E} \cdot \mathbf{g}_{ab}^1.$$

(20.22)

Since any kinetic equation describes the balance from the drift $-e\mathbf{E} \cdot \partial_k f$ and a proper collision integral which is a complicated integral over the two-particle Green's function (20.22) we obtain the momentum balance from (20.22)

$$\mathbf{E} = R\mathbf{j} + \frac{\delta E}{E}\mathbf{E}$$

(20.23)

where $\mathbf{j}$ is the current and R and $\frac{\delta E}{E}$ correspond to the collision integrals coming from the first and second term of (20.22) respectively. Therefore, we see that an effective electric field,

$$\tilde{\mathbf{E}} = \left(1 - \frac{\delta E}{E}\right)\mathbf{E} = R\mathbf{j}$$

(20.24)

is connected to the current. The conductivity $\mathbf{j} = \sigma\mathbf{E}$ is therefore

$$\sigma = \frac{1 - \frac{\delta E}{E}}{R}$$

(20.25)

where the relaxation field δE and the resistivity R become apparent.

If we now consider the effective field selfconsistently we derive the feedback effect. Therefore we observe that the deviation of the two-particle Green's function (20.22) is due to the effective field itself instead of due to the external field

$$g_{ab} = g_{ab}^0 + \tilde{\mathbf{E}} \cdot \mathbf{g}_{ab}^1.$$

(20.26)

The balance would then lead to

$$\mathbf{E} = R\mathbf{j} + \frac{\delta E}{E}\tilde{\mathbf{E}} \tag{20.27}$$

instead of (20.23). Now the effective field is again connected to the current as $\tilde{\mathbf{E}} = \gamma \mathbf{E} = R\mathbf{j}$ but with a yet-unknown factor γ when compared to (20.24). This factor is obtained if we compare with (20.27)

$$R\mathbf{j} = \mathbf{E} - \frac{\delta E}{E}\tilde{\mathbf{E}} = \left(1 - \gamma\frac{\delta E}{E}\right)\mathbf{E} = \gamma\mathbf{E} \tag{20.28}$$

from which one obtains $\gamma = 1/(1 - \frac{\delta E}{E})$ and finally from (20.27) the conductivity

$$\sigma = \frac{1}{R}\frac{1}{1 + \frac{\delta E}{E}}. \tag{20.29}$$

We see that the selfconsistency or modification on the three-particle level leads to a different relaxation effect (20.29) compared to (20.25).

Summarising, the relaxation effect accounts for the internal disturbance of the two-particle correlations due to an external field and includes quantum interference effects diminishing the conductivity up to a possible localisation (Morawetz, 2000c). This term unifies the picture of interference in the theory of dense plasmas where it is known as the Debye–Onsager relaxation effect (Kadomtsev, 1958; Klimontovich and Ebeling, 1972; Ebeling, 1976; Ebeling and Röpke, 1979; Röpke, 1988; Morawetz and Kremp, 1993; Esser and Röpke, 1998; Morawetz, 2000c) which was first derived within the theory of electrolytes (Debye and Hückel, 1923; Onsager, 1927; Falkenhagen, 1953; Kremp, Kraeft and Ebeling, 1966; Falkenhagen, Ebeling and Kraeft, 1971) with weak localisation corrections or Anderson localisation (Vollhardt and Wölfle, 1980; Kirkpatrick and Belitz, 1986). Technically, we will see this term emerge from the field-dependent collision integral (Morawetz, 2003).

The result of electrolytes was obtained by Debye (Debye and Hückel, 1923)

$$E^{\text{eff}} = E\left(1 - \frac{\kappa e^2}{6T}\right) \tag{20.30}$$

and Onsager corrected it by the result (Onsager, 1927)

$$E^{\text{eff}} = E\left(1 - \frac{\kappa e^2}{3(2 + \sqrt{2})T}\right). \tag{20.31}$$

Amazingly, the Onsager result is not to be reproduced by accounting for more correlations but by asymmetrically treating the correlations or alternatively assuming an infinitesimal friction with a background as illustrated in Appendix A.3.

20.4   Conductivity with electron–electron interaction

Let us calculate all conductivity contributions explicitly and consider an electron system interacting with impurities and with themselves being exposed to an external electric field. We assume that the field is weak enough and a stationary current has developed. The scattering with impurities will relax the distribution function with a relaxation time $\tau(k)$. The electron–electron interaction has no direct influence on the conductivity but changes the occupation relaxing the distribution towards the equilibrium one shifted by the mean-momentum $f_0(|\mathbf{k} - \bar{\mathbf{p}}|)$. Therefore this interaction leads to an indirect effect on the conductivity by the heated displaced distribution (Wingreen, Stanton and Wilkins, 1986).

The current in D-dimensions can be calculated and assumes the form

$$\mathbf{j} = -\frac{e}{m} g \int \frac{d\mathbf{k}}{(2\pi\hbar)^D} \mathbf{k} f(\mathbf{k}) = \frac{e}{m} g \int \frac{d\mathbf{k}}{(2\pi\hbar)^D} \mathbf{k}(\bar{\mathbf{p}} \cdot \partial_k) f_0(k) = -\frac{en}{m}\bar{\mathbf{p}} \qquad (20.32)$$

where we have used the fact that the deviation from symmetry is due to the mean current $\bar{\mathbf{p}}$ and introduced the density

$$n = g \int \frac{d\mathbf{k}}{(2\pi\hbar)^D} f_0(k). \qquad (20.33)$$

The possible spin degeneracy is denoted by g.

The linearised kinetic equation around an equilibrium f_0 then reads

$$-e\tilde{\mathbf{E}}(\mathbf{k}) \cdot \partial_k f_0 = -\frac{f(\mathbf{k}) - f_0(k)}{\tau(k)} - \frac{f(\mathbf{k}) - f_0(|\mathbf{k} - \bar{p}|)}{\tau_{ee}(k)}. \qquad (20.34)$$

Here we take into account that the field acting on the electrons is not the external field $\mathbf{E}$ but the one diminished by the relaxation effects condensed in a relaxation field $\delta E(k)$

$$\tilde{E}(k) = E\left(1 - \frac{\delta E(k)}{E}\right). \qquad (20.35)$$

This relaxation effect should not be mixed with the dielectric screening field created by electrostatic induced potentials but appears as a friction force due to polarisation by the surrounding media. Consequently, this effect appears only if the particles move.

The kinetic equation (20.34) yields the linearised solution

$$\delta f = f(\mathbf{k}) - f_0(k) = -\frac{\tau(k)}{\tau(k) + \tau_{ee}(k)}\bar{\mathbf{p}} \cdot \partial_k f_0(k) + \frac{\tau(k)\tau_{ee}(k)}{\tau(k) + \tau_{ee}(k)} e\tilde{\mathbf{E}}(k) \cdot \partial_k f_0(k). \qquad (20.36)$$

Using the solution (20.36) in the second equality of (20.32), we obtain the identity

$$g \int \frac{d\mathbf{k}}{(2\pi\hbar)^D} \mathbf{k} \frac{\tau_{ee}(k)\tau(k)}{\tau(k)+\tau_{ee}(k)} e(\tilde{\mathbf{E}}(k) \cdot \partial_k) f_0(k) = -g \int \frac{d\mathbf{k}}{(2\pi\hbar)^D} \mathbf{k} \frac{\tau_{ee}(k)}{\tau(k)+\tau_{ee}(k)} (\bar{\mathbf{p}} \cdot \partial_k) f_0(k).$$

(20.37)

Applying the angular integration; one can show that for any dimension D, this identity leads to

$$eEg \int \frac{d\mathbf{k}}{(2\pi\hbar)^D} \mathbf{k} \left(1 - \frac{\delta E(k)}{E}\right) \frac{\tau_{ee}(k)\tau(k)}{\tau(k)+\tau_{ee}(k)} (\mathbf{k}\cdot\partial_k) f_0(k)$$

$$= -\bar{\mathbf{p}}g \int \frac{d\mathbf{k}}{(2\pi\hbar)^D} \frac{\tau_{ee}(k)}{\tau(k)+\tau_{ee}(k)} (\mathbf{k}\cdot\partial_k) f_0(k).$$

(20.38)

With the help of this expression for the mean current $\bar{\mathbf{p}}$ in (20.32) we obtain, for the conductivity $\mathbf{j}=\sigma\mathbf{E}$,

$$\sigma = \frac{e^2 n}{m} \frac{\left\langle \frac{\tau_{ee}\tau}{\tau+\tau_{ee}} \left(1-\frac{\delta E}{E}\right)\right\rangle}{\left\langle \frac{\tau_{ee}}{\tau+\tau_{ee}}\right\rangle}$$

(20.39)

with the thermal averaging for the dimension D introduced as

$$\langle A \rangle = -\frac{1}{nD} \int \frac{d^D k}{(2\pi)^D} \mathbf{k} \cdot \partial_k f_0(k) A(k).$$

(20.40)

The conductivity (20.39) contains different special cases discussed in the literature. The adiabatic limit of (20.39) is reached if the electron–electron interaction is negligible $\tau_{ee} \to \infty$

$$\sigma_{\mathrm{ad}} = \frac{e^2 n}{m} \left\langle \tau \left(1-\frac{\delta E}{E}\right)\right\rangle.$$

(20.41)

In this limit 'all moments of the single-particle distribution are important since the nonequilibrium state of the electron system cannot be described by the total temperature' (Zubarev, Morozov and Röpke, 1997).

In the opposite isothermal limit we have a dominant electron–electron interaction $\tau_{ee} \to 0$ and obtain from (20.39)

$$\sigma_{\mathrm{th}} = \frac{e^2 n}{m} \frac{1-\langle \frac{\delta E}{E}\rangle}{\langle \frac{1}{\tau}\rangle}.$$

(20.42)

Here 'all moments of the single-particle distribution except the lowest one relax rapidly. The nonequilibrium state of the electron system can be described by the total energy or temperature and the total momentum' (Zubarev, Morozov and Röpke, 1997). A comprehensive discussion about these two different forms of conductivity and the related problem of convergence of the perturbation series can be found in (Huberman and Chester, 1975).

It is instructive to explicitly calculate both limits directly from the kinetic equation.

20.5   Isothermal conductivity

In this regime, the nonequilibrium distribution is entirely determined by a (time-dependent) mean current

$$f(\mathbf{k}) = f_0(|\mathbf{k} - \bar{p}(t)|) \approx f_0(k) - \bar{p}(t) \cdot \partial_k f_0(k). \tag{20.43}$$

The conductivity is then found from the balance equation; i.e. multiplying the kinetic equation with $\mathbf{k}$ and integrating leads to

$$\partial_t [n\bar{p}(t)] - ne\mathbf{E} = g \int \frac{d\mathbf{k}}{(2\pi\hbar)^D} \mathbf{k} I(k) \equiv \mathscr{F}_p + \mathscr{F}_E \tag{20.44}$$

with the forces derived in (20.45):

$$\mathscr{F}_E = -ne\mathbf{E} \left\langle \frac{\delta E}{E} \right\rangle, \qquad \mathscr{F}_p = -n\bar{p}(t) \left\langle \frac{1}{\tau} \right\rangle. \tag{20.45}$$

Before showing this we note the conductivity deduced from the stationary current $\mathbf{j} = ne\bar{p} = \sigma\mathbf{E}$ of (20.44)

$$\sigma = \frac{ne^2}{m} \frac{1}{\langle \frac{1}{\tau} \rangle} \left(1 - \left\langle \frac{\delta E}{E} \right\rangle \right). \tag{20.46}$$

Comparing this with (20.42) we see that the field-dependent collision integral leads to the relaxation factor. These represent quantum corrections to the conductivity.

20.5.1   Quasi two-dimensional example

We will now derive the forces (20.45) from the field-dependent collision integral in the Born approximation which reads

$$I(k) = \frac{2n_i s}{\hbar^2} \int \frac{d\mathbf{q}}{(2\pi\hbar)^2} V^2(q) \int_0^{t-t_0} dt' \cos\left[\frac{t'}{\hbar}\left(\epsilon_k - \epsilon_{k-q} - \frac{e\mathbf{E} \cdot \mathbf{q}t'}{2m}\right)\right] [f(\mathbf{k} - \mathbf{q} - e\mathbf{E}t') - f(\mathbf{k} - e\mathbf{E}t')],$$

$$\tag{20.47}$$

where the impurity distribution has been integrated out yielding the density n_i and we assume quadratic dispersion $\epsilon_k = k^2/2m$. Further, we do not consider short-time effects, $t_0 \to -\infty$.

Using (20.43) we obtain the forces on the right-hand side of the momentum balance in (20.44) up to a linear order in the external field or in the mean momenta

$$s \int \frac{d\mathbf{k}}{(2\pi\hbar)^2} \mathbf{k} I(k) \equiv \mathscr{F}_p + \mathscr{F}_{E1} + \mathscr{F}_{E2}. \tag{20.48}$$

The force proportional to the mean momenta is

$$\mathscr{F}_p = -\frac{2n_i s^2}{\hbar^2} \int \frac{d\mathbf{q}d\mathbf{k}}{(2\pi\hbar)^4} V^2(q)\mathbf{k} \int_0^\infty dt' \cos\left[\frac{t'}{\hbar}\left(\epsilon_k - \epsilon_{k-q}\right)\right](\bar{\mathbf{p}}) \cdot (\partial_{k-q} - \partial_k)f_0 \tag{20.49}$$

and the two forces linear in the field are

$$\mathscr{F}_{E1} = -\frac{2n_i s^2}{\hbar^2} \int \frac{d\mathbf{q}d\mathbf{k}}{(2\pi\hbar)^4} V^2(q)\mathbf{k} \int_0^\infty dt' \cos\left[\frac{t'}{\hbar}\left(\epsilon_k - \epsilon_{k-q}\right)\right](e\mathbf{E}t') \cdot (\partial_{k-q} - \partial_k)f_0. \tag{20.50}$$

and

$$\mathscr{F}_{E2} = \frac{2n_i s^2}{\hbar^2} \int \frac{d\mathbf{q}d\mathbf{k}}{(2\pi\hbar)^4} V^2(q)\mathbf{k} \left[f_0(\mathbf{k}-\mathbf{q}) - f_0(\mathbf{k})\right] \int_0^\infty dt' \cos\left[\frac{t'}{\hbar}\left(\epsilon_k - \epsilon_{k-q} - \frac{e\mathbf{E}\cdot\mathbf{q}t'}{2m}\right)\right]. \tag{20.51}$$

By substituting $\mathbf{k} - e\mathbf{E}t'/2 \to \mathbf{k}$, one can easily see that the force (20.51) is just $\mathscr{F}_{E2} = -\mathscr{F}_{E1}/2 + o(E^2)$ such that we have a net effect

$$\mathscr{F}_E = \mathscr{F}_{E1} + \mathscr{F}_{E2} = \frac{1}{2}\mathscr{F}_{E1}. \tag{20.52}$$

Further evaluation is done employing the Plemelj formula (8.5). The force (20.49) can be read-off directly to be

$$\mathscr{F}_p = -n\bar{\mathbf{p}}(t)\left\langle\frac{1}{\tau}\right\rangle \tag{20.53}$$

with the relaxation time

$$\frac{1}{\tau(k)} = \frac{2\pi n_i s}{\hbar} \int \frac{d\mathbf{q}}{(2\pi\hbar)^2} V^2(q)(1 - \cos\alpha)\delta(\epsilon_k - \epsilon_{k-q}) \tag{20.54}$$

and the angle between $\mathbf{k}$ and $\mathbf{k} - \mathbf{q}$ denoted by α.

We will now explicitly calculate the relaxation time for quasi-2D systems. There, the Coulomb interaction is $V(q) = e^2\hbar/2\epsilon_0 q$, see (10.57), and the screened one reads

$$V(q) = \frac{e^2\hbar}{2\epsilon_0(q + \hbar\kappa)} \tag{20.55}$$

with the inverse screening length

$$\kappa = \frac{e^2}{4\pi\epsilon_0}\frac{ms}{\hbar^2}. \tag{20.56}$$

Using this potential we obtain for the relaxation time (20.54)

$$\frac{1}{\tau(k)} = \frac{4\pi\hbar n_i}{ms}I_1\left(\frac{\hbar\kappa}{2k}\right) \tag{20.57}$$

with the integral

$$I_1(c) = c^2\left(\pi + \frac{2c}{1-c^2}\right) + \frac{2c^3(2-c^2)}{(c^2-1)^{3/2}}\arccos\frac{1}{c}.$$

The relaxation force $\mathscr{F}_E$ needs some more care in order to be calculated analytically. We read-off from (20.50) and (20.52)

$$\mathscr{F}_E = -4m^2 en_i s^2 \int\frac{d\mathbf{k}}{(2\pi\hbar)^2}\mathbf{E}\cdot\partial_k f_0 \int\frac{d\mathbf{q}}{(2\pi\hbar)^2}\mathbf{q}\frac{V^2(|\mathbf{q}|)}{2\mathbf{k}\cdot\mathbf{q}-q^2} \tag{20.58}$$

to be understood as the principal integration value. Employing symmetries by transforming $\mathbf{k} \to -\mathbf{k}$ and $\mathbf{q} \to -\mathbf{q}$ and introducing $y = q/k$ as well as $x = \cos(q,k)$, one finds

$$\mathscr{F}_E = e\frac{16\pi n_i}{s\kappa^2}\int\frac{d\mathbf{k}}{(2\pi\hbar)^2}\mathbf{k}(\mathbf{E}\cdot\partial_k)f_0(k)I_2\left(\frac{\hbar\kappa}{2k}\right) \tag{20.59}$$

with the integral

$$I_2(c) = \frac{c^2(2-6c^2+c^4)}{(c^2-1)^2} - \frac{3c^5\ln(c-\sqrt{c^2-1})}{(c^2-1)^{5/2}}. \tag{20.60}$$

From (20.59), the relaxation force follows immediately (20.45) with

$$\frac{\delta E(k)}{E} = -\frac{16\pi n_i}{s\kappa^2}I_2\left(\frac{\hbar\kappa}{2k}\right). \tag{20.61}$$

With (20.57) and (20.61) we have derived analytical expressions for the the forces (20.45) and therefore for the isothermal conductivity (20.46).

20.6   Adiabatic conductivity

Since, in the adiabatic regime, all moments of the distribution contribute, we will solve the kinetic equation for the linear deviation from the equilibrium, $f(\mathbf{k}) = f_0(k) + \delta f(\mathbf{k})$ being proportional to the field $\delta f(\mathbf{k}) = \mathbf{k} \cdot \mathbf{E} f_1(k)$. Then the collision integral (20.47) expands in first order $I(k) = I_0(k) + I_{\mathrm{E}}(k)$ of the external field. The zero order is easily rewritten in the relaxation time (20.54) such that the kinetic equation reads

$$-e\mathbf{E}(k) \cdot \partial_k f_0 = -\frac{\delta f(\mathbf{k})}{\tau(k)} + I_{\mathrm{E}}(k) \tag{20.62}$$

with the solution

$$\delta f(\mathbf{k}) = \tau(k)e\mathbf{E} \cdot \partial_k f_0(k) + \tau(k)I_{\mathrm{E}}. \tag{20.63}$$

The current (20.32) becomes therefore

$$\mathbf{j} = \frac{e^2 n}{m}\langle\tau\rangle\mathbf{E} - \frac{e}{m}g\int\frac{d\mathbf{k}}{(2\pi\hbar)^D}\mathbf{k}\tau(k)I_{\mathrm{E}}. \tag{20.64}$$

20.6.1   Quasi two-dimensional example

The last term of (20.64) can be treated analogously as was done when deriving (20.58) but an additional factor, $\tau(k)$ has to be considered. This leads instead of to (20.58), to

$$g\int\frac{d\mathbf{k}}{(2\pi\hbar)^D}\mathbf{k}\tau_k I_{\mathrm{E}}$$

$$= 4m^2 en_i s^2\int\frac{d\mathbf{k}}{(2\pi\hbar)^2}\mathbf{E}\cdot\partial_k f_0 \int\frac{d\mathbf{q}}{(2\pi\hbar)^2}\left[(\mathbf{k}-\mathbf{q})\tau(|\mathbf{k}-\mathbf{q}|) - \mathbf{k}\tau(k)\right]\frac{V^2(|\mathbf{q}|)}{2\mathbf{k}\cdot\mathbf{q}-q^2}. \tag{20.65}$$

A further exact treatment would lead to integrals like I_2 but with the additional functional arguments of $\tau(k)$. Since the momentum dependence of τ is thermally averaged, we neglect the momentum dependence of $\tau(|\mathbf{k}-\mathbf{q}|) \approx \tau(k)$ in (20.65) and obtain from (20.64) just the conductivity (20.41)

$$\sigma_{\mathrm{ad}} = \frac{e^2 n}{m}\left\langle\tau\left(1 - \frac{\delta E}{E}\right)\right\rangle. \tag{20.66}$$

with the relaxation field (20.61). This completes the derivation of the adiabatic conductivity.

20.7   Debye–Onsager relaxation effect

One can also analytically integrate the case of three dimensions. The statically screened result leads to (Morawetz, 2000c) and the following relaxation field:

$$\frac{\partial}{\partial t} < \mathbf{k}_a > -n_a e_a \mathbf{E} \left(1 + \frac{\delta E_a}{E} \right) = n_a e_a \mathbf{J} \, R(E) \tag{20.67}$$

with

$$\frac{\delta E_a}{E} = -\frac{e_a \pi}{6\kappa} \sum_b \frac{4 n_b e_b^2}{\mu_{ab}} \frac{\frac{e_b}{m_b} - \frac{e_a}{m_a}}{\left(\frac{T_b}{m_b} + \frac{T_a}{m_a} \right)^2} F(|x|)$$

$$F(x) = -\frac{3}{x^2} \left[3 - x + \frac{1}{1+x} - \frac{4}{x} \ln(1+x) \right] \tag{20.68}$$

and the classical field parameter

$$x = \frac{e}{\zeta} = \frac{E}{2 T_{ab} \kappa} \left(\frac{m_a}{m_a + m_b} e_b - \frac{m_b}{m_a + m_b} e_a \right). \tag{20.69}$$

We see that, for a plasma consisting of particles with equal charge-to-mass ratios, no relaxation field appears. The link to the known Debye–Onsager relaxation effect can be found if we assume that we have a plasma consisting of electrons (m_e, $e_e = e$) and ions with charge $e_i = eZ$ and temperatures $T_e = T_i = T$. Then (20.68) reduces to

$$\frac{\delta E_a}{E} = -\frac{\kappa e_a^2}{6T} \frac{Z(1 + \frac{m_e}{m_i} Z)}{(1+Z)(1 + \frac{m_e}{m_i})} F \left(\frac{eE}{T\kappa} \frac{Z(1 + \frac{m_e}{m_i} Z)}{1 + \frac{m_e}{m_i}} \right) = -\frac{e^2 \kappa_e}{6T} \begin{cases} \frac{1}{2} + o(E) \\ \frac{3\kappa T}{2eE} + o(E)^{-2} \end{cases} \text{ for } Z = 1. \tag{20.70}$$

This formula gives the classical relaxation effect for statically screened approximations up to any field strength and represents a result beyond the linear response. We see that in the case of single-charged heavy ions, the Debye result (20.30) is underestimated by a factor of two.

We like to remark that we neglect any field dependence on the screening ϵ^R itself here. As presented in (Morawetz and Jauho, 1994) a field-dependent screening function can be derived. However, this field dependence gives rise to a field dependence starting quadratically and will be not considered here.

The dynamically screened result can be given, as well as the analytically achieved one, and the linear relaxation field (20.68) takes the form

$$\frac{\delta E^{\mathrm{dyn}}}{E} = \frac{4e\pi\kappa}{3\sum_c \kappa_c^2 \sqrt{\frac{m_c}{T_c}}} \sum_b n_b e_b^2 \sqrt{\frac{m_a m_b}{T_a T_b}} \left(\frac{e_a}{T_a^{3/2}\sqrt{m_a}} - \frac{e_b}{T_b^{3/2}\sqrt{m_b}} \right) + o(E). \qquad (20.71)$$

The difference from (20.68) becomes more evident if we consider again only electrons and ions with equal temperatures

$$\frac{\delta E^{\mathrm{dyn}}}{E} = -\frac{\kappa e^2}{6T} \frac{2Z(1 + \sqrt{\frac{m_e}{m_i}} Z)}{(Z + \sqrt{\frac{m_e}{m_i}})} + o(E). \qquad (20.72)$$

The differences from (20.70) are obvious in the different mass dependence. This result overestimates the Debye result by a factor of two, in agreement with the result of the hierarchy (A.39).

The classical result beyond the linear response reads for (20.71) for single-charged ions and big mass differences

$$\frac{\delta E^{\mathrm{dyn}}}{E} = -\frac{\kappa e^2}{6T} \mathscr{F}\left[\frac{eE}{\kappa T} \right] = -\frac{e^2 \kappa_e}{6\,T} \begin{cases} 2 + o(E) \\ \frac{3\kappa T}{\sqrt{2}eE} + o(1/E)^2 \end{cases}$$

$$\mathscr{F}[x] = \frac{3}{x^3}\left(\frac{2\left(-2x + 3\left(-1 + \sqrt{1+x}\right)\right)}{\sqrt{1+x}} + \sqrt{2}\left(-\mathrm{artanh}\left(\frac{1}{\sqrt{2}}\right) + \mathrm{artanh}\left(\frac{\sqrt{1+x}}{\sqrt{2}}\right)\right) \right.$$

$$\left. + \frac{x^2 + \log(1-x^2)}{\sqrt{2}} \right) = 2 + o(E). \qquad (20.73)$$

This result will be compare with the statically screened result (20.70) and the hydrodynamical result (A.34) in Figure 20.1.

20.7.1   Thermally averaged dynamically screened result

There is an approximate treatment of the dynamical screening used in (Ebeling and Röpke, 1979) to replace the dynamical screening in the collision integral (20.15) which is $\epsilon^R(\omega, q)^{-2}$ by $(1 + \kappa^2 V_{aa}(q)/4\pi)^{-1}$. This represents a thermal averaging (Klimontovich, 1975) of $(\epsilon^R)^{-2}$. We obtain the relaxation effect of (20.68) and (20.70) but with a different field-function F

$$F^{\mathrm{dyn}}(x) = -\frac{3}{x^2}\left[2 - x - \frac{2}{x}\ln(1 + x) \right] = \begin{cases} 2 + o(x) \\ \frac{3}{x} + o(1/x)^2 \end{cases}. \qquad (20.74)$$

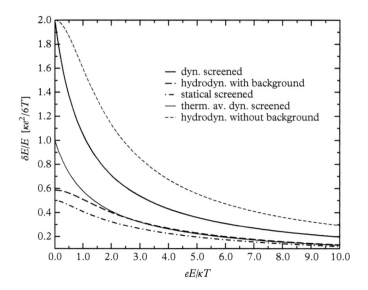

Figure 20.1: *Debye–Onsager relaxation effect vs. electric field for electrons and single charged ions. The hydrodynamical approximation (A.34) leads to the Onsager result (20.31) for small field strengths $2 - \sqrt{2}$, the statically screened result (20.68) or (20.70) to half the Debye result (20.30). The thermally averaged dynamical screening (20.74) approaches the Debye result too, while the full dynamically screened approximation (20.73) leads to twice the Debye result as the hydrodynamical result (A.39) without the background.*

Therefore the relaxation effect (20.70) in linear response for single charged ions takes the form of (20.30) and twice the statically screened result (20.70) and half of the dynamically screened result (20.73).

As we see from Figure 20.1 the different approximations lead to very different results. The statically screened result (20.70) underestimates the Debye result by factor 2 which is corrected by the thermal averaged treatment of the screening. If instead we calculate the complete dynamically screened result (20.72) we obtain twice the Debye result (20.30) and the thermally averaged screened result. However there is a completely different charge dependence. We have to observe that the perfectly symmetric treatment of screening does not reproduce the hydrodynamical result which is the Onsager result (20.31) for linear response.

20.7.2   Asymmetric dynamically screened result

We want now to proceed and ask under which assumptions the Onsager result (20.31) might be reproduced. Following the results from the hierarchy in Appendix A.3, we have consequently to treat the electrons (specie a) and ions (all other species) asymmetrically.

This we will perform in the same spirit as Onsager in that the ions have to be treated dynamical (as before) but the electrons are screened statically.

This means we consider as the potential, not the bare Coulomb one, but a statically screened Debye potential for specie a. The ions (all other species) will then form the dynamical screening. The modification is for the linear response result (20.72) where for electron-ion plasma $q = 1/(Z + 1)$ and

$$\frac{\delta E^{\mathrm{asy}}}{E} = \frac{\delta E^{\mathrm{dyn}}}{E} \frac{Zq}{\sqrt{q}+1} = -\frac{\kappa e^2}{3T} \frac{Zq}{\sqrt{q}+1} + o(E)$$

(20.75)

which agrees with (A.31) if we consider that the mobilities are very different $b_i/b_e \to 0$ in (A.32). We obtain the same result from the thermally averaged result (20.70) but (Morawetz, 1997)

$$F^{\mathrm{dyn}}_{\mathrm{asy}}(x) = 2F^{\mathrm{dyn}}(x) - \sqrt{2}F^{\mathrm{dyn}}(\sqrt{2}x) = \begin{cases} 2 - \sqrt{2} + o(x) \\ \\ \frac{3}{2x} + o(1/x)^2 \end{cases}$$

(20.76)

with F^{dyn} of (20.74). The linear response then leads exactly to the same result as from the dynamical screening (20.75), i.e. the Onsager result with the same charge dependence. The asymmetrically screened result approaches the hydrodynamical or Onsager result (20.31) rather well for small fields while it is too low at high fields. On the other hand, the thermally averaged symmetrically screened result (20.76), agrees with the hydro-dynamical approximation (A.34) in the low- and high-field limit. With this asymmetric screening we have given a further example of the asymmetric treatment of correlations derived in Chapter 11.5.

21

Kinetic Theory of Systems with SU(2) Structure

21.1 Transport in electric and magnetic fields

21.1.1 Basic notation

The dipole magnetic moment for a circular current surrounding an area $\mathbf{A}$ can be written

$$\mathbf{m} = \frac{1}{2}\int \mathbf{r} \times \mathbf{J} d^3 r = \frac{1}{2}\mathcal{J}\int \mathbf{r} \times d\mathbf{r} = \mathcal{J}\mathbf{A} \tag{21.1}$$

with the current density $\mathcal{J}$. For a moving point charge $\mathbf{J} = e\mathbf{v}\delta(\mathbf{r} - \mathbf{r}_e)$ the magnetic moment is therefore expressed by its angular momentum via the gyromagnetic ratio γ as

$$\mathbf{m} = \frac{1}{2}e\mathbf{r} \times \mathbf{v} = \frac{e}{2m}\mathbf{L} = \gamma\mathbf{L}. \tag{21.2}$$

The torque $d\mathbf{L}/dt$ is given by $\mathbf{m} \times \mathbf{B}$ such that one obtains the equation of motion

$$\dot{\mathbf{m}} = \gamma\dot{\mathbf{L}} = \gamma(\mathbf{m} \times \mathbf{B}). \tag{21.3}$$

One can consider the magnetic field as an effective one which reacts to damping mechanisms such that a small delay time δt appears

$$\mathbf{B}(t - \delta t) = \mathbf{B}(t) - \delta t \frac{\partial \mathbf{B}}{\partial \mathbf{m}}\dot{\mathbf{m}}. \tag{21.4}$$

Then (21.3) takes the form

$$\dot{\mathbf{m}} = \gamma\mathbf{m} \times \mathbf{B} - \frac{\mathbf{m}}{m} \times (A \cdot \dot{\mathbf{m}}) \tag{21.5}$$

where the damping tensor is introduced $A = \gamma\,\delta t m\partial\mathbf{B}/\partial\mathbf{m}$. Assuming $A \cdot \dot{\mathbf{m}} = \alpha\dot{\mathbf{m}}$ one finally obtains the Landau–Lifshitz–Gilbert equation (Landau and Lifshitz, 1935; Ament and Rado, 1955; Gilbert, 2004)

Interacting Systems far from Equilibrium. Klaus Morawetz, Oxford University Press (2018).
© Klaus Morawetz. DOI: 10.1093/oso/9780198797241.001.0001

$$\dot{\mathbf{m}} = \gamma \mathbf{m} \times \mathbf{B} - \frac{\alpha}{m}(\mathbf{m} \times \dot{\mathbf{m}}). \tag{21.6}$$

This equation is equivalent to the Landau–Lifshitz equation

$$\dot{\mathbf{m}} = \gamma^* \mathbf{m} \times \mathbf{B} - \frac{\lambda}{m}(\mathbf{m} \times (\mathbf{m} \times \mathbf{B})) \tag{21.7}$$

if one replaces in the lasst term the equation itself $\mathbf{m} \times \mathbf{B} = \frac{1}{\gamma^*}(\dot{\mathbf{m}} + \frac{\lambda}{m^2}(\mathbf{m} \times (\mathbf{m} \times \mathbf{B}))$ and set $\alpha = \lambda/\gamma m$ and $\gamma = \gamma^*(1 + \alpha^2)$.

21.1.2 Spin

If we consider the spin of the electron as an internal degree of freedom analogously to circular motion we can identify the angular momentum with the spin $\hat{\mathbf{L}} = \hat{\mathbf{S}} = \frac{\hbar}{2}\sigma$ and the Pauli matrices σ for the three components with the quantisation axes in the z-direction

$$\sigma = \mathbf{e}_x \begin{pmatrix} 0 & 1 \\ 1 & 0 \end{pmatrix} + \mathbf{e}_y \begin{pmatrix} 0 & -i \\ i & 0 \end{pmatrix} + \mathbf{e}_z \begin{pmatrix} 1 & 0 \\ 0 & -1 \end{pmatrix}. \tag{21.8}$$

The elementary magnetic moment as the product of the current and the enclosed area is then identified for the spin in terms of the Bohr magneton μ_B,

$$\hat{\mathbf{m}} = g\frac{e}{2m_e}\frac{\hbar}{2}\sigma = \frac{g}{2}\mu_B\sigma \tag{21.9}$$

with the anomalous gyromagnetic ratio $g \approx 2$ for electrons. If there are many fermions with densities $n_\pm$ of spins parallel/anti-parallel to the magnetic field, the total magnetisation is $m_z = g\mu_B s_z$ with the polarisation $s_z = (n_+ - n_-)/2$. We want to access the density and polarisation density distributions and employ therefore four Wigner functions

$$\hat{\rho}(\mathbf{r}, \mathbf{p}, t) = f(\mathbf{r}, \mathbf{p}, t) + \sigma \cdot \mathbf{g}(\mathbf{r}, \mathbf{p}, t) = \begin{pmatrix} f + g_z & g_x - ig_y \\ g_x + ig_y & f - g_z \end{pmatrix}. \tag{21.10}$$

The density and polarisation density are given by

$$\sum_p f(\mathbf{p}, \mathbf{r}, t) = n(\mathbf{r}, t), \qquad \sum_p \mathbf{g}(\mathbf{p}, \mathbf{r}, t) = \mathbf{s}(\mathbf{r}, t) \tag{21.11}$$

where we abbreviate $\sum_p = \int d^D p/(2\pi\hbar)^D$ for D dimensions in the following. The magnetisation (magnetic moment) density becomes $\mathbf{M}(\mathbf{r}, t) = \gamma \mathbf{s}(\mathbf{r}, t)$ with $\gamma = g\mu_B/\hbar$. The advantage is that we can describe any direction of magnetisation density created by microscopic correlations.

21.1.3   Conductivity for crossed electric and magnetic fields: orbital motion

We can calculate the magnetic moment and the conductivity of an electron moving in an electric and magnetic field by starting from the Newton equation

$$m_e \dot{\mathbf{v}} = e(\mathbf{v} \times \mathbf{B}) + e\mathbf{E} - m_e \frac{\mathbf{v}}{\tau} \tag{21.12}$$

where besides the Lorentz force we assume some damping τ due to friction or any dissipative process. Eq. (21.12) can be seen as the balance of the simple kinetic equation in relaxation-time approximation

$$\partial_t f + \left(e\mathbf{E} + \frac{e}{m} \mathbf{k} \times \mathbf{B} \right) \partial_{\mathbf{k}} f = \frac{f_0 - f}{\tau} \tag{21.13}$$

by multiplying with $\mathbf{k}/m$ and integrating. The corresponding density balance as a simple momentum integral is seen to be $\dot{n} = 0$.

The Newton equation (21.12) describes the Ohmic regime where $\mathbf{J} = ne\mathbf{v} = \sigma\mathbf{E}$ since we can Fourier transform the time dependence of $\mathbf{v}_t, \mathbf{E}_t$ where $\mathbf{B} = const$ into frequency and obtain from (21.12) the vector equation

$$\mathbf{J}(\omega) - \frac{\omega_c \tau}{1 - i\omega\tau} \mathbf{J}(\omega) \times \mathbf{B}_0 = \frac{\sigma_0}{1 - i\omega\tau} \mathbf{E}(\omega) \tag{21.14}$$

where we have introduced the static-free conductivity $\sigma_0 = ne^2\tau/m_e$, the Lamour frequency $\omega_c = eB/m_e$ and $\mathbf{B}_0 = \mathbf{B}/B$ in the direction of $\mathbf{B}$. The vector equation is readily solved with the result

$$\mathbf{J} = \sigma \left[\mathbf{E}_\omega + \alpha(\mathbf{E}_\omega \times \mathbf{B}_0) + \alpha^2 \mathbf{B}_0 (\mathbf{B}_0 \cdot \mathbf{E}_\omega) \right] \tag{21.15}$$

where we introduce the dynamical scalar conductivity

$$\sigma = \sigma_0 \frac{1 - i\omega\tau}{(1 - i\omega\tau)^2 + (\omega_c\tau)^2} = \frac{\bar{\sigma}}{1 + \bar{\sigma}^2 R^2 B^2} \tag{21.16}$$

and the abbreviation $\alpha = \omega_c\tau/(1 - i\omega\tau)$, $\bar{\sigma} = \sigma_0/(1 - i\omega\tau)$ and the Hall coefficient

$$R = -\frac{1}{ne}. \tag{21.17}$$

The Hall conductivity $\mathbf{J} \sim \sigma_H \mathbf{B} \times \mathbf{E}$ is therefore

$$\sigma_H = \frac{R\bar{\sigma}^2}{1 + R^2\bar{\sigma}^2 B^2}. \tag{21.18}$$

Of course, the last term in (21.15) is absent for electromagnetic fields since $\mathbf{E} \perp \mathbf{B}$ but we keep it for completeness since later we will obtain similar structures where the precession axis allows non-perpendicular effective fields. With the help of the inverse Fourier transform it is easy to see that the time-dependence of the current starting with $\mathbf{J}(0) = 0$ and driven by the electric field reads

$$
\mathbf{J}(t) = \sigma_0 \int_0^t \frac{d\bar{t}}{\tau} e^{-\frac{\bar{t}}{\tau}} \{\cos(\omega_c \bar{t}) \mathbf{E}(t - \bar{t}) + \sin(\omega_c \bar{t}) \mathbf{E}(t - \bar{t}) \times \mathbf{B}_0
$$
$$
+ [1 - \cos(\omega_c \bar{t})][\mathbf{E}(t - \bar{t}) \cdot \mathbf{B}_0]\mathbf{B}_0\} \tag{21.19}
$$

which illustrates nicely the threefold orbiting of the electrons with cyclotron frequency: in the direction of the electric and magnetic field, respectively, and in the direction perpendicular to the magnetic and electric field.

The stationary solutions for long times in (21.19) takes for $E(\infty) = E$ the form

$$
\mathbf{J}(\infty) = \frac{\sigma_0}{1 + \omega_c^2 \tau^2} \left[\mathbf{E} + \omega_c \tau \mathbf{E} \times \mathbf{B}_0 + \omega_c^2 \tau^2 (\mathbf{E} \cdot \mathbf{B}_0)\mathbf{B}_0\right] \tag{21.20}
$$

which agrees with (21.15) in the $\omega = 0$ limit.

The solution (21.15) represents the dynamical conductivity tensor

$$
\mathbf{J} = \hat{\sigma} \cdot \mathbf{E} \tag{21.21}
$$

which takes a form of a 3×3 matrix. Let's assume that the magnetic field is in z-direction, then we obtain for the conductivity tensor the canonical form

$$
\hat{\sigma} = \sigma \begin{pmatrix} 1 & \alpha & 0 \\ -\alpha & 1 & 0 \\ 0 & 0 & 1 + \alpha^2 \end{pmatrix} \equiv -i\omega\epsilon_0\hat{\epsilon}. \tag{21.22}
$$

Please note that we did not made any assumptions about the direction of the electric field. The inversion of the conductivity tensor provides the resistivity $\mathbf{E} = \hat{\rho}\mathbf{j}$ with

$$
\hat{\rho} = \frac{1}{\sigma(1 + \alpha^2)} \begin{pmatrix} 1 & -\alpha & 0 \\ \alpha & 1 & 0 \\ 0 & 0 & 1 \end{pmatrix}. \tag{21.23}
$$

Writing in components one obtains

$$
E_x = \rho \mathcal{J}_x + RB\mathcal{J}_y, \qquad E_y = \rho \mathcal{J}_y - RB\mathcal{J}_x, \qquad E_z = \rho \mathcal{J}_z \tag{21.24}
$$

with the scalar dynamical resistivity

$$\rho = \frac{1}{\bar\sigma} = \frac{1 - i\omega\tau}{\sigma_0} \tag{21.25}$$

and the Hall coefficient (21.17). This could have been read off from (21.14) also directly as $\mathbf{E} = \rho\mathbf{J} + R\mathbf{J} \times \mathbf{B}$.

21.2   Systems with electro-magnetic fields and spin-orbit coupling

21.2.1   Various spin-orbit coupling as SU(2) structure

Let us consider spin-polarised fermions which interact with an impurity potential V_i and themselves with $\hat{V}$, covered by the Hamiltonian

$$\hat{H} = \sum_i \Psi_i^+ \left[\frac{(\mathbf{p} - e\mathbf{A}(\mathbf{r}_i, t))^2}{2m_e} + e\Phi(\mathbf{r}_i, t) - \mu_B\sigma \cdot \mathbf{B} + \sigma \cdot \mathbf{b}(\mathbf{p}, \mathbf{r}_i, t) + \hat{V}_i(\mathbf{r}_i) \right] \Psi_i$$

$$+ \frac{1}{2} \sum_{ij} \Psi_i^+ \Psi_j^+ \hat{V}(\mathbf{r}_i - \mathbf{r}_j) \Psi_j \Psi_i \tag{21.26}$$

with the spinor $\Psi_i = (\psi_{i\uparrow}, \psi_{i\downarrow})$, the spin-orbit coupling $\mathbf{b}$ and the Zeeman term $\mu_B \cdot \mathbf{B}$. Any spin-orbit coupling used in different fields, say plasma systems, semiconductors, graphene or nuclear physics can be recast into the general form

$$H^{\text{s.o.}} = A(\mathbf{p})\sigma_x - B(\mathbf{p})\sigma_y + C(\mathbf{p})\sigma_z = \mathbf{b} \cdot \sigma \tag{21.27}$$

with a momentum-dependent spin-orbit vector $\mathbf{b}$ illustrated in Table 21.1 and which can become space and also time-dependent in nonequilibrium.

We further assume a two-particle interaction which has a scalar and a spin-dependent part

$$\hat{V} = V_0 + \sigma \cdot \mathbf{V} \tag{21.28}$$

where the latter is responsible for spin-flip reactions. The vector part of the potential describes e.g. spin-dependent scattering. The scattering by impurities consists of a vector potential from magnetic impurities and a scalar one from charged or neutral impurities

$$\hat{V}_i = V_{i0} + \sigma \cdot \mathbf{V}_i. \tag{21.29}$$

In this way we have included the Kondo model as a specific case which was solved exactly in equilibrium (Wiegmann, 1981) and at zero temperature. Here we will consider the nonequilibrium form of this model in the mean-field approximation including

Table 21.1 *Selected 2D and 3D systems with the Hamiltonian described by (21.27) taken from (Chen and Guo, 2009; Cserti and Dávid, 2006)*

2D system	$A(p)$	$B(p)$	$C(p)$
Rashba	$\beta_R p_y$	$\beta_R p_x$	
Dresselhaus [001]	$\beta_D p_x$	$\beta_D p_y$	
Dresselhaus [110]	βp_x	$-\beta p_x$	
Rashba–Dresselhaus	$\beta_R p_y - \beta_D p_x$	$\beta_R p_x - \beta_D p_y$	
cubic Rashba (hole)	$i\frac{\beta_R}{2}(p_-^3 - p_+^3)$	$\frac{\beta_R}{2}(p_-^3 + p_+^3)$	
cubic Dresselhaus	$\beta_D p_x p_y^2$	$\beta_D p_y p_x^2$	
Wurtzite type	$(\alpha + \beta p^2)p_y$	$(\alpha + \beta p^2)p_x$	
single-layer graphene	$v p_x$	$-v p_y$	
bilayer graphene	$\frac{p_-^2 + p_+^2}{4m_e}$	$\frac{p_-^2 - p_+^2}{4m_e i}$	
3Dsystem	$A(p)$	$B(p)$	$C(p)$
bulk Dresselhaus	$p_x(p_y^2 - p_z^2)$	$p_y(p_x^2 - p_z^2)$	$p_z(p_x^2 - p_y^2)$
Cooper pairs	Δ	0	$\frac{p^2}{2m} - \epsilon_F$
extrinsic $\beta = \frac{i}{\hbar}\lambda^2 V(p)$	$q_y p_z - q_z p_y$	$q_z p_x - q_x p_z$	$q_x p_y - q_y p_x$
neutrons in nuclei $\beta = iW_0\left(n_n + \frac{n_p}{2}\right)$	$q_z p_y - q_y p_z$	$q_x p_z - q_z p_x$	$q_y p_x - q_x p_y$

relaxation due to collisions. This model is applicable to any system with an SU(2) structure. In graphene we can consider the two sub-lattices described by the value of the z-component of the Pauli matrix (Kane and Mele, 2005; Scholz, Stauber and Schliemann, 2012). Since the two K-points are not coupled (Scholz, Stauber and Schliemann, 2012) we account for them by the degeneracy $g = 2$ and an additional factor two for spin degeneracy. For single-layer graphene the previous form of spin-orbit coupling can be considered as a pseudo-spin representing the linear dispersion for the kinetic energy $\mathbf{b} = v(p_x, p_y, 0)$ and the limit of $m \to \infty$.

21.3 Meanfield kinetic equations

21.3.1 Gradient expansion

We use the Wigner mixed representation (7.31) and expand the convolution up to second-order gradients, see Chapter 9.1.1. All Green's functions become 2×2 matrices in spin space. Matrix product terms $A \circ B$ appearing in the Kadanoff and Baym equation (7.28) can be written as

$$A \circ B \rightarrow e^{\frac{i}{2}(\partial_\Omega \partial_t^B - \partial_t^A \partial_\Omega^B - \partial_p^A \partial_r^B + \partial_r^A \partial_p^B)} AB. \tag{21.30}$$

The quasiclassical limit is obtained by keeping only the first gradients in space and time of the above gradient expansion

$$A \circ B \rightarrow AB + \frac{i}{2} \{A, B\}, \tag{21.31}$$

where curly brackets denote Poisson brackets, i.e. $\{A, B\} = \partial_\Omega A \partial_t B - \partial_t A \partial_\Omega B - \partial_p A \partial_r B + \partial_r A \partial_p B$. Therefore, in the lowest-order gradient approximation, we have the following rule to evaluate the commutators $[A, B]_-$:

$$[A \circ, B]_- \rightarrow [A, B]_- + \frac{i}{2} (\{A, B\} - \{B, A\})$$

$$= [A, B]_- + \frac{i}{2} \left([\partial_r A, \partial_p B]_+ - [\partial_p A, \partial_r B]_+ + [\partial_\Omega A, \partial_t B]_+ - [\partial_t A, \partial_\Omega B]_+ \right). \tag{21.32}$$

Please note that the quantum spin structure remains untouched even after gradient expansion due to the commutators.

In order to prevent ambiguous results for different choices of gauges, we need to formulate the theory in a gauge-invariant way. Under U(1), the local gauge for electromagnetic fields, we have introduced the gauge-invariant Fourier transform (20.1) with the vector and scalar potential $A_\mu = (\Phi(\mathbf{r}, t), \mathbf{A}(\mathbf{r}, t))$ in gradient expansion (20.3) leading to the rules for gauge-invariant formulation (20.1). This treatment ensures that one has included all orders of a constant electric field.

21.3.2 Mean field

We want to consider now the mean-field selfenergy for impurity interactions as well as spin-orbit couplings. A general four-point potential can be written

$$\langle x_1 x_2 | \hat{V} | x_1' x_2' \rangle = \sum_{p,p'} e^{-ip(x_1 - x_2) + ip'(x_1' - x_2')} \langle p | \hat{V} \left(\frac{x_1 + x_2}{2} - \frac{x_1' + x_2'}{2}, \frac{\frac{x_1 + x_2}{2} + \frac{x_1' + x_2'}{2}}{2} \right) | p' \rangle$$

$$= \hat{V}_{-p,p'} \delta \left(\frac{x_1 + x_2}{2} - \frac{x_1' + x_2'}{2} \right) \tag{21.33}$$

with

$$\hat{V}_{-p,p'} = \begin{cases} V_0(p' - p) \\ \sigma \cdot \mathbf{V}(p' - p) \\ \frac{i\lambda^2}{\hbar} \sigma \cdot (\mathbf{p} \times \mathbf{p}') V(p' - p) \end{cases} \tag{21.34}$$

for scalar, magnetic impurities (21.29), the two-particle interaction (21.28), and extrinsic spin-orbit coupling. Since the potential is time-local, the Hartree-mean field is the convolution with the Wigner function (21.10) and written with Fourier transform of difference coordinates

$$\hat{\Sigma}(p,r,t) = \sum_{R'qQ} e^{iq(r-R')} \hat{\rho}(Q+p,r',t) \hat{V}_{\frac{q-Q}{2},\frac{q+Q}{2}}. \tag{21.35}$$

Due to the occurring product of the potentials (21.28), (21.29) and the Wigner function (21.10) one has

$$\hat{V}\hat{\rho} = V_0 f + \mathbf{g}\cdot\mathbf{V} + \sigma\cdot[f\mathbf{V} + V_0\mathbf{g} + i(\mathbf{V}\times\mathbf{g})]. \tag{21.36}$$

The last term in (21.36) is absent since we work in symmetrized products as they appear on the left side of (7.26) from now on. Consequently the selfenergy possesses a scalar and a vector component

$$\hat{\Sigma}^H(\mathbf{p},\mathbf{r},t) = \Sigma_0(\mathbf{p},\mathbf{r},t) + \sigma\cdot\mathbf{\Sigma}^H(\mathbf{p},\mathbf{r},t). \tag{21.37}$$

The interaction between a conduction electron and the magnetic impurity $\sigma\cdot\mathbf{V}_i$ where the direction of $\mathbf{V}_i$ is the local magnetic field we assume randomly distributed on different sites within an angle θ_l from the $\mathbf{e}_z$ direction. The directional average (Chen and Hershfield, 1998) leads then to

$$\overline{\sum_p f\mathbf{V}} = |V|\frac{\sin\theta_l}{\theta_l}\mathbf{e}_z n(\mathbf{q},t) = \mathbf{V}(q)n(\mathbf{q},t)$$

$$\overline{\sum_p \mathbf{g}\cdot\mathbf{V}} = |V|\frac{\sin\theta_l}{\theta_l}\mathbf{e}_z\cdot\mathbf{s}(\mathbf{q},t) = \mathbf{V}(q)\cdot\mathbf{s}(\mathbf{q},t). \tag{21.38}$$

The angle θ_l allows us to describe different models. A completely random local magnetic field $\theta_l = \pi$ is used for magnetic impurities in a paramagnetic spacer layer and in a ferromagnetic layer one uses $\theta_l = \pi/4$. The latter one describes the randomly distributed orientation against the host magnet (Chen and Hershfield, 1998).

For impurity potentials the spatial convolution with the density and spin polarisation reads when Fourier transformed, $\mathbf{r} \to \mathbf{q}$,

$$\Sigma_0^{\text{imp}}(p,q,t) = n(q)V_0(q) + \mathbf{s}(q)\cdot\mathbf{V}(q)$$

$$\mathbf{\Sigma}^{\text{imp}}(p,q,t) = \mathbf{s}(q)V_0(q) + n(q)\mathbf{V}(q). \tag{21.39}$$

For extrinsic spin-orbit coupling we obtain

$$\Sigma_0^{\text{s.o.}}(\mathbf{p}, \mathbf{q}, t) = i\frac{\lambda^2}{\hbar^2} V(q) \left[m_e(\mathbf{d}_j(q) \times \mathbf{q})_j - \mathbf{s}(q) \cdot (\mathbf{p} \times \mathbf{q}) \right]$$

$$\boldsymbol{\Sigma}^{\text{s.o.}}(\mathbf{p}, \mathbf{q}, t) = i\frac{\lambda^2}{\hbar^2} V(q) \left[m_e(\mathbf{j}(q) \times \mathbf{q}) - n(q)(\mathbf{p} \times \mathbf{q}) \right]. \qquad (21.40)$$

We used the particle density and the spin polarisation (21.11) and introduced the current and spin current

$$\mathbf{j} = \sum_p \frac{\mathbf{p}}{m_e} g(\mathbf{p}, \mathbf{q}, t); \quad \mathbf{d}_i = \sum_p \frac{\mathbf{p}}{m_e} [\mathbf{g}(\mathbf{p}, \mathbf{q}, t)]_i; \quad d_{ij} = S_{ji}. \qquad (21.41)$$

Please note the summation over indices in the first line of (21.40) after cross products. Collecting these results the inverse retarded Green's function reads

$$\hat{G}_R^{-1}(\mathbf{k}, \Omega, \mathbf{r}, t) = \Omega - H - \boldsymbol{\sigma} \cdot \boldsymbol{\Sigma}(\mathbf{k}, \mathbf{r}, t) \qquad (21.42)$$

with the effective scalar Hamiltonian

$$H = \frac{k^2}{2m} + \Sigma_0(\mathbf{k}, \mathbf{r}, t) + e\Phi(\mathbf{r}, t) \qquad (21.43)$$

and $\mathbf{k} = \mathbf{p} - e\mathbf{A}(\mathbf{r}, t)$. We have summarised the Zeeman term, the intrinsic spin-orbit coupling, and the vector part of the Hartree–Fock selfenergy due to impurities and extrinsic spin-orbit coupling into an effective selfenergy

$$\boldsymbol{\Sigma} = \boldsymbol{\Sigma}^H(\mathbf{k}, \mathbf{r}, t) + \mathbf{b}(\mathbf{k}, \mathbf{r}, t) + \mu_B \mathbf{B} \qquad (21.44)$$

such that the effective Hamiltonian possesses the Pauli structure

$$\hat{H}_{\text{eff}} = H + \boldsymbol{\sigma} \cdot \boldsymbol{\Sigma}. \qquad (21.45)$$

Please note that one can consider (21.44) as an effective Zeeman term where the spin-orbit competes with the magnetic field leading to additional degeneracies in Landau levels (Shen, Ma, Xie and Zhang, 2004).

21.3.3 Commutators

Now we are ready to evaluate the commutators according to (21.32). In the following we drop the vector notation where it is obvious. We calculate first the commutator with the scalar parts of (21.42) where we use the gauge-invariant Green's function such that one has

$$\partial_p G = \partial_k \bar{G}$$

$$\partial_r G = \partial_r \bar{G} - e\nabla\Phi\partial_\omega\bar{G} - e(\nabla A_i)\partial_{k_i}\bar{G}$$

$$\partial_t G = \partial_t \bar{G} - e\partial_t A\partial_k\bar{G} - e\partial_t\Phi\partial_\omega\bar{G}$$

$$\partial_\Omega G = \partial_\omega\bar{G}. \tag{21.46}$$

Further one calculates

$$\partial_t H = e\dot{\Phi} - e\frac{k}{m_e}\dot{A} - e\dot{A}\partial_k\Sigma_0 + \dot{\Sigma}_0$$

$$\partial_p H = \frac{k}{m_e} + \partial_k\Sigma_0$$

$$\partial_r H = e\partial_r\Phi + \frac{1}{m_e}\left[k\times\partial_r\times(p-eA) + (k\cdot\partial_r)(p-eA)\right] + \partial_r\Sigma_0 - e\partial_r A_i\partial_{k_i}\Sigma_0$$

$$= e\nabla\Phi - \frac{k}{m_e}\times eB - \frac{e}{m_e}(k\cdot\nabla)A + \partial_r\Sigma_0 - e\partial_r A_i\partial_{k_i}\Sigma_0 \tag{21.47}$$

where we have used $\frac{1}{2}\nabla u^2 = u\times\nabla\times u + (u\cdot\nabla)u$. We obtain, for the commutator with the scalar part of (21.42), according to (21.32)

$$\frac{1}{i}[(\Omega - H)\circ, G^<]_- \rightarrow \left[\partial_t + (\dot{\Sigma}_0 + evE)\partial_\omega + v\partial_r + (eE + ev\times B - \partial_r\Sigma_0)\partial_k\right]\hat{G}^< \tag{21.48}$$

where the mean velocity of the particles is given by

$$v = \frac{k}{m_e} + \partial_k\Sigma_0 \tag{21.49}$$

and $E = -\partial_r\Phi - \dot{A}$ and $B = \partial_r\times A$.

In order to get the equation for the Wigner distribution we integrate over frequency and the second term on the right-hand side of (21.48) disappears. This term has the structure of the power supplied to the particles which is composed of the contribution by the electric field and the time change of the scalar field which feeds energy to the system. The first and third part of (21.48) together are the co-moving time derivative of a particle with velocity (21.49). The fourth term in front of the momentum derivative of $\bar{G}^<$ represents the forces exercised on the particles which appears as the Lorentz force and the negative gradient of the scalar part of the selfenergy which acts therefore like a potential.

Next we calculate the commutator (21.32) with the vector components of (21.42). Therefore we employ the relations

$$[\sigma \cdot \mathbf{A}, \hat{G}^<]_+ = 2\sigma \cdot \mathbf{A}\, G^< + 2\mathbf{A} \cdot \mathbf{G}^<$$

$$\left[\sigma \cdot \mathbf{A}, \hat{G}^<\right]_- = 2i\sigma \cdot (\mathbf{A} \times \mathbf{G}^<) \tag{21.50}$$

where $\hat{G}^< = G^< + \sigma \cdot \mathbf{G}^<$ and using (21.46) we obtain

$$- i[-\sigma \cdot \mathbf{\Sigma}\circ, \hat{G}^<]_- \to -2\sigma \cdot (\mathbf{\Sigma} \times \mathbf{G}^<)$$

$$+ \left[(\dot{\mathbf{\Sigma}} + eE\partial_k\mathbf{\Sigma})\partial_\omega + \partial_k\mathbf{\Sigma}\partial_r + (e\partial_k\mathbf{\Sigma} \times B - \partial_r\mathbf{\Sigma})\partial_k\right] \cdot (\sigma\, G^< + \mathbf{G}^<). \tag{21.51}$$

Please note that the vector notations for $\partial_k, \partial_r, B, E$ are suppressed in order to indicate that they form scalar products among each other and the vectors written explicitly among themselves, respectively. We recognise the same drift terms as for the scalar selfenergy components (21.48). Additionally, the vector selfenergy couples the scalar and spinor part of the Green's function by an analogous drift but controlled by the vector selfenergy instead of the scalar one.

21.3.4   Coupled kinetic equations

By integrating (21.48) over frequency and adding (21.51) we have the complete kinetic equation as required from the Kadanoff and Baym equation (7.26). In order to make it more transparent we separate the equation according to the occurring Pauli matrices. This is achieved by once forming the trace and once multiplying with σ and forming the trace. We finally obtain two coupled equations for the scalar and vector part of the Wigner distribution

$$(\partial_t + \mathscr{F}\partial_k + v\partial_r)f + \mathscr{A} \cdot \mathbf{g} = 0$$

$$(\partial_t + \mathscr{F}\partial_k + v\partial_r)\mathbf{g} + \mathscr{A}f = 2(\mathbf{\Sigma} \times \mathbf{g}) \tag{21.52}$$

which describes the drift and force of the scalar and vector part with the velocity (21.49) and the effective Lorentz force

$$\mathscr{F} = (e\mathbf{E} + e\mathbf{v} \times \mathbf{B} - \partial_r\Sigma_0). \tag{21.53}$$

The coupling between spinor parts is given by the vector drift

$$\mathscr{A}_i = (\partial_k\Sigma_i\partial_r - \partial_r\Sigma_i\partial_k + e(\partial_k\Sigma_i \times \mathbf{B}) \cdot \partial_k). \tag{21.54}$$

Remember that we subsumed in the vector selfenergy (21.44) the magnetic impurity meanfield, the spin-orbit coupling vector, and the Zeeman term.

The term (21.54) in the second parts on the left sides of (21.52) represent the coupling between the spin parts of the Wigner distribution. The vector part contains additionally the spin-rotation term on the right-hand side. These coupled meanfield kinetic equations including the magnetic and electric field, Zeeman coupling, and spin-orbit coupling are the final result of the section. On the right-hand side one has to additionally consider collision integrals which can be derived from the KB equation taking the selfenergy beyond the meanfield approximation. In the simplest way one approximates it by a relaxation time.

If one neglects the coupling of the scalar distribution f to the vector distribution $\mathbf{g}$ in the second equation of (21.52) one has the Eilenberger equation (Schwab, Dzierzawa, Gorini and Raimondi, 2006) extended here by magnetic and electric fields as well as selfenergy effects. Compared to (Meyerovich, 1989) we write both scalar and vector components and have included meanfield quasiparticle renormalisation and the vector selfenergy. The coupling of the vector equation to the scalar one has been neglected in (Leggett and Rice, 1968; Mineev, 2005) too but selfconsistent quasiparticle energies and Zeeman fields have been taken into account. In (Meyerovich, Stepaniants and Laloë, 1995; Golosov and Ruckenstein, 1995) only the transverse components have been considered and (Meyerovich, Stepaniants and Laloë, 1995) approximates two of the four degrees of freedom in (21.52). The same reduction of degrees of freedom at the beginning has been used by the projection technique in (Nacher, Tastevin and Laloë, 1991*b*,*a*) since the focus of all these papers had been on the proper collision integral instead. Selfconsistent quasiparticle equations have been presented in (Jeon and Mullin, 1989; Bashkin, daProvidencia and daProvidencia, 2000) which have been decoupled by Landau Fermi-liquid assumptions (Meyerovich and Musaelian, 1992) and variational approaches. The coupled kinetic equations have been derived without spin-orbit coupling terms and reduced vector selfenergies in (Williams, Nikuni and Clark, 2002) which disentangle the equation of moments. Here we present all these effects without the assumptions found in different places of the above-mentioned approaches.

21.3.5 Quasi stationary solution

The time-independent stationary solution should obey the stationary mean-field equation (21.52) since any collision integral is then zero yielding in the Fermi distribution. However, the arguments and functional dependence as well as spin structure of the solution are already determined from the stationary equation of the mean-field equation (21.52). We write them in formal notation

$$\mathscr{D}f + \mathscr{A} \cdot \mathbf{g} = 0$$

$$\mathscr{D}\mathbf{g} + \mathscr{A}f = 2(\mathbf{\Sigma} \times \mathbf{g}) \tag{21.55}$$

with $\mathscr{D} = \{\epsilon_k, ...\}$, $\mathscr{A}_i = \{\Sigma_i, ...\}$ and the Poisson bracket $\{a, b\} = \partial_k a \cdot \partial_r b - \partial_r a \cdot \partial_k b$. The electric field is given by a scalar potential $e\mathbf{E}(r) = -\nabla\Phi(r)$ such that we have the quasiparticle energy

$$\epsilon_k(r) = \frac{k^2}{2m_e} + \Sigma_0(k, r) + \Phi(r). \tag{21.56}$$

The choice of gauge is arbitrary since we have ensured that the kinetic equation is gauge invariant.

We rewrite (21.55) into one equation again by the spinor representation $\hat{\rho} = f + \sigma \cdot \mathbf{g}$ using the identity

$$\mathbf{c} \cdot \mathbf{g} + (\sigma \cdot \mathbf{c})f - 2\sigma \cdot (\sigma \times \mathbf{g}) = \sigma \cdot \frac{\mathbf{c} + 2i\sigma}{2}\hat{\rho} + \hat{\rho}\sigma \cdot \frac{\mathbf{c} - 2i\sigma}{2} \tag{21.57}$$

to arrive at

$$[\mathscr{D} + \sigma \cdot \mathscr{A}, \hat{\rho}]_+ + 2i[\sigma \cdot \mathbf{\Sigma}, \hat{\rho}]_- = 0 \tag{21.58}$$

which is equivalent to (21.55). Now we search for a solution which renders both the anti-commutator and commutator zero separately.

For the anti-commutator, the equation

$$(\mathscr{D} + \sigma \cdot \mathscr{A})\hat{\rho} = 0 \tag{21.59}$$

and correspondingly $\hat{\rho}(\mathscr{D} + \sigma \cdot \mathscr{A}) = 0$ are solved by any function of the argument

$$\hat{\rho}_0 \left[\epsilon_k(r) + \mathbf{\Sigma}(k, r) \cdot \sigma \right]. \tag{21.60}$$

Employing the relation

$$e^{\sigma \cdot \mathbf{\Sigma}} = \cosh |\mathbf{\Sigma}| + \sigma \cdot \mathbf{e} \sinh |\mathbf{\Sigma}| = \sum_{s=\pm} \hat{P}_s e^{s|\mathbf{\Sigma}|} \tag{21.61}$$

with the projectors $\hat{P}_\pm = \frac{1}{2}(1 \pm \mathbf{e} \cdot \sigma)$ and $\mathbf{e} = \mathbf{\Sigma}/|\mathbf{\Sigma}|$ we have to have the stationary solution in the form

$$\hat{\rho}_0 \left[\epsilon_k(r) + \mathbf{\Sigma}(k, r) \cdot \sigma \right] = \sum_{\pm} P_\pm \hat{\rho}_\pm (\epsilon_k \pm |\mathbf{\Sigma}|). \tag{21.62}$$

The demand of vanishing commutator in (21.58) works further down the still general possibility of distribution $\hat{\rho}_\pm = \bar{f}_\pm + \sigma \cdot \mathbf{g}_\pm$. In fact it demands

$$0 = [\sigma \cdot \mathbf{\Sigma}, \hat{\rho}]_- = [\sigma \cdot \mathbf{\Sigma}, \sigma \cdot \mathbf{g}]_- = i\sigma \cdot (\mathbf{\Sigma} \times \mathbf{g}) \tag{21.63}$$

which implies that $\mathbf{g} = \mathbf{e}g$ with the effective direction $\mathbf{e} = \mathbf{\Sigma}/|\mathbf{\Sigma}|$. Together with (21.62) we obtain the stationary solution of (21.58) and consequently of (21.55) to have the form (Morawetz, 2015a)

$$\hat{\rho}(\hat{\varepsilon}) = \sum_{\pm} \hat{P}_{\pm} f_{\pm} = \frac{f_+ + f_-}{2} + \sigma \cdot \mathbf{e} \, \frac{f_+ - f_-}{2} \equiv f + \sigma \cdot \rho \qquad (21.64)$$

with $f_{\pm} = \bar{f}_{\pm} + g_{\pm} = f_0(\epsilon_k(r) \pm |\boldsymbol{\Sigma}(k,r)|)$ and f_0 an unknown scalar function which is determined by the vanishing of the collision integral to be the Fermi–Dirac or Bose–Einstein distribution in equilibrium.

21.4 Normal and anomal currents

Due to the spin-orbit coupling (21.45) with (21.44), the current possesses a normal and anomaly part. Using $[\boldsymbol{\Sigma}(\hat{\mathbf{p}}), \hat{x}_j] = -i\hbar \partial_{p_j} \boldsymbol{\Sigma}(\hat{\mathbf{p}})$ we have

$$\hat{v}_j = \frac{i}{\hbar}[\hat{H}, \hat{x}_j] = \partial_{p_j}\epsilon + \partial_{p_j}\boldsymbol{\Sigma} \cdot \sigma \qquad (21.65)$$

if the single particle Hamiltonian is given by the quasiparticle energy $\epsilon(p)$. Together with the Wigner function (21.10) one has

$$\hat{\rho}\hat{v}_j = f\partial_{p_j}\epsilon + \mathbf{g} \cdot \partial_{p_j}\boldsymbol{\Sigma} + \sigma \cdot (\partial_{p_j}\epsilon\mathbf{g} + f\partial_{p_j}\boldsymbol{\Sigma} + i\partial_{p_j}\boldsymbol{\Sigma} \times \mathbf{g}) \qquad (21.66)$$

and the particle and pseudo-spin current densities read

$$\hat{j}_j = \sum_p [\hat{\rho}, v_j]_+ = 2\sum_p \left[f\partial_{p_j}\epsilon + \mathbf{g} \cdot \partial_{p_j}\boldsymbol{\Sigma} + \sigma \cdot (\partial_{p_j}\epsilon\mathbf{g} + f\partial_{p_j}\boldsymbol{\Sigma}) \right] = j_j + \sigma \cdot \mathbf{S}_j. \qquad (21.67)$$

The scalar part describes the particle current $\mathbf{j} = \mathbf{j}^n + \mathbf{j}^a$ consisting of a normal and anomalous current and the vector part describes the pseudo-spin current S_{ij} not to be confused with the polarisation $\mathbf{s}$.

The stationary solution allows one to learn about the seemingly cumbersome structure of the particle current (21.67) consisting of normal and anomalous parts. In fact both parts are necessary to guarantee the absence of particle currents in stationary spin-orbit coupled systems since both parts separately are nonzero and only their sum vanishes. To see this, we expand the normal particle current with (21.64) and $\boldsymbol{\Sigma} = \boldsymbol{\Sigma}_n + \mathbf{b}$ linear in the spin-orbit coupling $\mathbf{b}$ to get

$$j_i^n = \frac{1}{2}\sum_p \partial_{p_i}\epsilon \left[f(\epsilon_p + \Sigma) + f(\epsilon_p - \Sigma) \right] = \frac{1}{2}\sum_p \partial_{p_i}\epsilon \frac{\boldsymbol{\Sigma}_n \cdot \mathbf{b}}{\Sigma_n} \partial_\epsilon \left[f(\epsilon_p + \Sigma_n) - f(\epsilon_p - \Sigma_n) \right].$$

$$(21.68)$$

The anomalous current reads linearly in **b**

$$j_i^a = \frac{1}{2}\sum_p (\partial_{p_i}\mathbf{b}) \cdot \mathbf{e}\left[f(\epsilon_p + \Sigma) - f(\epsilon_p - \Sigma)\right] = \frac{1}{2}\sum_p \frac{\Sigma_n \cdot \partial_{p_i}\mathbf{b}}{\Sigma_n}\left[f(\epsilon_p + \Sigma) - f(\epsilon_p - \Sigma)\right].$$

(21.69)

Combining both currents one obtains

$$j_i = j_i^a + j_i^n = \frac{\Sigma_n}{\Sigma_n} \cdot \frac{1}{2}\sum_p \partial_{p_i}\left\{\mathbf{b}\left[f(\epsilon_p + \Sigma) - f(\epsilon_p - \Sigma)\right]\right\} = 0$$

(21.70)

as one should. This demonstrates the importance of the anomalous current. One can consider the spin-orbit coupling as a continuous current of normal quasiparticles compensated by the spin-induced one. Any disturbance and linear response will lead to interesting effects due to this disturbed balance such as the anomalous Hall and spin-Hall effects.

21.5   Linear response

21.5.1   Linearisation to external electric field

We will now consider the linearisation of kinetic equation (21.52) with respect to an external electric field. We Fourier transform the time $\partial_t \to -i\omega$ and the spatial coordinates $\partial_r \to i\mathbf{q}$. The distribution is linearised according to $\hat{\rho}(p,r,t) = f(p) + \delta f(p,r,t) + \sigma \cdot [\mathbf{g}(p) + \delta\mathbf{g}(p,r,t)]$ due to the external electric field perturbation $e\delta\mathbf{E} = e\mathbf{E}(r,t) = -\nabla\Phi$. The magnetic field consists of a constant and an induced part $B(r,t) = B + \delta B(r,t)$. Since the external electric field perturbation is produced due to an external potential U^{ext} one sees from the Maxwell equation that $\delta\dot{\mathbf{B}} = -\nabla \times \delta\mathbf{E} = 0$, which means that all terms linear in the induced magnetic field vanish. It is convenient to work in velocity variables instead of momentum defined ones, according to (21.49) and we use $\epsilon_p = p^2/2m_e + \Sigma_0$ for the quasiparticle energy. Further we assume a collision integral of the relaxation time approximation (Hakim, Mornas, Peter and Sivak, 1992)

$$-\frac{1}{2}[\hat{\tau}^{-1}, \delta\hat{\rho}^l]_+$$

(21.71)

with a vector and scalar part of relaxation times $\hat{\tau}^{-1} = \tau^{-1} + \sigma \cdot \tau^{-1}$ and

$$\tau = \frac{\tau^{-1}}{\tau^{-2} - |\tau^{-1}|^2}, \quad \tau = -\frac{\tau^{-1}}{\tau^{-2} - |\tau^{-1}|^2}.$$

(21.72)

Scalar relaxation is assumed, not towards the absolute equilibrium $f_0(\epsilon \pm |\Sigma| - \mu)$ characterised by the chemical potential μ, but towards a local one $f^l = f_0(\epsilon \pm |\Sigma| - \mu - \delta\mu)$. The latter one can be specified as

$$\delta n = \sum_p (f - f_0) = \sum_p (f - f^l + f^l - f_0) = \sum_p (f^l - f_0) = \partial_\mu n \, \delta\mu \qquad (21.73)$$

such that the density is conserved (Mermin, 1970; Das, 1975) as expressed in the step to the second line, see Chapter 18.1. Therefore the relaxation term becomes

$$-\frac{\delta\hat{\rho}^l}{\tau} = -\frac{\delta\hat{\rho}}{\tau} + \frac{\delta n}{\tau \partial_\mu n} \partial_\mu \hat{\rho}_0. \qquad (21.74)$$

In this way the density is conserved in the response function which could be extended to included more conservation laws (Morawetz, 2002, 2013).

In order to facilitate the vector notation we want to understand $q\partial_p = \mathbf{q} \cdot \partial_{\mathbf{p}}$ in the next section. We obtain from (21.52) with the velocity $v = \partial_p \epsilon_p$

$$\left[-i\omega + iqv + \frac{1}{\tau} + (v \times eB)\,\partial_p \right] \delta f + \left[iq\partial_p \Sigma_i + (\tau^{-1})_i - (\partial_p \Sigma_i \times eB)\,\partial_p \right] \delta g_i = S_0 \quad (21.75)$$

and

$$\left[-i\omega + iqv + \frac{1}{\tau} + (v \times eB)\,\partial_p \right] \delta g_i - 2(\Sigma \times \delta g)_i$$

$$+ \left[iq\partial_p \Sigma_i + (\tau^{-1})_i - (\partial_p \Sigma_i \times eB)\,\partial_p \right] \delta f = S_i \qquad (21.76)$$

with the source terms arising from the external field $\mathbf{q}\Phi = ie\mathbf{E}$ and the induced meanfield variations (21.39)

$$S_0 = iq\partial_p f\Phi + \frac{\delta n}{\tau \partial_\mu n} \partial_\mu f_0 + \left(iq + eB \times \partial_p^\Sigma \right) \delta\Sigma_0 \partial_p f + \left(iq - eB \times \partial_p^\Sigma \right) \delta\Sigma_i \partial_p g_i$$

$$S_i = iq\partial_p g_i\Phi + 2(\delta\Sigma \times g)_i + \frac{\delta n \partial_\mu g}{\tau \partial_\mu n} e_i + \left(iq + eB \times \partial_p^\Sigma \right) \delta\Sigma_0 \partial_p g_i + \left(iq - eB \times \partial_p^\Sigma \right) \delta\Sigma_i \partial_p f.$$

$$(21.77)$$

Due to the magnetic field we see that the sources get additional rotation terms coupled with the momentum-dependent derivation of meanfields which is present only with spin-orbit coupling.

21.5.2   Conductivities without magnetic fields

It is interesting to note that the coupled kinetic equation (21.52) for a homogeneous system neglecting magnetic fields

$$(\partial_t + eE\partial_p)f = 0, \qquad (\partial_t + eE\partial_p)\mathbf{g} = 2(\Sigma \times \mathbf{g}) \qquad (21.78)$$

allows for a finite conductivity even without collisions and a Hall effect without an external magnetic field. This is due to the interference between the two-fold splitting of the band (21.64) and will be the reason for the anomalous Hall effect. The linearised solution of (21.78) one gets from (21.76) with $E\partial_p = \mathbf{E}_\omega \cdot \partial_p$

$$\delta\mathbf{g}(\omega, p) = \frac{i\omega}{4|\Sigma|^2 - \omega^2} eE\partial_p\mathbf{g} - 2\frac{1}{4|\Sigma|^2 - \omega^2}\Sigma \times eE\partial_p\mathbf{g} - 4i\frac{1}{\omega(4|\Sigma|^2 - \omega^2)}\Sigma(\Sigma \cdot eE\partial_p\mathbf{g}).$$
(21.79)

Each term from (21.79) corresponds therefore to a specific precession motion analogously to the one seen in the conductivity of a charge in crossed electric and magnetic fields. If we replace $\omega_c = eB/m \leftrightarrow 2|\Sigma|$ one has just (21.19) which was the solution of the Newton equation of motion (21.12). It illustrates the threefold orbiting of the electrons.

According to (21.67) we have therefore the anomaly current

$$\delta j_\alpha^a = \sum_p \partial_{p_\alpha}\Sigma \cdot \delta\mathbf{g} = \sum_p \Sigma\partial_{p_\alpha}\mathbf{e} \cdot \delta\mathbf{g} + \sum_p (\mathbf{e} \cdot \partial_{p_\alpha}\Sigma)\mathbf{e} \cdot \delta\mathbf{g}$$
(21.80)

and note that $\partial_p\Sigma = \partial_p\mathbf{b}$. Writing $j_\alpha = \sigma_{\alpha\beta}E_\beta$ we obtain from the first part of (21.80) using (21.79) the interband and the anomalous Hall conductivity

$$\sigma_{\alpha\beta}^{\text{inter}} = 2e^2 \sum_p \frac{g}{1 - \frac{\omega^2}{4|\Sigma|^2}} \frac{i\omega}{2|\Sigma|} \partial_{p_\alpha}\mathbf{e} \cdot \partial_{p_\beta}\mathbf{e}$$

$$\sigma_{\alpha\beta}^{\text{Hall}} = 2e^2 \sum_p \frac{g}{1 - \frac{\omega^2}{4|\Sigma|^2}} \mathbf{e} \cdot (\partial_{p_\alpha}\mathbf{e} \times \partial_{p_\beta}\mathbf{e})$$
(21.81)

where the first part of (21.79) leads to the interband and the second part of (21.79) to the Hall conductivity. The latter one is exactly the expression one gets from the Kubo–Bastin–Streda formula (Streda and Jonckheere, 2010) using the Berry phase connection, see Chapter V. in (Morawetz, 2015*a*). The second part of (21.80) with (21.79) leads to the intraband conductivity

$$\sigma_{\alpha\beta}^{\text{intra}} = i2e^2 \sum_p \partial_{p_\alpha}\partial_{p_\beta}\Sigma\frac{g}{\omega}$$
(21.82)

where the first and the third part of (21.79) have contributed (Morawetz, 2016).

21.5.3 Collective modes

Multiplying the linearised kinetic equations (21.75) and (21.76) with powers of momentum and integrating one obtains a coupled hierarchy of moments. A large variety of treatments neglect certain Landau-liquid parameters (Mineev, 2004) based on the work of (Leggett, 1970) in order to close such system. A more advanced closing procedure was provided by (Mineev, 2005) where the energy dependence of δs was assumed to be factorised from space and direction $\mathbf{p}$ dependencies.

One does not need to follow these approximations but can solve the linearised equation exactly (Morawetz, 2015a) to provide the solution of the balance equations and the dispersion. Amazingly, this yields a quite involved and extensive structure with many more terms than are usually presented (Morawetz, 2015b). It is instructive to first have a look at the balance equation for the densities from (21.75) and (21.76)

$$\partial_t \delta n + \partial_{x_i} j_i + \tau^{-1} \cdot \delta \mathbf{s} = 0$$

$$\partial_t \delta \mathbf{s} + \partial_{x_i} \mathbf{S}_i + \tau^{-1} \delta n - 2 \sum_p \boldsymbol{\Sigma} \times \delta \mathbf{g} = 2 \delta \boldsymbol{\Sigma} \times \mathbf{s} \qquad (21.83)$$

where we Fourier transformed the wave vector q back to spatial coordinates x. We are now interested in the long wave-length limit $q \to 0$, which means we neglect any spatial derivative in (21.83). Alternatively, we might consider this as the spatially integrated values providing the change of number of particles and magnetic moment (21.11)

$$N = \int d^3 x\, n(x) = n_{q=0}, \quad \mathbf{m} = \gamma \int d^3 x\, \mathbf{s}(x) = \gamma \mathbf{s}_{q=0}. \qquad (21.84)$$

With the help of the first equation of (21.83) we get the closed equation for the magnetisation from (21.83)

$$i \omega \delta \mathbf{m} + \frac{i}{\omega} \left(\tau^{-1} + \frac{2}{\gamma} \mathbf{m} \times \mathbf{V} \right) (\tau^{-1} \cdot \delta \mathbf{m}) + 2(N\mathbf{V} + \mu_B \mathbf{B}) \times \delta \mathbf{m} = 2\gamma \sum_p \delta \mathbf{g}_{q=0} \times \mathbf{b}.$$

$$(21.85)$$

On the right-hand side, all terms are collected that are coming from the explicit knowledge of the solution $\delta \mathbf{g}$ needed to evaluate this sum over the momentum-dependent spin-orbit term $\mathbf{b}$. Please compare to the Landau–Lifshitz–Gilbert equation (21.6) to identify the microscopic sources of the different terms there.

The separation of the balance equation in the form (21.85) has the merit of already seeing the collective spin-mode structure. Since we have $\mathbf{m} = m\mathbf{e}_Z$ and $\mathbf{V} = V\mathbf{e}_Z$ the equation for the magnetisation becomes

$$\begin{pmatrix} -i\omega & 2(NV + \mu_B B) & 0 \\ -2(NV + \mu_B B) & -i\omega & 0 \\ 0 & 0 & -i\omega \end{pmatrix} \delta \mathbf{m} = 2 \sum_p \mathbf{b} \times \delta \mathbf{g}_{q=0} \qquad (21.86)$$

neglecting the quadratic terms of the vector relaxation times τ^{-1}.

Inverting (21.86) provides the solution of the magnetisation change provided we know the solution of δg on the right-hand side. Interestingly, this inversion is only possible for a nonzero determinant. The vanishing determinant provides therefore the two spin waves with some possible further modifications due to the spin-orbit coupling

$$\omega_{\text{spin}} = \pm 2|NV + \mu_B B| \tag{21.87}$$

which shows linear splitting due to the driving external magnetic field and the permanent magnetisation $\mathbf{V} = V\mathbf{e}_z$.

21.6   Response with magnetic fields

In order to solve (21.75) and (21.76) we use the same coordinate system as Bernstein (Bernstein, 1958). The magnetic field $\mathbf{B}$ points in the v_z direction and the q vector is in the $v_z - v_x$ plane with an angle Θ between v_x and q

$$\mathbf{q} = q\sin\Theta\mathbf{e}_x + q\cos\Theta\mathbf{e}_z. \tag{21.88}$$

For the velocity v we use polar coordinates around $\mathbf{B}$ with an azimuthal angle ϕ

$$\mathbf{v}(\phi) = w\cos\phi\mathbf{e}_x + w\sin\phi\mathbf{e}_y + u\mathbf{e}_z \tag{21.89}$$

and one gets

$$\frac{1}{\tau} - i\omega + i\mathbf{q}\mathbf{v} + \left(\mathbf{v} \times \frac{e\mathbf{B}}{m}\right)\partial_{\mathbf{v}} = \frac{1}{\tau} - i\omega + i\mathbf{q} \cdot \mathbf{v}(\phi) - \omega_c\partial_\phi$$

$$= \frac{1}{\tau} - i\omega + i\mathbf{q} \cdot \mathbf{v}(\omega_c t_\phi) - \partial_t \equiv -i\Omega_{t_\phi} - \partial_{t_\phi} \tag{21.90}$$

with the orbiting time

$$t_\phi = \phi/\omega_c. \tag{21.91}$$

We can write the equations (21.75) and (21.76) as

$$\left(-i\Omega_{t_\phi} - \partial_{t_\phi}\right)\delta f + \left(\frac{iq\partial_v\Sigma}{m_e} + \tau^{-1}\right) \cdot \delta g = S_0$$

$$\left(-i\Omega_{t_\phi} - \partial_{t_\phi}\right)\delta g_i + \left(\frac{iq\partial_v\Sigma_i}{m_e} + (\tau^{-1})_i\right)\delta f - 2(\boldsymbol{\Sigma} \times \delta g)_i = S_i. \tag{21.92}$$

where the corresponding right-hand sides are given by (21.77). Now we employ the identity

$$\mathcal{G} \cdot \delta\mathbf{g} + (\sigma \cdot \mathcal{G})\delta f - 2\sigma \cdot (\mathbf{\Sigma} \times \delta\mathbf{g}) = \sigma \cdot \frac{\mathcal{G} + 2i\mathbf{\Sigma}}{2}\delta\hat{F} + \delta\hat{F}\sigma \cdot \frac{\mathcal{G} - 2i\mathbf{\Sigma}}{2} \quad (21.93)$$

with $\mathcal{G} = iq\partial_v \mathbf{\Sigma}/m_e + \tau^{-1}$ and $\delta\hat{F} = \delta f + \sigma \cdot \delta\mathbf{g}$ which one proves with the help of $(\tau a)(\tau b) = a \cdot b + i\tau(a \times b)$. This allows us to rewrite (21.92) into

$$-\partial_{t_\phi}\delta F - i\Omega_{t_\phi}\delta F + \sigma \cdot \left(\frac{\mathcal{G}}{2} + i\mathbf{\Sigma}\right)\delta\hat{F} + \delta\hat{F}\sigma \cdot \left(\frac{\mathcal{G}}{2} - i\mathbf{\Sigma}\right) = \hat{S}_{p_\phi}(\omega) \quad (21.94)$$

where $\hat{S}_{p_\phi} = S_0 + \sigma \cdot \mathbf{S}$.

Please note that due to (21.91) the integration over the azimuthal angle is translated into the time integration about orbiting intervals. Equation (21.94) is readily solved as

$$\delta F = -\int\limits_\infty^t d\bar{t}\, e^{\frac{i}{i}\int\limits_{\bar{t}}^{t}\Omega_{t'}^+ dt'}\, e^{\sigma\int\limits_{\bar{t}}^{t}\left(\frac{\mathcal{G}_{t'}}{2} + i\mathbf{\Sigma}_{t'}\right)dt'}\, \hat{S}_{p_{\omega_c}\bar{t}}\, e^{\sigma\int\limits_{\bar{t}}^{t}\left(\frac{\mathcal{G}_{t'}}{2} - i\mathbf{\Sigma}_{t'}\right)dt'}$$

$$= \int\limits_{-\infty}^{0} dx\, e^{-i\int\limits_{0}^{x}\Omega_{t-y}^+ dy}\, e^{\sigma\int\limits_{0}^{x}\left(\frac{1}{2}\mathcal{G}_{t-y} + i\mathbf{\Sigma}_{t-y}\right)dy}\, \hat{S}_{p_{\omega_c}(t-x)}(\omega)e^{\sigma\int\limits_{0}^{x}\left(\frac{1}{2}\mathcal{G}_{t-y} - i\mathbf{\Sigma}_{t-y}\right)dy} \quad (21.95)$$

where we used $\omega^+ = \omega + i\tau^{-1}$ as before. The first exponent can be calculated explicitly with the definitions of (21.89) and (21.90)

$$i\int\limits_{0}^{x}\Omega_{t-y}^+ dy = i\omega^+ x - i\mathbf{q} \cdot \mathcal{R}_x \cdot \mathbf{v}(t) \quad (21.96)$$

with the matrix (Walter, Zwicknagel and Toepffer, 2005)

$$\mathcal{R}_x = \frac{1}{\omega_c}\begin{pmatrix} \sin\omega_c x & 1 - \cos\omega_c x & 0 \\ \cos\omega_c x - 1 & \sin\omega_c x & 0 \\ 0 & 0 & \omega_c x \end{pmatrix} \quad (21.97)$$

having the property $\mathcal{R}_{-x} = -\mathcal{R}_x^T$.

We employ the long-wave-length approximation and neglect the vector relaxation, $\mathcal{G} \approx 0$. The integration over an azimuthal angle $x = \phi/\omega_c$ is coupled to the momentum (velocity) arguments. The spin-orbit coupling provides a momentum-dependent $\boldsymbol{\Sigma}$ which couples basically to $q\partial_p\boldsymbol{\Sigma}$. Since

$$m_e q\partial_p = q\sin\Theta\left(\cos\phi\partial_w - \frac{\sin\phi}{w}\partial_\phi\right) + q\cos\Theta\partial_u \tag{21.98}$$

in the coordinates (21.88) and (21.89), it means we neglect higher than first-order derivatives in ϕ and $\partial_p\boldsymbol{\Sigma}$ when approximating $\boldsymbol{\Sigma}_{t-y} \approx \boldsymbol{\Sigma}_t$ in the exponent. We obtain

$$\delta\hat{F} = \delta f + \sigma\cdot\delta\mathbf{g} = \int_{-\infty}^{0} dx e^{i[q\mathscr{R}_x \mathbf{v}(t)-\omega^+ x]} e^{ix\sigma\cdot\boldsymbol{\Sigma}(t)}\hat{S}_{p_{\omega_c}(t-x)} e^{-ix\sigma\cdot\boldsymbol{\Sigma}(t)}. \tag{21.99}$$

To work it out further we use $e^{i\tau\cdot a} = \cos|a| + i\frac{\tau\cdot a}{|a|}\sin|a|$ to see that

$$e^{i\sigma\cdot\boldsymbol{\Sigma}x}(S_0 + \sigma\cdot\mathbf{S})e^{-i\sigma\cdot\boldsymbol{\Sigma}x} = S_0 + (\sigma\cdot\mathbf{S})\cos(2x|\boldsymbol{\Sigma}|)$$

$$+\ \sigma(\mathbf{S}\times\mathbf{e})\sin(2x|\boldsymbol{\Sigma}|) + (\sigma\cdot\mathbf{e})(\mathbf{S}\cdot\mathbf{e})(1-\cos(2x|\boldsymbol{\Sigma}|)) \tag{21.100}$$

with the direction $\mathbf{e} = \boldsymbol{\Sigma}/|\boldsymbol{\Sigma}|$ and (21.44).

The effect of a magnetic field is basically condensed in two places. First the phase term $\mathbf{q}\cdot\mathscr{R}_x\cdot\mathbf{p} = \mathbf{q}\cdot\mathbf{p} + o(B, q^2)$ and then we have $\bar{\omega} = \omega - \mathbf{p}\cdot\mathbf{q}/m + i/\tau$

$$\int_{-\infty}^{0} e^{-i(\omega x - \mathbf{q}\cdot\mathscr{R}_x\cdot\frac{\mathbf{p}}{m})}\begin{pmatrix}\cos 2|\boldsymbol{\Sigma}|x \\ 1-\cos 2|\boldsymbol{\Sigma}|x \\ -\sin 2|\boldsymbol{\Sigma}|x\end{pmatrix} = \frac{1}{2}\begin{pmatrix}\frac{i}{\bar{\omega}+2\boldsymbol{\Sigma}}+\frac{i}{\bar{\omega}-2\boldsymbol{\Sigma}} \\ \frac{2i}{\bar{\omega}}-\frac{i}{\bar{\omega}+2\boldsymbol{\Sigma}}-\frac{i}{\bar{\omega}-2\boldsymbol{\Sigma}} \\ \frac{1}{\bar{\omega}-2\boldsymbol{\Sigma}}-\frac{1}{\bar{\omega}+2\boldsymbol{\Sigma}}\end{pmatrix} + o(q^2) = \begin{pmatrix}\frac{i}{\bar{\omega}} \\ 0 \\ \frac{2\boldsymbol{\Sigma}}{\omega^2}\end{pmatrix} + o(B, \boldsymbol{\Sigma}^2).$$

$$\tag{21.101}$$

The magnetic-field-dependent phase factor $\mathscr{R}_x$ does play a role only in inhomogeneous systems with finite wave length. In the limit of a large wave-length this effect can be ignored. The *sin* and *cos* terms are results of the precession of spins around the effective direction $\mathbf{e} = \boldsymbol{\Sigma}/\boldsymbol{\Sigma}$ and can be considered as Rabi oscillations. For the limit of small $\boldsymbol{\Sigma}$ we can expand the *cos* and *sin* terms in first order and it leads to THz out-of-plane signals (Morawetz, K., 2013).

The second effect is the retardation in $t = \phi/\omega_c$ which means that the precession time in the arguments $S(t - x)$ contains important magnetic field effects. In fact, this retardation represents all kinds of normal-Hall effects as we will convince ourselves now.

21.6.1   Retardation subtleties by magnetic field

The magnetic field causes a retarding integral in Section 21.6 over the precession time $t = \phi/\omega_c$ coupled to any momentum by the representation in Bernstein coordinates

$$\frac{\mathbf{p}_\phi}{m} = (w\cos\phi, w\sin\phi, u). \tag{21.102}$$

This retardation is crucial for any kind of Hall effect. In order to get a handle on such expressions we concentrate first on the mean values of the scalar part δf. The general field-dependent solution provides the form

$$\langle A \rangle = \sum_{p_\phi} \int_{-\infty}^{0} dx\, e^{-i(\omega^+ x - q\mathscr{R}_x \frac{\mathbf{p}_\phi}{m})} A(\mathbf{p}_\phi) S_0(p_{\phi - x\omega_c}) \tag{21.103}$$

where S_0 is the scalar source term (21.77). The trick is to perform first a shift $\phi \to \phi + \omega_c x$ and integrate then about $p = p_\phi$. This has the effect that the retardation is only condensed in the momentum of variable A

$$P(x) = \mathbf{p}_{\phi + \omega_c x} = \mathbf{p}_\phi \cos(\omega_c x) + \mathbf{e}_z(\mathbf{e}_z \cdot \mathbf{p}_\phi)[1 - \cos(\omega_c x)] + \mathbf{e}_z \times \mathbf{p}_\phi \sin(\omega_c x)$$

$$= \mathbf{p}_\phi + \mathbf{e}_z \times \mathbf{p}_\phi \omega_c x + o(\omega_c^2). \tag{21.104}$$

and the exponent

$$\mathscr{R}_x \frac{\mathbf{p}_{\phi + \omega_c x}}{m} = \frac{\mathbf{p}_\phi}{m}\frac{\sin\omega_c x}{\omega_c} + \left(\mathbf{e}_z \times \frac{\mathbf{p}_\phi}{m}\right)\frac{1 - \cos\omega_c x}{\omega_c} = \frac{\mathbf{p}_\phi}{m}x + \omega_c\frac{x^2}{2}\mathbf{e}_z \times \frac{\mathbf{p}_\phi}{m} + o(\omega_c^2). \tag{21.105}$$

The phase effect leads to the first-order corrections in ω_c or alternatively in wave-length q

$$\langle A \rangle = \sum_p S_0(p)\left[1 - \frac{\omega_c}{2m}\mathbf{q}\cdot(\mathbf{e}_z \times \mathbf{p})\partial_\omega^2 + o(\omega_c^2 q^2)\right]\int_{-\infty}^{0} dx\, e^{-i(\omega^+ x - \frac{\mathbf{q}\cdot\mathbf{p}}{m})} A[P(x)]$$

$$= \sum_p S_0(p) A[P(i\partial_\omega)]\left[1 - \frac{\omega_c}{2m}\mathbf{q}\cdot(\mathbf{e}_z \times \mathbf{p})\partial_\omega^2 + o(\omega_c^2 q^2)\right]\frac{i}{\omega^+ - \frac{\mathbf{q}\cdot\mathbf{p}}{m}}. \tag{21.106}$$

where the integration variable x in the momentum (21.104) can be transformed into derivatives of ω if needed. Completely analogously we can perform any mean value over the vector part of the distribution δg. We have

$$\langle \mathbf{A} \rangle = \sum_p A(p) \delta \mathbf{g} = \sum_p \left[1 - \frac{\omega_c}{2m} \mathbf{q} \cdot (\mathbf{e}_z \times \mathbf{p}) \, \partial_\omega^2 + o(\omega_c^2 q^2) \right] \int_{-\infty}^{0} dx e^{i(\frac{q\mathbf{p}}{m} x - \omega^+ x + o(\omega_c, q^2))} A[P(x)]$$

$$\times [\mathbf{S} \cos(2\Sigma x) + \mathbf{e} \times \mathbf{S} \sin(2\Sigma x) + \mathbf{e}(\mathbf{e} \cdot \mathbf{S})(1 - \cos(2\Sigma x))] \tag{21.107}$$

where the arguments of $\mathbf{S}$, Σ and $\mathbf{e}$ are the momentum p and there is no retardation anymore. The exponent can be written in complete B-dependence with $\mathscr{R}_x$ of course. Then the x-integration over the *cos* and *sin* terms has to be performed numerically. Analytically we can proceed if we expand the phase effect in the orders of $\mathbf{q}$. We obtain, with the help of (21.101),

$$\langle \mathbf{A} \rangle = \sum_p A[P(i\partial_\omega)] \left[1 - \frac{\omega_c}{2m} \mathbf{q} \cdot (\mathbf{e}_z \times \mathbf{p}) \, \partial_\omega^2 + o(\omega_c^2) \right]$$

$$\times \left[[\mathbf{e} \times (\mathbf{S} \times \mathbf{e})] \frac{i}{2} \left(\frac{1}{\bar{\omega} + 2\Sigma} + \frac{1}{\bar{\omega} - 2\Sigma} \right) + \mathbf{e} \times \mathbf{S} \frac{1}{2} \left(\frac{1}{\bar{\omega} + 2\Sigma} - \frac{1}{\bar{\omega} - 2\Sigma} \right) + \mathbf{e}(\mathbf{e} \cdot \mathbf{S}) \frac{i}{\bar{\omega}} \right] \tag{21.108}$$

with $\bar{\omega} = \omega + \frac{i}{\tau} - \frac{\mathbf{pq}}{m}$ and (21.104). The formulas (21.106) and (21.108) establish the rules for calculating mean values with magnetic fields.

21.6.2   Classical-Hall effect

Now, we are in a position to see how the Hall effect is buried in the kinetic theory. Therefore, we neglect any mean-field and spin-orbit coupling for the moment such that the f and g distributions decouple and use the $q \to 0$ limit, i.e. a homogeneous situation. We obtain from (21.95) with (21.100) and (21.77)

$$\delta f = - \int_{-\infty}^{0} dx e^{-i\omega^+ x} e\mathbf{E} \cdot \partial_{\mathbf{p}_\phi} f(p_{\phi - \omega_c x}) \tag{21.109}$$

where we now pay special care to the retardation since this provides the Hall effect which was overseen in many treatments of magnetised plasmas.

After the shift of coordinates in the azimuthal angle ϕ, as outlined in Section 21.6.1, we can carry out the x-integration with the help of (21.104), (21.106) and (21.101):

$$\mathbf{J} = e \sum_{p_\phi} \frac{\mathbf{p}_\phi}{m_e} \delta f = -e^2 \sum_{p_\phi} \mathbf{E} \cdot \partial_{\mathbf{p}_\phi} f(p_\phi) \int_{-\infty}^{0} dx e^{-i\omega^+ x} \frac{\mathbf{p}_{\phi + \omega_c x}}{m}$$

$$= \sigma_0 \frac{1 - i\omega\tau}{(1 - i\omega\tau)^2 + (\omega_c\tau)^2} \left[\mathbf{E} + \frac{(\omega_c\tau)^2}{(1 - i\omega\tau)^2} (\mathbf{E} \cdot \mathbf{e}_z)\mathbf{e}_z + \frac{\omega_c\tau}{1 - i\omega\tau} \mathbf{E} \times \mathbf{e}_z \right] \tag{21.110}$$

which agrees with the elementary solution of (21.12).

In order to obtain all three precession terms we have used the complete form (21.104) and no expansion in ω_c.

21.6.3 Quantum-Hall effect

If we consider low temperatures such that the motion of electrons become quantised in Landau levels we have to use the quantum kinetic equation (10.15) and not the quasi-classical one. However, we can establish a simple *re-quantisation rule* which allows us to translate the previously discussed quasi-classical results into the quantum expressions. Therefore we recall the linearisation of the quantum-Vlasov equation, which is the quantum kinetic equation with only the mean field in operator form

$$\dot{\rho} - \frac{i}{\hbar}[\rho, H] = 0. \tag{21.111}$$

The perturbing Hamiltonian due, to external electric fields, is $\delta H = e\mathbf{E} \cdot \hat{\mathbf{x}}$ such that the linearisation in eigenstates E_n of the unperturbed Hamiltonian reads

$$\delta\rho_{nn'} = -e\mathbf{E} \cdot \mathbf{x}_{nn'} \frac{\rho_n - \rho_{n'}}{\hbar\omega - E_n + E_{n'}}. \tag{21.112}$$

One obtains the same result in the vector gauge since $[\rho, \delta H] = \frac{eEt}{m_e}[\rho, \mathbf{p}] = e\mathbf{E}[\rho, \mathbf{v}]$ and the same matrix elements appear. Now, we investigate the quasiclassical limit where the momentum states are proper representations. We chose

$$\langle n| = \langle p_1| = \langle p + \frac{q}{2}|, \quad |n\rangle' = |p_2\rangle = |-p + \frac{q}{2}\rangle \tag{21.113}$$

and have in the quasiclassical $q \to 0$ approximation

$$\mathbf{x}_{nn'}(\rho_n - \rho_{n'}) = \frac{\hbar}{i}\partial_q\delta(\mathbf{q})(\rho_{p+\frac{q}{2}} - \rho_{p-\frac{q}{2}}) \approx \frac{\hbar}{i}\partial_q\delta(\mathbf{q})\mathbf{q} \cdot \partial_p\rho_p = -\frac{\hbar}{i}\delta(\mathbf{q})\partial_p\rho_p \tag{21.114}$$

from which follows

$$\delta\rho \approx -i\hbar \frac{e\mathbf{E} \cdot \partial_p\rho}{\hbar\omega - \frac{\mathbf{p}\cdot\mathbf{q}}{m_e}}. \tag{21.115}$$

This is precisely the quasiclassical result we obtain from quasiclassical kinetic equations. Turning the argument around we see that we can re-quantise our quasiclassical results by applying the rule

$$\mathbf{E} \cdot \partial_p f \to \mathbf{E} \cdot \mathbf{v}_{nn'} \frac{f_n - f_{n'}}{E_{n'} - E_n}. \tag{21.116}$$

Let us exploit this in the normal-Hall conductivity. The calculation in 3D can be found in (Vliet and Vasilopoulos, 1988). Here, we represent the 2D calculation. We use the area density $1/A$ and re-normalise the level distribution $\sum_n f_n = 1$ to obtain for the static conductivity $\omega = 0$

$$\sigma_{\alpha\beta} = \frac{e^2 \hbar i}{A} \sum_{nn'} f_n (1 - f_{n'}) \frac{1 - e^{\beta(E_n - E_{n'})}}{(E_n - E_{n'})^2} v^\alpha_{nn'} v^\beta_{n'n} \tag{21.117}$$

which is nothing but the Kubo formula. Further evaluation for Landau levels has been performed by Vasilopoulos (Vasilopoulos, 1985; Vliet and Vasilopoulos, 1988). Therefore one chose the gauge $\mathbf{A} = (0, Bx, 0)$ and the corresponding energy levels are

$$E_n = \left(n + \frac{1}{2}\right) \hbar \omega_c + \frac{p_z^2}{2m_e} \tag{21.118}$$

where the last term is only in 3D. The wave functions read

$$|n\rangle = \frac{1}{\sqrt{A}} \phi_n(x + x_0) e^{i p_y y / \hbar} e^{i p_z z / \hbar} \tag{21.119}$$

with the harmonic oscillator functions ϕ_n, $x_0 = l^2 p_y / \hbar$, $l^2 = \hbar/eB$, and $A = L_y L_z$ where the corresponding z parts are absent in 2D. One easily obtains

$$v_{nn'} v_{n'n} = \frac{i \hbar \omega_c}{2m} \left[n \delta_{n',n-1} - (n+1)\delta_{n',n+1} \right] \delta(p_y - p'_y). \tag{21.120}$$

Introducing this into (21.117) and using

$$\sum_{p_y} = \frac{L_y}{2\pi \hbar} \int_{-\frac{\hbar L_x}{2l^2}}^{\frac{\hbar L_x}{2l^2}} dp_y = \frac{A}{2\pi l^2} \tag{21.121}$$

one arrives for $T \to 0$ at

$$\frac{e^2}{h} \sum_{n'} (n' + 1) f_{n'} (1 - f_{n'+1}) \left(1 - e^{-\beta \hbar \omega_c}\right) \to \frac{e^2}{h} (\bar{n} + 1) \tag{21.122}$$

with $\bar{n} \omega_c \le \epsilon_f \le (\bar{n} + 1)\omega_c$. This is von Klitzing's result of quantised Hall conductivity.

21.7 Spin-Hall effect, conductivity of graphene

21.7.1 Dirac (Weyl) dispersion

We can now reproduce the results for Dirac particles to the limit

$$\epsilon_\pm = \frac{p^2}{2m} + \Sigma_0 \pm |\mathbf{\Sigma}| \to \pm vp \tag{21.123}$$

which takes the form of single-layer graphene for $m \to \infty$ and a spin-orbit coupling $\mathbf{\Sigma} = \mathbf{b} = v\mathbf{p}$. This means

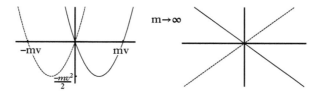

Please note that during this limit the bounded dispersion from below turns into an unbounded Luttinger-type of dispersion. Therefore this limit changes the structure of equations in an nontrivial way. In fact we will see that various limits cannot be interchanged with this infinite-mass limit.

In the limit of infinite mass, $\epsilon_p \to 0$ and therefore $\partial_{p_j}\epsilon \to 0$ we obtain that graphene can only possess an anomalous particle current (21.67) since the normal one would be of Drude type vanishing for $m \to \infty$. The normal pseudo-spin current, however, possesses a finite $m \to \infty$ limit which is unexpected. We will treat this in Section 21.7.4.

For graphene we have $\partial_{p_\alpha}\Sigma_i = v\delta_{\alpha,i}$ and the anomalous current (21.80) reads together with the linear response (21.79) to any time-dependent electric field (Katsnelson, 2006)

$$\delta j^a(t) = 2ev\sum_p \delta g = -2ev\sum_p \int_0^t d\bar{t}\, e^{-\frac{\bar{t}}{\tau}} \Bigg\{ \cos(2vp\bar{t})E_{t-\bar{t}}\partial_p \mathbf{g}$$

$$+ \sin(2vp\bar{t})\mathbf{p}_0 \times E_{t-\bar{t}}\partial_p\mathbf{g} + [1 - \cos(2vp\bar{t})] \left[\mathbf{p}_0 \cdot E_{t-\bar{t}}\partial_p\mathbf{g} \right] \mathbf{p}_0 \Bigg\} \tag{21.124}$$

with $\mathbf{p}_0 = \mathbf{p}/|\mathbf{p}|$. Each term from (21.79) corresponds therefore to a specific precession motion analogous to the one seen in the conductivity of a charge in crossed electric and magnetic fields. If we replace $\omega_c = eB/m \leftrightarrow 2|\mathbf{\Sigma}| = 2vp$ in (21.124) one has again just (21.19).

21.7.2 Conductivity

The interband (21.81) and intraband (21.82) conductivities provide both parts of the longitudinal conductivity. The zero-temperature limit of the interband part considered in (Morawetz, 2015a) reads for particles with quadratic dispersion

$$\sigma_{xx}^{\text{inter}} = \frac{e^2}{8\pi\hbar}\left\{\frac{4\epsilon_v\Sigma_n^2\tau_\omega/\hbar}{2\epsilon_v\mu + \Sigma_n^2} + \left(1 - \frac{4\Sigma_n^2\tau_\omega^2}{\hbar^2}\right)\arctan\left[\frac{4\epsilon_v\tau_\omega\hbar}{\hbar^2 + 4(2\epsilon_v\mu + \Sigma_n^2)\tau_\omega^2}\right]\right\} \quad (21.125)$$

with $\epsilon_v = mv^2/\hbar$. This conductivity represents a contribution in the direction of the applied electric field and is caused by collisional correlations. To translate (21.125) into the formula for graphene we perform the limit of infinite mass or $\epsilon_v \to \infty$. The order of limits becomes now essential. If we apply the limit of vanishing friction $\tau^{-1} \to 0$ before the infinite-mass limit we obtain $\sigma_{xx}^{\text{inter}} = 0$ and if we perform the static limit afterwards we obtain the negative result as opposed to the expected one. Since we translate the results of spin-orbit coupled systems to graphene with the help of the infinite-mass limit, the latter should be correctly performed first before any other specifications. We get for zero temperature

$$\sigma_{xx}^{\text{inter}} = \frac{e^2}{8\pi\hbar}\left[\begin{array}{c}\frac{2\Sigma_n^2\tau_\omega}{\hbar\mu} + \left(1 - \frac{4\Sigma_n^2\tau_\omega^2}{\hbar^2}\right)\text{arccot}\frac{2\mu\tau_\omega}{\hbar}, \;\; \mu > \Sigma \\ \\ \frac{2\Sigma_n\tau_\omega}{\hbar} + \left(1 - \frac{4\Sigma_n^2\tau_\omega^2}{\hbar^2}\right)\text{arccot}\frac{2\Sigma_n\tau_\omega}{\hbar}, \;\; \mu < \Sigma\end{array}\right]. \quad (21.126)$$

Since $\tau_\omega = \tau/(1 - i\omega\tau)$ the limits of infinite frequency $\omega \to \infty$, vanishing scattering $\tau \to \infty$, and vanishing density $\mu \to 0$ are not interchangeable. In fact, if we neglect the Zeeman field, $\Sigma_n = 0$, in (21.126) we get

$$\sigma_{xx}^{\text{inter}} = \frac{e^2}{16\hbar}\left[\begin{array}{c}-1 - i\frac{4\mu}{\pi\hbar\omega} + o\left(\frac{1}{\omega^2}\right) \\ \\ 1 - \frac{4}{\pi\hbar}\frac{\mu\tau}{\omega\tau + i} + o\left(\mu^2\right) \\ \\ -1 - i\frac{4\mu}{\pi\hbar\omega} + o\left(\mu^2\right) + o\left(\frac{1}{\tau}\right) \\ \\ 0 + o(\omega) + o\left(\frac{1}{\tau}\right) \\ \\ 1 + o\left(\frac{1}{\tau}\right) + o(\omega)\end{array}\right]. \quad (21.127)$$

Only the second limiting procedure leads to the right result. One obtains even a zero value if the static limit is used after vanishing scattering which is different from interchanging both limits. This illustrates the care one has to take when integrating zero-temperature values. In graphene the chiral nature of the charge carriers leads to a minimal finite conductivity even with the vanishing density of scatterers. If there are no charge carriers the field has to first create electron-hole pairs before they can be accelerated. Since the absolute value of the velocity is fixed, only the direction can change which provides an anomaly transport (Kao, Lewkowicz and Rosenstein, 2010). This remarkable feature of dissipation even in an ideal crystal is reached by various limiting procedures. The finite-temperature interband conductivity can be found in (Morawetz, 2016).

Intraband contribution We have seen that the interband contribution leads to the universal low-density conductivity at zero temperature and decreases with higher densities. Now we will discuss the intraband contribution which vanishes for zero temperatures and low densities. The finite-temperature expression of the intraband conductivity reads explicitly

$$\sigma_{xx}^{\text{intra}} = \frac{i\epsilon_0\omega_p^2}{\omega + \frac{i}{\tau}} = \frac{e^2 T\tau_\omega}{4\pi\hbar^2}\int\limits_{\frac{\Sigma_n}{T}}^{\infty}\frac{dx}{x}\left(\frac{\Sigma_n^2}{T^2} - x^2\right)\left[\frac{1}{\left(e^{\frac{x-\mu}{T}} + 1\right)\left(e^{\frac{-x+\mu}{T}} + 1\right)} + (\mu \leftrightarrow -\mu)\right]. \tag{21.128}$$

Consequently, this intraband contribution can be written in the form of the frequency-dependent Drude conductivity with a collective frequency $\omega_p[\mu, T, \Sigma_n]$ dependent on the chemical potential, temperature and effective Zeeman field. The intraband conductivity has a threshold at the effective Zeeman field. Higher temperatures mix both components and smooth the threshold of intraband conductivity.

Optical conductivity: comparison with experiment The optical conductivity $\sigma(\omega)$ is important to know if one wants to calculate the optical transparency (Stauber, Peres and Geim, 2008)

$$t(\omega) = \left[1 + \frac{\sigma(\omega)}{2\epsilon_0 c}\right]^{-2}. \tag{21.129}$$

If the effective Zeeman field becomes larger than the chemical potential (Fermi energy) then the conductivity is exclusively due to intraband transitions and independent density

$$\sigma_{xx}^{\text{inter}}(\Sigma_n > \mu) = \frac{e^2}{8\pi\hbar}\left[\frac{4\Sigma_n i}{\hbar(\omega + \frac{i}{\tau})} + \frac{i}{\pi}\left(1 + \frac{4\Sigma_n^2}{\hbar^2(\omega + \frac{i}{\tau})^2}\right)\ln\frac{2\Sigma_n - (\omega + \frac{i}{\tau})}{2\Sigma_n + (\omega + \frac{i}{\tau})}\right]. \tag{21.130}$$

The real part of the conductivity starts at the threshold $\hbar\omega = 2\Sigma_n$ accompanied by a minimum in the imaginary part. The finite scattering smears this step-like behaviour and leads in the strong scattering limit to a constant real conductivity of universal value. The astonishing fact is that not only a universal value appears for small densities due to the chiral nature of particles but that a whole universal optical conductivity appears independent of density and solely determined by strong effective Zeeman fields.

In Figure 21.1 we compare with the experimental data of (Li, Henriksen, Jiang, Hao, Martin, Kim, Stormer and Basov, 2008). We find the best fit with the help of fitting the effective Zeeman field and the relaxation time (Nomura and MacDonald, 2006) calculated in screened Coulomb interaction. The relaxation time is nine times smaller for unscreened potentials (Morawetz, 2016) which shows the importance of screening.

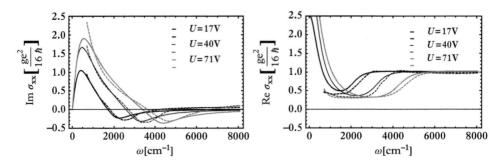

Figure 21.1: *Imaginary (left) and real (right) optical conductivity of graphene versus frequency for different applied voltages (Morawetz, 2016) compared with the experimental data (dashed) of (Li, Henriksen, Jiang, Hao, Martin, Kim, Stormer and Basov, 2008). The applied voltage induces a density $n = \frac{\epsilon \epsilon_0}{ed} U = 7.3 \times 10^{10} cm^{-2} V^{-1} U$ for graphene on typical SiO_2 substrates (Zhang, Tan, Stormer and Kim, 2005; Li, Henriksen, Jiang, Hao, Martin, Kim, Stormer and Basov, 2008). The density is linked to the chemical potential $n = \frac{g\mu^2}{4\pi \hbar^2 v^2} = 1010cm^{-2}(\mu/eV)^2$.*

21.7.3   Hall conductivity

Let us assume the magnetic field in the z-direction and the electric field in the x-direction. Then we have for graphene

$$\sigma_{xy}^{\text{Hall}} = \frac{e^2}{2\pi \hbar} \left(\frac{\Sigma_n}{T}\right) \left(\frac{T\tau_\omega}{\hbar}\right)^2 \int\limits_{\frac{\Sigma_n}{T}}^{\infty} dx \frac{\left(e^{\frac{-x-\mu}{T}} + 1\right)^{-1} - \left(e^{\frac{x-\mu}{T}} + 1\right)^{-1}}{1 + 4x^2 \left(\frac{T\tau_\omega}{\hbar}\right)^2}. \tag{21.131}$$

This Hall conductivity for particles with $p^2/2m$ dispersion at zero temperature and $\mu > \Sigma$ reads (Morawetz, 2015a)

$$\sigma_{xy}^{\text{Hall}} = \frac{e^2}{4\pi \hbar^2} \Sigma_n \tau_\omega \arctan \left[\frac{4\epsilon_v \tau_\omega/\hbar}{\hbar^2 + 4(2\epsilon_v \mu + \Sigma_n^2)\tau_\omega^2}\right] \tag{21.132}$$

with the energy $\epsilon_v = mv^2$ and the chemical potential (Fermi energy) μ. We see that the anomalous Hall effect vanishes with vanishing effective Zeeman field (21.44). The latter one being an external magnetic field or an effective magnetised domain.

The infinite-mass limit now reads

$$\sigma_{xy}^{\text{Hall}} = \frac{e^2}{4\pi \hbar^2} \tau_\omega \Sigma_n \begin{cases} \text{arccot} \frac{2\mu\tau_\omega}{\hbar}, & \mu > \Sigma_n \\ \text{arccot} \frac{\Sigma_n \tau_\omega}{\hbar}, & \mu < \Sigma_n \end{cases} \rightarrow \frac{e^2}{8\pi \hbar} \begin{cases} \frac{\Sigma_n}{\mu} + o(\tau_\omega^{-1}), & \mu > \Sigma_n \\ 1 + o(\tau_\omega^{-1}), & \mu < \Sigma_n \\ \frac{\Sigma_n \tau}{\hbar}\left(\pi - \frac{4\tau\mu}{\hbar} + o(\mu^2)\right), & \mu > \Sigma_n \end{cases} \tag{21.133}$$

which is nothing but the zero-temperature limit of (21.131). One sees that a step-like structure appears if the chemical potential exceeds the effective magnetic field Σ_n. Remarkably, for the larger effective Zeeman field Σ_n, we obtain a result independent of the chemical potential and therefore independent of the density. Even in the limit of vanishing scattering the universal value $\sigma_{xy} \to e^2/8\pi\hbar$ appears.

21.7.4 Pseudo-spin conductivity

Now we consider the pseudo-spin current according to (21.67)

$$\mathbf{S}_\alpha = 2 \sum_p \left(\partial_{p_\alpha} \epsilon \delta \mathbf{g} + \delta f \partial_{p_\alpha} \mathbf{b} \right) \tag{21.134}$$

where the first part represents the normal and the second part is the anomaly part. For graphene in the limit of infinite mass we would not expect a normal part since $\partial_{p_j}\epsilon = p_j/m \to 0$ at a first glance. However, there is a subtle problem here. This first normal part is

$$\mathbf{S}_\alpha^n = -\frac{e\tau}{m_e(1-i\omega\tau)}\mathbf{s}E_\alpha + \sigma_{\alpha\beta}E_\beta. \tag{21.135}$$

The normal spin-Hall coefficient consists analogously as the anomalous Hall effect of a symmetric and an asymmetric part ($\omega \to \omega + i/\tau$)

$$\left.\begin{array}{c}\sigma_{\alpha\beta}^{\mathrm{Hall}}\\[2mm]\sigma_{\alpha\beta}^{\mathrm{inter}}\end{array}\right\} = \frac{e}{m_e\omega}\sum_p \frac{p_\alpha g}{1-\frac{\omega^2}{4|\Sigma|^2}} \left\{\begin{array}{c}\frac{i\omega}{2|\Sigma|}\mathbf{e}\times\partial_{p_\beta}\mathbf{e}\\[2mm]i\partial_{p_\beta}\mathbf{e}\end{array}\right. \tag{21.136}$$

with the explicit integration in zero temperature and linear Rashba coupling (Morawetz, 2015a)

$$\sigma_{yx}^z = \frac{e}{8\pi\hbar}\left[1 - \frac{\hbar^2 + 4\Sigma_n^2\tau_\omega^2}{4\epsilon_v\tau_\omega\hbar}\arctan\left(\frac{4\hbar\epsilon_v\tau_\omega}{\hbar^2 + 4\tau_\omega^2(2\epsilon_v\mu + \Sigma_n^2)}\right)\right]$$

$$\sigma_{xx}^z = \frac{2}{\hbar}\Sigma_n\tau\sigma_{yx}^z \tag{21.137}$$

with $\tau_\omega = \tau/(1-i\omega\tau)$ and $\epsilon_v = mv^2$. Neglecting the selfenergy and using the static limit it is just the result of (Schliemann and Loss, 2004; Schliemann, 2006). For the Dresselhaus linear spin-orbit coupling one has (21.137) with the opposite sign.

For graphene we perform the limit of infinite mass of (21.137). Amazingly we obtain just the universal limit

$$\lim_{m\to\infty} \sigma_{yx}^z = \frac{e}{8\pi\hbar} \tag{21.138}$$

contrary to the expectation at the beginning of this chapter that these normal parts of pseudo-spin current should vanish. This presents a puzzle. The infinite-mass limit of the normal spin-Hall conductivity leads to a finite result though the quasiparticle velocity $\partial_{p_j}\epsilon$ vanishes and we would have expected no result from the normal current. One obtains such a zero effect if one performs the limit of vanishing density first and then the absence of collisions $\tau \to \infty$. The same is true if we first perform the limit of vanishing collisions and then the static limit.

The universal constant $e/8\pi\hbar$ has been first described by (Sinova et al., 2004) and raised an intensive discussion. It was shown that the vertex corrections cancel this constant (Inoue et al., 2004; Raimondi and Schwab, 2005). A suppression of Rashba spin-orbit coupling has been obtained due to disorder (Chalaev and Loss, 2005), or electron-electron interaction (Dimitrova, 2005) and found to disappear in the self-consistent Born approximation (Liu, Lei and Horing, 2006). The conclusion was that the two-dimensional Rashba spin-orbit coupling does not lead to a spin-Hall effect as soon as there are relaxation mechanisms present which damp the pseudo-spins towards a constant value. Beyond the mean field, in relaxation-time approximation by including vertex corrections (Zala, Narozhny and Aleiner, 2001) or treating collision integrals (Shytov, Mishchenko, Engel and Halperin, 2006; Scholz, Stauber and Schliemann, 2012) lead to vanishing spin-Hall current. The spin-Hall effect does not vanish with magnetic fields or spin-dependent scattering processes (Schliemann, 2006). This was discussed by kinetic theory (Gorini, Schwab, Dzierzawa and Raimondi, 2008) for a dirty and clean regime and the Kubo formula (Inoue, Kato, Ishikawa, Itoh, Bauer and Molenkamp, 2006) where it was shown that the Zeeman field gives a non-vanishing spin-Hall current. In (van den Berg, Raymond and Verga, 2011) magnetic impurities were treated in tight-binding approximation and the importance of the order of large scattering time and small frequencies were pointed out. The anomalous part as the second part of (21.134) is wave-length dependent and vanishes in the long-wave-length limit (Morawetz, 2016).

22

Relativistic Transport

We will see that due to many-particle effects, processes are possible, which can be interpreted as particle creation and annihilation due to in-medium one-meson exchange. This is connected with in-medium cross sections from the generalised derivation of collision integrals possessing complete crossing symmetries.

Since Walecka (1974, Serot and Walecka (1986)) has established his model of meson exchange, field theoretical models were successfully applied for the description of heavy-ion collisions at intermediate energies (Stöcker and Greiner, 1986; Bertsch and Gupta, 1988; Snider, 1990; Cubero, 1990). Thereby the derivation of transport equations starting from a microscopic model was paid great attention (Botermans and Malfliet, 1990; Danielewicz, 1990; Ivanov, 1987). A number of publications are devoted to the description of relativistic many-body theory of high-density matter (Wilets and et. al., 1976; Chin, 1977).

22.1 Model and basic equations

We consider relativistic systems with the metric

$$g_{\mu\nu} = \begin{pmatrix} 1 & 0 & 0 & 0 \\ 0 & -1 & 0 & 0 \\ 0 & 0 & -1 & 0 \\ 0 & 0 & 0 & -1 \end{pmatrix} \tag{22.1}$$

and the contra- and covariant four-momentum

$$p^{\mu} = \left(\frac{E}{c}, \mathbf{p}\right) = -\frac{\hbar}{i}\frac{\partial}{\partial x_{\mu}} = \left(-\frac{\hbar}{i}\frac{\partial}{\partial ct}, \frac{\hbar}{i}\frac{\partial}{\partial \mathbf{r}}\right) = -\frac{\hbar}{i}\partial^{\mu}$$

$$p_{\mu} = g_{\mu\nu}p^{\nu} = \left(-\frac{\hbar}{i}\frac{\partial}{\partial ct}, -\frac{\hbar}{i}\frac{\partial}{\partial \mathbf{r}}\right) = -\frac{\hbar}{i}\partial_{\mu} \tag{22.2}$$

Interacting Systems far from Equilibrium. Klaus Morawetz, Oxford University Press (2018).
© Klaus Morawetz. DOI: 10.1093/oso/9780198797241.001.0001

such that the relativistic energy balance reads

$$m^2 c^2 = p^\mu p_\mu = \frac{E^2}{c^2} - \mathbf{p}^2. \tag{22.3}$$

The description of families of particles with the same symmetries shall be illustrated now. Considering the isospin for protons and neutrons we might write a wave-function $\bar\Psi = (\Psi_p, \Psi_n)$ and creation operators for protons and neutrons as

$$\tau_+ \bar\Psi = \Psi_p \qquad \text{with } \tau_+ = \begin{pmatrix} 0 & 1 \\ 0 & 0 \end{pmatrix}$$

$$\tau_- \bar\Psi = \Psi_n \qquad \text{with } \tau_- = \begin{pmatrix} 0 & 0 \\ 1 & 0 \end{pmatrix}. \tag{22.4}$$

From this one can build three isotopic-spin matrices which are the Pauli matrices (21.8) with $\sigma_x = \tau_+ + \tau_-$, $\sigma_y = -i(\tau_+ - \tau_-)$ and $\sigma_z = \tau_+\tau_- - \tau_-\tau_+$ which obey $\sigma_i\sigma_j = \delta_{ij} + i\epsilon_{ijk}\sigma_k$. It is of advantage to use the doubling of matrices $\gamma_\mu = (\gamma_0, \boldsymbol{\gamma})$ into Dirac matrices with

$$\gamma_i = \begin{pmatrix} 0 & \sigma_i \\ \sigma_i & 0 \end{pmatrix}, \quad \gamma_0 = \begin{pmatrix} 1 & 0 \\ 0 & -1 \end{pmatrix}, \quad \gamma_5 = \begin{pmatrix} 0 & 1 \\ 1 & 0 \end{pmatrix} \tag{22.5}$$

and $[\gamma^\mu, \gamma^\nu]_+ = g^{\mu\nu}$ such that the baryons with the four-component wave-function ψ obey the Dirac equation $(i\gamma_\mu \partial^\mu - m_0)\psi = 0$ with their conjugate $\bar\psi = \psi^+\gamma^0$.

The three Pauli matrices as generators of the SU(2) symmetry allows us to describe e.g. three mesons which are represented usually by a complex field ϕ, $\Pi^+ = (\phi + \phi^*)/\sqrt{2}$, $\Pi^- = (\phi - \phi^*)/\sqrt{2}$, and $\Pi^0 = \phi_0$ called pseudoscalar. Their coupling to the baryons is realised by an additional γ^5 matrix : $\bar\psi\gamma^5\sigma_i\psi\phi_i$: obeying the required symmetries. The four γ matrices allow to describe the vector mesons. In general one can describe different mesons–baryon couplings by : $\bar\psi\Omega_i\Phi^i\psi$: illustrated in Table 22.1. With :: one denotes the normal ordering procedure to ensure the renormalisation of baryon density (Roman, 1969). The different meson couplings will create repulsive and attractive parts in the interaction of baryons as we will see in a minute. We will restrict this to the $\sigma - \omega$ model where we 'invent' a σ meson to describe the repulsion and take the ω meson for an attraction to fit to the experiments. Any more realistic description just extends the number of equations.

The Yukawa Lagrangian, which couples scalar and vector meson fields to fermion fields describes the simplest form of a model for nuclear matter (Serot and Walecka, 1986)

$$\mathcal{L} = -\bar\Psi\left(-i\gamma^\mu\partial_\mu + \kappa_0\right)\Psi + +\frac{1}{2}\partial_\mu\Phi\partial^\mu\Phi - \frac{1}{2}m_{\sigma_0}^2\Phi^2 + \, : g_\sigma\bar\Psi\Phi\Psi :$$

$$-\frac{1}{4}F_{\mu\nu}F^{\mu\nu} + \frac{1}{2}m_{\omega_0}^2\varphi_\lambda\varphi^\lambda + \, : g_\omega\bar\Psi\gamma_\lambda\Psi\varphi^\lambda : . \tag{22.6}$$

Table 22.1 *Some meson descriptions according to the required symmetries from (Brown, Puff and Wilets, 1970). Please note the artificial construction of the σ − ω model neglecting real mesons.*

Ω^i	i		meson	m[MeV]	spin^{parity}	isospin	charge conj.
$g_\Pi \gamma^5 \sigma_i$	1, 2, 3		Π	138	0^-	1	+
$g_\eta \gamma^5$	4		η	550	0^-	0	+
$g_\omega \gamma^\mu$	$5 + \mu$		ω	783	1^-	0	−
$g_\Phi \gamma^\mu$	$9 + \mu$		Φ	1020	1^-	0	−
$g_\rho \gamma^\mu \sigma_k$	$13 + 4(k-1) + \mu$		ρ	770	1^-	1	−
g_σ	25		σ		0^+	0	

Here m_{s_0}, m_{v_0} are the bare scalar and vector meson mass and κ_0 is the bare nucleon one respectively. The field tensor is $F^{\mu\nu} = \partial^\mu \phi^\nu - \partial^\nu \phi^\mu$. After renormalisation of the masses the coupling constants have to be chosen in such a way that, on one hand, the equilibrium ground state properties can be fitted, and on the other hand the scattering data can be reproduced (Snider, 1990; Ko, Li and Wang, 1987). Because of the renormalisability of the model Lagrangian, we can choose a standard procedure for the occurring divergent vacuum terms resulting in physical masses. For our used selfconsistent Hartree–Fock and RPA approximation this is well-known (Roman, 1969). In the following we use therefore the normalised fields and masses, which may be considered as an effective model.

From (22.6) the Lagrange equations of motion

$$\frac{\partial}{\partial x^\mu} \frac{\delta \mathscr{L}}{\partial (\partial_\mu \phi)} - \frac{\delta \mathscr{L}}{\delta \phi} = 0 \tag{22.7}$$

read

$$\left(\Box + m_\sigma^2\right) \Phi = g_\sigma \quad : \quad \overline{\Psi}\Psi \quad : \equiv j_\Psi \tag{22.8}$$

$$\left(\Box + m_\omega^2\right) \varphi_\nu = g_\omega \quad : \quad \overline{\Psi}\gamma_\nu\Psi \quad : \equiv j_{\Psi\nu} \tag{22.9}$$

$$\left(i\gamma^\mu \partial_\mu - \kappa_0\right) \Psi = g_\sigma \quad : \quad \Psi\Phi \quad : \quad - g_\omega \quad : \quad \gamma^\lambda \varphi_\lambda \Psi \quad : \equiv j_\varphi. \tag{22.10}$$

22.2 Coupled Green's functions for meson and baryons

With the help of the (retarded) Green's function $G_0^\sigma = (-k^2 + m_\sigma^2)^{-1}$ and analogously for G_0^ω of the meson equations (22.8, 22.9), the mesonic degree of freedom in Eq. (22.10) can be eliminated (Botermans and Malfliet, 1990; Brown, Puff and Wilets, 1970; Wilets and et. al., 1976)

$$\left(i\gamma^{\mu}\partial_{\mu}-\kappa\right)\Psi_{\alpha}=\left[g_{\sigma}^{2}\delta_{\alpha\beta}\delta_{\gamma\delta}G_{o}^{\sigma}-g_{\omega}^{2}\,(\gamma_{\mu})_{\alpha\beta}\,(G_{0}^{\omega})^{\mu\rho}\,(\gamma_{\rho})_{\gamma\delta}\right)\bar{\Psi}_{\gamma}\Psi_{\delta}\Psi_{\beta}. \tag{22.11}$$

Here $\alpha,\beta,\gamma,\delta$ indicate the spinor indices. One may construct from (22.11) a four-vector potential, which has the structure $U_{\omega}-U_{\sigma}$, which means the vector mesons act repulsively and the scalar mesons attractively.

The meanfields can be seen from averaging the equation (22.10) and factorising the right side. Then the averaged meson fields are needed which come from (22.8) and (22.9)

$$\langle\Phi\rangle=\frac{g_{\sigma}}{m_{\sigma}^{2}-p^{2}}\,\langle\bar{\Psi}\Psi\rangle=\frac{g_{\sigma}}{m_{\sigma}^{2}}\,n_{s}$$

$$\langle\varphi_{\nu}\rangle=\frac{g_{\omega}}{m_{\omega}^{2}-p^{2}}\,\langle\bar{\Psi}\gamma_{\nu}\Psi\rangle=\frac{g_{\omega}}{m_{\omega}^{2}}\,\delta_{\nu 0}n_{B} \tag{22.12}$$

for homogeneous systems $p^{2}\ll m^{2}$ and $j_{\psi_{l}}=0$. The scalar density n_{s} and the baryon density n_{B} will be calculated soon, when we are able to calculate the averaged baryon operators with the help of the Green's functions.

For the nucleon system we consider the Green's function (7.2) which become 4×4 matrices according to the Dirac spinors and for the meson fields ϕ we introduce the causal Green's function and correlation functions

$$id(1,2)=\left\langle T\varphi(1)\varphi^{*}(2)\right\rangle-\langle\varphi(1)\rangle\left\langle\varphi^{*}(2)\right\rangle$$

$$d^{<}(1,2)=-\left\langle\varphi^{*}(2)\varphi(1)\right\rangle+\left\langle\varphi^{*}(2)\right\rangle\langle\varphi(1)\rangle$$

$$d^{>}(1,2)=\left\langle\varphi(1)\varphi^{*}(2)\right\rangle-\langle\varphi(1)\rangle\left\langle\varphi^{*}(2)\right\rangle. \tag{22.13}$$

For reasons of legibility we will write only the scalar meson equations. The vector meson Green's functions carry additional spinor indices. All final results will be presented for both vector and scalar mesons.

Using the definition (7.2), (7.3) and (22.13), the equations of motion for the different Green's functions can be derived. From the equation of motion for the causal one reads, for instance

$$\left(i\gamma^{\mu}\,\partial_{\mu}-\kappa_{0}\right)G(1,2)=\delta(1-2)+\frac{1}{i}\left\langle Tj_{\varphi}(1)\bar{\psi}(2)\right\rangle \tag{22.14}$$

$$\left(\Box+m_{\sigma}^{2}\right)d(1,2)=-\delta\,(1-2)+\frac{1}{i}\left\langle Tj_{\psi}(1)\varphi^{*}(2)\right\rangle, \tag{22.15}$$

with j_{φ} and j_{ψ} from Eq. (22.8) and (22.10). All steps of developing Green's function from Chapter 7.1 can be repeated. The weakening of initial correlation reads, for the right side of (22.14),

$$\lim_{t\to-\infty}\left\langle j_{\varphi}(1)\bar{\psi}(2)\right\rangle=\left\langle\psi(1)\bar{\psi}(2)\right\rangle\left\{g_{\sigma}\,\langle\Phi(1)\rangle-g_{\omega}\gamma^{\lambda}\,\langle\varphi_{\lambda}(2)\rangle\right\}. \tag{22.16}$$

This is an asymptotic condition, which breaks the time-symmetry and provides irreversible evolution in nonequilibrium systems. Now we are able to enclose the equation of motion (22.14) resulting in the nonequilibrium Dyson equation (8.11) for the causal Green's function

$$\left(i\gamma^{\mu}\partial_{\mu} - m\right) G_{bc}(1,2) = \delta_{bc}(1-2) + \int_{-\infty}^{\infty} d3 \left\{\Sigma(1,3) G(3,2) - \Sigma^{<}(1,3)G^{>}(3,2)\right\}$$

(22.17)

and for the meson correlation functions (22.15) one has analogously

$$\left(\Box + m^2\right) d(1,2) = \delta(1-2) + \int d3 \left\{\Pi(1,3)d(3,2) - \Pi^{<}(1,3)d^{>}(3,2)\right\}.$$ (22.18)

To study the structure of the selfenergy Σ and the polarisation function Π explicitly, we want to use the variational technique of Chapter 11 developed by Schwinger (Roman, 1969). Therefore we introduce an infinitesimal meson generating flux j_{Φ} for scalar mesons and j_{λ} for vector mesons, as a new type of interaction

$$L_{\text{int}} = -j_{\Phi}\Phi - j_{\lambda}\varphi^{\lambda}.$$ (22.19)

In our presented formalism it is not necessary to introduce a generating functional for nucleons. It turns out that the infinitesimal interaction (22.19) is sufficient to obtain the Kadanoff and Baym equations (7.26) which are consequently valid for any density or correlations in the system.

Now we introduce an interaction picture with respect to the infinitesimal interaction (22.19) and the time-evolution operator (11.6) which reads

$$S = \hat{T}_c e^{-i\int \left(-j_{\Phi}\Phi - j_{\mu}\varphi^{\mu}\right)}.$$ (22.20)

Here $\hat{T}$ is the time-ordering operator. Now special relations can be established by variational technique

$$\frac{\delta S}{\delta j} = -i\,\hat{T}_c\,\varphi^*\,S$$ (22.21)

where j stand for j_{Φ} or j_{λ}. In a straightforward manner one can express all correlation functions (22.13) through variations with respect to j (Bezzerides and DuBois, 1972). The causal Green's function e.g. can be expressed as

$$i\frac{\delta\langle\varphi(1)\rangle}{\delta j(2)} = \langle T\varphi(1)\varphi(2)\rangle - \langle\varphi(1)\rangle\langle\varphi(2)\rangle.$$ (22.22)

With some manipulations, the introduced polarisation function (22.18) of mesons can be derived from (22.15) in the form

$$\Pi(1,2) = -g_\sigma^2 \int d3 d4 G(1,3)\Gamma(3,4,2)G(4,1^+) \tag{22.23}$$

where we have introduced the vertex function

$$\Gamma(3,4,2) = -\frac{1}{g_\sigma}\frac{\delta G^{-1}(3,4)}{\delta\langle\varphi(2)\rangle} = -i\delta(3-4)\,\delta(3-2) + \frac{1}{g_\sigma}\frac{\delta\Sigma(3,4)}{\delta\langle\varphi(2)\rangle} \tag{22.24}$$

which is quite easy to verify by means of (22.17).

Concerning the nucleons we find that the right-hand side of equation (22.14) can be expressed with the help of variations in such a way that the equation (22.14) takes the form

$$\left[i\gamma^\mu\delta_\mu - \kappa_0 + g_\sigma\langle\phi\rangle - g_\omega\langle\varphi_\lambda\rangle\gamma^\lambda\right]G(1,2) = \delta(1-2) + \left\{\frac{g_\sigma}{i}\frac{\delta}{\delta j_\phi(1)} - \frac{g_\omega}{i}\gamma_\lambda\frac{\delta}{\delta j_\lambda(1)}\right\}G(1,2). \tag{22.25}$$

After the introduction of the same vertex function (22.24), the structure of the selfenergy introduced in (22.17) can finally be written as

$$\Sigma(1,3) = -\int d\bar{1}d2 g_\sigma^2 G(1,\bar{1})\Gamma(\bar{1},3,2)d(2\,1). \tag{22.26}$$

Equations (22.17), (22.18), (22.23), (22.24) and (22.26) form a complete quantum statistical description of the many-particle system in nonequilibrium and serves as a starting point for any further approximated treatment. If we compare them with Chapter 11.3 we just recognise the Hedin equation.

In principle one has to admit that in the straightforward derivation of the above expressions one has to care about divergent contributions (vacuum polarisation graphs). As we have already pointed out after Eq. (22.6) and as it is discussed in several articles, e.g. (Poschenrieder and Weigel, 1988) and the citations therein, one may neglect such contributions for the calculation of many-body properties to get a first insight in to the structure of equations. The renormalisation scheme can be found in (Roman, 1969).

For the spectral property we need the retarded function (22.17)

$$\left(i\gamma^\mu\partial_\mu - \kappa\right)G^R(1,2) = \delta(1-2) + \int d\bar{1}\Sigma^R(1,\bar{1})G^R(\bar{1},2). \tag{22.27}$$

After the Wigner transformation, the variables are split into macroscopic and microscopic ones and it is assumed $\Sigma^R(X,p) \gg \left|\frac{\partial}{\partial x^\mu}\frac{\partial}{\partial p_\mu}\Sigma^R(X,p)\right|$. This means that $\Delta x^\mu \Delta p_\mu \gg 1$. Here the characteristic length is Δp, at which the selfenergy varies in four-momentum space, corresponding to the inverse space–time interaction range.

Therefore the approximation demands the shortness of the space–time interaction range when compared with the system's space–time inhomogeneity scale (Mrowczynski and Danielewicz, 1990). This is the physical background when we apply the gradient expansion, see Chapter 9.1.1, which now yields for (22.27)

$$\left(\gamma^\mu p_\mu - \kappa - \Sigma^R(p, X)\right) G^R(p, X) = 1. \tag{22.28}$$

To make further progress, we write the selfenergy in characteristic parts in terms of the γ–matrices,

$$\Sigma^R = \gamma^\mu \Sigma^R_\mu + \Sigma^R_I. \tag{22.29}$$

Introducing the following medium-dressed variables

$$\tilde{P}_\mu = p_\mu - Re\Sigma_\mu$$

$$\kappa = \kappa_0 + Re\Sigma_I$$

$$\tilde{\Gamma} = \left(\gamma_0\, p_0\, \eta + \frac{\Gamma}{2}\right)\left(\gamma\tilde{P} + \kappa\right) \tag{22.30}$$

with $\eta \to 0$, one obtains the complete spectral function from (22.28)

$$A = i\left\{G^R(\omega + i\eta) - G^A(\omega - i\eta)\right\} = \left(\gamma^\mu \tilde{P}_\mu + \kappa\right)\frac{2\tilde{\Gamma}}{\left(\tilde{P}^2 - \kappa^2\right)^2 + \tilde{\Gamma}^2}. \tag{22.31}$$

This derivation underlines the correct expression using $\epsilon\gamma^0$ instead of ϵ in the denominator. In the case of vanishing damping Γ we get the spectral function, which determines the quasiparticle energies in the quasiparticle picture

$$A = 2\pi \left\{\gamma^\mu \tilde{P}_\mu + \kappa\right\} \delta\left(\tilde{P}^2 - \kappa^2\right) \operatorname{sgn}\left(\tilde{P}_0\right). \tag{22.32}$$

The δ-function in (22.32) represents the defining equation for quasiparticle energies

$$\epsilon_\pm = Re\Sigma^R_0 \pm \sqrt{\left(\vec{p} - Re\vec{\Sigma}^R\right)^2 + (\kappa_o + Re\Sigma_I)^2}\ |_{\omega=\epsilon_\pm}. \tag{22.33}$$

This is a nonlinear equation indicated by the many-particle influence and immediately shows the particle/antiparticle symmetry. Therefore the single particle and antiparticle states become dressed by the medium and yield a natural generalisation of the Dirac–Brueckner Theory. It can be seen that (22.33) is determined only by the retarded selfenergy, which yields a mass off-shell behaviour of the quasiparticles. Please note the similarity of (22.33) with (12.32).

22.3   Equilibrium and saturation thermodynamic properties

At this point it is instructive to consider the meanfield approximation where the spectral function (22.32) is exact and the correlation function reads explicitly with the equilibrium (7.53) Fermi–Dirac distribution f

$$G^<(p, \omega) = A(p, \omega)f(\omega) = \pi (\gamma^0 \epsilon_+ - \gamma^l p_l + \kappa)\delta(\omega - \epsilon_+)f(\epsilon_+)$$

$$- \pi (\gamma^0 \epsilon_- - \gamma^l p_l + \kappa)\delta(\omega - \epsilon_-)[1 - \bar{f}(-\epsilon_-)] \qquad (22.34)$$

where we used (22.33) and the effective baryon mass $\kappa = \kappa_0 + \mathrm{Re}\Sigma_I$. Since the distribution should describe physical particles and antiparticles we use the Dirac interpretation in the quasiparticle picture. There an empty state of particles with negative energy equal to an antiparticle state with positive energy and we can write

$$f(-\omega) = 1 - \bar{f}(\omega), \qquad (22.35)$$

where $\bar{f}$ signs the antiparticle distribution (Bezzerides and DuBois, 1972). In equilibrium, this is identical with the fact that the chemical potential has to be chosen with opposite signs for particles and antiparticles, which follows immediately from the conserved baryon density (Serot and Walecka, 1986). For spin-polarised nucleons see (Mrowczynski and Heinz, 1994).

With (22.34) the meanfield selfenergies (22.12) take the explicit forms

$$\mathrm{Re}\Sigma_0^R = \frac{g_\omega^2}{m_\omega^2} n_B = \Delta, \qquad \mathrm{Re}\Sigma_I = -\frac{g_\sigma^2}{m_\sigma^2} n_s = s \qquad (22.36)$$

with the baryon and scalar density

$$n_B = \langle \bar{\Psi} \gamma^0 \Psi \rangle = \mathrm{Tr}\gamma^0 G^< = \frac{g}{2\pi^2} \int_0^\infty dp p^2 (f - \bar{f})$$

$$n_s = \langle \bar{\Psi} \Psi \rangle = \mathrm{Tr} G^< = \frac{g}{2\pi^2} \int_0^\infty dp p^2 \frac{\kappa}{\sqrt{p^2 + \kappa^2}} (f + \bar{f}) \qquad (22.37)$$

and spin-degeneracy g. We see that the effective baryon mass in the distribution functions turns out to obey a nonlinear equation

$$\kappa = \kappa_0 + s = \kappa_0 - \frac{g_\sigma^2}{m_\sigma^2} \frac{g}{2\pi^2} \int_0^\infty dp p^2 \frac{\kappa}{\sqrt{p^2 + \kappa^2}} (f + \bar{f}) \qquad (22.38)$$

and the effective chemical potential becomes $\mu = \mu_0 + \Delta$ in the distributions.

The pressure and total energy we can obtain directly from Green's functions. Multiplying (22.10) from the left with $\bar{\psi}$ one creates, on the right-hand side, the negative of the interacting part of the Lagrangian (22.6)

$$(i\gamma^0\partial_t + i\gamma^l\nabla_l - \kappa_0)\bar{\psi}\psi = -\mathscr{L}_{\text{int}}. \tag{22.39}$$

Averaging over the statistical operator means that we can write the mean-interaction energy

$$\langle V_\lambda\rangle = -\langle\mathscr{L}_\lambda\rangle = \text{Tr}\int\frac{dp^4}{(2\pi)^4}(\gamma^\mu p_\mu - \kappa_0)G^<_\lambda = \Delta n_B\lambda^2 + sn_s\lambda^2 \tag{22.40}$$

where we used the quasiparticle propagator (22.34) and assume a λ factor at each coupling constant g_i. This allows us to apply the charging formula (7.73) which means we get for the interacting part of the pressure

$$\mathscr{P}(\kappa,\mu) - \mathscr{P}_0(\kappa_0,\mu_0) = -\int_0^1\langle V_\lambda\rangle\frac{d\lambda}{\lambda} = -\frac{\Delta}{2}n_B - \frac{s}{2}n_s \tag{22.41}$$

where we used the notation

$$\mathscr{P}_0(\kappa,\mu) = \frac{g}{2\pi^2\beta}\int_0^\infty dpp^2\left\{\ln\left[e^{-\beta(\sqrt{p^2+\kappa^2}-\mu)} + 1\right] + \ln\left[e^{-\beta(\sqrt{p^2+\kappa^2}+\mu)} + 1\right]\right\}$$

$$= \mathscr{P}_0(\kappa_0,\mu) - sn_s + o(s^2) = \mathscr{P}_0(\kappa_0,\mu_0) - sn_s - \Delta n_B + o(s^2,\Delta^2) \tag{22.42}$$

and $\mu = \mu_0 + \Delta$ seen from the quasiparticle energies. The total energy can be calculated from the thermodynamic relation

$$E = \mu n_B - \frac{\partial[\beta\mathscr{P}(\kappa,\mu)]}{\partial\beta} = \mu n_B + E_{QP}(\kappa_0) + \mathscr{P}_0(\kappa\mu) - \mathscr{P}(\kappa,\mu) + \Delta n_B - \mu_0 n_B \tag{22.43}$$

with the quasiparticle energy

$$E_{\text{qp}}(\kappa) = \frac{g}{2\pi^2}\int_0^\infty dpp^2\sqrt{p^2+\kappa^2}(f+\bar{f}) = E_{\text{qp}}(\kappa_0) + sn_s. \tag{22.44}$$

Summarising, we have for the pressure and energy

$$\mathscr{P} = \mathscr{P}_0(\kappa,\mu) + \frac{\Delta}{2}n_B + \frac{s}{2}n_s, \qquad E = E_{QP}(\kappa) + \frac{\Delta}{2}n_B - \frac{s}{2}n_s. \tag{22.45}$$

The saturation effect can be seen from the zero-temperature result of (22.38)

$$\kappa = \kappa_0 - \frac{g_\sigma^2}{m_\sigma^2}\frac{g}{4\pi^2}\kappa\left(k_f\epsilon_f - \kappa^2\ln\frac{k_f+\epsilon_f}{\kappa}\right) \tag{22.46}$$

where $\epsilon_f = \sqrt{k_f^2 + \kappa^2}$ and one can show that mean-field approximation is thermodynamically consistent by proving

$$\mathscr{P} = n_B^2\frac{\partial}{\partial n_B}\left(\frac{E_{QP}}{n_B}\right). \tag{22.47}$$

The high-density expansion of the energy reads

$$E \approx n_B\left[\frac{g_\omega^2}{2m_\omega^2}n_B + \frac{3}{4}k_f + \frac{m_\sigma^2\kappa^2}{2g_\sigma^2 n_B} + o\left(n_B^{-5/3}\right)\right]. \tag{22.48}$$

The first part gives the repulsion by vector mesons, the second is the Fermi energy of massless baryons and the third term gives the scalar contribution which is damped due to the mass renormalisation (22.46).

The binding property shows up for the low-density expansion

$$E \approx n_B\left[\kappa + \frac{3k_f^2}{10\kappa} - \frac{3k_f^4}{56\kappa^3} + \frac{g_\omega^2}{2m_\omega^2}n_B - \frac{g_\sigma^2}{2m_\sigma^2}n_B + \frac{3g_\sigma^2 k_f^2}{10m_\sigma^2\kappa^2}n_B + o\left(n_B^2\right)\right]. \tag{22.49}$$

We recognise the rest mass as the first term and the nonrelativistic Fermi-gas energy as is the second term and its relativistic correction as third term. Then the nonrelativistic binding property of the repulsive vector meson and attractive scalar mesons appear. Further, the relativistic corrections are damped again by a small effective mass. For very high and very low density the systems are unbound.

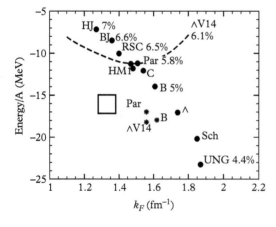

Figure 22.1: *Energy per particle vs Fermi wave vector for different realistic NN interactions (Dickhoff, 2016). The circles and stars indicate the minims of the saturation curves for the Brueckner–Hartree–Fock approximation with various interactions. The box identifies the experimental data which are reproduced by the relativistic meanfield Walecka model as well as a three-particle Skyrme interaction.*

With the help of the two degrees of freedom of the relativistic approach, the effective mass by the gap equation and the vector selfenergy, one can reproduce the nuclear binding energy as illustrated in Figure 22.1. Any nonrelativistic approach with two-body forces remain on the so-called Coester line (Delfino et al., 2005; Hassaneen et al., 2011) above the experimental values as illustrated in Figure 22.1. On the other side, a three-particle interaction resulting into a density-dependent Skyrme-like two-particle interaction allows us to describe the binding as well (Morawetz, 2000d). This means that either three-particle correlations or the relativistic gap equation for the effective mass are the responsible mechanisms for saturation.

22.4 Kinetic equations

Let us now consider the meson equations, which read for the retarded correlated function in operator notation

$$(\Box + m^2)\left(-d^R\right) = 1 - \Pi^R d^R, \tag{22.50}$$

where integration about inner variables is assumed. Moreover, we have for the correlation function

$$(\Box + m^2)(-d)^{\gtrless} = -\Pi^R d^{\gtrless} - \Pi^{\gtrless} d^A. \tag{22.51}$$

Combining (22.50) and (22.51) we obtain the following important relation:

$$-d^{\gtrless} = d^R \Pi^{\gtrless} d^A, \tag{22.52}$$

which presents the optical theorem. Following the same arguments as in deriving (22.28), one gets for the product $d^R d^A$ in the gradient expansion

$$d^R d^A = \frac{1}{\left(\omega^2 - E_d^2\right)^2 + \left(\frac{\Gamma_\pi}{2} + 2\varepsilon\omega\right)^2}. \tag{22.53}$$

Here the quasi-particle energy of mesons is introduced by

$$E_d^2 = \vec{p}^2 + m^2 - Re\Pi. \tag{22.54}$$

The spectral function of mesons in the case of nearly vanishing damping reads

$$B = i\left(d^R - d^A\right) \longrightarrow 2\pi\delta\left(\omega^2 - E_d^2\right)\text{sgn}(\omega) \qquad \text{for} \qquad \Gamma \to 0. \tag{22.55}$$

In a static case with vanishing damping one obtains from (22.53) the Fourier transformed Yukawa potential for the meson exchange

$$d^R d^A \approx \frac{1}{g^4 (4\pi)^4} V^2(p)$$

$$V(r) = g^2 \frac{e^{-r/r_0}}{r} \qquad r_0^{-2} = m^2 - Re\Pi. \qquad (22.56)$$

When Π is determined by (22.23) we have the possibility to construct a selfconsistent system including fluctuation phenomena.

As a first step of the approximation we neglect higher-vertex corrections in agreement with the high-density limit, see Chapter 11.3.1, and derive from (22.24) for the vertex function

$$\Gamma_{a\bar{a}d}\left(\bar{1}, \overset{=}{1}, 2\right) \approx \delta_{a\bar{a}}(1-\bar{1})\delta_{ad}(\bar{1}-2). \qquad (22.57)$$

In the framework of this relativistic random-phase approximation we get from (22.17), (22.18), (22.23), (22.26), (22.52) and (22.57) the complete set of equations

$$\left[G^{r^{-1}}, G^<\right](p, X) = G^<(p, X)\Sigma^>(p, X) - G^>(p, X)\Sigma^<(p, X)$$

$$\Sigma^{\gtrless}(xX) = -ig_0^2 \, G^{\gtrless}\left(X + \frac{x}{2}, X - \frac{x}{2}\right) d^{\lessgtr}\left(X - \frac{x}{2}, X + \frac{x}{2}\right)$$

$$d^{\gtrless}(p, X) = \frac{1}{\left|4\pi g_0^2\right|^2} V^2(p, X)\Pi^{\gtrless}(p, X)$$

$$\Pi^{\gtrless}(xX) = ig^2 \, \mathrm{Tr}\left\{G^{\gtrless}\left(X + \frac{x}{2}, X - \frac{x}{2}\right) G^{\lessgtr}\left(X - \frac{x}{2}, X + \frac{x}{2}\right)\right\}, \qquad (22.58)$$

where all variables mean four-vectors and p is the Fourier transform of x. It has to be stressed that the *dynamical* potential introduced by the optical theorem (22.53) now contains an infinitesimal sum of nucleon fluctuations by the polarisation function resulting in an effective meson exchange mass. Therefore this approximation can be considered as a modified first Born approximation including collective effects, especially density fluctuations. The set of equations (22.58) forms a closed set and determines the correlation function $G^{\gtrless}$.

We proceed with the Kadanoff and Baym ansatz (8.33) neglecting any nonlocalities and off-shell contributions. With the help of (22.32) and (22.33) we obtain for $G^<$

$$G^< = \frac{i\pi}{\sqrt{P_1^2 + \kappa^2}} \left\{ \left(\gamma^0 \epsilon_+ - \gamma^1 P_1 + \kappa\right) \, \mathrm{f}(\epsilon_+) \, \delta\,(\omega - \epsilon_+) \right.$$

$$\left. - \left(-\gamma^0 \epsilon_- - \gamma^1 P_1 + \kappa\right) \mathrm{f}(-\epsilon_-) \, \delta\,(\omega + \epsilon_-) \right\}. \qquad (22.59)$$

If we now introduce (22.35) and (22.59) into the kinetic equation (22.58) we have to perform the Poisson brackets on the left side carefully. After partial integration one can show that the renormalisation denominator, which arises from the spectral function (22.32) cancels exactly with the factors following from the gradient expansion on the left side so that the drift term takes the form

$$\frac{\partial}{\partial t} f + \nabla_r \epsilon_+ \nabla_p f - \nabla_p \epsilon_+ \nabla_r f$$

and analogously for antiparticle $\bar{f}$ with ϵ_-. We want to restrict this to a virtual meson exchange and neglect all density terms in the meson Green's function. After somewhat extensive but straightforward calculation using the set of equations (22.58) we finally arrive at the kinetic equation for the particle distribution, if we integrate both sides over positive frequencies. For convenience we use the following abbreviation of the *dynamical potential* (22.53) without static approximation

$$V^2(p, \bar{p}, p_1, \bar{p}_1) = \left(2(p\bar{p}_1)(\bar{p}p_1) + 2(pp_1)(\bar{p}\bar{p}_1) - 2\kappa^2[(p_1\bar{p}_1) + (p\bar{p})] + 4\kappa^4\right) V_\omega^2(\bar{p} - p)$$

$$- (p\bar{p} + \kappa^2)(p_1\bar{p}_1 + \kappa^2) V_\sigma^2(\bar{p} - p), \qquad (22.60)$$

with $V_\omega^2(p) = g_\omega^4/(p^2 - m_\omega^2 + \mathrm{Re}\Pi_\omega^R)$ and correspondingly for V_σ^2. As already mentioned following (22.26) no mixed terms between g_ω^2, g_σ^2 occur in the cross sections due to the additive behaviour of the different self energies. The kinetic equation is the Landau-Silin one (9.24). The scattering integral now contains eight processes (Morawetz and Kremp, 1995) which can be seen from the off-shell momentum (22.30) according to $\pm p_1, \pm \bar{p}, \pm \bar{p}_1$ when p is fixed.

The first process describes the elastic particle–particle scattering-in for the s-channel $p_1 + p \to \bar{p}_1 + \bar{p}$, for scattering-out

$$\int \frac{d\bar{p}^3}{(2\pi)^3} \frac{dp_1^3}{(2\pi)^3} \frac{d\bar{p}_1^3}{(2\pi)^3} (2\pi)^4 \delta^{(4)} (\bar{p}_1 + \bar{p} - p_1 - p) \frac{1}{|\epsilon_+ \bar{\epsilon}_1 \epsilon \bar{\epsilon}|}$$

$$\times V^2(p, \bar{p}, p_1, \bar{p}_1) \left\{ f_{\bar{p}} f_{\bar{p}_1} N\left(f_p f_{p_1} \right) - f_p f_{p_1} N\left(f_{\bar{p}} f_{\bar{p}_1} \right) \right\}. \qquad (22.61)$$

Here and in the following we used the Pauli blocking factors

$$N(\bar{f}) = 1 - \bar{f}, \quad N(\bar{f} f) = (1 - \bar{f})(1 - f), \quad N(\bar{f} f G) = (1 - \bar{f})(1 - f)(1 - G).$$

The next two processes are the crossing symmetric processes of (22.61) in the corresponding t-channel $p - \bar{p}_1 \to \bar{p} - p_1$ and u-channel $p - \bar{p} \to \bar{p}_1 - p_1$. They are obvious by the crossing symmetry and describe the elastic particle–antiparticle scattering as illustrated in Figure 22.2.

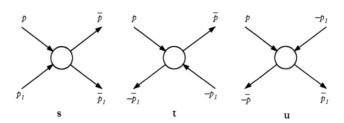

Figure 22.2: *The three elastic-scattering process of s-, t-, and u-channel respectively.*

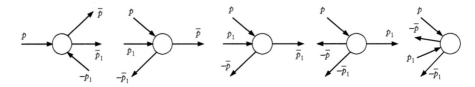

Figure 22.3: *The five inelastic scattering process only possible due to medium effects.*

The next five processes illustrated in Figure 22.3 are only possible due to the off-shell character of the quasi-particle energies (22.33) and are forbidden for free particles

$$
\int \frac{d\bar{p}^3}{(2\pi)^3} \frac{dp_1^3}{(2\pi)^3} \frac{d\bar{p}_1^3}{(2\pi)^3} (2\pi)^4 \frac{1}{|\epsilon_+ \bar{\epsilon}_1 \epsilon \bar{\epsilon}|}
$$

$$
\times [-V^2(p,\bar{p},-p1,\bar{p}1)\delta^{(4)}(\bar{p}_1 + p_1 + \bar{p} - p) \{f_p N (f_{\bar{p}}\bar{f}_{p_1}f_{\bar{p}_1}) - f_{\bar{p}}f_{\bar{p}_1}\bar{f}_{p_1} N(f_p)\}
$$

$$
- V^2(p,\bar{p},p_1,-\bar{p}_1)\delta^{(4)}(-\bar{p}_1 - p_1 + \bar{p} - p) \{f_p f_{p_1}\bar{f}_{\bar{p}_1} N(f_{\bar{p}}) - f_{\bar{p}} N (f_p f_{p_1}\bar{f}_{\bar{p}_1})\}
$$

$$
- V^2(p,\bar{p}_1,p_1,-\bar{p})\delta^{(4)}(-\bar{p} - p_1 + \bar{p}_1 - p) \{f_p f_{p_1}\bar{f}_{\bar{p}} N(f_{\bar{p}}) - \bar{f}_{p_1} N (f_p f_{p_1}\bar{f}_{\bar{p}})\}
$$

$$
- V^2(p,-\bar{p},-p_1,-\bar{p}_1)\delta^{(4)}(-\bar{p}_1 + p_1 - \bar{p} - p) \{f_{\bar{p}}\bar{f}_{\bar{p}}\bar{f}_{\bar{p}_1} N(\bar{f}_{p_1}) - \bar{f}_{p_1} N (f_{\bar{p}}\bar{f}_{\bar{p}}\bar{f}_{\bar{p}_1})\}
$$

$$
+ V^2(p,-\bar{p},p_1,-\bar{p}_1)\delta^{(4)}(-\bar{p}_1 - p_1 - \bar{p} - p)
$$

$$
\times \{f_p \bar{f}_{\bar{p}}f_{p_1}\bar{f}_{\bar{p}_1} - (1-f_p)(1-\bar{f}_{\bar{p}})(1-f_{p_1})(1-\bar{f}_{\bar{p}_1})\}]. \tag{22.62}
$$

This can be understood as particle creation and destruction by means of virtual meson exchange. It is important to recognise that these processes are special effects of many-particle surrounding and are forbidden by momentum and energy conserving δ-functions in the case of vanishing many-particle influence. Analysing the energy and momentum conserving δ-function one finds in connection with (22.33) that these processes occur if the c.m. energy $\sqrt{s}$ fulfils the condition

$$
\sqrt{s} = |p + p_1| = -2Re\Sigma_o^R. \tag{22.63}
$$

This means that, in a nucleus system, where the zero part of the selfenergy becomes negative, new reaction channels will be opened by many-particle influence. One may think on pseudo-scalar and pseudo-vector meson coupling, where the absence of mean-field terms of the vector mesons ensures the condition (22.63).

Further, it can be seen explicitly that the relativistic treatment yields some fore-factors (22.60) to the dynamical potentials (22.53). The dynamical potentials have to be determined self-consistently with the kinetic equation and the polarisation function (22.58). Consequently, we arrived at a generalised relativistic Lenard–Balescu equation. The dynamical potentials reflect the influence of virtual mesons and are written for each process with the influence of the dynamic density fluctuations by (22.53).

To complete, we give the in-medium cross sections derived from the elastic process (22.61). For abbreviation, we denote $Re\Sigma_o^R = \Delta$ and $\kappa = \kappa_0 + Re\Sigma_i^R$. The proton-proton cross section (10.162) can be identified from the $V^2/|\epsilon\epsilon_1\bar{\epsilon}\bar{\epsilon}_1|$ term of (22.60) and reads

$$
\frac{d\sigma}{d\Omega} = \frac{1}{32\pi^2} \left\{ \frac{2g_\omega^4}{s(m_\omega^2+t)^2} \left[\left(\Delta(\sqrt{s}+\Delta) + \kappa_0^2 - \frac{u}{2} \right)^2 + \left(\Delta(\sqrt{s}+\Delta) - \kappa_0^2 + \frac{s}{2} \right)^2 \right. \right.
$$

$$
\left. -4\kappa^2 \left(\Delta(\sqrt{s}+\Delta) - \frac{t}{2} \right) + 4\kappa^2 \left(4\kappa^2 - \kappa_0^2 \right) \right] - \frac{2g_\sigma^4}{s(m_\sigma^2+t)^2} \left(\Delta(\sqrt{s}+\Delta) + \kappa^2 + \kappa_0^2 - \frac{t}{2} \right)^2 \right\}. \quad (22.64)
$$

Here $s = (p+p_1)^2$, $t = (p-\bar{p}_1)^2$, and $u = (p-\bar{p})^2$ are the Mandelstam variables. This means that the in-medium cross sections are affected by the dressed Baryon mass and an additional term arises which is proportional to the vector part of the self energy.

23

Simulations of Heavy-Ion Reactions with Nonlocal Collisions

23.1 Scenario of low-energy heavy-ion reactions

Among all nonequilibrium systems investigated by physics, the features of Fermi systems met in nuclear physics are special. During a heavy-ion collision we can access a state of matter which gives insight into special aspects of nonequilibrium processes. The dominant features are that we have strong short-range interactions of typically 1 fm = 10^{-15} m and the product of the range of interaction with Fermi momentum is in the orders of one characteristic for a degenerate quantum system. Since the radius of typical nuclei is $R \approx 1.2A^{1/3}$ fm we see that the product of the radius with Fermi momentum is of few $\hbar$ indicating strong spatial inhomogeneity. It is useful to start by sketching out the scenario of a reaction with a projectile energy $E/A < 100$ MeV.

One can crudely talk about three stages of central reactions (Stöcker and Greiner, 1986; Bertsch and Gupta, 1988; Aichelin, 1991). In the first touch of two nuclei, nucleons at the front sides are highly compressed reaching about twice the normal density. In this stage one can expect that correlations between projectile and target nucleons develop. In the second stage, two penetrating streams of nucleons try to establish a local equilibrium, while the whole system expands and adiabatically cools down. In the last stage, the system may break into clusters if fragmentation is happening. The kinetic equation is primarily oriented on the second stage.

With increasing excitation energy, the theoretical description shifts from correlations (in the ground state) towards the dynamics of particles in the self-consistent mean field (for $E/A < 10$ MeV) and finally to dominant two-particle collisions (for $E/A > 10$ MeV). These different energy domains are naturally described within different theoretical approaches. A unified theory based on nonequilibrium Green's functions covers all these energy domains.

Interacting Systems far from Equilibrium. Klaus Morawetz, Oxford University Press (2018).
© Klaus Morawetz. DOI: 10.1093/oso/9780198797241.001.0001

23.1.1   Numerical simulations of heavy-ion reactions

Refined non-relativistic two-particle calculations using Brueckner theory (Köhler, 1975) and approximation beyond (Hjorth-Jensen, Muether, Polls and Osnes, 1996) are not able to reproduce the correct saturation of nuclear matter, see Figure 22.1. An agreement has been achieved within relativistic treatments (Walecka, 1974; Jaminon, Mahaux and Rochus, 1981; Horowitz and Serot, 1983; Serot and Walecka, 1986; Nuppenau, Mackellar and Lee, 1990; Gross-Boelting, Fuchs and Faessler, 1999). The major relativistic contribution comes from the effect of nucleons on mesons which results in effectively density-dependent binary nucleon-nucleon interaction (Bouyssy, Mathiot and Giai, 1987). This feature can be captured by three-nucleon interactions (Carlson, Pandharipande and Wiringa, 1983; Jackson, Rho and Krotscheck, 1983). On a phenomenological level one can account for the nonlinear density dependence by an effective potential, originally introduced by Skyrme (Skyrme, 1956, 1959).

Numerical simulations extensively used to interpret experimental data from heavy-ion reactions, are based either on the Boltzmann (BUU) equation or on the quantum molecular dynamics (QMD). Due to their quasiclassical character, they offer a transparent picture of the internal dynamics of reactions and allow one to link the spectrum of the detected particles with individual stages of reactions. They fail, however, to describe some energy and angular distributions of neutrons and protons in low- and mid-energy domains (Tõke and et. al., 1995; Baldwin and et. al., 1995; Skulski and et. al., 1996). Appreciable values of the collision delay and space displacements show that the nonlocal collisions should be accounted for. In this chapter we show that they improve the agreement between simulation results and experimental data.

The nonlocal collision has been implemented in (Morawetz, Špička, Lipavský, Kortemeyer, Kuhrts and Nebauer, 1999c) within the QMD and in (Morawetz, 2000a; Morawetz, Ploszajczak and Toneev, 2000b) within the BUU equation. The full set of corrections (13.28), however, represents a serious numerical demand. Two additional approximations were used to simplify the numerical treatment. Firstly, according to the approximations used within the Boltzmann equation, the medium effect on binary collision has been neglected, i.e. the Δ's are evaluated from the free-space T-matrix discussed in Chapter 14. Secondly, the scattering integral was rearranged into an instant but nonlocal form. As shown in (Lipavský, Špička and Morawetz, 1999), the instant approximation is compatible with the variational definition of the quasiparticle energy, i.e. it is consistent with the Skyrme approximation of the mean field. Moreover, the instant form allows one to employ computational methods developed within the theory of gases (Alexander, Garcia and Alder, 1995) similarly as it has been done in (Kortemeyer, Daffin and Bauer, 1996).

23.2   Instant nonlocal approximation

The non-instant collision corresponds to an event in which particles a and b of momentum k and p start to collide at time t being at positions r_a and $r_b = r_a + \Delta_2$. The collision ends at time $t + \Delta_t$ when the particles have momenta $k' = k - q + \Delta_K$ and $p' = p + q + \Delta_K$

being at positions $r'_a = r_a + \Delta_3$ and $r'_b = r_a + \Delta_4$. In the instant approximation, this non-instant event is replaced by an effective one in which particles make sudden jumps such that their asymptotic motion for $t \to \infty$ is exactly reproduced.

The first question is: to which time instant $\tilde{t}$ the collision should be attributed. Two different choices are important with respect to practical implementations. The central time, $\tilde{t} = t + \frac{1}{2}\Delta_t$ represents the most natural theoretical choice. Moreover, it maintains the time and space reversal symmetries which are vital for the evaluation of effective Δ's from scattering phase shifts, see (14.38). The closest-approach time, $\tilde{t} = t_{ca}$ defined as the time instant at which the approaching particles would reach the shortest distance, $\left| r_a + (t_{ca} - t)\frac{k}{m} - r_b - (t_{ca} - t)\frac{p}{m} \right| = $ min., violates this symmetry but it corresponds to the time instant easily identified within numerical simulation codes.

Initial and final states   Regardless of the choice of the time instant, the effective collision results from the extrapolation of the non-instant event into time $\tilde{t}$. Extrapolated coordinates

$$\tilde{r}_a = r_a + \frac{k}{m}(\tilde{t} - t), \qquad \tilde{r}_b = r_b + \frac{p}{m}(\tilde{t} - t),$$

$$\tilde{r}'_a = r'_a + \frac{k-q}{m}(\tilde{t} - t - \Delta_t), \quad \tilde{r}'_b = r'_b + \frac{p+q}{m}(\tilde{t} - t - \Delta_t) \tag{23.1}$$

and extrapolated momenta

$$\tilde{k} = k - \frac{\partial U}{\partial r_a}(\tilde{t} - t), \qquad \tilde{p} = p - \frac{\partial U}{\partial r_b}(\tilde{t} - t),$$

$$\tilde{k}' = k - q + \Delta_K - \frac{\partial U}{\partial r_a}(\tilde{t} - t - \Delta_t) \quad \tilde{p}' = p + q + \Delta_K - \frac{\partial U}{\partial r_b}(\tilde{t} - t - \Delta_t) \tag{23.2}$$

represent initial and final states of the effective event.

Conserving quantities   From the properties of the free-particle T-matrix (14.22) and extrapolation (23.2) one finds that the total momentum conserves in the effective event,

$$\tilde{k}' + \tilde{p}' = \tilde{k} + \tilde{p}. \tag{23.3}$$

In the extrapolation towards the instant event, the momentum gain is naturally cancelled by the effect of the mean-field force on particles during free motion. Indeed, within the free-particle T-matrix coupled to other particles only via the mean field, the momentum gain describes only the effect of forces during the collision.

Similarly, from (14.22) and (23.2), one finds that energy conservation has no energy gain,

$$\epsilon_a(\tilde{k}', \tilde{r}'_a, \tilde{t}) + \epsilon_b(\tilde{p}', \tilde{r}'_b, \tilde{t}) = \epsilon_a(\tilde{k}, \tilde{r}_a, \tilde{t}) + \epsilon_b(\tilde{p}, \tilde{r}_b, \tilde{t}). \tag{23.4}$$

Briefly, in the instant event the mean field has no time to pass any momentum and energy to the colliding pair.

Finally, in agreement with the continuity of the centre of mass motion, one finds from (14.31) and (23.2) that

$$\tilde{r}'_a + \tilde{r}'_b = \tilde{r}_a + \tilde{r}_b. \tag{23.5}$$

This relation reduces a number of independent displacements. One has to find only the initial relative displacement $\tilde{\Delta} = \tilde{\Delta}_2 = \tilde{r}_b - \tilde{r}_a$ and the final relative displacement $\tilde{\Delta}' = \tilde{\Delta}_4 - \tilde{\Delta}_3 = \tilde{r}'_b - \tilde{r}'_a$.

23.2.1   Displacements from the on-shell shifts

Now we take the central time $\tilde{t} = t + \frac{1}{2}\Delta_t$ and evaluate the relative displacements $\tilde{\Delta}$ and $\tilde{\Delta}'$. From extrapolation (23.2), (14.30) and (13.3) one finds

$$\tilde{\Delta} = -\frac{\partial\hat{\phi}}{\partial\kappa} - \frac{\kappa}{2\mu}\frac{\partial\hat{\phi}}{\partial\omega}, \tilde{\Delta}' = \frac{\partial\hat{\phi}}{\partial\kappa_f} + \frac{\kappa_f}{2\mu}\frac{\partial\hat{\phi}}{\partial\omega}, \tag{23.6}$$

where $\kappa = \frac{1}{2}(\tilde{k} - \tilde{p})$ and $\kappa_f = \kappa - q$ are initial and final momenta in the barycentric system, and $\mu = m/2$ is the relative mass.

Relations (23.6) can be written as on-shell derivatives

$$\tilde{\Delta} = -\frac{\partial\phi_{on}}{\partial\kappa}, \tilde{\Delta}' = \frac{\partial\phi_{on}}{\partial\kappa_f}, \tag{23.7}$$

of the on-shell scattering phase shift

$$\phi_{on}(\kappa, \kappa_f) = \hat{\phi}(\omega, \kappa, \kappa_f)\Big|_{\omega = \frac{1}{2}\left(\frac{\kappa^2}{2\mu} + \frac{\kappa_f^2}{2\mu}\right)}. \tag{23.8}$$

We note that it is customary to define the on-shell value with respect to the initial or final kinetic energy only, $\omega = \frac{\kappa^2}{2\mu}$ or $\omega = \frac{\kappa_f^2}{2\mu}$, see (Goldberger and Watson, 1964). The symmetric choice (23.8) has an advantage that it preserves the symmetry between the initial and final momenta needed to evaluate the derivatives with respect to the absolute value of the momentum.

To complete the evaluation of the relative displacements, one can use formulae derived in Chapter 14, substituting ϕ_{on} instead of $\hat{\phi}$. It is easy to check that this substitution works correctly. Since ϕ_{on} does not depend on ω, all contributions due to the collision delay drop out so that (14.22) gives no gains of energy and momentum and (14.31) reduces to the continuity of the centre-of-mass motion. Accordingly, it is sufficient to replace $\hat{\phi}'$ by

$$\phi'_{on}(|\kappa|, \cos\theta) = \phi_{on}(\kappa, \kappa_f)\Big|_{|\kappa_f| = |\kappa|} \tag{23.9}$$

in relations (14.40) and (14.41) and take $\tilde{\Delta} = \tilde{\Delta}_2 = \tilde{\Delta}_4 + \tilde{\Delta}_3$ and $\tilde{\Delta}' = \tilde{\Delta}_4 - \tilde{\Delta}_3$. As before, θ is the deflection angle connected to momenta, $\cos\theta = \frac{\kappa \kappa_f}{|\kappa||\kappa_f|}$.

The scattering phase shift ϕ'_{on} is observable in the scattering of isolated nucleons and usually only ϕ'_{on} is found in textbooks. The nonlocal corrections in the instant approximation can thus be evaluated directly from experimental values with no regard to models of the interaction potential.

Closest approach time   In simulation codes, two particles are selected for a collision if they meet at the point of closest approach, i.e. at $\tilde{t} = t_{ca}$. Naturally, their displacement at t_{ca} is different from the displacement $\tilde{\Delta}$ required within the centre-of-time picture. This mismatch can be corrected following the scheme of Thirring (1981).

The time which is required to travel from $\tilde{\Delta}$ to the point of closest approach is

$$\tilde{\Delta}_t = t_{ca} - t - \frac{1}{2}\Delta_t = \frac{\kappa \tilde{\Delta}}{|\kappa|} \frac{\mu}{|\kappa|}. \tag{23.10}$$

By $\tilde{\Delta}_t$ we have to transform the coordinates of particles. From (23.2) one finds that the relative displacements Δ_{ca} and Δ'_{ca} at t_{ca} before and after the re-displacement read

$$\Delta_{ca} = \tilde{r}_b + \frac{p}{m}\tilde{\Delta}_t - \tilde{r}_a - \frac{k}{m}\tilde{\Delta}_t = \tilde{\Delta} - \frac{\kappa}{\mu}\tilde{\Delta}_t,$$

$$\Delta'_{ca} = \tilde{r}'_b - \frac{p+q}{m}\tilde{\Delta}_t - \tilde{r}'_a + \frac{k-q}{m}\tilde{\Delta}_t = \tilde{\Delta}' + \frac{\kappa_f}{\mu}\tilde{\Delta}_t. \tag{23.11}$$

Displacements Δ_{ca} and Δ'_{ca} are the values incorporated into simulations codes.

Scattering styles   To explain how the nonlocal corrections are incorporated into the QMD or BUU simulation codes, it is necessary to clarify how the code deals with the collision without nonlocal corrections. Two particles are selected for a collision if their positions R_a and R_b are closer than the radius R_σ of the scattering cross section and their trajectories are at the point of closest approach. The deflection angle is then randomly selected, but the collision plane (given by initial and final relative momenta) is not correlated with the direction $R_b - R_a$. In this scheme called after Halbert (1981), the $4\pi I$ style (4π impact scattering), the displacements $R_b - R_a$ have no correlation with the transferred momentum, therefore their mean contribution to the collision flux vanishes.

In an alternative scattering style denoted RN by Halbert (repulsive in-plane scattering with no core), the collision plane is identified from the initial relative momentum and $R_b - R_a$. In the RN style, the displacement $R_b - R_a$ is strongly correlated with the transferred momentum because their components are orthogonal to the initial momentum and always point in the same direction having the same sign. The difference between these two schemes is convincingly demonstrated by the numerical results of Halbert.

For collisions of nucleons at small and medium energies, $\kappa^2/2\mu < 50$ MeV, the radius of the cross section, R_σ is much larger than the expected displacement Δ_{ca}. The simplest approximation is thus to neglect Δ_{ca} using the $4\pi I$ style with the final displacement only.

We found this approximation sufficient for heavy ion collisions, at least for tested cases we have not found any effect visible on the scale of numerical fluctuations. If it is of interest, one can include the initial displacement combining the $4\pi I$ style with the RN style in the share which reproduces the collision flux due to Δ_{ca}.[1]

The implementation of the final displacement Δ'_{ca} does not depend on the scattering style. Once the collision is selected, one simply shifts the particle b by $\frac{1}{2}\Delta'_{ca}$ and the particle a by $-\frac{1}{2}\Delta'_{ca}$. This scheme was used in (Morawetz, Špička, Lipavský, Kortemeyer, Kuhrts and Nebauer, 1999c).

23.2.2 Realistic displacements

From (23.11) one finds that the final displacement has two components,

$$\Delta'_{ca} = \frac{\partial \phi'_{on}}{\partial |\kappa|} n_{\parallel} + \frac{1}{|\kappa|} \frac{\partial \phi'_{on}}{\partial \cos \theta} n_{\perp}, \tag{23.12}$$

where $n_{\parallel}$ is the vector parallel to κ_f and $n_{\perp}$ is perpendicular to $n_{\parallel}$ while it stays in the collision plane defined by κ_f. As shown in Fig. 23.1, the parallel component is much larger than the orthogonal one. The displacements have been evaluated from the Bonn potential which correctly reproduces the experimental phase shifts.

The component parallel to the final relative momentum κ_f has a nearly constant value of 0.5 fm at energies close to the Fermi energy ~ 40 MeV. This displacement gives the dominant contribution. The orthogonal component reaches smaller values of the order of 0.2 fm. Moreover, the orthogonal component has both positive and negative signs so that its net contribution to the system dynamics is further reduced.

The numerical value of the initial displacement Δ_{ca} equals the orthogonal component of Δ'_{ca}. Indeed, at the point of closest approach, the initial displacement is orthogonal to the initial momentum. From the symmetry it thus follow that these values are equal. Numerical tests made within studies discussed below have shown that the initial displacement and the orthogonal component of the final displacement can be neglected.

The parallel displacement is identical to the correction obtained from Wigner's collision delay discussed by Danielewicz and Pratt (1996). The numerical results show that this correction is the dominant one for nuclear matter.

Perhaps we should comment on the model of classical hard spheres used in earlier simulations. For hard spheres, the displacement points in direction of the transferred momentum $\kappa - \kappa_f$ which determines a relation between the amplitudes of parallel and orthogonal components. Displacements evaluated from scattering phase shifts do not follow this relation which signals that the model of hard spheres is not suitable for nuclear matter.

---

[1] The scattering style RC (repulsive in-plane with hard-sphere hard core) proposed by Halbert (1981) for an incorporation of the collision flux which is smaller that the hard-sphere flux is not applicable to nuclear matter since its lowest value given by the RC style already overestimates the flux due the correlation between $R_b - R_a$ and the transferred momentum.

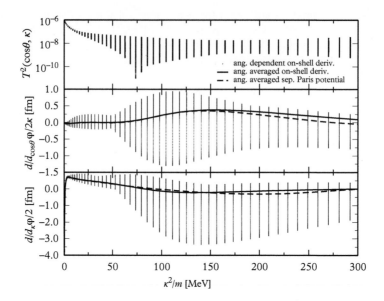

Figure 23.1: *The parallel and orthogonal components of the final displacement* Δ'_{on}. *The vertical lines composed of dots show the spread of components with deflection angle. The lines show the angle-averaged values which are used in the QMD code where the collisions are treated in the isotropic approximation. The amplitude of the T-matrix is presented (above) to indicate the weight of individual processes.*

23.3   Numerical simulation results

Till now there are only few implementations of nonlocal collisions. Nevertheless, these results show that the nonlocal corrections lead to a better agreement between the numerical simulations and experimental data.

Production of light clusters   The nonlocal collision within the above approximation has been used in (Morawetz, Špička, Lipavský, Kortemeyer, Kuhrts and Nebauer, 1999c) to explain the spectrum of light clusters emitted from the central reaction of ^{129}Xe→^{119}Sn at 50 MeV/A. The result of this QMD simulation presented in Fig. 23.2a shows the exclusive proton spectra subtracting the protons bound in clusters. The subtraction has been performed within a spanning tree model which is known to describe a production of a light-charged cluster in a reasonable agreement with the experimental data, Figs. 23.2b–f.

Within the local approximation, the distribution of high-energy protons is too low to meet the experimental values. As one can see, the inclusion of nonlocal collisions corrects this shortage of the QMD simulation. As demonstrated in Fig. 23.2, productions of other light clusters are rather insensitive to nonlocal corrections. This shows that the improvement of the proton production is not at the cost of worse results in other spectra.

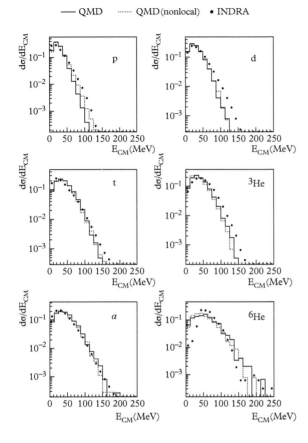

— QMD ⋯⋯ QMD(nonlocal) • INDRA

Figure 23.2: *The particle spectra for central collision of* $^{129}Xe\rightarrow$ ^{119}Sn *at 50 MeV/A with and without nonlocal corrections. The data are extracted from recent INDRA experiments. The nonlocal corrections bring the spectrum of the protons towards the experimental values leaving the clusters almost unchanged.*

The energy spectra of emitted particles does not allow one to trace down the microscopic mechanism responsible for their shapes. This is in part due to the integral character of the observed quantity in part due to the applied coalescence model by the spanning tree program. One may speculate that the high energy protons appear due to enhanced collision rate seen in the simulation (Morawetz, Špička, Lipavský, Kortemeyer, Kuhrts and Nebauer, 1999c). Later studies shown, however, offer the interpretation that this enhancement of spectra is due to more energy being transferred during the nonlocal collisions compared to the local collision scenario.

In order to trace down the microscopic mechanism we have to search for more direct observables which are not dependent on additional assumptions like cluster formation programmes or coalescence models. This will be the charge-velocity distribution.

Neck between target and projectile The influence of the nonlocal collisions on the reaction of heavy ions has been studied for the $^{181}_{73}Ta + ^{197}_{79}Au$ reaction at 33MeV (Morawetz, Lipavský, Normand, Cussol, Colin and Tâmain, 2001b). The nonlocal corrections have

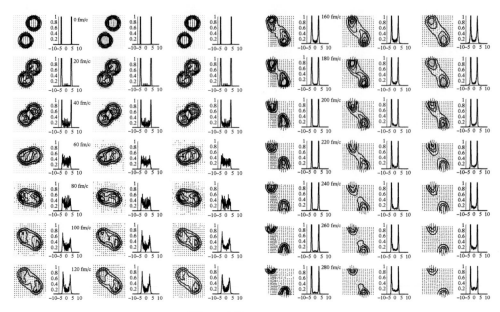

Figure 23.3: *The time evolution of $^{181}_{73}Ta + ^{197}_{79}Au$ collisions at $E_{lab}/A = 33$ MeV and 8 fm impact parameter in the local BUU (left), nonlocal BUU (middle), and the nonlocal BUU with quasiparticle renormalisation (right). Plots in the first column show the $(x-z)$-density cut of 30 fm × 30 fm where Ta as projectile comes from below. The mass momenta are shown by arrows. The corresponding second column gives the charge density distribution versus relative velocity in cm/ns where the target like distribution of Au is on the left and the projectile-like distributions of Ta on the right.*

been implemented into the BUU simulation code and compared with the treatment using the local collisions.

The time evolution of the particle density can be seen in the left (local) and middle (nonlocal) pictures of Fig. 23.3. Let us focus on the local picture (leftmost column) first. The *Ta* projectile approaches the *Au* target from below. At initial time, 0 fm/c, the two nuclei are well separated in two spheres of nuclear matter. The arrows indicate the mean velocities of local matter. At a time of about 20 fm/c, the two nuclei touch and start to build up a composed object. During the next time period from 20 – 200 fm this composed nucleus rotates then changes its shape and emits matter which can be seen as the spreading of small velocity arrows. At around 200 fm/c the composed nucleus breaks into projectile- and target-like fragments. At a short time period of 160–180 fm/c we see that a part of matter is located between the projectile- and target-like fragments, called the neck.

The time evolution of the reaction within the nonlocal scenario (middle figures) has identical gross features. Two differences can be noticed, however. There is higher emission of matter visible in particular at 60 fm/c. Secondly, the neck contains more matter and stays longer, till 220 fm/c compared to 180 fm/c in the local case.

The formation of a neck-like structure in peripheral heavy ion reactions is important for the production of light clusters (Stuttgé and et al., 1992; Lecolley and et al., 1995; Larochelle and et al., 1999; Dempsey and et al., 1996; Tõke and et al., 1995; Chen and et al., 1996; Montoya and et al., 1994; Casini and et al., 1993; Stefanini, 1995; Bocage and et al., 2000). It has been suggested that the neck instability can be important for the fast decomposition of matter and is probably neutron rich (Colonna, DiToro and Guarnera, 1995; Sobotka, Dempsey, Charity and Danielewicz, 1997; Colonna, DiToro, Guarnera, Maccarone, Zielinska-Pfabé and Wolter, 1998; Fabbri, Colonna and DiToro, 1998). The too-weak neck, seen in the local picture, has also been found from the Landau equation (Pethik and Ravenhall, 1987; Donangelo, Dorso and Marta, 1991*a*; Donangelo, Romanelli and Schifino, 1991*b*) or BUU equations (Kiderlen and Hofmann, 1994; Ayik, Colonna and Chomaz, 1995).

Except for the nonlocal picture, a sufficiently large neck has been achieved by the additional / inclusion of fluctuations in the Boltzmann (BUU) equation (Chattopadhyay, 1995, 1996) resulting in Boltzmann–Langevin pictures (Suraud et al., 1990; Ayik and Gregoire, 1990; Randrup and Remaud, 1990; Ayik et al., 1992; Colonna et al., 1993, 1994). The Boltzmann–Langevin equation has been derived assuming an additional coarse graining of phase space (Reinhard et al., 1992; Reinhard and Suraud, 1992). Fluctuations to the TDHF equation have been analysed before by Balian and Veneroni (1985), Flocard (1989) and tested in (Troudet and Vautherin, 1985).

Hydrodynamic picture The process of collision seen in Fig 23.3 can be characterised by a hydrodynamic motion of matter. The spatial motion is visible from the already discussed density (left columns)

$$n(r, t) = \int \frac{dp}{(2\pi)^3} f(p, r, t). \tag{23.13}$$

It is instructive to view the same process with respect to the hydrodynamic velocity,[2]

$$v(r, t) = \frac{1}{m \, n(r, t)} \int \frac{dp}{(2\pi)^3} p f(p, r, t). \tag{23.14}$$

From this we define the distribution of hydrodynamical velocities

$$F(\bar{v}, t) = \int dr \, n(r, t) \, \delta(\bar{v} - v_{\text{fiss}}(r, t)) \tag{23.15}$$

---

[2] The mean velocity $v(r, t)$ does not include the Fermi energy. In the case that we have a different repartitioning of Fermi energy during the collision than described in our kinetic equation, e.g., by dynamical cluster formation, there will be an ambiguity. Since the dynamical cluster formation is not described in our approach we might have an effect of Fermi energy on the mass velocity. This confirms the observation that BUU or nonlocal kinetic equations have too much stopping compared to the experiment when more central collisions are considered. For peripheral collisions we believe that this kinetic description is sufficient, which we will prove by proper association of experimental events to the maximum in the velocity distribution.

where $v_{\text{fiss}}(r, t)$ is the projection of $v(r, t)$ onto the fission line. This distribution is close to the experimentally observable charge density distribution (Lecolley and et al., 2000; Bocage and et al., 2000; Plagnol and et al., 2000) if one chooses the proton density for $n(r, t)$.

In Figure 23.3 this normalised charge distribution is plotted versus velocity (corresponding right panels). Until times of 160 fm/c there is no difference between local and nonlocal scenarios. After that time, an appreciably higher distribution appears at small velocities (mid-rapidity) for the nonlocal scenario (mid panel). Note that this difference appears at the instant when the neck formation starts to differ in local and nonlocal scenarios. From the experimental point of view, it is important that this enhanced mid-rapidity distribution does not vanish at a larger time. Accordingly, it can be observed. We will compare experimental data with simulations later in Section 23.4.1 after introducing quasiparticle renormalisation.

Transformation of the potential and kinetic energies   Before we will proceed and add additional corrections to the BUU code, we would like to understand the microscopic origin of enhanced neck formation. Galichet (1998) has shown that the enhanced neck formation can be created by increasing the cross section and correspondingly the number of collisions. This applies only to smaller-impact parameters, $b < 6$ fm. For more peripheral reactions, as in Fig. 23.3 with $b = 8$ fm, the increase of the cross section does not lead to a visible difference.

In Fig. 23.4, the time evolution of the collision rate is presented for 8 fm at impact parameter and scattering cross section, which is artificially doubled. One can see that the number of collisions are visibly enhanced by doubling the cross section while for the nonlocal scenario there is only a slight enhancement at the beginning, and after 200 fm/c even lower values appear compared to the local BUU result. The latter fact comes from the earlier decomposition of matter in the nonlocal scenario. Apparently, the small enhancement of collision rate at about 100 fm/c due to non-localities is much lower that the artificial enhancement obtained from the doubled scattering cross section.

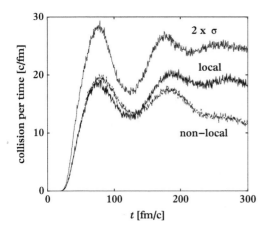

Figure 23.4: *The time evolution of the number of nucleon collisions for Ta + Au at $E_{lab}/A = 33$ MeV and different impact parameter in the BUU (thick black line), nonlocal kinetic model (broken line) and for the case of 8 fm impact parameter the local BUU with a doubled cross section (thin dark line).*

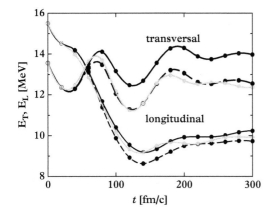

Figure 23.5: *The time evolution of the longitudinal (thin lines) and transverse energy (thick lines) including Fermi motion of nucleon collisions for Ta + Au at $E_{lab}/A = 33$ MeV and 8 fm impact parameter. The local BUU (black dashed line), nonlocal kinetic equation (solid line) and the local BUU but with twice the cross section (gre line) are compared.*

A different picture appears when one looks at transverse and longitudinal kinetic energy

$$E_T(t) = \int \frac{dp}{(2\pi)^3} dr \frac{p_x^2 + p_y^2}{2m} f(p, r, t)$$

$$E_L(t) = \int \frac{dp}{(2\pi)^3} dr \frac{p_z^2}{2m} f(p, r, t) \qquad (23.16)$$

shown in Fig. 23.5. Before the reaction, the transverse kinetic energy is exclusively due to the Fermi energy which requires finite velocities of nucleons even at zero temperature. In the equilibrium, the longitudinal energy would be half of the transverse one, its higher value in the beginning is due to the relative motion of the nuclei. Due to collisions, the equilibrium ratio is almost approached after 200 fm/c.

In the course of the reaction there is an oscillation between mean-field energy and transverse-kinetic energy due to an excited compressional mode. The simulation shows that the time evolution of the kinetic energy is not sensitive to the actual value of the cross section. On the other hand, the nonlocal scenario leads to an increase of the transverse and longitudinal energies of about 2 MeV and 1 MeV respectively. The nonlocal scenario thus leads to a different transformation between the potential and kinetic energies. This is likely the mechanism for the enhancement of high-energetic particles in Fig. 23.2. The increase of the longitudinal kinetic energy is connected to the enhanced formation of the neck.

23.4   Quasiparticle renormalisation

So far we have discussed the nonlocal shifts as if there were free classical particles. The interaction affects, however, the free motion of particles between individual collisions.

The dominant effect is due to mean-field forces which bind the nucleus together, accelerating particles close to the surface towards the centre. These forces are conveniently included via potentials of the Skyrme and Hartree type.

Beside forces, the interaction also modifies the velocity with which a particle of a given momentum propagates in the system. This effect is known as the mass renormalisation. A numerical implementation of the renormalised mass is rather involved since a plain use of the renormalised mass instead of the free one leads to incorrect currents. Within the Landau concept of quasiparticles, this problem is cured by the back flow, but it is not obvious how to implement the back flow within the BUU simulation scheme.

One can circumvent the problem of back flows using explicit zero-angle collisions to which one adds a nonlocal correction. This trick is based on the fact that the Pauli exclusion principle blocks a majority of phase space cell on the energy shell so that the zero-angle collisions dominate over the dissipative events. One can use these blocked events to simulate the renormalisation of the mass (Morawetz, Lipavský, Normand, Cussol, Colin and Tamain, 2001b).

Assuming that, at each blocked collision, there is a displacement D oriented along the difference momentum,

$$\Delta = |\Delta| \frac{k-p}{|k-p|}, \tag{23.17}$$

where k is a momentum of the particle while p belongs to its partner in the prohibited collision. In general, the displacement $|\Delta|$ is a function of k and p. For simplicity we take a constant value $|\Delta|$ and fit it to the mass renormalisation.

We fit the displacement Δ to zero temperature when all collisions are blocked by the Pauli exclusion principle. The mean velocity of the particle is then given by its free motion and the mean value of the displacements per the time unit,

$$v = \frac{k}{m} + \sum_a \int \frac{dp}{(2\pi\hbar)^3} \sigma \frac{|k-p|}{m} f_p |\Delta| \frac{k-p}{|k-p|}. \tag{23.18}$$

The mean value of displacements is proportional to the frequency of binary entertainments, i.e. it is the sum of integrals over distributions of protons and neutrons weighted with the scattering cross-section σ and their relative velocity to the observed particle.

With a good approximation the cross-section σ is independent of energy so that one can easily evaluate the integral in (23.18),

$$v = \frac{k}{m} + n\sigma|\Delta|\frac{k}{m} - n\sigma|\Delta|\langle v \rangle. \tag{23.19}$$

For the system at rest, $\langle v \rangle = 0$, we find the mass renormalisation,

$$\frac{m}{m^*} = 1 + n\sigma|\Delta|. \tag{23.20}$$

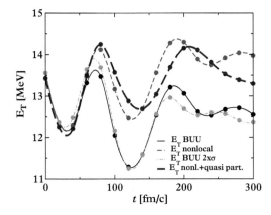

Figure 23.6: *The time evolution of the transverse energy including the Fermi motion for Ta + Au at $E_{lab}/A = 33$ MeV and 8 fm impact parameter in the local BUU (black line), the nonlocal BUU (dashed line), the local BUU with doubled cross section (dashed dotted line) and the nonlocal scenario with the quasiparticle renormalisation (long dashed line).*

Formula (23.20) allows us to fit $|\Delta|$ from known value of the effective mass. For m^* : $m = 3:4$, $\sigma = 40$ mb and $n = 0.16$ fm^{-3} one finds value $|\Delta| = 0.5$ fm. This value is very close to the nonlocal correction in dissipative collisions, see Fig. 23.1.

The quasiparticle velocity (23.19) relates to the quasiparticle energy by $v = \partial\epsilon_k/\partial k$. For the moving nuclear matter $\langle v \rangle \neq 0$ one finds

$$\epsilon_k = \frac{k^2}{2m} + n\sigma|\Delta|\frac{(k - m\langle v \rangle)^2}{2m}. \tag{23.21}$$

An approximation of this structure is commonly used in simple applications of the Landau concept of quasiparticles.

We see in Figure 23.3 (right panel) that the mid-rapidity distribution is yet more enhanced when one includes quasiparticle renormalisation. The detailed comparison of the time evolutions of the transverse energy for 8 fm impact parameter can be seen in Figure 23.6. We recognise that the transverse energies including quasiparticle renormalisation are similar to the nonlocal scenario and higher than in the local scenario.

The difference appears in the period of oscillation which corresponds to a giant resonance. This period becomes longer for the case of quasiparticle renormalisation. We can conclude that the compressibility decreases. In other words, the quasiparticle renormalisation leads to a softer equation of state. The nonlocal collision by itself does not have such strong effect on the compressibility.

23.4.1 Experimental charge density distribution

Recent INDRA observation shows the discussed enhancement of emitted matter in the mid-rapidity region (Bocage and et al., 2000; Plagnol and et al., 2000). The simulations can be compared to the experimental data of the $Ta + Au$ collision (Morawetz, Lipavský, Normand, Cussol, Colin and Tamain, 2001b). In Figure 23.7, the theoretical and experimental charge density distributions are compared. The experimental charge density distribution has been obtained using the procedure described in reference (Lecolley and et al., 2000). The data are represented by light grey points, the standard BUU calculation by the thin line and the nonlocal BUU with quasiparticle renormalisation calculation by the thick line. A reasonable agreement is found for the nonlocal scenario including quasiparticle renormalisation while simple BUU fails to reproduce mid-rapidity matter.

In summary, as documented by the improvement of the high-energy proton production and the midrapidity charge density distribution, the nonlocal treatment of the binary collisions brings a desirable contribution to the dynamics of heavy ion reactions. According to the experience from the theory of gases, one can also expect a vital role of non-localities in the search for the equation of state of nuclear matter. It is encouraging that the nonlocal corrections are easily incorporated into the BUU and QMD simulation codes and do not increase the computational time.

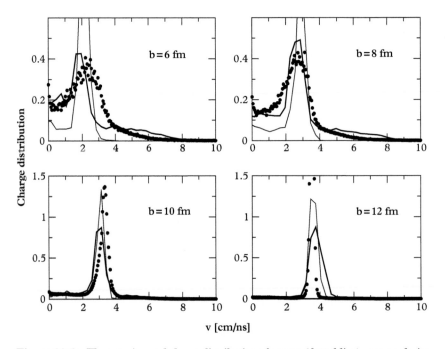

Figure 23.7: *The experimental charge distribution of matter (dotted line) versus velocity in comparison with in the BUU (thin solid line) and the nonlocal model with quasiparticle renormalisation (thick line), from (Morawetz, Lipavský, Normand, Cussol, Colin and Tamain, 2001b).*

Appendix A
Density-Operator Technique

A.1 BBGKY hierarchy

In the Schrödinger picture, the n-particle density operator is a superposition of state projections weighted with the probability ω_n

$$\hat{\rho}(t) = \sum_n \omega_n |\Psi_n(t)\rangle\langle\Psi_n(t)| \tag{A.1}$$

such that with the help of the Schrödinger equation the van–Neumann equation follows

$$\frac{d}{dt}\hat{\rho} = \frac{i}{\hbar}[\hat{\rho}, \hat{H}]. \tag{A.2}$$

This equation contains all the information and is reversible since from it follows for the entropy

$$\frac{d}{dt}\langle S\rangle = -k_B\frac{d}{dt}\langle\hat{\rho}\ln\hat{\rho}\rangle = -k_B\,\mathrm{Tr}(\ln\hat{\rho}-1)[\hat{\rho},\hat{H}] = 0. \tag{A.3}$$

Any approximate reduced information like single-, two-particle etc. density operator shows an entropy production due to the loss of information into higher-order terms. If we divide the N-particle Hamiltonian into an s-particle one, an N-s-particle one and the interactions between the s and N-s subsystems,

$$V_{s...N-s} = \sum_{i=1}^{s}\sum_{j=s+1}^{N} V_{ij} \tag{A.4}$$

we can trace-off from the van-Neumann equation the $s + 1...N$ particles to get for the s-particle density operator

$$\frac{d}{dt}\hat{\rho}_{1...s} = \frac{i}{\hbar}\,\mathrm{Tr}_{s+1...N}\left([\hat{\rho}, \hat{H}_{1...s}] + [\hat{\rho}, \hat{H}_{1...N-s}] + [\hat{\rho}, \hat{V}_{s...N-s}]\right). \tag{A.5}$$

The first term on the right-hand side is easily evaluated as

$$\mathrm{Tr}_{s+1...N}[\hat{\rho}, \hat{H}_{1...s}] = [\hat{\rho}_s, \hat{H}_{1...s}] \tag{A.6}$$

since the s-particle Hamiltonian is independent of the $s+1...N$ variables. From cyclic permutation one convinces oneself that the second therm on the right-hand side of (A.5) vanishes. Finally the last term becomes

$$\mathrm{Tr}_{s+1...N}[\hat{\rho}, \hat{V}_{s...N-s}] = (N-s)\,\mathrm{Tr}_{s+1}\sum_{i=1}^{s}[\hat{\rho}_{1...s+1}, \hat{V}_{i,s+1}] \tag{A.7}$$

such that we obtain from (A.5) the Born–Bogoliubov–Green–Kirkwood–Yvon hierarchy (BBGKY) first derived for classical systems,

$$\frac{\hbar}{i}\frac{\partial}{\partial t}\hat{\rho}_{1...s} = [\hat{\rho}_{1...s}, H_{1...s}] + (N-s)\,\mathrm{Tr}_{1...s+1}[\hat{\rho}_{1...s+1}, \hat{V}_{s,s+1}]. \tag{A.8}$$

This couples the one-particle to the two-particle density operator and so on and establishes a hierarchy of correlations which might by truncated at a specific level. As long as it remains exact it is reversible due to (A.3). A coupled set of equations of motion for the reduced density operators was first derived by Irving and Zwanzig (Irving and Zwanzig, 1951). The formal structure is similar to the BBGKY hierarchy (Balescu, 1978) for reduced distribution functions in classical statistical physics. The first to use this BBGKY hierarchy in deriving kinetic equations were Bogoliubov (Bogoliubov, 1946), Born and Green (Born and Green, 1946) and Kirkwood (Kirkwood, 1946).

A.2   Derivation of the Levinson equation and non-Markovian energy conservation

We want to consider the binary collision approximations, i.e. to neglect the influence of density operators higher than the two-particle one ρ_{12}. Then the equation for the two-particle density operator (A.8) is easily solved in the first Born approximation. To derive this expression we use

$$\frac{d}{d\lambda}e^{\lambda(\hat{A}+\hat{B})}e^{-\lambda\hat{A}} = e^{\lambda(\hat{A}+\hat{B})}\hat{B}e^{-\lambda\hat{A}} \tag{A.9}$$

and integrate from 0 to 1 to obtain the operator identity

$$e^{(\hat{A}+\hat{B})} = e^{\hat{A}} + \int_{0}^{1}e^{\lambda(\hat{A}+\hat{B})}\hat{B}e^{(1-\lambda)\hat{A}}. \tag{A.10}$$

Now we can solve (A.8)

$$\partial_t\hat{\rho}_{12} = \frac{i}{\hbar}[\hat{\rho}_{12}, \hat{H}_{12} + \hat{V}_{12}] = (\hat{L}^0 + \hat{L}^V)\hat{\rho}_{12} \tag{A.11}$$

as

$$\hat{\rho}_{12} = e^{(L^0+L^V)t}\rho_{12}^0 \tag{A.12}$$

and with the help of (A.10)

$$\rho_{12}(t) = \rho_1(t)\rho_2(t) + \frac{i}{\hbar}\int_{t_0}^{t}e^{-\frac{i}{\hbar}H_{12}^0(t-t')}[\rho_1(t')\rho_2(t'), V_{12}]e^{\frac{i}{\hbar}H_{12}^0(t-t')}dt' + o(V_{12}^2) \tag{A.13}$$

where at the artificial time t_o the two-particle density can be factorised into $\rho_{12}(t_0) = \rho_1(t_0)\rho_2(t_0)$. This is also known as a weakening of the initial correlations. The two-particle Hamiltonian was split into $H_{12} = H_1 + H_2 + V_{12}$. Substituting this equation back into the first equation of (A.8), one obtains the following Levinson kinetic equation (19.8) for the one-particle distribution function in momentum representation

$$\frac{\partial}{\partial t} \rho(p_1) = \frac{1}{\hbar^2} \int \frac{dp_2 \, dp_1' \, dp_2'}{(2\pi\hbar)^9} V(|\, p_1 - p_1' \,|)^2 (2\pi\hbar)^3 \, \delta(p_1 + p_2 - p_1' - p_2')$$

$$\times \int_0^\infty d\tau \left[\rho(p_1', t-\tau)\rho(p_2', t-\tau) - \rho(p_1, t-\tau)\rho(p_2, t-\tau) \right] 2\cos\left[\frac{1}{\hbar}(\epsilon_1 + \epsilon_2 - \epsilon_1' - \epsilon_2')\tau \right]. \quad (A.14)$$

Here we used the normalisation $\int \frac{dp_1}{(2\pi\hbar)^3} \rho(p_1) = n$ and $\epsilon = \frac{p^2}{2m}$. This Levinson equation is well known in the literature (Klimontovich, 1975; Morawetz and Kremp, 1994; Haug and Ell, 1992; Thoai and Haug, 1993) and plays an important role in considerations of high field problems (Seminozhenko, 1982; Jauho and Wilkins, 1984; Morawetz and Kremp, 1994). If one neglects the time retardation in the distribution functions, one obtains the usual Boltzmann collision integral by the identity $\int_0^\infty d\tau \cos(a\tau) = \pi\delta(a)$. The influence of retardation in Eq. (A.14) can be calculated numerically. This leads to a diminishing of the influence of collisions on the temporal development of the distribution function (Morawetz, Walke and Röpke, 1994). For transport problems in electric fields we have seen in chapter 20.3 and 20.7 that the linearisation of eq. (A.14) reproduces precisely the Debye–Onsager-relaxation effect, which diminishes the applied electric field through the moving screening cloud (Morawetz and Kremp, 1993). Beyond the linear response, the double count of the correlation leads to spurious time evolution as we discussed in Chapters 19.3 and 19.4. The missing Pauli blocking factors in (A.14) come from the classical form of the hierarchy. It can be reintroduced taking the proper symmetrisation and antisymmetrisations into account (McLennan, 1989).

It is now instructive to see the conservation laws. Multiplying Eq. (A.14) by $1, \mathbf{p}_1$ and integrating over p_1, one obtains the density (momentum) conservation

$$\frac{\partial}{\partial t} n = 0 \qquad \frac{\partial}{\partial t} \langle p_1 \rangle = 0$$

where the momentum-conserving δ distribution function is used in the manner known from treating the ordinary Boltzmann equations. The total energy conservation is more interesting. While the ordinary quantum-Boltzmann equation conserves only the kinetic energy it will be shown now that Eq. (A.14) leads to total energy conservation. This can be elaborated by multiplying Eq. (A.14) by $\frac{p_1^2}{2m}$ and integrating over p_1. The right-hand side of the resulting equation can be written as a time derivation and one finally obtains

$$\frac{\partial}{\partial t} \left(\langle \frac{p_1^2}{2m} \rangle + \langle V_{12} \rangle \right) = 0 \qquad (A.15)$$

with

$$\langle V_{12} \rangle = -\frac{1}{2\hbar} \int \frac{dp_1 \, dp_2 \, dp_1' \, dp_2'}{(2\pi\hbar)^{12}} V(|\, p_1 - p_1' \,|)^2 (2\pi\hbar)^3 \, \delta(p_1 + p_2 - p_1' - p_2')$$

$$\times \int_0^\infty d\tau \left[\rho(p_1', t-\tau)\rho(p_2', t-\tau) - \rho(p_1, t-\tau)\rho(p_2, t-\tau) \right] \sin\left[\frac{1}{\hbar}(\epsilon_1' + \epsilon_2' - \epsilon_1 - \epsilon_2)\tau \right]. \quad (A.16)$$

Now it remains to show that (A.16) is just the interaction part as we have already indicated. Therefore we calculate the mean correlation energy

$$\langle V_{12} \rangle = \frac{n^2}{2} \operatorname{Tr}_{12}(\rho_{12} V_{12}). \tag{A.17}$$

On the same level of approximation as used for the derivation of the kinetic equation we take the expression for the two-particle density operator (A.13) and obtain from Eq. (A.17) in momentum representation precisely the expression (A.16). Therefore it is proved that the non-Markovian collision integral (A.14) leads to complete energy conservation. Solving it numerically shows, however, that the correlation energy and kinetic energy increases are unlimited for larger time scales inhibiting a stationary solution. This comes due to internal double counts of correlations as discussed in Chapter 19.3. This shows that the two-time internal propagator causes the same problems as in the two-time Green's function. Therefore we have dealt within the book, from the beginning, with two-time functions in order to understand their internal structure.

A.3   Debye–Onsager relaxation effect from classical hierarchy

We will consider here an interesting example of a non-analyticity in two-particle correlations. We will show that assuming an infinitesimal friction with the background leads to another solution than without friction. Though the friction drops out of the result, it is necessary to assume in order to obtain the correct expression. This astonishing fact has been observed first by Onsager (1927) when calculating the relaxation effect which is the internal force created by the different velocity of the screening cloud surrounding a moving charge. We have considered this effect already in Chapter 20.3.

The starting point is the classical limit of the BBGKY hierarchy (A.8) (Bogoliubov, 1946; Klimontovich and Ebeling, 1962) where all commutators translate into Poisson brackets. The hierarchy reads for the one-particle distribution function F_a

$$\frac{\partial F_a}{\partial t} + \mathbf{v}\frac{\partial F_a}{\partial \mathbf{r}} + \frac{e_a}{m_a}\bar{\mathbf{E}}\frac{\partial F_a}{\partial \mathbf{v}} - S_a F_a = \sum_b \frac{n_b e_a e_b}{m_a}\frac{\partial}{\partial\mathbf{v}}\int d\mathbf{r}'\, d\mathbf{v}'\, F_{ab}(\mathbf{r},\mathbf{r}',\mathbf{v},\mathbf{v}')\frac{\partial}{\partial\mathbf{r}}\frac{1}{|\mathbf{r}-\mathbf{r}'|} \tag{A.18}$$

and the two-particle distribution function F_{ab}

$$\frac{\partial F_{ab}}{\partial t} + \mathbf{v}\frac{\partial F_{ab}}{\partial \mathbf{r}} + \mathbf{v}'\frac{\partial F_{ab}}{\partial \mathbf{r}'} + \frac{e_a}{m_a}\bar{\mathbf{E}}\frac{\partial F_{ab}}{\partial \mathbf{v}} + \frac{e_b}{m_b}\bar{\mathbf{E}}\frac{\partial F_{ab}}{\partial \mathbf{v}'} - S_a F_{ab} - S_b F_{ab}$$

$$= e_a e_b \frac{\partial}{\partial\mathbf{r}}\frac{1}{|\mathbf{r}-\mathbf{r}'|}\left(\frac{1}{m_a}\frac{\partial F_{ab}}{\partial\mathbf{v}} - \frac{1}{m_b}\frac{\partial F_{ab}}{\partial\mathbf{v}'}\right)$$

$$+ \sum_c n_c e_c \int d\mathbf{r}''\, d\mathbf{v}''\left(\frac{e_a}{m_a}\frac{\partial}{\partial\mathbf{r}}\frac{1}{|\mathbf{r}-\mathbf{r}''|}\cdot\frac{\partial F_{abc}}{\partial\mathbf{v}} + \frac{e_b}{m_b}\frac{\partial}{\partial\mathbf{r}'}\frac{1}{|\mathbf{r}'-\mathbf{r}''|}\cdot\frac{\partial F_{abc}}{\partial\mathbf{v}'}\right) \tag{A.19}$$

with the external field **E**. S_a describes a collision integral with some background which we will specify later. This hierarchy is truncated approximating that (Klimontovich and Ebeling, 1962)

$$F_{ab} = F_a F_b + g_{ab}$$

$$F_{abc} = F_a F_b F_c + F_a g_{bc} + F_b g_{ac} + F_c g_{ab} \qquad (A.20)$$

where $g_{ab}(\mathbf{r}_a, \mathbf{r}_b, \mathbf{v}_a, \mathbf{v}_b)$ is the two-particle correlation function.

Within the local equilibrium approximation we suppose a stationary (for example a local Maxwellian) distribution for the velocities in the one- and two-particle distribution functions

$$f_a(\mathbf{r}, \mathbf{v}, t) = n_a(\mathbf{r}, t) \left(\frac{m_a}{2\pi T} \right)^{3/2} e^{-\frac{m_a(\mathbf{v}-\mathbf{u}_a)^2}{2T}}$$

$$g_{ab}(\mathbf{r}, r', v, v', t) = F_{ab} - F_a F_b = h_{ab}(\mathbf{r}, r', t) \left(\frac{m_a m_b}{4\pi^2 T^2} \right)^{3/2} e^{-\frac{m_a(v-w_{ab})^2}{2T} - \frac{m_b(v'-w_{ba})^2}{2T}}. \qquad (A.21)$$

Here we have introduced the local one-particle density and the local average velocity

$$n_a(\mathbf{r}, t) = \int d\mathbf{v} F_a(\mathbf{r}, \mathbf{v}, t), \quad \mathbf{u}_a = \frac{1}{n_a} \int d\mathbf{v}\, \mathbf{v}\, F_a(\mathbf{r}, \mathbf{v}, t) \qquad (A.22)$$

as well as the pair-correlation function and the average pair velocity

$$h_{ab}(\mathbf{r}, \mathbf{r}', t) = \int d\mathbf{v}\, d\mathbf{v}'\, g_{ab}(\mathbf{r}, \mathbf{r}', \mathbf{v}, \mathbf{v}', t),$$

$$w_{ab}(\mathbf{r}, \mathbf{r}', t) = \frac{1}{h_{ab}} \int d\mathbf{v}\, d\mathbf{v}'\, \mathbf{v} g_{ab}(\mathbf{r}, \mathbf{r}', \mathbf{v}, \mathbf{v}', t). \qquad (A.23)$$

Further on, we suppose that the particles interact with some background (e.g. neutrals or electrolyte solvent) by the collision integrals S_a with the following properties

$$\int d\mathbf{v} S_a f_a = 0, \quad \int d\mathbf{v}\,\mathbf{v} S_a f_a = \frac{1}{b_a m_a} \rho_a \mathbf{u}_a,$$

$$\int d\mathbf{v}\,\mathbf{v} S_a g_{ab}(\mathbf{r}, r', v, v', t) = \frac{1}{b_a m_a} h_{ab} \mathbf{w}_{ab} \qquad (A.24)$$

where b_a is the mobility of particle of type a. This friction with a background serves here to couple the two-particle equations and will be considered infinitesimally small in the end. However, as we will demonstrate, this yields to a symmetry breaking in the system which leads basically to different results than neglecting this friction.

Fourier transform of the resulting two equations (A.20) into momentum space and assuming a homogeneous density $n(\mathbf{r}) = n$ we arrive at the coupled equation system

$$\frac{e_a}{T} \bar{\mathbf{E}}(\mathbf{u}_a - \mathbf{v}_a) f_a = \frac{\mathbf{u}_a}{b_a m_a} + \sum_b \frac{4\pi n_b e_a e_b}{T} \int \frac{d\mathbf{k}}{(2\pi)^3} \frac{i\mathbf{k} \cdot (\mathbf{v}_a - \mathbf{w}_{ab})}{k^2} f_a(\mathbf{v}_a - \mathbf{w}_{ab} + \mathbf{u}_a) h_{ab}(\mathbf{k}) \qquad (A.25)$$

and

$$i\mathbf{k}(v_a - v_b)g_{ab} - (e_a(\mathbf{v}_a - w_{ab}) + e_b(\mathbf{v}_b - w_{ba}))\frac{\bar{\mathbf{E}}}{T}g_{ab} = S_a g_{ab} + S_b g_{ab}$$

$$i e_a e_b \frac{4\pi}{k^2}\mathbf{k}(\mathbf{u}_a - \mathbf{v}_a + \mathbf{v}_b - \mathbf{u}_b)f_a f_b + i\int\frac{d\bar{\mathbf{k}}}{(2\pi)^3}\frac{4\pi e_a e_b}{T k^2}\bar{\mathbf{k}}(w_{ab} - \mathbf{v}_a + \mathbf{v}_b - w_{ba})g_{ab}(\mathbf{k} - \bar{\mathbf{k}})$$

$$- \sum_c n_c\int dv_c\frac{4\pi i e_c}{T k^2}\left[e_a\mathbf{k}(\mathbf{v}_a - \mathbf{u}_a)f_a g_{cb}(\mathbf{k}) - e_b\mathbf{k}\cdot(\mathbf{v}_b - \mathbf{u}_b)f_b g_{ac}(\mathbf{k})\right] \qquad (A.26)$$

with

$$\bar{\mathbf{E}} = \mathbf{E} - \sum_b n_b e_b\int d\mathbf{r}_b dv_b\frac{\partial}{\partial\mathbf{r}_b}\frac{1}{|\mathbf{r}_a - \mathbf{r}_b|}F_b. \qquad (A.27)$$

By multiplying the above equation system by $1, v_a, v_b$ and integrating over the velocities we obtain the Onsager equation (Onsager, 1927)

$$b_a\left[Th_{ab}(\mathbf{k})\left(1 + i\frac{e_a}{e}E_k\right) + e_a\Phi_b(-\mathbf{k})\right] = -b_b\left[Th_{ab}(\mathbf{k})\left(1 - i\frac{e_b}{e}E_k\right) + e_b\Phi_a(\mathbf{k})\right] \qquad (A.28)$$

with the abbreviation $E_k = \frac{e\mathbf{k}\cdot\bar{\mathbf{E}}}{k^2 T}$ and

$$k^2\Phi_a(\mathbf{k}) = 4\pi e_a + \sum_c n_c e_c h_{ac}(\mathbf{k}), \quad k^2\Phi_a(-\mathbf{k}) = 4\pi e_a + \sum_c n_c e_c h_{ca}(\mathbf{k}) \qquad (A.29)$$

for the two-particle correlation function h_{ab}.

Let us already remark here that the friction with a background described by the mobilities b couple the two sides of the Equation (A.28). If we had not considered this friction, $S_i = 0$, we would have obtained that the left- and right-hand sides of (A.28) vanish separately. This will lead essentially to a different result even for an infinitely small friction. There is no continuous transition between these two extreme cases pointing to a symmetry breaking in the two treatments. Let us first discuss the case with background friction.

A.3.1   With background friction

The system (A.28) for electrons, $e_e = e$, and ions, $e_i = -Ze$, with charge Z reads expanded

$$h_{ee} = -e\frac{\Phi_e(-\mathbf{k}) + \Phi_e(\mathbf{k})}{2T},$$

$$h_{ei} = -e\frac{\Phi_i(-\mathbf{k}) - Z\frac{b_i}{b_e}\Phi_e(\mathbf{k})}{T\left[1 + \frac{b_i}{b_e} + iE_k\left(1 + \frac{b_i}{b_e}Z\right)\right]}$$

$$h_{ie} = -e\frac{\Phi_i(\mathbf{k}) - Z\frac{b_i}{b_e}\Phi_e(-\mathbf{k})}{T\left[1 + \frac{b_i}{b_e} - iE_k\left(1 + \frac{b_i}{b_e}Z\right)\right]},$$

$$h_{ii} = Ze\frac{\Phi_i(-\mathbf{k}) + \Phi_i(\mathbf{k})}{2T}. \qquad (A.30)$$

This we can solve this together with (A.29). We calculate the effective field strength at the position of the electron in linear response and obtain the Onsager result (Onsager, 1927)

$$\frac{\delta E}{E} \mathbf{E} = -i \frac{\mathbf{E}}{E} \frac{1}{(2\pi)^2} \int\limits_0^\infty k^3\, dk \int\limits_{-1}^1 d(\cos\theta)\, \cos\theta\, \Phi_e(\mathbf{k}) = \mathbf{E} \frac{\kappa e^2}{3T} \frac{Zq}{\sqrt{q}+1} \tag{A.31}$$

with $\kappa^2 = \kappa_e^2(1+Z) = \frac{4\pi(e^2 n_e + Z^2 e^2 n_i)}{T}$ and

$$q = \frac{b_e + Zb_i}{(1+Z)(b_e+b_i)}. \tag{A.32}$$

For single-charged ions $Z = 1$ the influence of the mobilities drops out and we recover the Onsager result (20.31) (Onsager, 1927)

$$E^{\text{eff}} = E\left(1 - \frac{\kappa e^2}{3(2+\sqrt{2})T}\right). \tag{A.33}$$

Since this result is independent of the mobilities one could conclude that this is a universal limiting law. However we will express two doubts here. As one sees for charges $Z > 1$ the result (20.31) is approached only in the limit where the ion mobilities are much smaller than the electron mobilities $b_i/b_e \to 0$. This means of course that the electrons have different frictions due to the imagined background than the ions. In other words there is an explicit symmetry breaking mechanism included by assuming such collision integrals with the background. Therefore we will obtain another solution if we consider no friction.

For completeness, we want to recall the expression of the nonlinear Onsager result (Ortner, 1997; Morawetz, 1997) which is obtained from the limit $b_i/b_e \to 0$ of the system (A.30)

$$Th_{ee} + e\frac{\varphi_e(-\mathbf{k}) + \varphi_e(\mathbf{k})}{2} = 0, \qquad h_{ei}\left(T + i\,e\frac{\mathbf{kE}}{k^2}\right) + e\varphi_i(-\mathbf{k}) = o\left(\frac{b_i}{b_e}\right) = 0$$

$$-h_{ie}\left(T - i\,e\frac{\mathbf{kE}}{k^2}\right) - e\varphi_i(\mathbf{k}) = o\left(\frac{b_i}{b_e}\right) = 0, \qquad Th_{ii} - Ze\frac{\varphi_i(-\mathbf{k}) + \tilde{\varphi}_i(\mathbf{k})}{2} = 0. \tag{A.34}$$

One obtains (Falkenhagen, 1953; Ortner, 1997) the result for $Z = 1$

$$\delta E = -\frac{e^2\kappa_e}{3(1+\sqrt{2})\,T}\mathbf{E}\,F_H\left(\frac{eE}{T\kappa_e}\right) = -\frac{e^2\kappa_e}{6\,T}\mathbf{E}\begin{cases}2-\sqrt{2}+o(E) \\[2mm] \frac{3\kappa T}{2eE} + o(1/E)^2\end{cases} \tag{A.35}$$

with (Morawetz, 1997; Ortner, 1997)

$$F_H(\alpha) = \frac{3(1+\sqrt{2})}{\alpha^2}\left[\frac{1}{2}\sqrt{\alpha^2+2} - 1 + \frac{1}{\alpha}\arctan(\alpha) - \frac{1}{\alpha}\arctan\left(\frac{\alpha}{\sqrt{\alpha^2+2}}\right)\right]. \tag{A.36}$$

A.3.2   Without background

Now we reconsider the steps from (A.26) to (A.28) without friction with the background. We obtain that both sides of (A.28) vanish separately

$$Th_{ab}(\mathbf{k})\left(1 + i\frac{ea}{e}a\right) + e_a\Phi_b(-\mathbf{k}) = 0, \quad Th_{ab}(\mathbf{k})\left(1 - i\frac{eb}{e}a\right) + e_b\Phi_a(\mathbf{k}) = 0. \tag{A.37}$$

Both equations have identical solutions h_{ab} which can be easily verified using the symmetry $h_{ab}(\mathbf{k}) = h_{ba}(-\mathbf{k})$. Together with (A.29) we can solve for Φ_e and the relaxation field is obtained instead of (A.35) as

$$\delta\mathbf{E} = -\frac{e^2\kappa_e\sqrt{1+Z}}{6T}(Z+1)\mathbf{E}F_N\left(\frac{eE}{T\kappa_e}\right) \tag{A.38}$$

which takes for $Z = 1$

$$\frac{\delta E}{E} = -\frac{e^2\kappa_e}{6T}\begin{cases} 2 + o(E) \\ \frac{3\kappa T}{eE} + o(1/E)^2 \end{cases} \tag{A.39}$$

with

$$F_N(\alpha) = \frac{3}{(1+Z)\alpha^2}\left[\sqrt{4 + (1+Z)\alpha^2} + \frac{4}{\sqrt{1+Z}\alpha}\log\frac{2}{\sqrt{1+Z}\alpha + \sqrt{4 + (1+Z)\alpha^2}}\right]. \tag{A.40}$$

We see that the linear response result for $Z = 1$ is twice the Debye result (20.30)

$$E^{\text{eff}} = E\left(1 - \frac{\kappa e^2}{6T}\right). \tag{A.41}$$

For equal charged system $Z = -1$ which would coincide with a one-component plasma no relaxation effect appears as one would expect. In other words, in a perfectly symmetric mathematical two-component plasma there is another relaxation effect than in a system which distinguishes the components by a different treatment of friction. The Onsager result (A.35) does not vanish for the limit of one-component plasma $Z = -1$. This is due to the different treatment of ions and electrons there which assumes explicitly a two-component plasma. Therefore the limit $Z = -1$ does not work there.

This result is quite astonishing. One would expect that the limiting procedure which transforms the system (A.29) into (A.37) would also lead to a smooth transitions of the end results. However this is not the case. While the separate limit of $b_{e,i} \to \infty$ of (A.29) leads to (A.37) there is no possibility of transforminig the result (A.31) into the linear response result of (A.39). This underlines that due to even infinitesimally small friction assumed in obtaining (A.31) there occurs a symmetry breaking in the sense that the electrons and ions are not symmetrically treated anymore. This is derived in Chapter 11.5 and used to derive the Debye–Onsager relaxation effect in Chapter 20.7 and 20.7.2. The different results are compared in Figure 20.1.

Appendix B
Complex Time Path

On the complex time path one can develop a perturbative expansion for G^c in the same way as for the ground state or Matsubara's functions. Related techniques employ either resummation of diagrams (Abrikosov, Gorkov and Dzyaloshinski, 1963) or functional derivatives (Kadanoff and Baym, 1962; Semkat, Kremp and Bonitz, 2000). The time evolution from a given nonequilibrium initial condition is covered by the Keldysh formalism (Keldysh, 1964). The same equation of motion results from the older Schwinger's approach (Kadanoff and Baym, 1962) where an equilibrium state is supposed in a remote past and the nonequilibrium initial condition of the collision is prepared by unspecified external fields. There are a number of later modifications of both approaches (Korenman, 1966; Craig, 1968; Langreth and Wilkins, 1972; Langreth, 1976), it is merely a question of habit which formalism authors use. We prefer the so-called generalised Kadanoff and Baym formalism (GKB) which has been developed by Langreth and Wilkins as the continuation of Schwinger's original approach.

The basic idea of Schwinger's approach is outlined in Fig. B.1. In the remote past τ_0 the system is supposed to be in equilibrium where its grand canonical averaging operator is known to be

$$\hat{\rho} = \frac{1}{\text{Tr}\left(e^{\beta\mu\hat{N}}e^{-\beta\hat{H}}\right)} e^{\beta\mu\hat{N}}e^{-\beta\hat{H}}, \tag{B.1}$$

where $\hat{N}$ is the operator of the number of particles and $\beta = \frac{1}{k_B T}$ is the inverse temperature at τ_0. The time evolution of the field operator from τ_0 to any later time is given by Heisenberg's equation of motion

$$\hat{\psi}(1) = \hat{T}_- e^{-i\int_{t_1}^{\tau_0} d\tau \hat{H}(\tau)} \hat{\psi}_{a_1}(x_1, \tau_0) \hat{T}_+ e^{-i\int_{\tau_0}^{t_1} d\tau \hat{H}(\tau)}, \tag{B.2}$$

$$\hat{\psi}^\dagger(3) = \hat{T}_- e^{-i\int_{t_3}^{\tau_0} d\tau \hat{H}(\tau)} \hat{\psi}_{b_3}(x_3, \tau_0) \hat{T}_+ e^{-i\int_{\tau_0}^{t_3} d\tau \hat{H}(\tau)}. \tag{B.3}$$

In the time interval $\tau_1 < \tau < \tau_2$, the Hamiltonian $H(\tau)$ includes the 'unspecified' external fields. The Dyson's operators $\hat{T}_\pm$ provide (anti)causal time-ordering of the field operators in the exponential. With Matsubara's ordering operator $\hat{T}_\beta$ along the complex interval $(\tau_0, \tau_0 - i\beta)$ we can write the essential part of the grand canonical averaging operator in the form of the complex evolution

$$e^{-\beta\hat{H}} = \hat{T}_\beta e^{-i\int_{\tau_0}^{\tau_0 - i\beta} d\tau \hat{H}(\tau)}. \tag{B.4}$$

A compact notation is achieved if all evolution operators in (B.2)–(B.4) are joined to an evolution along the complex path c outlined in Fig. B.1. The corresponding Dyson's operator $\hat{T}_c$ acting

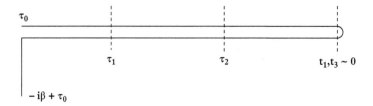

Figure B.1: *Complex time path for nonequilibrium perturbative expansion. The equilibrium state at $\tau_0 \to -\infty$ is disturbed by unspecified external fields in the interval $\tau_1 < t < \tau_2$ so that at τ_2 the system is already composed of two nuclei heading to their collision. The reaction of nuclei takes part at $t_{1,3} \sim 0$ during the time interval which is very short on the time scale of the path. The small down-turning appendix reflects the temperature $1/(k_B\beta)$ of the initial state.*

along path c is composed of $\hat{T}_\pm$ and $\hat{T}_\beta$. It orders operators according to the inequality $t_i <_c t_j$ which means that going from τ_0 to $\tau_0 - i\beta$ one reaches t_i before t_j.

As in the case of ground state Green's functions or Matsubara's functions for equilibrium, the perturbative expansion is formally simplest for the causal Green's function $G^c(1,3)$ defined with the help of Dyson's operator $\hat{T}_c$ as

$$G^c(1,3) = -i\mathrm{Tr}\left(\hat{T}_c \hat{\rho}\hat{\psi}(1)\hat{\psi}^\dagger(3)\right). \tag{B.5}$$

The desired correlation function $G^<$ is its particular element

$$G^<(1,3) = -iG^c(1,3) \quad \text{for} \quad t_1 <_c t_3. \tag{B.6}$$

Besides the so-far considered correlation function $G^<$ we have also a corresponding second correlation function

$$G^>(1,3) = \mathrm{Tr}\left(\hat{\rho}\hat{\psi}(1)\hat{\psi}^\dagger(3)\right), \tag{B.7}$$

which is in equilibrium connected to $G^<$ by the Kubo–Martin–Schwinger relation (Kadanoff and Baym, 1962) $g^<(\omega, k) = \exp(-\beta\omega)\, g^>(\omega, k)$. In nonequilibrium both correlation functions remain unconnected, therefore we have to deal with a two-function formalism. Therefore one extends the complex path according to Figure B.1 to real times forth and back leading to the Kadanoff and Baym equation of motions.

This can be derived as follows. Our intention was to express the two-particle Green's function in terms of the single-particle Green's function so that the first equation of the hierarchy (7.8) becomes a closed equation. To complete the set of functions, one has to introduce the anti-causal Green's function

$$G^a(1,2) = -\frac{1}{i}\Theta(t_2 - t_1)G^>(1,2) \pm \frac{1}{i}\Theta(t_1 - t_2)G^<(1,2) = i\langle \hat{T}^a \hat{\psi}_1 \hat{\psi}_2^\dagger\rangle, \tag{B.8}$$

where $\hat{T}^a\hat{A}\hat{B} = \hat{A}\hat{B}$ for $t_A < t_B$ and $\hat{T}^a\hat{A}\hat{B} = \mp\hat{B}\hat{A}$ for $t_B < t_A$.

It is possible to construct a 2×2 matrix from the causal, anti-causal and correlation functions and develop the perturbative expansion in terms of these matrices. Such an approach was adopted by Keldysh (1964) on the complex time path C and called the Keldysh contour, on which the time runs along the real-time axis from some initial time t_0 to ∞ and back. Here we follow this approach. The function

$$G(1, 2) = \frac{1}{i}\Theta_C(t_1 - t_2)G^>(1, 2) \mp \frac{1}{i}\Theta_C(t_2 - t_1)G^<(1, 2) = \frac{1}{i}\langle \hat{T}\hat{\psi}_1\hat{\psi}_2^\dagger\rangle, \tag{B.9}$$

is causal along the path C and the time-ordering operator acts according to the order of times on this path. The first equation of the hierarchy (7.8) extended to the Keldysh contour is formally identical to the causal form,

$$\left(i\frac{\partial}{\partial t_1} + \frac{\nabla_1^2}{2m} - U(1)\right) G(1, 2) = \delta(1 - 2) \mp i \int_C d3 V(1, 3)G_2(1, 3, 2, 3^+). \tag{B.10}$$

The superscript $+$ denotes $t_3^+ >_C t_3$ which means $t_3^+ > t_3$ on the forward branch and $t_3^+ < t_3$ on the backward branch with $t_3^+ - t_3 \to 0$, as it corresponds to the order of times on the path C.

A formally closed equation for the single-particle Green's function can be reached with the introduction of the selfenergy Σ defined via (7.9) in terms of which Eq. (B.10) reads (7.19). Inverting the differential operator we obtain the Dyson equation on the path

$$G = G^0 + G^0\Sigma G. \tag{B.11}$$

Being defined by the time-ordering operator $\hat{T}$, the path Green's function (or its selfenergy) can be expressed with the help of Feynman's diagrams or it can be developed using the technique of functional derivatives. In this part the complex time on the path is an advantage. As soon as one leaves the level of general relations and approaches implementations of the theory, the complex time becomes a drawback as it complicates the Fourier transformation—one of the most powerful mathematical tools for differential equations. Before an implementation it is thus necessary to dismantle the complex part and express all time integrations as simple integrals along the time axis.

Let us assume for the moment that the approximation of the selfenergy is specified and take Σ as a known function. Langreth and Wilkins (1972) formulated simple rules for how to handle the time integration of the product $Q = \Sigma G$. Let us assume an order of times on the path $t_1 <_C t_2$, then

$$Q(1, 2)_{t_1 <_C t_2} = \pm iQ^<(1, 2)$$

$$= \pm \int_{t_0}^{t_1} d3\ \Sigma^>(1, 3)G^<(3, 2) - \int_{t_1}^{t_2} d3\ \Sigma^<(1, 3)G^<(3, 2) \pm \int_{t_2}^{t_0} d3\ \Sigma^<(1, 3)G^>(3, 2). \tag{B.12}$$

Integrals going to ∞ and back cancel and the function $C^<$ depends exclusively on the previous occasions, as is required by the principle of causality. To make the structure of time integrals in Eq. (B.12) transparent, we express them via integrals from the initial time $t_0 \to -\infty$,

$$Q^<(1,2) = -i \int_{-\infty}^{t_1} dt_3 \Sigma^> G^< \mp i \left(\int_{-\infty}^{t_1} dt_3 - \int_{-\infty}^{t_2} dt_3 \right) \Sigma^< G^< + i \int_{-\infty}^{t_2} d\bar{t}_3 \Sigma^< G^> . \qquad \text{(B.13)}$$

The integrals with identical upper time limit can be joined. Moreover it is advantageous express the upper time limit via the Θ-function,

$$Q^<(1,2) = -i \int_{-\infty}^{t_1} d3 \, [\Sigma^> \pm \Sigma^<] G^< + i \int_{-\infty}^{t_2} d3 \, \Sigma^< [G^> \pm G^<]$$

$$= \int_{-\infty}^{\infty} d3 \, (-i)\Theta(t_1 - t_3)[\Sigma^>(1,3) \pm \Sigma^<(1,3)] G^<(3,2)$$

$$+ \int_{-\infty}^{\infty} d3 \, \Sigma^<(1,3)i\Theta(t_2 - t_3)[G^>(3,2) \pm G^<(3,2)]. \qquad \text{(B.14)}$$

Elements of the selfenergy in the first integral form the retarded function (7.15) and the elements of the Green's function in the second integral form the advanced function (7.16). Definitions of retarded and advanced functions are general. Any double-time function defined on the Keldysh time contour can be converted to retarded, advanced and less and greater functions linked by these definitions. We recognise the LW rules of Langreth and Wilkins (1972) how one obtains from a causal product $C = AB$ defined on the complex-time path into tractable relation on the real time axis (7.18) and (7.17). In other words the complex-path formalism is equivalent to the one we presented in Chapter 7.1.

Appendix C
Derived Optical Theorem

Beside the optical theorem (8.40) for the imaginary (anti-Hermitian) part of the T-matrix, there is a similar identity for the real (Hermitian) part of the T-matrix $[\mathscr{T} = \frac{1}{2}(\mathscr{T}^R + \mathscr{T}^A) \equiv \text{Re}\,\mathscr{T}^R]$

$$\mathscr{T}_{\text{ex}} = \mathscr{T}_{\text{sc}}^R \left(\mathscr{V}^{-1} - \mathscr{G}\right) \mathscr{T}_{\text{sc}}^A. \tag{C.1}$$

This identity follows directly from a simple identity for the non-anti-symmetrized T-matrix

$$\mathscr{T} = \frac{1}{2}\left(\mathscr{T}^R + \mathscr{T}^A\right) = \mathscr{T}^R \frac{1}{2}\left(\mathscr{T}_R^{-1} + \mathscr{T}_A^{-1}\right)\mathscr{T}^A = \mathscr{T}^R \left(\mathscr{V}^{-1} - \mathscr{G}\right)\mathscr{T}^A, \tag{C.2}$$

and the anti-symmetrisation.

The identity (C.1) has no straightforward physical meaning and there is no common name for it. As far as we know, the identity (C.1) itself is rarely used. Its energy derivative, however, appears in various forms in most of the papers dealing with quantum virial corrections. Such a derived identity has been presented by a number of authors under different names, we finds as the most instructive name 'the derived optical theorem' introduced by Zimmermann and Stolz (1985).

Let us take the energy derivative of (C.1),

$$\frac{\partial \mathscr{T}_{\text{ex}}}{\partial E} = \frac{\partial \mathscr{T}_{\text{sc}}^R}{\partial E}\left(\mathscr{V}^{-1} - \mathscr{G}\right)\mathscr{T}_{\text{sc}}^A + \mathscr{T}_{\text{sc}}^R\left(\mathscr{V}^{-1} - \mathscr{G}\right)\frac{\partial \mathscr{T}_{\text{sc}}^A}{\partial E} - \mathscr{T}_{\text{sc}}^R \frac{\partial \mathscr{G}}{\partial E}\mathscr{T}_{\text{sc}}^A. \tag{C.3}$$

To rearrange the first and second terms, we write

$$\mathscr{V}^{-1} - \mathscr{G} = \mathscr{V}^{-1} - \mathscr{G}^A + \frac{i}{2}\mathscr{A} = \mathscr{V}^{-1} - \mathscr{G}^R - \frac{i}{2}\mathscr{A}, \tag{C.4}$$

and use identities

$$\mathscr{T}_{\text{sc}}^R \left(\mathscr{V}^{-1} - \mathscr{G}^R\right) = \mathscr{U}_{\text{sc}},$$
$$\left(\mathscr{V}^{-1} - \mathscr{G}^A\right)\mathscr{T}_{\text{sc}}^A = \mathscr{U}_{\text{sc}}, \tag{C.5}$$

where $\mathscr{U}_{\text{sc}}$ is an operator of the anti-symmetrisation,

$$\mathscr{U}_{\text{sc}} = \left(1 - \delta_{a_1 a_2}\right)\delta(1{-}3)\delta(2{-}4) + \frac{1}{\sqrt{2}}\delta_{a_1 a_2}\left(\delta(1{-}3)\delta(2{-}4) - \delta(1{-}4)\delta(2{-}3)\right). \tag{C.6}$$

From definitions of $\mathscr{T}^{R,A}_{\text{ex,sc}}$ one finds

$$\mathscr{T}^{R}_{\text{sc}}\mathscr{U}_{\text{sc}} = \mathscr{T}^{R}_{\text{ex}}, \qquad \mathscr{U}_{\text{sc}}\mathscr{T}^{A}_{\text{sc}} = \mathscr{T}^{A}_{\text{ex}}. \tag{C.7}$$

Since $\mathscr{U}_{\text{sc}}$ does not depend on the energy E, from (C.7) directly follows

$$\frac{\partial \mathscr{T}^{R}_{\text{sc}}}{\partial E}\mathscr{U}_{\text{sc}} = \frac{\partial \mathscr{T}^{R}_{\text{ex}}}{\partial E},$$

$$\mathscr{U}_{\text{sc}}\frac{\partial \mathscr{T}^{A}_{\text{sc}}}{\partial E} = \frac{\partial \mathscr{T}^{A}_{\text{ex}}}{\partial E}. \tag{C.8}$$

With the help of (C.4)–(C.8), the first and the second terms of (C.3) are rearranged as

$$\frac{\partial \mathscr{T}^{R}_{\text{sc}}}{\partial E}\left(\mathscr{V}^{-1}-\mathscr{G}\right)\mathscr{T}^{A}_{\text{sc}}+\mathscr{T}^{R}_{\text{sc}}\left(\mathscr{V}^{-1}-\mathscr{G}\right)\frac{\partial \mathscr{T}^{A}_{\text{sc}}}{\partial E} = 2\frac{\partial \mathscr{T}_{\text{ex}}}{\partial E}+\frac{i}{2}\frac{\partial \mathscr{T}^{R}_{\text{sc}}}{\partial E}\mathscr{A}\,\mathscr{T}^{A}_{\text{sc}}-\frac{i}{2}\mathscr{T}^{R}_{\text{sc}}\mathscr{A}\frac{\partial \mathscr{T}^{A}_{\text{sc}}}{\partial E}. \tag{C.9}$$

Substituting (C.9) into (C.3) one immediately obtains the derived optical theorem

$$\frac{\partial \mathscr{T}_{\text{ex}}}{\partial E} - \mathscr{T}^{R}_{\text{sc}}\frac{\partial \mathscr{G}}{\partial E}\mathscr{T}^{A}_{\text{sc}} = -\frac{i}{2}\frac{\partial \mathscr{T}^{R}_{\text{sc}}}{\partial E}\mathscr{A}\,\mathscr{T}^{A}_{\text{sc}} + \frac{i}{2}\mathscr{T}^{R}_{\text{sc}}\mathscr{A}\frac{\partial \mathscr{T}^{A}_{\text{sc}}}{\partial E}. \tag{C.10}$$

For proofs of (C.10) see also (Zimmermann and Stolz, 1985; Schmidt, Röpke and Schulz, 1990; Morawetz and Röpke, 1995).

The derived optical theorem couples an energy derivative of the real part of the T-matrix with energy derivatives of the scattering phase shift. Accordingly, some of its forms appear in all of alternative approaches to the virial corrections. For the formulation of this identity used within the quantum theory of gases see e.g. (de Haan, 1990a).

Apparently, one finds a similar identity taking the derivative with respect to any other variable. In the energy conservation we will need the time modification

$$\frac{\partial \mathscr{T}_{\text{ex}}}{\partial t} - \mathscr{T}^{R}_{\text{sc}}\frac{\partial \mathscr{G}}{\partial t}\mathscr{T}^{A}_{\text{sc}} = -\frac{i}{2}\frac{\partial \mathscr{T}^{R}_{\text{sc}}}{\partial t}\mathscr{A}\,\mathscr{T}^{A}_{\text{sc}} + \frac{i}{2}\mathscr{T}^{R}_{\text{sc}}\mathscr{A}\frac{\partial \mathscr{T}^{A}_{\text{sc}}}{\partial t}. \tag{C.11}$$

Correlated observables in an extended quasiparticle picture

The derived optical theorem is particularly useful when one wants to link single-particle observables evaluated from the $\rho[f]$-functional with the non-instant correction to the scattering integral. Assume that we want to evaluate the observable o given by the function $o_a(k)$,

$$o = \int \frac{dk}{(2\pi)^3}o\rho. \tag{C.12}$$

According to the $\rho[f]$-functional (9.67), the observable can be decomposed into the free and the correlated part,

$$o^{\text{free}} = \sum_a \int \frac{dk}{(2\pi)^3} f o, \tag{C.13}$$

$$o^{\text{corr}} = \sum_a \int \frac{dk}{(2\pi)^3}(\rho - f)o = \sum_a \int \frac{dk}{(2\pi)^3} \frac{d\omega}{2\pi}\left(\sigma^<(1-f) - \sigma^> f\right) \frac{\wp'}{\omega - \varepsilon} o. \tag{C.14}$$

We will focus on the correlated part.

For the selfenergy we substitute from (15.100) so that (C.14) turns into a functional of the T-matrix

$$o^{\text{corr}} = \sum_{ab} \int \frac{dk\,dp\,dq}{(2\pi)^9} o_1\, |t_{\text{sc}}(\varepsilon_3 + \varepsilon_4, k, p, q)|^2\, \frac{\wp'}{\varepsilon_3 + \varepsilon_4 - \varepsilon_1 - \varepsilon_2}$$

$$\times\, [(1-f_1)(1-f_2)f_3 f_4 - f_1 f_2(1-f_3)(1-f_4)]. \tag{C.15}$$

To bring this formula into a representation of two-particle operators used above for the derived optical theorem we use a trivial identity

$$(1-f_1)(1-f_2)f_3 f_4 - f_1 f_2(1-f_3)(1-f_4) = (1-f_1-f_2)f_3 f_4 - f_1 f_2(1-f_3-f_4), \tag{C.16}$$

and split the correlated observable into two terms,

$$o^{\text{corr}} = \sum_{ab} \int \frac{dk\,dp\,dq}{(2\pi)^9} o_1\, |t_{\text{sc}}(\varepsilon_3 + \varepsilon_4, k, p, q)|^2\, \frac{\wp'}{\varepsilon_3 + \varepsilon_4 - \varepsilon_1 - \varepsilon_2}(1-f_1-f_2)f_3 f_4$$

$$- \sum_{ab} \int \frac{dk\,dp\,dq}{(2\pi)^9} o_1\, |t_{\text{sc}}(\varepsilon_3 + \varepsilon_4, k, p, q)|^2\, \frac{\wp'}{\varepsilon_3 + \varepsilon_4 - \varepsilon_1 - \varepsilon_2}f_1 f_2(1-f_3-f_4). \tag{C.17}$$

With the help of two dummy energy variables, E and E' we can now introduce the two-particle representation,

$$\mathcal{G}^<(E) = 2\pi\delta(E - \varepsilon_3 - \varepsilon_4)f_3 f_4, \quad \mathcal{G}^<(E') = 2\pi\delta(E' - \varepsilon_1 - \varepsilon_2)f_1 f_2,$$

$$\mathcal{A}(E) = 2\pi\delta(E - \varepsilon_3 - \varepsilon_4)(1 - f_3 - f_4), \quad \mathcal{A}(E') = 2\pi\delta(E' - \varepsilon_1 - \varepsilon_2)(1 - f_1 - f_2). \tag{C.18}$$

These two-particle functions do not include the off-shell corrections that can be neglected because we are already dealing with linear corrections in the limit of small scattering rates. From now on, all two-particle functions will only be in this lowest-order approximation. For convenience of notation, we do not use a special subscript to denote this approximation, but remind ourselves of this fact when necessary.

We associate the observable o with an operator $\mathcal{O}$ in the two-particle space. For instance, the correlated density n_c^{corr} is given by

$$\mathcal{O}_c = \delta_{ac}. \tag{C.19}$$

The operator $\mathcal{O}_c$ measures whether a particle of spin and isospin c is present in the two-particle interaction. The observable o is supposed to be a function of the phase-space point, i.e. the operator $\mathcal{O}$ commutes with $\mathcal{A}, \mathcal{G}^<$. For observables which conserve during the collision, the operator $\mathcal{O}$ commutes also with the T-matrix.

In terms of two-particle functions, the correlated contribution to an observable o reads

$$o^{\text{corr}} = \int \frac{dE\, dE'}{2\pi\, 2\pi} \frac{\wp'}{E-E'} \text{Tr}\left[\mathcal{T}_{\text{sc}}^R(E)\mathcal{G}^<(E)\,\mathcal{T}_{\text{sc}}^A(E)\mathcal{A}(E')\mathcal{O}\right]$$

$$-\int \frac{dE\, dE'}{2\pi\, 2\pi} \frac{\wp'}{E-E'} \text{Tr}\left[\mathcal{T}_{\text{sc}}^R(E)\mathcal{A}(E)\,\mathcal{T}_{\text{sc}}^A(E)\mathcal{G}^<(E')\mathcal{O}\right], \qquad (C.20)$$

where Tr denotes trace over two-particle states. The (C.20) is representation-free form of (C.17).

The second term can be simplified with the help of the optical theorem (8.40), $\mathcal{M}_{\text{ex}} = \mathcal{T}_{\text{sc}}^R \mathcal{A}\, \mathcal{T}_{\text{sc}}^A$. Using the Kramers–Kronig relation between the imaginary and real parts of analytic functions,

$$\int \frac{dE}{2\pi} \mathcal{M}_{\text{ex}}(E) \frac{\wp'}{E-E'} = \frac{\partial}{\partial E'} \int \frac{dE}{2\pi} \mathcal{M}_{\text{ex}}(E) \frac{\wp}{E-E'} = -\frac{\partial}{\partial E'} \mathcal{T}_{\text{ex}}(E'), \qquad (C.21)$$

the second term of (C.20) can be simplified to

$$-\int \frac{dE\, dE'}{2\pi\, 2\pi} \frac{\wp'}{E-E'} \text{Tr}\left[\mathcal{T}_{\text{sc}}^R(E)\mathcal{A}(E)\,\mathcal{T}_{\text{sc}}^A(E)\mathcal{G}^<(E')\mathcal{O}\right]$$

$$= \int \frac{dE'}{2\pi} \text{Tr}\left[\mathcal{G}^<(E')\mathcal{O}\frac{\partial}{\partial E'}\mathcal{T}_{\text{ex}}(E')\right]. \qquad (C.22)$$

In the first term of (C.20), the Kramers–Kronig relations allow one to integrate out dummy variable E',

$$\int \frac{dE'}{2\pi} \mathcal{A}(E') \frac{\wp'}{E-E'} = -\frac{\partial}{\partial E} \int \frac{dE'}{2\pi} \mathcal{A}(E') \frac{\wp}{E-E'} = -\frac{\partial}{\partial E}\mathcal{G}(E). \qquad (C.23)$$

The first term of (C.20) thus can be written in a form

$$\int \frac{dE\, dE'}{2\pi\, 2\pi} \frac{\wp'}{E-E'} \text{Tr}\left[\mathcal{T}_{\text{sc}}^R(E)\mathcal{G}^<(E)\,\mathcal{T}_{\text{sc}}^A(E)\mathcal{A}(E')\mathcal{O}\right]$$

$$= -\int \frac{dE}{2\pi} \text{Tr}\left[\mathcal{G}^<(E)\mathcal{T}_{\text{sc}}^A(E)\left(\frac{\partial}{\partial E}\mathcal{G}(E)\right)\mathcal{O}\,\mathcal{T}_{\text{sc}}^R(E)\right]. \qquad (C.24)$$

Finally, one can employ the symmetry of the T-matrix which holds in the absence of gradients. From $\mathcal{T}(12, 34) = \mathcal{T}(21, 43)$ it follows that the trace can be equivalently written with the inverted order of matrices

$$\text{Tr}\left[\mathcal{G}^<(E)\mathcal{T}_{\text{sc}}^A(E)\left(\frac{\partial}{\partial E}\mathcal{G}(E)\right)\mathcal{O}\,\mathcal{T}_{\text{sc}}^R(E)\right] = \text{Tr}\left[\mathcal{T}_{\text{sc}}^R(E)\left(\frac{\partial}{\partial E}\mathcal{G}(E)\right)\mathcal{O}\,\mathcal{T}_{\text{sc}}^A(E)\mathcal{G}^<(E)\right]. \qquad (C.25)$$

Making the rotation of matrices under the trace one arrives at

$$\int \frac{dE\, dE'}{2\pi\, 2\pi} \frac{\wp'}{E-E'} \mathrm{Tr}\left[\mathscr{T}_{\mathrm{sc}}^{R}(E)\mathscr{G}^{<}(E)\mathscr{T}_{\mathrm{sc}}^{A}(E)\mathscr{A}(E')\mathscr{O}\right]$$

$$= -\int \frac{dE}{2\pi}\mathrm{Tr}\left[\mathscr{G}^{<}(E)\mathscr{T}_{\mathrm{sc}}^{R}(E)\left(\frac{\partial}{\partial E}\mathscr{G}(E)\right)\mathscr{O}\mathscr{T}_{\mathrm{sc}}^{A}(E)\right]. \tag{C.26}$$

Now both terms can be collected. Renaming the integration variable in (C.22) as $E' \to E$, the correlated observable results

$$o^{\mathrm{corr}} = \int \frac{dE}{2\pi}\mathrm{Tr}\left[\mathscr{G}^{<}\left(\mathscr{O}\frac{\partial\mathscr{T}_{\mathrm{ex}}}{\partial E} - \mathscr{T}_{\mathrm{sc}}^{R}\frac{\partial\mathscr{G}}{\partial E}\mathscr{O}\mathscr{T}_{\mathrm{sc}}^{A}\right)\right], \tag{C.27}$$

where all functions depend on the energy E. Finally, we split this expression into two parts

$$o^{\mathrm{corr}} = \int \frac{dE}{2\pi}\mathrm{Tr}\left[\mathscr{G}^{<}\mathscr{O}\left(\frac{\partial\mathscr{T}_{\mathrm{ex}}}{\partial E} - \mathscr{T}_{\mathrm{sc}}^{R}\frac{\partial\mathscr{G}}{\partial E}\mathscr{T}_{\mathrm{sc}}^{A}\right)\right] + \int \frac{dE}{2\pi}\mathrm{Tr}\left[\mathscr{G}^{<}\mathscr{T}_{\mathrm{sc}}^{R}\frac{\partial\mathscr{G}}{\partial E}\left(\mathscr{T}_{\mathrm{sc}}^{A}\mathscr{O} - \mathscr{O}\mathscr{T}_{\mathrm{sc}}^{A}\right)\right]. \tag{C.28}$$

The second term vanishes if the observable quantity conserves during the collision.

In the first term of (C.28) we employ the derived optical theorem (C.10). The right-hand side of the optical theorem has the form of the anti-Hermitian part of the operator $\frac{\partial\mathscr{T}_{\mathrm{sc}}^{R}}{\partial E}\mathscr{A}\mathscr{T}_{\mathrm{sc}}^{A}$. Accordingly, the correlated contribution reads

$$o^{\mathrm{corr}} = \mathrm{Im}\int \frac{dE}{2\pi}\mathrm{Tr}\left[\mathscr{O}\mathscr{G}^{<}\frac{\partial\mathscr{T}_{\mathrm{sc}}^{R}}{\partial E}\mathscr{A}\mathscr{T}_{\mathrm{sc}}^{A}\right] + \int \frac{dE}{2\pi}\mathrm{Tr}\left[\mathscr{G}^{<}\mathscr{T}_{\mathrm{sc}}^{R}\frac{\partial\mathscr{G}}{\partial E}\left(\mathscr{T}_{\mathrm{sc}}^{A}\mathscr{O} - \mathscr{O}\mathscr{T}_{\mathrm{sc}}^{A}\right)\right]. \tag{C.29}$$

The first term of (C.29) includes the energy derivative of the T-matrix which in the momentum representation results in the collision delay Δ_t. The second term vanishes for observables which conserve during the collision. For instance, the number of particles of the spin and isospin c conserves, therefore the second term of (C.29) does not contribute to the correlated density.

For observables which conserve during collision, $\mathscr{O}\mathscr{T}_{\mathrm{sc}}^{A} - \mathscr{T}_{\mathrm{sc}}^{A}\mathscr{O} = 0$, we write the correlated contribution in the explicit representation

$$o^{\mathrm{corr}} = \mathrm{Im}\int \frac{dE}{2\pi}\mathrm{Tr}\left[\mathscr{O}\mathscr{G}^{<}\frac{\partial\mathscr{T}_{\mathrm{sc}}^{R}}{\partial E}\mathscr{A}\mathscr{T}_{\mathrm{sc}}^{A}\right]$$

$$= \sum_{ab}\int \frac{dk\,dp\,dq}{(2\pi)^{8}} o_{1}(1-f_{1}-f_{2})f_{3}f_{4}\delta(\varepsilon_{1}+\varepsilon_{2}-\varepsilon_{3}-\varepsilon_{4})\mathrm{Im}\frac{\partial t_{\mathrm{sc}}^{R}(E)}{\partial E}t_{\mathrm{sc}}^{A}(E)\Big|_{E=\varepsilon_{1}+\varepsilon_{2}}$$

$$= \sum_{ab}\int \frac{dk\,dp\,dq}{(2\pi)^{8}} o_{1}(1-f_{1}-f_{2})f_{3}f_{4}\delta(\varepsilon_{1}+\varepsilon_{2}-\varepsilon_{3}-\varepsilon_{4})\Delta_{t}\,|t_{\mathrm{sc}}(\varepsilon_{1}+\varepsilon_{2},k,p,q)|^{2}. \tag{C.30}$$

In the explicit representation, the density of particles c is given by the operator $o_{1} = \delta_{ac}$.

Correlated energy

Now we use the derived optical theorem to evaluate the correlated energy $\mathscr{E}^{\mathrm{corr}}$ from (D.16). Substituting the selfenergy from (15.100) we obtain

$$\mathscr{E}^{\mathrm{corr}} = \frac{1}{2} \int \frac{dE}{2\pi} \frac{d\omega}{2\pi} \mathrm{Tr}\left[(\omega + \varepsilon_1) \frac{\wp'}{\omega - \varepsilon_1} 2\pi \delta(E - \omega - \varepsilon_2) \right.$$

$$\left. \times \left[(1 - f_2)(1 - f_1) \mathscr{T}_{\mathrm{ex}}^{<}(E) - f_2 f_1 \mathscr{T}_{\mathrm{ex}}^{>}(E) \right] \right]. \tag{C.31}$$

Integrating out the energy ω yields

$$\mathscr{E}^{\mathrm{corr}} = \frac{1}{2} \int \frac{dE}{2\pi} \mathrm{Tr}\left[(E - \varepsilon_2 + \varepsilon_1) \frac{\wp'}{E - \varepsilon_2 - \varepsilon_1} \left[(1 - f_2)(1 - f_1) \mathscr{T}_{\mathrm{ex}}^{<}(E) - f_2 f_1 \mathscr{T}_{\mathrm{ex}}^{>}(E) \right] \right]. \tag{C.32}$$

The energy difference $\varepsilon_1 - \varepsilon_2$ found in the numerator of (C.32) is anti-symmetric with respect to the interchange of 1 and 2, while other constituents are symmetric. Accordingly, the contribution of $\varepsilon_1 - \varepsilon_2$ vanishes.

To turn the focus on the important variables we use the two-particle notation (C.18), e.g. $\mathscr{G}^{<}(E') = f_1 f_2 2\pi \delta(E' - \varepsilon_1 - \varepsilon_2)$, in terms of which (C.32) reads

$$\mathscr{E}^{\mathrm{corr}} = \frac{1}{2} \int \frac{dE}{2\pi} \frac{dE'}{2\pi} E \frac{\wp'}{E - E'} \mathrm{Tr}\left[\mathscr{G}^{>}(E') \mathscr{T}_{\mathrm{ex}}^{<}(E) - \mathscr{G}^{<}(E') \mathscr{T}_{\mathrm{ex}}^{>}(E) \right]. \tag{C.33}$$

Finally, we use (8.40) to express the 'greater' functions in terms of the spectral functions,

$$\mathscr{E}^{\mathrm{corr}} = \frac{1}{2} \int \frac{dE}{2\pi} \frac{dE'}{2\pi} E \frac{\wp'}{E - E'} \mathrm{Tr}\left[\mathscr{A}(E') \mathscr{T}_{\mathrm{ex}}^{<}(E) - \mathscr{G}^{<}(E') \mathscr{M}_{\mathrm{ex}}(E) \right], \tag{C.34}$$

to make the formula ready for the application of Kramers–Kronig relations and the derived optical theorem. With the help of Kramers–Kronig relation (C.23) and a modification of (C.21),

$$\int \frac{dE'}{2\pi} E \mathscr{M}_{\mathrm{ex}}(E) \frac{\wp'}{E - E'} = \mathscr{V}_{\mathrm{ex}} - \mathscr{T}_{\mathrm{ex}}(E') - E' \frac{\partial \mathscr{T}_{\mathrm{ex}}}{\partial E'}, \tag{C.35}$$

one can integrate out the energy,

$$\mathscr{E}^{\mathrm{corr}} = \frac{1}{2} \int \frac{dE}{2\pi} \mathrm{Tr}\left[(\mathscr{T}_{\mathrm{ex}}(E) - \mathscr{V}_{\mathrm{ex}}) \mathscr{G}^{<}(E) \right] - \frac{1}{2} \int \frac{dE}{2\pi} E \, \mathrm{Tr}\left[\mathscr{T}_{\mathrm{ex}}^{<}(E) \frac{\partial \mathscr{G}}{\partial E} - \frac{\partial \mathscr{T}_{\mathrm{ex}}}{\partial E} \mathscr{G}^{<}(E) \right]. \tag{C.36}$$

In the second term we substitute (10.127), $\mathscr{T}_{\mathrm{ex}}^{<} = \mathscr{T}_{\mathrm{sc}}^{R} \mathscr{G}^{<} \mathscr{T}_{\mathrm{sc}}^{A}$, and then employ the derived optical theorem (C.10) which links the derivatives of $\mathscr{T}_{\mathrm{ex}}$ with the collision delay,

$$\mathscr{E}^{\mathrm{corr}} = \frac{1}{2} \int \frac{dE}{2\pi} \mathrm{Tr}\left[(\mathscr{T}_{\mathrm{ex}}(E) - \mathscr{V}_{\mathrm{ex}}) \mathscr{G}^{<}(E) \right] + \frac{1}{2} \int \frac{dE}{2\pi} E \, \mathrm{Im} \, \mathrm{Tr}\left[\frac{\partial \mathscr{T}_{\mathrm{sc}}^{R}}{\partial E} \mathscr{A} \mathscr{T}_{\mathrm{sc}}^{A} \mathscr{G}^{<} \right]. \tag{C.37}$$

The first term is the quasiparticle correction $\mathscr{E}_{\mathrm{qp}}^{\mathrm{corr}}$ to the uncorrelated energy, the second term is the molecular energy $\mathscr{E}^{\mathrm{mol}}$ needed for (D.4)–(D.7).

Appendix D
Proof of Drift and Gain Compensation into the Rate of Quasiparticles

Here, in this appendix, we prove the compensation of energy and momentum gains with the corresponding derivatives of the drift into a complete derivative of the quasiparticle energy.

The proof of momentum gain compensation (15.51)

$$\sum_a \int \frac{dk}{(2\pi)^3} \varepsilon \frac{\partial f}{\partial r} + \mathscr{F}^{\text{gain}} = \frac{\partial \mathscr{E}^{\text{qp}}}{\partial r} - \mathscr{F}_K^{\text{ext}} \tag{D.1}$$

with the source term

$$\mathscr{F}_K^{\text{ext}} = \sum_a \int \frac{dk}{(2\pi)^3} f \frac{\partial \epsilon_k}{\partial r} \tag{D.2}$$

covering the momentum passed to particles by external fields. It closely parallels the proof of the energy gain compensation (15.54)

$$\sum_a \int \frac{dk}{(2\pi)^3} \varepsilon \frac{\partial f}{\partial t} - \mathscr{I}_{\text{gain}}^E = \frac{\partial \mathscr{E}^{\text{qp}}}{\partial t} - \mathscr{I}_E^{\text{ext}} \tag{D.3}$$

if one only replaces the spacial derivatives in (D.1) and (D.2) by the time derivatives. Therefore we will only outline here the proof with time derivatives.

In order to proove (D.3) we have first to derive what the quasiparticle energy looks like from the definition using the extended quasiparticle approximation. We use the general expression (7.38) to evaluate the energy density. We will drop the arguments space r and time t.

Limit of small scattering rates

Now we consider the limit of small scattering rates, $\gamma \to 0$, and decompose the energy into the zeroth- and first-order contribution in $\gamma \to 0$,

$$\mathscr{E} = \mathscr{E}^{\text{free}} + \mathscr{E}^{\text{corr}}. \tag{D.4}$$

The energy $\mathscr{E}^{\text{free}}$ corresponds to uncorrelated motion, i.e. it includes the mean field also. The energy $\mathscr{E}^{\text{corr}}$ describes the potential energy of binary correlations. We will write both parts as functionals of the quasiparticle distribution.

Besides (D.4) we have found the decomposition

$$\mathcal{E} = \mathcal{E}^{\mathrm{qp}} + \mathcal{E}^{\mathrm{mol}}. \tag{D.5}$$

as conserving from the kinetic theory. Both forms (D.4) and (D.5) are related as follows. The $\mathcal{E}^{\mathrm{qp}}$ is the quasiparticle renormalisation of the uncorrelated energy,

$$\mathcal{E}^{\mathrm{qp}} = \mathcal{E}^{\mathrm{free}} + \mathcal{E}_{\mathrm{qp}}^{\mathrm{corr}}. \tag{D.6}$$

The molecular energy originates exclusively from the correlated motion. The correlated energy thus is

$$\mathcal{E}^{\mathrm{corr}} = \mathcal{E}^{\mathrm{mol}} + \mathcal{E}_{\mathrm{qp}}^{\mathrm{corr}}. \tag{D.7}$$

The two decompositions are needed for technical reasons. The $\mathcal{E} = \mathcal{E}^{\mathrm{qp}} + \mathcal{E}^{\mathrm{mol}}$ reflects the chemical picture which is convenient for intuitive insight. The $\mathcal{E} = \mathcal{E}^{\mathrm{free}} + \mathcal{E}^{\mathrm{corr}}$ is necessary for a consistent implementation of the limit of small scattering rates.

Now we implement the limit of small scattering rates by substituting the extended quasiparticle approximation (9.62) into the energy density (7.38),

$$\mathcal{E} = \sum_a \int \frac{dk}{(2\pi)^3} \frac{d\omega}{2\pi} \frac{1}{2} \left(\omega + \frac{k^2}{2m} \right) \left[fz 2\pi \delta(\omega - \varepsilon) + \sigma^< \frac{\wp'}{\omega - \varepsilon} \right]. \tag{D.8}$$

The off-shell term proportional to $\sigma^<$ is linear in γ. The on-shell term proportional to fz is a mixture of zeroth and linear order in γ. Higher orders in γ are neglected.

First we use the explicit wave-function renormalisation (9.66) to rearrange (D.8) as

$$\mathcal{E} = \sum_a \int \frac{dk}{(2\pi)^3} \frac{1}{2} (\varepsilon + \epsilon_k) f + \sum_a \int \frac{dk}{(2\pi)^3} \frac{d\omega}{2\pi} \frac{1}{2} (\omega + \epsilon_k) \frac{\wp'}{\omega - \varepsilon} \left(\sigma^< (1 - f) - \sigma^> f \right)$$

$$- \sum_a \int \frac{dk}{(2\pi)^3} \frac{d\omega}{2\pi} \frac{1}{2} (\varepsilon - \omega) \frac{\wp'}{\omega - \varepsilon} (\sigma^> + \sigma^<) f. \tag{D.9}$$

The last term simplifies with the help of the Kramers–Kronig relation,

$$\int \frac{dk}{(2\pi)^3} \frac{d\omega}{2\pi} (\varepsilon - \omega) \frac{\wp'}{\omega - \varepsilon} (\sigma^> + \sigma^<) = \sigma^{\mathrm{reg}}, \tag{D.10}$$

and cancels with σ^{reg} from the quasiparticle energy, $\varepsilon = \epsilon + \sigma^{HF} + \sigma^{\mathrm{reg}}$, in the first term of (D.9). The energy density then reads

$$\mathcal{E} = \sum_a \int \frac{dk}{(2\pi)^3} \epsilon_k f + \frac{1}{2} \sum_a \int \frac{dk}{(2\pi)^3} \sigma^{HF} f$$

$$+ \frac{1}{2} \sum_a \int \frac{dk}{(2\pi)^3} \frac{d\omega}{2\pi} (\omega + \epsilon_k) \frac{\wp'}{\omega - \varepsilon} \left(\sigma^< (1 - f) - \sigma^> f \right). \tag{D.11}$$

Now we can separate orders in γ. The first term of (D.11) is the bare energy

$$\sum_a \int \frac{dk}{(2\pi)^3} \epsilon_k f = \text{Tr}\,[\epsilon_k f]. \tag{D.12}$$

This is of zeroth order in γ. The second term of (D.11) is the mean field given by selfenergy $\sigma^{HF} = \mathcal{V}_{ex}\rho$. To separate its zeroth order, we split the Wigner distribution ρ into the quasiparticle distribution f and the off-shell correction $\rho - f$,

$$\sum_a \int \frac{dk}{(2\pi)^3} \sigma^{HF} f = \text{Tr}\,[\mathcal{V}_{ex}ff] + \text{Tr}\,[\mathcal{V}_{ex}(\rho - f)f]. \tag{D.13}$$

The first term is the mean-field interaction of quasiparticles. The second term is more complex. Among other contributions, it includes a mean-field interaction between quasiparticles and effective molecules. The kinetic energy and the mean field cover the uncorrelated energy,

$$\mathcal{E}^{free} = \text{Tr}\,[\epsilon_k f] + \frac{1}{2}\text{Tr}\,[\mathcal{V}_{ex}ff]. \tag{D.14}$$

Now we collect contribution linear in γ which are made up of the second term of the mean-field (D.13) and the last term of the expansion (D.11). To join both terms we express $\frac{1}{2}\text{Tr}\,[\mathcal{V}_{ex}(\rho - f)f]$ from the $\rho[f]$-functional (9.67),

$$\frac{1}{2}\text{Tr}\,[\mathcal{V}_{ex}(\rho - f)f] = \frac{1}{2}\text{Tr}\left[\sigma^{HF}(\rho - f)\right] = \frac{1}{2}\sum_a \int \frac{dk}{(2\pi)^3}\frac{d\omega}{2\pi}\sigma^{HF}\frac{\wp'}{\omega - \varepsilon}\left(\sigma^<(1 - f) - \sigma^> f\right). \tag{D.15}$$

The last term of Eq. (D.11) combines with Eq. (D.15) into

$$\mathcal{E}^{corr} = \frac{1}{2}\sum_a \int \frac{dk}{(2\pi)^3}\frac{d\omega}{2\pi}(\omega + \varepsilon)\frac{\wp'}{\omega - \varepsilon}\left(\sigma^<(1 - f) - \sigma^> f\right). \tag{D.16}$$

The explicit $\sigma^{>,<}$ in (D.16) guarantees that the $\mathcal{E}_{corr}$ is linear in γ. Other ingredients can thus be treated within the lowest order approximation. In this sense, we have already used $\mathcal{V}_{ex}f = \sigma^{HF}$ in Eq. (D.15) and $\omega + \epsilon_k + \sigma^{HF} = \omega + \varepsilon$ in Eq. (D.16).

Molecular energy density

Formulae (D.14) and (D.16) define the decomposition (D.4). The actual physical content is more easily readable from decomposition (D.5) which is reached from (D.4) when one decomposes the correlated density into its molecular and quasiparticle parts, see Eq. (D.7). To this end we substitute into (D.16) for the selfenergy from (15.100) and rearrange it with the help of the Kramers–Kronig relations and the derived optical theorem. Due to its technical character this algebra is appended in Appendix C. Formula (C.37), resulting directly from Eq. (D.16), in explicit notation reads

$$\mathscr{E}^{\mathrm{corr}} = \frac{1}{2} \sum_{ab} \int \frac{dk\, dp\, dq}{(2\pi)^9} 2\pi \delta(\varepsilon_1 + \varepsilon_2 - \varepsilon_3 - \varepsilon_4)(\varepsilon_1 + \varepsilon_2)$$

$$\times f_1 f_2 (1 - f_3 - f_4) \mathrm{Im}\, \frac{\partial t_{\mathrm{sc}}^R(E, k, p, q)}{\partial E} \left. t_{\mathrm{sc}}^A(E, k, p, q) \right|_{E = \varepsilon_1 + \varepsilon_2}$$

$$+ \frac{1}{2} \sum_{ab} \int \frac{dk\, dp}{(2\pi)^6} f_1 f_2 \left[\mathscr{T}_{\mathrm{ex}}(\varepsilon_1 + \varepsilon_2, k, p, 0) - \mathscr{V}_{\mathrm{ex}}(k, p, 0) \right], \tag{D.17}$$

where $q = 0$ denotes the T-matrix of the zero momentum transfer channel.

The first term of (D.17) is the molecular energy $\mathscr{E}^{\mathrm{mol}}$. Indeed, the energy derivative of the T-matrix results in the collision delay so that (D.17) agrees with the molecular energy (15.80) found from the kinetic equation. The second term is thus the correlated contribution to the quasiparticle part,

$$\mathscr{E}_{\mathrm{qp}}^{\mathrm{corr}} = \frac{1}{2} \mathrm{Tr}\left[(\mathscr{T}_{\mathrm{ex}} - \mathscr{V}_{\mathrm{ex}}) ff \right]. \tag{D.18}$$

According to (D.6) and (D.14), the quasiparticle part of the energy density,

$$\mathscr{E}^{\mathrm{qp}} = \sum_a \int \frac{dk}{(2\pi)^3} f_a(k) \epsilon_a(k) + \frac{1}{2} \sum_{ab} \int \frac{dk\, dp}{(2\pi)^6} f_a(k) f_b(p) \mathscr{T}_{\mathrm{ex}}(\varepsilon_1 + \varepsilon_2, k, p, 0). \tag{D.19}$$

has the same structure as the uncorrelated energy (D.14), the bare interaction potential is, however, re-normalised to the T-matrix.

To prove that (D.19) solves (15.54), we take the time derivative of (D.19),

$$\frac{\partial \mathscr{E}^{\mathrm{qp}}}{\partial t} = \sum_a \int \frac{dk}{(2\pi)^3} \frac{\partial f_1}{\partial t} \left(\epsilon_1(k) + \sum_b \int \frac{dp}{(2\pi)^3} \mathscr{T}_{\mathrm{ex}} f_2 \right)$$

$$+ \frac{1}{2} \sum_{ab} \int \frac{dk\, dp}{(2\pi)^6} f_1 f_2 \left(\frac{\partial \mathscr{T}_{\mathrm{ex}}}{\partial t} + \frac{\partial \mathscr{T}_{\mathrm{ex}}}{\partial E} \frac{\partial(\varepsilon_1 + \varepsilon_2)}{\partial t} \right)_{E = \varepsilon_1 + \varepsilon_2} + \sum_a \int \frac{dk}{(2\pi)^3} \frac{\partial \epsilon_1(k)}{\partial t} f_1. \tag{D.20}$$

In the T-matrix part of the first term we have used the symmetry between the states 1 and 2. The proof will be complete when we manage to rearrange the right-hand side of (D.20) into $\mathscr{I}_{\mathrm{drift}} - \mathscr{I}_{\mathrm{gain}}$ found in relation (15.54).

Quasiparticle energy

The first term of Eq. (D.20) reminds the integral $\int \varepsilon \partial f / \partial t$ resulting from the energy balance in the kinetic equation. To link these two forms we need the quasiparticle energy, $\varepsilon_1 \equiv \varepsilon_a(\mathbf{k}, \mathbf{r}, t) = \epsilon_a(\mathbf{k}, \mathbf{r}, t) + \sigma_a(\varepsilon_1, \mathbf{k}, \mathbf{r}, t)$, where $\sigma_a(\varepsilon_1, \mathbf{k}, \mathbf{r}, t)$ is the on-shell value of the real part of the retarded selfenergy. Below we suppress space and time arguments.

The retarded selfenergy (10.129) has two parts, $\Sigma^R = \mathscr{T}_{\mathrm{ex}}^R G^< - \mathscr{T}_{\mathrm{ex}}^< G^A$, its real part in the Wigner representation reads

$$\sigma(\varepsilon_1, k) = \sum_b \int \frac{d\omega}{2\pi} \frac{dp}{(2\pi)^3} \mathscr{T}_{\mathrm{ex}}(\varepsilon_1 + \omega, k, p, 0) g_b^<(\omega, p) + \sigma^{(2)} \tag{D.21}$$

where we denote the second term,

$$\sigma^{(2)} = -\sum_b \int \frac{d\omega}{2\pi} \frac{dp}{(2\pi)^3} \mathcal{T}_{\text{ex}}^< (\varepsilon_1 + \omega, k, p, 0) g_b(\omega, p) = -\mathcal{T}_{\text{ex}}^< g. \tag{D.22}$$

The first term of (D.21) includes the mean field, which we separate as

$$\mathcal{T}_{\text{ex}} g^< = \mathcal{V}_{\text{ex}} g^< + (\mathcal{T}_{\text{ex}} - \mathcal{V}_{\text{ex}}) g^<. \tag{D.23}$$

Since the potential $\mathcal{V}_{\text{ex}}$ is independent of energy, the energy integration of the first term directly leads to the Wigner distribution. The second term is linear in $\gamma \to 0$, therefore its correlation function can be approximated as $g^< = f \, 2\pi \delta(\omega - \varepsilon)$ what also allows us to integrate out the energy

$$\mathcal{T}_{\text{ex}} g^< = \mathcal{V}_{\text{ex}} \rho + (\mathcal{T}_{\text{ex}} - \mathcal{V}_{\text{ex}}) f = \mathcal{T}_{\text{ex}} f + \mathcal{V}_{\text{ex}} (\rho - f). \tag{D.24}$$

The first term, $\sigma^{(1)} = \mathcal{T}_{\text{ex}} f$, is the dominant contribution. The second term, $\sigma^{(3)} = \mathcal{V}_{\text{ex}}(\rho - f)$, represents the mean-field interaction between quasiparticles and molecules. With these pieces of the selfenergy, the drift contribution (15.25) reads

$$\sum_a \int \frac{dk}{(2\pi)^3} \varepsilon \frac{\partial f}{\partial t} = \sum_a \int \frac{dk}{(2\pi)^3} \frac{\partial f_1}{\partial t} \left(\epsilon_k + \sigma^{(1)} + \sigma^{(2)} + \sigma^{(3)} \right). \tag{D.25}$$

Energy conservation

Having the drift term (D.25) and the mean-energy gain (15.53), we are furnished to verify that (D.20) is equivalent to (15.54), i.e. that the kinetic equation conserves the energy density given by the extended quasiparticle approximation. To this end we decompose (D.20) into $\mathscr{I}_{\text{drift}}$ and $\mathscr{I}_{\text{gain}}$.

The first term of the energy derivative (D.20) is the dominant part of $\mathscr{I}_{\text{drift}}$,

$$\mathscr{I}_{\text{drift}}^{(1)} = \sum_a \int \frac{dk}{(2\pi)^3} \frac{\partial f_1}{\partial t} \left(\epsilon_k + \sum_b \int \frac{dp}{(2\pi)^3} \mathcal{T}_{\text{ex}} f_2 \right) = \sum_a \int \frac{dk}{(2\pi)^3} \frac{\partial f_1}{\partial t} \left(\epsilon_k + \sigma^{(1)} \right). \tag{D.26}$$

The second term of (D.20),

$$\frac{1}{2} \sum_{ab} \int \frac{dk\,dp}{(2\pi)^6} f_1 f_2 \left(\frac{\partial \mathcal{T}_{\text{ex}}}{\partial t} + \frac{\partial \mathcal{T}_{\text{ex}}}{\partial E} \frac{\partial (\varepsilon_1 + \varepsilon_2)}{\partial t} \right)_{E = \varepsilon_1 + \varepsilon_2}. \tag{D.27}$$

should thus result in three distinct contributions: the mean energy gain $\mathscr{I}_{\text{gain}}$, the part of the selfenergy $\sigma^{(2)}$ and the mean-field interaction of quasiparticles and molecules $\sigma^{(3)}$.

To identify the mean-energy gain we substitute the time-derived optical theorem (C.11),

$$
\frac{\partial \mathscr{T}_{ex}(E, k, p, 0)}{\partial t} = \int \frac{dq}{(2\pi)^3} |t_{sc}(E, k, p, q)|^2 (1 - f_3 - f_4)
$$

$$
\times \left(-2\Delta_E \, 2\pi \delta(E - \varepsilon_3 - \varepsilon_4) + \frac{\wp'}{E - \varepsilon_3 - \varepsilon_4} \frac{\partial(\varepsilon_3 + \varepsilon_4)}{\partial t} \right)
$$

$$
- \int \frac{dq}{(2\pi)^3} |t_{sc}(E, k, p, q)|^2 \left(\frac{\partial f_3}{\partial t} + \frac{\partial f_4}{\partial t} \right) \frac{\wp}{E - \varepsilon_3 - \varepsilon_4}, \tag{D.28}
$$

into the first part of (D.27). The term proportional to Δ_E in (D.28) directly results into the mean energy gain $\mathscr{I}_{gain}^E$ of Eq. (15.53) and the last $\wp$-term of (D.28) we write as

$$
\mathscr{I}_{drift}^{(2)} = -\frac{1}{2} \sum_{ab} \int \frac{dE}{2\pi} \int \frac{dk\,dp\,dq}{(2\pi)^9} f_1 f_2 \, |t_{sc}(E, k, p, q)|^2
$$

$$
\times \left(\frac{\partial f_3}{\partial t} + \frac{\partial f_4}{\partial t} \right) \frac{\wp}{E - \varepsilon_3 - \varepsilon_4} 2\pi \delta(E - \varepsilon_1 - \varepsilon_2) \tag{D.29}
$$

since it will turn out to be the contribution $\sigma^{(2)}$ to the drift. We have introduced the on-shell energy condition via a dummy energy, $\int dE \, \delta(E - \varepsilon_1 - \varepsilon_2)$.

Collecting the remaining last part of (D.28) and the second part of (D.27), which is the sum of $\frac{\partial \mathscr{T}_{ex}}{\partial E}$-term of (D.20) and $\wp'$-term of (D.20), into

$$
\mathscr{I}_{drift}^{(3)} = \frac{1}{2} \sum_{ab} \int \frac{dE}{2\pi} \int \frac{dk\,dp}{(2\pi)^6} f_1 f_2 \frac{\partial \mathscr{T}_{ex}}{\partial E} \frac{\partial(\varepsilon_1 + \varepsilon_2)}{\partial t} 2\pi \delta(E - \varepsilon_1 - \varepsilon_2)
$$

$$
+ \int \frac{dE}{2\pi} \int \frac{dq}{(2\pi)^3} |t_{sc}(E, k, p, q)|^2 (1 - f_3 - f_4) \frac{\wp'}{E - \varepsilon_3 - \varepsilon_4} \frac{\partial(\varepsilon_3 + \varepsilon_4)}{\partial t} 2\pi \delta(E - \varepsilon_1 - \varepsilon_2) \tag{D.30}
$$

so that Eq. (D.20) reads

$$
\frac{\partial \mathscr{E}^{qp}}{\partial t} = -\mathscr{I}_{gain}^E + \mathscr{I}_{drift}^{(1)} + \mathscr{I}_{drift}^{(2)} + \mathscr{I}_{drift}^{(3)} + \sum_a \int \frac{dk}{(2\pi)^3} \frac{\partial \epsilon_1}{\partial t} f_1. \tag{D.31}
$$

We have already (D.26) and it remains to identify $\mathscr{I}_{drift}^{(2)}$ and $\mathscr{I}_{drift}^{(3)}$.

To rearrange $\mathscr{I}_{drift}^{(2)}$ of (D.29), we interchange states 1 and 2 with 3 and 4 by substitution (15.9)

$$
\mathscr{I}_{drift}^{(2)} = -\sum_a \int \frac{d\bar{k}}{(2\pi)^3} \frac{\partial f_1}{\partial t} \sum_b \int \frac{d\bar{p}}{(2\pi)^3} \frac{dE}{2\pi} \frac{\wp}{E - \varepsilon_1 - \varepsilon_2}
$$

$$
\times \int \frac{d\bar{q}}{(2\pi)^3} |t_{sc}(E, \bar{k}, \bar{p}, \bar{q})|^2 f_3 f_4 2\pi \delta(E - \varepsilon_3 - \varepsilon_4). \tag{D.32}
$$

The δ function together with distributions $f_3 f_4$ forms the two-particle correlation function,

$$\mathscr{G}^< = f_3 f_4 2\pi \delta(E - \varepsilon_3 - \varepsilon_4) \tag{D.33}$$

and the integral over the transferred momentum represents the matrix product in the equation of motion (10.127). Therefore

$$\int \frac{d\bar{q}}{(2\pi)^3} \left| t_{\text{sc}}(E, \bar{k}, \bar{p}, \bar{q}) \right|^2 \mathscr{G}^< = \mathscr{T}_{\text{ex}}^<(E, \bar{k}, \bar{p}, 0). \tag{D.34}$$

The integral over E and $\bar{p}$ in (D.32) represents a convolution of $\mathscr{T}_{\text{ex}}^<$ with the real part of the propagator $\text{Re}\, G^R = g(E - \varepsilon_1, \bar{p}) = \frac{\wp}{E - \varepsilon_1 - \varepsilon_2}$, accordingly

$$\sum_b \int \frac{d\bar{p}}{(2\pi)^3} \frac{dE}{2\pi} \frac{\wp}{E - \varepsilon_1 - \varepsilon_2} \mathscr{T}_{\text{ex}}^<(E, \bar{k}, \bar{p}, 0) = -\sigma^{(2)}(\varepsilon_1, \bar{k}) \tag{D.35}$$

such that

$$\mathscr{I}_{\text{drift}}^{(2)} = \sum_a \int \frac{d\bar{k}}{(2\pi)^3} \frac{\partial f_1}{\partial t} \sigma^{(2)}(\varepsilon_1, \bar{k}), \tag{D.36}$$

what we wished to prove.

To complete the proof of the energy conservation it remains to show that $\mathscr{I}_{\text{drift}}^3$ is the $\sigma^{(3)}$-contribution to the drift. We express $\frac{\partial \mathscr{T}_{\text{ex}}}{\partial E}$ from the derived optical theorem (C.10),

$$\frac{\partial \mathscr{T}_{\text{ex}}(E, k, p, 0)}{\partial E} = \int \frac{dq}{(2\pi)^3} \left| t_{\text{sc}}(E, k, p, q) \right|^2 (1 - f_3 - f_4)$$
$$\times \left(\Delta_t \, 2\pi \delta(E - \varepsilon_3 - \varepsilon_4) - \frac{\wp'}{E - \varepsilon_3 - \varepsilon_4} \right) \tag{D.37}$$

that the sum of all remaining contributions to (D.20) reads

$$\mathscr{I}_{\text{drift}}^{(3)} = \frac{1}{2} \sum_{ab} \int \frac{dE\, dk\, dp\, dq}{2\pi \, (2\pi)^9} \frac{\partial (\varepsilon_1 + \varepsilon_2)}{\partial t} 2\pi \delta(E - \varepsilon_1 - \varepsilon_2)$$
$$\times f_1 f_2 \left| t_{\text{sc}}(E, k, p, q) \right|^2 (1 - f_3 - f_4) 2\pi \delta(E - \varepsilon_3 - \varepsilon_4) \Delta_t$$
$$+ \frac{1}{2} \sum_{ab} \int \frac{dE\, dk\, dp\, dq}{2\pi \, (2\pi)^9} 2\pi \delta(E - \varepsilon_1 - \varepsilon_2) \left| t_{\text{sc}}(E, k, p, q) \right|^2$$
$$\times f_1 f_2 \frac{\partial (\varepsilon_3 + \varepsilon_4 - \varepsilon_1 - \varepsilon_2)}{\partial t} (1 - f_3 - f_4) \frac{\wp'}{E - \varepsilon_3 - \varepsilon_4}. \tag{D.38}$$

In the second term we have already introduced the dummy energy E and joined both $\wp'$-terms. The two integrals of (D.38) correspond to the two parts in (C.29) with 'observable' $o = \frac{\partial \varepsilon}{\partial t}$.

Formula (D.38) represents the correlated contribution to the mean value of the time derivative of the quasiparticle energy, i.e. $\mathscr{I}_{\text{drift}}^{(3)} = \left(\frac{\partial \varepsilon}{\partial t}\right)^{\text{corr}}$.

Using the relation for a correlated observable in the backward manner, from Eq. (C.14) we find

$$
\mathscr{I}_{\text{drift}}^{(3)} = \sum_a \int \frac{dk}{(2\pi)^3} (\rho_1 - f_1) \frac{\partial \varepsilon_1}{\partial t}
$$

$$
= \sum_a \int \frac{dk}{(2\pi)^3} (\rho_1 - f_1) \frac{\partial \varepsilon_1}{\partial t} + \sum_a \int \frac{dk}{(2\pi)^3} (\rho_1 - f_1) \frac{\partial \sigma_1^{HF}}{\partial t}. \tag{D.39}
$$

Since $\rho - f$ is linear in $\gamma \to 0$, in the second line we keep only zeroth order terms. The first term adds with the last term of Eq. (D.31) into the energy change $\mathscr{I}_E^{\text{ext}}$ defined by Eq. (D.2).

In zeroth order in $\gamma \to 0$ the mean-field selfenergy σ^{HF} can be approximated by its quasiparticle part

$$
\frac{\partial \sigma_1^{HF}}{\partial t} = \sum_b \int \frac{dp}{(2\pi)^3} \mathscr{V}_{\text{ex}} \frac{\partial f_2}{\partial t}. \tag{D.40}
$$

Finally, we substitute (D.40) into (D.39), and interchange variables 1 and 2 so that

$$
\mathscr{I}_{\text{drift}}^{(3)} - \sum_a \int \frac{dk}{(2\pi)^3} (\rho_1 - f_1) \frac{\partial \varepsilon_1}{\partial t} = \sum_{ab} \int \frac{dk\,dp}{(2\pi)^6} (\rho_1 - f_1) \mathscr{V}_{\text{ex}} \frac{\partial f_2}{\partial t}
$$

$$
= \sum_a \int \frac{dk}{(2\pi)^3} \frac{\partial f_1}{\partial t} \sum_b \int \frac{dp}{(2\pi)^3} (\rho_2 - f_2) \mathscr{V}_{\text{ex}} = \sum_a \int \frac{dk}{(2\pi)^3} \frac{\partial f_1}{\partial t} \sigma^{(3)}. \tag{D.41}
$$

Now we are ready to substitute $\mathscr{I}_{\text{drift}}^{(1)}$, $\mathscr{I}_{\text{drift}}^{(2)}$ and $\mathscr{I}_{\text{drift}}^{(3)}$ into Eq. (D.31). According to Eq. (D.25) they sum into the quasiparticle energy weighted with the time derivative of the quasiparticle distribution, and the correlated part of the external field energy feed $\mathscr{I}_E^{\text{ex}}$. The time derivative of the energy density (D.31) thus is

$$
\frac{\partial \mathscr{E}^{\text{qp}}}{\partial t} = -\mathscr{I}_{\text{gain}}^E + \mathscr{I}_E^{\text{ext}} + \sum_a \int \frac{dk}{(2\pi)^3} \varepsilon_1 \frac{\partial f_1}{\partial t} \tag{D.42}
$$

which is identical to Eqs. (15.54) and (D.3). The proof of the space gradient is analogous.

Appendix E
Separable Interactions

A separable expansion is always possible for local interactions provided that the potential range is finite ($\mathscr{V}(r) = 0$ for $r > R$). We follow the presentation of Koike et al. (1997) and start from the Lippmann–Schwinger equation with arbitrary energy ω (called off-shell if $\omega \neq \frac{p^2}{2m}$),

$$|\psi\rangle = |p\rangle + \mathscr{G}_0 \mathscr{V} |\psi\rangle. \tag{E.1}$$

We expand the wave function inside the potential range,

$$|\psi\rangle = \sum_i |B_i\rangle \xi_i, \tag{E.2}$$

and multiply (E.1) by $\langle B_j|\mathscr{V}$ from the left to obtain

$$\sum_i \langle B_j|\mathscr{V}|B_i\rangle \xi_i = \langle B_j|\mathscr{V}|p\rangle + \sum_i \langle B_j|\mathscr{V}\mathscr{G}_0\mathscr{V}|B_i\rangle \xi_i. \tag{E.3}$$

Abbreviating $\mu_{ji} = \langle B_j|\mathscr{V}|B_i\rangle$, $\zeta_j = \langle B_j|\mathscr{V}|p\rangle$ and introducing the form factors

$$\langle g_j| = \langle B_j|\mathscr{V} \qquad |g_i\rangle = \mathscr{V}|B_i\rangle \tag{E.4}$$

we can write (E.3) as a matrix equation

$$\sum_i (\mu_{ji} - \langle g_j|\mathscr{G}_0|g_i\rangle) \xi_i = \zeta_j \tag{E.5}$$

such that the expansion coefficients become

$$\xi_j = \sum_i \tau_{ij} \zeta_j \tag{E.6}$$

with $\tau_{ij}^{-1} = \mu_{ij} - \langle g_i|\mathscr{G}_0|g_j\rangle$. Inserting (E.6) into (E.2) we have

$$\mathscr{V}|\psi\rangle = \sum_{ij} |g_i\rangle \tau_{ij} \langle g_j|p\rangle \tag{E.7}$$

which shows that the effect of the potential on the scattering wave function can be expanded in separable form.

E.1 Yamaguchi form factor

We consider a rank-one separable potential

$$\langle \mathbf{p}|\mathscr{V}|\mathbf{q}\rangle = \lambda g(\mathbf{p})g^*(\mathbf{q}) \tag{E.8}$$

with the traditional form factor of Yamaguchi (1954)

$$g_p = \frac{1}{\beta^2 + p^2}. \tag{E.9}$$

Other formfactors can be found in (Morawetz, Schreiber, Schmidt, Ficker and Lipavský, 2005*b*). The strength λ is positive/negative for repulsive/attractive interactions. We can compute with the potential (E.8) the unrestricted two-particle scattering T-matrix (10.156)

$$\langle p_1 P|\mathscr{T}(\omega)|p_2 P'\rangle = \lambda \delta_{PP'} \frac{g_{p_1} g_{p_2}}{1 - \lambda \mathscr{J}(P,\omega)} \tag{E.10}$$

which is controlled by

$$\lambda \mathscr{J}\left(P, \Omega = \omega - \frac{P^2}{4m}\right) = \frac{\lambda}{2\pi^2} \int\limits_0^\infty dq q^2 \frac{g_q^2}{\Omega - \frac{q^2}{m} + i\eta}. \tag{E.11}$$

We will use a dimensionless scaled potential strength $\xi = m\lambda/2\pi^2\beta^3$ presently. The *T*-matrix possesses a pole, the bound-state energy $\Omega = -E_b$ is given by

$$1 = \lambda \mathscr{J}(-E_b) = -\frac{\pi \xi}{2(1 + \sqrt{2}y)^2} \tag{E.12}$$

with $y = \sqrt{mE_b/\beta^2}$. We obtain bound states if

$$\xi = -\frac{2(1 + \sqrt{2}y)^2}{\pi} < -\frac{2}{\pi}. \tag{E.13}$$

The two-particle Schrödinger equation in the centre-of-mass frame

$$\left(\Omega - \frac{p^2}{m}\right)\phi(\mathbf{p}) = \int \frac{d^3q}{(2\pi)^3} \bar{v}(\mathbf{p},\mathbf{q})\phi(\mathbf{q}) = \lambda g(\mathbf{p}) \int \frac{d^3q}{(2\pi)^3} g^*(\mathbf{q})\phi(\mathbf{q})$$

yields directly one bound state for $\Omega = -\frac{\alpha^2}{m} = -E_B$ with the wave function

$$\phi_b(\mathbf{p}) = C_b \frac{g(\mathbf{p})}{\alpha^2 + p^2}, \qquad \frac{1}{C_b^2} = \int \frac{d^3p}{(2\pi)^3} \frac{|g(\mathbf{p})|^2}{(\alpha^2 + p^2)^2}. \tag{E.14}$$

This bound-state wave function in real space for this form factor is

$$\phi_b(\mathbf{r}) = \int \frac{d^3p}{(2\pi)^3} e^{i\mathbf{p}\mathbf{r}} \phi_b(\mathbf{p}) = \sqrt{\frac{1}{2\pi}\left(\frac{1}{\alpha} + \frac{1}{\beta}\right)} \frac{e^{-\alpha r} - e^{-\beta r}}{\frac{1}{\alpha} - \frac{1}{\beta}}$$

where $\mathbf{r}$ is the vector from one particle to the other. From the bound-state wave function follows the bond length

$$\bar{r} = \int rd^3r\,|\phi_b(\mathbf{r})|^2 = \frac{1}{2\alpha} + \frac{1}{2\beta} + \frac{1}{\alpha + \beta}. \tag{E.15}$$

The bound-state parameter α can be obtained from the bound-state condition (E.13)

$$\alpha = \gamma\left(\sqrt{-\frac{\lambda}{\lambda_{c0}}} - 1\right),$$

but a bound state is only possible for (E.13) which we can rewrite

$$\lambda < -\lambda_{c0} = -\frac{8\pi\beta^3}{m}. \tag{E.16}$$

Assuming an incoming plane wave and $\epsilon = \frac{k^2}{m}$ one obtains the scattering state

$$\phi_s(\mathbf{p}) = (2\pi)^3\delta(\mathbf{p} - \mathbf{k}) + \tau\left(\frac{k^2}{m}\right)\frac{g(\mathbf{p})g^*(\mathbf{k})}{\frac{k^2}{m} - \frac{p^2}{m} + i\eta} \tag{E.17}$$

where $\eta \to +0$ guarantees an outgoing scattered wave and

$$\tau(\epsilon) = \lambda\left(1 - \lambda\int\frac{d^3q}{(2\pi)^3}\frac{|g(\mathbf{q})|^2}{\epsilon - \frac{\hbar^2q^2}{m} + i\eta}\right)^{-1}. \tag{E.18}$$

The scattering T-matrix for the potential (E.8) is $\langle\mathbf{p}|\hat{\mathcal{T}}(\epsilon)|\mathbf{k}\rangle = \tau(\epsilon)g(\mathbf{p})g^*(\mathbf{k})$ with $\epsilon = k^2/m$, what shows the relation of the T-matrix to the scattering wave function (E.17). The scattering wave function in real space

$$\phi_s(\mathbf{r}) = \int\frac{d^3p}{(2\pi)^3}e^{i\mathbf{p}\mathbf{r}}\phi_s(\mathbf{p}) = e^{i\mathbf{k}\mathbf{r}} - \frac{m}{4\pi}\tau\left(\frac{k^2}{m}\right)g^2(\mathbf{k})\frac{e^{ikr} - e^{-\beta r}}{r}$$

has the asymptotics far from the scattering centre and for low-energy scattering

$$\phi_s(\mathbf{r}) \overset{r\gg\beta^{-1}}{=} e^{i\mathbf{k}\mathbf{r}} + \frac{m}{4\pi}\left|\tau\left(\frac{k^2}{m}\right)\right|g^2(\mathbf{k})\frac{e^{ikr + i\varphi(k)}}{r} \overset{k=0}{=} 1 - \frac{a_0}{r}$$

which determines the scattering phase $\varphi(k)$ and the scattering length a_0.

In terms of the T-matrix, the phase shift is then given by

$$\tan\varphi = \left.\frac{\mathrm{Im}\,\mathscr{T}}{\mathrm{Re}\,\mathscr{T}}\right|_{\Omega=\frac{p^2}{m}} = \frac{\lambda\,\mathrm{Im}\,\mathscr{J}}{1-\lambda\,\mathrm{Re}\,\mathscr{J}} \tag{E.19}$$

and the parameter of the potential can be linked to the scattering length a_0 and the range of the potential r_0 via the small wave-vector expansion

$$p\cot\varphi = -\frac{1}{a_0} + \frac{r_0}{2}p^2 + \dots. \tag{E.20}$$

For these scattering states, $\Omega = \frac{p^2}{m} > 0$, we obtain for (E.11)

$$\lambda\mathscr{J} = \xi\beta^2\pi\frac{p^2-\beta^2-2ip\beta}{2(\beta^2+p^2)^2} \tag{E.21}$$

and the scattering phase shift becomes

$$\cot\varphi = \frac{\mathrm{Re}\,\tau\left(\frac{p^2}{m}\right)}{\mathrm{Im}\,\tau\left(\frac{p^2}{m}\right)} = \frac{(p^2+\beta^2)^2}{-\pi\xi p\beta^3} + \frac{p^2-\beta^2}{2\beta p} = -\frac{1}{a_0 p} + \frac{1}{2}r_0 p - \frac{1}{2}\frac{\lambda_{c0}}{\lambda}\frac{p^3}{\beta^3} \tag{E.22}$$

where we expand in p and compare with (E.20) to determine the potential parameters

$$a_0 = \frac{m}{4\pi\hbar^2}\tau(0) = \frac{2}{\beta}\left(\frac{\lambda_{c0}}{\lambda}+1\right)^{-1} = \frac{2}{\beta(\frac{2}{\pi\xi}+1)},$$

$$r_0 = \frac{1}{\beta}\left(1-\frac{4}{\xi\pi}\right) = \frac{1}{\beta}\left(1-2\frac{\lambda_{c0}}{\lambda}\right) \tag{E.23}$$

and r_0 is called the effective range of the interaction These interaction parameter can be fitted to properties of e.g. hydrogen, helium, lithium and potassium (Maennel, 2011). A unique value one gets from $\cos\varphi(0) = -\frac{a_0}{|a_0|}$ and $\cos\varphi(\infty) = -\frac{\lambda}{|\lambda|}$ and therefore

$$\varphi(0) - \varphi(\infty) = \begin{cases} 0 & \text{for } \lambda > -\lambda_{c0} \\ \pi & \text{for } \lambda < -\lambda_{c0} \end{cases} \tag{E.24}$$

which is an expression of the Levinson theorem (Negele, 1982, p. 357), i.e. the phase-shift difference (E.24) is π times the number of bound states, see Fig. E.1.

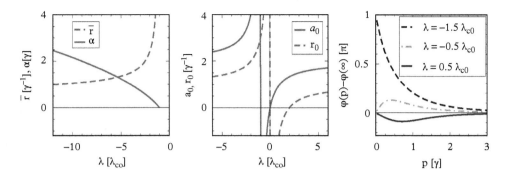

Figure E.1: *Parameters for the Yamaguchi potential. The scattering length and bond length diverge at $\lambda = -\lambda_{c0}$ where the bound state appears. A negative scattering phase for $\lambda > 0$ indicates a repulsive interaction and vice versa.*

References

Abramowitz, M. and Stegun, I. A. (1984). *Pocketbook of mathematical functions*. Verlag Harri Deutsch, Frankfurt/Main.

Abrikosov, A. A., Gorkov, L. P., and Dzyaloshinski, I. E. (1963). *Methods of Quantum Field Theory in Statistical Physics*. Prentice Hall, New York.

Abrikosov, A. A. and Khalatnikov, I. M. (1959). *Rep. Prog. Phys.*, **22**, 329.

Aichelin, J. (1991). *Phys. Rep.*, **202**, 235.

Aleksandrov, A. S., Grebenev, V. N., and Mazur, E. A. (1987). *Pisma Zh. Eksp. Teor. Fiz.*, **45**, 357. [JETP Lett. 34, 455].

Alexander, F. J., Garcia, A. L., and Alder, B. J. (1995). *Phys. Rev. Lett.*, **74**, 5212.

Alm, T., Röpke, G., Schnell, A., Kwong, N. H., and Köhler, H. S. (1996). *Phys. Rev. C*, **53**, 2181.

Ambegaokar, V. (1969). In *Superconductivity* (ed. R. D. Parks), Volume 1, Chapter 5, p. 259. Marcel Dekker, New York.

Ament, W. S. and Rado, G. T. (1955). *Phys. Rev.*, **97**, 1558.

Andersen, J. O. (2004). *Rev. Mod. Phys.*, **76**, 599.

Anderson, M. H., Ensher, J. R., Matthews, M. R., Wieman, C., and Cornell, E. A. (1995). *Science*, **269**, 198.

Aryasetiawan, Ferdi and Biermann, Silke (2009). *J. Phys.: Condens. Matter*, **21**, 0642332.

Axt, V. M. and Kuhn, T. (2004). *Rep. Prog. Phys.*, **67**, 433.

Ayik, S. and Boilley, D. (1992). *Phys. Lett. B*, **276**, 263. Errata ibd. 284 (1992) 482.

Ayik, S., Colonna, M., and Chomaz, Ph. (1995). *Phys. Lett. B*, **353**, 417.

Ayik, S. and Gregoire, C. (1990). *Nucl. Phys. A*, **513**, 187.

Ayik, S., Suraud, E., Belkacem, M., and Boilley, D. (1992). *Nucl. Phys. A*, **545**, 35.

Bala, Renu, Srivastava, Sunita, and Pathak, Kare Narain (2015). *European Phys. J. B*, **88**(10), 258.

Baldwin, S.P. and et. al. (1995). *Phys. Rev. Lett.*, **74**, 1299.

Balescu, R. (1975). *Equilibrium and Nonequilibrium Statistically Mechanics*. Wiley, New York.

Balescu, R. (1978). *Equilibrium and Nonequilibrium Statistically Mechanics (russ. trans.)*. Volume 2. MIR, Moskau.

Balian, R. and Veneroni, M. (1985). *Ann. of Phys.*, **164**, 334.

Banyai, L., Thoai, D. B. Tran, Remling, C., and Haug, H. (1992). *phys. stat. sol. (b)*, **173**, 149.

Bányai, L., Vu, Q. T., Mieck, B., and Haug, H. (1998). *Phys. Rev. Lett.*, **81**, 882.

Bardeen, J., Cooper, L. N., and Schrieffer, J. R. (1957). *Phys. Rev.*, **108**, 1175.

Bärwinkel, K. (1969*a*). *Z. Naturforsch*, 24a, 38.

Bärwinkel, K. (1969*b*). *Z. Naturforsch*, 24a, 22.

Bashkin, E. P., daProvidencia, C., and daProvidencia, J. (2000). *Phys. Rev. B*, **62**, 3968.

Baym, G. (1962). *Phys. Rev.*, **127**, 1391.

Baym, G., Blaizot, J. P., Holzmann, M., Laloë, F., and Vautherin, D. (2001). *Eur. Phys. J. B*, **24**, 107.

Baym, G. and Kadanoff, L. P. (1961). *Phys. Rev.*, **124**(2), 287.

Baym, G. and Pethick, Ch. (1991). *Landau Fermi-Liquid Theory*. Wiley, New York.

Beliaev, S. T. (1958). *Soviet. Phys. JETP*, 7, 289.

Bernstein, I. B. (1958). *Phys. Rev.*, **109**, 10.

Bertoncini, R. and Jauho, A. P. (1991). *Phys. Rev. B*, **44**(8), 3655.

Bertoncini, R., Kriman, A. M., and Ferry, D. K. (1989). *Phys. Rev. B*, **40**(5), 3371.

Bertsch, G. F., Bortignon, P. F., and Broglia, R. A. (1983). *Rev. of Mod. Phys.*, **55**, 287.

Bertsch, G. F. and Gupta, S. Das (1988). *Phys. Rep.*, **160**, 189.

Beth, G. E. and Uhlenbeck, E. (1937). *Physica*, **4**, 915.

Beyer, M., Röpke, G., and Sedrakian, A. (1996). *Phys. Lett. B*, **376**, 7.

Bezzerides, B. and DuBois, D. F. (1972). *Ann. Phys. (NY)*, **70**, 10.

Bijlsma, M. and Stoof, H. T. C. (1997). *Phys. Rev. A*, **55**, 498.

Birse, M. C. (2011). *Phil. Trans. R. Soc. A*, **369**, 2662.

Birse, M. C., Epelbaum, E., and Geglia, J. (2016). *Eur. Phys. J. A*, **52**, 26.

Birse, M. C., McGovern, J. A., and Richardson, K. R. (1999). *Phys. Lett. B*, **464**, 169.

Bishop, R. F., Strayer, M. R., and Irvine, J. M. (1974, Dec). *Phys. Rev. A*, **10**(6), 2423.

Bishop, R. F., Strayer, M. R., and Irvine, J. M. (1975). *J. of Low Temp. Phys.*, **20**(5–6), 573.

Blanter, Ya. M. and Büttiker, M. (2000). *Phys. Rep.*, **336**, 1.

Bocage, F. and et al. (2000). *Nucl. Phys. A*, **676**, 391.

Bogner, S. K., Schwenk, A., Kuo, T. T. S., and Brown, G. E. (2001). arXiv:nucl-th/0111042.

Bogner, S. K., Furnstahl, R. J., Ramannan, S., and Schwenk, A. (2007). *Nucl. Phys. A*, **784**, 79.

Bogner, S. K., Kuo, T. T. S., Schwenk, A., Entem, D. R:, and Machleidt, R. (2003). *Phys. Lett. B*, **576**, 265.

Bogoliubov, N. (1947). *J. Phys. (USSR)*, **11**, 23.

Bogoliubov, N. N. (1946). *J. Phys. (USSR)*, **10**, 256. transl. in *Studies in Statistical Mechanics*, Vol. 1, editors D. de Boer and G. E. Uhlenbeck (North-Holland, Amsterdam 1962).

Bogoliubov, N. N. and Gurov, K. P. (1947). *Zh. Eksp. Teor. Fiz.*, **17**, 614.

Bok, J. and Klein, J. (1968). *Phys. Rev. Lett.* **20**, 660.

Bollé, D. (1981). On Classical Time Delay (ed. H. Mitter and L. Pittner), In Volume XXIII, *Acta Physica Austriaca, Suppl.*, p. 587. Springer-Verlag, Wien.

Boltzmann, L. (1872). *Sitz.-Ber. Akad. Wiss. Wien.*, **66**, 275.

Boltzmann, L. (1893). *Phil. Mag.*, **35**, 161.

Boltzmann, L. (1895). *Phil. Mag.*, **51**, 414.

Bonitz, M. (1998). *Quantum Kinetic Theory.* Teubner, Stuttgart.

Bonitz, M. and et. al. (1996). *J. Phys.: Condens. Matter*, **8**, 6057.

Bonitz, M., Kwong, N. H., Semkat, D., and Kremp, D. (1999). *Contrib. Plasma Phys.*, **39**, 37.

Born, M. and Green, H. S. (1946). *Proc. R. Soc. London A*, **188**, 10.

Bornath, Th., Kremp, D., Kraeft, W. D., and Schlanges, M. (1996). *Phys. Rev. E*, **54**, 3274.

Bose, D. (1924). *Zeitschrift für Physik*, **26**(1), 178.

Botermans, W. and Malfliet, R. (1988). *Phys. Lett. B*, **215**, 617.

Botermans, W. and Malfliet, R. (1990). *Phys. Rep.*, **198**(3), 115.

Bothe, W. and Gentner, W. (1937). *Z. Phys.*, **106**, 236.

Bouyssy, A., Mathiot, J. F., and Giai, N. Van (1987). *Phys. Rev. C*, **36**, 380.

Braaten, E. and Radescu, E. (2002). *Phys. Rev. A*, **66**, 063601.

Bradley, C. C., Sackett, C. A., Tollett, J. J., and Hulet, R. G. (1995). *Phys. Rev. Lett.*, **75**, 1687.

Brieva, F. A. and Rook, J. R. (1977). *Nucl. Phys. A*, **291**, 299.

Brown, W. D., Puff, R. D., and Wilets, L. (1970). *Phys. Rev. C*, **2**(2), 331.

Budd, H. F. and Vannimenus, J. (1973). *Phys. Rev. Lett*, **31**, 1218.

Burbury, S. H. (1890). *Phil. Mag.*, **30**, 301.

Büttiker, M. (1992). *Phys. Rev. B*, **46**, 12485.

Calzetta, E. and Hu, B. L. (1988). *Phys. Rev. D*, **37**, 2878.

Camalet, S., Kohler, S., and Hänggi, P. (2004). *Phys. Rev. B*, **70**, 155326.
Carlson, J., Pandharipande, V. R., and Wiringa, R. B. (1983). *Nucl. Phys. A*, **401**, 59.
Casini, G. and *et al.* (1993). *Phys. Rev. Lett.*, **71**, 2567.
Chalaev, Oleg and Loss, Daniel (2005, Jun). *Phys. Rev. B*, **71**, 245318.
Chapman, S. (1937). *Nature*, **139**, 931.
Chapman, S. and Cowling, T. C. (1939). *The mathematical Theory of Nonuniform Gases*. Cambridge University Press, Cambridge.
Chapman, S. and Cowling, T. G. (1990). *The Mathematical Theory of Non-uniform Gases*. Cambrigde University Press, Cambridge. Third edition Chapter 16.
Chattopadhyay, S. (1995). *Phys. Rev. C*, **52**, R480.
Chattopadhyay, S. (1996). *Phys. Rev. C*, **53**, R1065.
Chen, Jian and Hershfield, Selman (1998, Jan). *Phys. Rev. B*, **57**, 1097.
Chen, Q., Stajic, J., Tan, S., and Levin, K. (2005). *Phys. Rep.*, **412**, 1.
Chen, S. L. and et al. (1996). *Phys. Rev. C*, **54**, 2214.
Chen, Tsung-Wei and Guo, Guang-Yu (2009). *Phys. Rev. B*, **79**(12), 125301.
Chin, S. A. (1977). *Annals of Physics*, **108**, 301.
Cohen, E. (1962). *Fundamental Problems in Statistical Mechanics*. Nort-Holland, Amsterdam.
Cohen, E. G. D. and Dorfman, J. R. (1965). *Phys. Rev. Lett.*, **16**, 24.
Cohen, E. G. D. and Dorfman, J. R. (1972). *Phys. Rev. A*, **1**, 776.
Colonna, M., Burgio, G. F., Chomaz, P., DiToro, M., and Randrup, J. (1993). *Phys. Rev. C*, **47**, 1395.
Colonna, M., Chomaz, Ph., and Randrup, J. (1994). *Nucl. Phys. A*, **567**, 637.
Colonna, M., DiToro, M., and Guarnera, A. (1995). *Nucl. Phys. A*, **589**, 160.
Colonna, M., DiToro, M., Guarnera, A., Maccarone, S., Zielinska-Pfabé, M., and Wolter, H. H. (1998). *Nucl. Phys. A*, **642**, 449.
Courteille, P., Freeland, R. S., Heinzen, D. J., van Abeelen, F. A., and Verhaar, B. J. (1998). *Phys. Rev. Lett.*, **81**, 69.
Craig, R. A. (1966a). *Ann. Phys.*, **40**, 434.
Craig, R. A. (1966b). *Ann. Phys.*, **40**, 416.
Craig, R. A. (1968). *J. Math. Phys.*, **9**, 605.
Cserti, József and Dávid, Gyula (2006, Nov). *Phys. Rev. B*, **74**, 172305.
Cubero, M. (1990). Ph.D. thesis, TU Darmstadt. GSI report GSI-90-17,Darmstadt.
Dahm, T. and Tewordt, L. (1995). *Phys. Rev. B*, **52**, 1297.
Daniel, E. and Vosko, S. H. (1960). *Phys. Rev.*, **120**, 2041.
Danielewicz, P. (1984a). *Ann. Phys. (NY)*, **152**, 305.
Danielewicz, P. (1984b). *Ann. Phys. (NY)*, **152**, 239.
Danielewicz, P. (1990). *Ann. Phys. (NY)*, **197**, 154.
Danielewicz, P. (2000). *Nucl. Phys. A*, **673**, 375.
Danielewicz, P. and Bertsch, G. F. (1991). *Nucl. Phys. A*, **533**, 712.
Danielewicz, P. and Pratt, S. (1996). *Phys. Rev. C*, **53**, 249.
Das, A. K. (1975). *J. Phys. F*, **5**, 2035.
DasSarma, S. (1986). *Phys. Rev. B*, **33**, 5401.
DasSarma, S. and Hwang, E. H. (1999). *Phys. Rev. Lett.*, **83**, 164.
Davis, K. B., Mewes, M. O., Andrews, M. R., van Druten, N. J., Durfee, D. S., Kurn, D. M., and Ketterle, W. (1995). *Phys. Rev. Lett.*, **75**, 3969.
de Gennes, P. G. (1966). *Superconductivity of Metals and Alloys*. Benjamin, New York.
de Haan, H. (1990a). *Physica A*, **165**, 224.
de Haan, H. (1991). *Physica A*, **170**, 571.

de Haan, M. (1990*b*). *Physica A*, **164**, 373.

de Jong, F. and Malfliet, R. (1991). *Phys. Rev. C*, **44**(3), 998.

Debye, P. and Hückel, E. (1923). *Phys. Zeitsch.*, **15**, 305.

Deisz, J. J. and Slife, T. (2009). *Phys. Rev. B*, **80**, 094516.

Delfino, A., Malheiro, M., Timóteo, V. S., and Martins, J. S. Sá (2005). *Brazilian Journal of Physics*, **35**, 190.

Dellafiore, A., Matera, F., and Brink, D. M. (1995). *Phys. Rev. A*, **51**, 914.

Dempsey, J. F. and et al. (1996). *Phys. Rev. C*, **54**, 1710.

Dew-Huges, D. (2001). *Low Temp. Phys.*, **27**, 713.

Dickhoff, W. H. (2016). *Journal of Physics: Conference Series*, **702**, 012013.

Dimitrova, Ol'ga V. (2005). *Phys. Rev. B*, **71**, 245327.

DiToro, M., Kolomietz, V. M., and Larionov, A. B. (1999). *Phys. Rev. C*, **59**, 3099.

Donangelo, R., Dorso, C. O., and Marta, H. D. (1991*a*). *Phys. Lett. B*, **263**, 19.

Donangelo, R., Romanelli, A., and Schifino, A. C. Sicardi (1991*b*). *Phys. Lett. B*, **263**, 342.

Dorfman, J. R. and Cohen, E. G. (1967). *J. Math. Phys.*, **8**, 282.

Ebeling, W. (1976). *Ann. Phys.*, **33**, 5.

Ebeling, W. and Röpke, G. (1979). *Ann. Phys. (Leipzig)*, **36**, 429.

Eckstein, Martin, Kollar, Marcus, and Werner, Philipp (2009, Jul). *Phys. Rev. Lett.*, **103**, 056403.

Einstein, A. (1925, January). *Sitzungsberichte der preußischen Akademie der Wissenschaften, physikalisch-mathematische Klasse*, 3.

Elliott, R. J., Krumhansl, J. A., and Leath, P. L. (1974). *Rev. Mod. Phys.*, **46**, 465.

ElSayed, K., Schuster, S., Haug, H., Herzel, F., and Henneberger, K. (1994). *Phys. Rev. B*, **49**, 7337.

Elze, H. Th. and et. al., (1987). *Mod. Phys. Lett. A*, **2**, 451.

Enskog, D. (1917). *Kinetische Theorie der Vorgänge in mäßig verdünnten Gasen*. Almqvist & Wiksells, Uppsala.

Enskog, D. (1972). In *Kinetic theory* (ed. S. Brush), Volume 3. Pergamon Press, New York. orig.: *Kungl Svenska Vet. Akad. Handl.* **63**(3), 3 (1921).

Ernst, M. H. (1983). In *Nonequilibrium Phenomena I* (ed. J. L. Lebowitz and E. W. Montroll), Chapter 3, p. 52. North-Holland Publishing Company, North-Holland.

Esser, A. and Röpke, G. (1998). *Phys. Rev. E*, **58**, 2446.

Evans, D. J. and Morriss, G. P. (1990). *Statistical Mechanics of Nonequilibrium Liquids*. Academic Press, London.

Fabbri, G., Colonna, M., and DiToro, M. (1998). *Phys. Rev. C*, **58**, 3508.

Falkenhagen, H. (1953). *Elektrolyte*. S. Hirzel Verlag, Leipzig.

Falkenhagen, H., Ebeling, W., and Kraeft, W. D. (1971). Equilibrium properties of ionized dilute electrolytes. In *Ionic Interaction* (ed. Petrucci), Chapter 1, p. 1. Academic Press, New York and London.

Farid, B. (1999). *Phil. Mag. B*, **79**, 1097.

Farina, J. E. G. (1983). In *The International Encyclopedia of Physical Chemistry and Chemical Physics* (ed. R. McWeeny), Volume 4. Pergamon, Oxford.

Fetter, A. L. and Walecka, J. D. (1971). *Quantum Theory of Many-Particle Systems*. McGraw-Hill Publ., New York.

Feynman, R. P. (1954). *Phys. Rev.*, **94**, 262.

Feynman, R. P. and Cohen, M. (1956). *Phys. Rev.*, **102**, 1189.

Flesch, A., Cramer, M., McCulloch, I. P., Schollwöck, U., and Eisert, J. (2008). *Phys. Rev. A*, **78**, 033608.

Flocard, H. (1989). *Ann. of Phys.*, **191**, 382.

Flügge, S. (1994). *Practical Quantum Mechanics*. Springer, Berlin.

Friedmann, B. and Pandharipande, V. R. (1981). *Phys. Lett. B*, **100**, 205.

Fröhlich, H (1950). *Phys. Rev.*, **79**, 845.

Fuhrmann, U., Morawetz, K., and Walke, R. (1998). *Phys. Rev. C*, **58**, 1473.

Fulde, P. (1991). *Electron Correlations in Molecules and Solids*. Springer Series in Solid-State Sciences, Berlin.

Galichet, E. (1998). Ph.D. thesis, Insitut de Physique Nucléaire de Lyon.

Galitskii, V. M. (1958). *Zh. Eksp. Teor. Fiz.*, **34**, 151. [Sov. Phys.–JETP 7, 104 (1958)].

Gartner, P., Banyai, L., and Haug, H. (1999). *Phys. Phys. B*, **60**, 14234.

Gell-Mann, Murray and Brueckner, Keith A. (1957). *Phys. Rev.*, **106**, 364.

Gervais, G., Thoennessen, M., and Ormand, W. E. (1998). *Phys. Rev. C*, **58**, 1377(R).

Giaever, I (1960). *Phys. Rev. Lett.*, **5**, 147.

Gilbert, T. L. (2004). *IEEE Transactions on Magnetics*, **40**(6), 3443.

Glyde, H. R. (1994). *Excitations in liquid and solid helium*. Clarendon Press, Oxford.

Glyde, H. R. and Hernadi, S. I. (1983). *Phys. Rev. B*, **28**, 141.

Gold, A. and Dolgopolov, V. T. (1986). *Phys. Rev. B*, **33**, 1076.

Goldberger, M. L. and Watson, K. M. (1964). *Collision Theory*. Wiley, New York.

Goldman, R. and Frieman, E. A. (1967). *J. Math. Phys.*, **8**, 1410.

Golosov, D. I. and Ruckenstein, A. E. (1995). *Phys. Rev. Lett.*, **74**, 1613.

Gori-Giorgi, Paola and Ziesche, Paul (2002, Dec). *Phys. Rev. B*, **66**, 235116.

Gorini, C., Schwab, P., Dzierzawa, M., and Raimondi, R. (2008). *Phys. Rev. B*, 78, 125327.

Gorkov, L. P. (1958). *Zh. Eksp. Fiz.*, **34**, 735. *Sov. Phys. JETP*, **7** (1958) 505.

Gradshteyn, I. S. and Ryzhik, I. M. (2014). *Table of Integrals, Series, and Products*. 2014 Elsevier Inc.

Green, H. S. (1952). *The Molecular Theory of Fluids*. North-Holland, Amsterdam.

Grimaldi, C., Pietronero, L., and Strässler, S. (1995*a*). *Phys. Rev. Lett.*, **75**(6), 1158.

Grimaldi, C., Pietronero, L., and Strässler, S. (1995*b*). *Phys. Rev. B*, **52**, 10530.

Groot, De, a. van Leeuwen, W., and van Weert, Ch. G. (1980). *Relativistic Kinetic Theory*. North Holland Publishing Company, Amsterdam.

Gross-Boelting, T., Fuchs, C., and Faessler, A. (1999). *Nucl. Phys. A*, **648**, 105.

Gulian, A. M. and Zharkov, G. F. (1999). *Nonequilibrium Electrons and Phonons in Superconductors*. Kluwer Academic/Plenum Publishers, New York.

Hakim, R., Mornas, L., Peter, P., and Sivak, H. D. (1992). *Phys. Rev. D*, **46**, 4603.

Halbert, E. C. (1981). *Phys. Rev. C*, **23**, 295.

Hassaneen, Kh. S. A., Abo-Elsebaa, H. M., Sultan, E. A., and Mansour, H. M. M. (2011). *Annals of Physics*, **326**, 566.

Hasselmann, N., Ledowski, S., and Kopietz, P. (2004). *Phys. Rev. A*, **70**, 063621.

Haug, H. and Ell, C. (1992). *Phys. Rev. B*, **46**(4), 2126.

Haug, H. and Jauho, A. P. (1996). *Quantum Kinetics in Transport and Optics of Semiconductors*. Springer, Berlin Heidelberg.

Hauge, E. H. and Støvneng, J. A. (1989). *Rev. Mod. Phys.*, **61**(4), 917.

Haussmann, R. (1993). *Z. Phys. B*, **91**, 291.

Haussmann, R. (1994). *Phys. Rev. B*, **49**, 12975.

He, Y., Chien, C. C., Chen, Q., and Levin, K. (2007). *Phys. Rev. B*, **76**(22), 224516.

Hedin, Lars (1965). *Phys. Rev.*, **139**, A796.

Heidenberger, J. and Plessas, W. (1984). *Phys. Rev. C*, **30**, 1822.

Henning, P. A. (1995). *Phys. Rep.*, **253**(5&6), 235.

Henning, P. A., Nakamura, K., and Yamanaka, Y. (1996). *Int. J. Mod. Phys. B*, **10**, 1599.

Henning, P. A. and Umezawa, H. (1994). *Nucl. Phys. B*, **417**, 463.

Hergert, H., Bogner, S. K., Morris, T. D., Schwenk, A., and Tsukiyama, K. (2016). *Physics Reports*, **621**, 165. Memorial Volume in Honor of Gerald E. Brown.

Hertz, J. A., Levin, K., and Beal-Monod, M. T. (1976). *Sol. State Comm.*, **18**, 803.

Hicks, L. D. and Dresselhaus, M. S. (1993). *Phys. Rev. B*, **47**, 12727.

Hirschfelder, J. O., Curtiss, Ch. F., and Bird, R. B. (1964). *Molecular Theory of Gases and Liquids*. Wiley, New York. Chapters 6.4a and 9.3.

Hjorth-Jensen, M., Muether, H., Polls, A., and Osnes, E. (1996). *J. Phys. G*, **22**, 321.

Hodges, C., Smith, H., and Wilkins, J. W. (1971). *Phys. Rev. B*, **4**, 302.

Holzmann, M. and Baym, G. (2003). *Phys. Rev. Lett.*, **90**, 040402.

Holzmann, M., Grüter, P., and Laloë, F. (1999). *Eur. Phys. J. B*, **10**, 739.

Horowitz, C. J. and Serot, B. D. (1983). *Nucl. Phys. A*, **399**, 529.

Huang, K. (1964). *Statistische Mechanik II*. Bibliogr. Inst., Mannheim.

Huang, K. (1995). In *Bose–Einstein Condensation* (ed. A. Griffin, D. W. Snoke, and S. Stringari), p. 31. Cambridge University Press, Cambridge.

Huang, K. and Yang, C. N. (1957). *Phys. Rev.*, **105**, 767.

Huang, K., Yang, C. N., and Luttinger, J. M. (1957). *Phys. Rev.*, **105**, 776.

Huber, R., Kübler, C., Tübel, S., Leitenstorfer, A., Vu, Q. T., Haug, H., Köhler, F., and Amann, M.-C. (2005). *Phys. Rev. Lett.*, **94**, 027401.

Huber, R., Tauser, F., Brodschelm, A., Bichler, M., Abstreiter, G., and Leitenstorfer, A. (2001). *Nature*, **414**, 286.

Huber, R., Tauser, F., Brodschelm, A., and Leitenstorfer, A. (2002). *phys. stat. sol. (b)*, **234**, 207.

Huberman, M. and Chester, G. V. (1975). *Adv. Phys.*, **24**, 489.

Hugenholtz, N. M. and Pines, D. (1959). *Phys. Rev.*, **116**, 486.

Iannaccone, G. and Pellegrini, B. (1994). *Phys. Rev. B*, **49**, 16548.

Ikeda, M.A., Ogasawara, A., and Sugihara, M. (1992). *Phys. Lett. A*, **170**(4), 319.

Inoue, Jun-Ichiro, Bauer, Gerrit E. W., and Molenkamp, Laurens W. (2004). *Phys. Rev. B*, **70**, 041303.

Inoue, J. I., Kato, T., Ishikawa, Y., Itoh, H., Bauer, G. E. W., and Molenkamp, L. W. (2006). *Phys. Rev. Let.*, **97**(4), 046604.

Inouye, S., Andrews, M. R., Stenger, J., Miesner, H.-J., Stamper-Kurn, D. M., and Ketterle, W. (1998). *Nature*, **392**, 151.

Irving, J. H. and Zwanzig, R. W. (1951). *J. Chem. Phys.*, **19**, 1173.

Itzykson, Cl. and Zuber, J. B. (1980). *Quantum field theory*. McGraw-Hill, New York.

Itzykson, C. and Zuber, J. B. (1990). *Quantum field theory*. McGraw-Hill, New York.

Ivanov, Yu. B. (1987). *Nucl. Phys. A*, **474**, 669.

Ivanov, Yu. B., Knoll, J., and Voskresensky, D. N. (2000). *Nucl. Phys. A*, **672**, 313.

Jäckle, J. (1978). *Einführung in die Transporttheorie*. Friedr. Vieweg and Sohn, Braunschweig.

Jackson, A. D., Rho, M., and Krotscheck, E. (1983). *Nucl. Phys. A*, **407**, 495.

Jaminon, M., Mahaux, C., and Rochus, P. (1981). *Nucl. Phys. A*, **365**, 371.

Janis, V. (2001). *Phys. Rev. B*, **64**, 115115.

Janis, V. (2009). *J. Phys: Condens. Matter (London)*, **21**(48), 485501.

Jauho, A. P. (1991). In *Quantum Transport in Semiconductors* (ed. D. Ferry and C. Jacoboni), Chapter 7, p. 141. Plenum Press, New York.

Jauho, A. P. and Wilkins, J. W. (1984). *Phys. Rev. B*, **29**(4), 1919.

Jeon, J. W. and Mullin, W. J. (1989). *Phys. Rev. Lett.*, **62**, 2691.

Jiang, W., Wang, H., Zhao, S., and Wang, Y. (2009). *J. Phys. D: Appl. Phys.*, **42**, 102005.

Jones, W. and March, N. H. (1973). *Theoretical Solid State Physics*. Volume 2. Dover, New York.

Kadanoff, L. P. and Baym, G. (1962). *Quantum Statistical Mechanics*. Benjamin, New York.

Kadanoff, L. P. and Martin, P. C. (1961). *Phys. Rev.*, **124**(3), 670.

Kadomtsev, B. B. (1958). *Zh. Eksp. Teor. Fiz.*, **33**, 151. Sov. Phys. -JETP 33,117(1958).

Kakehashi, Y. (2002). *Phys. Rev. B*, **66**, 104428.

Kakehashi, Y. (2004). Quantum statistical approaches to correlations. In *Nonequilibrium Physics at Short Time Scales - Formation of Correlations* (ed. K. Morawetz), p. 3. Springer, Berlin.

Kamerdzhiev, S. P. (1969). *Yad. Fiz.*, **9**, 324.

Kane, C. L. and Mele, E. J. (2005). *Phys. Rev. Lett.*, **95**(22), 226801.

Kao, H. C., Lewkowicz, M., and Rosenstein, B. (2010). *Phys. Rev. B*, **82**, 035406.

Katsnelson, M. I. (2006). *Eur. Phys. J. B*, **51**, 157.

Kawasaki, K. and Oppenheim, I. (1965). *Phys. Rev.* **139**, A649.

Keil, W. (1988). *Phys. Rev. D*, **38**, 152.

Keldysh, L. V. (1964). *Zh. exp. teor. fiz.*, **47**, 1515.

Keller, Marcus, Metzner, Walter, and Schollwöck, Ulrich (1999). *Phys. Rev. B*, **60**, 3499.

Khan, F. S., Davies, J. H., and Wilkins, J. W. (1987). *Phys, Rev. B*, **36**(5), 2578.

Kiderlen, D. and Hofmann, H. (1994). *Phys. Lett. B*, **332**, 8.

Kira, M. and Koch, S. W. (2004). *Phys. Rev. Lett.*, **93**, 076402.

Kirkpatrick, T. R. and Belitz, D. (1986). *Phys. Rev. B*, **34**, 2168.

Kirkwood, J. G. (1946). *J. Chem. Phys.*, **14**, 180.

Kirzhnitz, D. A., Lozovik, Yu. E., and Shpatakovskaya, G. V. (1975). *Usp. Fiz. Nauk*, **117**, 3.

Klimontovich, Yu. L. (1975). *Kinetic theory of nonideal gases and nonideal plasmas*. Academic Press, New York.

Klimontovich, Y. L. and Ebeling, W. (1962). *Jh. Eksp. Teor. Fiz.*, **43**, 146.

Klimontovich, Y. L. and Ebeling, W. (1972). *Jh. Eksp. Teor. Fiz.*, **63**(3), 904.

Kneur, J. L., Neveu, A., and Pinto, M. B. (2004). *Phys. Rev. A*, **69**, 053624.

Ko, C. M., Li, Q., and Wang, R. (1987). *Phys. Rev. Lett.*, **59**(10), 1084.

Köhler, H. S. (1975). *Phys. Rep.*, **18**, 217.

Köhler, H. S. (1992). *Phys. Rev. C*, **46**, 1687.

Köhler, H. S. (1993). Spetral functions. In *Advances in Nuclear Dynamics* (ed. B. Back, W. Bauer, and J. Harris). World Scientific, Singapore.

Köhler, H. S. (1995a). *Phys. Rev. C*, **51**, 3232.

Köhler, H. S. (1995b). *Nucl. Phys. A*, **583**, 339.

Köhler, H. S. (1995c). *Phys. Rev. C*, **51**, 3232.

Köhler, H. S. (1996). *Phys. Rev. E*, **53**, 3145.

Köhler, H. S., Kwong, N. H., and Yousif, H. A. (1999). *Comp. Phys. Comm.*, **123**, 123.

Köhler, H. S. and Malfliet, R. (1993). *Phys. Rev. C*, **48**, 1034.

Köhler, H. S. and Morawetz, K. (2001). *Phys. Rev. C*, **64**, 024613.

Kohler, S., Lehmann, J., and Hänggi, P. (2005). *Phys. Rep.*, **406**, 379.

Koike, Y., Parke, W. C., Maximon, L. C., and Lehman, D. R. (1997). *Few-Body Systems*, **23**, 53.

Kopietz, Peter, Bartosch, Lorenz, and Schütz, Florian (2010). *Introduction to the Functional Renormalization Group*. Springer, Berlin. Lecture Notes in Physics 798, ISBN 978-3-642-05094-7.

Korenman, V. (1966). *Ann. Phys.*, **39**, 72.

Kortemeyer, G., Daffin, F., and Bauer, W. (1996). *Phys. Lett. B*, **374**, 25.

Koster, G. F. and Slater, J. C. (1954). *Phys. Rev.*, **95**, 1167.

Kraeft, W. D., Kremp, D., Ebeling, W., and Röpke, G. (1986). *Quantum Statistics of Charged Particle Systems*. Akademie Verlag, Berlin.

Kremp, D., Bonitz, M., Kraeft, W.D., and Schlanges, M. (1997). *Ann. of Phys.*, **258**, 320.

Kremp, D., Kraeft, D., and Ebeling, W. (1966). *Ann. Phys. (Leipzig)*, **18**, 246.

Kremp, D., Kraeft, W. D., and Lambert, A. D. J. (1984). *Physica A*, **127**, 72.

Kremp, D., Schlanges, M., and Bornath, Th. (1985). *J. Stat. Phys.*, **41**, 661.

Kulik, O. (1961). *Zh. Eksp. Teor. Fiz.*, **40**, 1343. [Sov. Phys. JETP 13, 946].

Kumagai, K. I., Nozaki, K., and Matsuda, Y. (2001). *Phys. Rev. B*, **63**, 144502.

Kwong, N. Hang and Bonitz, M. (2000). *Phys. Rev. Lett.*, **84**, 1768.

Lacombe, M., Loiseau, B., Richard, J. M., Mau, R. Vinh, Côté, J., Pirès, P., and de Tourreil, R. (1980). *Phys. Rev. C*, **21**, 861.

Laloë, F. (1989). *J. Phys. (Paris)*, **50**, 1851.

Laloë, F. and Mullin, W. J. (1990). *J. Stat. Phys.*, **59**, 725.

Landau, L. D. (1957*a*). *Soviet Phys. JETP*, **3**, 920.

Landau, L. D. (1957*b*). *Soviet Phys. JETP*, **5**, 101.

Landau, L. D. and Lifschitz, E. M. (1971). *Lehrbuch der Theoretischen Physik, Band VI*. Akademie Verlag, Berlin.

Landau, L. D. and Lifschitz, E. M. (1980). *Lehrbuch der Theoretischen Physik, Band IX*. Akademie Verlag, Berlin.

Landau, L. D. and Lifschitz, E. M. (1984). *Lehrbuch der Theoretischen Physik, Band V*. Akademie Verlag, Berlin.

Landau, L. D. and Lifshitz, E. M. (1935). *Phys. Z. Sowjet.*, **8**, 153.

Landsman, N. P. and van Weert, Ch. G. (1987). *Phys. Rep.*, **145**, 141.

Langreth, D. C. (1976). In *Linear and Nonlinear Electron Transport in Solids* (ed. J. T. Devreese and E. van Boren). Plenum, New York.

Langreth, D. C. and Wilkins, J. W. (1972). *Phys. Rev. B*, **6**(9), 3189.

Larochelle, Y. and et al. (1999). *Phys. Rev. C*, **59**, 565.

Lecolley, J. F. and et al. (2000). *Nucl. Inst. and Meth. A*, **441**, 517.

Lecolley, J. F. and et al. (1995). *Phys. Lett. B*, **354**, 202.

Lee, D., Fujita, S., and Wu, F. (1970). *Phys. Rev. A*, **2**, 854.

Leggett, A. J. (1970). *J. Phys. C*, **3**, 448.

Leggett, A. J. (2001). *Rev. Mod. Phys.*, **73**, 307.

Leggett, A. J. (2006). *Quantum Liquids*. Oxford University Press, New York.

Leggett, A. J. and Rice, M. J. (1968). *Phys. Rev. Lett.*, **20**, 586. erratum *Phys. Rev. Lett.* **21**, 506 (1968).

Leitenstorfer, A. (2003). *Nature*, **426**, 23.

Leng, Xia, Jin, Fan, Wei, Min, and Ma, Yuchen (2016). *Wiley Interdisciplinary Reviews: Computational Molecular Science*, **6**(5), 532.

Leoni, S., Døssing, T., and Herskind, B. (1996). *Phys. Rev. Lett.*, **76**, 4484.

Levinson, I. B. (1965). *Fiz. Tverd. Tela Leningrad*, **6**, 2113.

Levinson, I. B. (1969). *Zh. Eksp. Teor. Fiz.*, **57**(2), 660. [Sov. Phys.–JETP 30, 362 (1970)].

Li, Z. Q., Henriksen, E. A., Jiang, Z., Hao, Z., Martin, M. C., Kim, P., Stormer, H. L., and Basov, D. N. (2008). *Nature Physics*, **4**, 532.

Lifschitz, E. and Pitaevsky, L. P. (1981). *Physical Kinetics*. Akademie Verlag, Berlin.

Lipavský, P. (2008). *Phys. Rev. B*, **78**, 214506.

Lipavský, P., Koláček, J., Mareš, J. J., and Morawetz, K. (2001*a*). *Phys. Rev. B*, **65**, 012507.

Lipavský, P., Koláček, J., Morawetz, K., and Brandt, E. H. (2002*a*). *Phys. Rev. B*, **65**, 144511.

Lipavský, P., Koláček, J., Morawetz, K., and Brandt, E. H. (2002*b*). *Phys. Rev. B*, **66**, 134525.

Lipavský, P., Koláček, J., Morawetz, K., Brandt, E. H., and Yang, T. J. (2007). *Bernoulli potential in superconductors*. Springer, Berlin. Lecture Notes in Physics 733.

Lipavský, P., Morawetz, K., Koláček, J., Mareš, J. J., Brandt, E. H., and Schreiber, M. (2004*a*). *Phys. Rev. B*, **69**, 024524.

Lipavský, P., Morawetz, K., Koláček, J., Mareš, J. J., Brandt, E. H., and Schreiber, M. (2004*b*). *Phys. Rev. B*, **70**, 104518.

Lipavský, P., Morawetz, K., Koláček, J., Mareš, J. J., Brandt, E. H., and Schreiber, M. (2005). *Phys. Rev. B*, **71**, 024526.

Lipavský, P., Morawetz, K., Šopík, B., and Männel, M. (2014). *Europ. Phys. J. B*, **87**(1), 8.

Lipavský, P., Morawetz, K., and Špička, V. (2001*b*). *Kinetic equation for strongly interacting dense Fermi systems*. Volume 26,1, Annales de Physique. EDP Sciences, Paris.

Lipavský, P. and Špička, V. (1994). *Phys. Rev. B*, **50**, 13981.

Lipavský, P., Špička, V., and Morawetz, K. (1999). *Phys. Rev. E*, **59**, R1291.

Lipavský, P., Špička, V., and Velický, B. (1986). *Phys. Rev. B*, **34**, 6933.

Littlewood, P. B. (1990). *Phys. Rev. B*, **42**, 10075.

Liu, S. Y., Lei, X. L., and Horing, Norman J. M. (2006, Jan). *Phys. Rev. B*, **73**(3), 035323.

Lloyd-Hughes, J., Castro-Camus, E., Fraser, M. D., Jagadish, C., and Johnston, M. B. (2004). *Phys. Rev. B*, **70**, 235330.

London, F. (1938). *Phys. Rev.*, **54**, 947.

Loos, D. (1990*a*). *J. Stat. Phys.*, **59**, 691.

Loos, D. (1990*b*). *J. Stat. Phys.*, **61**, 467.

Lu, Y. M., Fan, X. W., Chen, L. C;, Guan, Z. P., and Yang, B. J. (1994). *Thin Solid Films*, **249**, 207.

Luo, J. and Bickers, N. E. (1993). *Phys. Rev. B*, **47**, 12153.

Luttinger, J. M. (1960). *Phys. Rev.*, **119**, 1153.

Machida, Y., Sakai, S., Izawa, K., Okuyama, H., and Watanabe, T. (2011). *Phys. Rev. Lett.*, **106**, 107002.

Machleidt, R. (1989). *Adv. Nucl. Phys.*, **19**, 189.

Machleidt, R. (1993). In *Computational Nuclear Physics* (ed. K. Langanke, J. A. Maruhn, and S. E. Koonin), Volume 2. Springer, New York.

Macke, W. (1950). *Z. Naturforschg. A*, **5**, 192.

Maennel, M. (2011). *Condensation phenomena in interacting Fermi and Bose gases*. Ph.D. thesis, Chemnitz University of Technology.

Mahan, G. D. (1987). *Phys. Rep.*, **145**(5), 251.

Malfliet, R. (1983). *Nucl. Phys. A*, **420**, 621.

Malfliet, R. (1992). *Nucl. Phys. A*, **545**, 3c.

Maly, J., Jankó, B., and Levin, K. (1999). *Phys. Rev. B*, **59**(2), 1354.

Männel, M., Morawetz, K., and Lipavský, P. (2010). *New J. Phys.*, **12**, 033013.

Männel, M., Morawetz, K., and Lipavský, P. (2013). *Phys. Rev. A*, **87**, 053617.

Manzke, G. and Kremp, D. (1979). *Physica A*, **97**, 153.

March, N. H. and Tosi, M. P. (2002). *Introduction to liquid state physics*. World Scientific, New Jersey.

Martin, A. Ph. (1981). In *Time Delay of Quantum Scattering Processes* (ed. H. Mitter and L. Pittner), Volume XXIII, Acta Physica Austriaca, Suppl., p. 157. Springer-Verlag, Wien.

Martin, P. C. and Schwinger, J. (1959). *Phys. Rev.*, **115**, 1342.

Matsubara, T. (1955). *Prog. Theor. Phys.*, **14**, 351.

McLennan, J. A. (1989). *Introduction to Nonequilibrium Statistical Mechanics*. Prentice-Hall, Englewood Cliffs.

Mermin, N. D. (1970). *Phys. Rev. B*, **1**, 2362.

Mesiah, A. (1961). *Quantum mechanics*. Volume 2. North-Holland Publ. Co., Amsterdam.

Meyerovich, A. E. (1989). *Phys. Rev. B*, **39**, 9318.

Meyerovich, A. E. and Musaelian, K. A. (1992). *J. Low Temp. Phys.*, **89**, 781.

Meyerovich, A. E., Stepaniants, S., and Laloë, F. (1995). *Phys. Rev. B*, **52**, 6808.

Migdal, A. B. (1958). *Zh. Eksp. Teor. Fiz.*, **34**, 1438. [Sov. Phys. JETP 34, 996].

Mineev, V. P. (2004). *Phys. Rev. B*, **69**, 144429.

Mineev, V. P. (2005). *Phys. Rev. B*, **72**, 144418.

Molinari, L. G. (2005). *Phys. Rev. B*, **71**, 113102.

Monthoux, P. (2003). *Phys. Rev. B*, **68**, 064408.

Monthoux, P. and Scalapino, D. J. (1994). *Phys. Rev. Lett.*, **72**, 1874.

Montoya, C. P. and *et al.* (1994). *Phys. Rev. Lett.*, **73**, 3070.

Morawetz, K. (1994). *Phys. Rev. E*, **50**, 4625.

Morawetz, K. (1995). *Phys. Lett. A*, **199**, 241.

Morawetz, K. (1997). *Contrib. to Plasma Physics*, **37**, 195. errat. ibid 37 (1997) 4.

Morawetz, K. (2000*a*). *Phys. Rev. C*, **62**, 44606.

Morawetz, K. (2000*b*). *Phys. Rev. E.*, **61**, 2555.

Morawetz, K. (2000*c*). *Phys. Rev. E*, **62**, 6135. errat. ibid 69 (2004) 029902.

Morawetz, K. (2000*d*). *Phys. Rev. C*, **63**, 014609.

Morawetz, K. (2002). *Phys. Rev. B*, **66**, 075125. errat. ibid 88 (2013) 039905(E).

Morawetz, K. (2003). *Phys. Rev. B*, **67**, 115125.

Morawetz, K. (2010). *Phys. Rev. B*, **82**, 092501.

Morawetz, K. (2011). *J. Stat. Phys.*, **143**, 482.

Morawetz, K. (2013). *Phys. Rev. E*, **88**, 022148.

Morawetz, K. (2014). *Phys. Rev. B*, **90**, 075303.

Morawetz, K. (2015*a*, Dec). *Phys. Rev. B*, **92**, 245425. errat. ibid 93 (2016) 239904(E).

Morawetz, K. (2015*b*, Dec). *Phys. Rev. B*, **92**, 245426.

Morawetz, K. (2016). *Phys. Rev. B*, **94**, 165415.

Morawetz, K., Bonitz, M., Morozov, V. G., Röpke, G., and Kremp, D. (2001*a*). *Phys. Rev. E*, **63**, R020102.

Morawetz, K., Fuhrmann, U., and Walke, R. (1999*a*). *Nucl. Phys. A*, **649**, 348.

Morawetz, K., Fuhrmann, U., and Walke, R. (2000*a*). Collective modes in asymmetric nuclei. In *Isospin Effects in Nuclei* (ed. B. A. Li and U. Schroeder), Chapter 7. World Scientific, Singapore. correct version: nucl-th/0001032.

Morawetz, K., Gemming, S., Luschtinetz, R., Kunze, T., Lipavský, P., Seifert, G., and Eng, L. M. (2008). *Phys. Rev. B*, **79**, 085405.

Morawetz, K. and Jauho, A. P. (1994). *Phys. Rev. E*, **50**, 474.

Morawetz, K. and Koehler, S. (1999). *Eur. Phys. J. A*, **4**, 291.

Morawetz, K. and Kremp, D. (1993). *Phys. Lett. A*, **173**, 317.

Morawetz, K. and Kremp, D. (1994). *Phys. Plasmas*, **1**(2), 225.

Morawetz, K. and Kremp, D. (1995). *Z. Phys. A*, **352**, 265.

Morawetz, K., Lipavský, P., J.Koláček, Brandt, E. H., and Schreiber, M. (2007*a*). *Int. J. Mod. Phys. B*, **21**, 2348.

Morawetz, K., Lipavský, P., Normand, J., Cussol, D., Colin, J., and Tamain, B. (2001*b*). *Phys. Rev. C*, **63**, 034619.

Morawetz, K., Lipavský, P., and Schreiber, M. (2005*a*). *Phys. Rev. B*, **72**, 233203.

Morawetz, K., Lipavský, P., and Špička, V. (2001*c*). *Ann. of Phys.*, **294**, 134.

Morawetz, K., Lipavský, P., Špička, V., and Kwong, N.-H. (1999*b*). *Phys. Rev. C*, **59**, 3052.

Morawetz, K., Männel, M., and Schreiber, M. (2007*b*). *Phys. Rev. B*, **76**, 075116.

Morawetz, K., Ploszajczak, M., and Toneev, V. D. (2000*b*). *Phys. Rev. C*, **62**, 064602.

Morawetz, K. and Röpke, G. (1995). *Phys. Rev. E*, **51**, 4246.

Morawetz, K. and Röpke, G. (1996). *Z. Phys. A*, **355**, 287.

Morawetz, K., Schreiber, M., Schmidt, B., Ficker, A., and Lipavský, P. (2005*b*). *Phys. Rev. B*, 72, 014301.

Morawetz, K., Špička, V., and Lipavský, P. (1998). *Phys. Lett. A*, 246, 311.

Morawetz, K., Špička, V., Lipavský, P., Kortemeyer, G., Kuhrts, Ch., and Nebauer, R. (1999*c*). *Phys. Rev. Lett.*, 82, 3767.

Morawetz, K., Vogt, M., Fuhrmann, U., Lipavský, P., and Špicka, V. (1999*d*). *Phys. Rev. C*, 60, 054601.

Morawetz, K., Walke, R., and Röpke, G. (1994). *Phys. Lett. A*, 190, 96.

Morawetz, K. (2013). *EPL*, 104(2), 27005.

Mori, H. and Ono, S. (1952). *Prog. Theor. Phys.*, 8, 327.

Morozov, V. G. and Röpke, G. (1995). *Physica A*, 221, 511.

Morris, T. D. and Brown, J. B. (1971). *Physica*, 55, 760.

Mrowczynski, S. and Danielewicz, P. (1990). *Nucl. Phys. B*, 342, 345.

Mrowczynski, S. and Heinz, U. (1994). *Ann. Phys.*, 229, 1.

Nacher, P. J., Tastevin, G., and Laloë, F. (1989). *J. Phys. (Paris)*, 50, 1907.

Nacher, P. J., Tastevin, G., and Laloë, F. (1991*a*). *Ann. Phys. (Leipzig)*, 503, 149.

Nacher, P. J., Tastevin, G., and Laloë, F. (1991*b*). *Journal de Physique I*, 1, 181.

Negele, J. W. (1982). *Rev. Mod. Phys.*, 54, 913.

Nomura, Kentaro and MacDonald, Allan H. (2006). *Phys. Rev. Lett.*, 96, 256602.

Nuppenau, C., Mackellar, A. D., and Lee, Y. J. (1990). *Nucl. Phys. A*, 511, 525.

Onsager, L. (1927). *Phys. Zeitsch.*, 8, 277.

Ortner, J. (1997). *Phys. Rev. E*, 56, 6193.

Pao, Chien-Hua and Bickers, N. E. (1994). *Phys–. Rev. Lett.*, 72, 1870.

Pao, Chien-Hua and Bickers, N. E. (1995). *Phys. Rev. B*, 51(22), 16310.

Park, T., Sidorov, V. A:, Lee, H., Fisk, Z., and Thompson, J. D. (2005). *Phys. Rev. B*, 72, 060410(R).

Pauli, W. (1958). In *Encyclopaedia of Physics* (ed. S. Flügge), p. 60. Springer, Berlin.

Pavlyukh, Y. and Hübner, W. (2007). *Journal of Mathematical Physics*, 48(5), 052109.

Perdew, John P. and Wang, Yue (1992). *Phys. Rev. B*, 45, 13244.

Pethick, C. J. and Smith, H. (2001). *Bose–Einstein Condensation in Dilute Gases*. Cambridge University Press, Oxford.

Pethick, C. J. and Smith, H. (2004). *Bose–Einstein Condensation in Dilute Gases*. Cambridge University Press, Cambridge.

Pethik, C. J. and Ravenhall, D. G. (1987). *Nucl. Phys. A*, 471, 19c.

Pines, D. and Nozieres, P. (1966). *The Theory of Quantum Liquids*. Volume 1. Benjamin, New York.

Plagnol, E. and et al. (2000). *Phys. Rev. C*, 61, 014606.

Popyrin, S. L. (1998). *Dokl. Phys.*, 43, 671.

Poschenrieder, P. and Weigel, M. K. (1988). *Phys. Rev. C*, 38(1), 471.

Prange, R. E. (1960). In *Proc. of Int. Spring School of Physics, Naples*. Academic Press of Japan.

Prange, R. E. and Kadanoff, L. P. (1964). *Phys. Rev.*, 134, A566.

Prange, R. E. and Sachs, A. (1967). *Phys. Rev.*, 158, 672.

Prigogine, I. (1962). *Nonequilibrium Statistical Mechanics*. Wiley, New York.

Prokof'ev, N., Ruebenacker, O., and Svistunov, B. (2004). *Phys. Rev. A*, 69, 053625.

Puff, H. (1979). *Näherungsverfahren in der Theorie von Vielteilchensystemen mit direkter Paarwechselwirkung*. Volume Preprint 79-1. Zentralinstitut für Elektronenphysik, Berlin.

Raimondi, Roberto and Schwab, Peter (2005). *Phys. Rev. B*, 71, 033311.

Rainwater, J. C. and Snider, R. F. (1976). *J. Chem. Phys.*, 65, 4958.

Rammer, J. and Smith, H. (1986). *Rev. Mod. Phys.*, **58**, 323.

Randrup, J. and Remaud, B. (1990). *Nucl. Phys. A*, **514**, 339.

Rees, H. D. (1969). *J. Phys. Chem. Solids*, **30**, 643.

Reinhard, P. G. and Suraud, E. (1992). *Ann. of Phys.*, **216**, 98.

Reinhard, P. G., Suraud, E., and Ayik, S. (1992). *Ann. of Phys.*, **213**, 204.

Rey, M., Strass, M., Kohler, S., Hänggi, P., and Sols, F. (2007). *Phys. Rev. B*, **76**, 085337.

Rickayzen, G. (1969). *J. Phys. C*, **2**, 1334.

Roberts, J. L., Claussen, N. R., Burke, J. P., Greene, C. H., Cornell, E. A., and Wieman, C. E. (1998). *Phys. Rev. Lett.*, **81**, 5109.

Roman, P. (1969). *Introduction to Quantum Field Theory.* John Wiley & son, New York.

Röpke, G. (1987). *Statistische Mechanik für das Nichtgleichgewicht.* Verlag der Wissenschaften, Berlin.

Röpke, G. (1988). *Phys.Rev.A*, **38**(6), 3001.

Röpke, G. (1994). *Ann. Physik (Leipzig)*, **506**, 145.

Röpke, G. (1995). *Z. Phys. B*, **99**, 83.

Röpke, G., Münchow, L., and Schulz, H. (1982*a*). *Nucl. Phys. A*, **379**, 536.

Röpke, G., Münchow, L., and Schulz, H. (1982*b*). *Phys. Lett. B*, **110**, 21.

Röpke, G., Münchow, L., and Schulz, H. (1982*c*). *Nucl. Phys. A*, **379**, 536.

Röpke, G., Seifert, T., Stolz, H., and Zimmermann, R. (1980). *phys. stat. sol. (b)*, **100**, 215.

Schindlmayr, Arno and Godby, R. W. (1998). *Phys. Rev. Lett.*, **80**, 1702.

Schlanges, M. and Bornath, Th. (1987). *Wiss. Zeitschrift d. Universität Rostock*, **36**, 65.

Schliemann, J. (2006). *Int. J. of Mod. Phys. B*, **20**(9), 1015.

Schliemann, J. and Loss, D. (2004). *Phys. Rev. B*, **69**(16), 165315.

Schmidt, M. and Röpke, G. (1987). *phys. stat. sol. (b)*, **139**, 441.

Schmidt, M., Röpke, G., and Schulz, H. (1990). *Ann. Phys. (NY)*, **202**, 57.

Scholz, Andreas, Stauber, Tobias, and Schliemann, John (2012). *Phys. Rev. B*, **86**, 195424.

Schram, P. P. J. M. (1991). *Kinetic Theory of Gases and Plasmas.* Kluwer Academic Publishers, Dordrecht.

Schwab, P., Dzierzawa, M., Gorini, C., and Raimondi, R. (2006). *Phys. Rev. B*, **74**, 155316.

Schwinger, J. (1961). *J. Math. Phys.*, **2**, 407.

Seminozhenko, V. P. (1982). *Phys. Rep.*, **91**(3), 103.

Semkat, D., Kremp, D., and Bonitz, M. (1999). *Phys. Rev. E*, **59**, 1557.

Semkat, D., Kremp, D., and Bonitz, M. (2000). *J. Math. Phys.*, **41**, 7458.

Serene, J. W. and Rainer, D. (1983). *Phys. Rep.*, **101**, 221.

Serot, B. D. and Walecka, J. D. (1986). The relativistic nuclear many-body problem. In *Adavances of Nuclear Physics* (ed. J. Negele and E. Vogt), Volume 16. Plenum Press, New York.

Shashkin, A. A., Kravchenko, S. V., Dolgopolov, V. T., and Klapwijk, T. M. (2002). *Phys. Rev. B*, **66**, 073303.

Shen, S. Q., Ma, M., Xie, X. C., and Zhang, F. C. (2004). *Phys. Rev. Lett.*, **92**(25), 256603.

Sherman, E. Ya., Lemmens, P., Busse, B., Oosawa, A., and Tanaka, H. (2003). *Phys. Rev. Lett.*, **91**, 057201.

Shi, H. and Griffin, A. (1998). *Phys. Rep.*, **304**, 1.

Shytov, A. V., Mishchenko, E. G., Engel, H.-A., and Halperin, B. I. (2006). *Phys. Rev. B*, **73**, 075316.

Sinova, Jairo, Culcer, Dimitrie, Niu, Q., Sinitsyn, N. A., Jungwirth, T., and MacDonald, A. H. (2004, Mar). *Phys. Rev. Lett.*, **92**, 126603.

Skulski, W. and et. al. (1996). *Phys. Rev. C*, **53**, R2594.

Skyrme, T. H. R. (1956). *Phil. Mag.*, **1**, 1043.

Skyrme, T. H. R. (1959). *Nucl. Phys.*, **9**, 615.

Smith, H. and Hojgaard-Jensen, H. (1989). *Transport Phenomena*. Clarendon Press, Oxford.

Snider, R. F. (1960). *J. Chem. Phys.*, **32**, 1051.

Snider, R. F. (1964). *J. Math. Phys.*, **5**, 1580.

Snider, R. F. (1990). *J. Stat. Phys.*, **61**, 443.

Snider, R. F. (1991). *J. Stat. Phys.*, **63**, 707.

Snider, R. F. (1995). *J. Stat. Phys.*, **80**, 1085.

Snider, R. F., Mullin, W. J., and Laloë, F. (1995). *Physica A*, **218**, 155.

Snider, R. F. and Sanctuary, B. C. (1971). *J. Chem. Phys.*, **55**, 1555.

Sobotka, L. G., Dempsey, J. F., Charity, R. J., and Danielewicz, P. (1997). *Phys. Rev. C*, **55**, 2109.

Šopík, B., Lipavský, P., Männel, M., Morawetz, K., and Matlock, P. (2011). *Phys. Rev. B*, **84**, 094529.

Soven, P. (1967). *Phys. Rev.*, **156**(3), 809.

Špička, V. and Lipavský, P. (1994). *Phys. Rev. Lett.*, **73**, 3439.

Špička, V. and Lipavský, P. (1995). *Phys. Rev. B*, **52**(20), 14615.

Špička, V., Lipavský, P., and Morawetz, K. (1997*a*). *Phys. Rev. B*, **55**, 5095.

Špička, V., Lipavský, P., and Morawetz, K. (1997*b*). *Phys. Rev. B*, **55**, 5084.

Špička, V., Lipavský, P., and Morawetz, K. (1998). *Phys. Lett. A*, **240**, 160.

Špička, V., Lipavský, P., and Velický, B. (1992). *Phys. Rev. B*, **46**, 9408.

Špička, V., Morawetz, K., and Lipavský, P. (2001). *Phys. Rev. E*, **64**, 046107.

Stauber, T., Peres, N. M. R., and Geim, A. K. (2008). *Phys. Rev. B*, 78, 085432.

Stefanini, A. A. (1995). *Z. Phys. A*, **351**, 167.

Stein, H., Porthun, C., and Röpke, G. (1998). *Eur. Phys. J. B*, **2**, 393.

Steinhauer, J., Ozeri, R., Katz, N., and Davidson, N. (2002). *Phys. Rev. Lett.*, **88**, 120407.

Stenger, J., Inouye, S., Chikkatur, A. P., Stamper-Kurn, D. M., Pritchard, D. E., and Ketterle, W. (1999). *Phys. Rev. Lett.*, **82**, 4569.

Stern, F. (1980). *Phys. Rev. Lett.*, **44**, 1469.

Stöcker, H. and Greiner, W. (1986). *Phys. Rep.*, **137**, 277.

Stoleriu, L., Chakraborty, P., Hauser, A., Stancu, A., and Enachescu, C. (2011). *Phys. Rev. B*, **84**, 134102.

Stolz, H. and Zimmermann, R. (1979). *phys. stat. sol. (b)*, **94**, 135.

Stone, M. B., Zaliznyak, I. A., Hong, T., Broholm, C. L., and Reich, D. H. (2006). *Nature*, **440**, 187.

Streda, P. and Jonckheere, T. (2010). *Phys. Rev. B*, **82**(11), 113303.

Stuttgé, L. and et al. (1992). *Nucl. Phys. A*, **539**, 511.

Suraud, E., Ayik, S., Stryjewski, J., and Belkacem, M. (1990). *Nucl. Phys. A*, **519**, 171c.

Syassen, N., Bauer, D. M., Lettner, M., Volz, T., Dietze, D., Garca-Ripoll, J. J., Cirac, J. I., Rempe, G., and Drr, S. (2008). *Science*, **320**(5881), 1329.

Takahashi, Y. (1957). *Il Nuovo Cim.*, **VI**, 371.

Tastevin, G., Nacher, P. J., and Laloë, F. (1989). *J. Phys. (Paris)*, **50**, 1879.

Taylor, J. R. (1972). *Scattering Theory of Nonrelativistic Collisions*. Wiley, New York.

Theobald, W., Häßner, R., Wülker, C., and Sauerbrey, R. (1996). *Phys. Rev. Lett.*, **77**, 298.

Thirring, W. (1981). In *Classical Scattering Theory* (ed. H. Mitter and L. Pittner), Volume XXIII, Acta Physica Austriaca, Suppl., p. 3. Springer-Verlag, Wien.

Thoai, D. B. Tran and Haug, H. (1993). *Z. Phys. B*, **91**, 199. errat. ibid B92, 532 (1993).

Thomas, M. W. and Snider, R. F. (1970). *J. Stat. Phys.*, **2**, 61.

Thouless, D. J. (1960). *Ann. Phys.*, **10**, 553.

Tinkham, M. (1966). *Introduction to Superconductivity*. McGraw Hill, New York.

Tohyama, M. (1987). *Phys. Rev. C*, **86**, 187.

Tõke, J. and et. al. (1995). *Phys. Rev. Lett.*, **75**, 2920.

Tõke, J. and et al. (1995). *Nucl. Phys. A*, **583**, 519.

Trotzky, S., Chen, Y. A., Flesch, A., McCulloch, I. P., Schollwöck, U., Eisert, J., and Bloch, I. (2012). *Nature Physics*, **8**, 325.

Troudet, T. and Vautherin, D. (1985). *Phys. Rev. C*, **31**, 278.

Umezawa, H. (1993). *Advanced Field Theory: Micro, Macro and Thermal Physics*. American Institut of Physics, New York.

Vagov, A. V., Shanenko, A. A., Milosevic, M. V., Axt, V. M., and Peeters, F. M. (2012). *Phys. Rev. B*, **85**, 014502.

van Beijeren, H. and Ernst, M. H. (1979). *J. Stat. Phys.*, **21**, 125.

van den Berg, T. L., Raymond, L., and Verga, A. (2011, Dec). *Phys. Rev. B*, **84**, 245210.

Vasilopoulos, P. (1985). *Phys. Rev. B*, **32**, 771.

Velicky, B. (1969). *Phys. Rev.*, **184**, 614.

Velický, B., Kalvová, A., and Špička, V. (2008). *Phys. Rev. B*, **77**(4), 041201.

Velický, B., Kirkpatrick, S., and Ehrenreich, H. (1968). *Phys. Rev.*, **175**, 747.

Venkatasubramanian, R. (2000). *Phys. Rev. B*, **61**, 3091.

Venkatasubramanian, R., Silvola, E., Colpitts, T., and O'Quinn, B. (2001). *Nature*, **413**, 597.

Vliet, Carolyn M. Van and Vasilopoulos, P. (1988). *Journal of Physics and Chemistry of Solids*, **49**(6), 639.

Vollhardt, D. and Wölfle, P. (1980). *Phys. Rev. B*, **22**, 4666.

Volovik, G. E. (2003). *The Universe in a Helium Droplet*. Oxford University Press, Oxford.

Vu, Q. T. and Haug, H. (2000). *Phys. Phys. B*, **62**, 7179.

Waldmann, L. (1957). *Z. Naturforsch. A*, **12**, 660.

Waldmann, L. (1958). *Z. Naturforsch. A*, **13**, 609.

Waldmann, L. (1960). *Z. Naturforsch. A*, **15**, 19.

Walecka, J. D. (1974). *Ann. of Phys.*, **83**, 491.

Walter, M., Zwicknagel, G., and Toepffer, C. (2005). *Eur. Phys. J. D*, **35**, 527.

Wang, Yue and Perdew, John P. (1991). *Phys. Rev. B*, **44**, 13298.

Ward, J. C. (1950). *Phys. Rev.*, **78**, 182.

Weinberg, S. (1990). *Phys. Lett. B*, **251**, 288.

Weinstock, J. (1963a). *Phys. Rev*, **132**, 454.

Weinstock, J. (1963b). *Phys. Rev*, **132**, 470.

Weinstock, J. (1965). *Phys. Rev. A*, **140**, 460.

Wiegmann, P. B. (1981). *J. Phys. C: Solid State Phys.*, **14**, 1463.

Wigner, E. P. (1955). *Phys. Rev.*, **98**, 145.

Wild, W. (1960). *Z. Phys. A*, **158**(3), 322.

Wilets, L. and et. al. (1976). *Phys. Rev. C*, **14**(6), 2269.

Williams, J. E., Nikuni, T., and Clark, Charles W. (2002). *Phys. Rev. Lett.*, **88**, 230405.

Wingreen, N. S., Stanton, C. J., and Wilkins, J. W. (1986). *Phys. Rev. Lett.*, **57**(8), 1084.

Wojciechowski, R. J. (1999). *Physica B: Condensed Matter*, **259**, 498.

Yamaguchi, Y. (1954). *Phys. Rev.*, **95**, 1628.

Yanase, Yoichi, Jujo, Takanobu, Nomura, Takuji, Ikeda, Hiroaki, Hotta, Takashi, and Yamada, Kosaku (2003). *Physics Reports*, **387**(1–4), 1.

Zala, Gábor, Narozhny, B. N., and Aleiner, I. L. (2001, Nov). *Phys. Rev. B*, **64**, 214204.

Zeman, J., Šmíd, V., Krištofik, J., and Mareš, J. J. (1993). *Phil. Mag. B*, **67**, 49.

Zhang, Yuanbo, Tan, Yan-Wen, Stormer, Horst L., and Kim, Philip (2005). *Nature*, **438**, 201.

Ziesche, P. (2007). *phys. stat. sol. (b)*, **244**, 2022.

Ziesche, P. (2010). *Ann. Phys. (Berlin)*, **522**, 739.

Zimmermann, R., Kilimann, K., Kraeft, W., Kremp, D., and Röpke, G. (1978). *phys. stat. sol. (b)*, **90**, 175.

Zimmermann, R. and Stolz, H. (1985). *phys. stat. sol. (b)*, **131**, 151.

Zinn-Justin, J. (2007). *Phase Transitions and Renormalization Group*. Oxford University Press, Oxford.

Zubarev, D. N. (1971). *Nonequilibrium Statistical Thermodynamics (russ.)*. NAUKA, Moscow. German translation: Akademie Verlag, Berlin, 1976.

Zubarev, D. N., Morozov, V., and Röpke, G. (1996). *Statistical Mechanics of Nonequilibrium Processes*. Volume 1. Akademie Verlag, Berlin.

Zubarev, D. N., Morozov, V., and Röpke, G. (1997). *Statistical Mechanics of Nonequilibrium Processes*. Volume 2. Akademie Verlag, Berlin.

Zwicknagel, G. (1999). *Contrib. Plasma Phys.*, **39**, 155.

Zwicknagel, G., Toepffer, C., and Reinhard, P. G. (1995). In *Physics of strongly coupled plasmas* (ed. W. D. Kraeft and M. Schlanges), Singapore, p. 45. World Scientific.

Index